The Reduction of Iron Ores

L. von Bogdandy · H.-J. Engell

The Reduction of Iron Ores

Scientific Basis and Technology

Revised and Translated Edition of
"Die Reduktion der Eisenerze"

With 381 Figures and 54 Tables

1971
Springer-Verlag Berlin Heidelberg GmbH

Prof. Dr.-Ing. Ludwig von Bogdandy
Member of Managing Committee,
Hüttenwerk Oberhausen AG, Oberhausen

Prof. Dr. rer. nat. Hans-Jürgen Engell
Director, Max-Planck-Institut für Eisenforschung,
Düsseldorf

The translation of Chapters 1 to 4 was procured by arrangement with the Trade and Industrial Translation Centre Ltd., London. These four chapters were translated by Commander E. R. Dawson of the University of Dundee, Scotland, and edited by Mr. F. B. Traice of the Swinden Laboratories of British Steel Corporation. – Chapter 5 has been translated and edited jointly by Messrs. R. Sewell and A. J. H. Leadley of the Swinden Laboratories of British Steel Corporation.

DK 669.051
669.046.46: 669.162.12
669.162.263.23

ISBN 978-3-662-10402-6 ISBN 978-3-662-10400-2 (eBook)
DOI 10.1007/978-3-662-10400-2

Originally published by Springer-Verlag, Berlin/Heidelberg and Verlag Stahleisen m.b.H., Düsseldorf in 1971
Softcover reprint of the hardcover 1st edition 1971

Preface to the English Edition

For the English edition the book was revised by the authors, taking into account a number of suggestions of the readers of the German edition. Some of the most important publications in the field of iron ore reduction, which appeared since 1967, have been used to bring the manuscript as far as possible up to date. The kind assistance of Dr. K. BOHNENKAMP of the Max-Planck-Institut für Eisenforschung, Düsseldorf, was much appreciated in this respect. Chapters 2.9 and 2.10, dealing with the reduction of molten oxide slags by solid carbon and with the contribution of the water-gas reaction to iron ore reduction, have been added for the English edition. Chapter 2.9 has been completely revised with the kind assistance of Dr. H. J. GRABKE, Stuttgart. Dipl.-Ing. J. LODDE contributed to this edition by revising the bibliography. Owing to the rapid development of the blast furnace it was necessary to revise Chapter 5 considerably. In this field many valuable suggestions have been made by Dipl.-Ing. G. LANGE and Dipl.-Ing. P. HEINRICH.

Furthermore, Chapters 3 and 4 have been thoroughly revised by Dr.-Ing. E. FÖRSTER and Dr.-Ing. U. SCHIERLOH.

Last, but not least, we have to thank our translators for their excellent work.

Oberhausen and Düsseldorf,
November 1970

L. von Bogdandy **H.-J. Engell**

Preface to the German Edition

In 1838 ROBERT WILHELM VON BUNSEN climbed up the blast furnace of the Veckerhagen "Herrschaftliche" Ironworks and lowered into the burden a tube made from gun barrels welded together and covered with fireproof material. With the aid of lead pipes and glass tubes, bits of rubber tubing and an air pump, BUNSEN drew from the furnace shaft, via this tube, gas samples which he analyzed, finding carbon dioxide and monoxide, hydrogen and methane. He proved that it would be well worthwhile to collect the gas and use it to heat the Henschel steam-engine which drove the air blower, and which had previously required 35 lb. of beechwood per hour. BUNSEN himself said: "Even if the experiments mentioned cannot be regarded in any sense as exhausting the subject, they may still be not without value as a preliminary report, since they embody the most important features which can serve as a basis for practical investigations into this subject, which is of such importance for the whole of metallurgy".

No-one who reflects on the then state of affairs in science and engineering and on their subsequent evolution can dispute this opinion. The methods and aims of metallurgical research are shown quite clearly by this example. The aim is to improve the degree of utilization of the raw materials employed, and to increase manufacturing efficiency: the method consists in developing suitable methods of making measurements (here it was the method of gas analysis), combined with a scientifically based appraisal of the experimental results. These methods, these aims – coupled with theoretical penetration into the mechanism of the process, and with the development of better designs and more efficient auxiliary plant for the blast furnace – these are the foundation of the developments of the ensuing century. Countless engineers and research-workers have concerned themselves with the blast-furnace process, with the result that the blast furnace of today must be regarded as one of the most efficient devices used in metallurgy.

In this monograph, the thermodynamic principles and the basis in reaction kinetics for the reduction of ores are recounted first, and then the laws of gas flow and heat transfer in packed columns are reviewed. Using this background material, we try to describe scientifically the technology of the blast-furnace process and to compute limits for its performance. In particular, we discuss in detail: the preparation of the burden from the standpoint of process technology and reaction kinetics; the significance of the chemical and physical properties of the materials used, and the methods for determining these properties; and the behaviour of alternative fuels. The conditions in the "dry" part of the blast furnace where the reactions between gases and solids predominate are discussed in particular detail.

By means of this theoretical "tool-kit" we can in many cases distinguish the essential from the fortuitous; we can calculate in advance the effect of particular properties of the materials used, and the influence of the operational parameters, on the manufacturing costs and on the operational results, i.e., on the manufacturing performance and the specific fuel consumption. Hence a basis can be constructed for the optimization and automation of blast-furnace operation, and we can judge whether improvements will be possible in the future, and what these may be.

Apart from this statement of problems, which is mainly directed towards improving the blast-furnace process itself (the process which today is of primary importance), we also discuss the problem of "direct reduction", i.e., reduction without using a blast furnace. For several decades attempts have been made to reduce iron ores to the metal in some apparatus other than the blast furnace, obtaining the product either in the form of molten pig iron or in the solid phase, as sponge iron or metallized pellets, for example. It is true that no method of this sort has yet achieved the performance of the blast furnace. However, the increasing availability of economically priced gaseous and liquid fuels, especially of natural gas, in the Middle East and in Central and South America, for instance, makes it increasingly important to find a method for the reduction of iron ores using these fuels instead of coke. The reduction of iron ores without a blast furnace could be the key to the problem of providing the national economies of these countries with supplies of iron on extremely economic terms by using their indigenous raw materials and fuels. And iron is, of course, essential for building up national industries and increasing the general prosperity.

We hope that this monograph may be of assistance to both research workers and practising engineers in solving these and other problems and that it will stimulate them to improve and cheapen iron production by making well-conceived modifications of the methods of operation. But we also address ourselves to students, hoping to enable them to become more closely acquainted with the fundamentals of iron-ore reduction and their application to technology, and to acquire an insight into the often quite complicated relationships.

It was not intended to compile a handbook that would take all opinions, views and researches into account. Rather have we selected from the literature those articles which we regarded as indicating a way forward, which seemed to us to illustrate the way of thinking and working which we have presented, and which were related to our own observations, calculations and evaluations. This means that our choice of references was subjective, and it should also be added that on occasion some omissions may have occured.

The second-named of the two authors is responsible for Chapters 1 and 2, the first-named for Chapters 3 to 5.

We must thank our co-workers who have helped in the compilation of references, in the calculations, the textual presentation, and in the proof-reading. In particular, we should like to record the energetic assistance rendered by Dr.-Ing. G. Gross and Dr.-Ing. U. Pückoff of Dortmund, Dr.-Ing. H. D. Pantke, Dipl.-Ing. H. Singer and Dr.-Ing. D. Terlaak of Oberhausen, and Dr.-Ing. W. Pluschkell of Stuttgart. We have also to thank a large number of professional colleagues in the iron-making industry and at the Universities and Research Institutes for communicating unpublished results to us and for fruitful discussions. Dipl.-Ing. H. Kegel of Düsseldorf has contributed to the success of the project by many informative and clarificatory conversations. The instigation to write this monograph came from Professor H. Schenck, Dr.-Ing., Dr.-Ing. E. h., and to him we should like to express our sincere thanks for his encouragement.

Oberhausen and Stuttgart,
October 1966

L. von Bogdandy **H.-J. Engell**

Contents

Frequently Used Symbols

Symbol	*Meaning*	*Unit*
a	thermal diffusivity	$cm^2\ sec^{-1}$
c_i (or n_i)	concentration of the component i	mole cm^{-3}
d	a characteristic length; or particle diameter	cm
g	gravitational constant	cm sec^{-2}
i	running subscript	
j	flux of material	mole $cm^{-2}\ sec^{-1}$ or mole sec^{-1}
k_n $(n = 1, 2, 3\ldots)$	rate constant of phase-boundary reaction	mole $cm^{-2}\ sec^{-1}\ atm^{-1}$
k_v	rate constant of phase-boundary reaction with reference to volume of ore	mole $cm^{-3}\ sec^{-1}\ atm^{-1}$
l	diffusion path length; pore length; or boundary-layer thickness	cm
m	mass	g
m, n	stoichiometric numbers	
n	number	
n_i'	concentration of component i	cm^{-3}
p	gas pressure	atm or dyne cm^{-2}
Δp	pressure drop	dyne cm^{-2}
p_i	partial pressure of component i	atm
q	heat flow	kcal $m^{-2}\ h^{-1}$ or cal $cm^{-2}\ sec^{-1}$
s	concentration per unit area	cm^{-2}
t	time	sec
u	gas velocity (empty-tube)	cm sec^{-1}
u_F	descent rate of material particles in a flowing medium	cm sec^{-1}
u_{cr}	critical gas-velocity in a charge column	cm sec^{-1}
u_L	gas velocity at the transition from a static bed to a fluidized bed	cm sec^{-1}
x	space coordinate	cm
y	mole-fraction of point defects, e.g. of vacancies in wüstite	
z	electric charge number	
A	activation energy	cal $mole^{-1}$
B	mobility	cm $sec^{-1}\ dyne^{-1}$ or $cm^2\ sec^{-1}\ cal^{-1}$
C	volume fraction	
D	vessel diameter	cm

Symbol	*Meaning*	*Unit*
D_i	coefficient of diffusion of the *i*-th component	$cm^2\ sec^{-1}$
E	replacement ratio	e.g., (kg coke)/(kg heavy oil)
F	cross-sectional area of vessel; or surface area of charge	cm^2
F	Faraday constant	A sec/g equivalent
ΔG	free enthalpy of reaction	cal $mole^{-1}$
H	charge height	cm
ΔH	enthalpy of reaction	cal $mole^{-1}$
H_u	lower calorific value	kcal/m^3 at NTP, kcal/kg
K	force	dyne
K	equilibrium constant	
K_h	heating coke requirement	kg/ton Fe
L	throughput capacity	tons Fe/day
M	molecular weight	g $mole^{-1}$
M_C	carbon content of charge	g cm^{-3}
M_{Fe}	Fe-content of charge	g cm^{-3}
N_{Gr}	Grashof number	
N_L	Loschmidt number	$mole^{-1}$
N_{Nu}	Nusselt number	
N_{Pr}	Prandtl number	
N_{Re}	Reynolds number	
N_{Sc}	Schmidt number	
N_{Sh}	Sherwood number	
Q	heat requirement	kcal
R	general gas constant	cal/mole °K
ΔS	entropy of reaction	cal/mole °K
T	temperature	°C, °K
V	potential difference	V
V	gas-flow rate	mole sec^{-1}, $m^3\ sec^{-1}$ at NTP
V_{gas}	specific gas-throughput	m^3/kg Fe at NTP
V_R	volume of reaction space (often the same as the charge volume)	cm^3
Z	number of collisions	$cm^{-2}\ sec^{-1}$
$\mathfrak{F}$	electric field strength	V cm^{-1}
$\mathfrak{R}$	degree of reduction	
α	heat transfer number	cal $cm^{-2}\ sec^{-1}$ $°K^{-1}$
α	adsorption coefficient	
β	material transfer number	cm sec^{-1}
γ	porosity	
ε	voids fraction of a (uniform) charge	
ε_k	voids fraction for the component with least particle diameter	
ε_m	voids fraction for a multi-granular charge	
η	dynamic viscosity	g $cm^{-1}\ sec^{-1}$
η	electro-chemical potential	cal $mole^{-1}$, erg $mole^{-1}$, or eV $mole^{-1}$

Symbol	*Meaning*	*Unit*
η_{H_2}	degree of chemical utilization of H_2, given by the ratio of concentrations in the spent gas: $C_{H_2O}/(C_{H_2O} + C_{H_2})$, corrected for the amount of H_2O introduced in the charge	
η_{CO}	degree of chemical utilization of CO	
Θ	gas temperature	°C
Θ	degree of covering; angle of contact	
$\varkappa$	electrical conductivity	$ohm^{-1}\ cm^{-1}$
λ	thermal conductivity	$cal\ cm^{-1}\ sec^{-1}\ °K^{-1}$
$\bar{\lambda}$	mean free path	cm
μ	chemical potential	$cal\ mole^{-1}$
$\nu = \eta/\varrho$	kinematic viscosity	$cm^2\ sec^{-1}$
ν	rate of desorption	$cm^{-2}\ sec^{-1}$
ϱ	density	$g\ cm^{-3}$
σ	collision diameter	cm
σ	boundary or surface tension, free boundary or surface energy	$dyne\ cm^{-1}$
τ	holding time in reaction zone; half-life	sec
φ	electric potential	V
φ	particle form factor	
ψ	resistance parameter	
ξ	labyrinth factor	
ξ	thickness of layer of compact, solid, reaction products	cm
+ −	charges of lattice ions	
• ′	charges of lattice defects	

Thermodynamic Tables

Synopsis

Abbreviations and Symbols

c	crystalline
g	gaseous
G_T^0	molar free enthalpy at $P = 1$ atm and T°C
G_{25}^0	molar free enthalpy at $P = 1$ atm and 25°C
H_T^0	molar enthalpy at $P = 1$ atm and T°C
H_{25}^0	molar enthalpy at $P = 1$ atm and 25°C
l	liquid
M	atomic weight of an element or molecular weight of a pure substance
S_T^0	molar entropy at $P = 1$ atm and T°C
S_{25}^0	molar entropy at $P = 1$ atm and 25°C
T	absolute temperature in °K or temperature in °C
ΔG_T^0	free enthalpy of reaction
ΔH_T^0	enthalpy of formation
ΔS_T^0	entropy of formation

(ΔG_T^0, ΔH_T^0, ΔS_T^0:) at $P = 1$ atm and T°C for a chemical reaction of the type $\alpha A + \beta B = \gamma C$

Methane CH_4

Molecular weight $M = 16.04$			Entropy $S^0_{25} = 44.50$ cal/°C mole			Enthalpy of formation ΔH^0_T Entropy of formation ΔS^0_T Free enthalpy ΔG^0_T of the reaction $C_{(c)} + 2\,H_{2(g)} = CH_{4(g)}$		
T °C	T °K	State at T °C	$H^0_T - H^0_{25}$ cal/mole	$S^0_T - S^0_{25}$ cal/°C mole	$G^0_T - G^0_{25}$ cal/mole	ΔH^0_T cal/mole	ΔS^0_T cal/°C mole	ΔG^0_T cal/mole
25	298	g	0	0	0	−17800	−19.3	−12100
100	373	g	680	1.9	− 3370	−18430	−20.9	−10870
200	473	g	1690	4.3	− 8130	−19120	−22.6	− 8440
300	573	g	2830	6.5	−13150	−19740	−23.7	− 6140
400	673	g	4120	8.6	−18370	−20270	−24.6	− 3720
500	773	g	5690	10.6	−23640	−20570	−25.2	− 1070
600	873	g	7070	12.4	−29380	−21070	−25.7	1330
700	973	g	8710	14.2	−35160	−21350	−26.0	3910
800	1073	g	10450	15.9	−41130	−21570	−26.2	6470
900	1173	g	12270	17.6	−47250	−21740	−26.3	9140
1000	1273	g	14180	19.1	−53530	−21870	−26.4	11710
1100	1373	g	16160	20.6	−59970	−21930	−26.4	14340
1200	1473	g	18180	22.0	−66560	−21960	−26.5	17070
Relative error			1%	1%	1%	<3%	<3%	>10%

According to VALLET: Publ. de IRSID Serie B, No. 26, 1955.

Carbon Monoxide CO

Molecular weight $M = 28.01$			Entropy $S^0_{25} = 47.30$ cal/°C mole			Enthalpy of formation ΔH^0_T Entropy of formation ΔS^0_T Free enthalpy ΔG^0_T of the reaction $C_{(c)} + \frac{1}{2} O_{2(g)} = CO_{(g)}$		
T°C	T°K	State at T°C	$H^0_T - H^0_{25}$ cal/mole	$S^0_T - S^0_{25}$ cal/°C mole	$G^0_T - G^0_{25}$ cal/mole	ΔH^0_T cal/mole	ΔS^0_T cal/°C mole	ΔG^0_T cal/mole
25	298	g	0	0	0	−26400	21.4	−32800
100	373	g	550	1.6	− 3600	−26310	21.7	−34410
200	473	g	1230	3.2	− 8560	−26300	21.7	−36540
300	573	g	1940	4.6	− 13700	−26320	21.7	−38780
400	673	g	2680	5.8	− 18940	−26390	21.6	−40940
500	773	g	3420	6.8	− 24310	−26490	21.5	−43100
600	873	g	4190	7.7	− 29760	−26600	21.4	−45230
700	973	g	4970	8.6	− 35320	−26720	21.2	−47380
800	1073	g	5770	9.4	− 40920	−26840	21.1	−49480
900	1173	g	6580	10.1	− 46630	−26990	12.0	−51590
1000	1273	g	7400	10.7	− 52370	−27130	20.8	−53650
1100	1373	g	8290	11.4	− 58200	−27270	20.7	−55710
1200	1473	g	9070	12.0	− 64140	−27420	20.6	−57810
1300	1573	g	9910	12.5	− 70100	−27580	20.6	−59890
1400	1673	g	10760	13.0	− 76100	−27740	20.5	−61980
1500	1773	g	11610	13.6	− 82200	−27890	20.4	−64010
1600	1873	g	12470	14.0	− 88290	−28070	20.3	−66040
1700	1973	g	13330	14.5	− 94470	−28260	20.2	−68080
1800	2073	g	14200	14.9	−100600	−28450	20.1	−70030
1900	2173	g	15070	15.3	−106800	−28650	20.0	−72020
2000	2273	g	15950	15.7	−113070	−28840	19.9	−74000
2500	2773	g	20350	17.4	−145050			
3000	3273	g	24800	18.9	−177850			
3500	3773	g	29280	20.2	−211340			
4000	4273	g	33790	21.3	−245370			
4500	4773	g	38320	22.3	−279930			
Relative error			1%	1%	0.5%	<3%	<3%	<3%

According to VALLET: Publ. de IRSID Serie B, No. 26, 1955.

Carbon Dioxide CO_2

Molecular weight $M = 44.01$			Entropy $S^0_{25} = 51.06$ cal/°C mole			Enthalpy of formation ΔH^0_T, Entropy of formation ΔS^0_T, Free enthalpy ΔG^0_T of the reaction $C_{(c)} + O_{2(g)} = CO_{2(g)}$		
T°C	T°K	State at T°C	$H^0_T - H^0_{25}$ cal/mole	$S^0_T - S^0_{25}$ cal/°C mole	$G^0_T - G^0_{25}$ cal/mole	ΔH^0_T cal/mole	ΔS^0_T cal/°C mole	ΔG^0_T cal/mole
25	298	g	0	0	0	−94000	0.7	−94260
100	373	g	700	2.0	− 3820	−94050	0.7	−94270
200	473	g	1710	4.4	− 9320	−94080	0.6	−94370
300	573	g	2790	6.5	− 14990	−94120	0.6	−94440
400	673	g	3940	8.4	− 20840	−94160	0.5	−94490
500	773	g	5130	10.0	− 26870	−94200	0.4	−94530
600	873	g	6370	11.5	− 33060	−94210	0.4	−94570
700	973	g	7640	12.9	− 39320	−94290	0.3	−94610
800	1073	g	8950	14.2	− 45860	−94330	0.3	−94680
900	1173	g	10290	15.4	− 52440	−94370	0.3	−94700
1000	1273	g	11640	16.5	− 59150	−94400	0.2	−94710
1100	1373	g	13020	17.5	− 65950	−94420	0.2	−94720
1200	1473	g	14400	18.5	− 72860	−94450	0.2	−94740
1300	1573	g	15810	19.4	− 79820	−94480	0.2	−94750
1400	1673	g	17230	20.3	− 86910	−94510	0.2	−94760
1500	1773	g	18650	21.1	− 94130	−94550	0.1	−94790
1600	1873	g	20100	21.9	−101350	−94590	0.1	−94800
1700	1973	g	21530	22.7	−198720	−94620	0.1	−94820
1800	2073	g	22930	23.4	−116070	−94690	0.0	−94780
1900	2173	g	24460	24.0	−123450	−94750	0.0	−94750
2000	2273	g	25930	24.7	−130970	−94810	0.0	−94740
2500	2773	g	33340	27.7	−169710			
3000	3273	g	40860	30.2	−209790			
3500	3773	g	48430	32.5	−251650			
Relative error			1.5%	1.5%	1%	<3%	>10%	<3%

According to Vallet: Publ. de IRSID Serie B, No. 26, 1955.

Water Vapour H_2O

Molecular weight $M = 18.02$		Entropy $S^0_{25} = 45.10$ cal/°C mole				Enthalpy of formation ΔH^0_T Entropy of formation ΔS^0_T Free enthalpy ΔG^0_T of the reaction $H_{2(g)} + \frac{1}{2} O_{2(g)} = H_2O_{(g)}$		
T°C	T°K	State at T°C	$H^0_T - H^0_{25}$ cal/mole	$S^0_T - S^0_{25}$ cal/°C mole	$G^0_T - G^0_{25}$ cal/mole	ΔH^0_T cal/mole	ΔS^0_T cal/°C mole	ΔG^0_T cal/mole
25	298	g	0	0	0	−57700	−10.6	−54600
100	373	g	610	1.3	− 3260	−57980	−11.6	−53660
200	473	g	1440	3.7	− 8220	−58210	−11.7	−52670
300	573	g	2290	5.4	− 13190	−58430	−12.1	−51480
400	673	g	3170	6.8	− 18300	−58640	−12.5	−50230
500	773	g	4080	8.0	− 23560	−58830	−12.8	−48960
600	873	g	5020	9.2	− 28920	−59010	−13.0	−47670
700	973	g	5980	10.3	− 34400	−59170	−13.1	−46350
800	1073	g	6970	11.2	− 39980	−59320	−13.3	−45080
900	1173	g	7990	12.1	− 45670	−59470	−13.4	−43740
1000	1273	g	9030	13.0	− 51440	−59590	−13.5	−42400
1100	1373	g	10100	13.8	− 57310	−59720	−13.6	−41100
1200	1473	g	11190	14.6	− 63250	−59810	−13.7	−39700
1300	1573	g	12300	15.3	− 69250	−59910	−13.7	−38350
1400	1673	g	13430	16.0	− 75330	−60090	−13.8	−36970
1500	1773	g	14590	16.7	− 81530	−60080	−13.8	−35620
1600	1873	g	15760	17.3	− 87740	−60170	−13.8	−34260
1700	1973	g	16950	17.9	− 94070	−60230	−13.8	−32930
1800	2073	g	18150	18.6	−100380	−60310	−13.9	−31570
1900	2173	g	19360	19.1	−106750	−60370	−13.9	−30190
2000	2273	g	20590	19.7	−113220	−60440	−13.9	−28850
2500	2773	g	26860	22.3	−146490	−60760	−14.0	−22040
Relative error			0.5%	0.5%	0.4%	<3%	<3%	3−10%

According to VALLET: Publ. de IRSID Serie B, No. 26, 1955.

Calcium Carbonate $CaCO_3$

Molecular weight $M = 100.09$			Entropy $S^0_{25} = 22.2$ cal/°C mole			Enthalpy of formation ΔH^0_T, Entropy of formation ΔS^0_T, Free enthalpy ΔG^0_T of the reaction $CaO_{(c)} + CO_{2(g)} = CaCO_{3(c)}$		
T°C	T°K	State at T°C	$H^0_T - H^0_{25}$ cal/mole	$S^0_T - S^0_{25}$ cal/°C mole	$G^0_T - G^0_{25}$ cal/mole	ΔH^0_T cal/mole	ΔS^0_T cal/°C mole	ΔG^0_T cal/mole
25	298	c	0	0	0	−42500	−38.4	−31070
100	373	c	1630	4.7	− 1780	−42360	−38.1	−28130
200	473	c	3960	10.3	− 4780	−42170	−37.5	−24420
300	573	c	6500	15.2	− 8290	−41880	−36.9	−20700
400	673	c	9160	19.5	−12260	−41560	−36.4	−17030
500	773	c	11910	23.3	−16620	−41200	−36.0	−13400
600	873	c	14730	26.7	−21350	−40840	−35.5	− 9830
700	973	c	17650	29.9	−26400	−40440	−35.1	− 6290
800	1073	c	20630	32.8	−31770	−40000	−34.7	− 2820
900	1173	c	23710	35.5	−37400	−39530	−34.3	+ 650
Relative error			2%	2%	5%	<3%	<3%	<3%

According to VALLET: Publ. de IRSID Serie B, No. 26, 1955.

Wollastonite $CaSiO_3$

Transition point ($\beta \rightarrow \alpha$): 1190°C
Melting point: 1544°C

Molecular weight $M = 116.14$			Entropy $S^0_{25} = 19.6$ cal/°C mole			Enthalpy of formation ΔH^0_T, Entropy of formation ΔS^0_T, Free enthalpy ΔG^0_T of the reaction $CaO_{(c)} + SiO_{2(c)} = CaSiO_{3(c)}$		
T°C	T°K	State at T°C	$H^0_T - H^0_{25}$ cal/mole	$S^0_T - S^0_{25}$ cal/°C mole	$G^0_T - G^0_{25}$ cal/mole	ΔH^0_T cal/mole	ΔS^0_T cal/°C mole	ΔG^0_T cal/mole
25	298	c	0	0	0	−21300	0.1	−21330
100	373	c	1690	4.9	− 1590	−21300	0.1	−21330
200	473	c	4110	10.7	− 4360	−21310	0.1	−21340
300	573	c	6690	15.6	− 7650	−21340	0.0	−21350
400	673	c	9400	20.0	−11400	−21380	−0.1	−21350
500	773	c	12230	23.9	−15560	−21410	−0.1	−21320
600	873	c	15110	27.4	−20080	−21760	−0.5	−21290
700	973	c	18020	30.6	−24970	−21700	−0.5	−21260
800	1073	c	20970	33.5	−30140	−21770	−0.4	−21350
900	1173	c	23980	36.2	−35600	−21570	−0.3	−21180
1000	1273	c	27050	38.7	−41300	−21490	−0.3	−21160
1100	1373	c	30160	41.0	−47240	−21390	−0.2	−21130
Relative error			0.5%	0.5%	1%	<3%	>10%	<3%

According to Vallet: Publ. de IRSID Serie B, No. 26, 1955.

1127	1400	c				−21190		−20280
1227	1500	c				−21440		−19930
1327	1600	c				−21370		−19820
1427	1700	c				−21270		−19720
1527	1800	c				−21150		−19640

According to Elliott, Gleiser, and Ramakrishna: Thermochemistry for Steelmaking, Vol. 2, Reading (Mass.)/London 1963.

Enthalpy of formation: D. R. Torgeson and T. G. Sahama: J. Amer. Chem. Soc. 70 (1948) p. 2156.

Entropy of formation: S. S. Todd: J. Amer. Chem. Soc. 73 (1951) p. 3277.

High-temperature enthalpy and entropy increment: K. K. Kelley: U.S. Bur. Mines Bull. (1949) p. 476.

Dicalcium Silicate Ca_2SiO_4

Transition point ($\beta \rightarrow \alpha'$): 697°C Heat of transition: 440 cal/mole
Transition point ($\alpha' \rightarrow \alpha$): 1437°C Heat of transition: 3390 cal/mole
Melting point: 2130°C

Molecular weight $M = 172.25$			Enthalpy of formation ΔH_T^0 Free enthalpy ΔG_T^0 of the reaction $2\,CaO_{(c)} + SiO_{2(c)} \rightarrow Ca_2SiO_4O_{(c,\beta)}$	
T°C	T°K	State at T°C	ΔH_T^0 cal/mole	ΔG_T^0 cal/mole
25.15	298.15	c	−30190 (±230)	−30640
697	970	c	−30110	−31500
697	970	c	−29670	−31500
927	1200	c	−29150	−31850
1027	1300	c	−28910	−32110
1127	1400	c	−29590	−32360
1227	1500	c	−28170	−32625
1327	1600	c	−27670	−32950
1427	1700	c	−27040	−33300
1437	1710	c	−26970	−33360
1437	1710	c	−23580	−33360
1527	1800	c	−23190	−33860
1727	2000	c	−22400	−34800

According to ELLIOTT, GLEISER, and RAMAKRISHNA: Thermochemistry for Steelmaking, Vol. 2, Reading (Mass.)/London 1963.

Enthalpy of formation: E. G. KING: J. Amer. Chem. Soc. 73 (1951) p. 656.

Entropy of formation: S. S. TODD: J. Amer. Chem. Soc. 73 (1951) p. 3277.

High-temperature enthalpy, entropy increment, and heat of transition: J. P. COUGHLIN and C. J. O'BRIEN: J. Phys. Chem. 61 (1957) p. 767.

Dicalcium Ferrite $Ca_2Fe_2O_5$

Melting point: 1477°C Heat of melting: 36110 cal/mole

Molecular weight $M = 271.86$			Enthalpy of formation ΔH_T^0 Free enthalpy ΔG_T^0 of the reaction $2\,CaO_{(c)} + Fe_2O_{3(c,\gamma)} \rightarrow Ca_2Fe_2O_{5(c,l)}$	
T°C	T°K	State at T°C	ΔH_T^0 cal/mole	ΔG_T^0 cal/mole
25.15	298.15	c	− 7400 (±1000)	− 8772
927	1200	c	− 9290	−12090
1027	1300	c	− 9440	−12330
1127	1400	c	− 9630	−12530
1227	1500	c	− 9860	−12725
1327	1600	c	−10160	−12930
1427	1700	c	−10550	−13050
1477	1750	c	−10850	−13180
1477	1750	l	+25260	−13170
1527	1800	l	+25740	−14240

According to ELLIOTT, GLEISER, and RAMAKRISHNA: Thermochemistry for Steelmaking, Vol. 2, Reading (Mass.)/London 1963.

Enthalpy of formation: E. S. NEWMAN and R. H. HOFFMAN: J. Res. Natl. Bur. Standards 56 (1956) p. 313.

Entropy of formation: E. G. KING: J. Amer. Chem. Soc. 76 (1954) p. 5849.

High-temperature enthalpy and entropy increment: K. R. BONNICKSON: J. Amer. Chem. Soc. 76 (1954) p. 1480.

Enthalpy of iron(III)-oxide: J. P. COUGHLIN: J. Amer. Chem. Soc. 73 (1951) p. 3891.

Iron(II)-oxide (Wüstite) FeO

Melting point: 1377°C Heat of melting: 7490 cal/mole

Molecular weight $M = 71.85$			Entropy $S^0_{25} = 13.4$ cal/°C mole			Enthalpy of formation ΔH^0_T Entropy of formation ΔS^0_T Free enthalpy ΔG^0_T of the reaction $0.95\,Fe_{(c)} + \frac{1}{2} O_{2(g)} = Fe_{0.95}O_{(c,l)}$		
T°C	T°K	State at T°C	$H^0_T - H^0_{25}$ cal/mole	$S^0_T - S^0_{25}$ cal/°C mole	$G^0_T - G^0_{25}$ cal/mole	ΔH^0_T cal/mole	ΔS^0_T cal/°C mole	ΔG^0_T cal/mole
25	298	c	0	0	0	−63700	−17.6	−58400
100	373	c	900	2.7	− 1000	−63400	−17.0	−57100
200	473	c	2200	5.8	− 2800	−63200	−16.4	−55500
300	573	c	3500	8.3	− 4900	−63000	−15.9	−53800
400	673	c	4800	10.5	− 7200	−62800	−15.7	−52300
500	773	c	6200	12.4	− 9700	−62700	−15.5	−50700
600	873	c	7500	14.0	−12300	−62700	−15.5	−49200
700	973	c	8900	15.5	−15200	−62800	−16.1	−47600
800	1073	c	10300	16.9	−18100	−63200	−16.0	−46100
900	1173	c	11700	18.2	−21200	−63200	−16.0	−44500
Relative error			1%	1%	7%	<3%	3−10%	<3%

According to Vallet: Publ. de IRSID Serie B, No. 26, 1955.

T°C	T°K	State at T°C	$H^0_T - H^0_{25}$	$S^0_T - S^0_{25}$	$G^0_T - G^0_{25}$	ΔH^0_T	ΔS^0_T	ΔG^0_T
927	1200	c				−63300		−44450
1027	1300	c				−63100		−42900
1127	1400	c				−62900		−41400
1227	1500	c				−62700		−39850
1327	1600	c				−62500		−38300
1377	1650	c				−62350		−37550
1377	1650	l				−54850		−37550

According to Elliott and Gleiser: Thermochemistry for Steelmaking, Vol. 1, Reading (Mass.)/London 1960.

Enthalpy of formation: G. L. Humphrey, E. G. King, and K. K. Kelley: Bur. Mines Rept. Investigations (1952) p. 4870.

Free enthalpy of the reaction: L. S. Darken and R. W. Gurry: J. Amer. Chem. Soc. 67 (1945) p. 1398. — R. Schenk, Th. Dingmann, P. H. Kirscht, and H. Wesselkock: Z. anorg. Chem. 182 (1929) p. 97. — N. A. Gokcen: J. Metals 8 (1956) p. 1558.

Iron(II, III)-oxide (Magnetite) Fe_3O_4

1. Phase changes of the metal (Fe)

Curie-point:	769°C	—	
Transition point ($\alpha \rightarrow \gamma$):	911°C	Heat of transition:	215 cal/mole
Transition point ($\gamma \rightarrow \delta$):	1392°C	Heat of transition:	270 cal/mole
Melting point:	1536°C	Heat of melting:	3700 cal/mole

2. Phase changes of the oxide (Fe_3O_4)

Curie-point:	627°C	—	
Melting point:	1597°C	Heat of melting:	3300 cal/mole

Molecular weight $M = 231.55$			Entropy $S^0_{25} = 35.0$ cal/°C mole			Enthalpy of formation ΔH^0_T, Entropy of formation ΔS^0_T, Free enthalpy ΔG^0_T of the reaction $3\ Fe_{(c)} + 2\ O_{2(g)} = Fe_3O_{4(c,l)}$		
T°C	T°K	State at T°C	$H^0_T - H^0_{25}$ cal/mole	$S^0_T - S^0_{25}$ cal/°C mole	$G^0_T - G^0_{25}$ cal/mole	ΔH^0_T cal/mole	ΔS^0_T cal/°C mole	ΔG^0_T cal/mole
25	298	c	0	0	0	−267000	−82.5	−242420
100	373	c	2830	8.1	− 2830	−266670	−81.7	−236210
200	473	c	6960	18.0	− 7670	−266000	−79.9	−228230
300	573	c	11490	26.7	−13440	−265100	−78.2	−220320
400	673	c	16510	34.7	−19960	−263980	−76.5	−212500
500	773	c	21800	42.2	−27410	−262690	−74.8	−204890
600	873	c	27230	48.8	−35460	−261660	−73.5	−197470
700	973	c	32750	54.7	−44140	−261300	−72.9	−190350
800	1073	c	38370	60.3	−53420	−261440	−73.0	−183140
Relative error			2%	2%	8%	3−10%	<3%	<3%

According to Vallet: Publ. de IRSID Serie B, No. 26, 1955.

827	1100	c				−260500		−181800
911	1184	c				−260500		−175700
911	1184	c				−261200		−175700
927	1200	c				−261100		−174600
1027	1300	c				−260600		−167400
1127	1400	c				−260100		−160300
1227	1500	c				−259600		−153200
1327	1600	c				−259100		−146100
1392	1665	c				−258800		−141500
1392	1665	c				−259600		−141500
1427	1700	c				−259600		−139000
1527	1800	c				−259500		−131900
1536	1809	c				−259500		−131300
1536	1809	c				−270600		−131300
1597	1870	c				−270600		−126600
1597	1870	*l*				−237600		−126600

According to Elliott and Gleiser: Thermochemistry for Steelmaking, Vol. 1, Reading (Mass.)/London 1960.

Source of data: J. P. Coughlin, E. G. King, and K. R. Bonnickson: J. Amer. Chem. Soc. 73 (1951) p. 3891.

Iron(III)-oxide (Hematite) Fe_2O_3

1. Phase changes of the metal (Fe)

Curie-point:	769°C	—	
Transition point ($\alpha \rightarrow \gamma$):	911°C	Heat of transition:	215 cal/mole
Transition point ($\gamma \rightarrow \delta$):	1392°C	Heat of transition:	270 cal/mole
Melting point:	1536°C	Heat of melting:	3700 cal/mole

2. Phase changes of the oxide (Fe_2O_3)

—	677°C	Heat:	160 cal/mole
Curie-point:	777°C	—	
Decomposition point:	1457°C	Heat of decomposition: unknown	

Molecular weight $M = 159.70$ Entropy $S^0_{25} = 21.5$ cal/°C mole

Enthalpy of formation ΔH^0_T
Entropy of formation ΔS^0_T
Free enthalpy ΔG^0_T of the reaction
$2\,Fe_{(c)} + \frac{3}{2}\,O_{2(g)} = Fe_2O_{3(c)}$

T°C	T°K	State at T°C	$H^0_T - H^0_{25}$ cal/mole	$S^0_T - S^0_{25}$ cal/°C mole	$G^0_T - G^0_{25}$ cal/mole	ΔH^0_T cal/mole	ΔS^0_T cal/°C mole	ΔG^0_T cal/mole
25	298	c	0	0	0	—196500	—65.0	—177140
100	373	c	1960	5.6	— 1750	—196280	—62.5	—171470
200	473	c	4840	12.5	— 4840	—195780	—63.3	—165850
300	573	c	8000	18.6	— 8570	—195210	—62.1	—159630
400	673	c	11430	24.1	—12860	—194520	—61.0	—153470
500	773	c	15060	29.2	—17690	—193690	—60.0	—147330
600	873	c	18770	33.7	—23090	—193020	—59.2	—141350
700	973	c	22510	37.7	—28720	—193860	—58.9	—135590
800	1073	c	26270	41.4	—34850	—193060	—59.0	—129740
Relative error			2%	2%	6%	<3%	<3%	<3%

According to VALLET: Publ. de IRSID Serie B, No. 26, 1955

T°C	T°K	State at T°C	$H^0_T - H^0_{25}$ cal/mole	$S^0_T - S^0_{25}$ cal/°C mole	$G^0_T - G^0_{25}$ cal/mole	ΔH^0_T cal/mole	ΔS^0_T cal/°C mole	ΔG^0_T cal/mole
827	1100	c				—192400		—128100
911	1184	c				—192300		—123200
911	1184	c				—192700		—123200
927	1200	c				—192700		—122200
1027	1300	c				—192300		—116400
1127	1400	c				—191900		—110600
1227	1500	c				—191500		—104800
1327	1600	c				—191100		— 99000
1392	1665	c				—190700		— 95300
1392	1665	c				—191300		— 95300
1427	1700	c				—191200		— 93300

According to ELLIOTT and GLEISER: Thermochemistry for Steelmaking, Vol. 1, Reading (Mass.)/London 1960.

Enthalpy of formation: L. S. DARKEN and R. W. GURRY: J. Amer. Chem. Soc. 68 (1946) p. 799. — W. A. ROTH and F. WIENERT: Arch. Eisenhüttenwes. 7 (1934) p. 460.

Free enthalpy of the reaction: J. SMILTENS: J. Amer. Chem. Soc. 79 (1957) p. 4887.

Cristobalite SiO_2

Transition point ($\alpha \rightarrow \beta$) 250°C Heat of transition: 200 cal/mole
Melting point: 1713°C Heat of melting: 3100 cal/mole

Molecular weight $M = 60.06$			Entropy $S^0_{25} = 10.19$ cal/°C mole			Enthalpy of formation ΔH^0_T Entropy of formation ΔS^0_T Free enthalpy ΔG^0_T of the reaction $Si_{(c)} + O_{2(g)} = SiO_{2(c)}$		
T°C	T°K	State at T°C	$H^0_T - H^0_{25}$ cal/mole	$S^0_T - S^0_{25}$ cal/°C mole	$G^0_T - G^0_{25}$ cal/mole	ΔH^0_T cal/mole	ΔS^0_T cal/°C mole	ΔG^0_T cal/mole
25	298	c	0	0	0	−205000	−43.3	−192090
100	373	c	890	2.6	− 830	−205020	−43.5	−188800
200	473	c	2150	5.7	− 2320	−205020	−43.3	−184540
250	523	c	2910	7.2	− 3130	−204920	−43.1	−182330
250	523	c	3110	7.5	− 3130	−204720	−42.7	−182330
300	573	c	3890	8.9	− 4030	−204600	−42.6	−180200
400	673	c	5430	11.8	− 6350	−204420	−41.9	−176240
500	773	c	7030	13.6	− 8340	−204210	−42.0	−171750
600	873	c	8650	15.6	−10830	−204010	−41.8	−167560
700	973	c	10290	17.4	−13500	−203810	−41.6	−163380
800	1073	c	11940	19.0	−16340	−203630	−41.3	−159280
900	1173	c	13610	20.5	−19350	−203440	−41.2	−155170
1000	1273	c	15330	21.9	−22480			
1100	1373	c	17050	23.2	−25760			
1200	1473	c	18770	24.4	−29170			
1300	1573	c	20520	25.6	−32680			
1400	1673	c	22270	26.6	−36310			
1500	1773	c	24050	27.7	−40040			
1600	1873	c	25840	28.7	−43860			
1700	1973	c	27630	29.6	−47820			
Relative error			0.5%	0.5%	1.5%	<3%	<3%	<3%

According to VALLET: Publ. de IRSID Serie B, No. 26, 1955.

Free enthalpy ΔG^0_T of the reaction $Si_{(c)} + O_{2(g)} = SiO_{2(c)}$:

$$\Delta G^0_T = -(222800 - 47.62\,T) \pm 2000 \text{ cal}, \quad 1300 \text{ to } 1600°\text{C}$$

(H. F. RAMSTAD and F. D. RICHARDSON: Trans. Metallurg. Soc. AIME 221 (1961) p. 1021–1028);

$$\Delta G^0_T = -(226500 - 47.50\,T)$$

(J. CHIPMAN and R. BASCHITZ: Trans. Metallurg. Soc. AIME 227 (1963) p. 473);

$$\Delta G^0_T = -(212100 - 40.48\,T) \pm 2900 \text{ cal}, \quad 1227 \text{ to } 1412°\text{C}$$

$$\Delta G^0_T = -(224200 - 47.66\,T) \pm 2900 \text{ cal}, \quad 1412 \text{ to } 1447°\text{C}$$

(M. F. ANCEY-MORET, M. OLETTE, and F. D. RICHARDSON: Mém. Sci. Rev. Métallurg. 61 (1964) p. 169–176);

$$\Delta G^0_T = -(221000 - 46.2\,T) \pm 500 \text{ cal}, \quad 1450 \text{ to } 1620°\text{C}$$

(K. SCHWERDTFEGER and H.-J. ENGELL: Trans. Metallurg. Soc. AIME 233 (1965) p. 1327–1332).

Fayalite Fe_2SiO_4

Melting point: 1217°C Heat of melting: 22030 cal/mole

Molecular weight $M = 203.76$

Entropy $S^0_{25} = 35.4$ cal/°C mole

Enthalpy of formation ΔH^0_T
Entropy of formation ΔS^0_T
Free enthalpy ΔG^0_T of the reaction
$2\,FeO_{(c)} + SiO_{2(c)} = Fe_2SiO_{4(c,l)}$

T°C	T°K	State at T°C	$H^0_T - H^0_{25}$ cal/mole	$S^0_T - S^0_{25}$ cal/°C mole	$G^0_T - G^0_{25}$ cal/mole	ΔH^0_T cal/mole	ΔS^0_T cal/°C mole	ΔG^0_T cal/mole
25	298	c	0	0	0	−10900	−1.4	−10480
100	373	c	2600	7.5	− 2800	−11000	−1.9	−10290
200	473	c	6350	16.5	− 7510	−11160	−2.2	−10100
300	573	c	10400	24.2	−13030	−11180	−2.3	− 9870
400	673	c	14700	31.2	−19580	−11140	−2.1	− 9610
500	773	c	19200	37.4	−26540	−10940	−2.0	− 9400
600	873	c	23700	42.9	−33710	−11560	−2.2	− 9600
700	973	c	28200	47.8	−41770	−11020	−2.1	− 8980
800	1073	c	32850	52.3	−50130	−10920	−2.0	− 8750
900	1173	c	37550	56.6	−59180	−10740	−1.7	− 8720*
1000	1273	c	42400	60.5	−68490			
1100	1373	c	47300	64.2	−78190			
Relative error			2%	2%	4.5%	<3%	>10%	3−10%

According to Vallet: Publ. de IRSID Serie B, No. 26, 1955.

T°C	T°K	State at T°C	$H^0_T - H^0_{25}$ cal/mole	$S^0_T - S^0_{25}$ cal/°C mole	$G^0_T - G^0_{25}$ cal/mole	ΔH^0_T cal/mole	ΔS^0_T cal/°C mole	ΔG^0_T cal/mole
1127	1400	c				− 8625		−1610
1217	1490	c				− 8400		−1185
1217	1490	l				+13630		−1185
1227	1500	l				+13740		−1275
1327	1600	l				+14690		−2270
1377	1650	l				+15140		−2745
1427	1700	l				+ 458		−2925
1527	1800	l				+ 1065		−3110

According to Elliott, Gleiser, and Ramakrishna: Thermochemistry for Steelmaking, Vol. 2, Reading (Mass.)/London 1963.

Enthalpy and entropy of formation: E. G. King: J. Amer. Chem. Soc. 74 (1952) p. 4446.

High-temperature enthalpy, entropy increment, and heat of melting: R. L. Orr: J. Amer. Chem. Soc. 75 (1953) p. 528.

* −6000 cal/mole according to the decomposition curves of R. Schenck et al., cf. Stahl u. Eisen 52 (1932) p. 731–732.

Magnesium Ferrite $MgFe_2O_4$

Transition point ($\beta \to \gamma$): 957°C   Heat of transition: 350 cal/mole
Melting point: ≈1720°C

Molecular weight $M = 200.02$			Enthalpy of formation ΔH_T^0 Free enthalpy ΔG_T^0 of the reaction $MgO_{(c)} + Fe_2O_{3(c)} \to MgFe_2O_{4(c)}$	
T°C	T°K	State at T°C	ΔH_T^0 cal/mole	ΔG_T^0 cal/mole
25.15	298.15	c	−2500	−2575
927	1200	c	−3530	−3030
957	1230	c	−3562	−3055
957	1230	c	−3212	−3055
1027	1300	c	−3430	−3000
1127	1400	c	−3720	−2950
1227	1500	c	−3880	−2870
1327	1600	c	−3940	−2820
1427	1700	c	−3950	−2760
1527	1800	c	−4030	−2680

According to Elliott, Gleiser, and Ramakrishna: Thermochemistry for Steelmaking, Vol. 2, Reading (Mass.)/London 1963.

Enthalpy of formation: B. E. Levin: Izv. Akad. Nauk USSR, Ser. Fiz. 18 (1954) p. 519.

Entropy of formation: E. G. King: J. Amer. Chem. Soc. 76 (1954) p. 5849.

High-temperature enthalpy, entropy increment, and heat of transition: K. R. Bonnickson: J. Amer. Chem. Soc. 76 (1954) p. 1480.

Enthalpy of iron(III)-oxide: J. P. Coughlin: J. Amer. Chem. Soc. 73 (1951) p. 3891.

Amphibole $MgSiO_3$

Melting point: 1525°C Heat of melting: 14700 cal/mole

Molecular weight $M = 100.38$			Entropy $S^0_{25} = 16.2$ cal/°C mole			Enthalpy of formation ΔH^0_T Entropy of formation ΔS^0_T Free enthalpy ΔG^0_T of the reaction $MgO_{(c)} + SiO_{2(c)} = MgSiO_{3(c)}$		
T°C	T°K	State at T°C	$H^0_T - H^0_{25}$ cal/mole	$S^0_T - S^0_{25}$ cal/°C mole	$G^0_T - G^0_{25}$ cal/mole	ΔH^0_T cal/mole	ΔS^0_T cal/°C mole	ΔG^0_T cal/mole
25	298	c	0	0	0	−8700	−0.2	−8640
100	373	c	1610	4.6	− 1330	−8680	−0.2	−8620
200	473	c	3930	10.2	− 3720	−8620	−0.1	−8570
300	573	c	6430	15.0	− 6610	−8650	−0.1	−8610
400	673	c	9090	19.3	− 9960	−8620	0.0	−8600
500	773	c	11860	23.1	−13720	−8620	0.0	−8590
600	873	c	14710	26.6	−17830	−8940	0.0	−8580
700	973	c	17600	29.8	−22280	−8840	0.0	−8550
800	1073	c	20550	32.6	−27030	−8740	0.0	−8520
900	1173	c	23550	35.3	−32040	−8660	0.0	−8500
1000	1273	c	26590	37.8	−37320	−8560	0.0	−8500
1100	1373	c	29680	40.1	−42850	−8430	0.0	−8490
Relative error			0.5%	0.5%	1.5%	<3%	>10%	<3%

According to Vallet: Publ. de IRSID, Serie B, No. 26, 1955

1127	1400	c				−8860		−7610
1227	1500	c				−8820		−7530
1327	1600	c				−8800		−7440
1427	1700	c				−8790		−7310
1525	1798	c				−8755		−7120
1525	1798	*l*				+5945		−7120

According to Elliott, Gleiser, and Ramakrishna: Thermochemistry for Steelmaking, Vol. 2, Reading (Mass.)/London 1963.

Enthalpy of formation: D. R. Torgeson and T. G. Sahama: J. Amer. Chem. Soc. 70 (1948) p. 2156.

Entropy of formation: K. K. Kelley: U.S. Bur. Mines Bull. (1950) p. 477.

High-temperature enthalpy and entropy increment: K. K. Kelley: U.S. Bur. Mines Bull. (1949) p. 476.

Heat of melting: U.S. Bur. Standards Circular No. 500 (1952).

Forsterite Mg_2SiO_4

Melting point: 1890°C

Molecular weight $M = 140.73$			Enthalpy of formation ΔH_T^0 Free enthalpy ΔG_T^0 of the reaction $2\,MgO_{(c)} + SiO_{2(c)} \rightarrow Mg_2SiO_{4(c)}$	
T°C	T°K	State at T°C	ΔH_T^0 cal/mole	ΔG_T^0 cal/mole
25.15	298.15	c	−15120 (±500)	−15020
927	1200	c	−14670	−14680
1027	1300	c	−14530	−14710
1127	1400	c	−14320	−14610
1227	1500	c	−14070	−14750
1327	1600	c	−13820	−14810
1427	1700	c	−13570	−14910
1527	1800	c	−13330	−14990
1727	2000	c	−12880	−15020

According to Elliott, Gleiser, and Ramakrishna: Thermochemistry for Steelmaking, Vol. 2, Reading (Mass.)/London 1963.

Enthalpy of formation: D. R. Torgeson and T. G. Sahama: J. Amer. Chem. Soc. 70 (1948) p. 2156.

Entropy of formation: K. K. Kelley: U.S. Bur. Mines Bull. (1950) p. 477.

High-temperature enthalpy and entropy increment: R. L. Orr: J. Amer. Chem. Soc. 75 (1953) p. 528.

Diopside $CaMg(SiO_3)_2$

Melting point: 1391.5°C

Molecular weight $M = 216.58$			Enthalpy of formation ΔH_T^0 Free enthalpy ΔG_T^0 of the reaction $CaO_{(c)} + MgO_{(c)} + 2\,SiO_{2(c)} \rightarrow CaMg(SiO_3)_2$	
T°C	T°K	State at T°C	ΔH_T^0 cal/mole	ΔG_T^0 cal/mole
25.15	298.15	c	−35250 (±220)	−35820
927	1200	c	−35010	−36490
1027	1300	c	−34950	−36630
1127	1400	c	−34830	−36760
1227	1500	c	−34660	−36895
1327	1600	c	−34440	−37050

According to Elliott, Gleiser, and Ramakrishna: Thermochemistry for Steelmaking, Vol. 2, Reading (Mass.)/London 1963.

Enthalpy of formation: K. J. Neuvonen: Bull. Conim. Geol. Finlande 158 (1952) p. 50.

Entropy of formation: E. G. King: J. Amer. Chem. Soc. 79 (1957) p. 5437.

High-temperature entropy and enthalpy increment: K. K. Kelley: U.S. Bur. Mines Bull. (1949) p. 476.

Melting point: U.S. Bur. Standards Circular No. 500 (1952).

1. Fundamentals

1.1. Equilibria

1.1.1. The Iron-Oxygen System

A general view of the solid, thermodynamically stable phases which occur between 400 and 1400°C in the iron-oxygen system is given in the binary diagram Fig. 1, based on the data of DARKEN and GURRY[1], PHILLIPS and MUAN[2], and SALMON[3]. The curves correspond closely to

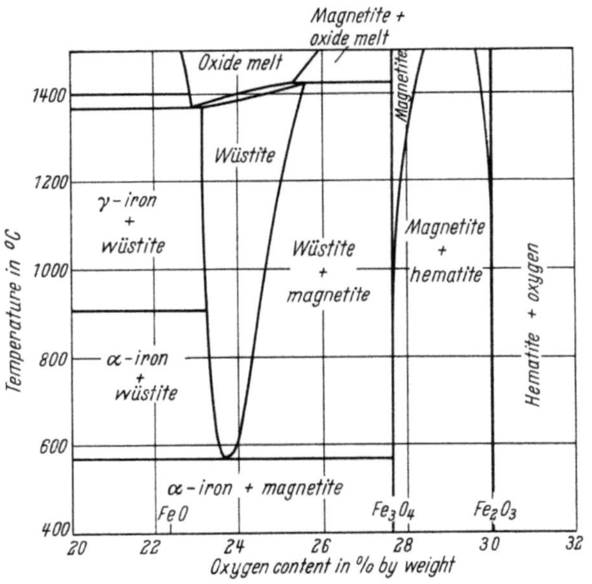

Fig. 1. Iron-oxygen system (section)

those given by HANSEN[4] and by ELLIOTT, GLEISER and RAMAKRISHNA[5]. The solubility of oxygen in solid α- and γ-iron is extremely small and is disregarded in the diagram. As determined by SIFFERLEN[6] the solubility is less than 0.01% O_2 by weight. The oxygen content has therefore no effect on the transition temperatures of the solid iron modifications. Molten iron has a much greater ability to dissolve oxygen. Figure 2 shows the saturation values of the solubility of oxygen in pure iron for the region

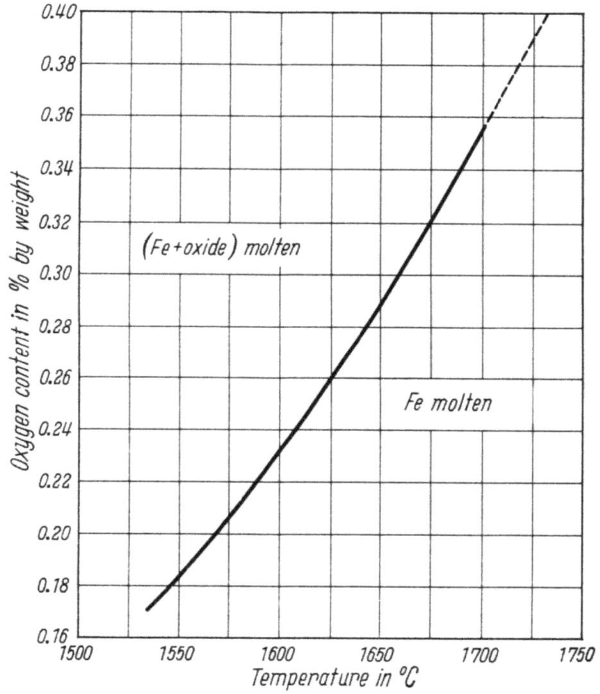

Fig. 2. Solubility of oxygen in molten iron in equilibrium with a pure ferrous oxide slag, according to ELLIOTT[5])

from 1535°C, the melting point of pure iron, to about 1700°C, according to the data of ELLIOTT et al.[5]).

Iron forms with oxygen the three stable, solid compounds $Fe_{1-y}O$, Fe_3O_4 and Fe_2O_3. To distinguish it from the cubic γ-Fe_2O_3 modification, rhombohedric* hematite is also called α-Fe_2O_3.

To describe the lattice structure of the oxides, the deviations from the ideal structure have to be considered.

In this book a notation suggested by SCHOTTKY is used to designate *point defects*; viz., a symbol between two vertical lines denotes the lattice position which the ion written before this symbol occupies: e.g.,

* Of Al_2O_3 type.

1. $Fe^{3+}|Fe^{2+}|^{\bullet}$ = a trivalent iron ion in the lattice position of a divalent iron ion;
2. $Mn^{2+}|Fe^{2+}|$ = a divalent manganese ion in the lattice position of a divalent iron ion;
3. $Fe^{2+}|Fe^{2+}|$ = a divalent iron ion in the normal lattice position;
4. $Mn^{3+}|Mn^{4+}|'$ = a trivalent manganese ion in the lattice position of a tetravalent manganese ion;
5. $Mn^{2+}|Mn^{3+}|'$ = a divalent manganese ion in the lattice position of a trivalent manganese ion.

The charges of the lattice ions are denoted by $^+$ and $^-$, the charges of the point defects relative to the undisturbed lattice (when the points are occupied as they should ideally be) are denoted by $^{\bullet}$ and $'$.

Correspondingly we use the following notation:

6. $|Fe^{2+}|''$ = an iron ion vacancy (an unoccupied iron ion lattice position);
7. $Fe^{\bullet\bullet}$ = an iron ion in interstitial position;
8. e' = a free electron (conduction electron);
9. $|e|^{\bullet}$ = an electron defect point (defect electron).

In wüstite, with which we shall be very much concerned in this book, the defects 1 and 9 are identical, but the notation 1 stresses the material form, while 9 stresses the state of the electronic charge.

Associated with the above, we also use the notation

10. $FeO \lrcorner$ = an oxide molecule in the half-crystal position (Fig. 3).

Wüstite $Fe_{1-y}O$ may contain between 23.1 and 25.6% O; it does not therefore attain the stoichiometric composition of Fe/O = 1, which is about 22.3%. Wüstite crystallizes in the cubic NaCl lattice. The oxygen sub-lattice is largely fully occupied, while, on the other hand, a number

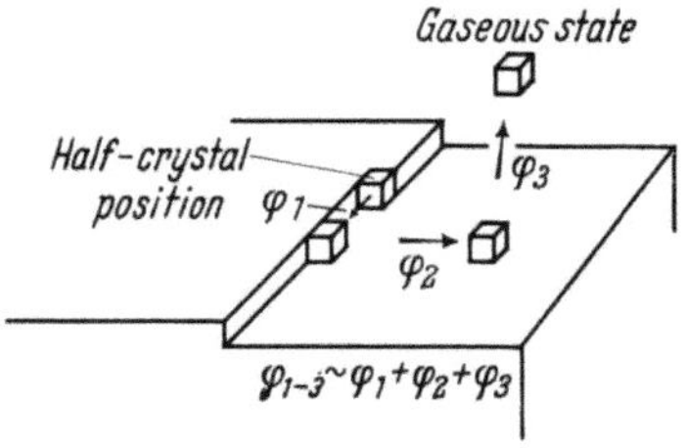

Fig. 3. Diagram of a crystal surface, according to STRANSKI[6a])
φ = total internal work function

of lattice points in the iron sub-lattice are vacant[7]). According to ROTH[8]) the wüstite lattice shows certain deviations from the NaCl type. ROTH's experiments on neutron diffraction indicate that in some regions of the lattice there is superimposed a structure of iron ion vacancies and iron ions in inter-lattice positions, giving a spinel-type structure.

The *oxygen excess* of wüstite is to be construed, from the lattice structure, as an *iron deficiency*. In the formula for wüstite $Fe_{1-y}O$ the letter y denotes the proportion of the vacant iron ion lattice positions relative to the available number of iron lattice-points:

$$y = n(|Fe^{2+}|'')/\{n(|Fe^{2+}|'') + n(Fe^{2+}|Fe^{2+}|) + n(Fe^{3+}|Fe^{2+}|^{\bullet})\}$$
$$= N(|Fe^{2+}|''). \quad (1)$$

$N(|Fe^{2+}|'')$ is the mole fraction of the iron ion vacancies. Because of the iron deficiency an appropriate proportion of trivalent iron ions must be present in the lattice, since otherwise the condition for electron neutrality

$$\sum z_i N_i = 0, \quad (2)$$

where z_i is the charge number for ions of type i, and $i = Fe^{2+}, Fe^{3+}, O^{2-}$, would not be satisfied. (2) implies that the mole fraction $N(Fe^{3+}|Fe^{2+}|^{\bullet})$ of the iron ions which carry this excess positive charge is given by

$$N(Fe^{3+}|Fe^{2+}|^{\bullet}) = N(|e|^{\bullet}) \quad (3)$$
$$= n(Fe^{3+}|Fe^{2+}|^{\bullet})/\{n(Fe^{3+}|Fe^{2+}|^{\bullet}) + n(Fe^{2+}|Fe^{2+}|) + n(|Fe^{2+}|'')\}.$$

The positive excess charges of the metal ions can migrate through the lattice by interchange of charge between the divalent and trivalent iron ions. They are called *positive holes* and are the charge carriers of the electrical conductivity of wüstite. The iron ion vacancies, on the other hand, enable the metal ions to move through the lattice: an iron ion can jump to a neighbouring empty lattice position and thus advance by one lattice interval. In this process the iron ion vacancy has migrated in the opposite direction through one lattice interval. Figure 4 illustrates this process.

The migration of a vacancy by such interchanges of position with neighbouring iron ions across a wüstite crystal corresponds to a migration of an iron ion in the opposite direction through the crystal. If j denotes the flux of material, measured, e.g., in mole $cm^{-2}\,sec^{-1}$, then

$$j(|Fe^{2+}|'') = -j(Fe). \quad (4)$$

The self-diffusion coefficient $D^*(Fe)$ of the iron in wüstite and the diffusion coefficient $D(|Fe^{2+}|'')$ of the vacancies are related by the equation

$$y\,D(|Fe^{2+}|'') = (1 - y)\,D^*(Fe) \quad (5)$$

given by WAGNER[9]). From measurements of the self-diffusion coefficient of iron in wüstite carried out by HIMMEL, MEHL, and BIRCHENALL[10]), it can with (5) be calculated that the diffusion coefficient of the vacancies is largely independent of the concentration y of the vacancies and can be expressed by

$$\left.\begin{aligned} D(|Fe^{2+}|'') &= D^0(|Fe^{2+}|'')\exp(-\Delta H^*/RT), \\ D^0(|Fe^{2+}|'') &= 4.5 \cdot 10^{-2}\ cm^2\,sec^{-1}, \\ \Delta H^* &= 27\,800\ cal\ mole^{-1}. \end{aligned}\right\} \quad (6)$$

Measurements of the transport number of iron ions in wüstite[12]) and measurements of the oxidation of iron[11]) yield negligibly different values of the diffusion coefficients.

Within its domain of homogeneity, the composition of wüstite varies with the activity of the iron or with the partial pressure of the oxygen in the gas phase, with which the oxide is in equilibrium. The iron deficiency

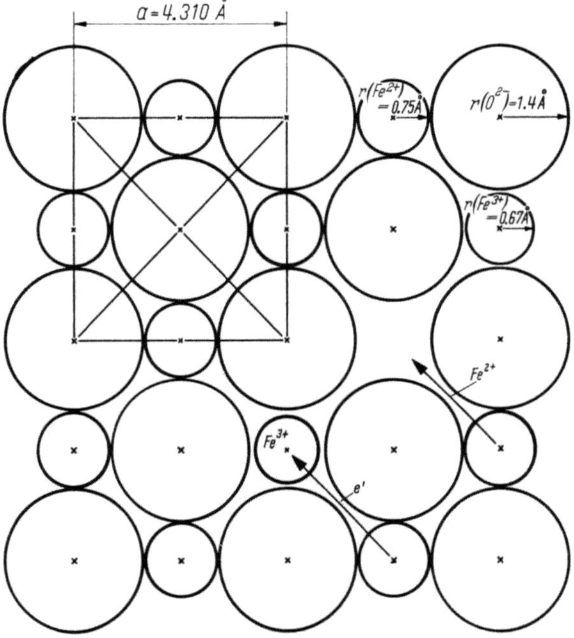

Fig. 4. (100)-plane of wüstite with an iron ion vacancy and two trivalent iron ions

i. e. the mole fraction of iron ion vacancies y may conveniently be taken as a concentration measure to describe the composition of the wüstite. The relation between this number y and the partial pressure $p(O_2)$ of the oxygen can be calculated by means of the *thermodynamics of ordered mixed phases* developed by WAGNER and SCHOTTKY[13]) and WAGNER[14]), if a suitable model is set up for the deviations of the actual crystal from the perfectly ordered lattice structure. In agreement with the fact[7]) that the iron deficiency produces a particular concentration in respect of vacant iron ion lattice positions, it may be assumed that in a hypothetical, stoichiometric wüstite a certain number of iron ions quit their lattice positions and are built in at positions which would remain vacant in the ideal wüstite lattice. Let us denote these interstitial ions by $Fe^{\bullet\bullet}$. The formation of interstitial ions takes place according to the reaction

$$Fe^{2+}\,|Fe^{2+}| \rightleftharpoons Fe^{\bullet\bullet} + |Fe^{2+}|''. \tag{6a}$$

In stoichiometric wüstite ($y = 0$) the mole fractions of the interstitial ions $Fe^{\bullet\bullet}$ and the vacancies $|Fe^{2+}|''$ must be equal:

$$N(Fe^{\bullet\bullet}) = N(|Fe^{2+}|'') \equiv \alpha \quad \text{for} \quad y = 0. \tag{7}$$

α is called the *degree of ion disorder*. Electron disorder may occur as well as ion disorder. This may be conceived, in chemical terms, as follows: an electron is separated from a divalent iron ion, for example, and moves freely through the lattice, and may therefore be called a conduction electron. If we denote the conduction electrons by the symbol e', then the relation

$$Fe^{2+}\,|Fe^{2+}| \rightleftharpoons Fe^{3+}\,|Fe^{2+}|^{\bullet} + e',$$

or with the previously used notation for positive holes $|e|^{\bullet}$ for the third positive charge of the iron ions:

$$Fe^{2+}\,|Fe^{2+}| \rightleftharpoons e' + |e|^{\bullet}, \tag{8}$$

holds for this process. Similar to (7) we also have

$$N(e') = N(|e|^{\bullet}) \equiv \beta \quad \text{for} \quad y = 0. \tag{9}$$

If we now regard an actual crystal as an ideal solution of vacancies, interstitial ions, conduction electrons and positive holes in a hypothetical, well-ordered FeO-lattice as the solvent, then the law of mass-action can be applied to (6a) and (8). If the concentration of the lattice-position ions is taken to be constant, it follows that

$$\begin{aligned} N(Fe^{\bullet\bullet})\, N(|Fe^{2+}|'') &= \alpha^2, \\ N(e')\, N(|e|^{\bullet}) &= \beta^2. \end{aligned} \tag{10}$$

The oxide comes into equilibrium with the gas phase according to the reactions

$$Fe^{2+}\,|Fe^{2+}| + \tfrac{1}{2}\,O_2 \rightleftharpoons |Fe^{2+}|'' + 2\,|e|^{\bullet} + FeO \tag{11}$$

and

$$Fe^{\bullet\bullet} + 2e' + \tfrac{1}{2}\,O_2 \rightleftharpoons FeO. \tag{12}$$

Since by (10) the concentrations of electrons and positive holes and also of interstitial ions and vacancies are related, (11) suffices to describe this equilibrium. If the concentration of iron ions in the normal lattice positions $Fe^{2+}\,|Fe^{2+}|$ is again assumed constant, the law of mass-action applied to (11) gives

$$K_y = p^{1/2}(O_2)/N(|Fe^{2+}|'')\, N^2(|e|^{\bullet}). \tag{13}$$

For $y \neq 0$ the general relation (2) for electron neutrality is to be used instead of (7) and (9). In the present case it is useful and customary to sum only over the ionic and electronic point defects, leaving out the lattice elements in the normal lattice positions, since their state of charge has been adopted as the zero-charge datum. A defect of a divalent positive iron ion, i.e., an iron ion vacancy, is thus regarded as two units negative, a trivalent iron ion, i.e., a positive hole, is counted as one unit positive, and so on. Hence we have

$$N(|e|^\bullet) + 2N(\mathrm{Fe}^{\bullet\bullet}) - N(e') - 2N(|\mathrm{Fe}^{2+}|'') = 0. \quad (14)$$

From (10), (13), (14) and

$$y = N(|\mathrm{Fe}^{2+}|'') - N(\mathrm{Fe}^{\bullet\bullet}) \quad (15)$$

it follows that

$$p^{1/2}(\mathrm{O}_2) = K_y\left(y + \sqrt{\beta^2 + y^2}\right)^2 \left(y/2 + \sqrt{\alpha^2 + y^2/4}\right). \quad (16)$$

In the region of larger deviations from the stoichiometric composition, we may take $y \gg \beta$ and $y \gg \alpha$. The concentration of conduction electrons may then be neglected compared with that of the positive holes, and so

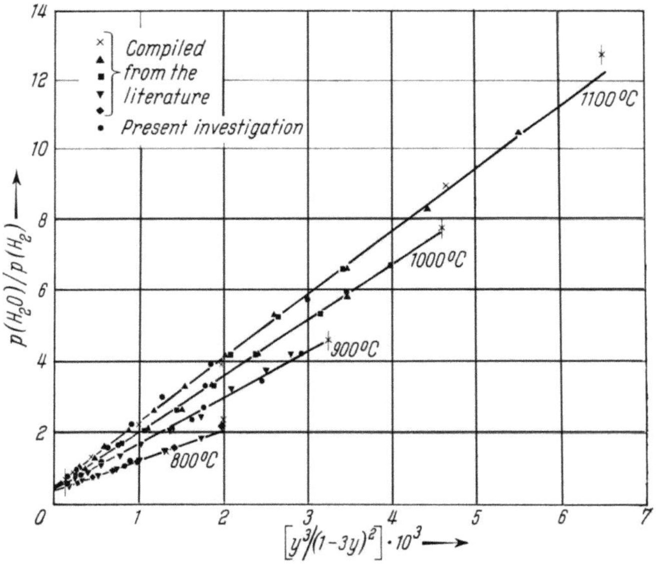

Fig. 5. Dependence of the vacancy concentration y on the oxygen pressure $p_0 \sim p(\mathrm{H_2O})/p(\mathrm{H_2})$ according to formula (18), for wüstite $\mathrm{Fe}_{1-y}\mathrm{O}$, according to the measurements and the compilation of BOHNENKAMP and RIECKE[18a])

can that of the interstitial ions compared with that of the vacancies. On these assumptions (16) can be approximated by

$$y^3 \approx p^{1/2}(\mathrm{O}_2)/4K_y. \quad (17)$$

With higher concentrations of vacancies, the reduction of the $\mathrm{Fe}^{2+}|\mathrm{Fe}^{2+}|$ by the formation of vacancies and positive holes has to be taken into account. Writing (11) in the form

$$3\,Fe^{2+}\,|Fe^{2+}| + \tfrac{1}{2}\,O_2 \rightleftharpoons |Fe^{2+}|'' + 2\,Fe^{3+}\,|Fe^{2+}|^{\bullet} + FeO,$$

$$FeO = Fe^{2+}\,|Fe^{2+}| + O^{2-}\,|O^{2-}|,$$

and taking into account that

$$N(Fe^{2+}\,|Fe^{2+}|) = 1 - 3y$$

must hold, we get by applying the law of mass-action

$$y^3/(1 - 3y)^2 = p^{1/2}(O_2)/4K_y'. \qquad (18)$$

This expression can also be derived statistically[15]).

Table 1 gives the values of K_y and K_y' calculated from the formulae (17) and (18), together with the corresponding references. It can be seen that both constants are weakly dependent on the vacancy concentration.

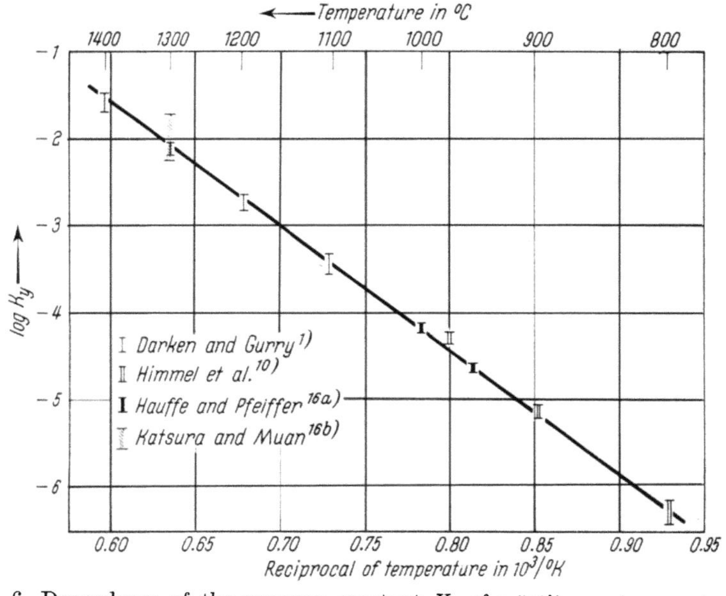

Fig. 6. Dependence of the vacancy constant K_y of wüstite on temperature

Formula (18) describes the relation between y and $p^{1/2}(O_2)$ or between y and $p(H_2O)/p(H_2) = K\,p^{1/2}(O_2)$ quite satisfactorily, as may be seen from Fig. 5, which gives the results of new measurements and of a literature review, both by Bohnenkamp and Riecke[18a]). Figure 6 shows the graph of $\log K_y$ against the reciprocal of the absolute temperature, and in this diagram the length of the vertical strokes takes into account the variation of K_y with y. A closer agreement between the theory and the observed values throughout the whole range of y-values is obtained on the hypothesis that vacancies never occur at neighbouring lattice-points[16]).

Table 1. *Values of the constants for the vacancies balance in wüstite calculated from formulae* (17) *and* (18)

Author	T (°C)	y	lg K_y	lg K'_y
KATSURA and MUAN[16b])	1300	0.037	−1.74	−1.84
		0.069	−2.15	−2.35
		0.097	−2.21	−2.51
		0.121	−2.16	−2.55
		0.143	−2.07	−2.556
		0.162	−1.96	−2.54
DARKEN and GURRY[1])	1100	0.0483	−3.30	−3.43
		0.0810	−3.55	−3.79
		0.1190	−3.47	−3.85
	1200	0.0526	−2.68	−2.83
		0.0817	−2.83	−3.07
		0.1210	−2.76	−3.15
		0.1395	−2.66	−3.13
	1300	0.0558	−2.11	−2.28
		0.0837	−2.22	−2.47
		0.1232	−2.15	−2.55
		0.1420	−2.04	−2.52
	1400	0.0579	−1.60	−1.77
		0.0867	−1.71	−1.97
		0.1256	−1.61	−2.02
		0.1443	−1.50	−1.99
HIMMEL, MEHL, and BIRCHENALL[10])	800	0.057	−6.17	−6.33
		0.065	−6.26	−6.45
		0.074	−6.32	−6.61
		0.082	−6.36	−6.61
		0.091	−6.37	−6.65
		0.099	−6.35	−6.66
	897	0.057	−5.06	−5.22
		0.065	−5.12	−5.31
		0.074	−5.16	−5.38
		0.082	−5.18	−5.43
		0.091	−5.21	−5.49
		0.099	−5.20	−5.51
		0.107	−5.20	−5.53
	983	0.057	−4.21	−4.37
		0.065	−4.24	−4.43
		0.074	−4.27	−4.49
		0.082	−4.29	−4.54
		0.091	−4.32	−4.60
		0.099	−4.32	−4.63
		0.107	−4.32	−4.65
		0.1145	−4.32	−4.69
		0.1205	−4.33	−4.72

Table 1 (continuation)

Author	T (°C)	y	$\lg K_y$	$\lg K'_y$
HAUFFE and PFEIFFER[16a])	950	0.072	−4.63	−4.74
		0.080	−4.60	−4.84
		0.093	−4.59	−4.88
		0.111	−4.59	−4.91
	1000	0.074	−4.21	−4.50
		0.084	−4.20	−4.45
		0.097	−4.19	−4.49
		0.0115	−4.18	−4.55

The deviation seen in Fig. 5 between the observed values and the derived formula (18) is due to the fact that, when the content of vacancies is small, the assumption that y is large compared with the degrees of disorder, α and β, is no longer justified. Since α and β cannot be determined for wüstite, because to do this measurements of the relation between y and $p(O_2)$ as $y \to 0$ would be required, a representation by the more exact formula (16) is not possible. As a proof of how well this formula describes the equilibrium of an oxide with a gas atmosphere when there is a deficiency of the metal, in Fig. 7 a curve calculated from (16) taking $\alpha = 1 \cdot 10^{-4}$, $\beta = 1 \cdot 10^{-2}$, and $K_y = 250$ is juxtaposed with the measured values obtained by DAVIES and RICHARDSON[16c]) for MnO at 1650°C.

That the model developed above is applicable to the disorder equilibrium of MnO is confirmed by more recent measurements of the oxidation of Mn in CO/CO_2 atmospheres[16d]). From the measured rate constants these authors determine the self-diffusion coefficients of manganese in MnO.

Figure 8 shows the wüstite *phase domain*. The phase boundaries are drawn to correspond with the results obtained by an electrochemical method[17]) and with the gravimetric and analytic determinations of the equilibria by MARION[18]), CIRILLI[19]), FOSTER and WELCH[20]), and DARKEN and GURRY[1]). In the diagram are plotted the curves of equal oxygen pressure, calculated by means of the formula

$$\log K_y = 4.6 - 8750/T\,(°\mathrm{K}) \tag{19}$$

which is derived from Fig. 6.

According to GEIGER, LEVIN and WAGNER[20a]) at temperatures $\leqq 850$°C the relation

$$y \sim p^{1/3.6}\,(O_2)$$

holds. BRANSKY and TANNHAUSER[20b]) obtain the same result. Both authors use the electrical conductivity of the wüstite specimen as the indicator of the concentration of positive holes. A reduced dependence of the conductivity on the oxygen pressure in the neighbourhood of the phase

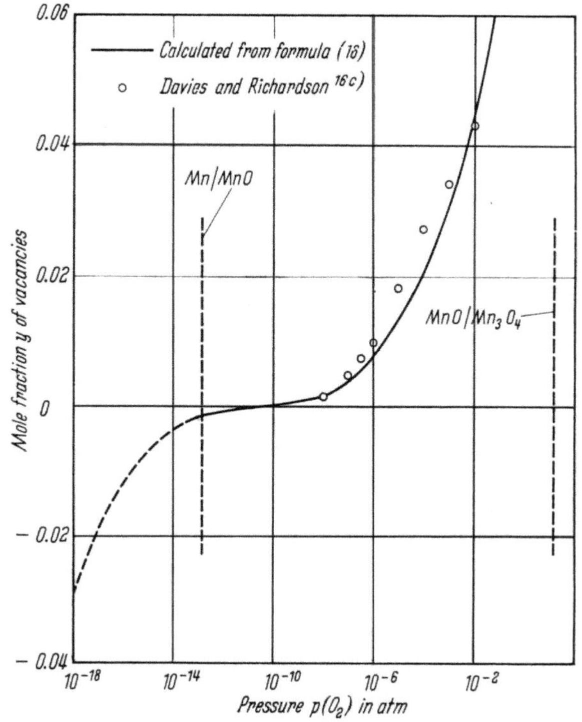

Fig. 7. Relation between y and $p(O_2)$ for $Mn_{1-y}O$ at 1650°C

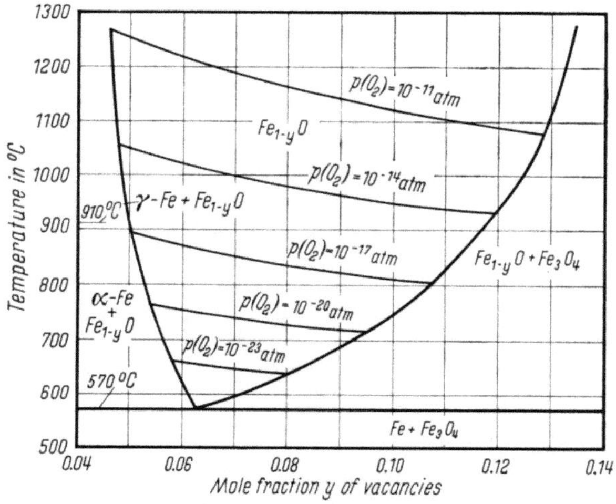

Fig. 8. The wüstite phase domain in the iron-oxygen system with the oxygen isobars

boundary Fe/FeO is in both cases interpreted as due to an arrangement of the vacancies. In the model we have just developed this effect is included for $y \leqq \beta$ without the need for any additional hypothesis.

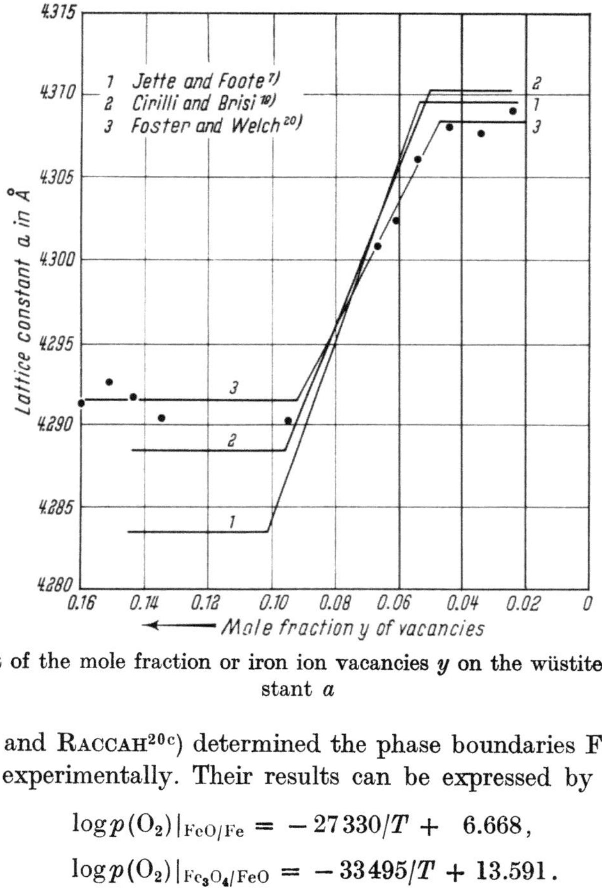

Fig. 9. Effect of the mole fraction or iron ion vacancies y on the wüstite lattice constant a

VALLET and RACCAH[20c] determined the phase boundaries FeO/Fe and Fe_3O_4/FeO experimentally. Their results can be expressed by

$$\text{a)} \qquad \log p(O_2)|_{FeO/Fe} = -27330/T + 6.668,$$

$$\text{b)} \qquad \log p(O_2)|_{Fe_3O_4/FeO} = -33495/T + 13.591.$$

In the opinion of these authors the relation between the oxygen pressure and the disorder cannot be regarded as continuous over the whole temperature range.

The lattice constant a of wüstite varies approximately linearly with the vacancy content. Figure 9 shows the variation of a with y according to a compilation by FOSTER and WELCH[20]). The density of wüstite decreases with increasing vacancy content from 5.728 to 5.613 g/cm³ [7]). At 570°C the equilibrium curves of $Fe/Fe_{1-y}O$ and $Fe_{1-y}O/Fe_3O_4$ intersect. Below this temperature, therefore, wüstite must decompose into iron and magnetite. The decomposition of wüstite has been particularly investigated by FISCHER et al.[21]) and more recently by ILSCHNER and MLITZKE[21a]). According to these authors three stages can be distinguished:

I. single-stage decomposition,

II. separation of Fe_3O_4 leaving behind a metastable matrix,

III. simultaneous, quasi-eutectoid formation of Fe and Fe_3O_4.

The changes taking place during stage II are determined not by diffusion in the matrix but by jumps of ions at the coherent phase boundary

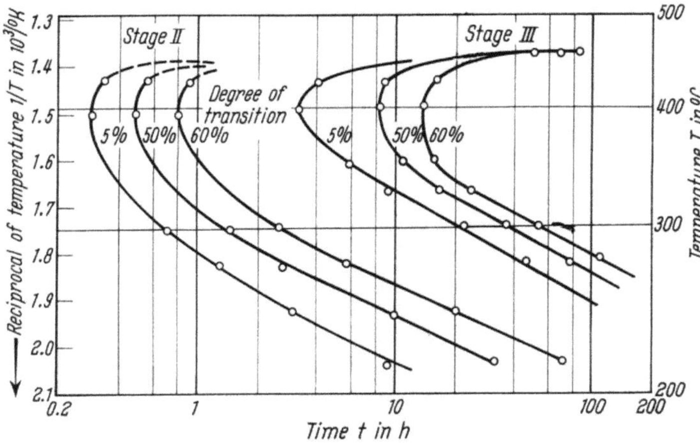

Fig. 10. The transition kinetics of wüstite in stages II and III, represented as a *TTT*-diagram

Fe_3O_4/FeO. For stage III a similar process at the phase boundary is taken to be the factor determining the rate. Figure 10 shows the results of the investigation of the course of the reaction.

Magnetite Fe_3O_4 crystallizes in the cubic spinel lattice. Since several of the oxide compounds which occur in ores and sinters belong to this class of compounds, we shall discuss the spinels in rather greater detail.

All the *spinels* have this in common, that in the lattice the oxygen forms a cubic, close-packed structure. The unit cell of a spinel contains 32 oxygen ions; these enclose 64 tetrahedral and 32 octahedral lattice vacancies, which can be occupied by the metal ions. According to the valency of the metal ions contained in the spinel, we distinguish the 2-3-spinels, of the formula $Me^{2+}Me_2^{3+}O_4$, and the 2-4-spinels, of the formula $Me_2^{2+}Me^{4+}O_4$. In both classes, 24 metal ions belong to the 32 oxygen ions of the unit cell. In the limiting case of an ideally ordered lattice they occupy 8 of the 64 tetrahedral vacancies and 16 of the 32 octahedral vacancies. The occupation of the vacancies is not random, but determined by a fixed law.

As regards the occupation of the octahedral and tetrahedral positions by the metal ions, there are two extreme cases. If we denote the ions in the octahedral positions by underlining them, there are the possibilities

$$Me^{2+}\underline{Me}_2^{3+}O_4 \quad \text{and} \quad Me^{4+}\underline{Me}_2^{2+}O_4,$$

in which all the metal ion lattice-points of the same coordination are occupied by ions of the same charge. The metal ions of different valencies may be of the same element or of different elements. This structure is called *normal spinel*. $Zn\underline{Fe}_2O_4$, for example, is of this type[22]). The other extreme case is described by

$$Me^{3+}\underline{Me^{3+}}\underline{Me^{2+}}O_4 \quad \text{and} \quad Me^{2+}\underline{Me^{4+}}\underline{Me^{2+}}O_4.$$

Here the octahedral positions are occupied by metal ions of different valencies. This type, whose most familiar representative is magnetite, $Fe^{3+}\underline{Fe^{3+}}\underline{Fe^{2+}}O_4$, is called *inverse spinel*.

In *mixed crystals* of spinels, one of which has the normal structure, the other the inverse structure, intermediate cases between the two types can occur. Many pure spinels are likewise to be regarded as mixed crystals of a normal and an inverse spinel of the same chemical composition, e.g., magnesium ferrite, $Fe^{3+}_{1-y}Mg^{2+}_{y}\underline{Fe}^{3+}_{1+y}\underline{Mg}^{2+}_{1-y}O_4$ [22]). However, such mixed crystals can also be regarded as disordered normal or inverse spinels, particularly if the deviations from one or other of the extreme cases are slight.

Besides this kind of disorder, another deviation from the ideal case occurs in spinels, in that octahedral or tetrahedral vacancies are occupied which would be unoccupied in the ideal crystal, and correspondingly a number of lattice positions which would be occupied in the ideal crystal remain vacant.

Both kinds of disorder play an important part in the diffusion of metal ions in the spinel lattice[23]). Because of the high order of the oxygen sublattice the diffusion coefficient for oxygen remains, in general, small compared with the diffusion coefficient for the metal ions[24]). This is also true of magnetite[24, 25]).

Electrons can very readily be interchanged between divalent and trivalent iron ions in octahedral positions. According to KRÖGER[26]) it can be supposed that the third valency electrons of the N divalent iron ions in the octahedral positions are distributed uniformly over the $2N$ occupied octahedral positions. The third valency electrons of the Fe^{2+} thus occupy a conduction band of the crystal, up to exactly one half (the energy state of the free, movable, conduction electrons is called the conduction band). The easy interchange of charge between the Fe^{2+} and Fe^{3+} results in the high electrical conductivity of the oxide, about 10^2 ohm^{-1} cm^{-1}.

Like most spinels, at higher temperatures magnetite can accept an excess of the trivalent metal in the oxide. A review of the oxide systems which have been well studied in this respect shows, however, that the phase width of the spinel becomes appreciable only if it is the sesquioxide Me_2O_3 itself which appears as the next phase with a higher oxygen content, and not other compounds of the sesquioxide with the oxide of the divalent metal.

A comparison of magnetite with γ-Fe_2O_3 enables a statement to be made about the lattice defects which arise in deviations from the stoichiometric composition of magnetite. This compound, γ-Fe_2O_3, has the same lattice structure as Fe_3O_4, but in each unit cell with 32 oxygen ions, 8/3 of the octahedral positions which are occupied by iron ions in magnetite remain vacant[27]). It can therefore be expected that, when magnetite takes in an excess of oxygen, iron ion vacancies in the octahedral positions will occur. This mechanism of taking in oxygen has also been demonstrated experimentally[28]).

According to SCHMALZRIED and WAGNER[29]) the concentration of cation vacancies in magnetite should be proportional to the fourth power of the activity of the FeO and therefore proportional to $p^{2/3}(O_2)$. A corresponding relation between the self-diffusion coefficient of the iron, $D^*(Fe)$, in magnetite and the oxygen pressure is also to be expected, if certain

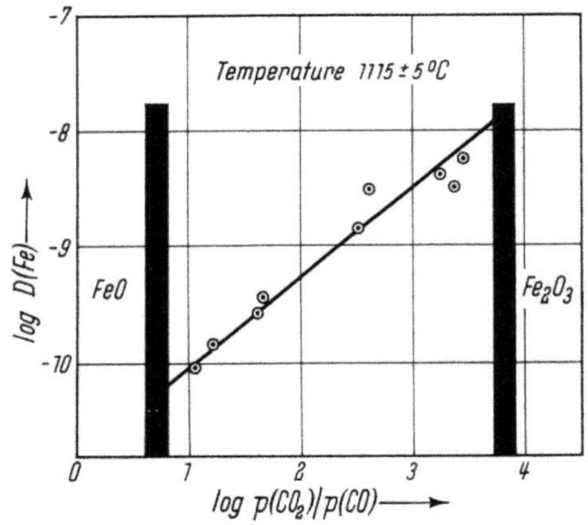

Fig. 11. Dependence of the self-diffusion coefficient of Fe in Fe_3O_4 on the oxygen partial pressure, according to SCHMALZRIED[24])

simple hypotheses are made regarding the process of mass transport. Experimentally, however, it is found that

$$D^*(Fe) \sim p^{0.4 \pm 0.05}(O_2),$$

as shown in Fig. 11[24]).

The deduction from the structure of γ-Fe_2O_3 about the disorder in magnetite with an oxygen excess is supported by the investigations carried out by HÄGG and SÖDERHOLM[30]) into mixed crystals of magnesium-aluminium spinel $MgAl_2O_4$ and Al_2O_3. In this system we can follow the step-by-step replacement of three Mg^{2+} ions by two Al^{3+} ions which results from the formation of vacancies in the octahedral positions of the metal ion sub-lattice. Complete replacement of all the magnesium of the spinel by aluminium would lead in this series of mixed crystals to γ-Al_2O_3, which has the same cubic structure as γ-Fe_2O_3.

Magnetite is *ferrimagnetic* with a Curie temperature of 627°C. This magnetic property is the prerequisite for the magnetic separation of magnetite ore. In magnetic roasting, for example in calcining spathic iron ores, magnetite is often required as the end-product of the roasting process, in order to make possible the separation of the roast into an iron-rich fraction and a gangue fraction[31]).

By oxidation of magnetite two different oxide phases with the composition Fe_2O_3 can be obtained. One of these phases, γ-Fe_2O_3, has already been mentioned because of the similarity of its lattice structure with the magnetite lattice; it is a spinel with an increased concentration of vacancies in the octahedral positions. It is noteworthy that this modification occurs in large quantities in the brown fume of oxygen-blown steel-making furnaces[31a]). Its production is possible only under quite specific conditions. According to DE BOER and SELWOOD[32]) the oxidation temperature must be below 400°C, since above this temperature the γ-Fe_2O_3 changes into hematite, the α-Fe_2O_3. DAVID and WELCH[33]) were able to obtain γ-Fe_2O_3 by the oxidation of magnetite at temperatures between 180 and 500°C if the magnetite, by reason of its production, contained water and before oxidation had been dried at temperatures between 250 and 850°C in a stream of argon. Non-dried magnetite containing water, and magnetite free from water from the outset, according to DAVID and WELCH, yielded on oxidation with oxygen only the rhombohedric α-phase hematite.

DAVID and WELCH came to the conclusion that at high concentrations of vacancies in the cation sub-lattice in γ-Fe_2O_3 the cubic spinel lattice could be stabilized only in the presence of water. The water can, perhaps, be regarded as dissociated. The hydrogen ions then occupy the cation vacancies, the hydroxyl ions could take the places of the oxygen ions. VERWEY[34]) had previously pointed out the importance of water for the stabilization of γ-Fe_2O_3. Similar circumstances are said to hold for γ-Al_2O_3 [35]).

FINCH and SINHA[36]) observed on calcining polished hematite crystals in air at temperatures of about 500°C the formation of thin, lamellar efflorescences, which in electron diffraction photographs yield different reflections from those by α- or γ-Fe_2O_3, and which, as a new phase, have been designated β-Fe_2O_3. On the other hand, on calcining polished hematite crystals at 700°C in air γ-oxide lamellae appear. FINCH and SINHA[36]) regard the β-form as an intermediate stage between α- and γ-Fe_2O_3 and, in contrast to DE BOER and SELWOOD[32]) and most other investigators of this phase, come to the conclusion that the α-γ-transformation may be not strictly monotropic in the direction $\gamma \rightarrow \alpha$, and that the γ-phase is a stable modification in particular temperature ranges. FEITKNECHT[38a]) has investigated the oxidation of Fe_3O_4 to γ-Fe_2O_3 and to hematite.

SCHRADER and BÜTTNER[37]) also represent the view that γ-Fe_2O_3 may be stable without stabilizing impurities. They describe the lattice of this oxide as an arrangement of three spinel unit cells mounted one on top of another. The same authors were able to identify a further modification of Fe_2O_3.

γ-Fe_2O_3, like magnetite, is ferrimagnetic. Its formation is therefore sought during ore-dressing in order, among other things, to be able to carry out finally a magnetic concentration[31, 38]).

The arrangement of the oxygen ions is very similar in FeO, Fe_3O_4, and γ-Fe_2O_3. The mole volumes, relative to 1 mole iron content and to 1 mole oxygen content, for the 4 iron oxides, together with their molecular weights and densities are shown in Table 2.

Table 2. *Molecular weights, densities, and relative mole volumes of the iron oxides*

	Molecular weight	Density (g/cm³)	Relative mole volumes V, relative to: 1 atom Fe V^*	Relative mole volumes V, relative to: 1 atom O V'^*
Fe	55.85	7.86** at 20°C	1	—
$Fe_{0.95}O$	69.05	5.73*** at 20°C	1.8	1
α-Fe_2O_3	159.7	5.355**	2.1	0.82
		5.26[170])	2.1	0.83
γ-Fe_2O_3	159.7	4.4[170])	2.6	1.0
Fe_3O_4	231.55	5.1**	2.1	0.93

* $V = V_M(Fe_nO_m)/[n \cdot V_M(Fe)]$ and $V' = V_M(Fe_nO_m)/[m \cdot V_M(Fe_{0.95}O)]$.
** LANDOLT-BÖRNSTEIN[76]), 5th ed., I, Suppl. I, Suppl. III, Part 1.
*** JETTE and FOOTE[7]), p. 27.

Since iron is the diffusing component in FeO and in Fe_3O_4, the transformation of these oxides one into the other takes place during reduction by the incorporation of additional iron. Since here too the mole volumes, relative to equal oxygen content, do not change essentially, an oriented growth of the layers of these oxides can take place, and the transformation leads to dense layers free from cracks and pores. On the other hand, the lattice of α-Fe_2O_3 differs appreciably from that of magnetite, and consequently in the reduction of hematite, reduction products with greater porosity are formed. The greater resulting surface area has the result that reduction of hematite proceeds more rapidly than that of magnetite or wüstite*.

It was observed by MEYER, RAUSCH and OTTO[38b]) and also by WATANABE[38c]) that some sorts of sinter containing hematite disintegrate at relatively low temperatures in the blast furnace, and at between 300 and 700°C in reduction experiments with H_2, CO and CO/CO_2 mixtures.

WATANABE's experiments show that the stability of the sinter is at a minimum at 500°C, a minimum which was particularly marked in the case of reduction with CO. As the proportion of CO_2 in the reduction gas rose, a decreasing tendency to sinter breakdown was established.

Observations in the hot-stage microscope showed that cracks arise during the reduction of hematite to magnetite. The reason for this may be the dependence of the morphology on temperature, and the transformation stresses during the reduction of hematite to magnetite.

From measurements of the conductivity in pure and doped hematite GEIGER and WAGNER[38d]) infer that in hematite below 850°C it is the oxygen vacancies, and above 850°C the electron defects, which preponderate and determine the conductivity.

Table 2a gives the values of the lattice parameter a (in Å) for iron and its oxides as functions of the temperature T (in °C).

Table 2a. *Lattice parameter a* (in Å) *for iron and its oxides as functions of the temperature T* (in °C), *according to* GORTON, BITSIANES *and* JOSEPH[38e])

α-Fe	$a = 2.8658 + 3.747 \cdot 10^{-5}T + 8.59 \cdot 10^{-9}T^2$
γ-Fe	$a = 3.5832 + 7.00 \cdot 10^{-5}T$
Fe_3O_4	$a = 8.3939 + 8.46 \cdot 10^{-5}T + 5.52 \cdot 10^{-8}T^2$
$Fe_{0.924}O$	$a = 4.2942 + 6.10 \cdot 10^{-5}T$
Fe_2O_3 (hexagonal)	$a = 5.0298 + 5.30 \cdot 10^{-5}T + 1.12 \cdot 10^{-8}T^2$ $c = 13.753 + 12.84 \cdot 10^{-5}T + 1.13 \cdot 10^{-8}T^2$
$Fe_2O_3 \cdot H_2O$ (orthorhombic)	$a = 4.574 + 5.70 \cdot 10^{-5}T + 1.33 \cdot 10^{-7}T^2$ $b = 9.953 + 11.00 \cdot 10^{-5}T$ $c = 3.020 + 6.00 \cdot 10^{-6}T + 1.60 \cdot 10^{-7}T^2$

The thermodynamic constants of the iron oxides are presented in the thermodynamic tables (pp. 10–12) according to the data of VALLET[39])

* The roughening of the surface during the removal of oxygen is discussed in detail in section 1.2.7.2. Similarity of lattice structure and mole volumes is not an essential condition for the formation of dense, imporous layers, as is shown by the growth of imporous iron films on wüstite[248]).

and Elliott et al.[5]). The values differ but little from those given by Kubaschewsky and Evans[40]), Kelley[41]), Landolt-Börnstein[42]), and the Hütte pocketbook[43]).

A general view of the *regions of thermodynamic stability* of the iron oxides is given in Fig. 12 which has been prepared using mean values for ΔH and ΔS from the thermodynamic tables, pp. 10–12. Particularly

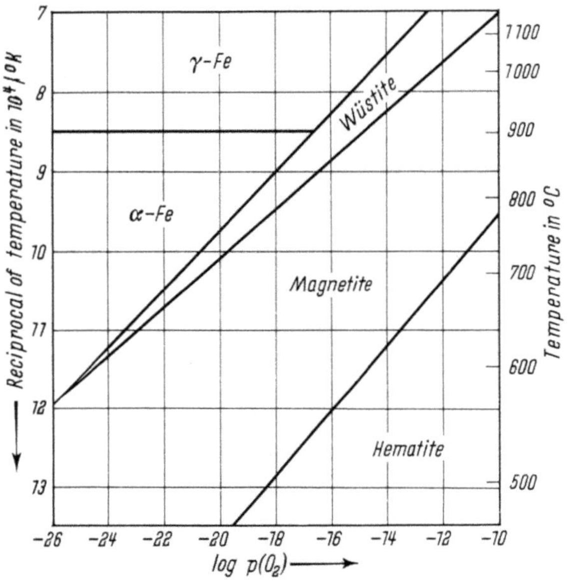

Fig. 12. Thermodynamic regions of stability of iron, wüstite, magnetite and hematite

striking is the narrow stability region of wüstite, which at 1100°C amounts to only about $2\frac{1}{2}$ powers of 10 for the oxygen pressure, decreases continuously as the temperature falls, and disappears at 570°C.

1.1.2. The Iron-Oxygen-Hydrogen System and the Iron-Oxygen-Carbon System

The equilibria between iron, oxygen, and hydrogen are obtained by combining the equilibria for iron-wüstite-oxygen, wüstite-magnetite-oxygen, and magnetite-hematite-oxygen according to the thermodynamic tables and Fig. 12 with the equilibrium

$$\left.\begin{aligned} &H_2 + \tfrac{1}{2}\,O_2 \rightleftharpoons H_2O, \\ p^{1/2}(O_2) &= p(H_2O)/\{K(H_2O)\,p(H_2)\}, \\ RT\log K(H_2O) &= -\Delta G^0(H_2O), \\ \log K(H_2O) &= -2.93 + 13020/T\ (^\circ K). \end{aligned}\right\} \quad (20)$$

The values for $\Delta G^0(H_2O)$ can be taken from the table on p. 5; the formula given for $\log K(H_2O)$ is valid between 600 and 1100°C.

However, since small errors in the values used for ΔG^0 greatly affect the shape of the equilibrium curves, it is better to go back to direct measurements for the equilibria. The shape of the equilibria lines is shown in Fig. 13, which is based on the compilation by GIVAUDON[44]), compared with the values given by KUN LI[45]). The lines of constant iron deficiency for wüstite have been computed using formulae (19) and (20). The course of the equilibria lines above 1300°C has been given by DARKEN and GURRY[1]).

The equilibria of iron and the iron oxides with a mixture of the gases CO and CO_2 and with solid carbon are obtained in a similar way, by taking into account, besides the relations corresponding to (20), the Boudouard equilibrium

$$CO_2 + C \rightleftharpoons 2\,CO. \tag{21}$$

For this equilibrium we have

$$K(C) = p^2(CO)/\{p(CO_2)\,a_c\}, \tag{21a}$$

and with the numerical values from the table on pp. 3–4, for 600 to 1100°C,

$$\log K(C) = 9.11 - 8840/T\ (°K). \tag{21b}$$

Based on the data in the literature[44, 45]), Fig. 14 shows the equilibria for the system iron-oxygen-carbon, which are important for the reduction process. This equilibrium diagram is generally known as the Baur-Glaessner diagram. Figure 15 shows this diagram in the form in which it was originally given by BAUER and GLAESSNER in 1903[46]).

From Fig. 14 it is to be inferred that, at temperatures above 710°C, all the iron oxides can be reduced by mixtures of CO/CO_2 at a total pressure of 1 atm which are in equilibrium with carbon, and can therefore be reduced by carbon itself. At lower temperatures only those mixtures have a reducing action on wüstite which are supersaturated with carbon and which therefore, according to (21), must react towards carbon deposition. Blast-furnace gas in general has a composition such that, on cooling below a temperature of 650 to 800°C, it runs into carbon deposition, but it acts reducingly at all temperatures on wüstite or magnetite and hematite. But the decomposition of CO, which, according to (21) is to be expected, in general fails to occur.

The systems C-O_2 and H_2-O_2 are connected through the *water-gas reaction:*

$$CO_2 + H_2 \rightleftharpoons CO + H_2O,$$

$$K(W) = p(CO)\,p(H_2O)/\{p(CO_2)\,p(H_2)\}. \tag{22}$$

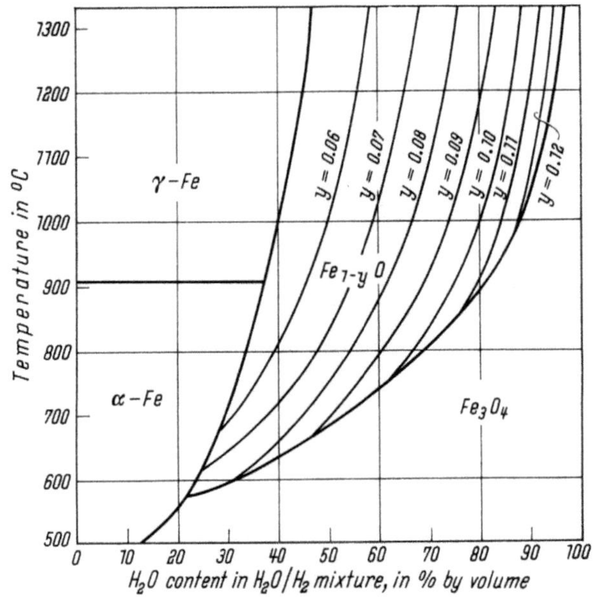

Fig. 13. Equilibria between iron, wüstite, magnetite and a hydrogen and water-vapour mixture

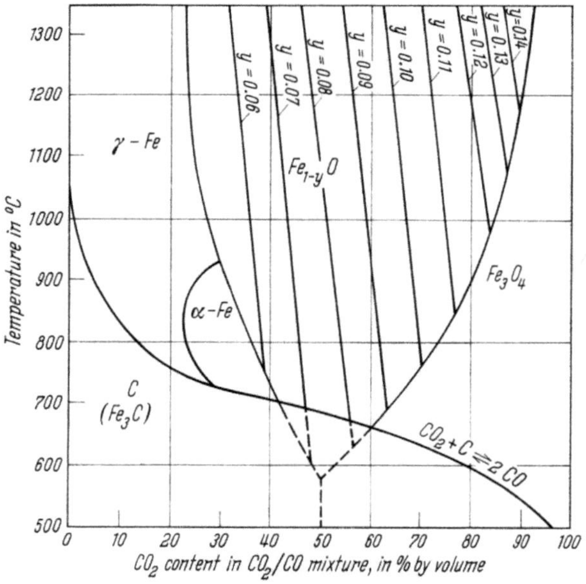

Fig. 14. Equilibria between iron, wüstite, magnetite and carbon monoxide, carbon dioxide, and carbon. $p(CO) + p(CO_2) = 1$ atm

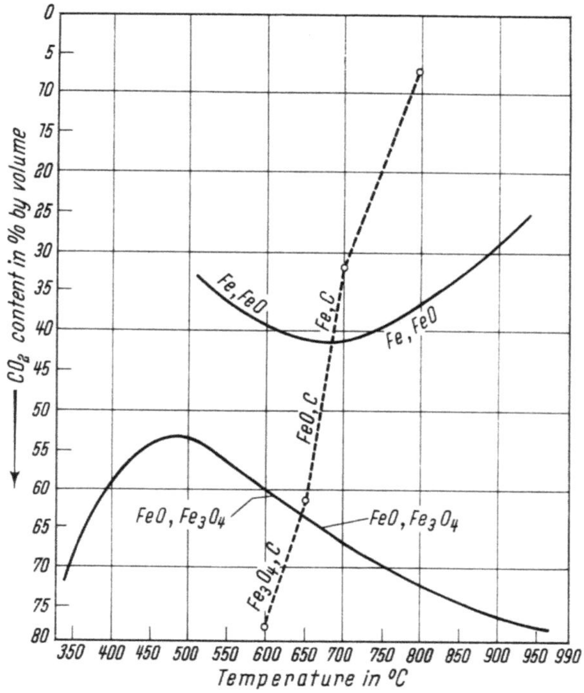

Fig. 15. Original form of the Baur-Glaessner diagram

For accurate calculations the values of $K(\mathrm{W})$ must be found from the values of ΔG^0 for the components, as given in the thermodynamic tables. For rough calculations the value given by

$$\log K(\mathrm{W}) = 1.6 - 1800/T\,(^\circ\mathrm{K}) \quad \text{for} \quad 600^\circ\mathrm{C} < T < 1100^\circ\mathrm{C}$$

is good enough; at 820°C $K(\mathrm{W}) = 1$.

1.1.3. Reduction Equilibria of Mixed Crystals with Iron Oxides and of Compounds Containing Iron Oxides

All the iron oxides can form mixed crystals with other oxides. The mixed crystals of wüstite with MnO and MgO are of importance in the reduction of ores. The variation of activity of wüstite in these mixtures is shown in Fig. 16[47]). Wüstite can also form mixed crystals with CaO; at 1100°C CaO dissolves maximally 15% FeO, and wüstite maximally 28% CaO [59]).

Fe_3O_4 forms mixed crystals with a number of cubic spinels. Fig. 17 shows the phase diagram for the binary system Fe_3O_4-Mn_3O_4. Below 1160°C it has a miscibility gap, since the low-temperature form of Mn_3O_4 has a tetragonal lattice, and only the high-temperature form is a cubic

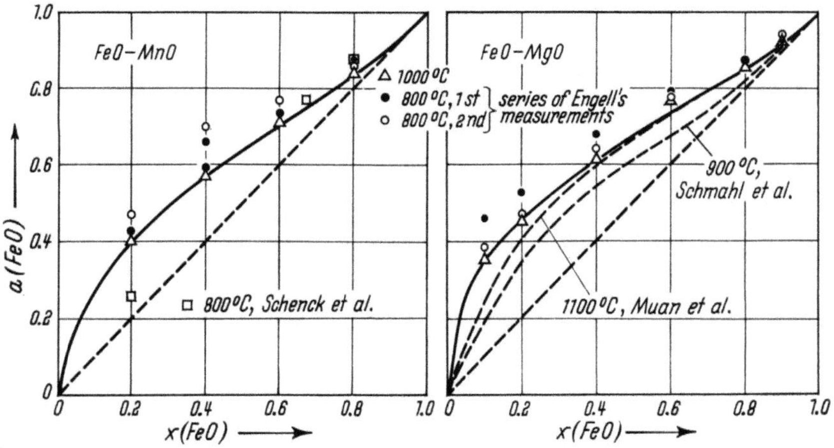

Fig. 16. Activity of wüstite in FeO-MnO and FeO-MgO mixed crystals. Gas equilibrium at different temperatures

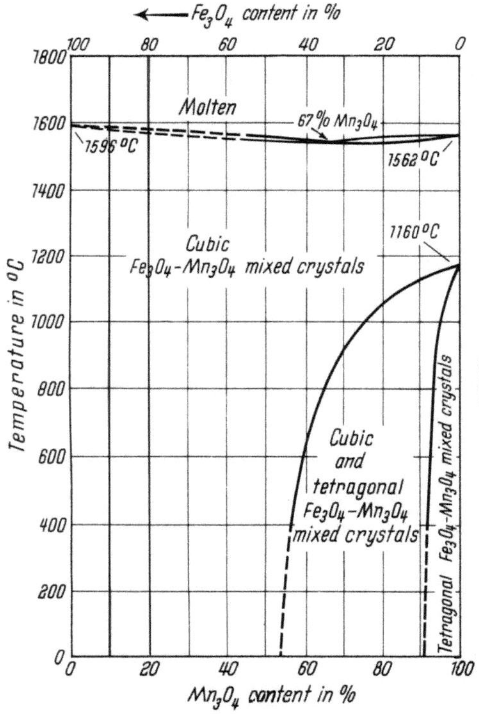

Fig. 17. Fe_3O_4-Mn_3O_4 system

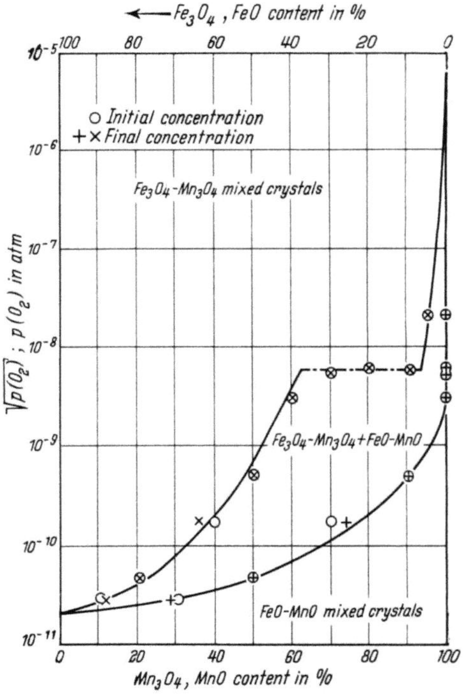

Fig. 18. Equilibrium diagram for Fe_3O_4-Mn_3O_4/FeO-MnO/oxygen at 700°C

spinel. Figure 18[48]) shows the equilibria between Fe_3O_4-Mn_3O_4 mixed crystals. From the diagram can be read off what composition of gases is necessary in order to reduce a mixed crystal of a given composition, and which FeO-MnO mixed crystals will form during the reaction. In ores, sinters, and pellets there also occur the mixed crystals of magnetite with iron-aluminium spinel $FeAl_2O_4$ [49]) and magnesium ferrite $MgFe_2O_4$ [50]). At 1350°C and at a pressure of 0.2 atm of oxygen Fe_2O_3 forms mixed crystals with Al_2O_3 in the ranges of concentration of Al_2O_3 from 0 to 15% and from 85 to 100% by weight, and also, at the same oxygen pressure and between 1330 and 1420°C, it forms the compound $Fe_2O_3 \cdot Al_2O_3$ [51]).

Up to 5% chromium oxide can be dissolved by wüstite at 1350°C; further it forms with the wüstite the compound $FeCr_2O_4$, and this, with magnetite, forms mixed crystals in all proportions. There is also continuous miscibility between Fe_2O_3 and Cr_2O_3 [52]). For the reduction of ores containing titanium the system FeO-TiO_2-Fe_2O_3 is important[53, 54]). For this system and for all other polyphase systems a knowledge of which may be required in individual cases, reference may be made to the appropriate volumes of tables[55, 56]). The thermodynamic characteristics of a number of compounds which contain iron oxides or which are of importance for the reduction of ores are given in the tables on pp. 1–17.

1.1.4. The System FeO-CaO-SiO_2 and Equilibria in Sinter Formation

Because of the particular importance which the mixtures and reaction products of the ternary system FeO-CaO-SiO_2 have in connection with the formation of sinter, this system will be discussed in some detail. Figure 19 [57]) shows sections for the temperatures 1130 and 1285°C, which are important in the sintering process, and for 700 and 980°C, since it is in this range mainly that gaseous reduction of iron ore takes place. There are more recent investigations at still higher temperatures[58]).

At 1285°C – depending on the composition of the sinter – the phases $CaSiO_3$, $3\,CaO \cdot 2\,SiO_2$, and Ca_2SiO_4 may form. When the temperature falls below 1210°C fayalite can crystallize out from the residual melt.

With decreasing temperature, $CaSiO_3$ incorporates increasing amounts of iron into its lattice; likewise the solubility of Fe_2SiO_4 in Ca_2SiO_4 increases and reaches its maximum value of about 10% at 1230°C. The solution of iron silicate in dicalcium silicate is of particular significance inasmuch as it can inhibit the phase changes of the Ca_2SiO_4, which is an advantage as regards the stability of the sinter. Wollastonite, $CaSiO_3$, achieves its greatest iron content with about 70% of "$FeSiO_3$" at 970°C. On further cooling, these mixed crystals decompose again with separation of hedenbergite, $CaFeSi_2O_6$.

At 1130°C the region of homogeneous melts is already closely confined between the saturation surfaces of the phases Fe_2SiO_4, $CaFeSiO_4$, and also (Ca, Fe)SiO_3. From about 1120°C downwards the liquidus curves of fayalite and iron-calcium-olivine touch one another; a series of mixed

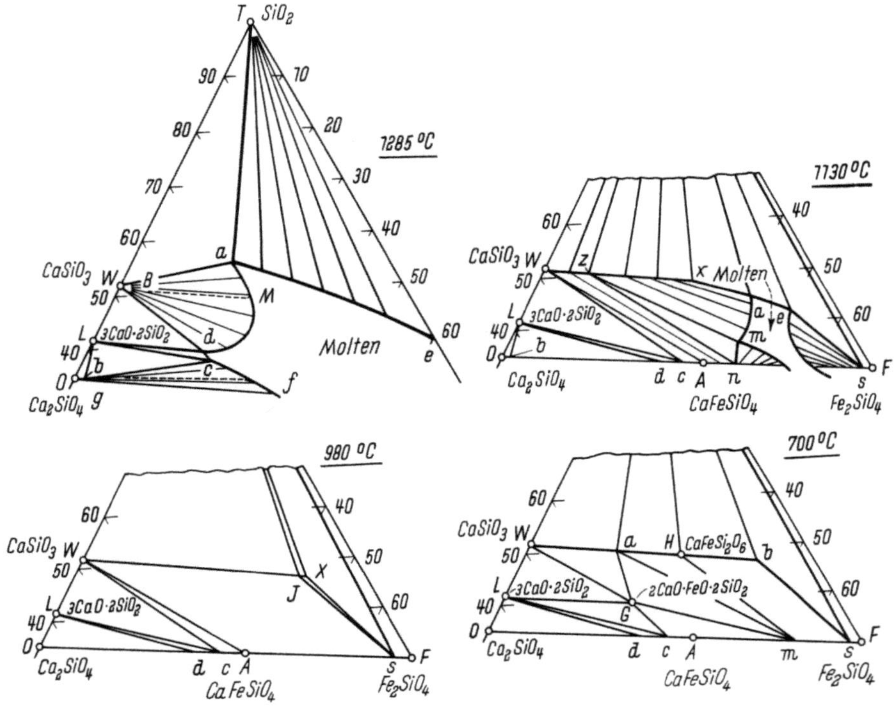

Fig. 19. Sections **from the ternary** system FeO-CaO-SiO_2, according to Bowen et al.[57])

crystals now forms, ranging in composition from Fe_2SiO_4 to about $CaFeSiO_4$. In the diagram for 700°C the further phase 2 CaO · FeO · 2 SiO_2 can be recognized.

The thermodynamic characteristics for the compounds Fe_2SiO_4, Ca_2SiO_4, and $CaSiO_3$ are given in the tables, pp. 1–17. Since dolomite may occasionally be used in the sinter, some silicates containing magnesium are included in the tables.

With higher admixtures of limestone or with increased lime content due to locality, in addition to the silicates discussed, compounds of the system FeO · Fe_2O_3 · CaO can also occur; this system has been particularly investigated by Phillips and Muan[59]) in recent years. Braun and Kwestroo[60]) have reported on further compounds in this system. The most important compounds in this system are the two calcium ferrites,

$CaFe_2O_4$ and $Ca_2Fe_2O_5$. The thermodynamic characteristics of dicalcium ferrite are given in the table on p. 9.

According to investigations by HASS, BITSIANES and JOSEPH[61a]), at temperatures between 1185 and 1195°C the ferrites $CaO \cdot 2\,Fe_2O_3$,

Fig. 20. General view of the compounds in the system FeO-CaO-Fe_2O_3, according to NYQUIST[61])

$C_2F = 2\,CaO \cdot Fe_2O_3$
$CF = CaO \cdot Fe_2O_3$
$CF_2 = CaO \cdot 2\,Fe_2O_3$
$C_4WF_4 = 4\,CaO \cdot FeO \cdot 4\,Fe_2O_3$
$C_3WF_7 = 3\,CaO \cdot FeO \cdot 7\,Fe_2O_3$
$CWF = CaO \cdot FeO \cdot Fe_2O_3$
$CW_3F = CaO \cdot 3\,FeO \cdot Fe_2O_3$

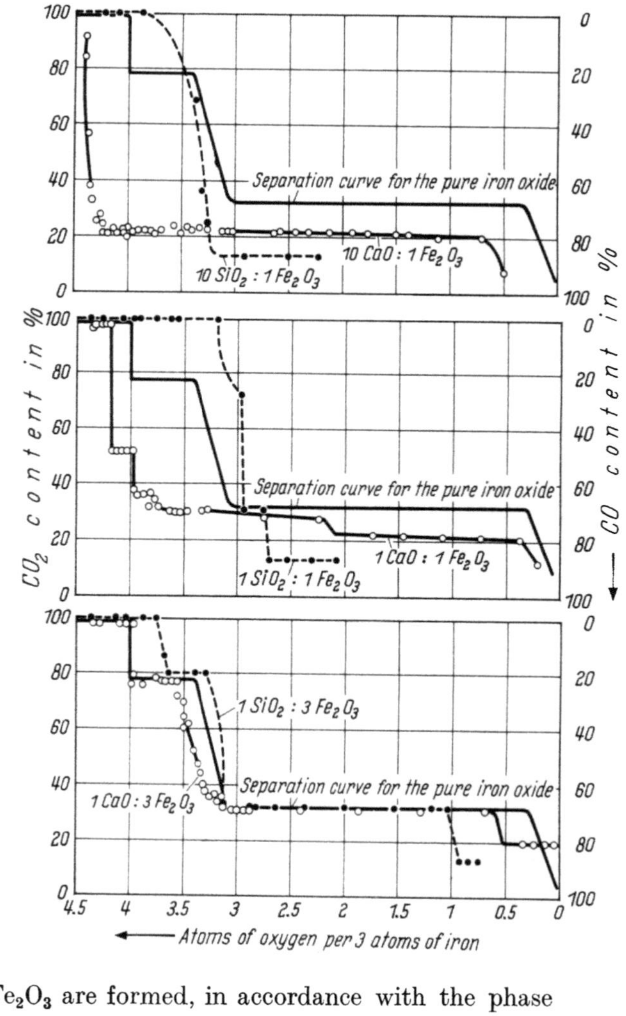

Fig. 21. Equilibria for the separation of oxygen at 900°C from mixtures of iron(III)-oxide with limestone or silica, according to SCHENCK et al.[62])

$CaO \cdot Fe_2O_3$ and $2\,CaO \cdot Fe_2O_3$ are formed, in accordance with the phase diagram, by a reaction in the solid state between hematite and limestone. Under reducing conditions, or in a reaction between limestone and magnetite in an inert atmosphere, only one ferrite phase is formed, which the authors call $CaO \cdot FeO \cdot Fe_2O_3$. The rate of formation of the ferrite phases in these solid-state reaction experiments is low and is presumably controlled by the diffusion in the layers formed; the diffusion coefficient of the migrating components appears to be greatest in $CaO \cdot 2\,Fe_2O_3$ and least in $CaO \cdot Fe_2O_3$. The reported rates of formation are not sufficiently high to allow any appreciable formation of ferrite by a solid-state reaction in the industrial sintering process; here the reaction must undoubtedly take place with participation of the molten components. The lowest temperature of melting in the system $CaO \cdot Fe_2O_3$ is somewhat above 1200°C. In the ternary system $CaO \cdot FeO \cdot Fe_2O_3$, and therefore under the reducing conditions of the industrial sintering process, the lowest eutectic temperatures correspondingly lie below 1200°C. EDSTRÖM[61b]) regards the compound $CaO \cdot 2\,Fe_2O_3$ as the most important binding phase for basic sinter.

For monocalcium ferrite it may be calculated, using the decomposition curves of SCHENCK et al.[62]) at 900°C, with the ΔG^0 value given by ELLIOTT[5]) for the dicalcium ferrite, that:

$$CaO + Fe_2O_3 \rightleftharpoons CaO \cdot Fe_2O_3, \qquad \Delta G \approx -10\ \text{kcal}.$$

Figure 20 gives a general view of the compounds appearing in the $FeO \cdot Fe_2O_3 \cdot CaO$ ternary system.

By its combination with CaO the activity of Fe_2O_3 in dicalcium ferrite is decreased so strongly that it is iron which appears directly as its reduction product.

In general, the activity of wüstite is reduced by the formation of a compound and hence the CO_2/CO or H_2O/H_2 ratio for the reduction to the next lower oxidation stage is decreased. Since the composition of the gas must be less than this ratio, if reduction is to take place, the formation of a compound affects the best possible utilization of the reduction gas. This shift of the equilibrium can also affect the kinetics of the reduction, since, as will be shown in section 2.2, the rate of gas diffusion and of the phase-boundary reaction in general depends linearly on the difference between the actual composition of the gas and the gas concentration for equilibrium. The change in the equilibrium CO_2/CO ratio through the combination of wüstite with CaO and with SiO_2 may be seen from Fig. 21[62]). We see that when either SiO_2 or CaO is added, the magnetite stage disappears from the decomposition curve. The addition of SiO_2 results in the formation of fayalite during the reduction of the Fe_2O_3, since through the combination with silica the FeO-activity is reduced so markedly that magnetite $FeO \cdot Fe_2O_3$ is no longer able to form. Correspondingly in the wüstite stage, the formation of the silicate makes itself apparent by a deterioration in the gas utilization. The first stage in the decomposition curve for mixtures of CaO with Fe_2O_3 in the proportion 1 : 1 is to be attributed to the reaction

$$2\,(CaO \cdot Fe_2O_3) + CO \rightleftharpoons 2\,CaO \cdot Fe_2O_3 + 2\,FeO + CO_2,$$

the second stage to the reduction of the wüstite produced in the first stage, and finally the last stage to the reaction

$$2\,CaO \cdot Fe_2O_3 + 3\,CO \rightleftharpoons 2\,CaO + 2\,Fe + 3\,CO_2 .$$

In the mixture with 1 CaO and 3 Fe_2O_3 the compounds richer in iron oxides also make themselves noticeable in corresponding stages.

The change in gas utilization due to the formation of compounds can be calculated if the free enthalpy of the reaction between the iron oxide and the reaction partner, ΔG^0, is known. If the CO_2/CO or H_2O/H_2 ratio for the equilibrium of the pure oxide with its next lower oxidation stage is denoted by K^*, and the corresponding ratio for the oxide combination by K, then

$$n(O) \log(K/K^*) = \Delta G^0/(2.3RT),$$

where $n(O)$ denotes the number of oxygen atoms which are removed from one molecule of the compound in the reduction. For $n(O) = 1$ and at 900°C, the ratio CO_2/CO or H_2O/H_2 will be reduced by the formation of a compound with $\Delta G^0 = -5.4$ kcal by a factor of 10, according to this formula.

The equilibria reactions given above show which compounds can occur in sinters that consist of iron oxides with lesser amounts of limestone and silica. Sinters poor in iron are discussed in section 5.2. Since the distribution of the individual components is not constant and has local variations, and since other constituents, such as alumina in particular, may be present, other compounds can be formed, and often the phases formed are not related to the overall composition of the sinter. The high rates of heating up and cooling off in sintering[63]), and also the short time for which the reaction temperature is maintained as a rule, limit the diffusion paths and hence also the reaction zones to very small neighbourhoods round the crystals or grains of the admixture, and inhibit the recrystallization equilibria-reactions during cooling. Table 3 gives an outline of the variation in the amounts of the phases in the sinter as the basicity CaO/SiO_2 increases, according to KNEPPER[66]).

Figure 22, taken from an investigation by NYQUIST[65]), is largely in agreement with the data of Table 3.

Apart from these phases, mixed crystals of various kinds appear in the sinter, e.g., Fe_3O_4 with up to 8% CaO, and Fe_2O_3 with about 1% CaO [57, 59, 61, 66]). The type of mineral phase occurring is independent of the absolute amount of gangue which the ore contains, but the amounts of these phases are obviously not independent of this. A uniform reaction throughout the sinter is promoted by fine grinding. The simultaneous occurrence of iron-bearing silicates and calcium ferrites in industrial sinters shows that genuine equilibrium is not attained. Since the thermodynamic

Table 3. *Effect of basicity on amount and type of phase in the sinter*[64])

Basicity B	$0 \to 1.0$	$1.0 \to 1.8$	$1.8 \to$
Phases	Increasing amount nonwater-etching silicates	Decreasing amount non-water-etching silicates	
		Increasing amount calcium ferrites	Increasing amount calcium ferrites
		Increasing amount β-dicalcium silicate	Decreasing proportion β-dicalcium silicate

Non-water-etching silicates: Glasses with FeO, Fe_2O_3, CaO, Al_2O_3, and SiO_2; fayalite Fe_2SiO_4; wollastonite $CaSiO_3$; (Ca, Fe)SiO_4; anorthite $CaO \cdot Al_2O_3 \cdot 2\,SiO_2$.

Calcium ferrite: $CaO \cdot xFe_2O_3$; $CaFe_2O_4 = CaO \cdot Fe_2O_3$; $Ca_2Fe_2O_5 = 2\,CaO \cdot Fe_2O_3$; $4\,CaO \cdot FeO \cdot 4\,Fe_2O_3$; $2\,CaO \cdot FeO \cdot 7\,Fe_2O_3$.

data for the phases appearing in the reaction are not all known, it is only possible to make an estimate for the actual equilibrium situation. Thus the free reaction enthalpy for the double-decomposition

$$\tfrac{1}{2}\,O_2 + Fe_2SiO_4 + 2\,CaFe_2O_4 \rightarrow Ca_2SiO_4 + 3\,Fe_2O_3$$

at 1000°C amounts to about -39 kcal. This reaction would therefore have to take place in the given direction only, even if the activity of the

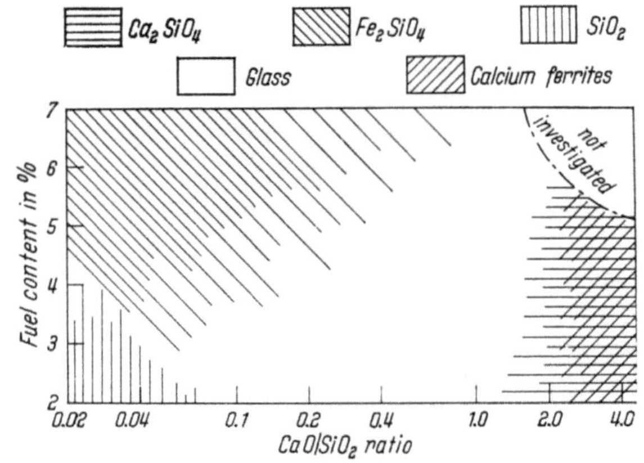

Fig. 22. Phase analysis of a magnetite sinter with 66% Fe, according to NYQUIST[65])

iron oxide in the silicate phase were reduced by the formation of mixed crystals, for example with dicalcium silicate, by a factor of 10^7. A similar statement holds regarding the activity of limestone in the ferrite. Under reducing conditions the fayalite would have to interact with the calcium ferrite in the sinter according to the formula

$$Fe_2SiO_4 + 2\,CaFe_2O_4 \rightleftharpoons 2\,FeO + 2\,Fe_2O_3 + Ca_2SiO_4; \quad \Delta G \approx -5\,\text{kcal}.$$

The frequently observed[64, 67]) minimum of the strength of sinter for basicities between 1 and 1.5 is to be attributed to the decrease in the amount of glass-like silicates, which serve as the binding medium, and the increase in the amount of β-Ca_2SiO_4, the phase changes of which during cooling affect the strength. The needle-like form of the precipitated calcium ferrite, which often binds neighbouring grains of ore together, causes the increase in strength of the sinter with higher basicity.

1.2. Fundamentals of Reduction Kinetics

1.2.1. Ore Reduction as a Sequence of Reactions

As a basis for our considerations we assume first that the reducing gas is CO or H_2 and that the ore is a porous lump of iron oxide. This lump of ore consists of individual *granules*; the *macro-pores* are the larger gaps between the granules, the *micro-pores* are the smaller gaps within the granules.

For a transformation according to the reaction

$$Fe_nO_m + m\,CO \rightarrow n\,Fe + m\,CO_2 \tag{1}$$

or

$$Fe_nO_m + m\,H_2 \rightarrow n\,Fe + m\,H_2O, \tag{2}$$

a number of *subsidiary processes* must take place. These, shown diagrammatically in Fig. 23, are as follows.

The reducing gas flows through the spaces between the lumps of ore. A *boundary layer* is formed round the individual lumps. The interchange of matter between the flowing gas-phase and the surface of the oxide takes place by transport of the reacting gas through the boundary layer. As the reduction progresses, the reducing gas diffuses through the macro-pores or micro-pores of the lump of ore or of the reaction products in the lump, and the oxidized gas diffuses outwards by the same path. The actual chemical reaction takes place at the phase boundary between the oxide and the gas; it includes the adsorption of the reducing gas, the separation of oxygen from the oxide lattice, the formation and growth of the nuclei of the reaction products, magnetite, wüstite, or iron, and finally the desorption of the oxidized gas-molecules from the surface of the solids. The further growth of the layers of the reaction products may involve reactions in the solid state and diffusion processes in the solid partners in the reaction, and these too are therefore included within the whole group of reactions.

Processes of this sort, made up of individual subsidiary processes taking place successively, are called *reaction sequences.* To each subsidiary process in a reaction sequence belongs a "sub-equilibrium". The deviation from this equilibrium may be regarded as the driving force in the subreaction, but of course, there is not necessarily a linear relation between the deviation from equilibrium and the rate of the reaction.

A measure for the driving force of a diffusion process for molecules of a type i is the gradient $d\mu_i/dx$ of the chemical potential, along the diffusion path; this is related to the gradient of a concentration or a partial pressure. dc_i/dx or dp_i/dx. The difference in composition at the ends of the diffusion path can therefore be taken here as a measure of the deviation from the

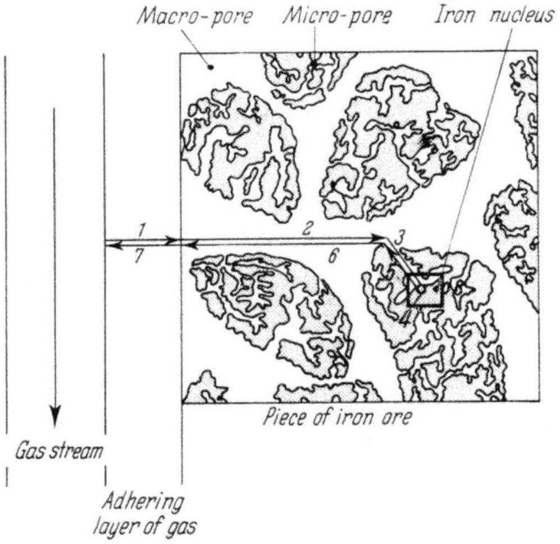

Fig. 23. Outline of the reduction mechanism for porous iron ores[106])

1 Diffusion of the hydrogen across the boundary layer, *2* Diffusion of the hydrogen through the macro-pores in the ore, *3* Diffusion of the hydrogen through the micro-pores to the reaction position, *4* Phase-boundary reaction, *5* Diffusion of the water vapour through the micro-pores, *6* Diffusion of the water vapour through the macro-pores, *7* Diffusion of the water vapour across the boundary layer. Migration of Fe^{2+} and $2\ominus$ to the iron nucleus

equilibrium. The rate of chemical reactions is also, in general, determined by the concentrations of the reacting partners. The superimposition of the forward and reverse reactions has the effect that, in reactions in which mixed phases participate, when the equilibrium concentrations are attained, the overall rate of transformation is zero.

Consider a sufficiently small time interval within which the total rates of the overall reactions (1) and (2), and also the concentrations of the intermediate materials such as the adsorbed gas molecules do not vary appreciably compared with the total transformation. In this time all the sub-reactions of the reaction sequence must proceed with the equivalent molar rate as the overall reaction. Such a state is called a *stationary* or *steady state*. With the steady reaction rate are associated definite deviations from equilibrium in the sub-processes. The situation can be represented diagrammatically by the block diagram in Fig. 24, which is laid out like an electric circuit. Here the driving-forces for the whole reaction and for the separate sub-reactions are taken to be the changes of the thermo-

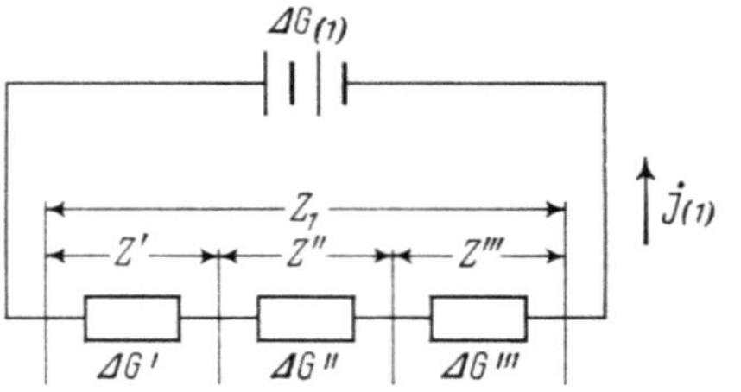

Fig. 24. Block diagram of a reaction sequence

dynamic potential G of the system in the course of the overall reaction represented by Eqs. (1) or (2) and in the course of the separate sub-reactions, and are denoted by $\Delta G_{(1)}$ or $\Delta G'$, $\Delta G''$, etc. $Z_{(1)}$ or Z', Z'', etc. are the resistances to the reactions and provide the connection between the driving-force ΔG of the reaction and the rate of the reaction $j_{(1)}$. It should again be stressed that only in special cases is there a linear relation between ΔG, Z, and j, such as would correspond to Ohm's law in the electrical analogy. By definition, in the steady state* the same reaction "current" j flows through all the "resistances" Z. But the deviations, ΔG, from equilibrium in the individual sub-reactions are different, depending in each case on the value of the resistance to the particular sub-reaction. If the resistance to the reaction in the i-th sub-process, Z_i, predominates over the sum of all the other resistances to reactions, then ΔG_i will be comparable with the driving-force $\Delta G_{(1)}$ for the overall reaction. In this case, the reaction rate j will practically be determined by Z_i alone, and the i-th sub-process will be said to be the *rate-determining process*. It frequently happens in a reaction sequence that one sub-process has a predominating influence on the reaction rate. In the reduction of ores — where

* Non-stationary states can also be considered by means of a similar block-diagram if "storages" — which would be represented by capacitances in the electrical analogy — are included in the circuit. The amounts of the materials stored in these storages would allow for time variation of the effective concentrations in the individual sub-reactions. In the steady state the "charges" in the storages, and hence also ΔG for the individual sub-reactions, remain constant, and j has the same value at each point of the circuit.

the temperature, composition of the reducing gas, rate of gas flow, the porous structure and size of the pieces of ore are all variables — various sub-processes may be rate-determinative.

It is an important task to establish which sub-process is rate-determinative in each case, since this is a prerequisite for influencing the reduction.

1.2.2. Gas Diffusion in the Boundary Layer and Pores

We consider a gas mixture of two components 1 and 2, say CO and CO_2. If local differences in the partial pressures p_1 and p_2 of the two gases or in their concentrations c_1 and c_2 (mole cm^{-3}) occur in the mixture, then diffusion of both components takes place. The rates of diffusion, j_1, j_2, of the gases 1 and 2, that is, the amount of gas (in moles) crossing an imaginary surface of area 1 cm^2 (normal to the direction of flow), are given by

$$\left.\begin{aligned} j_1 &= -D_{12}\frac{dc_1}{dx} = -D_{12}\frac{dp_1}{dx}\frac{1}{RT}, \\ j_2 &= -D_{21}\frac{dc_2}{dx} = -D_{21}\frac{dp_2}{dx}\frac{1}{RT}. \end{aligned}\right\} \tag{3}$$

If the total pressure and the temperature are the same at all points of the system, then

$$c_1 + c_2 = c \quad \text{and} \quad p_1 + p_2 = p.$$

These equations imply that

$$\frac{dc_1}{dx} = -\frac{dc_2}{dx} \quad \text{and} \quad \frac{dp_1}{dx} = -\frac{dp_2}{dx}. \tag{4}$$

These conditions can hold in the steady state only if also

$$j_1 = -j_2. \tag{5}$$

This condition also follows in the present case from the stoichiometry of the phase-boundary reaction. (3), (4), (5) imply that

$$D_{12} = D_{21}, \tag{6}$$

that is, the coefficient of diffusion of gas 1 in the mixture with gas 2 is equal to the coefficient of diffusion of gas 2 in the same mixture.

Numerical values of the diffusion coefficients can be calculated using the methods of the kinetic theory of gases. A first rough approximation is provided by an argument using the mean free path, thus:

Suppose that a coordinate system has been set up at an arbitrary point in the space domain containing the gases, and that this domain contains gas 1 in the concentration $n_1' = c_1 N_L$, where N_L is the Loschmidt number. The motion of the molecules being equally distributed, about one-third of the molecules will be moving in the direction of each axis, one-sixth in the positive direction, one-sixth in the negative direction. In 1 second an imaginary surface of area 1 cm² perpendicular to one direction of motion will be crossed by all those molecules moving towards it which are contained in a volume element whose base is the unit area and whose height is equal to the mean velocity $\bar{v}_1$ of the particles (in cm/sec). The number of these particles is $n_1' \bar{v}_1/6$. Only this number of particles can reach the surface in the chosen time of 1 second.

In a gas of homogeneous composition as many molecules will cross the surface in the one direction as in the opposite direction. But if, in the direction normal to the surface considered, there is a gradient, dn_1'/dx, in the concentration of the gas, then the flux of particles in the one direction is

$$\vec{j}_1 = \frac{\bar{v}_1}{6}\left(n_1' - x_0 \frac{dn_1'}{dx}\right)$$

and in the opposite direction

$$\overleftarrow{j}_1 = \frac{\bar{v}_1}{6}\left(n_1' + x_0 \frac{dn_1'}{dx}\right),$$

where x_0 denotes an element of length still to be determined.

Since the particles in a gas can move rectilinearly with constant velocity only between two collisions with other molecules, x_0 can here be taken to be the mean free path, λ. Consequently the flux of particles in the positive x-direction is given by

$$j_1 = \vec{j}_1 - \overleftarrow{j}_1 \approx -\frac{\lambda \bar{v}_1}{3} \frac{dn_1'}{dx}$$

or

$$j_1 \approx -\frac{\lambda \bar{v}_1}{3} \frac{dc_1}{dx} \tag{7}$$

for the diffusion flow in mole cm⁻² sec⁻¹. (3) and (7) imply that

$$D_{12} \approx \lambda \bar{v}_1/3.$$

If the actual (random) distribution of the directions of motion of the molecules is taken into account, we find that

$$D_{12} \approx \lambda \bar{v}_1/2.$$

Hence D_{12} can be computed if we take for the mean molecular velocity

$$\bar{v}_1 = \left(\frac{8RT}{\pi M_1}\right)^{1/2} \tag{8}$$

and for the mean free path

$$\lambda_1 = 1/\left[N_L \pi \left(c_1 \sigma_1^2 \sqrt{2} + c_2 \sigma_2^2 \sqrt{1 + M_1/M_2}\right)\right], \tag{9}$$

where M_1, M_2 are the molecular weights of the gases 1 and 2, and σ_1, σ_2 are the collision diameters of the molecules of these gases.

However, by applying the laws of momentum transfer[68]) in the calculation of the diffusion coefficients, we obtain the approximate expression

$$D_{12} = D_{21} \approx \frac{3}{8(c_1 + c_2)\sigma_{12}^2 N_L}\left(\frac{RT(M_1 + M_2)}{2\pi M_1 M_2}\right)^{1/2}, \tag{10}$$

where $\sigma_{12} = (\sigma_1 + \sigma_2)/2$ is the mean collision diameter of the molecules. Taking

$$c_1 + c_2 = P/RT,$$

we get

$$D_{12} = D_{21} \approx \frac{3}{8\sigma_{12}^2 P N_L} \left(\frac{M_1 + M_2}{2\pi M_1 M_2}\right)^{1/2} R T^{3/2} \tag{11a}$$

$$\approx 1.86 \cdot 10^{-3} \frac{T^{3/2}}{P\sigma_{12}^2} \left(\frac{M_1 + M_2}{M_1 M_2}\right)^{1/2}, \tag{11b}$$

where P is in atm, σ in Å.

The formula (10) is derived by means of the simplifying assumption that the molecules are hard spheres. Because of this limitation, (10) fails, in particular, to give correctly the relation between the diffusion coefficients and the temperature, as careful measurements show[69]).

To overcome this difficulty various other models have been used for the interaction of molecules in the gas domain. Figure 25 shows the variation of the forces of interaction with the distance between the molecules for

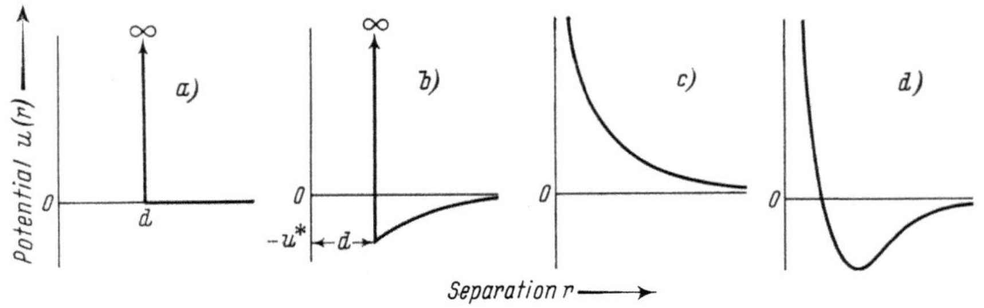

Fig. 25. Intermolecular potential functions for various models of the interaction of gas molecules
a) Billiard-ball model b) Sutherland—van-der-Waals model
c) Molecules as particles with mutual attraction proportional to $1/r$
d) As in c), but taking into account the repulsive forces (Lennard-Jones model)

rigid elastic spheres and for three other models. Formula (10) can be adapted to all the models by choosing a suitable temperature-dependent value for the effective diameter σ_{12} of interaction. For the *Sutherland—van-der-Waals model* (Fig. 25b)

$$\sigma_{\text{eff}} = \sigma'_\infty \sqrt{1 + S_{12}/T}; \tag{12}$$

S_{12} is the Sutherland constant and its numerical values are given in the literature[70]).

If, corresponding to Fig. 25c, we regard the molecules as particles which interact on one another with an energy E which is inversely proportional to some power of the distance r between them, i.e., if

$$E = E_{12}\, r^{-\nu},$$

then

$$\sigma_{\text{eff}} = \sigma_\infty \left(\frac{E_{12}}{2RT}\right)^{1/(\nu-1)}. \tag{13}$$

Formula (10) gives as the relation between the diffusion coefficients and the temperature

$$D_{12} \sim T^{1/2}.$$

However, in most observations on the relation between diffusion coefficients and temperature, it is not the concentration c but the pressure P which is held constant. In this case, by formula (11a) or (11b),

$$D_{12} \sim T^{3/2} \quad (P = \text{const}). \tag{14}$$

On the other hand, the use of formula (13) leads to

$$D_{12} \sim T^{3/2+2/(\nu-1)} \quad (P = \text{const}). \tag{15}$$

For rigid spheres with $\nu = \infty$, (15) becomes the same as (14). From measurements by VON OBERMAYER[69]), CHAPMAN and COWLING[71]) have calculated the values of the repulsion exponent ν for a number of binary mixtures of gases; these values range between 5 and 9. Formula (15) then gives

$$D_{12} \sim T^n, \quad 1.75 < n < 2 \quad (P = \text{const}). \tag{16}$$

HIRSCHFELDER[71a]) gives a different method of calculating the diffusion coefficients in binary mixtures of gases, leading to

$$D_{12} = 0.93 \cdot 10^{-3} \frac{T^{3/2}}{P\,\sigma_{12}^2\,W} \left(\frac{M_1 + M_2}{M_1 M_2}\right)^{1/2},$$

where the pressure is in atmospheres, the collision diameter σ_{12} in Å, the molecular weights in g/mole. The collision diameter σ_{12} is obtained additively from the collision cross-sections of the components, the latter being determined by HIRSCHFELDER from measurements of the internal friction. The collision integrals W are functions of the relative critical temperature of the gas mixture, $T/T_{12\,\text{crit}}$, where

$$T_{12\,\text{crit}} = 0.75 \sqrt{T_{1\,\text{crit}}\, T_{2\,\text{crit}}}.$$

$T/T_{12\,\text{crit}}$	W
2.0	0.540
2.4	0.506
3.0	0.475
4.0	0.442
5.0	0.421
6	0.386
8	0.386
10	0.371
50	0.288

Gas	T_{crit} (°K)	σ_0 (Å)
H_2	33.4	2.97
H_2O	647.5	~ 3.8
CO	133.1	3.59
CO_2	304.3	4.00
N_2	126.2	3.68

New expressions for the exact calculation of the temperature-dependence of the diffusion coefficients have been given by JOST[72]) and the results compared with the experimental values. The energy-distance relation according to the Lennard-Jones model, represented in Fig. 25d, has also been treated quantitatively by JOST.

The diffusion coefficients for binary mixtures of gases important in the reduction of ores, and the corresponding mean free paths, are given in Table 4. Formulae (9), (11b), and (13) were used for the calculations; the values used for σ_{12} (293°K) and ν were either taken from the literature[72]) or were estimated from the values for similar pairs of gases.

Table 4. *Diffusion coefficients* D (in cm²/sec) *and mean free path* λ (in cm) *in binary mixtures of gases* ($P = 1$ atm)

Temperature °C	Temperature °K	$D_T(CO/N_2)$	$D_T(H_2/N_2)$	$D_T(H_2/H_2O)$	$D_T(CO/CO_2)$
500	773	1.09	3.68	4.62	0.95
700	973	1.65	5.51	6.91	1.51
900	1173	2.30	7.64	9.58	2.20
1100	1373	3.05	10.06	12.62	3.01
Temperature °C	**Temperature °K**	$\lambda_T(CO/N_2)$	$\lambda_T(H_2/N_2)$	$\lambda_T(H_2/H_2O)$	$\lambda_T(CO/CO_2)$
500	773	$2.64 \cdot 10^{-5}$	$3.34 \cdot 10^{-5}$	$3.57 \cdot 10^{-5}$	$2.63 \cdot 10^{-5}$
700	973	$3.58 \cdot 10^{-5}$	$4.44 \cdot 10^{-5}$	$4.76 \cdot 10^{-5}$	$3.71 \cdot 10^{-5}$
900	1173	$4.52 \cdot 10^{-5}$	$5.62 \cdot 10^{-5}$	$6.01 \cdot 10^{-5}$	$4.91 \cdot 10^{-5}$
1100	1373	$5.53 \cdot 10^{-5}$	$6.84 \cdot 10^{-5}$	$7.32 \cdot 10^{-5}$	$6.22 \cdot 10^{-5}$

Thus far we have dealt with diffusion in gas mixtures of two components. But as a rule there are more than two gases present during industrial processes. Here we shall discuss only the case where, in addition to the reducing gas and the oxidized component – e.g., H_2 and H_2O or CO and CO_2 – there is also present one inert gas, such as nitrogen or argon, which does not take part in the reaction.

In the reduction of a sample of ore in such a mixture of gases it is again to be stipulated that the gas pressure shall be the same at all points, and this implies that

$$\frac{dc_1}{dx} + \frac{dc_2}{dx} + \frac{dc_3}{dx} = 0. \tag{17}$$

Using the subscript 1 for the reducing gas, 2 for the oxidized gas component, and 3 for the inert gas, we have further for the diffusion flows in the steady state

$$j_1 = -j_2, \quad j_3 = 0, \tag{18}$$

since j_1 and j_2 must be equal and opposite because the reaction is stoichiometric and in the steady state the diffusion flow of the inert gas must be zero because this gas is neither consumed nor generated.

The diffusion flow of each of the three components will be affected not only by the concentration gradient of this gas itself but also by the concentration gradients of the other two components. The nature of this coupling can be made clear qualitatively using the ideas of the kinetic theory of gases, thus:

Imagine a plane perpendicular to the x-axis at the point $x = 0$ in the gas domain. Let the inert gas have the constant concentration c_3, but the other two gases have variable concentrations. At a short distance x_0 from the reference plane the concentration of the first component is on the one side

$$\left[c_1 + x_0 \frac{dc_1}{dx}\right]_{x=+x_0}$$

and on the other side

$$\left[c_1 - x_0 \frac{dc_1}{dx}\right]_{x=-x_0}.$$

From (17), and (18),

$$\frac{dc_2}{dx} = -\frac{dc_1}{dx},$$

and therefore the concentrations of the second gas component on the two sides are

$$\left[c_2 - x_0 \frac{dc_1}{dx}\right]_{x=+x_0} \quad \text{and} \quad \left[c_2 + x_0 \frac{dc_1}{dx}\right]_{x=-x_0},$$

The number of collisions which the molecules of gas 3 situated at $x = 0$ encounter in the positive x-direction is

$$\vec{Z}_3 = \vec{Z}_{33} + \vec{Z}_{32} + \vec{Z}_{31}, \tag{19}$$

where the first subscript indicates the molecule struck, and the second the striking molecule. Each number of collisions, Z_{3i}, is proportional to the concentration of the struck molecules at $x = 0$ and to the concentration of the striking molecules at $x = -x_0$. It also depends on the probability of the collisions, which, assuming that the molecules are rigid spheres, is proportional to the mean value of the cross-sections of the participating molecules. If the mean values of the diameters of the molecules are again denoted by

$$\sigma_{32} = (\sigma_3 + \sigma_2)/2 \quad \text{and} \quad \sigma_{31} = (\sigma_3 + \sigma_1)/2,$$

then

$$\overleftarrow{Z}_3 \sim \sigma_{31}^2\left(c_1 - x_0 \frac{dc_1}{dx}\right) + \sigma_{32}^2\left(c_2 + x_0 \frac{dc_1}{dx}\right) + \sigma_3^2 c_3.$$

For the number of collisions in the opposite direction

$$\overleftarrow{Z}_3 \sim \sigma_{31}^2\left(c_1 + x_0\frac{dc_1}{dx}\right) + \sigma_{32}^2\left(c_2 - x_0\frac{dc_1}{dx}\right) + \sigma_3^2\, c_3 .$$

Hence for the difference in the number of collisions in the positive and negative x-directions we get

$$Z_3 = \overrightarrow{Z}_3 - \overleftarrow{Z}_3 \sim (\sigma_{32}^2 - \sigma_{31}^2)\frac{dc_1}{dx}. \tag{20}$$

This difference in the number of collisions vanishes only if the molecules of gases 1 and 2 have the same diameter σ. If Z_3 is not zero, then the gradients of the concentrations of gases 1 and 2 exert a dynamic force on gas 3 and cause a diffusion flow of this component. This effect has been demonstrated experimentally[73]). In the steady state of a reduction process, the flow of the inert gas must be zero, in accordance with (18). This is the case if a concentration gradient of the inert gas is built up which cancels out the effect of the other two gases. A detailed treatment of diffusion in ternary systems is given by HAASE[74]), based on the thermodynamics of irreversible processes. He comes to the conclusion that diffusion in ternary mixtures can be treated using just one diffusion coefficient, in the same way as in binary mixtures, only if the three sorts of molecules are closely similar. This corresponds to the point made earlier about the connection with the collision cross-sections, and is in agreement with HELLUND's experimental results[73]) on the one hand, and with the investigation of the diffusion of the very similar types of molecules, toluene and chlorbenzene in brombenzene, in the liquid phase[75]).

On the basis of the general Stefan-Maxwell theory of diffusion in mixtures of several gases SHERWOOD[75a]) has treated the diffusion of two gases in a mixture with a third gas which, in the steady state of the transport process, does not itself diffuse. The integration of the resulting differential equation has been carried out by GILLILAND[75b]). According to this, the quotient of the partial pressure, p_3 and p_3', of the inert gas at the ends of a diffusion path of length x, when $j_1 = -j_2$, is given by

$$\log\frac{p_3}{p_3'} = \frac{RT}{P}\, x \frac{j_1}{D_{13}}\left(1 - \frac{D_{13}}{D_{23}}\right).$$

In agreement with (20), the transport vanishes, i.e., $p_3 = p_3'$, if

$$D_{13} = D_{23} \quad \text{or if} \quad j_1 = 0,$$

which would correspond to

$$\sigma_{13} = \sigma_{23} \quad \text{or} \quad \frac{dc_1}{dx} = 0.$$

The ratios of the binary diffusion coefficients of CO/N_2, CO_2/N_2, and CO/CO_2 lie between 1 and 1.4. An appreciable transport, and therefore a noticeable difference in the partial pressures of the inert gas, the nitrogen, in the free gas domain and at the reacting ore surface is therefore not to be expected in normal blast-furnace operation, at any rate not if the hydrogen content of the reducing gas remains small. The diffusion coefficients of binary mixtures with hydrogen, on the other hand, are greater by a factor of about 4 than those of other binary mixtures of gases which are of importance for the reduction process. In reduction by gases having a higher hydrogen content then, the fact that the partial pressure of the inert gas is affected by the diffusion of the hydrogen to the ore surface and by the diffusion of water vapour back into the free gas domain must be taken into account; the partial pressure of the inert gas at the reacting oxide surface will, in fact, necessarily be less than its value in the free gas domain, because the inert gas is transported away by the water vapour.

For the diffusion flux j through the boundary layer, of apparent thickness l, which forms on the surface of the lumps of ore, it is convenient to define a material-transport coefficient β:

$$j = \beta \, \Delta p / RT, \qquad \beta = D/l. \tag{21}$$

Δp is the difference of the partial pressure of the diffusing component at the surface of the solid and in the gas space.

This concept has the advantage that it is meaningful not only for diffusion through a laminar boundary layer but also formally for the interchange of material even when turbulence occurs, although the boundary layer for the flow then ceases to exist in its original sense.

The *Sherwood number* N_{Sh} is made up of the material-transport coefficient, the diffusion coefficient D, and a characteristic length d — the diameter of a sphere in the flow, say:

$$N_{\mathrm{Sh}} = \beta \, d/D = d/l. \tag{22}$$

The *Reynolds number* N_{Re}, which characterizes the flow, is defined by

$$N_{\mathrm{Re}} = u \, d/\nu, \tag{23}$$

where u is the flow velocity, and ν is the kinematic viscosity, which is defined by $\nu = \eta/\varrho$. (If the viscosity η is measured in g cm^{-1} sec^{-1}, and the density ϱ in g cm^{-3}, then the kinematic viscosity is in cm^2 sec^{-1}.)

The *Schmidt number* N_{Sc} is defined by

$$N_{\mathrm{Sc}} = \nu/D. \tag{24}$$

These three numbers are connected by the relation[76])

$$N_{\mathrm{Sh}} = C + C' N_{\mathrm{Re}}^m N_{\mathrm{Sc}}^n, \tag{25}$$

where C and C' are constants. According to KRISCHER and LOOS[77]), for a sphere in the flow, $C = 2$. So, if $u = 0$, the Sherwood number also has the value 2. C' has a value of about 0.6 [76, 77]). For laminar flows, $m = 1/2$

and $n = 1/3$ [76]). For turbulent flow, the exponent m is between 0.5 and 0.8, while $n = 1/3$ still[77]). It follows that the material-transport coefficient β is roughly proportional to the square-root of the flow velocity u.

For pure gases the Schmidt number has values, dependent on the pressure, temperature, and type of gas, between 0.6 and 0.8 [68, 72]). For mixtures of gases the relation between viscosity and the mixing-ratio can be expressed by SUTHERLAND and THIESEN's formula

$$\eta_{12} = \eta_1[1 + a_{12} N_2/a_{11} N_1]^{-1} + \eta_2[1 + a_{21} N_1/a_{22} N_2]^{-1}, \qquad (26)$$

where N_1, N_2 are the mole fractions of the two gases, and the a_{jk} are empirical constants. The value of η_{12} given by (26) may increase or decrease monotonically or may show a flat maximum. Thus at 14.7°C, the viscosity of hydrogen is $89.3 \cdot 10^{-6}$, that of carbon dioxide is $146.8 \cdot 10^{-6}$, and the viscosity of a mixture of the two gases with 48.4% H_2 is[78]) $148.5 \cdot 10^{-6}$ g cm^{-1} sec^{-1}. A theoretical treatment of viscosity in a mixture of gases is given by BIRD, HIRSCHFELDER and CURTISS[78a]).

The inter-diffusion coefficient D_{12} for CO_2/H_2 mixtures, on the other hand, is largely independent of the mixing-ratio, and at normal temperature and pressure amounts to 0.550 cm^2 sec^{-1}. In accordance with formulae (10), (11a, b) this value lies between the diffusion coefficients of the pure gases, which amount to $D(H_2) = 1.24$ and $D(CO_2) = 0.104$.

Accordingly, the Schmidt number for mixtures of gases cannot have a constant value independent of the type and concentration of the gas mixture. Generally the Schmidt number takes smaller values for gas mixtures than for pure gases. This difference is the more pronounced the greater the differences in the diffusion coefficients and densities of the constituent gases in the mixture. For a 1 : 1 mixture of hydrogen and carbon dioxide at 20°C the Schmidt number is 0.25. As the proportion of the heavier gas increases, the Schmidt number decreases, since η is almost constant between 50 and 100% CO_2, ϱ increases, and D_{12}, as stated above, remains constant. In H_2/N_2 mixtures similar circumstances hold.

The dependence of ν_{12} on the temperature can be expressed by a formula which is similar to the Sutherland formula for the temperature-dependence of D_{12} and which is found by combining (11a) with (12):

$$\nu = \nu_0 T^{3/2} [1 + S_\nu/T]^{-1}.$$

Values of the viscosity for binary mixtures and the corresponding Sutherland constant S_ν are given in collections of tables[78]). For pure gases ANDRUSSOV[78b]) has discussed the dependence of viscosity, diffusion, and heat conductivity on temperature. A detailed guide to the literature on the viscosity of binary mixtures of gases is given by STAUDTE[79]).

In the diffusion through the pores of the lumps of ore or of the reduction products the restriction of the cross-section available for the diffusion of the gas must be taken into account. Further, a factor must be introduced to cater for the orientation of the pores relative to the diffusion path and also for the branching and cross-linking of the pores. This is achieved by using a porous-diffusion coefficient

$$D_p = D \gamma \xi, \tag{27}$$

where γ is the relative volume of pores, i.e., the volume of the pores relative to the whole volume of the lump of ore, and ξ is the labyrinth factor, which depends on the structure of the pores and can generally be estimated only empirically.

Substitution of D_P in Eq. (3) gives the diffusion flow j for porous diffusion, with j, for example, in mole sec^{-1} cm^{-2} relative to the exterior surface area of the lump of ore.

The gas molecules collide not only with one another but with the surfaces of the solid bodies, e.g., with the walls of the pores. If the semi-diameter of the pores $d/2$ is less than $\bar{\lambda}$, the mean free path of the gas molecules, then collisions with the walls of the pores will be more frequent than collisions of the gas molecules with one another. The assumptions made in deriving formula (10), (11a, b) are then no longer valid. In this case it is the *Knudsen diffusion coefficient* $D_{K,i}$ which is relevant. Replacement of $\bar{\lambda}$ by $d/2$ in formula (7) yields, together with $j = -D(dc/dx)$ and formula (8)

$$D_{K,i} = \frac{d}{6}\left(\frac{8RT}{\pi M_i}\right)^{1/2}. \tag{28}$$

Comparison with (11a) shows that, in contrast to the normal coefficient of gas diffusion, the Knudsen diffusion coefficient in pores with $d/2 < \bar{\lambda}$ is independent of the total pressure and of the effective cross-section of the molecules.

The mean free path in a mixture of gases is given, to a first approximation, by the formula (9), according to the kinetic theory of gases. Numerical values of $\bar{\lambda}$ are given in Table 4 (p. 54), these being calculated by means of (9) and (13).

If the rate-determining sub-process of a reaction is the diffusion of the gases, then, in accordance with section 1.2.1, it may be assumed that the reaction at the phase boundaries takes place in the immediate neighbourhood of the equilibrium conditions. If c_1^0 and c_2^0 denote the concentrations of the reducing gas and the oxidized gas respectively in the free gas space, and if c_1^* and c_2^* denote the equilibrium concentrations at the surface c^* of the oxide, then

$$\begin{aligned} j_1 &= -D_{12}\frac{dc_1}{dx} = -\frac{D_{12}}{l}(c_1^0 - c_1^*), \\ j_2 &= -D_{12}\frac{dc_2}{dx} = -\frac{D_{12}}{l}(c_2^0 - c_2^*). \end{aligned} \tag{29}$$

Here l denotes the length of the diffusion interval, i.e., the length of the pores or the apparent thickness of the diffusion layer. The coordinate system is chosen so that the positive x-axis extends from the reacting oxide surface perpendicularly, along the longitudinal axis of the pore, regarded as cylindrical, and into the free gas domain. The auxiliary conditions to be taken into account with (29) are

$$j_1 = -j_2; \quad c_1^0 + c_2^0 = c_1^* + c_2^* = c, \quad p_1^0 + p_2^0 = p_1^* + p_2^* = P. \tag{30}$$

Introducing the equilibrium constant K

$$K = c_1^*/c_2^* = p_1^*/p_2^*,$$

we obtain from (29) and (30)

$$j_1 = -j_2 = -\frac{D_{12}}{l} c_1^0 \frac{1}{1+K}\left(1 - K\frac{c_2^0}{c_1^0}\right). \tag{31}$$

So the diffusion flow is directly proportional to the concentration c_1^0, or to the partial pressure $p_1^0 = c_1^0 RT$, of the reducing gas in the free gas space, provided that the product $K\, c_2^0/c_1^0$ is much less than 1 or is constant.

If c_1^0 or p_1^0 changes due to a variation in the ratio of partial pressures p_1^0/p_2^0, then in general it is not to be expected that the proportionality relation

$$-j_1 \sim c_1^0$$

will hold. If

$$p_1^0 + p_2^0 = p_1^* + p_2^* = P,$$

then

$$j_1 = -j_2 = -\frac{D_{12}}{l}\frac{p_1^0}{RT}\left(1 - \frac{K}{1+K}\frac{P}{p_1^0}\right). \tag{31a}$$

This expression again shows that a linear relation between the diffusion flow and the partial pressure of the reducing gas exists only if the second term in the brackets is small compared with 1. Since the partial pressure p_1^0 cannot exceed the total pressure P, this will be the case only if the equilibrium constant K is small compared with 1. This occurs only for the reduction of Fe_2O_3 to Fe_3O_4 and for the reduction of Fe_3O_4 to FeO at high temperatures. The equilibrium constant for the latter process increases as the temperature decreases; for reduction with either hydrogen or with carbon monoxide it is already 0.44 at about 800°C. The temperature dependence of the diffusion flow will be determined by the temperature dependence of D_{12} in accordance with (16), but in addition by the effect of temperature on the equilibrium constant K. Figure 26 shows the variation of the expression in brackets in Eq. (31a) with temperature for $K = K(\text{Fe/FeO})$ and $K = K(\text{FeO}/Fe_3O_4)$ and with CO as the reducing gas for two values of $P/p^0(\text{CO})$ in each case. Since, from (11a), the diffusion coefficient D_{12} is inversely proportional to the total pressure P, i.e.,

$$D_{12} = D'_{12}/P,$$

(31 a) implies

$$j_1 = -\frac{D'_{12}}{RTl}\left(\frac{p_1^0}{P} - \frac{K}{1+K}\right). \tag{31 b}$$

In a change of the partial pressure p_1^0 due to a change in the total pressure P, therefore, the diffusion flow remains constant.

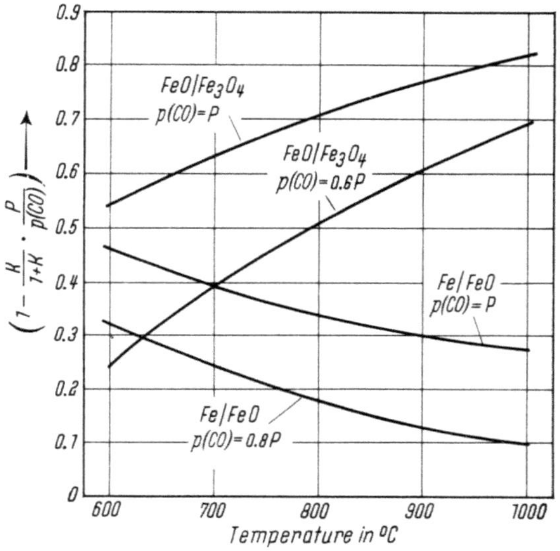

Fig. 26. Dependence on temperature of the gas composition for the reduction of iron oxides

1.2.3. Adsorption

The phase-boundary reaction of ore reduction takes place in sub-processes. The molecules of the reducing gas first impinge on the surfaces of the solid bodies. But only those molecules can react with the oxide which are not immediately reflected back into the gas space, because the phase-boundary reaction requires a certain time to take place. Some of the gas molecules do remain for some time on the surfaces because they become attached to them by physical or chemical forces. This binding process is called *adsorption*. In the range of temperatures in which reduction of ores takes place at a measurable rate, only those molecules remain for an appreciable time on the oxide surface which are attached by chemical forces. The physically adsorbed molecules remain only for about 10^{-7} sec, as will be shown. In the chemically adsorbed molecules individual molecular bonds may be loosened or even dissolved. The newly formed bonds between the chemically adsorbed molecules or atoms and the surface structural elements of the solid bodies may, on the other hand, be so strong that, in the subsequent desorption of the gas, atoms from the solid body's lattice may pass into the gas space by forming compounds with the gas. The separation of oxygen from an ore by H_2 or CO by the formation of H_2O or CO_2 may be imagined to take place atomically in this way. The greatest possible rate of reaction between gas and oxide is limited by the number of gas molecules which impinge on unit surface area in a unit of time. This number of impacts Z_i is given by

$$Z_i = n_i \bar{v}_i/4 = c_i N_L \bar{v}_i/4 = p_i N_L \bar{v}_i/4RT.$$

By (8), the mean velocity $\bar{v}_i$ of a gas molecule is

$$\bar{v}_i = (8RT/\pi M_i)^{1/2},$$

and therefore the number of impacts is

$$Z_i = p_i N_L/(2\pi RTM_i)^{1/2}. \tag{32}$$

For hydrogen, for example, at 1000°C and 1 atm, Eq. (32) gives about $5 \cdot 10^{23}$ impacts per cm^2 per second. Comparison with the measured values shown in Fig. 76 shows that the number of impacts is greater than the measured reaction rates by a factor of about 10^4.

An incident molecule interacts with the solid surface and may be adsorbed. Two groups of forces of interaction are distinguished, according to their type and magnitude; and from this classification adsorption may be subdivided into physical or chemical adsorption. In *physical adsorption* the forces acting are the same as those between individual particles of a gas. In *chemical adsorption*, on the other hand, the interaction is based on the same phenomena as is the chemical bond. Naturally, transitional types between the two sorts of adsorbent-adsorbate bond may occur.

The two main groups of interaction in chemical adsorption are known as *homopolar binding* and *ionic binding*. While in physical adsorption the energy of interaction seldom exceeds a few kilo-calories per mole, in chemical adsorption processes values of the heat of adsorption ΔH of -200 kcal/mole are no rarity; they frequently exceed the heat of formation of the corresponding solid compound which can be formed by the reaction of the gas with the adsorbent.

The amount of energy interchanged between the adsorbent and the adsorbate can lead, in metals, to the separation of individual metal atoms from the lattice structure and their incorporation into the layer of chemical adsorption[80]). With this process the boundary between chemical adsorption and subsequent formation of a new phase compound at the surface of the adsorbent has already been reached.

The high values of the heat of adsorption have the effect that, in the desorption of the gas, atoms can be torn out from the metal lattice with the gas, as may be checked by means of the field ion microscope[81]).

A good general picture of the interaction in physical adsorption and chemical adsorption can be gained from diagrams showing the operative forces as a function of the separation between the adsorbent and adsorbate.

As an example, Fig. 27, based on DE BOER's data[82]), shows such a diagram for the adsorption of hydrogen molecules, and the chemical

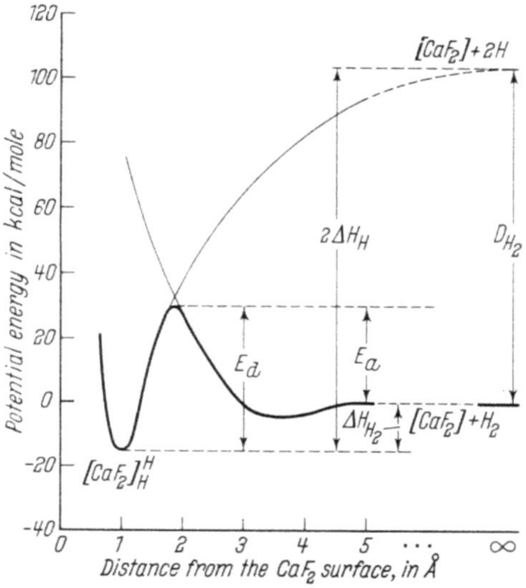

Fig. 27. Potential-distance diagram for the adsorption of hydrogen in calcium fluoride, according to DE BOER[82])

E_a, E_d = activation energy for adsorption and desorption of H_2 as H-atoms, ΔH_H, ΔH_{H_2} = adsorption energy for atomic and molecular hydrogen, D_{H_2} = dissociation energy of the hydrogen molecule

adsorption of hydrogen atoms, on calcium fluoride. The activation energy E_a for the chemical adsorption of hydrogen and the activation energy E_d of desorption were determined experimentally. It was also taken into account that the chemical adsorption of atomic hydrogen takes place without activation energy. The first, shallow potential-minimum is the equilibrium position for physical adsorption. The potential barrier between the first and the second minimum corresponds to the dissociation of the physically adsorbed hydrogen molecules, the atoms passing into the chemically adsorbed state.

Because of the behaviour as regards energy, exemplified by Fig. 27, the amount of gas adsorbed at constant pressure varies with temperature as shown in Fig. 28 for the adsorption of hydrogen on zinc oxide[83]). At low temperatures the gas is attached in the first, shallow energy minimum associated with physical adsorption. As the temperature rises, the amount of physically adsorbed gas decreases sharply because the stability of the attachment is slight. But as the temperature rises, it also becomes increasingly possible for the gas to surmount the potential barrier between physical and chemical adsorption, that is, between the first and second energy minimum. The renewed rise in the amount adsorbed above its minimum at around 200°K is to be attributed to this process. It is also the case for chemical adsorption that, in equilibrium, the amount of gas attached at the surface must decrease as the temperature rises. However,

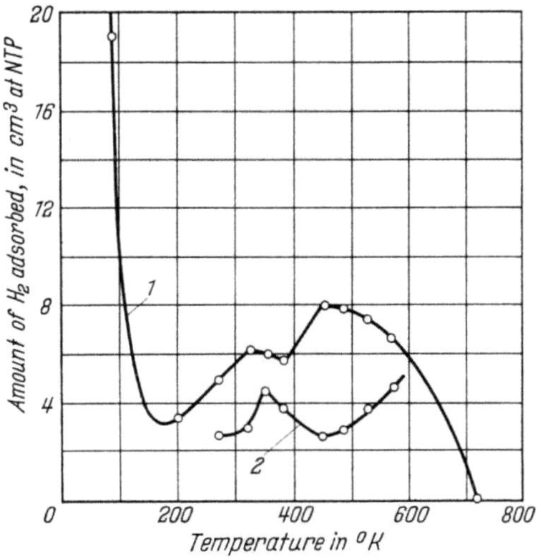

Fig. 28. Adsorption isobars for hydrogen on zinc oxide at 700 Torr, according to TAYLOR[83])

Curve *1* after 1000 min, Curve *2* after 5 min

the part of curve *1* between 200 and 450°K does not correspond to an equilibrium. Due to the slower rate of chemical adsorption in this temperature range, an equilibrium state is not set up, as may also be seen by comparing curves *1* and *2*.

In the temperature range which is of importance in the reduction of ores, it is a question of very small concentrations so far as physical adsorption is concerned; on the other hand, the rate of the physical adsorption is very high. An estimate may be arrived at from the following argument, which goes back to LANGMUIR[84]).

If we call the ratio of the actual surface-concentration to its maximum possible value the *covering ratio*, Θ, then we can suppose that the rate of adsorption is proportional to the free surface, $1 - \Theta$, and that the rate of desorption is proportional to the covered surface, Θ. Hence

$$\frac{d\Theta}{dt} = k_1(1 - \Theta) - k_2\,\Theta. \qquad (33)$$

Integrating with respect to time and taking $\Theta = 0$ for $t = 0$, we have

$$\Theta = [k_1/(k_1 + k_2)]\,[1 - \exp\{-(k_1 + k_2)\,t\}].$$

As $t \to \infty$, $\Theta \to \Theta^*$, its equilibrium value; therefore

$$\Theta^* = k_1/(k_1 + k_2),$$

and so, putting $k_1 + k_2 = k$, we have

$$\log[\Theta^*/(\Theta^* - \Theta)] = k\,t. \tag{34}$$

If s denotes the surface concentration in particles per cm^2, and if s_1 denotes the corresponding value when $\Theta = 1$, then

$$\frac{d\Theta}{dt} = \frac{1}{s_1}\,\frac{ds}{dt}.$$

The number of molecules adsorbed in unit time is given by the expression $\alpha\,Z(1 - \Theta)$, $Z(1 - \Theta)$ is the number of gas molecules impinging on the free surface and α is the *adsorption coefficient*, which is always less than 1. The number of molecules desorbed in unit time is equal to $\nu_1\Theta$, where ν_1 denotes the rate of desorption for $\Theta = 1$, i.e., for total covering. Hence

$$\frac{d\Theta}{dt} = \frac{1}{s_1}\,\frac{ds}{dt} = \frac{\alpha\,Z}{s_1}\,(1 - \Theta) - \frac{\nu_1}{s_1}\,\Theta. \tag{35}$$

Comparison of (35) and (33) gives

$$(\alpha\,Z + \nu_1)/s_1 = k = k_1 + k_2 \quad \text{and} \quad k_1 = \alpha\,Z/s_1; \qquad k_2 = \nu_1/s_1.$$

From (34) the time at which $\Theta = \Theta^*/2$ is given by

$$t|_{\Theta=\Theta^*/2} = (s_1 \log 2)/(\alpha\,Z + \nu_1) < s_1/\nu_1. \tag{36}$$

The expression s_1/ν_1 is equal to the mean time interval during which a gas molecule remains in the adsorption layer. For physical adsorption this time interval τ has been measured or calculated on various occasions[85, 86]). CLAUSINGS's measurements[87]) show that τ decreases sharply in numerical value as the temperature rises. At ambient air temperatures its value is between 10^{-7} and 10^{-5} sec.

For $\Theta = 1 - \Theta = 1/2$, the adsorption coefficient α can also be calculated. Using the value given earlier for Z, the number of impacts, for hydrogen at 1 atm and 1000°C, α is found to be between 10^{-3} and 10^{-1}. From α and Z we again get the maximum rate of interchange between the gas phase and the adsorption layer. It is so large that certainly physical adsorption cannot be the process determining the time for the reduction of ore.

It follows from (33) that for equilibrium, with $d\Theta/dt = 0$,

$$\Theta = \alpha\,Z/(\nu_1 + \alpha\,Z) = \frac{\alpha\,p_1\,N_L\,\bar{v}_1}{\alpha\,p_1\,N_L\,\bar{v}_1 + 4\,RT\,\nu_1}$$

$$= \frac{\alpha\,p_1\,N_L}{\alpha\,p_1\,N_L + (2\,RT\,\pi\,M_1)^{1/2}\,\nu_1}, \tag{37}$$

taking into account the relation between the number of impacts Z and the partial pressure of the impinging gas, as found from (32). Formula (37) is called the *Langmuir adsorption isotherm*. According to this equation, Θ increases proportionally to p_i for low gas pressures, but at higher pressures the rate of increase is smaller.

In the literature a non-linear dependence of the reaction rate on the partial pressure of the reducing gas has been described at various times; this dependence can be expressed by an equation similar to (37) and therefore it can be interpreted by assuming a proportionality between the reaction rate and the covering ratio Θ.

1.2.4. Chemical Adsorption

Physical adsorption depends only slightly on the chemical properties of the adsorbent and adsorbate, and is conditioned much more by physical quantities, such as the boiling-point and vapour pressure, mole volumes, van der Waals' constants, polarizability of the adsorbate, and the surface area and porous structure of the adsorbent.

In chemical adsorption, on the other hand, it is the chemical forces between the gas and the solid bodies which are effective. Hence the interaction and the coverage of the surface of a solid depend strongly on the chemical nature of the reagents, and particularly on their capability and tendency to exchange electrons or to share electron orbits. In the first case *ionic*, in the second, *covalent surface binding* occur. Transitional cases between these two types of binding occur. Information about the nature of the binding can be obtained by the following methods[88, 89]:

General: surface concentration s, $s = f(p)$; $s = f(T)$; $s = f(t)$.

Polar component of the binding: surface potential $\Delta\chi$; electron work-function φ_e; (electrical conductivity $\varkappa$).

Non-polar component of the binding: infra-red absorption.

Distribution of the charge-carriers: electrical conductivity $\varkappa$; magnetic susceptibility; electron-spin resonance.

Because of their different electronic structure, metals and semi-conducting oxides must be discussed separately as adsorbents. With metals, the rate of adsorption and the surface deposition attained is often far greater than with semi-conductors, but the activation energy considerably less. Ehrlich[90] interprets this difference between chemical adsorption on metals and semi-conductors by supposing that the adsorption points on the surface of semi-conductors are widely separated and that their potential curves do not overlap, so that the dissociation of a molecule which is tied to one place is not assisted by the attractive forces at the next place. Semi-conductors therefore have, in general, a high activation energy for the desorption of adsorbed atomic hydrogen and are poor catalysts for the recombination of atomic hydrogen.

Through investigations of the diffraction of slow electrons (by the LEED method – Low-Energy Electron Diffraction), it has recently been possible to show that the chemically adsorbed gas atoms form on the metal surface superstructures which vary with the covering ratio. Since the binding energy of the ad-atoms must depend on the structure, this finding implies that a gradual change of the constants for adsorption equilibrium is to be expected. But these gradations appear to be too shallow or blurred to be detected by the previous methods. The results of LEED measurements cannot yet be interpreted unequivocally, because of the double diffraction of electrons which occurs. They show, however, the sort of effect which the crystal structure of the adsorbing material can have on the surface reaction.

Regions of the metal close to the oxide may serve as surfaces for the initial chemical adsorption of the gas; then the gas particles adsorbed on the metal can, by surface diffusion, get to the oxide and react with it. The boundary of the three phases, oxide/metal/gas, is certainly important as a place of chemical adsorption and surface reaction.

For ore reduction the chemical adsorption of the reducing gas on semi-conducting oxides is also to be considered. To understand this process we have to go into the *electronic structure* of semi-conductors. In *metals* a certain number of electrons occupy incompletely filled energy bands. These electrons can move freely in the crystal lattice. In *semi-conductors* all the electrons are in completely filled bands at absolute zero and therefore cannot move. The upper filled band, *the valence band*, is separated from the first empty band, the *conduction band*, by an energy gap. When thermal energy is acquired, a certain number of electrons, depending on the width of this forbidden band, are able to leave the valence band and pass over into the conduction band, and are then free to move through the lattice as *conduction electrons*. The electron vacancies thus formed are called *holes*; the holes, too, can move freely through the lattice. In Fig. 29a these relations are represented schematically by means of the band model and the relations between the chemical and electrochemical potential of the defect positions μ and $\eta = \mu + z F \varphi$, where φ denotes the electric potential.

By incorporation of impurity atoms or by deviations from the stoichiometric composition of the semi-conductor, electron levels in the forbidden band can arise. The electrons contained in them do not move. If such an impurity level can give up electrons to the conduction band, it is called a *donor level*, see Fig. 29b. Impurity levels which accept electrons and can therefore produce holes are called *acceptors*.

Now such impurity levels can also arise on the surface of a semi-conductor. On pure surfaces they are produced by the formation of an unsaturated valency by the surface disturbance of the lattice periodicity[91]). On a pure germanium surface, for example, there are acceptor levels.

Impurity levels are also formed on semi-conductor surfaces through gas adsorption. Even the adsorbed gas particles themselves may appear as donors or acceptors, if they react with the solid by interchange of electrons and thus transform into positively or negatively charged ions[92]). The

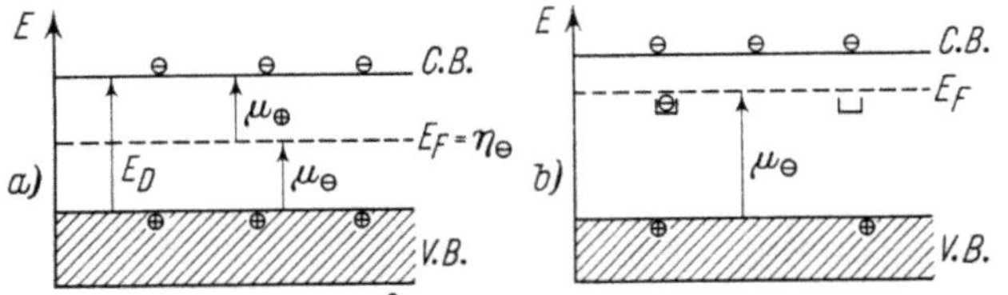

$$\ominus + \oplus \rightleftharpoons 0 \;; \quad n \times p = n_i^2 ; \quad \mu_\ominus + \mu_\oplus = \mu_i = E_D ; \quad \eta_\ominus = \mu_\ominus - e\varphi = E_F ;$$
$$\varphi = 0 : \eta_\ominus = \mu_i - \mu_\oplus = \mu_\ominus$$

Fig. 29. Energy levels of electrons in semi-conductors, in the form of a band model, and the chemical and electro-chemical potential μ and η of electrons $\ominus$ and defect electrons (or "holes") $\oplus$

E_F = Fermi potential, E_D = width of forbidden band, V.B. = valence band, C.B. = conduction band

a) Intrinsic semi-conductor b) Semi-conductor with n-type or donor impurities

formation of surface levels is also to be expected in covalent chemical adsorption[93]). By the linkage of the bonds the charge exchange levels of the surface elements of the solid are shifted, so that they can accept additional electrons or give up electrons to the semi-conductor.

In *insulators* also, i.e., semi-conductors with a very wide forbidden band, during chemical adsorption surface levels may form and exchange charge. In this case, acceptor levels must lie sufficiently close to the valence band and donor levels to the conduction band.

In chemical adsorption, with the passage of electrons between the semi-conductor and adsorbate and the formation of chemically adsorbed ions, the exchange of charge is always one part of the rate-determining process. In covalent chemical adsorption, the transfer of charge from the surface level formed may also belong to the rate-determining process. It may, however, be only a consequential process which is unimportant as regards the production of the surface linkage. In this case, of course, the energy exchange in the transfer of charge from the surface level formed must be small. Equivalently, the charge-transfer reaction takes place incompletely or the amount of charge transferred remains small. In this case, the formation of, and transfer of charge from, the level, and the associated changes in the electronic structure and properties of the semi-conductor, are to be regarded only as indicators of the chemical adsorption taking place and not as its prerequisites.

Figure 30 shows diagrammatically the variation of the chemical potential of the electrons and holes, $\mu_\ominus$ and $\mu_\oplus$, and also the electrochemical potential of the electrons $\eta_\ominus$ in a semi-conductor during the formation of acceptors by a chemical adsorption process. The negative counter-charges corresponding to the positive surface charges arising during the chemical ad-

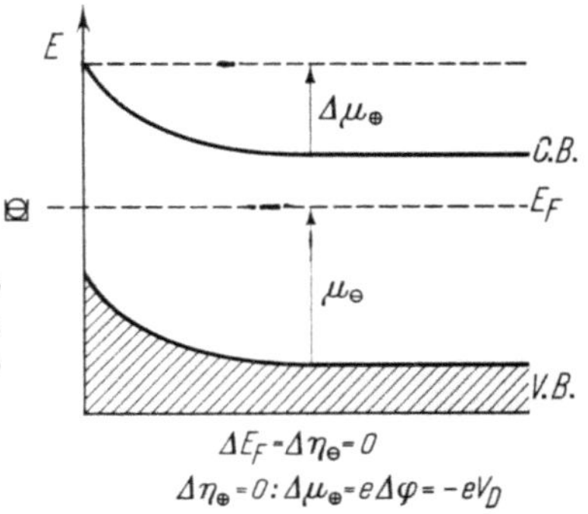

Fig. 30. Effect of chemical adsorption at the surface of a semi-conductor, with donors being formed on the energy bands

Symbols as in Fig. 29

sorption are disposed as space charges in the interior of the semi-conductor. The space charge zone may be up to 10000 Å thick, depending on the concentration of charge-carriers in the semi-conductor. By this enrichment in charge-carriers the chemical potentials of the electrons and holes in the surface are changed. The equilibrium conditions

$$\Delta\eta_\ominus = \Delta\mu_\ominus + F\Delta\varphi = 0 \quad (\varphi \text{ is the interior electrical potential})$$

therefore entail the existence of an electrical potential difference V_D between the surface and interior of the semi-conductor. The space charge density, concentration of the charged surface levels, and the potential difference V_D are connected with one another through *Boltzmann's* and *Poisson's equations*. The mathematical treatment is similar to that for the diffuse double-layer at electrodes in aqueous electrolytes. Various solutions of the equations have been given[93a]). The change in the concentration of charge-carriers in the boundary layer causes a change in the surface conductivity $\Delta\varkappa$. Figure 31 shows the calculated relationship between the potential difference V_D and the change in conductivity $\Delta\varkappa$ for weakly n-conducting germanium[94]).

For estimating the participation of free charge-carriers in chemical adsorption processes at the surface of semi-conductors, resistance measurements are more suitable than they are with metals, since the enrichment or impoverishment of carriers extends up to 10^{-4} cm into the interior and therefore can produce an appreciable effect even on samples up to 0.1 mm thick, if these layers are arranged as the current path parallel to the higher resistance of the interior of the semi-conductor. As an example, Fig. 32 shows the effect of the adsorption of water vapour on a single

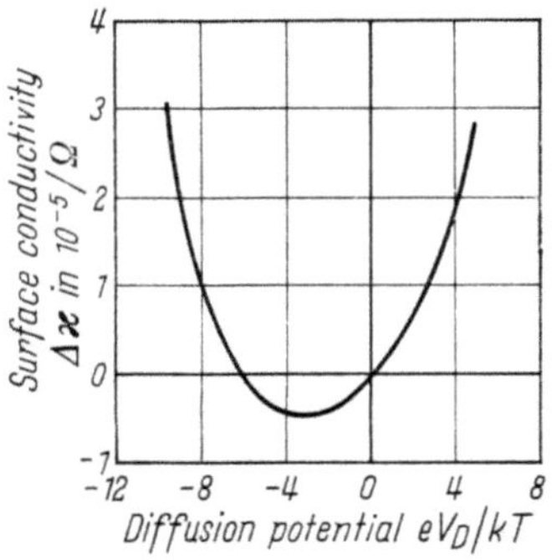

Fig. 31. Dependence of the surface conductivity $\Delta_\varkappa$ on the diffusion potential eV_D/kT for n-conducting germanium, according to SCHRIEFFER[94])

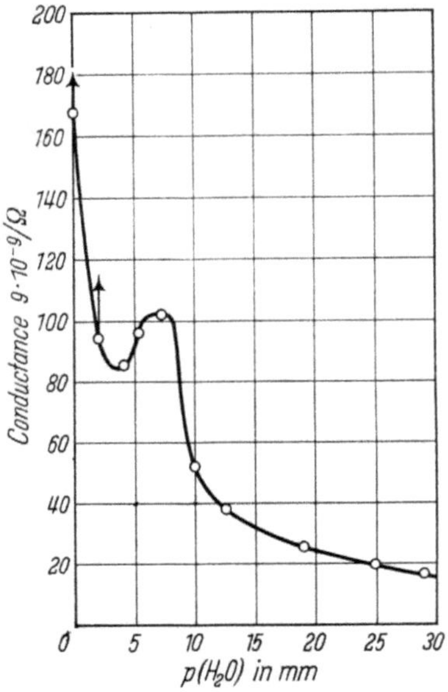

Fig. 32. Effect of water-vapour pressure on the electrical conductivity of Cu_2O at room temperatures, according to BRAUER[95])

crystal of Cu_2O on the electrical conductivity of the oxide[95]). With porous samples, such as sintered materials, gas can penetrate into the interior of the sample and by causing an impoverishment of charge-carriers at the surface of the crystals produce barrier layers between the crystallites. In this way an impoverishment in charge-carriers produces, in particular, a strong effect on the conductivity[96]). The formation of such barrier-layers causes a frequency dispersion in the value of the resistance to alternating current[97]).

The change in the electron work-function or in the surface potential $\Delta\chi$ is related to V_D by

$$\Delta\chi = V_D + d_{M-A}\,\mathfrak{F}(0).$$

The additive term $d_{M-A}\mathfrak{F}(0)$ takes into account the field $\mathfrak{F}(0)$ between the surface of the semi-conductor and the "centre of gravity" of the charges of the chemically adsorbed, charged gas particles; d_{M-A} is the separation of the charges.

In most cases the rate of chemical adsorption in semi-conductors can be expressed by the *Elovich equation*

$$ds/dt = a \exp(-b\,s). \tag{38}$$

This relation evidently implies that the activation energy of the process increases with the surface concentration s. This dependence may be explained, e.g., by the frequently observed linear relation between the covering-ratio and the adsorption energy, if it is assumed in addition that the activation energy increases linearly with falling adsorption energy[98]. Figure 33 shows this relation diagrammatically in the form of distance-potential curves. A different interpretation has been argued by TAYLOR[99].

TAYLOR shows that the Elovich equation can be deduced from the hypothesis that chemical adsorption takes place only at particular points on the surface and further that in the course of the chemical adsorption these sites of adsorption are reproduced by a bi-molecular reaction.

A further model is based on the assumption that a surface impoverishment of the charge-carriers necessary for the ionization of the adsorbed gas particles, or the associated potential difference V_D, is responsible for the increase in the activation energy.

To derive the Elovich equation from the assumption of a relation between the activation energy and V_D, a linear relation must be stipulated

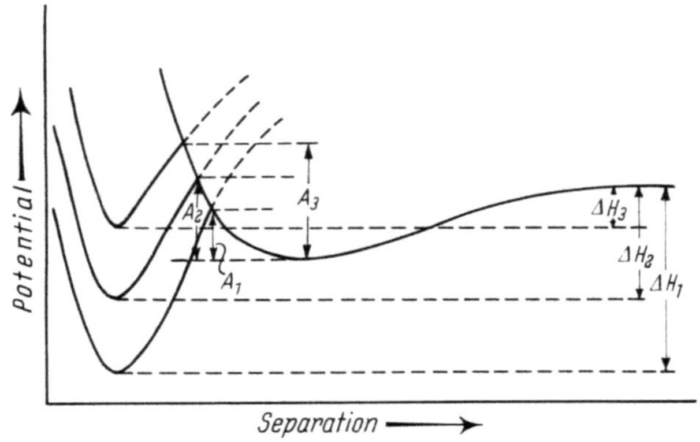

Fig. 33. Dependence of the activation energy A and the heat of reaction for chemical adsorption on the covering-ratio (diagrammatic)

between the potential difference V_D, and the number s of atoms or molecules chemically adsorbed when charge is taken up. But $V_D \sim s^2$ for an impoverishment of carriers and $V_D \sim \log s$ for an enrichment[100]. A time relation of the form

$$ds/dt \sim \exp(-V_D)$$

does not therefore lead to the Elovich equation.

Since the potential threshold which has to be overcome in the passage of charge between the adsorbent and the adsorbate is situated in space round the surface of the semi-conductor, it is affected by the field strength prevailing here, $\mathfrak{F}(0)$, and the separation of the physically adsorbed gas particles, d_{M-A}, from the semi-conductor surface. Since $\mathfrak{F}(0) \sim s$ always holds[100]), the Elovich equation can be interpreted as arising from the assumptions $ds/dt \sim \exp(-E_A)$ and $E_A = E_A^0 - d_{M-A}\mathfrak{F}(0)$.

Of course, this argument holds without further assumptions only if the chemically adsorbed gas particles themselves are the surface level. For homopolar chemical adsorption with the formation of levels in the semi-conductor surface, d_{M-A} must be given a different meaning. The transfer of charge for such surface levels is also an activated process, as measurements of the rate of transfer of charge, by means of the field effect, show[101]). Thus KINGSTON[102]) observed the half-life periods for the charge-transfer process of the surface levels in germanium to have values of up to one second. These measurements and other observations as well[100]) show that the half-life periods for charge transfer by surface levels have no unique value but that a spectrum of half-life periods occurs, and this points *a priori* to a heterogeneity of the surface.

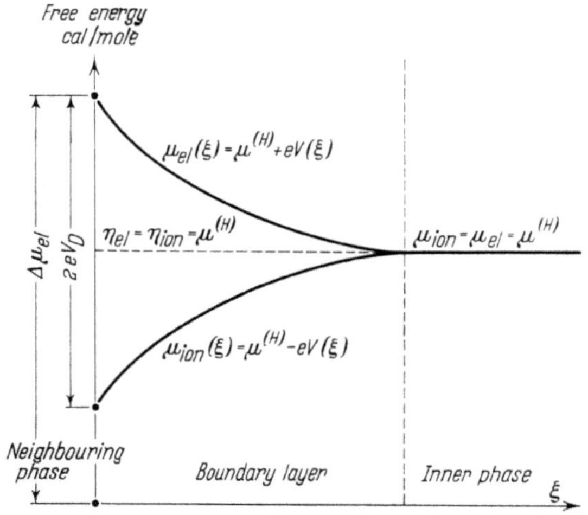

Fig. 34. Chemical and electrochemical potentials of ionic and electronic point defects in a boundary layer

Of particular importance for the phase-boundary reaction in the reduction of ore are the electric fields which are built up through the enrichment of charges in the semi-conductor surface by chemical adsorption. Figure 34 shows diagrammatically the variation of the chemical, electrical, and electrochemical potentials in a chemical adsorption boundary layer when all the equilibrium reactions have completely set in.

In the steady state of the reaction these electric fields are superimposed on the gradient of the chemical potential of the components of the oxide and act, like the gradients of the chemical potential, as the driving force for the transport processes. These transport processes often influence the actual chemical adsorption, are difficult to separate from it, and hinder the evaluation of adsorption measurements.

Chemical adsorption, the phase-boundary reaction, and incorporation into the lattice, are successive processes. It is particularly difficult, and scarcely possible, to separate them from one another if the adsorbed gas is one of the components from which the adsorbing solid has been formed. This is the case for the interaction of oxides with gases containing oxygen, such as O_2, CO/CO_2, or H_2/H_2O. These processes are of fundamental importance in the reduction of ores. The investigations of GARNER et al.[102b]), HAUFFE et al.[102c]), and WINTER[102d]) show that in these reactions too there is interaction between the electrons of the solid and the adsorbed gas, and that the electronic structure or the Fermi potential of the solid affects the course of the reaction.

1.2.5. The Phase-Boundary Reaction

The phase-boundary reaction in the reduction of ores, or the interface reaction in EUCKEN's sense[102a]), is the separation of oxygen from the oxide lattice by the chemically adsorbed reducing gas with the formation of CO_2 or H_2O. The rate of this reaction depends on the surface concentrations at the reacting phase boundary. It is found experimentally that, above 550°C, the rate of separation of oxygen is proportional to the pressure of the H_2 or CO gas. At higher pressures, it is often found that the actual gas pressure has no effect and so it must then be other processes which determine the reaction rate.

On the other hand, from the fundamental laws of chemical kinetics, it must be supposed that the activity, $a(O)$, of the second partner in the reaction, the oxygen in the oxide phase, affects the reaction rate v_1, i.e.,

$$v_1 = f[p(H_2), a(O)].$$

For the exchange of oxygen between the oxides of iron and mixtures of H_2/H_2O or CO/CO_2 this influence of the activity of the oxygen has been incontestably proved both directly[103, 104]) and indirectly[105]). The results discussed in section 2.6 for the reduction of mixed oxides can also be regarded as providing additional confirmation of this fact.

The most important stage in the reduction of iron oxides and the one which definitely determines the reaction rate and the degree of chemical utilization of the gas, is the separation of oxygen from wüstite. Even if. in industrial processes or laboratory experiments, higher oxides are put in, in the reduction from magnetite to wüstite[105]) and from wüstite to iron. separation of oxygen from this phase, wüstite, takes place as the phase-boundary chemical reaction above 570°C. Consequently this reaction will be discussed in special detail here.

In order to be able to study the exchange of oxygen between the oxide and the gas phase without disturbance by the deposition of the reaction products on the surface of the wüstite and also to exclude all effects from the formation and growth of the nuclei of a new phase, the reaction must be carried out so that the formation of the metal phase is avoided. This was achieved in GRABKE's investigations[103]) by studying the exchange of oxygen between $^{14}CO_2$ and ^{12}CO on wüstite as the catalyst, according to

$$\left.\begin{aligned} {}^{12}CO_2 &\rightleftharpoons {}^{12}CO + O^{ads}, \\ O^{ads} + {}^{14}CO &\underset{v_2}{\overset{v_1}{\rightleftharpoons}} {}^{14}CO_2. \end{aligned}\right\} \qquad (39)$$

The adsorbed oxygen is in equilibrium with the oxygen in the oxide lattice and so the oxygen activity of the oxide must affect the rate of reaction. The oxygen activity, $a(O)$, of the oxide is determined through the equilibrium set up. Following GRABKE, we put

$$a(O) = p(CO_2)/p(CO).$$

With this method of carrying out the experiment, the ratio of the partial pressures — both very small — of the $^{14}CO_2$ and the ^{14}CO differs little from the oxygen activity $a(O)$ of the oxide and the gas phase:

$$a^*(O) = p(^{14}CO_2)/p(^{14}CO).$$

For the rate of establishment of equilibrium we put

$$v = v_1 - v_2 = -\frac{dn(^{14}CO)}{dt} = k_1' \, p(^{14}CO) - k_2' \, p(^{14}CO_2),$$

and the experimental results can be expressed, in agreement with this, by

$$v = k_2' \, p(^{14}CO) \, [a(O) - a^*(O)], \qquad (40)$$

where

$$k_2' = k_2 \, a^{-m}(O) \qquad (40\text{a})$$

with $m \approx 0.6$ to 1 for wüstite and $m \approx 0.5$ for magnetite; the agreement can be seen from Fig. 35. Similar results have been obtained by LANDLER and KOMAREK[103a]) for the reduction of wüstite within the single phase domain.

The exchange of oxygen between wüstite and CO/CO_2 mixtures at 900 to 1000°C is therefore governed by

$$v = k_2\, p(^{14}CO)\,[a^{1/3}(O) - a^*(O)\, a^{-2/3}(O)]. \tag{41}$$

It is possible to explain this result by assuming for the reaction a mechanism based on the idea of the wüstite defects as expounded in section 1.1.

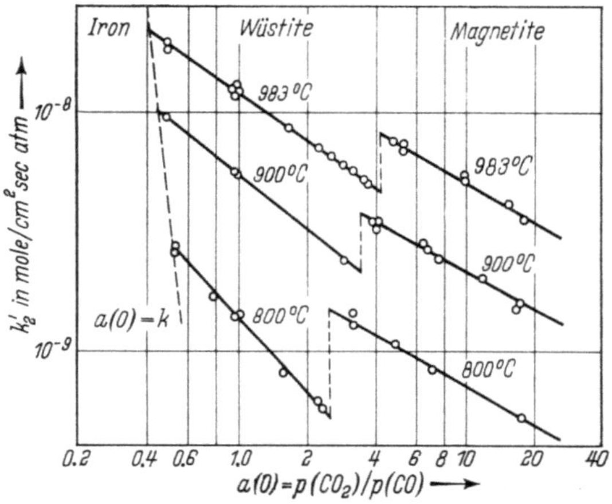

Fig. 35. Dependence of the oxygen activity $a(O) = p(CO_2)/p(CO)$ on the oxygen exchange reaction constant k_2' in wüstite and magnetite at three test temperatures, according to GRABKE[103])

Let us suppose that equilibrium for chemical adsorption of the ^{14}CO is established:

$$^{14}CO \rightleftharpoons {}^{14}CO^{(ads)},$$

$$(^{14}CO^{(ads)}) = K_1\, p(^{14}CO).$$

The adsorbed oxygen ions at the surface of the wüstite are in equilibrium with the iron ion vacancies $|Fe^{2+}|''$ and the iron ions in the normal lattice sites:

$$|Fe^{2+}|'' + FeO \rightleftharpoons O^{2-(ads)} + Fe^{2+}\,|Fe^{2+}|. \tag{42}$$

If y denotes the mole fraction of the iron ion vacancies, this gives

$$(O^{2-(ads)}) = K_2'(|Fe^{2+}|'') = K_2\, y. \tag{42a}$$

The rate-determining process is the forward and reverse reaction represented by

$$O^{2-(ads)} + {}^{14}CO^{(ads)} \underset{v_2}{\overset{v_1}{\rightleftharpoons}} {}^{14}CO_2^{2-(ads)}.$$

The adsorbed $^{14}CO_2$-ion is in equilibrium with the $^{14}CO_2$ of the gas phase and the holes (electron deficiencies):

$$^{14}CO_2 \rightleftharpoons {}^{14}CO_2^{2-(ads)} + 2\,|e|^{\bullet},$$
$$({}^{14}CO_2^{2-(ads)}) = K_3'\, p({}^{14}CO_2)\, (|e|^{\bullet})^{-2},$$

and if

$$(|e|^{\bullet}) = 2y,$$
$$({}^{14}CO_2^{2-(ads)}) = K_3\, p({}^{14}CO_2)\, y^{-2}.$$

We then have for the reaction rate

$$v_1 = k_1''(O^{2-(ads)})\,({}^{14}CO^{(ads)}), \quad v_2 = k_2''({}^{14}CO_2^{2-(ads)})$$

and with the given equilibrium conditions

$$v_1 = k_1'''\, p({}^{14}CO)\, y, \quad v_2 = k_2'''\, p({}^{14}CO_2)\, y^{-2}.$$

Taking into account the establishment of the equilibrium between the wüstite and an oxygen-containing gas phase:

$$\left.\begin{aligned} &\tfrac{1}{2}\, O_2 + Fe^{2+}\,|Fe^{2+}| \rightleftharpoons FeO\,\text{ℸ} + |Fe^{2+}|'' + 2\,|e|^{\bullet} \\ \text{and} \quad &a(O) = (|Fe^{2+}|'')\,(|e|^{\bullet})^2/K_y' = y^3/K_y \end{aligned}\right\} \qquad (43)$$

we get finally

$$v_1 = k_1''''\, p({}^{14}CO)\, a^{1/3}(O), \quad v_2 = k_2''''\, p({}^{14}CO_2)\, a^{-2/3}(O),$$

and hence, with the equilibrium conditions

$$v_1 = v_2 \quad \text{and} \quad p({}^{14}CO_2)/p({}^{14}CO) = a^*(O),$$

we get the Eq. (41), which was found experimentally. This equation also holds if the oxide activity of the whole gas phase is greater or less than that of the oxide surface. In this case an oxidation or reduction of the wüstite surface takes place and

$$v = v_1 - v_2 = k_2\, p({}^{14}CO)\, [a^{1/3}(O) - a^*(O)\, a^{-2/3}(O)]. \qquad (41)$$

In the surface of the oxide, disorders may also occur which, for geometrical and spatial reasons, do not appear in the interior of the crystal. In the case of wüstite, oxygen ion vacancies and iron ions in interstitial positions, in particular, are to be thought of in this connection. The ratio of metal to oxygen in the surface could also assume other values, and this again affects the electron deficiencies. Such possibilities, which have been discussed elsewhere[106]), ought perhaps to be particularly borne in mind when discussing reduction or oxidation at lower temperatures. The effect of such disorders might further change the relation, formulated in Eq. (41), between the reaction rate and $a(O)$.

The special feature of GRABKE's investigations is that the work was carried out with continuous equilibrium between the oxide and the gas phase. This condition does not, of course, hold in the reduction of ores. This difference does not arise in STOTZ's investigations[104]), in which the rate of establishing the binary equilibrium between CO_2/CO mixtures and wüstite was measured in accordance with formula (43). STOTZ used as the indicator for the progress of the reaction the holes $|e|^\bullet$ and their effect on the conductivity of extremely thin films of wüstite. Since the rate constant, in cm sec^{-1}, for the phase-boundary reaction is small compared with D/x, where D is the coefficient of diffusion of the vacancies, and x is half the thickness of the wüstite film, the change in conductivity follows practically instantaneously the progress of the phase-boundary reaction.

With this method of observation STOTZ was able to confirm GRABKE's results. In these measurements, too, any disturbance of the reaction due to the formation and growth of nuclei of the metal phase was excluded, since no separation of iron took place. The further measurements made by STOTZ in connection with the reduction of wüstite to iron will be discussed in the section on the formation of nuclei.

In measurements on the reduction of magnetite to wüstite, carried out by ULRICH[48]), the separation of oxygen from the wüstite was again the rate-determining sub-process. The wüstite covered the magnetite in a layer completely free from pores. The iron set free during the separation of oxygen diffused via the vacancies through the wüstite to the magnetite and changed it into wüstite:

$$FeO + |Fe^{2+}|'' + 2\,|e|^\bullet + H_2 \rightarrow Fe^{2+}\,|Fe^{2+}| + H_2O,$$

$$Fe_3O_4 + Fe^{2+}\,|Fe^{2+}| \rightarrow 4\,FeO + |Fe^{2+}|'' + 2\,|e|^\bullet.$$

Since the vacancy diffusion and the phase-boundary reaction (through which the vacancies are filled) are coupled by these reactions, the vacancy concentration y also changes with the thickness of the wüstite layer during the course of the reaction. This circumstance will be discussed in more detail in section 2.4. From his analysis of the course of the reaction, and by substituting for the rate equation

$$v = k_3'\,p(H_2) - k_4'\,p(H_2O),$$

ULRICH arrives at the result, analogous to (40), that

$$v = k_4'\,p(H_2)\,[a(O) - a^*(O)], \tag{40a}$$

where again $a^*(O)$ denotes the oxygen activity of the gas phase, and $a(O)$ denotes the oxygen activity of the oxide surface. In contrast to the results with CO and CO_2, the rate constant k_4' here is independent of the oxygen activity of the oxide.

By using the equilibrium relation (43), we can re-write (40a) as

$$v = k_4''\,p(H_2)\,(y^3 - y^{*3}), \tag{44}$$

where y^* denotes the vacancy concentration which would be set up in the wüstite in the case of equilibrium with the gas mixture used for the reduction.

From (44) the reaction rates measured at different times should be related to the corresponding vacancy concentrations in the wüstite by the equation

$$v(t)/v(t = 0) = [y^3(t) - y^{*3}]/[y^3(t = 0) - y^{*3}].$$

Figure 36 shows that at 800°C and for gas mixtures of several different ratios this relation does hold for the reaction between wüstite and CO/CO_2.

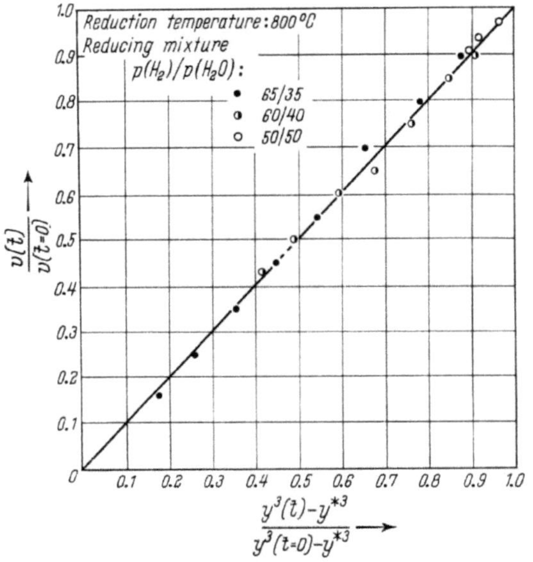

Fig. 36. Ratio of reduction rate v at time t to reduction rate at $t = 0$ as a function of the defect concentration y in the wüstite surface, according to Ulrich[49])

The results of this work thus give different relations between the reaction rates and $a(\mathrm{O})$ or y from those deduced from Grabke's[103]) and Stotz's[104]) observations. More recent investigations of Bohnenkamp and Riecke[18a]) show in accordance with Grabke's results for the reaction of wüstite with H_2O/H_2, that the exponent n in the relation $k \sim a^n(\mathrm{O})$ depends on temperature. Further, the pre-treatment of the oxide affects n.

The sequence

$$\mathrm{FeO}\urcorner + |\mathrm{Fe}^{2+}|'' \rightleftharpoons \mathrm{Fe}^{2+}\,|\mathrm{Fe}^{2+}| + \mathrm{O}^{2-(\mathrm{ads})}, \tag{45a}$$

$$2\,|e|^{\bullet} + \mathrm{O}^{2-(\mathrm{ads})} + \mathrm{H}^{(\mathrm{ads})} \rightleftharpoons \mathrm{OH}^{(\mathrm{ads})}, \tag{45b}$$

$$\mathrm{OH}^{(\mathrm{ads})} + \mathrm{H}^{(\mathrm{ads})} \underset{v_4}{\overset{v_3}{\rightleftharpoons}} \mathrm{H_2O} \tag{45c}$$

would be in agreement with Ulrich's experimental results, for instance, if we assume that the first two sub-processes take place in the immediate vicinity of equilibrium, but that the third sub-process is rate-determining for the forward and reverse reactions. Then, from (45c), we have

$$v_3 = k_3'''(H^{(ads)})\,(OH^{(ads)}). \tag{46}$$

Assuming that the adsorbed hydrogen atoms are in equilibrium with the gas phase in accordance with

$$H_2 \rightleftharpoons 2\,H^{(ads)},$$

we then have

$$(H^{(ads)}) = K_1\,p^{1/2}(H_2). \tag{47}$$

(45b) and (47) then imply

$$(OH^{(ads)}) = K_2\,p^{1/2}\,(H_2)\,(O^{2-(ads)})\,(H^{(ads)})$$

and using (45a)

$$(OH^{(ads)}) = K_3\,p^{1/2}(H_2)\,y\,(|e|^{\bullet})^2. \tag{48}$$

Substituting from (47) and (48) into (46), and using

$$(|e|^{\bullet}) = 2y \quad \text{and} \quad v = k_2''\,p(H_2)\,y^3 - k_4''\,p(H_2),$$

we obtain the above formula (44).

The mechanism described is speculative, but it is supported by the fact that, according to EISCHENS[107]) and WICKE[108]), when hydrogen is adsorbed on zinc oxide, covalent and ionogenically attached H-atoms and surface-OH groups occur.

The reactive oxygen ions, $O^{2-(ads)}$, in the surface of the oxide lattice should be determined, through (42), by the chemical disorder. But, apart from the reaction (42), reactive surface elements should also arise through other lattice defects such as grain boundaries or dislocations. One effect of these disorder centres on the rate of reduction has recently been demonstrated[109]).

The actual value of y or $a(O)$ in (41) and (42) depends, on the one hand, on the rate of oxygen removal and, on the other hand, on the transport of vacancies in the wüstite. As soon as separation of iron at the wüstite surface has begun, $a(O)$ and y attain stationary values. The rate equations for reduction by CO or H_2 can then be written in the simpler forms

$$v = k_1\,p(CO) - k_2\,p(CO_2), \tag{40b}$$

$$v = k_3\,p(H_2) - k_4\,p(H_2O). \tag{40c}$$

But now it is by no means obvious that from measurement of the rate constants for the phase-boundary reaction within the wüstite phase domain we can deduce their values during the reduction[109a, b]). Even if the iron which is formed is porous, and if the surfaces on which the phase-boundary reaction takes place – the interface between the porous iron and the wüstite situated below it – do not differ greatly in area from the wüstite surface on which the rate of oxygen removal has been measured when there is no iron separation, these reaction rates may still be very different. And, in actual fact, the removal of oxygen is apparently very much quicker if the wüstite surface is in contact with iron than previously, even though the removal of the oxygen excess takes place from the wüstite phase. This may be attributable to catalysis of the phase-boundary reaction at the point of contact of the iron and wüstite, or perhaps to the joint action of the iron and wüstite as "mixed catalysts". An analysis of the kinetics of growth of iron nuclei on wüstite clearly shows the predominant role of the three-phase boundary "iron/wüstite/gas" for the reaction. Apart from this, the formulae (40b) and (40c) describe the dependence on the partial pressures during the reduction of iron oxide to metal, in most cases.

If p^* denotes the gas pressure when there is equilibrium between the oxide and the metal, then

$$v = 0 \quad \text{when} \quad p(H_2) = p^*(H_2),$$

where $p^*(H_2)$ is a function of $p(H_2O)$ such that

$$p^*(H_2)/p(H_2O) = K(H_2O).$$

Hence it follows that

$$k_3\, p^*(H_2) = k_4\, p(H_2O),$$

and so

$$\left.\begin{aligned} v &= k_3[p(H_2) - p^*(H_2)], \\ v &= k_1[p(CO) - p^*(CO)] \\ \text{or} \quad v &= k_4[p^*(H_2O) - p(H_2O)], \\ v &= k_2[p^*(CO_2) - p(CO_2)]. \end{aligned}\right\} \tag{49}$$

The pairs of constants k_3, k_4 and k_1, k_2 are each related by

$$k_3/k_4 = K(H_2O) \quad \text{and} \quad k_1/k_2 = K(CO_2)$$

so that we also have, e.g.,

$$v = k_3[p^*(H_2O) - p(H_2O)]/K(H_2O).$$

According to Edmiston and Grace[109c]), in the reduction of wüstite by H_2/H_2O mixtures,

$$v \sim \frac{\Delta G - \Delta G_0}{RT} \quad \text{when} \quad \frac{\Delta G - \Delta G_0}{RT} \ll 1.$$

But since

$$\Delta G = RT \log p(H_2)/p(H_2O) \quad \text{and} \quad \Delta G_0 = RT \log p^*(H_2)/p(H_2O)$$

and

$$e^x \approx 1 + x \quad \text{for} \quad x < 1,$$

this does not contradict the formulae (49).

In the reduction by hydrogen of *hematite* to iron below the eutectoid decomposition temperature for wüstite an increased inhibition by the water vapour has, in general, been found; see, e.g.[111]). According to McKewan[110]) the observed results can be described by a reaction-rate constant k depending on the partial pressures according to

$$k = [a\,p(H_2) - b\,p(H_2O)]/[1 + c\,p(H_2O)/p(H_2)]. \tag{50}$$

The mechanism of the reduction of hematite to magnetite has been investigated by Geiger and Wagner[111a]). To do this they evaluated the linear, initial variation of the removal of oxygen from pure hematite and from mixed crystals of hematite with TiO_2 and CaO, basing their work on measurements of the electrical conductivity of these oxides as a function of the oxygen pressure and temperature. The observed results can be described by the expression

$$v = k'\,p(CO)\,\Gamma(|O|\,O) - k''\,p(CO_2)\,\Gamma(|O^{2-}|^{\bullet\bullet})\,a^2(e^-),$$

where $\Gamma(|O|\,O)$ and $\Gamma(|O^{2-}|^{\bullet\bullet})$ denote the surface concentrations of oxygen ions in the lattice positions and of oxygen ion vacancies respectively, and $a(e^-)$ denotes the activity of the electrons. The latter is increased by the incorporation of TiO_2 and reduced by the incorporation of CaO.

By reduction experiments with naturally found single crystals and synthetically prepared, dense hematite preparations, Gray and Henderson[111b]) were able to prove that the reduction of hematite to magnetite below 575°C is a topochemical process. The orientation relations here observed between hematite and magnetite are in agreement with the results of Meyer, Rausch and Ottow[38b]) and may perhaps be adduced to explain the disintegration of hematite during reduction at 500°C. Gray and Henderson, however, represent the view that it is unnecessary to take the topochemical nature of the reaction into account in the rate equation.

1.2.6. Formation of Nuclei

The genesis of both solid and liquid phases is initiated by the formation of nuclei. The critical nucleus is that volume of the newly formed phase for which the probability of its dissolution is equal to the probability of its further stepwise growth.

The theoretical treatment of nuclei-forming processes was first carried out, on the basis of work by Gibbs and Thomson, by Volmer et al. in 1925 [112]).

They calculate statistically the probability of the appearance in the *supersaturated mother-phase* of aggregates of 2 to n_i^* moles, n_i^* being the number of moles in the *critical nucleus*. From the kinetic theory of gases the rate of growth of such aggregates of molecules is deduced.

The numerical values of the energies involved can be obtained from the following chain of reasoning.

In the formation of a new phase from a supersaturated mother-phase the amount of energy liberated is $n_i \Delta \mu_i$, where n_i is the mole number of the substance i in the nucleus formed, and $\Delta \mu_i$ is the difference between the chemical potential of the substance in the nucleus phase and in the mother-phase. At the same time a new interface has to be formed, for which an amount of energy $\sigma_i F_K$ has to be expended, where σ_i is the surface tension, and F_K is the surface area of the nucleus. The change in the Gibbs free enthalpy in the formation of the nucleus is therefore

$$\Delta G_K = n_i \Delta \mu_i + \sigma_i F_K. \tag{1}$$

For a nucleus in the form of a spherical droplet of radius r we thus have

$$\Delta G_K = \frac{4}{3} r^3 \pi \frac{\Delta \mu_i}{V_M} + 4\pi r^2 \sigma_i; \tag{2}$$

V_M is the mole volume of the nucleus phase.

Since $\Delta \mu_i$ is negative and σ_i always positive, the two terms on the right-hand side of (2) have opposite signs. Putting

$$\frac{4}{3} \pi \frac{\Delta \mu_i}{V_M} = a \quad \text{and} \quad 4\pi \sigma_i = b,$$

we obtain the curves shown in Fig. 37 for the variation of the volume component $a r_i^3$, the surface component $b r_i^2$, and the work of nucleus formation ΔG_K with the radius r of the nucleus.

The energy that has to be expended in forming the nucleus at first increases with the radius; its maximum value, found by differentiating (2), is

$$\Delta G_K^* = -16\pi \sigma_i^2 V_M^2 / 3\Delta \mu_i^2; \tag{3}$$

and the radius of the nucleus corresponding to this maximum is found, by substituting (3) into (2), to be

$$r^* = -2\sigma_i V_M / \Delta \mu_i. \tag{4}$$

The quantity $\Delta \mu_i$ may be expressed in the form

$$\Delta \mu_i = RT \log[p_i(r = \infty)/p_i(r)], \tag{5}$$

where $p_i(r)$ is the vapour pressure of the mother-phase in which a droplet of the nucleus phase of radius r can remain in equilibrium, and $p_i(r = \infty)$ is the vapour pressure of the mother-phase in which the nucleus phase with a plane surface can remain in equilibrium.

Spherical nuclei with a radius greater than r^* can freely increase further in size, since, for $r > r^*$, the function $\partial G_K/\partial r_i$ and therefore also $\partial G_K/\partial n_i$ is negative.

The number of critical nuclei per unit volume is given by

$$z^* = n_i' \exp[-\Delta G_K^*/RT], \tag{6}$$

where n_i' is the number of molecules of the substance i in unit volume of the mother-phase.

The formation of solid reaction products in the reduction of metal oxides is a process in which the formation of nuclei must take place. Since, however, it is here a question of the formation of nuclei of a crystallized, solid phase with a surface which is partly in contact with the solid mother-phase and partly in contact with the gas phase, the above relations have to be modified.

If we assume that the surface tension, $\sigma(K, G)$, between nucleus and gas phase, and the surface tension, $\sigma(K, M)$, between nucleus and mother-phase, are isotropic, i.e., they have constant values in all directions, then

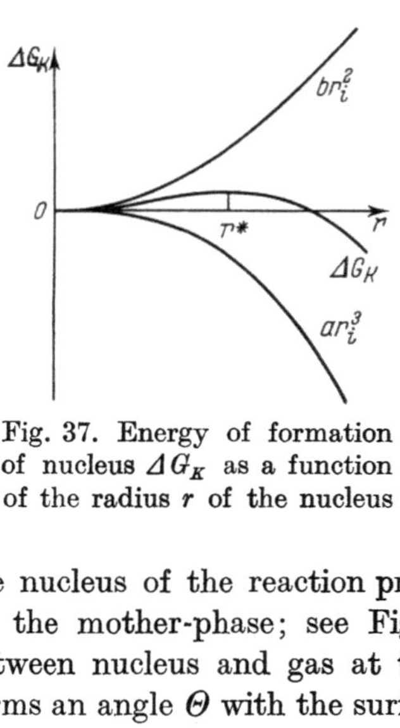

Fig. 37. Energy of formation of nucleus ΔG_K as a function of the radius r of the nucleus

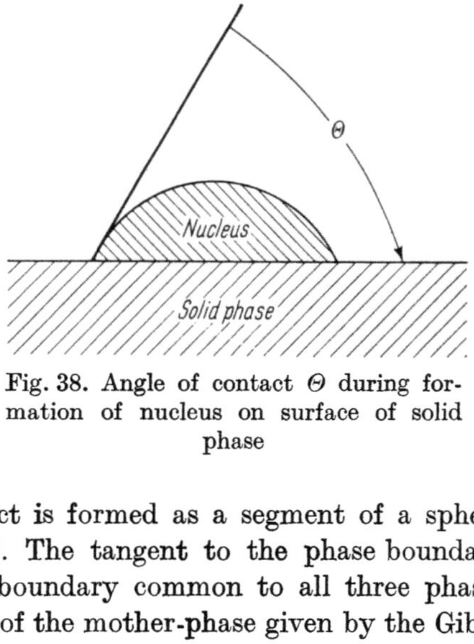

Fig. 38. Angle of contact Θ during formation of nucleus on surface of solid phase

the nucleus of the reaction product is formed as a segment of a sphere on the mother-phase; see Fig. 38. The tangent to the phase boundary between nucleus and gas at the boundary common to all three phases forms an angle Θ with the surface of the mother-phase given by the Gibbs relation

$$\sigma(M, G) = \sigma(K, M) + \sigma(K, G)\cos\Theta, \tag{7}$$

where $\sigma\ (M, G)$ is the surface tension between the mother phase and the gas phase.

The work which has to be done in forming the critical nucleus is

$$-\Delta G_K^* = 64\pi\,\sigma^3(K, G)\,V_M^2/[3\Delta\mu_i(2 + \cos\Theta)\,(1 - \cos\Theta)^2]. \tag{8}$$

The rate of formation of nuclei, J, is given by the number of nuclei of critical size which in unit time pass over into super-critical size through the acquisition of additional lattice elements and subsequently increase rapidly to macroscopic crystals. Hence

$$J = Z\,\omega\,z^{*\prime}, \tag{9}$$

where $z^{*\prime}$ is the number of critical nuclei on unit area of the mother-phase, and ω is the number of collisions of ad-atoms with one critical nucleus.

An ad-atom is to be understood to mean either a surface atom or an adsorbed atom of the nucleus substance on the surface of the mother-phase. The other factor, Z, in (9) is the disequilibrium factor introduced by BECKER and DÖRING[113]) and ZELDOVICH[114]); it takes into account the fact that nuclei of critical size can disappear by increasing to super-critical size and by decreasing to sub-critical size through the disassociation of atoms:

$$Z = V_M\,\Delta\mu_i^2/[8\pi\,\sigma(RT\,\sigma)^{1/2}] = [\Delta G_K^*/3\pi\,RT\,n_i^{*2}]^{1/2}. \tag{10}$$

The solution of Eq. (9) with (10) is [115])

$$J = C\,n_{\mathrm{ad}}\exp[-(\Delta G_{\mathrm{sd}}^* + \Delta G_K^*)/RT]$$

$$C = 2\pi\,r^*\,a\sin\Theta\,\bar{\omega}\,n_0[\Delta G^{*\prime}/4\pi\,RT\,n^{*2}]^{1/2}, \tag{11}$$

$$\Delta G_K^{*\prime} = \Delta G_K^* + \Delta G_q,$$

where

n_{ad} = surface concentration of ad-atoms on the surface of the mother-phase (mole cm^{-2}),
ΔG_{sd}^* = free enthalpy of the activated state of surface diffusion of the ad-atoms,
a = jump distance of surface diffusion,
n_0 = number of adsorption points on the surface of the mother-phase (mole cm^{-2}),
$\bar{\omega}$ = probability of transition for the activated complex, and
ΔG_q = an energy contribution to account for the distribution of nuclei on the surface: $\Delta G_q \approx RT\log(z^{*\prime}/n_0)$.

In this case, in the expression (8) for ΔG_K^* we have to put

$$\Delta\mu_i = RT\log a_{i,\mathrm{ad}}, \tag{12}$$

where $a_{i,\mathrm{ad}}$ is the activity of the adsorbed structural element of the substance of the nucleus. If it is a question of the formation of nuclei of a metal Me during the reduction of the oxide MeO, then

$$a_{\mathrm{Me,ad}} = a_{\mathrm{Me(MeO)}}\,.$$

In this case we have also to put

$$n_{\mathrm{ad}} = \mathrm{const}\cdot a_{\mathrm{Me(MeO)}}\,.$$

The relations also change if the nucleus has crystal surfaces with very different surface tension. In this case the nucleus is bounded by flat lattice planes and does not take the form, as previously assumed, of a segment of a sphere. The factor $64\pi/3$ in the formula (8) must then be replaced by some constant K whose value depends on the shape of the nucleus[116]). Moreover, in formula (2) a term must be included for each surface which appears.

The rate of formation of nuclei is therefore determined by the crystallographic form of the nucleus, and this in turn depends on the free surface energy of the individual crystal surfaces. According to VOLMER[112]), for a crystal in equilibrium with its vapour the relation

$$\sigma_k/r_k = [RT/2\,V_M]\log[p(r)/p(r=\infty)] = \text{const} \tag{13}$$

must hold, where σ_k is the surface tension of the k-th surface, and r_k is the central distance of this surface.

Further, according to GIBBS, the equilibrium form of the crystal is determined by the condition that

$$(\sum F_k\,\sigma_k)_{V=\text{const}}$$

shall be a minimum. From these conditions and a knowledge of the surface-tensions of the individual crystal surfaces, the habit of nuclei can be determined uniquely. This can also be done by STRANSKI's method[117]), using the energy φ of separation of lattice elements in the individual lattice planes, a quantity more amenable to theoretical treatment. We have

$$kT\log[p(r)/p(r=\infty)] = \varphi_{1/2} - \varphi_r, \tag{14}$$

where φ_r is the energy of separation from the uppermost lattice plane of the equilibrium form of the crystal, and $\varphi_{1/2}$ is the energy of separation from a half-crystal position.

The adsorption of foreign substances, and therefore of molecules of the reducing gas in particular, can alter considerably the energy of separation and so also the habit of crystals[118]).

1.2.7. Transport Processes in the Solid Phase

1.2.7.1. The Fundamentals of Diffusion in Iron Oxides

If, say, hematite Fe_2O_3 is reduced to iron, then in the course of the removal of the oxygen all the thermodynamically stable oxides of iron appear either successively or side by side. In many investigations indications have also been found of the occurrence of unstable decomposition products, such as wüstite at temperatures of reaction below 570°C[119]). Apart from such special cases, during the course of the reduction process, those phases which can co-exist with one another, i.e., Fe/FeO, FeO/Fe_3O_4, Fe_3O_4/Fe_2O_3, are always in direct contact with one another. Which of the reduction products at any given instant during the reduction lie on the surface of the oxide and are in contact with the gas phase depends not only on the corresponding equilibria but is also determined by the rates of removal of oxygen on the one hand and by the diffusion processes in the oxides on the other. For the reduction of magnetite ULRICH[120]) has described these circumstances in detail.

If the reduction products arise as non-porous layers so that the reducing gas no longer has direct access to the unreduced core of the oxide grains, then diffusion in these reaction products is essential for the progress of reduction. However, the overall rate of reduction can be affected by diffusion only in the outermost of the layers of metal or lower oxides which are being formed. The diffusion in the inner layer is a subsequent and subordinate process which has no direct effect on the course of the reaction.

In wüstite and magnetite it is almost exclusively the iron which diffuses[121]), the oxygen being in comparison immovable. The wüstite layer is transformed by the gas into iron:

$$FeO + H_2 \rightarrow Fe + H_2O.$$

The wüstite increases at the expense of the magnetite layer through the process

$$Fe_3O_4 + Fe \rightarrow 4\,FeO.$$

The magnetite is formed from the decomposition of hematite:

$$4\,Fe_2O_3 + Fe \rightarrow 3\,Fe_3O_4.$$

For the thicknesses ξ of the oxide layers to remain steady, the reaction rate dn_0/dt and the diffusion flows j(Fe) in the individual layers must be related by

$$dn_0/dt = 12j(\text{Fe in } Fe_3O_4) = 3j(\text{Fe in FeO}).$$

The driving force for these flows is the gradient of the chemical potential or the activity of the iron between the separate phase boundaries, i.e., in the oxide layers. If the magnetite layer is in equilibrium with the hematite and wüstite, and the wüstite layer with the iron and magnetite, then the iron activities on both sides of these oxide layers are determined. The two conditions given above can therefore both be satisfied only if the appropriate steady layer thicknesses are set up.

The thicknesses of the inner oxide layers are thus determined by the rate of reduction, not *vice versa*. On the other hand, the thickness of the outermost layer of reduction products increases regularly with the degree of reduction. If the solid-state diffusion in this layer requires an appreciable gradient in the concentration or activity, then the rate of reduction will depend on the thickness of this layer. The same thing is also basically true if the gas diffusion in the pores of the outermost reaction layer determines the reaction rate. Like diffusion in gases, diffusion in the solid state is governed by Fick's law, but the simple formula, (3) of section 1.2.2, holds only if the coefficient of diffusion is independent of the concentration and the activity of the diffusing substance depends linearly on its concentration. To be able to apply formula (3) it must also be known what substance is diffusing in the particular case in question, since its diffusion coefficient and its concentration have to be put in. The effect of the electric fields in the boundary zones of the oxide through chemical adsorption must also be taken into account.

For the transport of material, therefore, the general transport equation is to be applied:

$$j_i = -B_i\, c_i(\operatorname{grad}\mu_i + z_i\, F \operatorname{grad}\varphi), \tag{1}$$

where j_i is the flux, in mole cm^{-2} sec^{-1}, of the migratable lattice component, B_i and c_i are its mobility and concentration, and μ_i and z_i are its chemical potential and charge number. F is the Faraday constant and $\operatorname{grad}\varphi$ the local electric field strength.

Formula (1) can be applied directly if it is known which component of the lattice has the preponderating mobility. Diffusion, particularly in the lattice of a compound, takes place predominantly through the migration of lattice defects, such as the vacancies discussed in section 1.1.1, and of ions over interstitial positions. Those components having sublattices in which such defects occur are therefore the most liable to migrate. By way of illustration the transport in the wüstite layer will be discussed in greater detail[122]).

In wüstite the iron ion vacancies $|Fe^{2+}|''$ are the important defects. They make possible the diffusion of iron through the oxide lattice, this diffusion taking place, in the reduction of magnetite and hematite, through the outermost wüstite layer inwards from the wüstite/gas or wüstite/iron phase boundary to the wüstite/magnetite phase boundary. The iron ion vacancies $|Fe^{2+}|''$ have a double negative charge compared with the undisturbed lattice. Further, an intrinsic value of the chemical potential has to be ascribed to them, the significance of which will become apparent later.

The iron migrates via vacancies through the lattice in the form of Fe^{2+}-ions. To maintain electronic neutrality, a stream of electron deficiencies, or holes, $|e|^\bullet$, flows through the wüstite layer at the same time; chemically, these can be regarded as the third positive charge of Fe^{3+}-ions. This excess of charge moves by the interchange of charge between the Fe^{2+} and the Fe^{3+} ions of the lattice. The mobility of the holes $|e|^\bullet$ is considerably greater than that of ionic defects. In the steady state of the transport process we have

$$j(\mathrm{Fe}) = j(\mathrm{Fe}^{2+}) = -j(|\mathrm{Fe}^{2+}|'') = -\tfrac{1}{2} j(|e|^\bullet). \tag{2}$$

From (1),

$$j(|\mathrm{Fe}^{2+}|'') = -B(|\mathrm{Fe}^{2+}|'')\, c(|\mathrm{Fe}^{2+}|'')\, [\operatorname{grad}\mu(|\mathrm{Fe}^{2+}|'') - 2F \operatorname{grad}\varphi],$$

$$j(|e|^\bullet) = -B(|e|^\bullet)\, c(|e|^\bullet)\, [\operatorname{grad}\mu(|e|^\bullet) + F \operatorname{grad}\varphi].$$

Putting

$$c(|\mathrm{Fe}^{2+}|'') = y/V_M,$$

where y is the mole fraction of the vacancies, and V_M is the mole volume of the FeO, and also putting

$$j(|e|^\bullet) = 2j(|Fe^{2+}|''),$$

grad φ can be eliminated. Using the condition for electronic neutrality

$$c(|Fe^{2+}|'') = \tfrac{1}{2}c(|e|^\bullet),$$

we get

$$j(|Fe^{2+}|'') = -\frac{B(|e|^\bullet)\,B(|Fe^{2+}|'')}{B(|e|^\bullet) + 2B(|Fe^{2+}|'')}\,\frac{y}{V_M}\,[\operatorname{grad}\mu(|Fe^{2+}|'') + 2\operatorname{grad}\mu(|e|^\bullet)]. \tag{3}$$

The gradients in the square brackets of the chemical potentials of the ion and electron defects are formed from the gradients of the concentrations of these defects.

Under ideal conditions we have:

$$\mu_i = \bar{\mu}_i + RT\log c_i,$$

and so

$$\operatorname{grad}\mu_i = RT\operatorname{grad}(\log c_i).$$

Since

$$c(|e|^\bullet) = 2c(|Fe^{2+}|'') = 2y/V_M,$$

we obtain, with (2),

$$j(\mathrm{Fe}) = -j(|Fe^{2+}|'') = \frac{B(|e|^\bullet)\,B(|Fe^{2+}|'')}{B(|e|^\bullet) + 2B(|Fe^{2+}|'')}\,\frac{3}{V_m}\,RT\operatorname{grad}y.$$

The Nernst-Einstein relation

$$B_i RT = D_i,$$

and the fact that $B(|e|^\bullet) \gg B(|Fe^{2+}|'')$ give

$$j(\mathrm{Fe}) = D(|Fe^{2+}|'')\,\frac{3}{V_M}\operatorname{grad}y. \tag{4}$$

Between the metallic iron Fe(Me) and the defects in the wüstite lattice the reaction

$$\mathrm{Fe(Me)} + |Fe^{2+}|'' + 2\,|e|^\bullet \rightleftharpoons Fe^{2+}\,|Fe^{2+}|$$

can take place, where, as usual, $Fe^{2+}\,|Fe^{2+}|$ denotes a divalent metal ion in the wüstite in the normal lattice position. Since the concentration of iron ions in the lattice positions is nearly constant, from the corresponding equilibrium relation the expression in square brackets in (3) becomes

$$[\operatorname{grad}\mu(|Fe^{2+}|'') + 2\operatorname{grad}\mu(|e|^\bullet)] = -\operatorname{grad}\mu[\mathrm{Fe(Me)}]. \tag{5}$$

This relation can be regarded as a definition of $\mu(|Fe^{2+}|'')$ and $\mu(|e|^\bullet)$. Since in metal oxides Me_mO_n, regarded as ordered mixed phases, the Duhem-Margules equation simplifies to

$$m\,\mu(\mathrm{Me}) + \tfrac{1}{2}n\,\mu(O_2) = \Delta G(Me_mO_n),$$

we have further, for FeO,

$$\operatorname{grad}\mu[\mathrm{Fe(Me)}] = -\tfrac{1}{2}\operatorname{grad}\mu(O_2). \tag{6}$$

Hence formula (3) simplifies, with $B(|e|^\bullet) \gg B(|Fe^{2+}|'')$, to

$$j(\mathrm{Fe}) = -j(|\mathrm{Fe}^{2+}|'') = -RT\,B(|\mathrm{Fe}^{2+}|'')\,[y/V_M]\,\mathrm{grad}\log a[\mathrm{Fe(Me)}]$$
$$= RT\,B(|\mathrm{Fe}^{2+}|'')\,[y/V_M]\,\tfrac{1}{2}\,\mathrm{grad}\log a(\mathrm{O_2}),$$

where $a[\mathrm{Fe(Me)}]$ and $a(\mathrm{O_2})$ denote the activities of metallic iron and oxygen in the oxide.

Finally, using the Nernst-Einstein equation again and a relation given by WAGNER[123])

$$y\,D(|\mathrm{Fe}^{2+}|'') = (1-y)\,D^*(\mathrm{Fe}),$$

where $D^*(\mathrm{Fe})$ is the self-diffusion coefficient of iron in wüstite, we have, for $(1-y) \approx 1$,

$$j(\mathrm{Fe}) = -[D^*(\mathrm{Fe})/V_M]\,\mathrm{grad}\log a[\mathrm{Fe(Me)}] = [D^*(\mathrm{Fe})/2\,V_M]\,\mathrm{grad}\log a(\mathrm{O_2}). \quad (7)$$

Formula (4) is particularly suitable for the treatment of diffusion, since the coefficient of diffusion of the vacancies $D(|\mathrm{Fe}^{2+}|'')$ may be regarded as independent of y and therefore of the position in question, as Fig. 39 shows. However, this formula assumes that grad$y$ and $D(|\mathrm{Fe}^{2+}|'')$ are

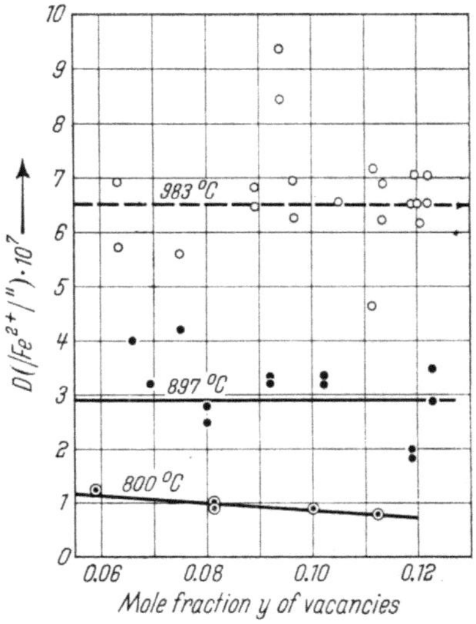

Fig. 39. Diffusion coefficients of iron ion vacancies, calculated from the self-diffusion coefficients of iron in wüstite; according to measurements of HIMMEL et al.[121])

known. In the case of wüstite the first quantity can be measured directly[122, 124]), the second can be calculated from measurements of the self-diffusion coefficient of iron in wüstite.

To use formula (7), only the experimentally more accessible value of $D^*(\mathrm{Fe})$ need be known. For the oxygen activity, which also appears in (7), the values for equilibrium at the phase boundaries of the oxide layer in question are generally known. Since formula (7) is independent of the actual structure defects, it is suitable when nothing more specific is known about the transport mechanism in the lattice. It must be remembered when using it that $D^*(\mathrm{Me})$ is a function of the defect concentration and therefore of $a(\mathrm{O}_2)$ or $a(\mathrm{Me})$.

Integrating (7) and putting ξ = thickness of the layer, we have

$$j(\mathrm{Fe}) = \frac{1}{2V_M\xi} \int_{a(\mathrm{O_2};\ \mathrm{FeO/Fe_3O_4})}^{a(\mathrm{O_2};\ \mathrm{FeO/gas})} D^*(\mathrm{Fe})\, d[\log a(\mathrm{O}_2)]. \tag{8}$$

Assuming ideal thermodynamic behaviour, (5) and (6) imply

$$-d[\log a(\mathrm{O}_2)] = 2(1 + |z|)\, d(\log y),$$

where $|z|$ is the absolute value of the charge of the defects capable of migration. Taking

$$D^*(\mathrm{Fe}) = y\, D(|\mathrm{Fe}^{2+}|'')$$

and carrying out the integration, we get

$$j(\mathrm{Fe})\,\xi V_M = [1 + |z(|\mathrm{Fe}^{2+}|'')|]\,[1 - y(\mathrm{FeO/gas})/y(\mathrm{FeO/Fe_3O_4})] \times$$
$$\times [D^*(\mathrm{Fe{:}FeO{:}FeO/Fe_3O_4})].\dagger \tag{9}$$

A corresponding formula holds for the transport of material in magnetite, where again it is primarily the iron ions which migrate. The defect concentrations, c, are known less accurately, but those, c', for equilibrium with wüstite are very much smaller than those, c'', for equilibrium with hematite. So we have

$$j(\mathrm{Fe\ in\ Fe_3O_4}) = \frac{1 + |z(|\mathrm{Fe}^{2+}|'')|}{\xi V_M(\mathrm{Fe_3O_4})} D^*(\mathrm{Fe{:}Fe_3O_4{:}Fe_3O_4/Fe_2O_3})\,(1 - c'/c'') \tag{10}$$

or, approximately,

$$j(\mathrm{Fe\ in\ Fe_3O_4}) \approx \frac{3}{\xi V_M(\mathrm{Fe_3O_4})} D^*(\mathrm{Fe{:}Fe_3O_4{:}Fe_3O_4/Fe_2O_3}). \tag{11}$$

For the growth, controlled by diffusion, of layers of reduction products the formula

$$dm/dt = k'/m \tag{12}$$

is often given, m being the amount of oxygen removed, in g cm^{-2}. k' is related to the self-diffusion coefficients of the predominatingly mobile components of the oxide by

$$k' = D^*\,\varrho^2 M(\mathrm{O})/\nu M(\mathrm{ox}), \tag{13}$$

† $D^*(\mathrm{Fe{:}FeO{:}FeO/Fe_3O_4})$ is the self-diffusion coefficient of iron in wüstite which is in equilibrium with $\mathrm{Fe_3O_4}$.

where

$M(\mathrm{O})$ = atomic weight of oxygen,
$M(\mathrm{ox})$ = molecular weight of the oxide,
ϱ = density of the oxide,
ν = number of moles of reduction product per mole of oxygen removed.

In the steady state of the reduction process of hematite with the formation of compact, non-porous layers of wüstite and magnetite, the diffusion flows of iron in the layers satisfy

$$j(\mathrm{Fe\ in\ FeO}) = 4j(\mathrm{Fe\ in\ Fe_3O_4}).$$

Taking

$$1 - y(\mathrm{FeO/gas})/y(\mathrm{FeO/Fe_3O_4}) \approx 2/3,$$

$$V_M(\mathrm{Fe_3O_4}) \approx 4\,V_M(\mathrm{FeO}),$$

and

$$1 + |z| = 3$$

for both oxides, formulae (9) and (11) yield for the ratio of the thicknesses of the magnetite and wüstite layers

$$\xi(\mathrm{Fe_3O_4})/\xi(\mathrm{FeO}) \approx 2D^*(\mathrm{Fe\ in\ Fe_3O_4})/3D^*(\mathrm{Fe\ in\ FeO}), \quad (14)$$

where the self-diffusion coefficients to be substituted are those for the

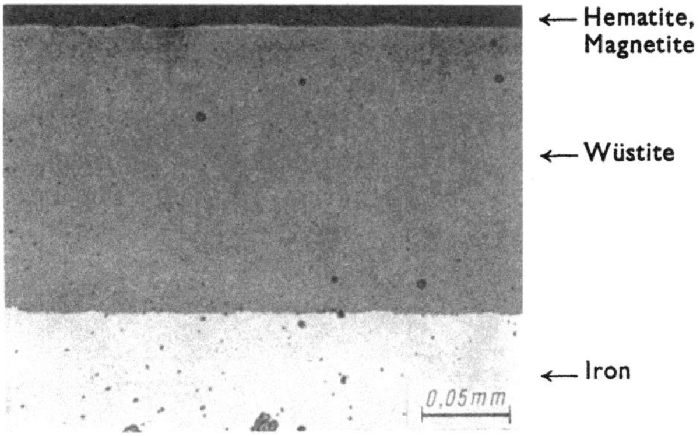

Fig. 40. Oxide layer formed on iron by oxidation at 850°C in air (200 : 1)

highest occurring concentrations of defects. According to the measurements of HIMMEL, MEHL, and BIRCHENALL[121]),

$$D^*(\mathrm{Fe\ in\ FeO}) = 4.5 \cdot 10^{-2}\, y(\mathrm{FeO/Fe_3O_4}) \exp(-27800/RT),$$

$$D^*(\mathrm{Fe\ in\ Fe_3O_4}) = 5.2 \exp(-55000/RT).$$

With these values it follows that, for 900°C,

$$\xi(\mathrm{Fe_3O_4})/\xi(\mathrm{FeO}) \approx 10^{-2}.$$

A ratio of thickness of layers of the same order of magnitude is also found experimentally when iron is oxidized in air or oxygen. Figure 40 shows, under 200 × magnification, the oxide scale formed on soft iron at 850°C in air; the outer, darker edge represents the layer of Fe_3O_4 and Fe_2O_3.

The diffusion coefficient of oxygen in hematite between 900 and 1250°C has been measured by HAGEL[124a]). He obtained the result

$$D(\mathrm{O}) = 2.04 \exp(-77900\ [\mathrm{cal/mole}]/RT) \pm 25\%\ [\mathrm{cm^2/sec}].$$

1.2.7.2. The Stability of the Surface Structure of the Oxide during Reduction

Measurements of the *porosity* and of the *distribution of sizes of pore diameters* in lumps of ore being reduced have shown[125]) that the porous structure changes markedly during the course of the reduction, as may be seen from Fig. 41. This phenomenon is quite understandable since the

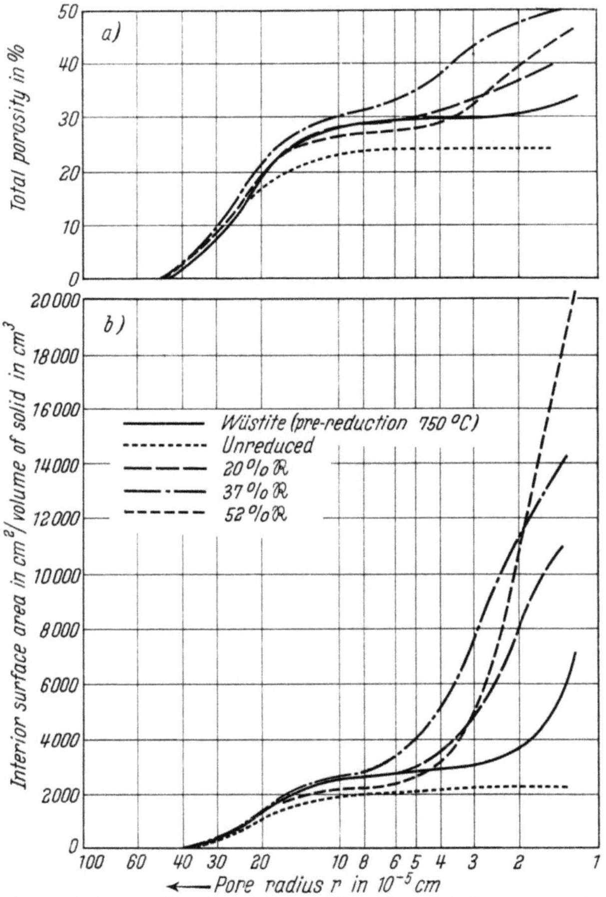

Fig. 41. Porosity and internal surface area as functions of the pore radius, for pellets of 3 to 5 mm diameter at 750 to 780°C and at various stages of reduction

metal formed has a smaller molar volume than that of the oxide. Consequently an expansion of the pores originally present would be expected. If, however, the metal separates as a porous sponge on the surface of the oxide grains and on the walls of the pores, then the original pores can also become smaller, and the dominating feature will be a new system of pores in the metal sponge, which may have an entirely different distribution

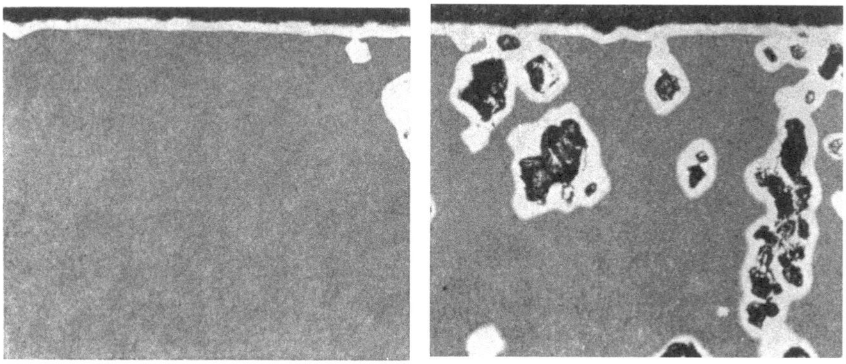

a) b)

Fig. 42. Cross-section of wüstite specimens with various concentrations of vacancies and varying degrees of pore formation after reduction at 800°C in a hydrogen/water-vapour mixture with about 2.5% hydrogen and total pressure of 700 Torr approximately (about 350 : 1, original 500 :1)

	Mole fraction of iron ion vacancies:	Reduction time:
a)	$y = 0.05$	1245 min
b)	$y = 0.063$	1326 min
c)	$y = 0.1$	108 min

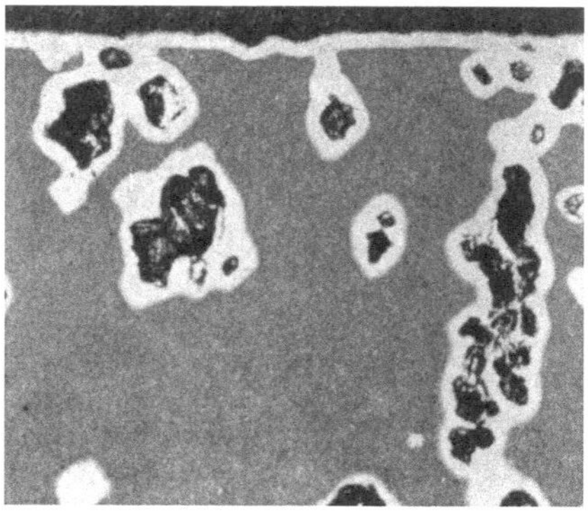

c)

of sizes. This is also the case if the iron nuclei, orientated into the interior of the wüstite grains, grow inwards. New pores are then continually being formed, in which the process of reduction, formation of nuclei, growth, formation of new pores, is repeated. Due to the continual formation of new FeO-surfaces capable of reacting, this mechanism leads to the highest rates of reduction. As will be shown later, the morphology of the formation of the iron nuclei can also produce a continuous film of iron which strongly inhibits the reaction.

Microscopic examination[126]) of compact and originally non-porous oxide specimens shows that through the decomposition of the oxide during the reduction a considerable additional porosity can arise in the oxide, the surface of which becomes roughened and segmented. Figure 42 shows wüstite specimens, which before reduction began had a smoothly polished surface and were free from pores, after reduction with hydrogen at 800°C. The clearly visible hollows are polished pores which, starting from the surface, increase inwards into the oxide. Even single-crystals of wüstite show this phenomenon.

As shown in Fig. 43, the initial content of vacancies, y, in the wüstite greatly influences the formation of pores and hence the course of the

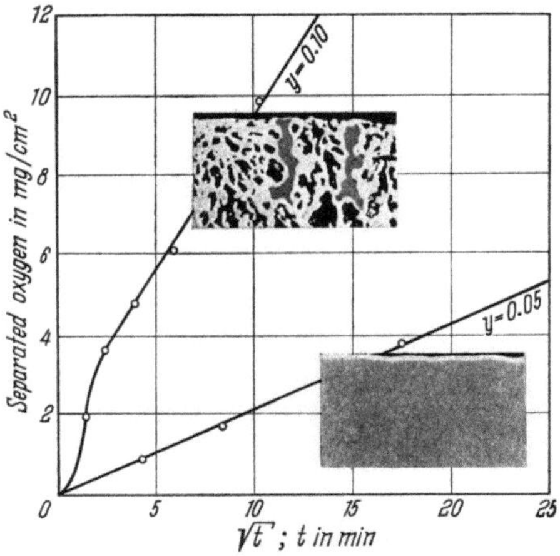

Fig. 43. Reduction of wüstite $Fe_{1-y}O$ with different vacancy content y. $T = 800°C$, $p(H_2) = 679.5$ Torr, $p(H_2O) = 17.5$ Torr

reduction. The conditions prevailing during the oxidation of metal alloys[127, 128]) lead us to imagine the reduction process occurring as follows.

At the outset of the reduction, the vacancy content in the interior of the wüstite is greater than that which would correspond to equilibrium with the iron. The removal of oxygen produces a filling up of the iron ion vacancies at the oxide surface:

$$FeO + |Fe^{2+}|'' + 2\,|e|^{\bullet} + H_2 \rightleftharpoons Fe^{2+}\,|Fe^{2+}| + H_2O\,.$$

$|Fe^{2+}|''$ is an iron ion vacancy, $|e|^{\bullet}$ an electron deficiency or the excess positive charge of a trivalent iron ion, and $Fe^{2+}\,|Fe^{2+}|$ a divalent iron ion in the normal lattice position for wüstite. The surface reduction in y, the mole fraction of iron ion vacancies, sets in motion a diffusion of vacancies and electron defects from the interior of the oxide towards the surface. Figure 44 shows diagrammatically the curves of equal concentration of vacancies in an oxide with a wavy surface, and the corresponding

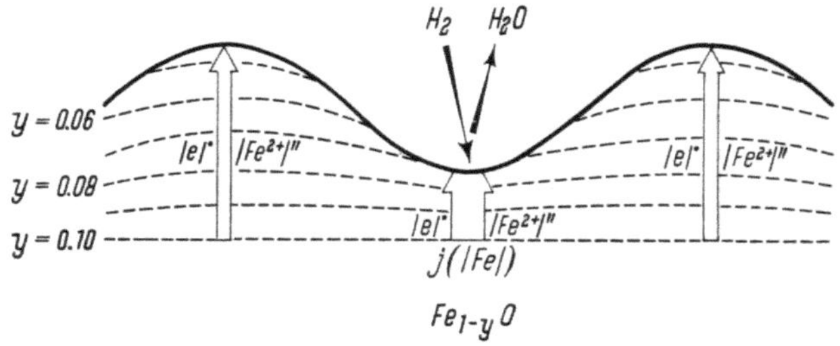

Fig. 44. Diagrammatic representation of flow of material and the formation of surface roughness in the reduction of wüstite

local diffusion flows $j(|Fe^{2+}|'')$, the magnitude of these being indicated by the thickness of the arrows. Since

$$j(|Fe^{2+}|'') \sim dy/dx,$$

the diffusion flow towards a depression of the surface has a higher value than that towards an elevated part of the surface. On the other hand, the rate of surface reaction is, by Eq. (44) of section 1.2.5, proportional to the cube of the local vacancy content. At the depressions, because of the increased diffusion flow towards the surface a higher concentration of vacancies is set up there and consequently a higher rate of oxygen removal, which results in a more rapid excavation of oxide from the depressions. A slight surface roughness, which is inevitably present, will thus be magnified by the interplay of the diffusion of vacancies and the phase-boundary reaction. A flat surface is therefore unstable if the vacancies can diffuse. In particular, sharp depressions, and therefore pores, expand rapidly into the interior of the oxide, since they are extending themselves in the direction opposite to the diffusion flow and hence support a high, local vacancy gradient.

The enlargement of surface area during the reduction of wüstite can take place for only so long as it takes the vacancy content in the whole specimen to assume the Fe/FeO equilibrium value. The time τ for a stationary value of the surface area to be set up should therefore be related to the coefficient of diffusion of the vacancies $D(|Fe^{2+}|'')$ and the thickness d of the specimen by

$$\tfrac{1}{2}d \approx [D(|Fe^{2+}|'')\,\tau]^{1/2}.$$

According to measurements by KOHL[129]), this condition does actually hold.

During the reduction of magnetite with formation of a covering layer of wüstite through which the liberated iron diffuses to the magnetite and converts it into wüstite, there exists a permanent flow of vacancies through the wüstite layer, and the wüstite-gas phase-boundary must continually grow rougher or have pores forming in the wüstite. This effect has actually been observed by ULRICH[130]), both directly by microscopic examination and indirectly from the analysis of the time variation of oxygen removal. However, this phenomenon occurs only at the higher rates of reduction, that is to say, when there is an appreciable gradient in the vacancy concentration and the oxygen activity at the wüstite surface is influenced by the diffusion of the vacancies.

1.2.7.3. Transport Processes in the Reduction of Mixed Oxides and Oxide Compounds

In the reduction of mixed oxides and oxide compounds the solid material, in general, is resolved into three components: into *oxygen*, which goes over into the gas phase in some combined form; into a *metallic phase*, which differs in composition from the molar ratio of the metals in the oxide phase; and into an *oxide phase*, with a correspondingly changed composition. The molar ratios of the metals in the metallic phase and in the oxide phase could be approximately the same only if removal of the oxygen takes place so quickly that an approximation to the equilibrium distribution between the solid phases can not take place. Equality of these molar ratios has to be considered only if the differences in the oxygen equilibrium pressures of the participating pure metals and oxides are small and if the rate of reduction is high. In the reduction of mixed crystals and compounds of the iron oxides with MnO, MgO, CaO, Al_2O_3, SiO_2, etc., which occur in the gangue or as additions in the reduction of iron ores, the first of these two conditions is not satisfied. All these oxides, when in equilibrium with their metal phase or with a lower oxide, have so small an oxygen pressure that they themselves cannot be reduced appreciably or only at high temperatures. At lower temperatures, then, the complex oxide will be split up into a metallic phase rich in iron, an oxide phase impoverished in iron, and oxygen. With limited formation of mixed crystals and with compounds which have only a limited width of phase domain, progress of the reduction requires the separation of a second oxide phase from the oxides; e.g., the separation of calcium oxide in the reduction of dicalcium ferrite $2\,CaO \cdot Fe_2O_3$:

$$2\,CaO \cdot Fe_2O_3 + 3\,H_2 \rightarrow 2\,Fe + 2\,CaO + 3\,H_2O\,.$$

Decomposition of the solid phase requires diffusion in the solid state. As in the previously discussed transport processes in iron oxides, it is again the degree of disorder in the crystals which determines the diffusion coefficients of the components in the case of mixed oxides and oxide compounds. Since in many cases reduction can take place only to the extent that decomposition through solid-state diffusion occurs, i.e., the solid-state diffusion is the rate-determining sub-process, the degree of disorder and transport phenomena are of particular importance in the reduction of mixed crystals and oxide compounds. Unfortunately, little is known about the degree of disorder and the transport mechanism in the phases which are important in this connection. In mixed crystals of wüstite with MnO and MgO the concentration of iron ion vacancies y should theoretically, under ideal conditions, follow the relation[131])

$$y \sim N^{2/3}(\mathrm{FeO})\, p^{1/6}(\mathrm{O_2}). \tag{15}$$

The relation found experimentally between y and $N(\mathrm{FeO})$, the mole fraction of FeO, is shown in Fig. 45; the curves plotted correspond to formula (15)[132]).

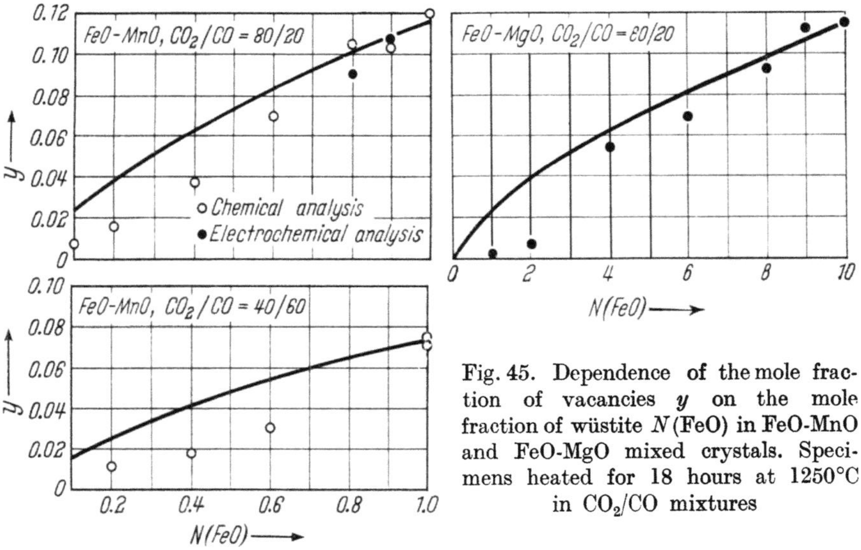

Fig. 45. Dependence of the mole fraction of vacancies y on the mole fraction of wüstite $N(\mathrm{FeO})$ in FeO-MnO and FeO-MgO mixed crystals. Specimens heated for 18 hours at 1250°C in CO_2/CO mixtures

In FeO-MgO mixed crystals BRYNESTAD and FLOOD[15a]) have determined the relation between the disorder and the mole fraction of the FeO and the partial pressure of the oxygen. In Fig. 46 the relative content of trivalent iron

$$N(\mathrm{Fe^{3+}}) = n(\mathrm{Fe^{3+}})/[n(\mathrm{Fe^{2+}}) + n(\mathrm{Fe^{3+}})]$$

found at 1400°C has been plotted against the mole fraction of MgO for oxygen pressures of $2 \cdot 10^{-7}$, $4 \cdot 10^{-6}$ and $9 \cdot 10^{-4}$ atmospheres. The continuous curves have been calculated from a thermodynamical formula derived by assuming association of vacancies with pairs of holes.

The defect-reduction with decreasing wüstite content naturally also affects the electrical conductivity of the oxide and the coefficients of diffusion of the iron in the mixed crystals. In the FeO-MnO series of mixed

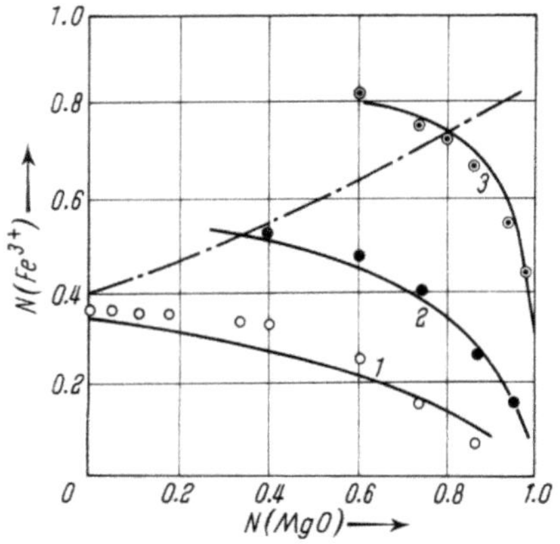

Fig. 46. Dependence of the relative content of trivalent iron on the mole fraction of MgO at 1400°C and oxygen pressures of $2 \cdot 10^{-7}$ (curve *1*), $4 \cdot 10^{-6}$ (curve *2*), and $9 \cdot 10^{-4}$ atm (curve *3*) for mixed crystals of the FeO-MgO series

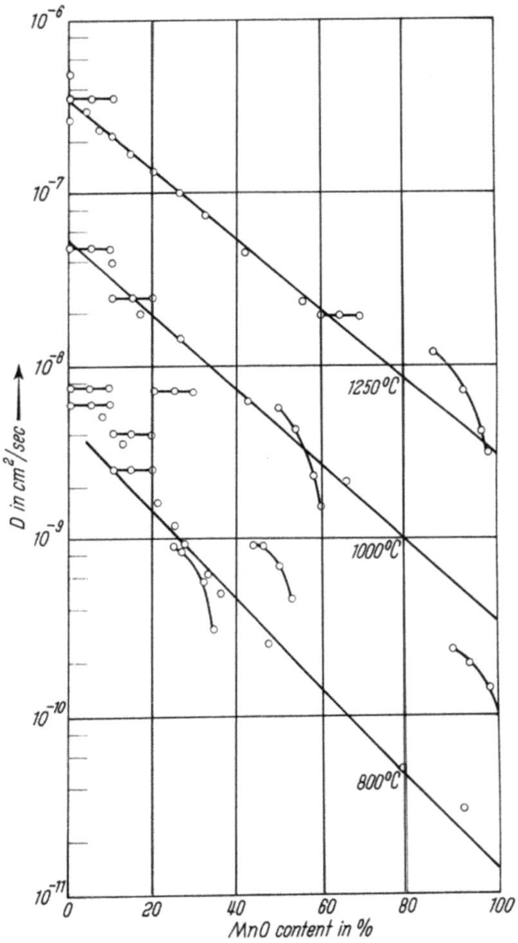

Fig. 47. Interdiffusion coefficient of metal ions in FeO-MnO mixed crystals, according to LIBANATI et al.[132a]

crystals the electrical conductivity at room temperature decreases abruptly at $N(\mathrm{FeO}) \approx 0.6$, because, once the concentration of electron deficiencies has been reduced by the addition of MnO, the adjacency of Fe^{2+} and Fe^{3+} ions in the lattice, which is necessary for the conduction process, becomes more improbable due to the incorporation of manganese ions into the metal-ion sub-lattice.

The change in the diffusion coefficient of the metal in the oxide, which, according to the relation

$$(1 - y)\, D^*(\mathrm{Me}) = y\, D(|\mathrm{Me}^{2+}|''),$$

must be produced by a change in y, assuming the diffusion coefficient of the vacancies, $D(|\mathrm{Me}^{2+}|'')$, to be constant, was determined experimentally by LIBANATI, MANENC, and PHILIBERT[132a]). The results of this investigation are shown in Fig. 47.

WÜNSCH and VASILOS[132b]) determined the interdiffusion coefficients in mixed crystals of MgO with 0 to 15% FeO, NiO, or CoO. The results are given in Table 5.

Table 5. *The diffusion coefficient* $D = D_0 \exp(-Q/RT)$ *in mixed oxides, according to* WÜNSCH *and* VASILOS[132b])

Mixed oxide	MgO/FeO	MgO/NiO	MgO/CoO
Q (kcal/mole)	41.6	48.3	47.4
D_0 (cm²/sec)	$8.8 \cdot 10^{-5}$	$1.8 \cdot 10^{-5}$	$5.8 \cdot 10^{-5}$

With a content of more than 10% of FeO, CoO or NiO the interdiffusion coefficient in this system, too, depends on the concentration[132c]).

MÜLLER and SCHMALZRIED[133]) have investigated defects and diffusion in cobalt ferrite. In crystals of the composition $\mathrm{Co}_{1-x}\mathrm{Fe}_{2-x}\mathrm{O}_4$ the diffusion coefficients of cobalt and iron have a minimum at $x \approx 0$ and increase slowly for $x < 0$ but rapidly in the region $x > 0$, see Table 6. This points

Table 6. *The self-diffusion coefficients of iron and cobalt in* $\mathrm{Co}_{1-x}\mathrm{Fe}_{2-x}\mathrm{O}_4$

		x								
		−0.21	−0.19	−0.16	−0.11	0.0	0.063	0.19	0.26	0.34
$D(\mathrm{Co}) \cdot 10^{10}$ (cm²/sec)	1180°C	—	—	2.5	1.6	0.44	1.1	11	—	36
	1130°C	—	1.5	—	0.5	0.35	—	14	19	—
	1080°C	0.4	—	—	—	0.27	—	2.7	—	—
$D(\mathrm{Fe}) \cdot 10^{10}$ (cm²/sec)	1180°C	—	—	0.27	—	0.21	—	—	—	25
	1130°C	—	0.22	—	—	0.1	—	—	20	—
	1080°C	0.035	—	—	—	—	—	5.8	—	—

to a diffusion of the vacancies over the octahedral positions for cobalt in the region $x > 0$ and for iron in the region $x < 0$. In the region $x > 0$ the iron diffuses over the interstitial positions, cobalt does so in the region $x < 0$. The diffusion coefficient also depends on the partial pressure of the oxygen, since with variable $p(\mathrm{O}_2)$ but a constant value of x the ratio of metal to oxygen changes, and hence so does the defect order. As seen in Fig. 48, the coefficient of diffusion of cobalt has, for $x = 0.063$, a minimum at $p(O_2) \approx 10^{-2}$ atm. When this pressure is exceeded, there is a transition from diffusion via the vacancies to diffusion via the interstitial positions. The detailed derivation of these relationships is contained in

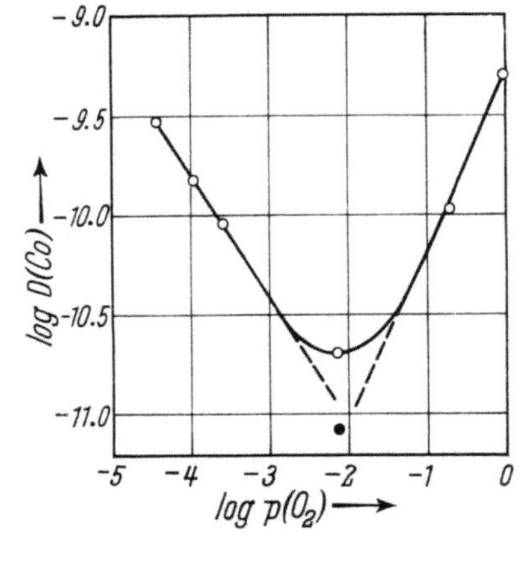

Fig. 48. Coefficient of self-diffusion of cobalt in cobalt ferrite $Co_{1-x}Fe_{2-x}O_4$ as a function of the partial pressure of oxygen at 1170°C and $x = 0.063$, according to SCHMALZRIED[133])

the cited publication[133]) and a previous report on defects in ternary oxide compounds[134]).

Figure 49 shows the values of the coefficients of diffusion of the metallic components of a number of oxide compounds containing iron and also of calcium in CaO. The results clearly can be appointed to one of three different groups: diffusion of iron in magnetite shows a high value of D_0 and a low value of Q; diffusion of Fe and Zn in ferrites other than magnetite shows high values of both D_0 and Q; and diffusion of Ca in CaO shows low values of both constants. The diffusion of Ca in $CaO \cdot Fe_2O_3$ belongs to the second group at higher temperatures and to the third at lower temperatures.

1.2.8. "Direct" Reduction: the Solid-State Reaction between Carbon and Iron Oxides

The Boudouard equilibrium and the equilibria between the iron oxides and CO/CO_2 mixtures represented in Fig. 14 of section 1.1.2 in principle permit a reaction between the iron oxides and solid carbon with the formation of iron and a CO/CO_2 mixture. According to the equilibrium diagram this reaction is possible with $p(CO) + p(CO_2) = 1$ atm at temperatures above 710°C. However, we have to enquire how the kinetics of this reaction are set up and what rate of reaction may be expected.

YUN[135]) investigated the reaction of mixtures of finely powdered graphite and hematite under a vacuum of $5 \cdot 10^{-4}$ Torr. At temperatures

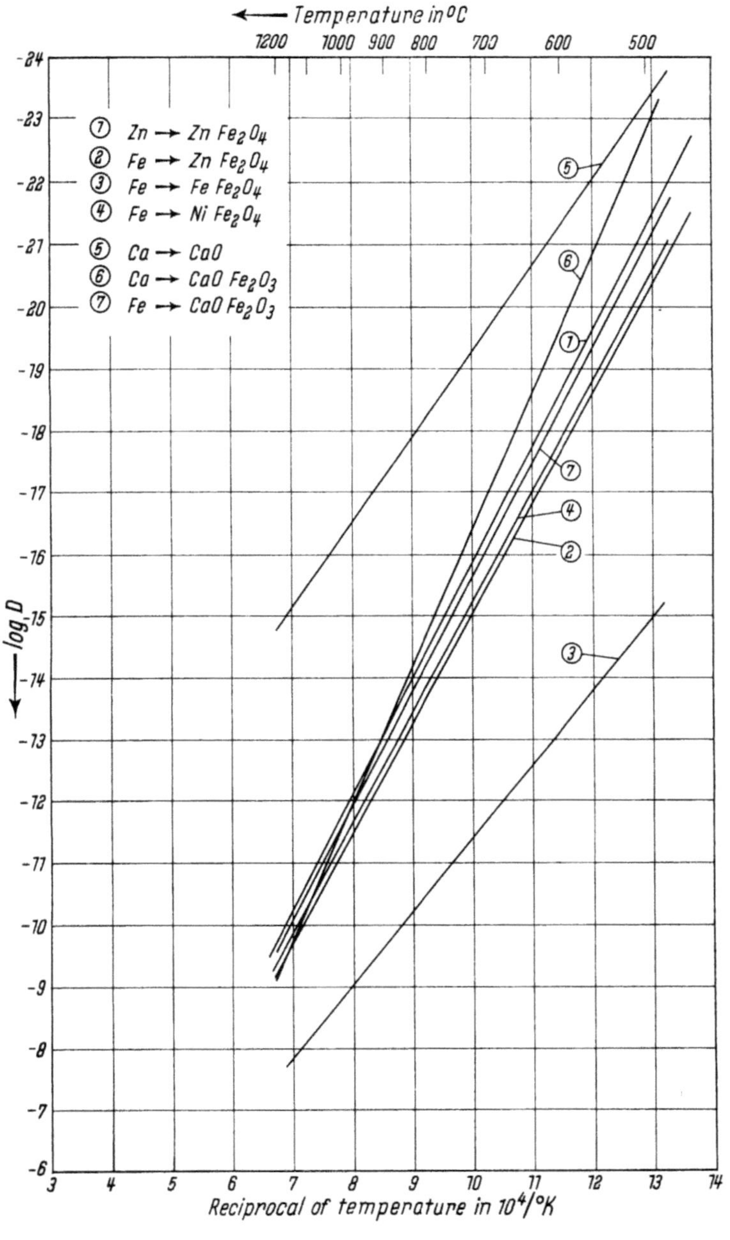

Fig. 49. Dependence on temperature of the coefficients of self-diffusion of the metallic components in oxide compounds, according to LINDNER[134a,b]) and HIMMEL et al.[121])

up to 900°C the reaction took place very slowly, and in 18 hours only Fe_3O_4 and FeO had formed but no iron. Only at higher temperatures was an appreciable transformation observed. YUN compared the course of the reaction with a formula derived by JANDER[136]) for solid-state reactions in powder mixtures and came to the conclusion that the rate of reaction is determined by the diffusion of the iron ions within the oxide phase. A deduction made by BAUKLOH and DURRER[137]) from corresponding in-

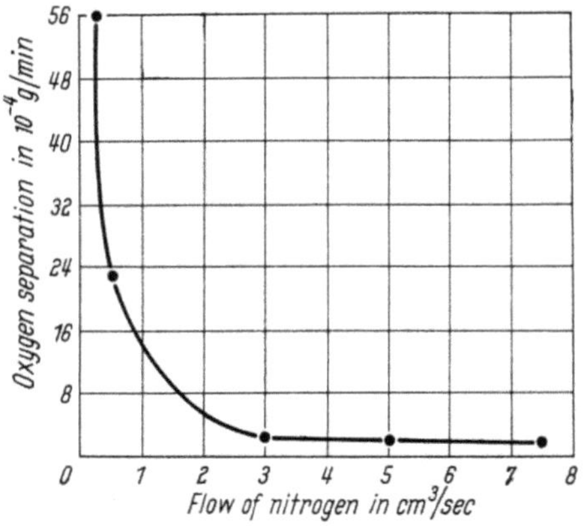

Fig. 50. Effect of nitrogen stream on the reaction rate between carbon and iron oxides, according to BALDWIN[138])
Degree of reduction: 6%

vestigations, that the carbon diffuses in the iron oxide, is perhaps only of historical interest. A comparison of the experimental results of YUN[135]) and BAUKLOH[137]) shows that the rate of reaction distinctly increases when the gas pressure above the powdered mixture increases. Moreover, in the pores of the mixture the gas pressure must be appreciably greater than measurement in the gas region indicates. In experiments of similar type in which a stream of nitrogen was passed through the mixture of carbon and oxide BALDWIN[138]) observed a marked decrease in the reaction rate as the flow of nitrogen was increased; see Fig. 50. It must be concluded that at least at temperatures from 900°C onwards the reaction proceeds, with the gas phase playing a decisive role, according to

$$\left.\begin{array}{llll} Fe_nO_m + m\,C & \rightarrow n\,Fe + m\,CO & \text{the starting reaction,} \\ \rightarrow m\,CO + Fe_nO_m & \rightarrow n\,Fe + m\,CO_2 & \text{reduction by gas,} \\ m\,CO_2 + m\,C & \rightarrow 2m\,CO & \text{the Boudouard reaction.} \end{array}\right\} \quad (1)$$

A reaction in the solid state between carbon and the oxides assumes in addition that there is continuous contact between the ore and the reducing medium; but this contact cannot be maintained during the course of the reaction because of the progressive removal of the products of the reaction. The consequence would be a marked decrease in the rate of reduction with the degree of reduction, as BALDWIN[138]) actually observed between 700 and 800°C. At higher temperatures, however, after an initial decrease in its speed, the reaction accelerates again, a behaviour which can be attributed to the setting up of the chain of reactions (1) and which confirms the importance of the gas phase for the progress of the reaction.

KOHL and MARINČEK[138a]) obtained similar results. Since metals catalyze the Boudouard reaction, and may support the transport of carbon to the oxide, an increase in reaction rate coincides with the onset of separation of the iron. Figure 50a shows that the steep increase takes place at a degree of reduction of 5%. This degree of reduction corresponds exactly

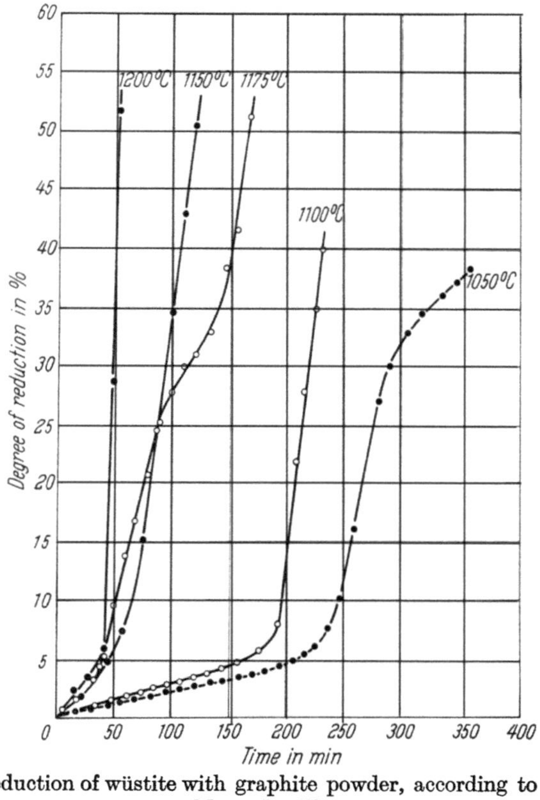

Fig. 50a. Reduction of wüstite with graphite powder, according to KOHL and MARINČEK[138a])

to the oxygen excess in the specimen which must become separated before the separation of the metal begins. That the Boudouard reaction participates in the chain of reactions is also supported by the measured activation energy during the course of the reaction.

The results adduced are in good agreement with experiments on the low-temperature carbonization of briquettes of coal and ore[139]). These experiments showed further that when nitrogen was passed over, and in low-temperature carbonization at temperatures between 700 and 1100°C, the Boudouard equilibrium was set up approximately, and the reaction between the gas and the oxide was rate-determining for the reduction.

A comparison of the observed values in all the experiments mentioned, whether carried out in vacuum or under nitrogen[135, 136, 138, 139]), with the rate of reduction of similar powdered oxides[140]) in CO or H_2, proves that the direct, solid-state reaction of carbon and ore – which is sometimes regarded as the actual mechanism of the direct reduction – is of no importance for the progress of industrial reduction processes. Rather, the direct reduction takes place in accordance with formula (1) in the form of a chain of reactions consisting in reduction by gas and the Boudouard reaction; in both blast and cupola furnaces and in reduction retorts, the first step of the chain, the formation of reducing gas, is replaced by the injection of reducing gas or its formation from oxygen and coke.

2. Results of Experimental Investigations of the Kinetics of Reduction

2.1. Methods and Apparatus for Measurements of the Course of the Reaction

The determination of the progress with time of the reduction of oxides depends in every case on measuring the progress of oxygen removal from the oxides or measuring the value of some quantity which is uniquely determined by the extent of oxygen removal. Methods of investigation entailing *discontinuous* working – in which, for example, the oxide specimen is removed from the reactor after a certain time, is cooled and examined (e.g., weighed) – are justified only in special cases, and will be mentioned here only in passing. The desideratum is *continuous* observation of the chosen quantity, because in a discontinuous method the course of the reaction is disturbed, and changes in the test material – by its being heated up and cooled, and by being taken out of the reaction gas into the atmosphere – are bound to occur.

The measurements may be concerned with the changes in either one of the *gas phases* or one of the *solid phases* participating in the reaction. The test system may be either *open* or *closed*. In the first case, a reaction gas whose composition does not vary in time flows into the system and leaves the test apparatus after reacting with the oxides. If the composition of the gas changes only inappreciably on passing through the apparatus, the system has the properties of a differential reactor. The composition of the gas participating in the reaction can then be regarded as constant at all times and at all points, and this simplifies the evaluation of the measured data. To ensure only differentially small changes in the composition of the gas, the rate of flow must be high or the dimensions of the oxide specimens must be small.

In a *closed* system there is no exchange of material across the boundary of the system during the course of an experiment; the amounts or concentrations of the participants in the reaction vary within the system. A method of this sort is being used, for example, if the gas pressure changes during the reaction, a result which can be achieved by freezing out the CO_2 or H_2O formed during the reduction of ore[141]). If the partial pressure of the component arising during the course of the reaction, in this case that of the water vapour, is kept sufficiently constant by such measures, and if we work with so small a turnover that the total pressure changes only insignificantly, then our closed system also may be regarded as a differential reactor.

In the choice and classification of experimental methods we must also consider whether the investigation is intended only to expose the fundamental relationships or whether in addition it is desired to simulate a reactor on the *technological* scale. In the latter case the choice of a differential reactor is impossible, because the requirement for the highest possible thermal and chemical efficiency in a full-scale process contradicts the conditions of a differential reactor.

2.1.1. Gravimetric Methods

The most obvious and most frequently used method of investigation is the measurement of changes in the *weight of the specimen*, caused by the removal of oxygen. If discontinuous methods are excluded, then for the gravimetric measurement of the rate of reduction we need for weighing the specimens a balance contained in a reaction vessel which is largely or entirely gas-tight.

A balance which is entirely inside the reaction-gas chamber is GULBRANSEN's *microbalance*[142]), Fig. 51. This was developed for oxidation experi-

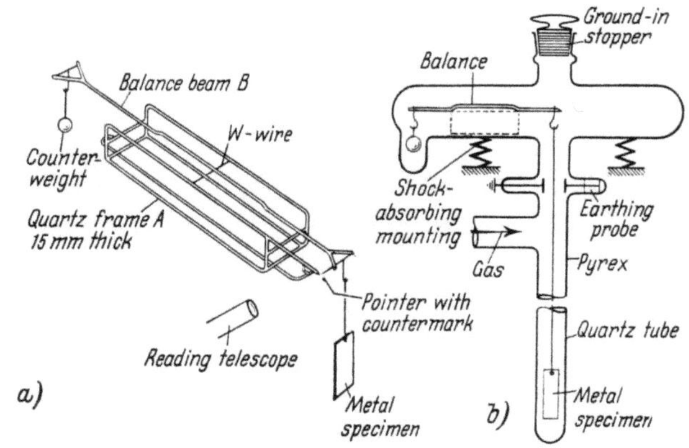

Fig. 51. Quartz micro-balance, according to GULBRANSEN[142])

a) Components of balance: length of beam 15 cm; weight of beam 1 g for test specimens of total weight 0.7 g

b) Arrangement of balance in the oxidation apparatus. The balance chamber of the glass apparatus is on a shock-absorbing mounting ments, and has proved successful in a large number of forms, particularly in the U.S.A. As shown in Fig. 51, the beam of the balance, made of quartz, rests on a tungsten wire fastened to a quartz frame. A tungsten pointer is attached to the quartz balance-arm, and its deflection from a quartz point on the frame of the balance is observed through a distant-reading telescope. Such a balance can weigh up to about 1 gram and its sensitivity is between 10^{-7} and 10^{-6} g. It is therefore suitable for measuring the slightest changes in very small specimens.

There are a number of variations on this type of balance, in which the load capacity and range of measurement are increased by using resetting devices such as an elastic suspension and magnetic weight-compensation[143]) or inductive resetting. GAST[144]) has developed a balance with elastic suspension and inductive resetting. The balance can take loads of 1 to 2 grams,

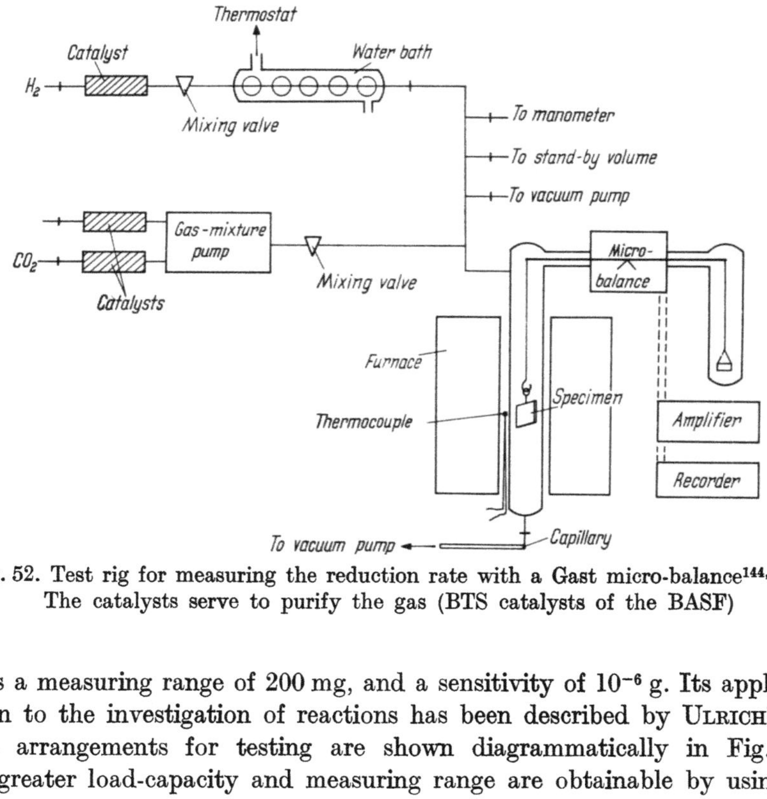

Fig. 52. Test rig for measuring the reduction rate with a Gast micro-balance[144, 145]). The catalysts serve to purify the gas (BTS catalysts of the BASF)

has a measuring range of 200 mg, and a sensitivity of 10^{-6} g. Its application to the investigation of reactions has been described by ULRICH[145]); his arrangements for testing are shown diagrammatically in Fig. 52. A greater load-capacity and measuring range are obtainable by using a normal analytical balance housed in a *gas-tight casing* connected to the apparatus and flushed out by the same gas[146]). The gas-tight casing is not needed if the measurements are made only *at atmospheric pressure.*

In this case the specimen is suspended on the balance by a fine wire or quartz fibre passed through a capillary tube, through which part of the reaction gas is passed[147, 148]).

For specimens up to a few grams in weight McBAIN's quartz spiral balance is also very suitable; this has been applied, e.g., by HAUFFE and PFEIFFER[149]) to oxidation experiments, and by QUETS, WADSWORTH and LEWIS[150]) to reduction experiments. Depending on the number and diameter of the coils, this balance gives a ratio of sensitivity to measuring range of about 10^{-5}. Figure 53 shows diagrammatically an apparatus in-

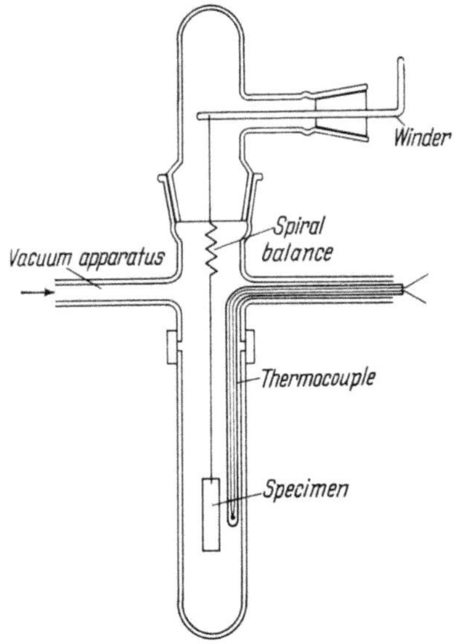

Fig. 53. Quartz spiral balance

corporating a quartz spiral balance, which, as may be seen, is entirely enclosed in the gas space. In a somewhat similar way the quartz spiral can be attached to one end of a quartz balance beam, the other end of which carries the suspended specimen, and the deflection of which is observed through a distant-reading telescope[151]).

The gravimetric methods so far mentioned are particularly, or only, suitable for dealing with individual specimens of the oxides under observation. For measurements on static *bulk material* the retort apparatus shown in Fig. 322, section 5.3.6, has been developed[152]). A reaction tube of heat-resistant steel hangs freely in a tubular furnace from a balance, reading to 1 g. The tube is filled with 350 to 450 g of the ore sample. The gas is passed in and out through flexible tubes attached to the lower end of the tube. The incoming gas first flows through the copper Raschig rings to be heated up to the reaction temperature before coming into contact with the pellets.

In the reduction of material in bulk a differentially small change in the composition of the gas flowing towards and from the material can be achieved only with the highest rates of gas flow, as the discussion in section 5.3 shows. SUBAT[153]) has given an analysis of the positionally dependent degree of reduction when using limited rates of flow.

2.1.2. Gas Analysis and Volumetric Methods

In open systems the course of the reaction can be followed by continuous analysis of the exhaust gas. If the apparatus is at the same time to operate as a *differential reactor*, i.e., if the changes in composition of the gas phase on passing through the apparatus are to be differentially small, then a

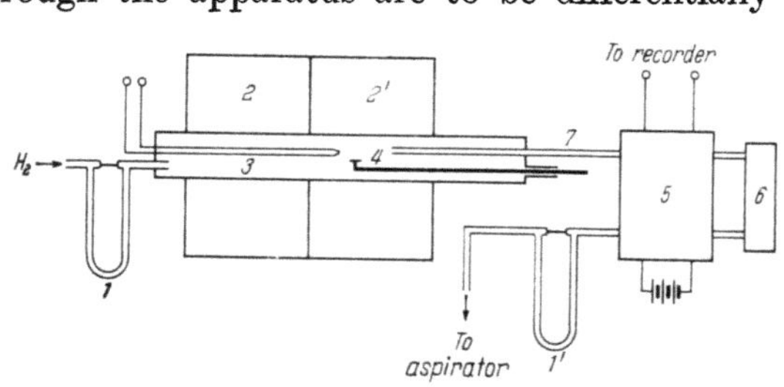

Fig. 54. Test rig for the determination of the rate of reduction by measurement of thermal conductivity and the difference in reducing gas flow-rate between inlet and outlet, according to HOCKINGS[154]) (explanation in text)

high order of accuracy of analysis will be required. The problem becomes even more acute if the partial pressure of the gaseous end-products in the reaction gas is to be *varied*, as is often necessary to elucidate the kinetics of the reaction.

HOCKINGS[154]) has described a test rig, see Fig. 54, in which the rate of reaction is determined by measurement of the thermal conductivity of the exhaust gas. The exhaust gas flow of H_2 and H_2O passes through one arm of the measuring cell *5*, is dried in the absorber *6*, and then passes through the comparison arm of the measuring cell. The deflection of the measuring instrument is proportional to the water-vapour content of the gas. To check the result, the change in the rate of flow of the gas between inlet and outlet is also determined by the flowmeters *1* and *1'*. It is proportional to the amount of H_2 transformed in the reactor. Both observed values give the rate of reaction as a function of time, and the degree of reaction has to be obtained by integration of the curves obtained. EZZ and WILD[155]) have used the change in the rate of flow of the reaction gas when the water vapour formed is frozen out to follow the course of the reaction in the dispersed-particle reduction apparatus shown in Fig. 55.

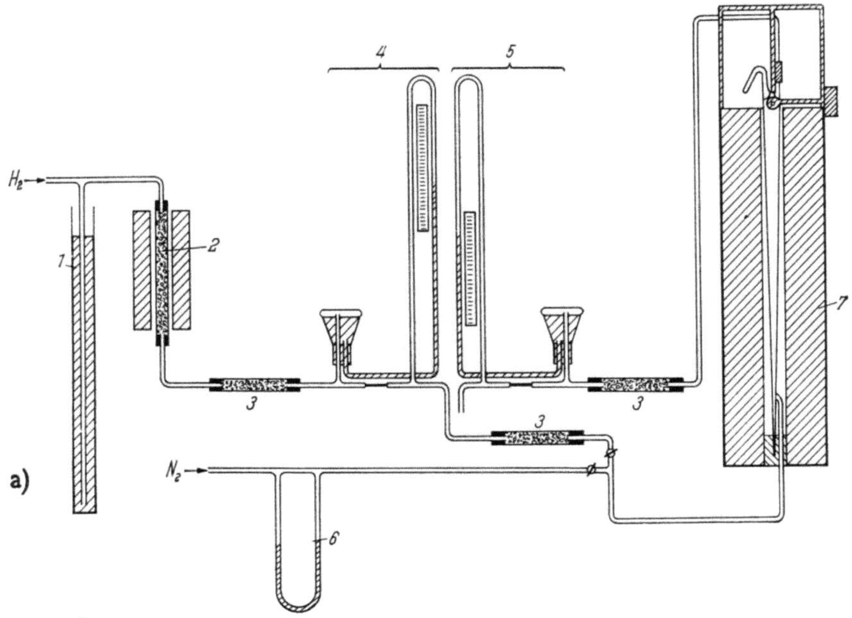

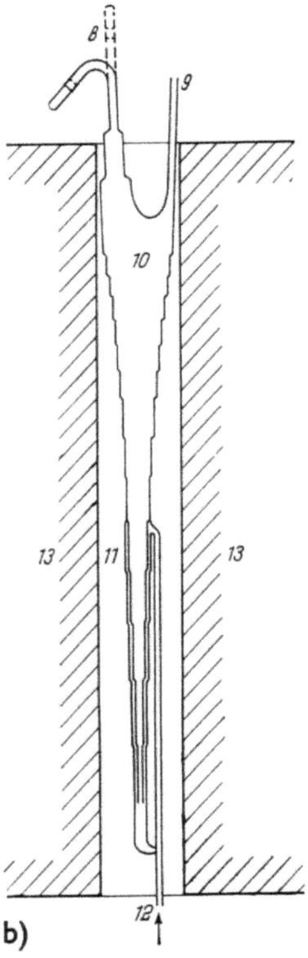

Fig. 55. Test rig for measuring the reduction of oxides in a dispersed-particle reactor, according to Ezz and Wild[155])

a) General view

1 Flow-regulator, *2* Catalyst for gas purification, *3* Drying-tubes, *4* and *5* Flow-meters on the input and output sides, *6* Flow-meter for N_2-admixture, *7* Dispersed-particle reactor

b) Detailed view of the dispersed-particle reactor

8 Arrangement for introduction of the ore specimen, *9* Gas outlet, *10* Reduction chamber, *11* Gas preheater, *12* Gas inlet, *13* Furnace

For the measurement of the rate of reduction of nickel oxide by hydrogen RAHMEL[156]) used a closed apparatus in which a uniform circulation of the gas is maintained by temperature differences. A condensation and saturation chamber is included in this circuit, by means of which the water-vapour pressure is kept constant. In this the water vapour formed during the reaction is condensed. The total pressure, and hence also all the partial pressures, in the apparatus are kept constant by a pressure column of mercury in a burette. The change in volume read off from the burette is equal to the amount of hydrogen used, and from this the degree of reaction can be calculated.

A further development of this principle by HEDDEN and LEHMANN[157]) is shown in Fig. 56. In this equipment the gas is circulated by a pump. The operation of the burette takes place automatically by means of a manometer with an electrical contactor and a servomotor, M_1. The move-

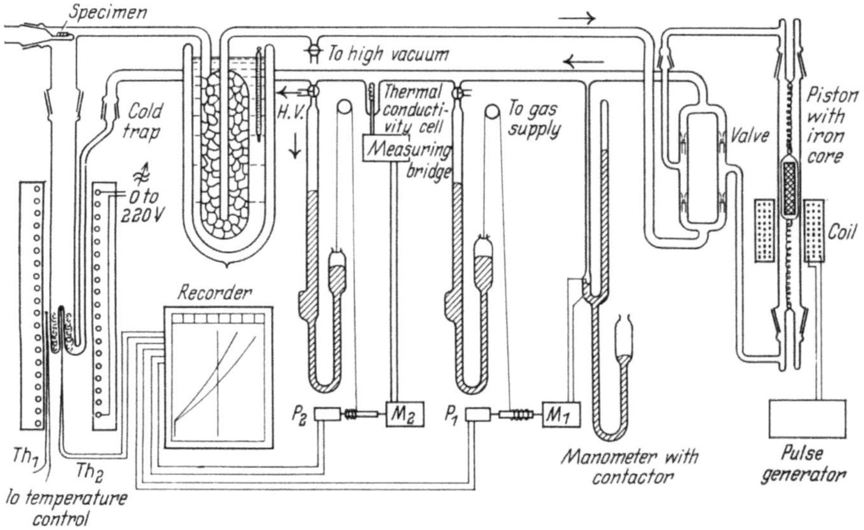

Fig. 56. Gas-volumetric measuring apparatus, according to HEDDEN et al.[157])

ment of this motor is proportional to the conversion achieved and therefore it can be used through the potentiometer p_1 to record the progress of the reaction. The second servomotor M_2, the corresponding burette, and the thermal-conductivity cell, are needed if carbon monoxide is used as the reducing gas.

An extension of the range of measurement of circulation apparatus of this sort is possible without increasing the amount of mercury needed, if extra gas is drawn from a reserve volume[158]).

The possibility has already been mentioned of freezing out the water vapour and carbon dioxide formed and following the progress of the reaction by means of the change in the total pressure[141]).

2.1.3. Special Methods

For elucidating the mechanism of the reaction, special methods of investigation have been successfully employed in some cases. GRABKE[159]) investigated the exchange of oxygen between mixtures of CO and CO_2 and

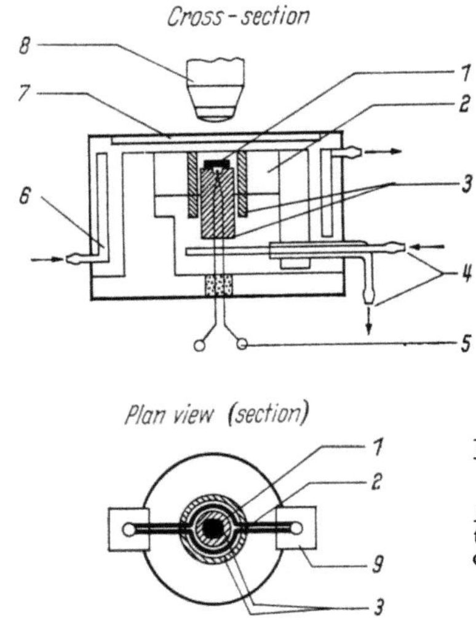

Fig. 57. Hot-stage microscope for direct observation of reduction processes

1 Specimen, *2* Platinum heating-band, *3* Alumina tube with insert, *4* Gas inlet and outlet, *5* Thermocouple, *6* Cooling system, *7* Quartz disc, *8* Objective, *9* Contact jaws

the oxides by following radiochemically the formation of ^{14}CO from added $^{14}CO_2$. This method can also be used for measurements of the exchange of oxygen between oxides and H_2/H_2O mixtures, if HD is added to the gas and the formation of HDO is followed by means of a mass spectrometer. The initial phase of reduction up to the formation of nuclei of the metal phase has been investigated in silver sulphide by measuring the potential of a solid electrolyte cell, and in wüstite by following the electrical conductivity of the oxide[160, 161]). Figure 57 shows an arrangement enabling the process of formation of the metal phase to be observed microscopically on a hot-stage. In section 2.5 some photographs taken with this apparatus are reproduced, showing the surface of wüstite specimens during the course of reduction by a mixture of hydrogen and water vapour[162]).

Measurement of the electrical conductivity of wires of oxide – made by oxidizing iron wires – can also be used to follow the whole course of the reduction[163]), since the conductivity increases with the thickness of the layer of iron formed. However, the relation between the conductivity and the degree of reduction is linear only if the structure of the spongy iron formed does not change during the course of the experiment.

2.1.4. Preparation and Investigation of Oxide Specimens

The purpose of laboratory experiments is usually to obtain information about the behaviour during reduction of the raw materials charged – ore, pellets, or sinter – in the industrial processes for making iron. For this reason such experiments are often carried out with these materials. But to establish the laws governing the sub-processes *synthetic* oxide specimens are more suitable, since by using these the experimental conditions can be set up more precisely and can be varied more easily.

To investigate the reaction at the phase boundary, compact, non-porous oxide specimens with well-defined surfaces are to be preferred. Specimens of wüstite or magnetite which satisfy these requirements can be obtained by oxidizing iron sheets in a CO_2/CO or H_2O/H_2 mixture of suitable composition. Iron oxidizes in H_2O/H_2 mixtures about 40 times faster than in CO_2/CO mixtures, but the production of higher H_2O partial pressures in the gas mixture is possible only with special equipment. By oxidizing iron in air, non-porous, dense hematite specimens can be obtained.

In the reduction of these dense hematite specimens, however, attention must be paid to the fact that cracks and pores do form in the original material and the reaction products[164]). In the reduction of magnetite[145]) and of wüstite[158]), too, the surface area may change during the course of the reaction.

Dense wüstite specimens can be prepared also by melting iron oxides, e.g., Fe_2O_3, in an iron crucible, provided it is possible to avoid the formation of cracks during solidification.

Oxide specimens with greater or lesser porosity can be obtained by sintering pellets or tablets pressed or moulded in the wet state. The porosity can be reduced to less than 5% by volume[147]). Specimens with a greater and definite porosity can be obtained by adding to the pressings ammonium carbonate, ammonium chloride or naphthalene; the additive must then evaporate off at a definite temperature during sintering.

To determine the porosity of natural or synthetic specimens of ores measurements of the volume of the pores, the distribution of size of pores and the internal surface area must be made[165]). The volume of the pores is calculated from the measured apparent density and the specific weight of the oxide. To determine the distribution by size of the pores, the *pore spectrometer* has been developed. By means of this instrument the volume of mercury which can be forced into the pores of the specimen is determined as a function of the pressure. The pressure P which has to be applied is needed to overcome the capillary forces and is therefore uniquely related to the boundary surface tension σ and the smallest area of cross-section through which the mercury has to flow to fill the corresponding volume. The radius r of this smallest cross-section is given by

$$r = \frac{2\sigma \cos\Theta}{P},$$

where Θ is the angle of contact of the mercury in the capillary. SCHENCK[166]) has described a pore spectrometer based on this principle, which has often been applied. The method gives satisfactory values for pore radii greater than 10^{-5} cm.

Measurement of the interior surface area is nowadays carried out almost exclusively by measuring the adsorption of an inert gas such as N_2 or Ar near its condensation point, following BRUNAUER, EMMETT and TELLER[167]). HAUL and DÜMBGEN[168]) have described an easily handled adsorption apparatus for this purpose. Measurements of adsorption and desorption can also be used to determine the distribution of pore sizes with radii less than 10^{-5} cm[169]).

Microscopic examination of the oxide specimens before and after reduction is indispensable. The original materials should be examined as regards the structure of the phases, the structure and distribution of pores, grain size and surface structure, by observation of the outer surface and cross-sections. Microscopic examination of outer surfaces and cross-sections after various degrees of reaction, and particularly of specimens withdrawn shortly after the beginning of the reaction, enable conclusions to be drawn regarding the change in the outer surfaces, the form of separation of the reaction products, the occurrence, type and structure of intermediate products, and changes in the pore structure. Information on microscopic techniques and the preparation of specimens for examination is given in the literature[170]).

2.2. Heat and Mass Transfer across the Gas-Flow Boundary-Layer

Being a countercurrent reactor, the blast furnace and other reduction plants require the heating and reduction of the ore to proceed in parallel, and the same applies to the other solid materials charged. The first sub-process of both processes is the interchange of material and heat between the charge and the gas across the flow boundary-layer. Because of the cross-relations between the two processes, it is useful to treat them together, starting with the heat transfer, since the relevant characteristic numbers have been more thoroughly investigated.

The *coefficient of heat transfer* α specifies the quantity of heat in cal cm^{-2} which is transferred in unit time with a temperature difference of 1°C between the gas and the surface of the solid. The heat flow, j_Q in cal cm^{-2} sec^{-1}, is therefore given by

$$j_Q = \alpha(\Theta - T), \tag{1}$$

where

Θ = temperature of the gas, and
T = temperature of the surface of the solid.

If λ_g is the coefficient of *thermal conductivity* of the gas, and if d is a characteristic length for the body in the flow (characterizing its size compared with geometrically similar bodies), then the *Nusselt number* N_{Nu} is related to the heat transfer coefficient α by

$$N_{\mathrm{Nu}} = \alpha\, d/\lambda_g .$$

Like its analogue, the *Sherwood number* in the transfer of material (see Eq. (22), section 1.2.2), the Nusselt number depends on the Reynolds number

$$N_{\mathrm{Re}} = u\, d/\nu,$$

where

u = velocity of the gas, and
ν = kinematic viscosity of the gas.

The Nusselt number also depends on the *Prandtl number*, N_{Pr},

$$N_{\mathrm{Pr}} = \nu/a, \quad \text{where} \quad a = \lambda_g/c_p\,\varrho_g,$$

which is the analogue of the *Schmidt number* in the transfer of material. Here c_p is the specific heat at constant pressure, in cal g^{-1} °C^{-1}, and a is the thermometric conductivity, or diffusivity, in cm^2 sec^{-1}. According to the kinetic theory of gases, the expression

$$\frac{\lambda_g}{\eta\, c_v} = \frac{a}{\nu}\,\frac{c_p}{c_v} = \frac{1}{N_{\mathrm{Pr}}}\,\frac{c_p}{c_v}$$

should be a constant, equal to 1.8 for a monatomic gas; here c_v is the specific heat of the gas at constant volume, in cal g^{-1} °C^{-1}. The ratio c_p/c_v depends on the structure and number of atoms in the gas molecule; for a monatomic gas it is equal to 1.66, for a diatomic gas 1.4, for a triatomic gas 1.3.

SCHRAMM and SCHÄFER[170a]) have given for the thermal conductivity λ of binary mixtures of gases the expression

$$\lambda = \lambda_1\left[1 + \frac{n_1'}{n_2'}\left(\frac{\sigma_{12}}{\sigma_{11}}\right)^2\left(\frac{M_1 + M_2}{2M_2}\right)^{1/2}\right]^{-1} + \lambda_2\left[1 + \frac{n_1'}{n_2'}\left(\frac{\sigma_{12}}{\sigma_{22}}\right)^2\left(\frac{M_1 + M_2}{2M_1}\right)^{1/2}\right]^{-1}.$$

This formula may be compared with Eq. (26) in section 1.2.2. Here

λ = thermal conductivity of the mixture,
λ_1, λ_2 = thermal conductivities of the gases 1 and 2,
n_1', n_2' = number of molecules per cm^3 for gases 1 and 2,
M_1, M_2 = molecular weights of gases 1 and 2,
σ_{11}, σ_{22} = effective diameter of molecules of gases 1 and 2,
σ_{12} = effective diameter in collisions of a molecule 1 with a molecule 2.

This formula is in excellent agreement with experiment.

The kinetic theory of gases shows that

$$\lambda_1 = n_1' c_{v1} \bar{v}_1 \bar{\lambda}_1/2N_L,$$

where

c_{v1} = molar specific heat at constant volume for the gas of type 1,
$\bar{v}_1$ = $(8RT/\pi M_1)^{1/2}$ = mean (kinetic theory) velocity of molecules of gas 1,
N_L = Loschmidt number,
$\bar{\lambda}_1$ = mean free path of molecules of gas 1 in a mixture of gases 1 and 2 [see Eq. (9), section 1.2.2]: in fact,

$$\bar{\lambda}_1 = 1/[n_1' \pi \sigma_{11}^2 \sqrt{2} + n_2' \pi \sigma_{22}^2 (1 + M_1/M_2)^{1/2}].$$

A similar formula for λ_2 holds for molecules of gas 2.

From the kinetic theory of gases we have, as in section 1.2.2, the following relations:

$$\sigma_{11}^2 = \sigma_{11\infty}^2 (1 + S_1/T),$$

$$\sigma_{12}^2 = [(\sigma_{11\infty} + \sigma_{22\infty})/2]^2 [1 + (S_1 S_2)^{1/2}/T],$$

where $\sigma_{11\infty}$ is the effective diameter for molecules of gas 1 as $T \to \infty$, etc., and S_1, S_2 are the Sutherland constants for molecules of gases 1 and 2.

For a *single, free body in a gas stream* we have the relation[171])

$$N_{\mathrm{Nu}} = C + C'(N_{\mathrm{Re}})^m (N_{\mathrm{Pr}})^{1/3}. \qquad (2)$$

For a *metal sphere in a free air stream* at values of the Reynolds number exceeding 100, observations[172]) give the relation

$$N_{\mathrm{Nu}} \approx 0.37 (N_{\mathrm{Re}})^{0.6}. \qquad (3)$$

Because the Prandtl numbers for air and blast-furnace gas are similar, this formula can also be used for the reduction process by CO_2/CO mixtures[173]).

For *loosely packed material* a corresponding relation between powers of the characteristic numbers is arrived at by taking into account the "void ratio" ε for the material; this is defined as the ratio of the volume of the unoccupied space in the loosely packed material to the total volume of its container.

At low rates of flow — which, however, do not often feature in practice, not in shaft furnaces, for example — an appreciable part of the heat is transferred through natural convection of the gases in the empty spaces between the grains. These conditions may be assumed to hold, for example, in the loosely packed material in rotary kilns. The effect of free convection is embodied in the *Grashof number*, N_{Gr}:

$$N_{Gr} = \frac{g(d/2)^3 \Delta T}{\nu^2 T_m}, \tag{4}$$

where

g = gravitational constant (cm/sec^2),
ΔT = difference in temperature between the surface of the material and the gas (°C),
T_m = mean absolute temperature for the surface of the material and the gas (°K).

According to JESCHAR[174]), the following relation holds for a *bed of loosely packed material consisting of equal sized spheres through which a gas flows*, when $N_{Pr} = 0.70$ and at small Grashof numbers:

$$N'_{Nu} = 2\frac{\varepsilon}{1-\varepsilon} + (N'_{Re})^{0.5} + 0.005 N'_{Re},$$

where (5)

$$N'_{Nu} = N_{Nu}\frac{\varepsilon}{1-\varepsilon}, \qquad N'_{Re} = N_{Re}\frac{1}{1-\varepsilon},$$

and here the Reynolds number for the gas velocity in the empty reaction tube is to be substituted.

N_{Pr} is approximately the same for almost all gases, and since only the power $(N_{Pr})^{1/3}$ occurs, it can be taken into the constants.

The upper curve in Fig. 58 shows the good agreement between Eq. (5) and a large number of observed values. JESCHAR[174]) carried out supplementary

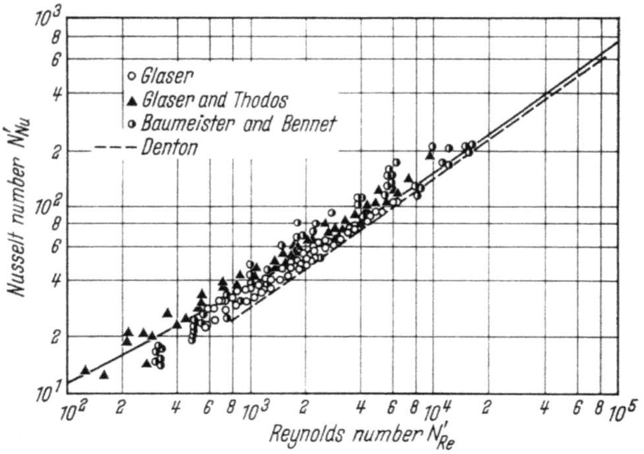

Fig. 58. Heat transfer in beds of spheres, according to JESCHAR[174])

measurements with a sphere electrically heated internally, shown in Fig. 59, whose surface temperature was observed, and which was placed with loosely packed material, consisting of spheres of equal size, in the apparatus shown in Fig. 60. The Reynolds number N_{Re} varied between 3 and 40000. These measurements, too, were in agreement with Eq. (5).

According to JESCHAR[174]) formula (5) can also be applied to beds with materials of several grain sizes, if d in the expressions for the Reynolds

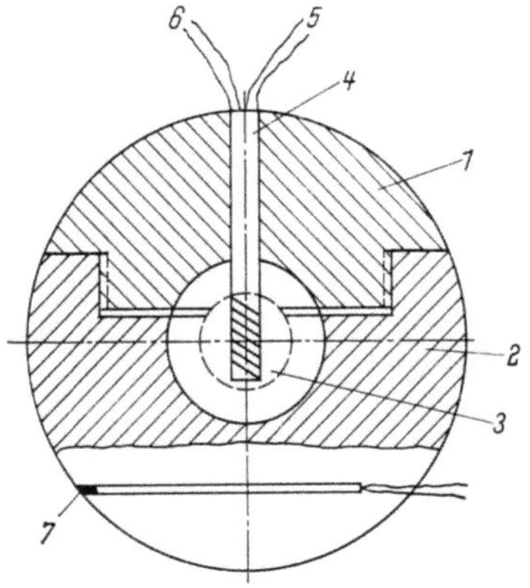

Fig. 59. Section through the measuring sphere, according to JESCHAR[174])

1 and *2* joined by a fine-thread screw, *3* Heating-coil, 0.2 mm diameter, *4* Small porcelain tube, *5* Current cable, 0.3 mm diameter, *6* Voltage cable, 0.1 mm diameter, *7* Soldered joint of thermocouple, 0.1 mm diameter

number N_{Re} and the Nusselt number N_{Nu} is replaced by a "packing diameter" d' given by

$$d' = [V_1/Vd_1 + V_2/Vd_2 + \cdots]^{-1},$$

where V_i and d_i are the volume and diameter for the grains of the i-th size.

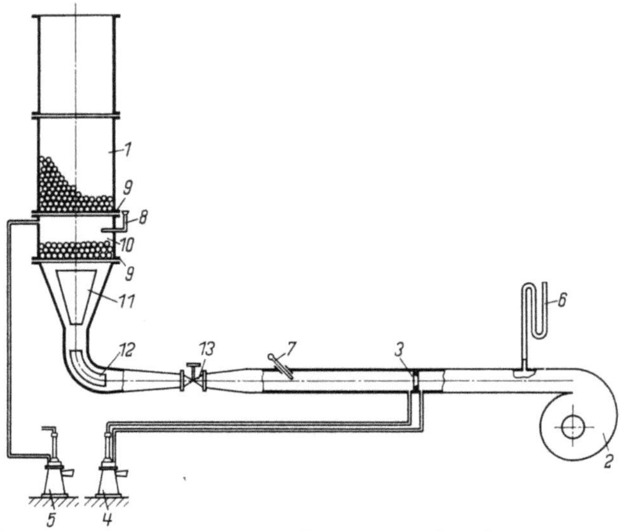

Fig. 60. Diagrammatic representation of apparatus for the measurement of heat transfer, according to JESCHAR[174])

1 Packed column, *2* Blower, *3* Orifice plate, *4* and *5* Micro-manometers, *6* U-tube, *7* and *8* Mercury thermometers, *9* Grid, *10* Packed material of uniform size, *11* and *12* Guides, *13* Valve

POLTHIER[174b)] has discussed heat transfer and pressure loss in loosely packed columns of irregularly shaped particles.

Eq. (5) can also be applied to loosely packed material consisting of different substances with different coefficients of heat transfer if α is replaced by

$$\alpha = \sum \alpha_i F_i / \sum F_i,$$

where α_i is the coefficient of heat transfer and F_i is the surface area for the i-th substance in the material.

The temperature distribution in a bed of spheres heated by hot gases can be calculated exactly from the heat-conduction formula[174c)]. The heat transfer between the gas and the solid bodies, the dependence of the thermal conductivity on temperature, the specific heats and densities of the materials are all taken into account in the boundary conditions.

BEER[175)] has determined the heat-transfer coefficient between a free air stream and coke spheres and pellets and also between air and beds of pellets and pieces of sinter, hematite ore, and coke.

Since the measurements of the coefficients in beds of material were carried out by a calorimetric method and the pieces had no one definite surface area, BEER could only determine a mean coefficient of heat transfer, α_Σ, relative to the volume and mean temperature T_i of the pieces. The relation

$$j'_Q = \alpha_\Sigma(\Theta - T_i) \qquad (\text{cal cm}^{-3}\,\text{sec}^{-1})$$

holds.

When comparatively large pieces, of low thermal conductivity, are heated rapidly there is a time lag before the temperature within a piece becomes equalized. The effect of porosity on the thermal conductivity can be taken into account by means of a formula given by EUCKEN[174a)]:

$$\lambda_M = \lambda_Z(1 - \gamma_p)/(1 + 0.5\gamma_p), \tag{6}$$

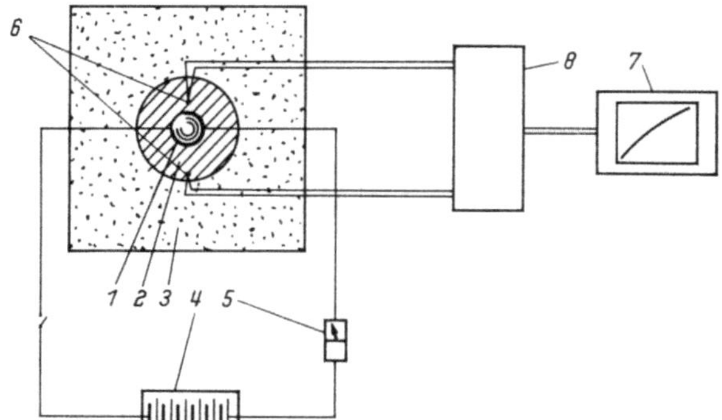

Fig. 61. Apparatus for measurement of thermal conductivity in porous bodies, according to BEER[175a)]

1 Heat source, *2* Test body, *3* Insulating material, *4* Heating battery, *5* Wattmeter, *6* Thermocouple, *7* Compensated recorder, *8* Thermocouple commutator switch

where

λ_M = thermal conductivity of the porous body,
λ_Z = thermal conductivity of the compact material of the body,
γ_p = ratio of the volume of the pores to that of the body.

BEER[175a]) has tested the applicability of Eq. (6) to unreduced and reduced ore fragments using the apparatus shown in Fig. 61. Figures 62

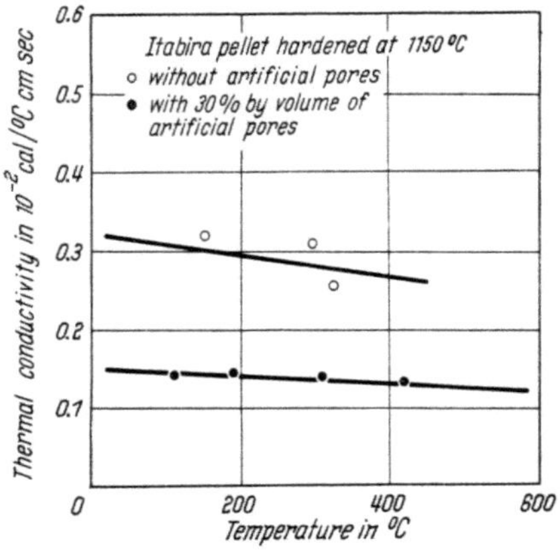

Fig. 62. Effect of porosity on thermal conductivity, according to BEER[175a])

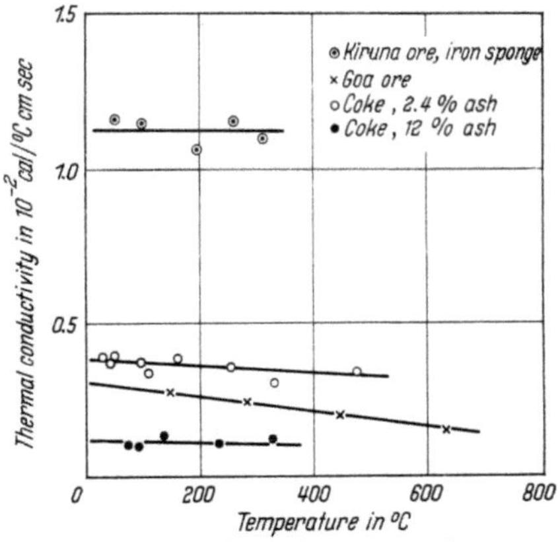

Fig. 63. Thermal conductivity of spongy iron made from Kiruna ore, of Goa ore, and of coke of different ash contents, as a function of temperature; according to BEER[175a])

and 63 show some of the results; it is clear from Fig. 62 that the porosity has a large influence, in accordance with Eq. (6).

By separate measurements on *pellets* and *spheres of coke* BEER was able to show, using a thermocouple method, that for ore and coke the relation

$$\frac{1}{\alpha_\Sigma} = \frac{1}{\alpha_v} + \frac{r_0}{14\lambda_c} \tag{7}$$

holds, where α_v is the heat transfer coefficient relative to the volume and surface temperature, r_0 is the equivalent radius for the fragment, and λ_c is the thermal conductivity of the solid. α_v and α are connected by the relation

$$\alpha_v = 3\alpha/r_0 .$$

Formula (7) makes it understandable that the heat intake is determined by the conduction of heat into the interior of the pieces when the thermal conductivity is low but by the heat transfer from the gas to the surface of the solid when the thermal conductivity is high. The effect of the second term in (7) particularly preponderates for large and porous pieces of material.

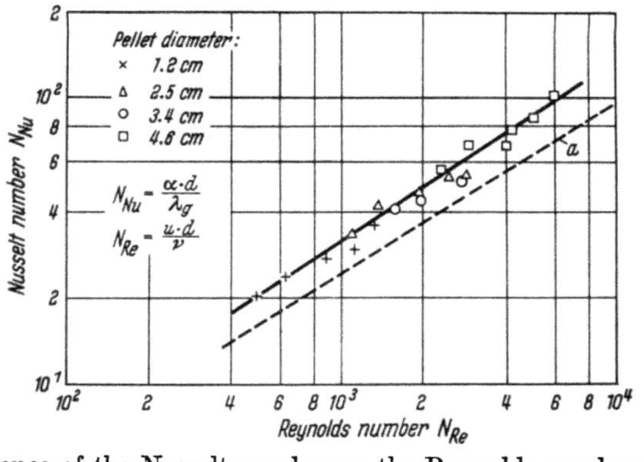

Fig. 64. Dependence of the Nusselt number on the Reynolds number for an ore pellet in a free air stream, according to BEER[175]); curve a according to McADAMS[172])

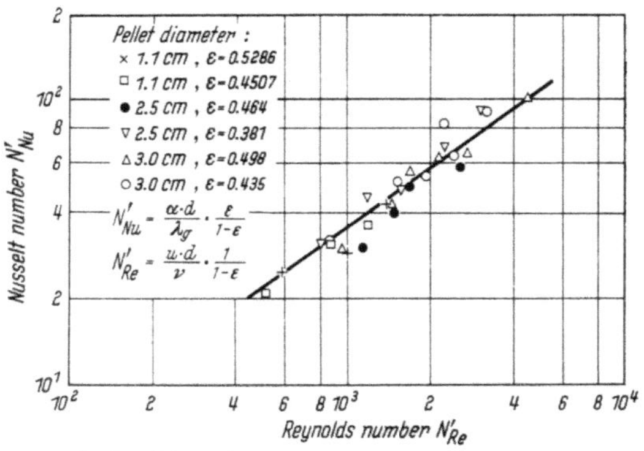

Fig. 65. Dependence of the Nusselt number on the Reynolds number for beds of pellets, according to BEER[175])

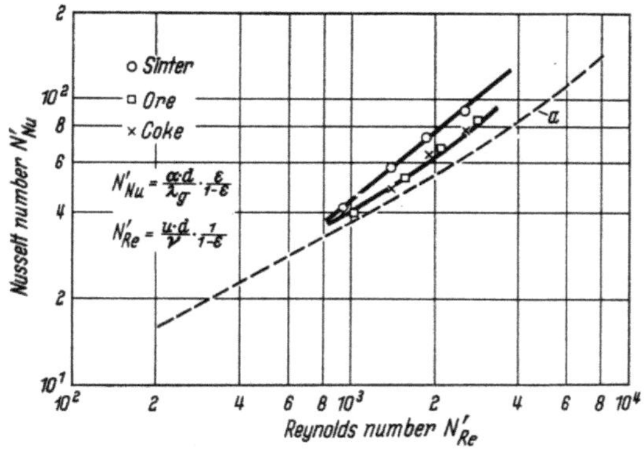

Fig. 66. Dependence of the Nusselt number on the Reynolds number for beds of sinter, ore, and coke, according to BEER[175]); curve a from JESCHAR's data for beds of metal spheres[174])

The Nusselt numbers calculated from the α-values converted in this way are shown in Figs. 64 to 66 as a function of the Reynolds number.

For *pellets in a free gas stream,*

$$N_{\mathrm{Nu}} = 0.36(N_{\mathrm{Re}})^{0.65}$$

and for *beds of pellets*

$$N'_{\mathrm{Nu}} = 0.28(N'_{\mathrm{Re}})^{0.7}$$

and for *beds of pieces of ore, sinter and coke*

$$N'_{\mathrm{Nu}} \approx 0.30(N'_{\mathrm{Re}})^{0.7}.$$

With Reynolds numbers between 500 and 5000 the values of Nusselt numbers calculated by this formula differ little from those given by (5). Working backwards from the relation between the Nusselt and Reynolds numbers the values of α for burden materials can be calculated, and these with Eq. (7) yield the *overall heat-transfer coefficient* α_Σ governing the taking up of heat by a blast-furnace charge.

BEER also carried out measurements on moist burden components. He established that the rate of heating is not affected to any important extent by the evaporation of the water. This fact is attributable to two effects working in opposite directions: it is true that the heat of evaporation must be additionally transferred to the solid material, but on the other hand during the evaporation the surface temperature of the solid remains at the boiling-point of water with the result that the difference in temperature between gas and surface, and hence also the heat flow, is increased. The heat-transfer coefficients for dry and damp burden materials are the same.

From the results of his measurements BEER reached the conclusion that in sized burdens with pieces smaller than 2.5 cm the temperature is equalized between gas and furnace burden almost completely within a few minutes of the furnace's being charged. See also section 4.6.

Comparison of the relations between the Nusselt and Reynolds numbers with the relation between the Sherwood and Reynolds numbers, as given by formula (25) of section 1.2.2, makes it possible to convert the Nusselt number into the Sherwood number and thus to determine the material-transport coefficient β from the heat-transfer coefficient α. Since the constant C' must have the same numerical value if the geometrical conditions are the same, we get

$$N_{\mathrm{Sh}} = N_{\mathrm{Nu}}[N_{\mathrm{Sc}}/N_{\mathrm{Pr}}]^{1/3} = N_{\mathrm{Nu}}[a/D]^{1/3} \tag{8}$$

and hence, since

$N_{\mathrm{Sh}} = \beta\, d/D$, $N_{\mathrm{Nu}} = \alpha\, d/\lambda_g$, $N_{\mathrm{Sc}} = \nu/D$, $N_{\mathrm{Pr}} = \nu/a$, and $a = \lambda_g/c_p\, \varrho_g$,

we find

$$\beta = \alpha[D/a]^{2/3}\,[c_p\, \varrho_g]^{-1}. \tag{9}$$

The *material-transport coefficient* β has the dimension cm sec^{-1}.

According to the kinetic theory of gases, heat conduction and material transport in gases are connected with internal friction[175b, c]. The coefficient of diffusion D and the thermometric conductivity a are related by the equation used above

$$D/a = N_{Pr}/N_{Sc}. \tag{10}$$

The *Schmidt number* N_{Sc}, as seen in section 1.2.2, for pure gases has values between 0.6 and 0.8. For diatomic gases, $N_{Pr} \approx 0.5$. So, for pure gases, we may take $D/a \approx 0.7$. This value also holds quite well for mixtures of gases of similar molecular structure and molecular weight. As indicated on p. 58, for gases which do not have these similarities, N_{Sc} has smaller values, and therefore D/a is different. From data in the literature[176, 177], D/a for mixtures of CO_2 and H_2 at 0°C and 1 atm is calculated to have the following values:

H_2 (% by volume)	D/a
25	2.74
50	1.46
75	0.81

If the value of the Nusselt number is known, the material-transport coefficient β may be calculated from

$$\beta = N_{Nu}(a/D)^{1/3}(D/d). \tag{8a}$$

(8a) also implies that, with $N_{Nu} = C'(N_{Re})^m$,

$$\beta = C'[u^m D/\nu^m d^{1-m}][a/D]^{1/3}. \tag{8b}$$

For beds of pellets, sinter or ore the index m has the value 0.7 according to Beer[175] and the constant C' is about 0.3. Hence

$$\beta = C'' u^{0.7}/d^{0.3}. \tag{8c}$$

Thus, for constant empty-tube velocity, the material-transport coefficient decreases as the size of the pieces increases. Also, for constant Reynolds number and therefore for constant Nusselt number, by formula (8a), β decreases as the piece size increases.

The maximal gas-flow rate through a static bed is the critical velocity which just preserves the stability of the static bed without fluidization. By formula (5) of section 3.1, this value is given by

$$u_{cr} \sim L^{0.58},$$

where $L = 2d\Theta$, and Θ is a factor taking into consideration the geometrical shape of the pieces.

Substituting this expression into (8c), we get

$$\beta_{cr} \approx C''' d^{0.28}.$$

At the equilibrium limit, then, β increases slowly with increasing piece diameter.

For a temperature of 900°C and a piece diameter of 2.5 cm, and also for a Reynolds number of 2000, which, according to HANSEN[178]), prevails at this temperature in blast furnaces, the value of the material-transport coefficient calculated [by formula (8a)] from BEER's measurements[175]) is about 250 cm sec^{-1} for H_2/H_2O mixtures and about 60 cm sec^{-1} for CO/CO_2 mixtures.

Comparison of this value with the measured reaction rates enables an estimate to be made of the effect of the diffusion in the boundary layer of the flow on the overall rate of reaction[175d]). According to section 1.2.2, the diffusion flux in the boundary layer of the flow is given by

$$j = \beta(c^0 - c') = \beta(p^0 - p')/RT,$$

where

c^0 = concentration of the reducing component of the gas in the free gas space,
c' = its concentration at the surface of the pellets,
p^0, p' = corresponding partial pressures.

If it is the reaction at the phase boundary which is rate-determining, then the rate of reduction v,

$$v = -\frac{dn_0}{dt},$$

measured e.g., in mole cm^{-2} sec^{-1}, is given by

$$v = k(p' - p^*),$$

where p^* is the equilibrium pressure of the reducing gas.

With the condition for the steady state of the reaction, $v = j$, it follows that

$$\frac{p^0 - p'}{p' - p^*} = \frac{\Delta p \text{ in boundary layer}}{\Delta p \text{ at phase boundary}} = \frac{k\,RT}{\beta}.$$

With pellets which are easily reduced it is to be supposed that the actual reacting surface area is greater than the geometrical surface area of the pellets; let us assume a factor of 10 to take this effect into account. If we take for k, the reaction rate constant, the comparatively high value given by McKEWAN[179]) of $4 \cdot 10^{-5}$ mole cm^{-2} sec^{-1} atm^{-1} at 900°C, then, with the assumed ratio of reacting to geometrical area, we find for reduction by hydrogen

$$\left.\frac{\Delta p \text{ in boundary layer}}{\Delta p \text{ at phase boundary}}\right|_{\max} \approx 0.15.$$

So large a decrease in the partial pressure of the reducing component of the gas in the boundary layer of the flow is not to be expected in blast-furnace operation. Reduction experiments on industrial ores[180]), carried out with the equipment shown in Fig. 322, section 5.3.6, give at 900°C with a gas mixture of 40% CO and 60% N_2 for a 40%-complete reaction, a maximum reaction rate of 0.6% per minute for ore pieces of size 2.5 cm. From this the reduction rate constant, reduced to the geometrical surface area, comes out to be $1.2 \cdot 10^{-5}$ mole cm^{-2} sec^{-1} atm^{-1}. This value gives 0.02 for the ratio of the decrements of partial pressure in the boundary layer of the flow and at the surface. This means that, in blast-furnace operation, in the lower region of the shaft and at comparatively high temperatures and also with ores of good reducibility, about 2% of the partial pressure of the reducing gas in the boundary layer of the flow is needed for the material transport to the gas and is lost for the reaction.

The dependence of the material-transport coefficient β on the temperature is determined by the diffusion constant D, which is approximately proportional to $T^{1.75}$. By formula (8b) it follows that $\beta \sim T^{1.2}$. The other sub-processes in reduction depend on temperature to a far greater extent. For this reason it is only at high temperatures that the material transport can be expected to have an appreciable effect on the overall rate of reduction.

In general, the effect of material transport from gas to solids can be disregarded in the further discussion of the kinetics of reduction, although it can certainly be demonstrated in laboratory experiments. Thus the reduction of a pellet of ore which is suspended asymmetrically in the reaction tube takes place at a rate which depends on the *local* gas flow[180a]).

2.3. The Diffusion of Gas and Solids in the Layer of Reaction Products

When the reducing gas has passed the boundary layer of the gas on the surface of the ore and has reached the surface of the solids, it converts the ore into its reduction products at this surface. In the general case – and particularly if the pieces are large and the ore is only slightly porous – the degree of reduction at a given time decreases from the outer surface of the ore inwards. Therefore the end products of the reduction appear first on the outer surface of the pieces of ore; and, depending on the composition of the gas and the temperature, these end-products may be either *iron* or *lower oxides*.

In a number of publications[181–187]) it has been conjectured or demonstrated that the diffusion of the gas in the porous end-products of the reaction exerts an influence on *the rate of reduction* or may even be the sub-process which determines the rate of reduction of the ore. The porosity γ and the labyrinth factor ξ, the thickness of the layer Δr, the temperature and composition of the gas at the boundary of the layer, all affect the rate of this process. The diffusion flow, in mole sec^{-1}, in a layer of a spherical pellet is given by

$$j_i = -4\pi r^2 D_i^P \frac{dc_i}{dr}, \tag{1}$$

where $D_i^P = D_i \gamma \xi$ is the coefficient of *porous diffusion* of the gas.

Let c_i' and c_i'' denote the concentrations of the reducing gas at the outer and inner surfaces of the layer of reduction products, and let r be the radius of the piece of ore.

Since the thickness of the layer Δr increases with time or with the degree of completeness of the reduction, j_i, c_i', and c_i'' will in general also change with the degree of completeness.

Integration of formula (1) gives

$$j_i(1/r_0 - 1/r') = 4\pi D_i^P (c_i' - c_i''), \tag{2}$$

where r_0 is the outer semi-diameter of the piece of ore, and r' is the semi-diameter of the unreduced core.

Let d_0 denote the oxygen content of the ore in mole cm^{-3}. Then

$$j_i = +4\pi (r')^2 d_0 \frac{dr'}{dt} \tag{3}$$

and with (2) this gives

$$\frac{r'}{r_0}(r_0 - r')\frac{dr'}{dt} = -\frac{D_i^P}{d_0}(c_i' - c_i''). \tag{4}$$

The diffusion in the porous layer of the reaction end-products will be *rate-determining*, if the reaction constant of removal of oxygen at the phase boundary is so high that at the inner surface of the layer there is practically equilibrium between the reducing gas, the ore, and the reduction products. In this case,

$$c_i' = c_i^0, \quad c_i'' = c_i^*,$$

where c_i^0 is the concentration in the free gas space, and c_i^* is the equilibrium concentration.

The relative thickness of the reduced layer is $(r_0 - r')/r_0 = \Delta r/r_0$ and is related to the degree of reduction $\mathfrak{R}$ by

$$\Delta r/r_0 = 1 - (1 - \mathfrak{R})^{1/3}.$$

Substitution of this expression in (4) gives[182, 184])

$$\dot{\mathfrak{R}} = \frac{d\mathfrak{R}}{dt} = \frac{3 D_i^P (c_i^0 - c_i^*)}{r_0^2 d_0[(1 - \mathfrak{R})^{-1/3} - 1]}. \tag{5}$$

It should be observed that in this derivation it has been assumed that the outer radius r_0 of a piece of ore remains unchanged throughout the reduction. This is possible only if the porosity of the layer corresponds to the difference in the molar volumes of the ore and the reduction product.

A variation of the porous structure with time can be taken into account by assuming that the labyrinth factor ξ depends on the degree of reduction[182].

Integration of formula (5) gives[184, 185, 186], for the ideal case where $D_i^P = \text{const}$,

$$\frac{1}{2} - \frac{\Re}{3} - \frac{(1-\Re)^{2/3}}{2} = \frac{D_i^P (c_i^0 - c_i^*)}{r_0^2\, d_0}\, t. \qquad (6)$$

For specimens of oxides in plate form, with sides large compared with their thickness r_0, the cross-section F of the layer can be regarded as constant at all points and at all times. If the gas diffusion through the pores is rate-determining, then for specimens of this shape

$$\frac{dn(\mathrm{O})}{dt} = -\dot{j}_i = F\,\frac{D_i^P (c_i^0 - c_i^*)}{l}. \qquad (2\text{a})$$

The thickness l of the layer is equal to $n(\mathrm{O})/F\, d_0$, where $n(\mathrm{O})$ is the amount of oxygen removed, d_0 is again the oxygen content of the ore specimen, in moles of oxygen per cm^3, and F is the surface area of the specimen.

Hence it follows that

$$\frac{dl}{dt} = \frac{D_i^P (c_i^0 - c_i^*)}{d_0\, l}. \qquad (4\text{a})$$

The degree of reduction, $\Re$, is here equal to the ratio of the thickness of the layer to half the thickness of the sample, $r_0/2$. Hence

$$\dot{\Re} = \frac{d\Re}{dt} = \frac{4 D_i^P (c_i^0 - c_i^*)}{r_0^2\, d_0\, \Re} \qquad (5\text{a})$$

and

$$\Re = \left[\frac{8 D_i^P (c_i^0 - c_i^*)}{r_0^2\, d_0}\, t\right]^{1/2}. \qquad (6\text{a})$$

From the formulae (5) and (5a), when the time for reduction is determined by diffusion through pores, the dependence of the rate of reduction, $\dot{\Re}$, on the temperature is determined by the dependence of the diffusion coefficients on temperature; $\dot{\Re}$ must therefore be approximately proportional to $T^{1.75}$, if $c^* \ll c^0$. Otherwise, the influence of T upon c^* must be considered, see section 1.2.2.

The *effect of temperature* has been checked by reduction experiments on Malmberget pellets of 3 to 31 mm diameter using mixtures of hydrogen and water vapour[182]. In Fig. 67 the apparent activation energies for various degrees of reduction are plotted against the size of pellet. We see that, for a 10% degree of reduction or $\Re = 0.1$ with pellets of about 20 mm, and for $\Re > 0.5$ onward with pellets of about 12 mm, the apparent activation energy for the diffusion in this temperature range reaches the value of about 1.2 kcal/mole. Figure 68 shows that the corrected rate of reduction $v' = v/(c^0 - c^*)$ at temperatures above 600°C follows the variation in the diffusion coefficients as regards its dependence on the temperature, but at lower temperatures reaches higher values of the apparent activation energy. It has been shown from photomicrographs that the removal of oxygen takes place in a narrow transitional zone between the skin of iron and the unreduced core of the ore.

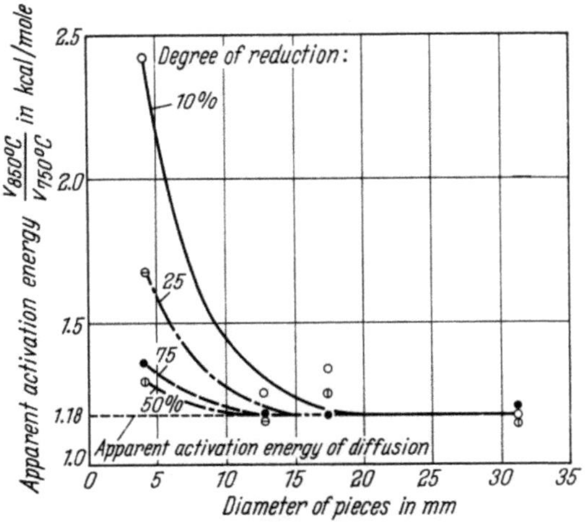

Fig. 67. Apparent activation energy in relation to size of pieces for different degrees of reduction[182])

Charge: 1000 g Fe_2O_3-pellets. Flow rate: 24 cm/sec (empty-tube velocity at 800°C)

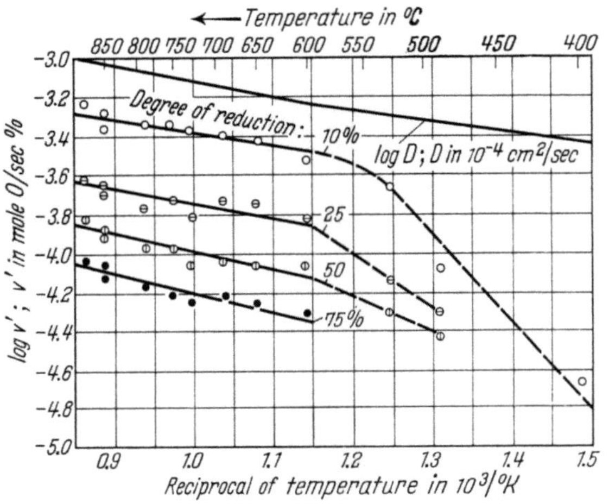

Fig. 68. Rate of reduction v and diffusion coefficient D in relation to the temperature T[182])

Charge: 1000 g Fe_2O_3-pellets (31 mm diameter). Gas flow: 60 litres H_2/min

Figure 69 shows an observed reduction curve and one calculated from formula (5). In the calculation it was assumed that the labyrinth factor ξ, contained in D^P, changes with the degree of reduction $\mathfrak{R}$ according to the relation[182])

$$\xi = 0.10(1 + 2.33\,\mathfrak{R}).$$

The absolute value of the labyrinth factor according to this lies between 0.1 and 0.3 and thus agrees well with the values determined experimentally

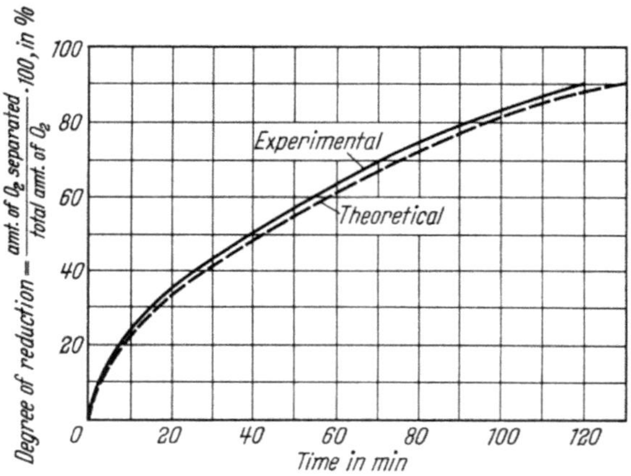

Fig. 69. Comparison of experimental and theoretical progress of reduction at 650°C[182])
Charge: 1000 g Fe_2O_3-pellets (31 mm diameter). Gas flow: 60 litres H_2/min

by WOODS[187]). Also the dependence to be expected from formula (5) of the rate of reduction on the partial pressures in the gas phase has been found by the authors to be confirmed[182]). KAWASAKI, SANSKRAINTE and WALSH[184]) have carried out (similar) measurements with pellets of 1.5 to 4.4 cm diameter and plates of oxides about 80 cm² in area and 0.69 to 1.24 cm thick at temperatures between 870 and 1200°C, using CO or H_2 as the reducing medium. In this investigation the progress of the reduction agreed with formula (6) if the improbably high values for the product of the porosity and the labyrinth factor of 1 for the CO and 0.7 for the hydrogen were used. OLSSON and MCKEWAN[187b]) determined the effect which the diffusion of the gas through the spongy iron exerts on the reduction of hematite. From the results they calculate the following values for the product of the porosity γ and the labyrinth factor ξ:

T (°C)	$\gamma\,\xi$
800	0.18
900	0.23
1000	0.24

From similar measurements by WARNER[187c] the following values are obtained:

T (°C)	$\gamma\,\xi$
650	0.28
750	0.33
850	0.39
950	0.51

The difference between these sets of values may be due to differences in the initial porosity.

It is noteworthy that the authors[184] found that the total pressure in the range 1 to 2 atm had no effect on the rate of reduction. Since, in fact, with increase of total pressure the factor $(c^0 - c^*)$ increases linearly in formula (5), while the diffusion coefficient D is, by formula (11a) of section 1.2.2, proportional to $1/P$, even this negative finding is a confirmation of the mechanism of diffusion through the pores.

An exact determination of the effect of the porous structure on the limiting pressure at which this phenomenon appears has been undertaken by HERR[187d]. HERR further gives an estimate of the labyrinth factor in specimens of platelike shape.

Following his treatment,

$$O_i = \pi\, n\, l\, d = V f,$$

$$F \gamma = \tfrac{1}{4} \pi\, d^2.$$

Combining them and using $V/F = L$ and the assumption $\xi = L/l$, we can write

$$\xi = \frac{L}{l} = \frac{4 n \gamma}{f d}.$$

In these formulae,

O_i = internal surface area of the specimen (cm²),
n = number of pores,
l = effective length of pores, from boundary into the interior (cm),
d = mean diameter of pores (cm),
f = specific surface area of the pores (cm²/cm³),
$2L$ = thickness of plate specimen (cm),
F = surface area of specimen (cm),
V = volume of specimen (cm³),
γ = fraction of the total surface area of specimen covered by pores.

From the measured porosity and the labyrinth factor calculated from his formulae, HERR determined from his reduction experiments a value for the coefficient of gas diffusion, $D(H_2/H_2O)$, which agreed with that calculated from the kinetic theory of gases to within a factor of 3. The comparison shows the approximate character of the method of calculation.

In the reduction of magnetite by hydrogen WIBERG[188]) and EDSTRÖM[189]) observed the formation of non-porous layers of iron on the oxide, as shown in Fig. 70. When 30% of CO was added to the reducing gas, these layers were burst open and had the appearance shown in Fig. 71. EDSTRÖM came

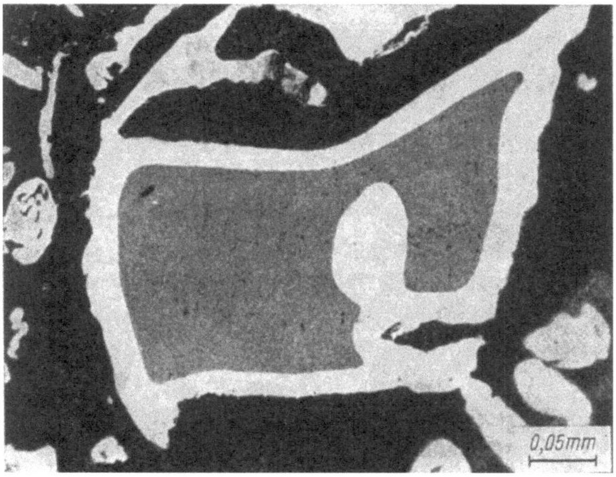

Fig. 70. Structure of the metal phase in the reduction of magnetite by hydrogen at 1000°C, according to WIBERG[188])

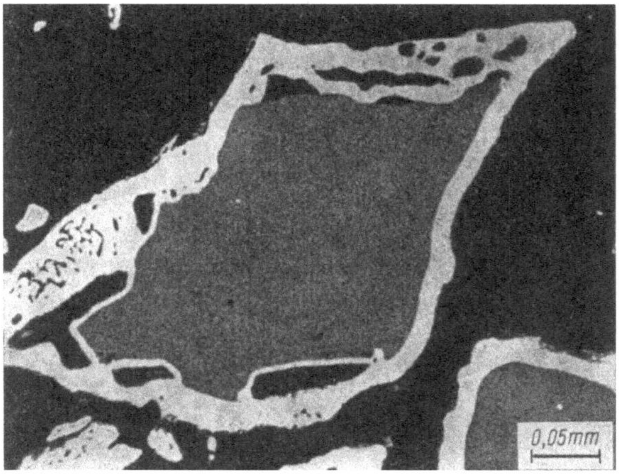

Fig. 71. The bursting of the iron layer in the reduction of magnetite in a mixture of hydrogen and carbon monoxide at 1000°C, according to WIBERG[188])

to the conclusion that the CO caused a carburization of the iron

$$2\,CO \rightarrow [C] + CO_2$$

and the carbon then reacted with the oxide

$$[C] + FeO \rightarrow Fe + CO.$$

The CO gas bursts the layer of metal open. BOHNENKAMP and RIECKE[190]) established by determining the kinetics of the reduction that, e.g., at 890°C the bursting of the layer is perceptible for an activity of the carbon in the iron of 0.2 to 0.3. From these values it may be calculated that at 890°C the critical partial pressure of the CO at the iron-wüstite phase-boundary, which must be exceeded for the layer to be burst open, is 4 to 5 atm.

In a blast-furnace shaft the activity of the carbon in the iron, and consequently the partial pressure of CO at the iron-wüstite phase-boundary, is determined by the composition of the reducing gas. Assuming equal activity of the carbon in the gas phase and in the metal phase, and using the values measured by SCHÜRMANN et al.[191]) for the temperatures and composition of the gas in the blast-furnace shaft, the following partial pressures of the CO at the wüstite-iron phase-boundary are obtained:

T (°C)	p(CO) (atm)
650	0.3
800	0.3
950	0.8
1100	8.4

It may be conjectured that, the higher the temperature, the lower the CO pressure at which the iron layers burst open. From the above values it can be estimated that covering layers of iron will be formed only at temperatures below 1000°C. With other reduction aggregates produced by means of *carbon monoxide* similar conditions hold for the bursting of the covering layer of iron. In reduction by *hydrogen*, bursting of the covering layer of iron by the water vapour in a similar process is not to be expected, since the water-vapour pressure at the Fe/FeO phase-boundary can never reach more than a fraction of the pressure outside. In the reduction of *nickel oxide*, in contrast, the equilibria and the pressures are very different. In this case, when $p(H_2) = 1$ atm, water-vapour pressures of 3 atm and more arise in the equilibrium between the oxide and the metal, and these very clearly burst the metal layer[192]).

In investigations of the reduction of wüstite by *hydrogen*[193]), sometimes covering layers of iron were formed, and sometimes porous layers of spongy metal. From these observations it may be assumed that the form of the metal layer depends on impurities in the oxide and thus on the effect of foreign nuclei.

The formation of covering, non-porous metal films greatly affects the kinetics of reduction (see Fig. 72), since in this case exchange of material between the oxide and the gas phase can take place only by solid diffusion through the growing metal layer. It is to be assumed that the particles capable of migration are oxygen atoms. If it is the diffusion of oxygen which determines the rate of reduction, v, then this rate for specimens

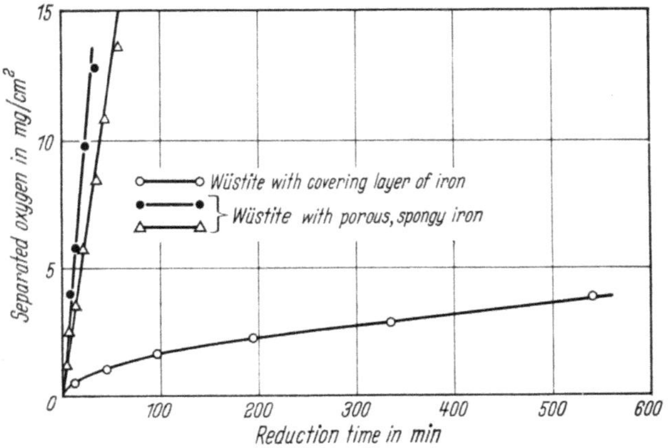

Fig. 72. Progress of reduction with the formation of porous, spongy iron, and with the formation of a covering layer of iron[193]). Reduction at 800°C in hydrogen/water-vapour with 2.7% H_2O. Total pressure: 680 Torr

with plane surfaces and small thickness is given by

$$\frac{dn(\mathrm{O})}{dt} = v = -j(\mathrm{O\ in\ Fe}) = D(\mathrm{O;\ Fe})\frac{dc(\mathrm{O})}{dx}$$

and if $c^0(\mathrm{O})$ = concentration of the oxygen in the iron at the phase-boundary with the gas, and $c^*(\mathrm{O})$ = the saturation concentration of oxygen in iron in equilibrium with wüstite, then

$$\frac{dn(\mathrm{O})}{dt} = D(\mathrm{O;\ Fe})\frac{c^*(\mathrm{O}) - c^0(\mathrm{O})}{\xi}. \tag{7}$$

The thickness ξ of the iron film formed is related to the amount of oxygen liberated, $n(\mathrm{O})$, in mole cm^{-2}, by

$$\xi = n(\mathrm{O})\, V_M(\mathrm{Fe}),$$

where $V_M(\mathrm{Fe})$ is the mole volume of the iron, in cm^3 mole^{-1}.

If we assume from now on that

$$c^*(\mathrm{O}) \gg c^0(\mathrm{O}),$$

then the amount of oxygen removed, $m(\mathrm{O})$, in g cm^{-2}, satisfies

$$\frac{dm(\mathrm{O})}{dt} = D(\mathrm{O;\ Fe})\, c^*(\mathrm{O})\frac{M^2(\mathrm{O})}{M(\mathrm{Fe})}\,\frac{\varrho(\mathrm{Fe})}{m(\mathrm{O})}, \tag{8}$$

where $\varrho(\mathrm{Fe})$ is the density of the iron, and M the molecular weights.

The concentration in (8) is to be expressed in mole cm^{-3}, but if we express the concentration in g cm^{-3}, then

$$\frac{dm(\mathrm{O})}{dt} = D(\mathrm{O};\ \mathrm{Fe})\, c^*(\mathrm{O}) \frac{M(\mathrm{O})}{M(\mathrm{Fe})} \frac{\varrho(\mathrm{Fe})}{m(\mathrm{O})} \tag{8a}$$

or, in integrated form,

$$m(\mathrm{O}) = \left[2D(\mathrm{O};\ \mathrm{Fe})\, c^*(\mathrm{O}) \frac{M(\mathrm{O})}{M(\mathrm{Fe})} \varrho(\mathrm{Fe})\, t\right]^{1/2}. \tag{9}$$

Figure 73 shows that the linear relation, to be expected according to (9), between the amount of oxygen removed and the square root of the time, in the formation of a covering layer of iron, is indeed observed experi-

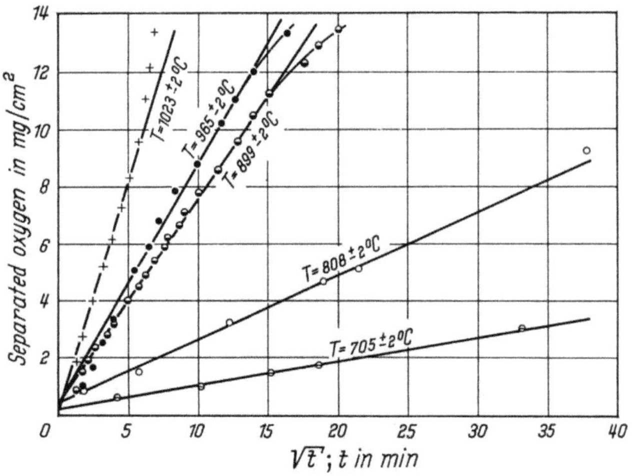

Fig. 73. Progress of reduction of wüstite in a H_2O/H_2 mixture with 2.5% water vapour and a total pressure of about 700 Torr at various temperatures[193]) (formation of dense iron layers)

mentally[193]). From the slope of the curves in Fig. 73 the permeability of the iron to oxygen, $D(\mathrm{O};\ \mathrm{Fe})\, c(\mathrm{O})$ can be calculated. The results are shown in Fig. 74. The values agree well with those calculated from measurements of the internal oxidation of iron alloys[194, 195]). If we assume that the diffusion coefficient of oxygen in iron corresponds to that of nitrogen, the values shown in Fig. 73 yield for the solubility of oxygen in iron a value of about $1.5 \cdot 10^{-4}$% by weight at 890°C and about $3 \cdot 10^{-3}$% by weight at 1100°C.

When covering layers of iron are formed, a relation analogous to (5) holds for the rate of reduction:

$$\dot{\mathfrak{R}} = \frac{3k}{r_0^2} \frac{1}{[(1-\mathfrak{R})^{-1/3} - (1-\nu\,\mathfrak{R})^{-1/3}]}, \tag{6b}$$

where

$$k = \frac{D(\text{O}; \text{Fe})\,[c^*(\text{O}) - c^0(\text{O})]}{d_0}\,\frac{V_M(\text{FeO})}{V_M(\text{Fe})} \tag{6c}$$

and

$$\nu = 1 - V_M(\text{Fe})/V_M(\text{FeO}).$$

V_M(Fe) and V_M(FeO) are the mole volumes of the metal phase and the oxide phase.

The derivation of this formula takes into account the fact that the outer dimensions of the grains vary during the reduction, because of the

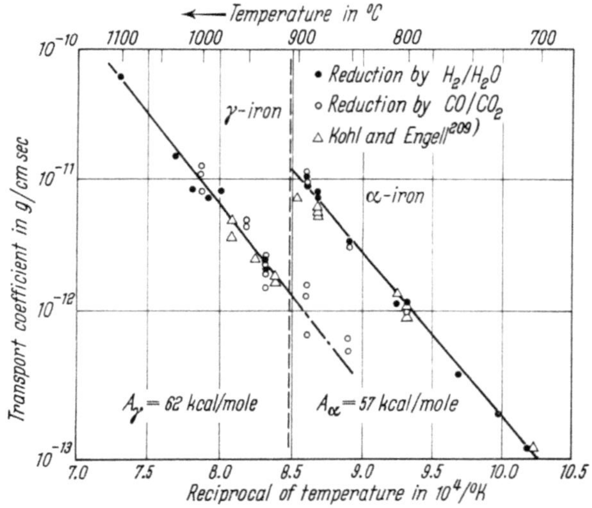

Fig. 74. Dependence of the transport coefficient $c_0 D_0$ on the temperature, between 700 and 1100°C

difference in the mole volumes of iron and wüstite. It can therefore also be used, on making suitable changes of notation, for the reduction when porous iron layers are formed, if the outer radius r_0 changes appreciably during the course of the reduction.

The constant k in formula (6b) is some orders of magnitude smaller than its equivalent for rate-determining diffusion of gas through a porous layer, which is $D_i^P(c_i^0 - c_i^*)/d_0$, as can be readily seen by comparison with formula (5). Therefore, the formation of non-porous iron films greatly reduces the rate of reduction, as shown by Fig. 104, section 2.6.

In the reduction of *hematite* no covering layers of iron are formed; the observations[196–199] agree on this point. This can be attributed to the porosity of the magnetite and wüstite which is formed, or to the lattice defects which arise in the reduction of hematite to the lower oxides.

Even the magnetite which is formed in the reduction of hematite is not free from penetrating cracks and pores[198]). It is therefore also not surprising that HEDDEN et al.[199]) observed, during the reduction of hematite to iron, layers of magnetite which were considerably thicker than would correspond to the value calculated from the coefficient of diffusion in magnetite.

2.4. Investigations of the Phase-Boundary Reaction

The basic laws for the phase-boundary reaction were discussed in section 1.2.5. At temperatures above 550°C, it is found experimentally almost without exception that for the rate of reduction v, in mole $cm^{-2}\,sec^{-1}$, the relation

$$-\frac{dn(O)}{dt} = v = k_3[p(H_2) - p^*(H_2)] \qquad (1)$$

holds for reduction by *hydrogen*, and

$$-\frac{dn(O)}{dt} = v = k_1[p(CO) - p^*(CO)] \qquad (2)$$

for reduction by *carbon monoxide*.

If k_4 and k_2 are the rate constants for the backward reaction, i.e., for oxidation by H_2O and CO_2 respectively, then

$$-\frac{dn(O)}{dt} = v = k_3\, p(H_2) - k_4\, p(H_2O) \qquad (3)$$

and

$$-\frac{dn(O)}{dt} = v = k_1\, p(CO) - k_2\, p(CO_2). \qquad (4)$$

The forward and backward reactions are therefore of the first order relative to the gaseous partner in the reaction[200–204, 145]). It has not been satisfactorily demonstrated that the phase-boundary reaction was rate-determining in all these investigations. Moreover, the area of the actual reacting surface is not known. Sometimes for the dependence on the oxygen activity of the oxide phase at which the phase-boundary reaction with the gas proceeds a first order of reaction was found, sometimes a fractional order of reaction[145, 206]). It is to be expected, in principle, that the value of the rate constant k for the phase-boundary reaction will depend on which of the iron oxides is reacting with the gas.

For the variation in time of the phase-boundary reaction we therefore have

$$-\Delta n(O) = F\,k\,t,$$

where

$-\Delta n(O)$ = amount of oxygen removed (mole),
F = area of the reacting surface,
t = duration of the test.

For the degree of reduction, $\mathfrak{R}$, we get

$$\mathfrak{R} = -\frac{\Delta n(\mathrm{O})}{n^0(\mathrm{O})} = \frac{F}{n^0(\mathrm{O})} k\,t = F\,k''\,t, \tag{5}$$

where $n^0(\mathrm{O})$ is the amount of oxygen contained in the ore specimen at $t = 0$; it is equal to the product of the oxygen content d_0 in mole cm^{-3} and the volume V of the specimen.

Depending in each case on the geometrical form of the ore, the surface area F changes with the degree of reduction $\mathfrak{R}$. For flat specimens with a plane surface and small thickness, $dn(\mathrm{O})/dt$, and so V too, remain constant over a wide range of $\mathfrak{R}$. This behaviour was observed by LEWIS[208]) and by KOHL[209]) in the reduction of iron oxides when spongy iron is formed, and is illustrated in Figs. 72 and 75.

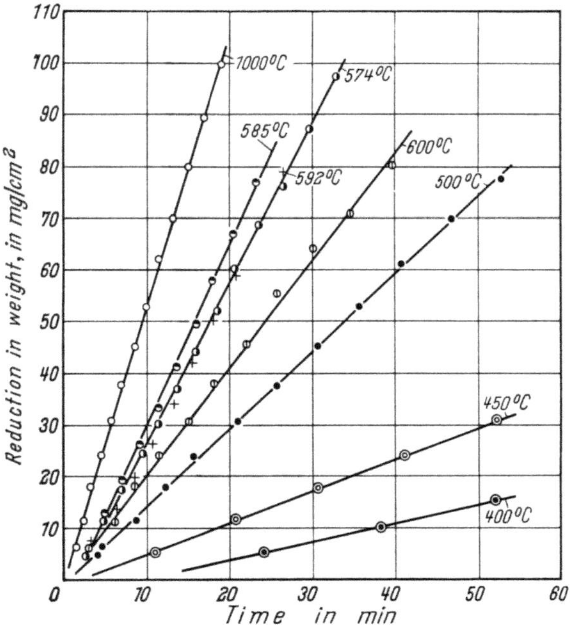

Fig. 75. The linear time law of the phase-boundary reaction in the reduction of oxide plates, according to LEWIS et al.[208]). $p(H_2) = 0.86$ atm

When the phase-boundary reaction is rate-determining, it is true for all shapes of specimen that the reacting surface moves inwards into the interior of the specimen linearly with time.

If r_0 is the initial radius of a sphere of oxide, and r_i is the radius of the unreduced part at time t, then

$$-\frac{dr_i}{dt} = \frac{k}{d_0}$$

and

$$\Re = 1 - (r_i^3/r_0^3).$$

These equations imply that

$$r_i = r_0 - k\, t/d_0$$

and also

$$d_0\, r_0 [1 - (1 - \Re)^{1/3}] = k\, t; \tag{6}$$

this expression for the duration of reduction of spheres of oxide was first obtained by JANDER[209a]) and later by McKEWAN[210]).

Differentiating (6), we have

$$\dot{\Re} = \frac{d\Re}{dt} = \frac{3k}{r_0\, d_0} (1 - \Re)^{2/3}. \tag{7}$$

The expression in square brackets on the left in (6) is the relative thickness of the reduced layer:

$$[1 - (1 - \Re)^{1/3}] = (r_0 - r_i)/r_0 = \Delta r/r_0. \tag{8}$$

The thickness of the reduced spherical shell, i.e., the quantity $[1 - (1 - \Re)^{1/3}]$, increases linearly with the duration of the experiment.

In deriving formulae (6) to (8) it has been assumed that the reaction at the phase boundary between the oxides and the gas is rate-determining. If the ore is largely free from pores before the reaction starts, as was the case in McKEWAN's researches[200, 210]), a sharply defined reaction front must form between the reduced and unreduced zones in the piece of ore, and the reduced zone must form a shell round the unreduced core.

The phase-boundary reaction may, however, be rate-determining even though the piece of ore has a high porosity so that the reducing gas can diffuse without hindrance to the whole of the interior surface of the piece of ore. Microscopically this circumstance can be recognized by the fact that a formation of the solid reaction products appears on the oxide cores across the whole cross-section of the piece of ore. The formulae (6) to (8) can be applied in this case too, except that r_0 now denotes the radius of the grains of ore. The rate of reduction will thus be independent of the radius of the piece of ore itself[211]).

The observed values of k_3 and k_4 according to Eq. (3), for the reduction of hematite, magnetite, and wüstite by *hydrogen*, are plotted logarithmically against the reciprocal of the temperature in Fig. 76. All the rate constants have been reduced to the units mole cm^{-2} sec^{-1} atm^{-1}. To convert them into the units cm sec^{-1}, they must be multiplied by $RT = 82.06$ (cm^3 atm $mole^{-1}$ $°K^{-1}$) $\cdot\, T$ (°K). Since different units are sometimes used by the original authors, and sometimes the rates are given relative to the concentrations instead of the partial pressures of the gas, the values for both k_0 and A in the formula

$$k = k_0 \exp[-A/RT],$$

as given here, may differ from those given in the references. This also explains the occasional divergence between the diagrams, tables, and text in the various references among themselves.

For the reduction of hematite to iron by pure hydrogen below 550°C, according to McKewan[210] the following relation holds:

$$k_3 = 6.5 \cdot 10^{-2} \exp[-14780/RT];$$

in the same temperature range, Hockings[212] gives, for two different hematites,

$$k_3 = 1.4 \cdot 10^{-2} \exp[-11790/RT],$$
$$k_3 = 2.1 \cdot 10^{-2} \exp[-11710/RT].$$

For the reduction of hematite to iron above 550°C McKewan[200] gives the values

$$k_3 = 2.3 \cdot 10^{-2} \exp[-14700/RT],$$
$$k_3 = 1.4 \cdot 10^{-2} \exp[-13880/RT],$$
$$k_4 = 0.33 \cdot 10^{-2} \exp[-9660/RT].$$

For the reduction of magnetite to iron by hydrogen, according to the data of Quets et al.[208]

$$k_3 = 2.0 \cdot 10^{-2} \exp[-14600/RT],$$

and according to McKewan[213]

$$k_3 = 1.3 \cdot 10^{-2} \exp[-13600/RT].$$

The results of the measurements by Hedden et al.[214] on the reduction of hematite and magnetite to iron by hydrogen, so far as they were carried out at a pressure of 1 atm of hydrogen and without the addition of water-vapour, are also plotted in Fig. 76. k_4 was not calculated, since here a different order of reaction was found.

For the reduction of wüstite to iron by hydrogen Knacke[203] found

$$k_3 = 6.4 \cdot 10^{-4} \exp[-13240/RT].$$

The values which Kohl[205] found in the reduction of wüstite by dry hydrogen in the temperature range 950 to 1200°C can be represented by

$$k_3 = 2 \cdot 10^{-3} \exp[-12700/RT].$$

In all the above 10 equations, as already stated, the k are expressed in the units mole cm^{-2} sec^{-1} atm^{-1}.

In all the measurements so far mentioned reduction down to iron was carried out, and it is to be supposed that, after a short working-up time, at temperatures above 600°C the oxygen removal takes place at the boundary of the three phases gas/wüstite/iron. Apart from the supersaturation which is necessary for the growth of the metal phase at some distance from the metal nuclei and whose exact amount is unknown, it may therefore be assumed, as a first approximation, that in these measurements the reducing gas is reacting with iron-saturated wüstite, the

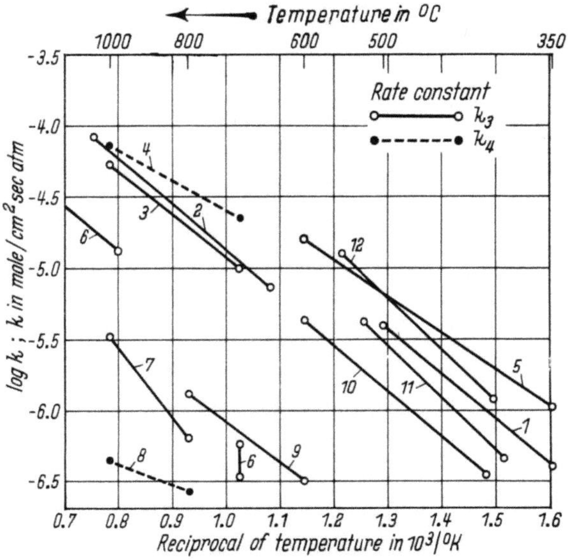

Fig. 76. Rate constants k_3 and k_4 for the reduction of iron oxides with hydrogen *1* to *4* according to McKewan[200, 210, 213]), *5* according to Hockings[212]), *6* according to Kohl[105, 209]), *7* and *8* according to Ulrich[145]), *9* according to Knacke[203]), *10* according to Quets et al.[208]), *11* reduction of magnetite by H_2, according to Hedden and Lehmann[214]), *12* reduction of hematite by H_2, according to Hedden and Lehmann[214])

metal phase being present at the oxide surface. The agreement between most of the results and the very slight difference between the reduction of hematite and magnetite is therefore not surprising. It is noticeable, however, that the rates of reduction measured by Knacke and by Kohl for wüstite to iron are considerably less. Mainly the values of the pre-exponential factor are smaller in these investigations. This difference suggests the hypothesis that in the reduction of the higher oxides effects appear which are absent in the reduction of wüstite. We should particularly bear in mind here the increase in surface area which is observed[145]) in the reduction of magnetite to iron and to wüstite; this increase comes about through the diffusion of vacancies through the wüstite and the effect of the concentration of vacancies on the phase-boundary reaction, cf. section 1.2, p. 78.

This phenomenon also occurs in the reduction of wüstite if there is a high content of vacancies at the beginning of the experiment[209]), but here the excess of vacancies at the start of the experiment is removed, and the surface, which is initially roughened, may be levelled up again as the reduction proceeds. Figure 77 shows a cross-sectioned magnetite specimen with a wüstite layer and surface roughening.

According to later investigations by McKEWAN[215], however, it is to be assumed that in his experiments the layers of wüstite and magnetite

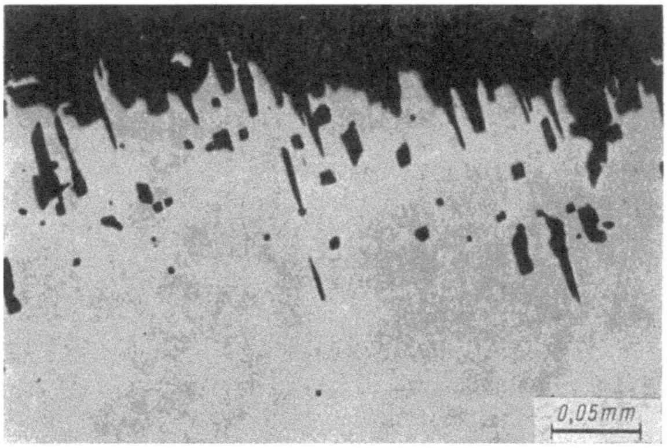

Fig. 77. Surface roughening in the reduction of magnetite to wüstite by hydrogen (200 : 1)[145]. Reduction at 1000°C. Composition of gas $H_2/H_2O = 1$. Total pressure $P = 31$ Torr

formed on the Fe_2O_3 were porous, and that the gas could reach the hematite by diffusing through the pores. McKEWAN established by photomicrographs, in fact, that in his experiments layers of wüstite and magnetite were formed of considerable thickness, which increased during the experiment. At 900°C and in pure hydrogen at 1 atm, for example, the wüstite layer increased at a rate of about $2 \cdot 10^{-4}$ cm sec^{-1}, and the magnetite at a rate of the same order of magnitude, see Fig. 78. By the end of the experiment the layers were therefore more than 10^{-2} cm thick. But from the thickness of these layers, the values of the diffusion coefficients for the vacancies, $D(|Fe^{2+}|'')$, and the equilibrium concentrations of the vacancies at the wüstite/magnetite phase-boundary, y_i, and at the wüstite/iron boundary, y_a, it can be calculated from the relation

$$\frac{d\xi(\mathrm{FeO})}{dt} = j(\mathrm{Fe};\ \mathrm{FeO})\ 4\,V_M(\mathrm{FeO}) = \frac{12\,D(|\mathrm{Fe}^{2+}|'')}{\xi}\,(y_i - y_a)$$

that the movement of the wüstite/magnetite phase-boundary through solid diffusion could take place only at a rate of about $5 \cdot 10^{-5}$ cm sec^{-1}. On the other hand, McKEWAN calculated from Fig. 78 a rate of movement of 0.28 mm min^{-1} or $5 \cdot 10^{-4}$ cm sec^{-1}. Even more extreme ratios are found for the movement of the magnetite/hematite phase-boundary, since the diffusion coefficient of iron in magnetite, $D(\mathrm{Fe};\ Fe_3O_4)$, is substantially smaller than the diffusion coefficient of iron in wüstite, $D(\mathrm{Fe}; \mathrm{FeO})$. Moreover, the layers of wüstite and of magnetite should assume a stationary thickness during the reduction of hematite to iron if the material transport takes place only by solid diffusion. The thicknesses of these layers are found from the relations

$$v = 3j(\mathrm{Fe};\ \mathrm{FeO}) = 12j(\mathrm{Fe};\ \mathrm{Fe_3O_4}),$$

see section 1.2.7.1, and with formula (4), p. 88 for wüstite we get

$$\xi(\mathrm{FeO}) = \frac{9\Delta y}{V_M(\mathrm{FeO})} \frac{D(|\mathrm{Fe}^{2+}|'')}{k_3[p(\mathrm{H_2}) - p^*(\mathrm{H_2})]}$$

and with formula (11), p. 90 for magnetite we get

$$\xi(\mathrm{Fe_3O_4}) \approx \frac{36}{V_M(\mathrm{Fe_3O_4})} \frac{D^*(\mathrm{Fe};\mathrm{Fe_3O_4})}{k_3[p(\mathrm{H_2}) - p^*(\mathrm{H_2})]}.$$

With the numerical values given on p. 91 for the coefficients of diffusion

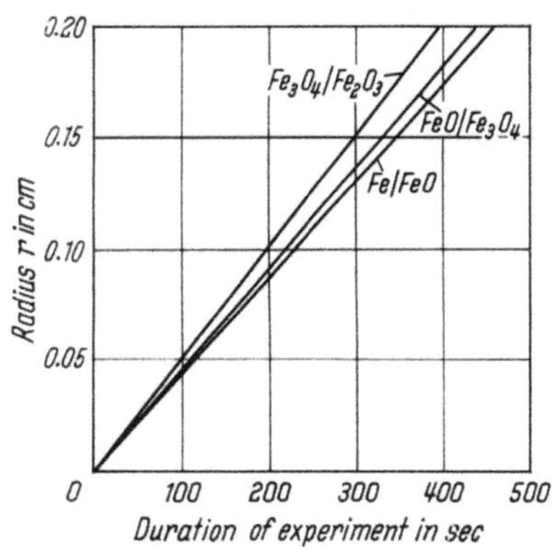

Fig. 78. Migration of the phase boundaries Fe/FeO, FeO/Fe_3O_4 and Fe_3O_4/Fe_2O_3 in the reduction of dense hematite pellets in pure hydrogen at 1 atm and 900°C, according to McKewan[215])

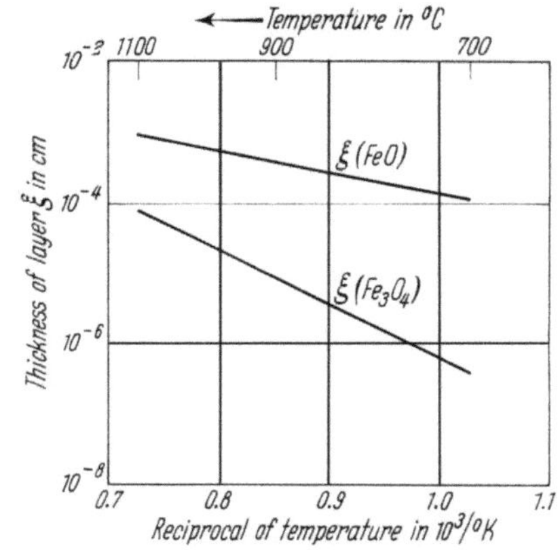

Fig. 79. Calculated thickness of layers of FeO and Fe_3O_4 in the reduction of hematite. Diffusion coefficients according to Himmel et al.[10]). Reaction rate constants k_3, according to McKewan[200, 210, 213])

and the rate constants k_3 given by McKewan we find that, at 800 to 1100°C,

$$\xi(\mathrm{FeO}) \approx \frac{0.19}{p(\mathrm{H_2}) - p^*(\mathrm{H_2})} \exp(-13900/RT),$$

where a constant mean value of 0.07 has been assumed for the difference Δy in the concentrations of vacancies at the phase boundaries of the wüstite. The corresponding relation for magnetite is

$$\xi(\mathrm{Fe_3O_4}) = \frac{315}{p(\mathrm{H_2}) - p^*(\mathrm{H_2})} \exp(-41000/RT).$$

Numerical values for the thicknesses ξ of the layers when $p^*(H_2) = 0$ and $p(H_2) = 1$, i.e., for pure hydrogen at 1 atm, are shown in Fig. 79. Comparison of the calculated values with the experimental values as in Fig. 78 gives:

	Calculated	Measured, with $t = 400$ sec
$\xi(FeO)$	$4 \cdot 10^{-4}$ cm	$1 \cdot 10^{-2}$ cm
$\xi(Fe_3O_4)$	$8 \cdot 10^{-6}$ cm	$2 \cdot 10^{-2}$ cm

The differences are particularly clear for the magnetite layer. The observed and calculated values found for a mixture of 70% hydrogen and 30% water vapour [$p(H_2) - p^*(H_2) = 0.2$] are:

	Calculated	Measured, with $t = 400$ sec
$\xi(FeO)$	$2 \cdot 10^{-3}$ cm	$9 \cdot 10^{-2}$ cm
$\xi(Fe_3O_4)$	$4 \cdot 10^{-5}$ cm	$2 \cdot 10^{-2}$ cm

PHILBROOK[228a]), too, came to the conclusion that in McKEWAN's measurements the rate of reduction was not determined only by the reaction at a clear-cut phase boundary. He showed that formula (6) also holds if the material transport in the interior of the pellet – which is necessary for the formation of intermediate products such as Fe_3O_4 and FeO – takes place by diffusion of the gas in the pores. This is intelligible, since the essential prerequisite for McKEWAN's formula is that the rate of reaction be proportional to $(1 - \mathfrak{R})^{2/3}$. For this it is not necessary to assume, as McKEWAN did, that the removal of oxygen takes place only at one surface, namely, the phase-boundary FeO/spongy-iron/gas, whose area is proportional to $(1 - \mathfrak{R})^{2/3}$. Instead, oxygen removal may take place at arbitrary positions anywhere in the whole pellet, provided only that the sum of all the reactions at any time is proportional to $(1 - \mathfrak{R})^{2/3}$. The next most simple picture of the situation after McKEWAN's is that of a pellet consisting of porous layers of intermediate and end-products of the reaction, with the removal of oxygen taking place in reaction zones in the neighbourhood of the surfaces of contact of the phases, the thickness of the zones being constant in time. The spatial extension of these binary reaction zones is determined by the ratio of the coefficients of porous diffusion to the product of the surface area of the phase rich in oxygen and the rate constant of the phase-boundary reaction. If these values are constant, then, as McKEWAN found experimentally, the boundaries of the individual zones move inwards at a rate which is constant at all times. This result is intelligible only if the increase in the oxygen activity of the diffusing gas in the pellet takes place predominantly in the binary-phase regions, and the increase in thickness of the single-phase layers therefore exerts no significant influence on the oxygen activity in the reaction zones.

For this a relatively high porosity of the single-phase layers is necessary. The relative thickness of the single-phase layers depends on the ratios of the corresponding coefficients of porous diffusion and the rates of the phase-boundary reactions.

The initial porosity affects the porosity in the layers of intermediate products, and consequently it affects the extension of the zones of two-phase reactions, and thus it affects the reactions in each of these zones. The sum of these reactions is equal to the rate of reduction.

The adduced considerations show further that the increment of the reaction rate with temperature in the case described here, which clearly corresponds roughly to McKewan's observations, must lie between that for the gas diffusion and that for the phase-boundary reaction, since both processes affect one another. With higher initial porosity and lower temperatures of reduction, the topology of the reaction must change in the direction of an extension of the zones of binary-phase reactions. In the limiting case this degenerates into a successive reduction of the whole pellet into the intermediate products FeO and Fe_3O_4, i.e., the thickness of the reaction zones will be equal to the semi-diameter of the pellet. With very slight porosity in the initial material and perhaps also in the zones of binary-phase reactions, the layers of the intermediate products and of the reaction zones shrink together. If the thickness of the two-phase layer tends to zero, the increment of reaction-rate with temperature reaches the value of the activation energy of the phase-boundary reaction at the FeO/spongy-iron/gas phase-boundary. Only in this case is McKewan's suggested mechanism for the reaction properly applicable in its strict interpretation. The thickness of the Fe_3O_4 and FeO layers must then assume the values which can be calculated from the coefficients of solid diffusion.

Figure 76 also shows the rate constants for the reduction of magnetite to wüstite according to Ulrich's data[145]):

$$k_3 = 2.8 \cdot 10^{-2} \exp[-22900/RT],$$
$$k_4 = 7.1 \cdot 10^{-6} \exp[-\ 7000/RT],$$

both measured in mole $cm^{-2}\,sec^{-1}\,atm^{-1}$. The corresponding measured values hold for the beginning of the reaction, i.e., when only very thin layers of wüstite have formed. The removal of oxygen is therefore taking place here from wüstite which is in equilibrium with magnetite.

The only gaseous reducing media which come into consideration for industrial processes are carbon monoxide and hydrogen. In industry *carbon monoxide* has hitherto played the predominant role; nevertheless, most of the laboratory tests have been carried out with *hydrogen*. In many of the investigations carried out with CO, it cannot be unequivocally decided that the phase-boundary reaction was rate-determining. They cannot therefore be invoked to compare the rates of the phase-boundary reaction for reduction by H_2 and by CO.

Bohnenkamp and Riecke[18a] investigated the rate of oxidation and of reduction of wüstite within the homogeneous phase field of the oxide. Their results for the reaction with CO/CO_2 agree excellently with Grabke's values[206] shown in Fig. 80.

Ulrich[145] determined the rate constants for the phase-boundary reaction, using H_2 and CO as the reducing media for the reduction of magnetite to wüstite. In these experiments the removal of oxygen took place from wüstite which was in equilibrium with magnetite. Figure 80

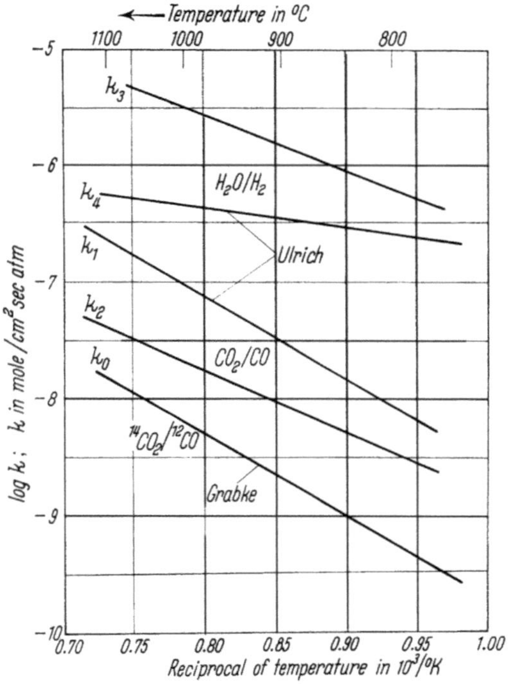

Fig. 80. Rate constants for the separation of oxygen from wüstite which is in equilibrium with magnetite, according to measurements by Ulrich[145]) and Grabke[206])

shows the rate constants k_3 and k_4 for the forward and reverse reactions with H_2, and k_1, k_2 for CO, according to formulae (3) and (4). We see that, at 900°C, k_3 and k_1 or k_4 and k_2 differ by a factor of about 40. The same difference in the rate constants k_3 and k_1 can be calculated from measurements[216]) of the oxidation of iron to wüstite in mixtures of CO_2/CO and of H_2O/H_2. The absolute values of the rate constants determined from the oxidation experiments cannot be compared with those in Fig. 80, because the various values given in the literature [206, 216–218] for the rate constants for the oxidation of iron in CO_2/CO mixtures do not even agree among themselves, and we therefore waive making any comparison. On the other hand, GRABKE's calculated values of the rate constants for the phase-boundary reaction obtained from measurements[206] of the rate of oxygen exchange between $^{14}CO_2$ and ^{12}CO are shown in Fig. 80. They depend on temperature in the same way as the value of k_1 from ULRICH's measurements, but are about one order of magnitude smaller than ULRICH's data.

From the shape of the curves in Fig. 80 we find

$$k_1 = 4.6 \cdot 10^{-2} \exp[-33200/RT],$$

$$k_2 = 3.2 \cdot 10^{-4} \exp[-24400/RT].$$

The corresponding relations for k_3 and k_4 have already been given on p. 144.

The exchange of oxygen between wüstite and H_2/H_2O as well as between this oxide and CO/CO_2 was re-investigated by BOHNENKAMP and RIECKE[18a]. They avoided any possible influence of the metal phase upon the reaction by changing the oxygen content of the wüstite within its region of stability only. Extrapolation of their results to a wüstite in equilibrium with iron yields:

$$k_1 = 2.8 \cdot 10^{-3} \exp[-28100/RT],$$

$$k_2 = 3.6 \cdot 10^{-2} \exp[-32200/RT],$$

$$k_3 = 9.5 \cdot 10^{-2} \exp[-28500/RT],$$

$$k_4 = 3.1 \cdot 10^{-2} \exp[-24600/RT].$$

The appreciably higher activation energies of these reaction constants as compared to all other values mentioned above may be due to an effect of the metal phase, which was present in the experimental work mentioned earlier. This question needs further investigation.

As before, the constants have the units mole cm^{-2} sec^{-1} atm^{-1}. In contrast to the results of GRABKE[206], BOHNENKAMP and RIECKE[18a] found with decreasing temperature a decrease of the values of the exponent m in formula (40a), section 1.2.5, $k_2' = k_2\, a^{-m}(O)$. It should be mentioned further that $a(O)$ is normalised here to the equilibrium between wüstite and iron, i.e., $a(O) = 1$ at this two-phase equilibrium.

The effect of the partial and total pressure of the reducing gas will be discussed in section 2.5.

2.5. The Interaction of Diffusion and the Phase-Boundary Reaction

2.5.1. The Phase-Boundary Reaction and Diffusion in the End-Products of the Reaction

In sections 2.3 and 2.4 we have discussed how the reaction depends on time if either the diffusion in the end-products of the reaction or the phase-boundary reaction is the slowest, and therefore the time-determining, process. If, however, both sub-processes have comparable rate constants, then both of them affect the rate of reaction equally, and their mutual interaction must be taken into account.

Since the resistance to the reaction due to diffusion increases as the thickness of the outer layer of the piece of ore, which has already reacted, increases, the relative influence of the two sub-processes on the kinetics of the reaction changes with the time of reaction t or with the degree of completion of the reaction, $\mathfrak{R}$; moreover, their relative influence also depends on the radius, r_0, of the ore pieces.

The dependence on time which results from the superimposed influence of the phase-boundary reaction and of the diffusion in the end-products of the reaction has been derived by WEI-KAO LU[219]) and by SETH and ROSS[220]). To do this, formula (5) of section 2.3 in the form

$$\dot{\mathfrak{R}} = \frac{3D^P(p^0 - p')}{r_0^2 d_0[(1-\mathfrak{R})^{-1/3} - 1]RT} \tag{1}$$

and formula (6) of section 2.4 in the form

$$\dot{\mathfrak{R}} = 3k_3(1-\mathfrak{R})^{2/3}(p' - p^*)/r_0 d_0 \tag{2}$$

are combined. p^0, p', p^* denote respectively the partial pressures of the reducing component at the surface of the pieces, at the phase boundary between the unreduced core of the ore and the end-products of the reaction, and the equilibrium pressure. (1) and (2) are combined by eliminating p'. This gives

$$\dot{\mathfrak{R}} = \frac{3D^P k_3(1-\mathfrak{R})^{2/3}(p^0 - p^*)}{d_0 r_0 D^P + d_0 r_0^2 k_3[(1-\mathfrak{R})^{1/3} - (1-\mathfrak{R})^{2/3}]RT}\,. \tag{3}$$

This formula includes the expression (5) of section 2.3 as a limiting case when $d_0 D^P \ll r_0 k_3$, i.e., when inhibition due to the diffusion preponderates; likewise, as we should expect, it also includes the expression (6) in section 2.4 as a limiting case when $d_0 D^P \gg r_0 k_3$, i.e., when inhibition due to the phase-boundary reaction predominates.

Integration of (3) with respect to time gives

$$k_3(p^0 - p^*)\,t/r_0 d_0 = [1 - (1-\mathfrak{R})^{1/3}] + RT\,k_3\,r_0[\tfrac{1}{2} - \tfrac{1}{3}\mathfrak{R} - \tfrac{1}{2}(1-\mathfrak{R})^{2/3}]/D^P. \tag{4}$$

A comparison of experimental results with the complete Eqs. (3) and (4) has not up to now been attempted. Figure 81 shows a calculated curve for the variation in the reaction rate with the degree of reduction according to Eq. (3); this holds at 900°C, assuming the values:

$$D^P = 3.5\ \text{cm}^2\ \text{sec}^{-1}, \quad d_0 = 1.5\ \text{g cm}^{-3}, \quad r_0 = 5\ \text{cm}, \quad p(H_2) = 1\ \text{atm},$$
$$k_3 = 2.3 \cdot 10^{-5}\ \text{g cm}^{-2}\ \text{sec}^{-1}\ \text{atm}^{-1}.$$

The curves for the cases where only the diffusion is time-determining, and where only the phase-boundary reaction is time-determining, have also been

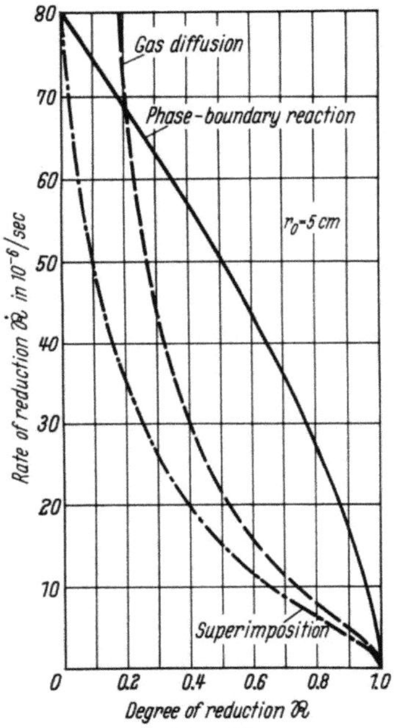

Fig. 81

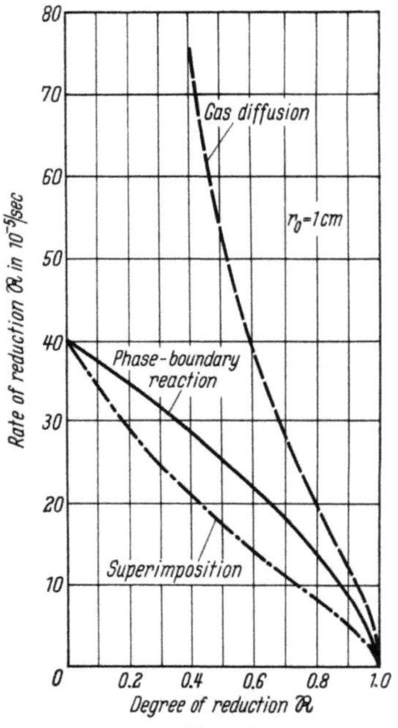

Fig. 82

Fig. 81/82. Rate of reduction as a function of the degree of completion of the reduction for $r_0 = 5$ cm and $r_0 = 1$ cm

plotted. We see that the shape of the combined curve is determined more by the phase-boundary reaction when the degree of reduction is small, but more by the diffusion when the reaction is more nearly complete. The curves in Fig. 82 have been calculated for the same values except that r_0 is now 1 cm. Even for pieces of ore of this small size the influence of diffusion on the rate of reduction is still clearly recognizable.

If we define the transition from rate-determining diffusion to rate-determining phase-boundary reaction to take place when exactly one half of the available partial pressure $p^0 - p^*$ lies in the layer of end-products of the reaction and the other half is available for the phase-boundary reaction, so that

$$p^0 - p' = p' - p^*,$$

then it follows from formulae (1) and (2) that the *critical degree of reduction* $\Re^*$ at which this transition occurs[221]) satisfies

$$(1 - \Re^*)^{1/3} - (1 - \Re^*)^{2/3} = D^P/RT\,k_3\,r_0 = \Theta = 1/RT\,N_{\mathrm{Da}}. \quad (5)$$

where $N_{\mathrm{Da}} = k_3\,r_0/D^P$ is the *Damköhler number.*

Figure 83 shows the critical degree of reduction $\Re^*$ plotted against the characteristic Θ. The range of $\Re$ in which the diffusion in the end-products

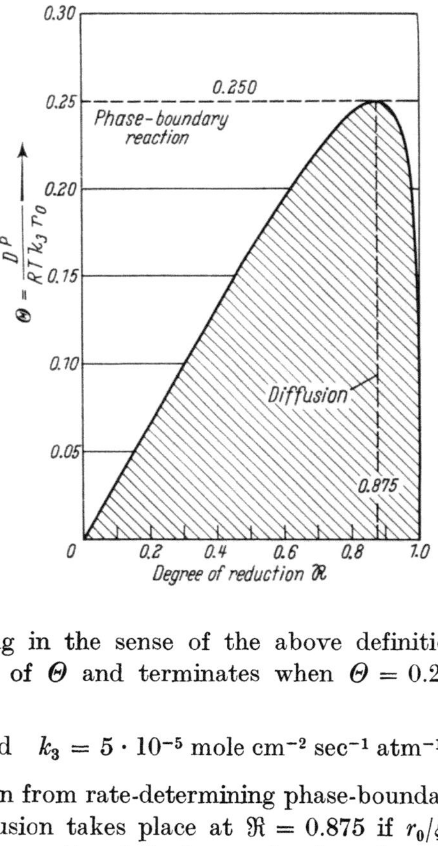

Fig. 83. The region of gas diffusion and of the phase-boundary reaction, according to PLUSCHKELL[221])

of the reaction is rate-determining in the sense of the above definition decreases with increasing values of Θ and terminates when $\Theta = 0.25$. With the values[222])

$$D(H_2/H_2O) = 7.6\ \mathrm{cm^2\,sec^{-1}} \quad \text{and} \quad k_3 = 5 \cdot 10^{-5}\ \mathrm{mole\ cm^{-2}\,sec^{-1}\,atm^{-1}}$$

which hold at 900°C, the transition from rate-determining phase-boundary reaction to rate-determining diffusion takes place at $\Re = 0.875$ if $r_0/\xi\,\gamma$ is greater than 6.25, γ being the porosity of the layer of end-products of the reaction, ξ the labyrinth factor for the system of pores in this layer, and r_0 the radius of the ore pieces. With $r_0/\xi\,\gamma > 12.5$, the domain of rate-determining diffusion in the sense of the above definition is already attained with a degree of reduction $\Re$ of 0.33. With $\gamma \approx 0.5$ and $\xi \approx 0.2$ [223])

there would be, for the beginning of rate-determining diffusion, a *critical radius for the ore pieces*, namely, $r_0^* = 0.6$ cm for $\mathfrak{R} = 0.825$, and $r_0^* = 1.3$ cm for $\mathfrak{R} = 0.33$. These values hold for pieces of ore which are largely free from pores, such as were used by McKEWAN in his researches, and for hydrogen as the reducing medium. With porous ores, the product of the labyrinth factor and the porosity certainly increases, but the reacting surface also increases to a more or less broad reaction zone extending into the interior, so that the total transformation from the phase-boundary reaction also increases. The two effects oppose one another. As a rough guide we may therefore take the above values of r_0 to hold also for porous pieces of ore. In reduction by carbon monoxide the diffusion coefficients are smaller by a factor of 4, but, according to McKEWAN's investigations, the rate constant for the phase-boundary reaction decreases by a factor of 1/40, and so the transition into the diffusion domain should occur first with larger-sized pieces.

The partial pressures were eliminated in the derivation of formula (5); also, by (11a) in section 1.2.3, the coefficient of diffusion is inversely proportional to the total pressure P. Therefore, at constant values of the porosity γ,

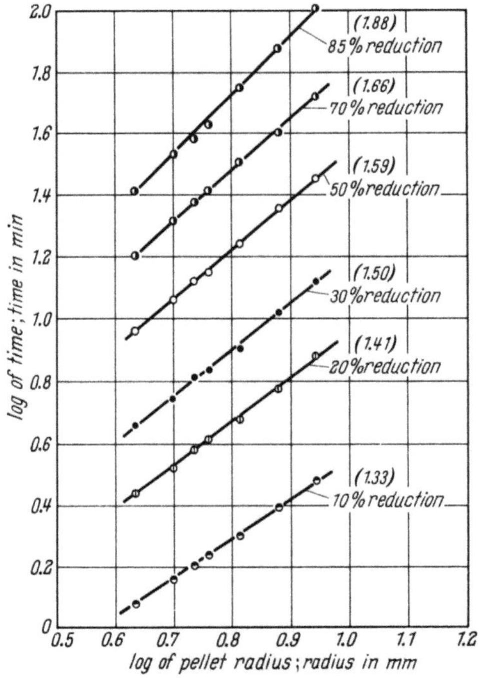

Fig. 84. Effect of the phase-boundary reaction and of gas diffusion on the rate of reduction; evaluation by SETH and ROSS[136])

labyrinth factor ξ, and radius of pieces r_0, the characteristic Θ decreases, according to formula (5), as the total pressure of the gas phase increases, provided that the partial pressures of the reaction gases H_2 and H_2O or CO and CO_2 remain constant or increase with the total pressure. The transition to the diffusion domain should therefore take place earlier the greater the total pressure.

Seth and Ross evaluated their own reduction experiments[186]) by plotting the logarithm of the time of reduction t corresponding to a fixed degree of reduction $\Re$ against the logarithm of the pellet radius, r_0. Figure 84 shows a family of such curves for various $\Re$; the gradient of a curve increases with the degree of reduction to which the curve corresponds. When the phase-boundary reaction is rate-determining, this gradient should, by formula (6) of section 2.4, have the value 1, and by formula (6) of section 2.3, when diffusion is rate-determining, the gradient should have the value 2. The same conclusion was reached by other investigators[224]). Seth and Ross find that, for $\Re = 0.1$, $t \sim r_0^{1.3}$, and for $\Re = 0.86$, $t \sim r_0^{1.88}$. This finding agrees qualitatively with the formulae (4) and (5). It is to be noted, with this representation, that the labyrinth factor ξ and the porosity γ vary with the degree of reduction[220, 224]).

Even when the end-products of the reaction form a non-porous layer, the phase-boundary reaction and the diffusion can influence each other. In the reduction of magnetite to wüstite a non-porous FeO-layer is formed[225]). The layer grows by the removal of oxygen at the wüstite surface, diffusion of the liberated iron via the vacancies and positive holes to the wüstite-magnetite phase-boundary, and transformation of the magnetite into wüstite:

$$FeO\,\llcorner + H_2 + |Fe^{2+}|'' + 2\,|e|^{\bullet} \rightarrow Fe^{2+}\,|Fe^{2+}| + H_2O,$$

$$Fe_3O_4 + Fe^{2+}\,|Fe^{2+}| \rightleftharpoons 4\,FeO + |Fe^{2+}|'' + 2\,|e|^{\bullet}.$$

The diffusion rate j and the rate v of the phase-boundary reaction must be the same [as regards the factor 3, see section 1.2.7.1, Eq. (4)]:

$$v = j(|Fe^{2+}|'') = -3D(|Fe^{2+}|'')\frac{dc(|Fe^{2+}|'')}{dx} = \frac{3D(|Fe^{2+}|'')}{V_M(FeO)}\,\frac{y'' - y'}{\xi}, \qquad (6)$$

where

$D(|Fe^{2+}|'')$ = coefficient of diffusion of the vacancies,
$V_M(FeO)$ = mole volume of the wüstite,
ξ = thickness of the layer,
y' = mole fraction of vacancies at the wüstite surface,
y'' = mole fraction of vacancies at the wüstite/magnetite phase-boundary.

The value of y'' is constant, since there is equilibrium between the two oxide phases, but y' must decrease with increasing ξ so that (6) may hold. By formula (44) of section 1.2.5, at 800°C v is proportional to $a(O)$ or $(y')^3$, and it is from the fact that v depends on $a(O)$ that the coupling between the phase-boundary reaction and the diffusion arises. Figure 85 shows the variation of y in a layer of wüstite which was formed in the reduction of magnetite[225]), the values of y being calculated from formula (6) using the measured values of v, the known coefficients of diffusion of the vacancies, and the equilibrium value of y''. The separate points plotted are measured values of y, determined by an electrochemical method[226, 227]).

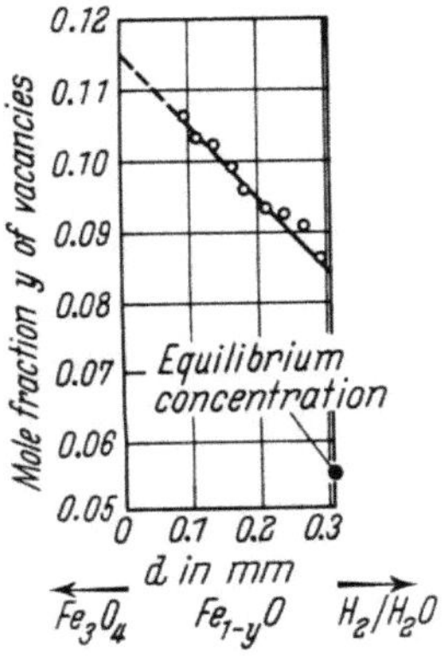

Fig. 85. Variation of the mole fraction y of vacancies in a dense wüstite layer growing on magnetite during the reduction of the latter. Reduction at 900°C by H_2/H_2O, $p(H_2)/p(H_2O) = 60/40$

So far we have dealt only with the interaction of the surface reaction and the diffusion of gas in the pores of the reduced shell of the ore or pellet. JESCHAR[228]) gives a complete picture of the kinetics of reduction in dense ore, taking into account the influence of the diffusion of the gas through the boundary layer and in the pores of the reduction products formed, and also the effect of the phase-boundary reaction. He starts from the equation

$$j_i = \beta[c_i(r_0) - c_i^0]$$

for the material transport of the reducing gas i, with concentration c_i^0 in the gas space, to a sphere of ore of radius r_0. He also uses the formulae (1) of section 2.3 and (1) of section 2.4. To simplify the computations, the Fourier number, $N_{\mathrm{Fo}} = D^P t/r_0^2$, is used as a dimensionless measure of the time of reaction t, the Biot number, $N_{\mathrm{Bi}} = \beta r_0/D^P$, is used as a dimensionless measure of the material transport, and the Damköhler number, $N_{\mathrm{Da}} = k r_0/D^P$, is introduced as a dimensionless, reduced measure of the phase-boundary reaction. The ratio k/β, which is also dimensionless, is called the Traustel number, N_{Tr}. The ratio $\zeta = r/r_0$ of the radius of the still unreduced sphere of ore to its initial radius r_0 serves as a dimensionless, reduced measure of the progress of the reaction. ζ is related to the degree of reduction $\Re$, by the equation $\Re = 1 - \zeta^3$. Combining and solving the given differential equations for the flow of material j_i and the rate of reaction v_i we obtain for the progress of the reaction the equation

$$\frac{r_0^2}{D^P}\frac{d\zeta}{dt} = \frac{d\zeta}{dN_{Fo}} = \frac{c_i^0}{d_0}\,\frac{1}{\zeta^2(1/N_{Bi}-1)+\zeta+1/N_{Da}}.$$

Integrating and putting $\zeta = 1$ for $N_{Fo} = 0$ we get

$$\frac{1}{N_{Da}}[1-(1-\Re)^{1/3}] + \frac{1}{2}[1-(1-\Re)^{2/3}] - \frac{1}{3}(1-1/N_{Bi})\,\Re$$
$$= \frac{c_i^0}{d_0}N_{Fo} = \frac{c_i^0\,D^P}{d_0 r_0^2}\,t.$$

For complete reduction of the sphere of ore, i.e., $\Re = 1$, we have the simple relation

$$[c_i^0\,N_{Fo}/d_0]_{\Re=1} = 1/6 + 1/(3N_{Bi}) + 1/N_{Da}.$$

From this equation the expressions can be derived which correspond to uninhibited material transport across the boundary layer and to negligible inhibition of the phase-boundary reaction. In the first of these cases, since $3N_{Bi} \gg N_{Da}$, the term in $1/N_{Bi}$ disappears from the right-hand side of the equation above; and in the second case, since $N_{Da} \gg 3$, the term $1/N_{Da}$ drops out. Moreover, the equations given by JESCHAR contain as special cases the formulae derived earlier in this text for the progress of the reduction when inhibition by diffusion or by the phase-boundary reaction is dominant, and for the superposition of the two sub-processes.

PHILBROOK et al.[228a]) have also given a treatment of the kinetics of reduction, which again takes into account the diffusion of the gas across the boundary layer and in the pores of the reduction products formed, as well as the phase-boundary reaction.

These authors show further that, even when the transport processes exert an appreciable influence, the advance of the reaction front may depend linearly on the time. Such a dependence on time was regarded by McKEWAN[222, 235]) as proving that the phase-boundary reaction was rate-determining.

2.5.2. Interaction of the Phase-Boundary Reaction with Diffusion in the Reacting Layers

In the reduction of pieces of ore of appreciable porosity, the removal of oxygen from the oxide takes place in a zone which may extend to varying degrees into the pieces of ore. In this case the phase-boundary reaction and the diffusion in the pores of the reacting oxides interact with one another.

On the theory of *gas diffusion in the pores with a simultaneous reaction at their walls* a good deal of fundamental work has already appeared which, in particular, goes back to THIELE[229]) and ZELDOVICH[230]). The theory leads to the result, which has been often verified experimentally, that the temperature increment of the reaction is reduced by a half if the porous diffusion is rate-determining and furthermore a large part of the internal surface of the solid participates in the reaction.

JOST[231]) discusses the case of a solid, where it is not the whole of the internal surface of the solid which participates, but instead the gas-solid reaction proceeds zonally, advancing inwards from the outside. His calculations are based on MORAWIETZ's researches[232]) into the reaction

$$6\,FeS + 4\,SO_2 \rightarrow 2\,Fe_3O_4 + 5\,S_2 .$$

THIELE's theory[229]) was developed for reactions at the surface of a porous catalyst, that is, for the case where the solid remains unchanged after the reaction. WICKE and his co-workers[233]) have shown, however, that THIELE's theory can also describe the gasification of coal and coke. The theory can also be applied to the reduction of porous ores[234]).

A piece of ore is assumed to be of spherical form, as represented in Fig. 86. In the free gas space, i.e., for $x > r_0$, the reducing gas has the

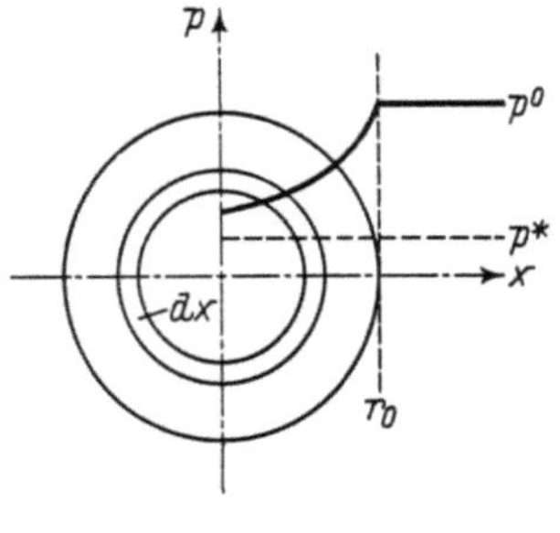

Fig. 86. Local variation in partial pressure of the reducing gas in the reduction of porous pieces of ore

partial pressure p^0. The reaction between gas and solid is of first order relative to the gas pressures of the reducing and oxidizing components of the gas and takes place without change of volume. For the reaction dv in a spherical shell of thickness dx we have

$$dv = 4\pi\, x^2\, dx\, k_v[p(x) - p^*], \tag{7}$$

where k_v is the rate constant relative to unit volume:

$$k_v = k_3\, F/V, \tag{8}$$

V being the volume of the piece of ore, and F the area of the reacting surface.

The reducing gas diffuses through the spherical shell into the interior of the piece. The diffusion flow j' is given by†

$$j' = -4\pi x^2 D^P \frac{1}{RT} \frac{d}{dx} [p(x) - p^*]. \tag{9}$$

The gas diffuses through the pores of the reacting solid and the diffusion flow therefore changes with position: in fact,

$$\frac{dj'}{dx} dx = -dv. \tag{10}$$

Differentiating (9) and substituting the result, together with (7), into the expression (10), we get

$$RT \frac{k_v}{D^P} \Delta p = \frac{\partial^2 \Delta p}{\partial x^2} + \frac{2}{x} \frac{\partial \Delta p}{\partial x}, \tag{11}$$

where

$$\Delta p = p(x) - p^*.$$

The boundary conditions must be taken into account:

$$\frac{\partial \Delta p}{\partial x} = 0$$

when $x = 0$ (i.e., at the centre of the piece of ore) and

$$\Delta p = \Delta p^0 = p^0 - p^*$$

when $x = r_0$.

With these conditions the differential equation (11) has the solution

$$\frac{\Delta p}{\Delta p^0} = \frac{r_0}{x} \frac{\sinh[(k_v RT/D^P)^{1/2} x]}{\sinh[(k_v RT/D^P)^{1/2} r_0]}. \tag{12}$$

The total reaction in the piece of ore is given by

$$v = \int_0^{r_0} dv = \int_0^{r_0} k_v \Delta p \, 4\pi x^2 \, dx. \tag{13}$$

Carrying out the integration we get

$$v = 4\pi r_0^2 \left[\frac{k_v RT}{D^P}\right]^{1/2} \left[\frac{D^P}{RT}\right] \Delta p \left\{\coth\left[r_0 \left(\frac{k_v RT}{D^P}\right)^{1/2}\right] - \frac{1}{r_0}\left(\frac{k_v RT}{D^P}\right)^{1/2}\right\}. \tag{14}$$

Following Thiele, we define a degree of utilization, η, by

$$\eta = v/v_{\max} = \frac{3}{\varphi}(\coth\varphi - 1/\varphi), \tag{15}$$

where

$$\varphi = r_0 (k_v RT/D^P)^{1/2}. \tag{16}$$

For small values of k_v and r_0 the degree of utilization tends to 1:

$$\lim_{\varphi \to 0} \eta = 1.$$

† Here j' has the dimension mole · sec^{-1}.

In this case the reaction in the piece of ore is not influenced by the diffusion in the macro-pores, the gas pressure is p^0 independently of position, and the overall reaction will be controlled by the phase-boundary reaction, provided that the limitations mentioned later do not apply. The overall reaction will be independent of the size of the ore pieces.

For large values of the modulus φ, i.e., for large rate constants of the phase-boundary reaction, large radii r_0 of pieces, and a small coefficient of diffusion in the pores, D^P, η tends to $3/\varphi$:

$$\lim_{\varphi\to\infty} \eta = \frac{3}{r_0[k_v RT/D^P]^{1/2}}.$$

Putting $v = 4\pi r_0^3 d_0 \dot{\mathfrak{R}}/3$ we get for these two cases

$$\dot{\mathfrak{R}} = \frac{3}{r_0 d_0} \Delta p [k_v D^P/RT]^{1/2} \quad \text{as} \quad \varphi \to \infty, \tag{17a}$$

$$\dot{\mathfrak{R}} = \frac{k_v}{d_0} \Delta p \quad \text{as} \quad \varphi \to 0. \tag{17b}$$

Thus, for large pieces of ore, the rate of reduction should be inversely proportional to the radius r_0 of the pieces and should be proportional to the square root of the product of the coefficient of porous diffusion, D^P, with the rate constant k_v for the reaction, relative to the volume. For small pieces, the reduction rate should, on the contrary, be independent

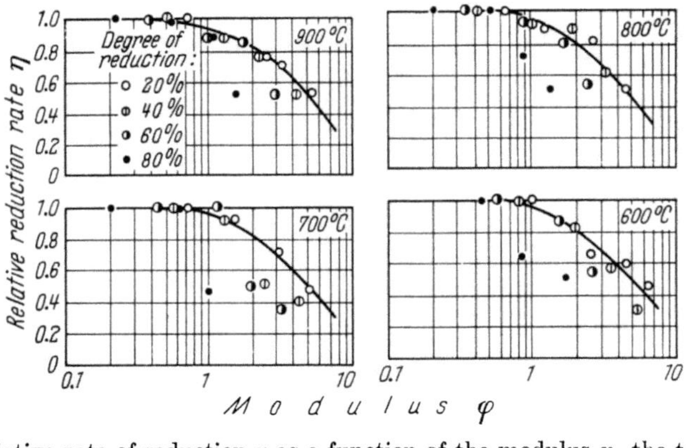

Fig. 87. Relative rate of reduction η as a function of the modulus φ, the temperature, and the degree of reduction[234])

of the size of the pieces and be determined by the rate of the phase-boundary reaction only.

With the smallest pieces of ore used in the investigation[234]) the rate of reduction was indeed independent of the radius r_0, in accordance with (17b); this experimental value was used to calculate the value of k_v from (17b).

To evaluate the reduction experiments[234]) in porous pieces of ore according to the above formulae it was assumed first that the diffusion in the outer shell of a piece of ore which has already reacted, does not affect the rate of reaction. For the degree of reduction, $\mathfrak{R}$, we put

$$\mathfrak{R} = \frac{r_0^3(t=0) - r_0^3(t)}{r_0^3(t=0)}.$$

D^P was then calculated from the measured rates of reduction, $\dot{\mathfrak{R}}$, from formula (17a) for $\mathfrak{R} = 0.2, 0.4, 0.6, 0.9$. Table 7 [234]) shows the results and the labyrinth factor ξ calculated from the values obtained for the porous diffusion coefficients, the porosity γ, and the known values of the coefficient D of the gas diffusion.

In Fig. 87 the values are plotted of the modulus φ, calculated by working backwards from the above values for D^P, against the degree of utilization, η, obtained from the observed values.

Table 7. *Evaluation of reduction experiments by means of formula* (17a)

°C	$D(H_2/H_2O)$	γ (%)	$\xi = D^P/\gamma D$
600	5.94	63	0.056
700	7.19	56	0.20
800	8.58	50	0.40
900	10.04	47	0.38

We see from Fig. 87 that the theory reproduces the experimental results correctly with a reasonable dispersion. The values which deviate markedly downwards originate from measurements on pellets of 30 mm diameter. For these, $\dot{\mathfrak{R}}$ decreases with r_0 more rapidly than formula (17a) suggests. Clearly, the diffusion in the layer of iron which has already reacted is here exerting its influence. In this case, by formula (1) of the present section, $\dot{\mathfrak{R}} \sim 1/r_0^2$.

However, whether the phase-boundary reaction alone determines the rate of reduction of small-sized pieces must be regarded as doubtful, judging by measurements of the corresponding *temperature increment*[224]). Figure 88 shows observed values from which a temperature increment of 8 kcal/mole can be read off. The marked decrease in the internal surface area during the reaction, which, according to Fig. 41 of section 1.2.7.2, is connected with the considerable alteration in the pore geometry, may, however, explain this value of the temperature increment, which is too small for the constant of the phase-boundary reaction. Indeed, if the rates of reduction are reduced to equal reacting areas, then a temperature increment of 18.5 kcal/mole is obtained. As may be seen from Fig. 89, however, the rate of reduction still does not show the expected linear dependence on the total pressure of the hydrogen, while, according to Fig. 90, the reaction rate is proportional to the partial pressure of the hydrogen[125]). For hydrogen pressures between 1 and 40 atm McKewan[235]), too, found marked deviations from a linear relation between reaction rate and gas pressure during the reduction of magnetite at temperatures between 350 and 500°C.

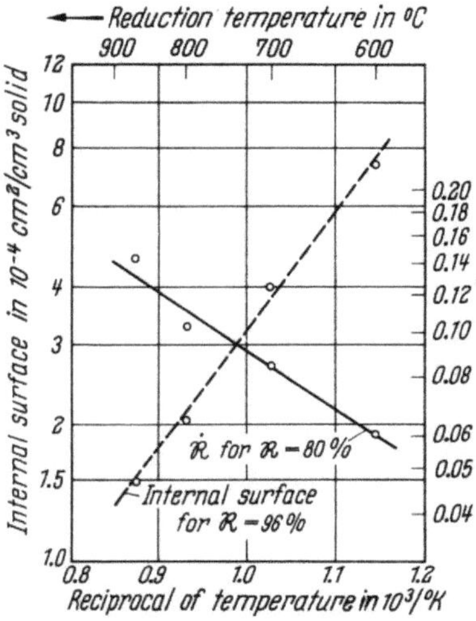

Fig. 88. Relation between the rate of reduction, internal surface area, and temperature of reduction[125])

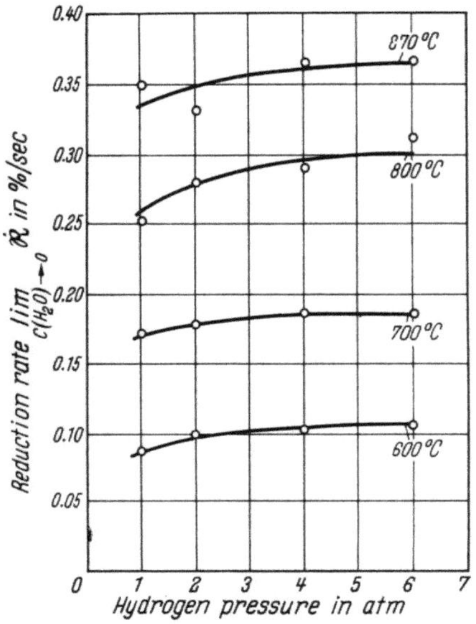

Fig. 89. Rate of reduction for wüstite pellets between 3 and 5 mm diameter, in hydrogen without inhibition by water vapour, between 20 and 70% reduction, with increasing gas pressure[125])

It has been shown in section 2.4 that, according to McKewan's observations on the reduction of hematite, it must be assumed that there is gaseous diffusion in the pores of the layers of Fe_3O_4 and FeO. Here, as in the experiments on porous ores[125]), the effect of the gaseous diffusion must therefore be taken into account in discussing the dependence on pressure.

The diffusion, like the phase-boundary reaction, takes place under a constant total pressure at a rate which is proportional to the partial pressure of the reacting or diffusing gas component. It is therefore evident why – at least at temperatures above 500°C – almost without exception, proportionality between the partial pressure and rate of reaction at constant total pressure was found.

If, however, the mole fraction of the reducing component of the gas is kept constant and the total pressure P varies, then the relation $D \sim 1/P$ must be taken into account. The rate of the phase-boundary reaction, on the other hand, depends only on the partial pressure and not on the total pressure, i.e., the rate constant k is independent of the pressure. By formula (16) the modulus φ must therefore increase with the total pressure P.

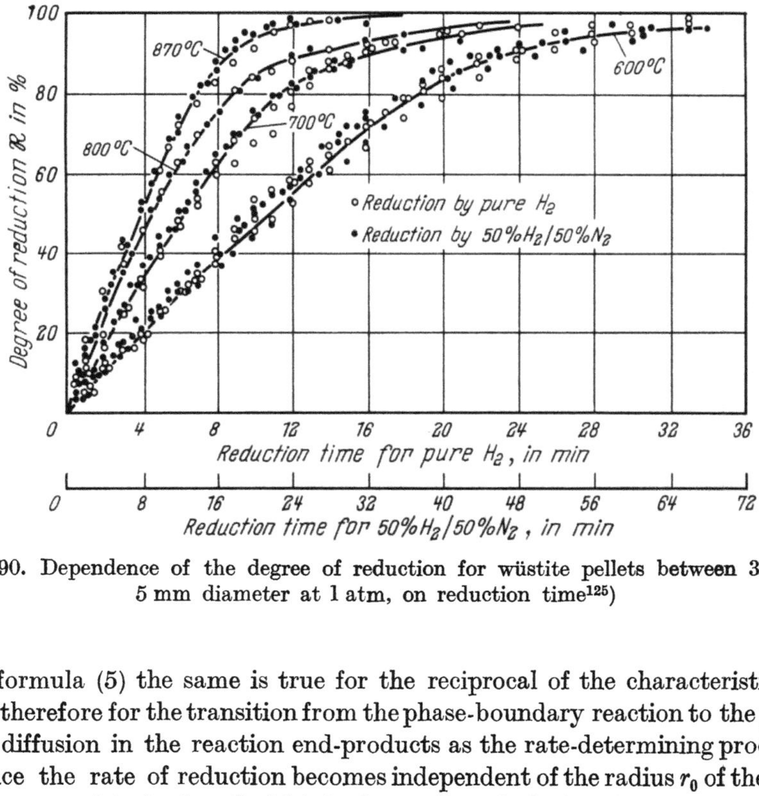

Fig. 90. Dependence of the degree of reduction for wüstite pellets between 3 and 5 mm diameter at 1 atm, on reduction time[125])

By formula (5) the same is true for the reciprocal of the characteristic Θ and therefore for the transition from the phase-boundary reaction to the porous diffusion in the reaction end-products as the rate-determining process. Hence the rate of reduction becomes independent of the radius r_0 of the ore pieces at a critical value r_0^* which is the smaller, the higher the pressure of the gas, and indeed we must have $r_0^* \sim 1/\sqrt{P}$ according to (16), or $r_0^* \sim 1/P$ according to (5).

On the other hand it has been proved[237]) in this connection that, in accordance with Fig. 41 of section 1.2.7.2, the pores which form during the reduction have for the most part diameters which have precisely the order of magnitude of the mean free path of the gas molecules at the temperature of the experiments and at a total pressure of 1 atm. At pressures above about 1 atm the diffusion in the pores should therefore take place through the normal mechanism of gas diffusion. When diffusion is rate-determining, the total gas pressure has, for the reason explained, no influence on the rate of reduction. With smaller total pressures, however, the mean free path of the gas molecules is greater than the diameters of the pores, so that now transport through the pores takes place by Knudsen flow. But by formula (28) of section 1.2.2, the *Knudsen diffusion coefficient* D_K is independent of the total pressure. If, then, the Knudsen flow in the pores is rate-determining, the rate of reduction at small gas-pressures must be proportional to the partial pressure and the total pressure, as is indeed shown also by the experimental results.

By correlating the pore spectrum and the limiting pressures at which the influence of the total pressure on the rate of reduction disappears, the validity of this model can be confirmed[237a]), as Fig. 90a shows.

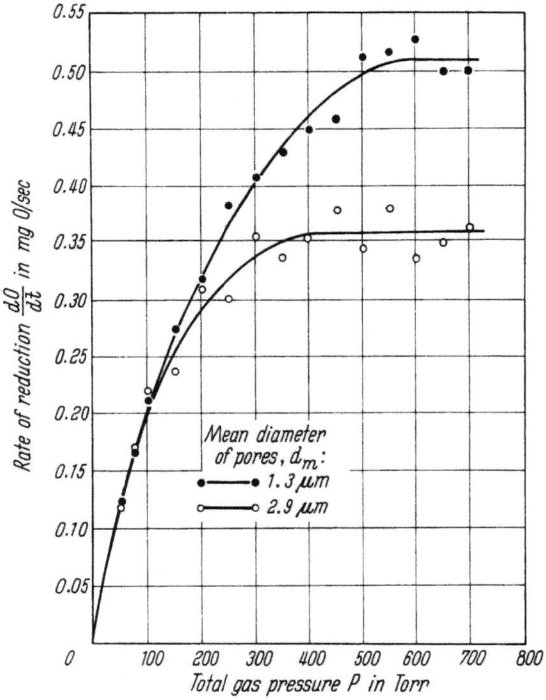

Fig. 90a. Dependence of the reduction rate of porous wüstite specimens on the total gas pressure, with a 50% degree of reduction

We must therefore assume that diffusion has a determining influence on the rate of reduction of porous ores even for piece sizes at which it is no longer possible to observe that piece size has any effect on that rate. Additional evidence for this is the fact that the ratio of the rates of reduction in hydrogen and in carbon monoxide, for values of r_0 between 3 and 20 mm, is independent of the piece size and has a value between 1.5 and 5 [224]). This ratio corresponds roughly to the ratio of the two diffusion coefficients, $D(H_2/H_2O)/D(CO/CO_2)$.

In contrast, when the phase-boundary reaction is rate-determining, a ratio of the reduction rates in H_2 and in CO of about 40 is said to be found[225]). The varying effects which added amounts of *hydrogen* in the reducing gas have[238]) on the reduction rate of different ores may be explained without difficulty by the different contributions of diffusion and phase-boundary reaction to the resistance to the reaction and the variation of these with the degree of reduction. The marked effect of a small addition of hydrogen, and the slight effect of a further increase in the hydrogen content, can also be interpreted by this mechanism: the first added amounts increase the rate of the phase-boundary reaction particularly, so that the diffusion becomes rate-determining, and then further additions of hydrogen can no longer produce any marked effect. In this connection the *water-gas reaction*, which leads to a regeneration of the used hydrogen, must also be taken into account, see section 2.10.

2.5.3. Consideration of the Combined Effects of the Phase-Boundary Reaction, Gas Flow, Diffusion in the Reacting Layers, and Diffusion in the End-Products of the Reaction

The progress of the reduction of iron oxides has been considered in the previous sections from the point of view of the rate-determining subprocess or processes. In industrial processes the effect of ore size, the porosity of the ores, the partial pressures in the gas phase, and the temperature of reaction, are particularly important. Figures 91 to 96 show diagrammatically the variation of the partial pressure of the reducing gas component (subscript 1) and of the oxidizing component (subscript 2) or the oxygen activity $a(O) \sim p_2/p_1$ in the pieces of ore. For both components an asterisk denotes the equilibrium value, p^*. The diagrams also show how the reduction rate depends on the diameter d of the pieces of ore, on the partial pressure p, and on the temperature T.

Figure 91 holds for the case where the material transport across the boundary layer controls the rate of reaction; this case is scarcely of much significance for the forced draught of industrial processes. The relation shown between the rate of reduction $\dot{\mathfrak{R}}$ and the diameter $2r_0$ of the pieces is obtained from formula (8c) of section 2.2, and the dependence of the reduction rate on the partial and total gas pressure and on the temperature is obtained from the fundamental laws of diffusion of gases, see section 1.2.2.

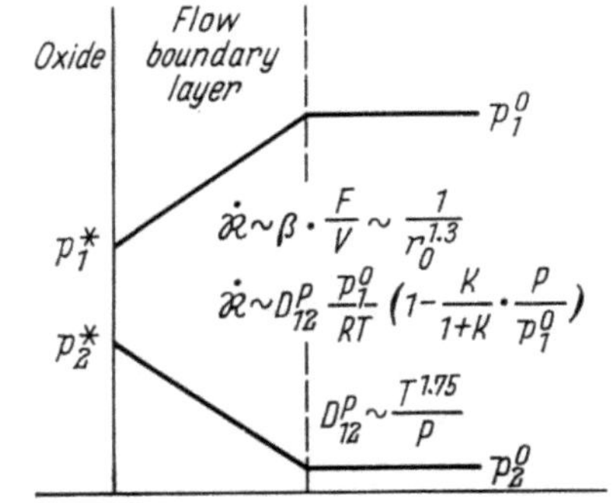

Fig. 91. Reduction of ores: gaseous diffusion in the flow boundary-layer is rate-determining

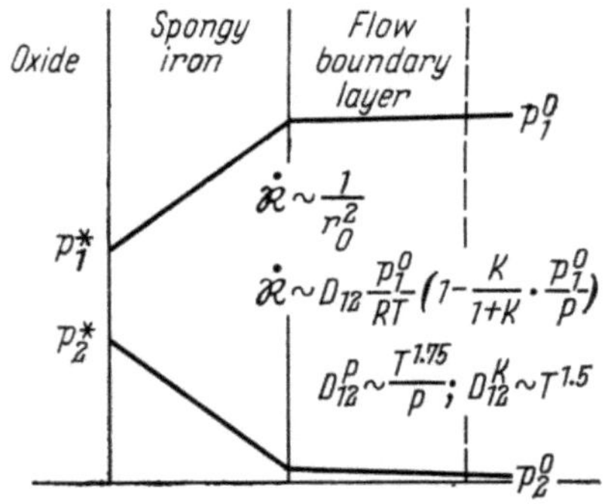

Fig. 92. Reduction of ores: gaseous diffusion in the porous reduction-products is rate-determining

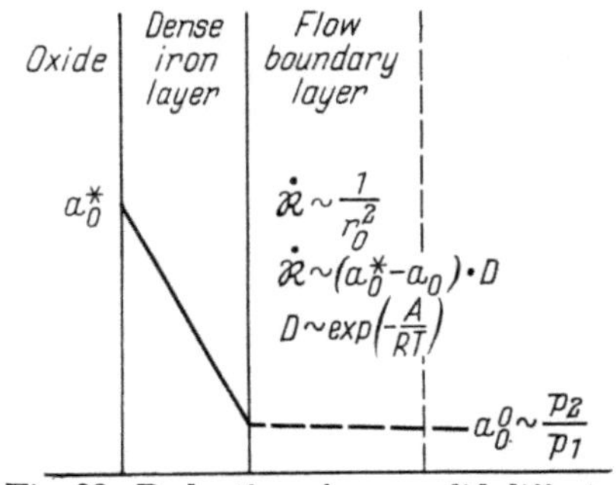

Fig. 93. Reduction of ores: solid diffusion in non-porous reduction-products is rate-determining

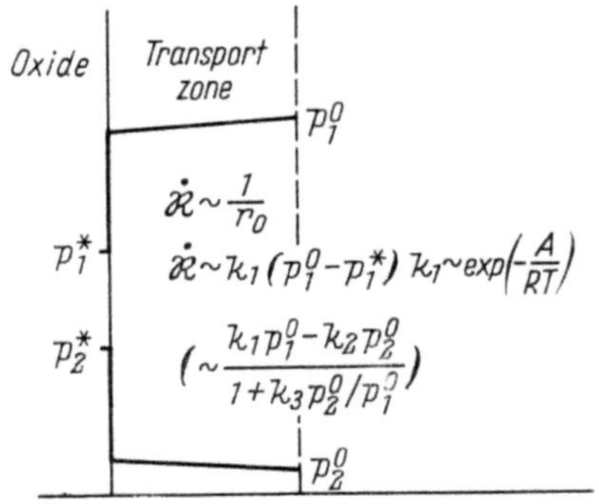

Fig. 94. Reduction of ores: phase-boundary reaction at the outer surface is rate-determining

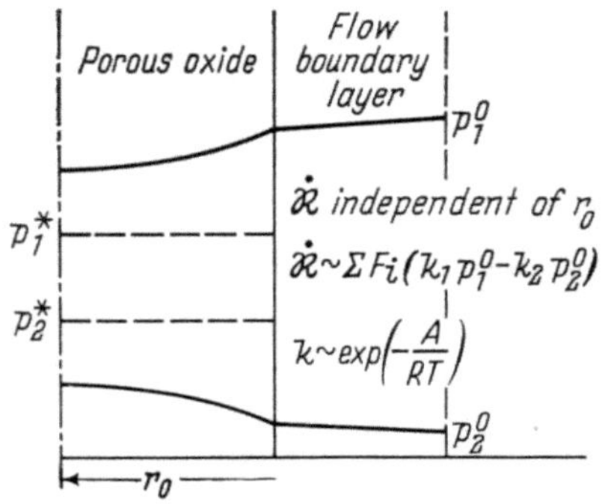

Fig. 95. Reduction of ores: phase-boundary reaction at the internal surface of porous ore is rate-determining

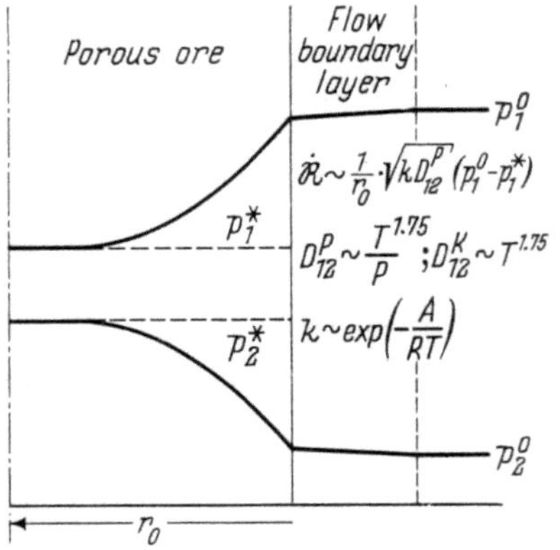

Fig. 96. Reduction of ores: diffusion and phase-boundary reaction in porous ore is rate-determining

Figure 92 shows the circumstances when gas diffusion in the pores of the spongy iron formed is rate-determining, in accordance with the formulae of section 2.3. As shown there, this case arises for large values of the ore diameter $2r_0$. The Knudsen diffusion coefficient, $D_{K,12}$, which is the one appropriate for diffusion in pores of diameter less than the mean free path of the gas molecules, is independent of the total pressure P.

Figure 93 represents the case where solid diffusion in the non-porous layers of the end-products of the reaction is rate-determining. Here, as for diffusion in the pores of spongy iron, the relation $\dot{\mathfrak{R}} \sim 1/(2r_0)^2$ holds, but the dependence of the reduction rate on the temperature is considerably greater.

When the phase-boundary reaction at or near to the external surface is rate-determining, Fig. 94, the decrease of partial pressure in the transport zone is slight. This zone consists of the boundary layer and the layer of end-products of the reaction. The relations between the reduction rate and the diameter d of the ore pieces, the partial pressures p, and the temperature T have been derived in section 1.2.5 and 2.4. The formula included in brackets holds when the reaction is markedly inhibited by water vapour, a phenomenon frequently observed below 550°C.

The case represented by Fig. 94 can occur with small-sized pieces of ores which are either non-porous or have very few pores. With very porous ores of small size the same circumstances hold at the surface of the grains which constitute each piece. The variation of the partial pressure in such a piece of ore is shown in Fig. 95. $\sum F_i$ denotes the internal surface area of the piece, which is accessible to the gas. In these circumstances the reduction rate is independent of the ore diameter $2r_0$.

With lower porosity or larger ore diameter, the diffusion in the pores of the reacting ore affects the reduction rate. In this case the relations shown in Fig. 96 hold; these come from formula (17a) in section 2.5.2. The reduction rate is then inversely proportional to the ore diameter $2r_0$. The temperature dependence will be chiefly determined, because of the relation $\dot{\mathfrak{R}} \sim \sqrt{k D}$, by the rate constant k for the phase-boundary reaction, but because of the square-root sign it corresponds to only half the apparent activation energy of the phase-boundary reaction.

In Fig. 97 [240]) the most important of the cases treated in Figs. 92, 93, 95 and 96 are collected together, and the transitions between the individual modes of behaviour are also taken into account. The structures which arise and the variation in the partial pressures or concentrations are shown in the upper part of the diagram, and in the lower part the graphs and formulae for the dependence of the reduction rate on the ore diameter $2r_0$.

The dependence of the reduction rate on the degree of reduction is also determined by the rate-determining sub-process. If the phase-boundary reaction is rate-determining, then for spherical pieces of ore

$$\dot{\mathfrak{R}} \sim (1 - \mathfrak{R})^{2/3},$$

and if gaseous diffusion in the pores of the end-products of the reaction is rate-determining, then

$$\dot{\mathfrak{R}} \sim [(1 - \mathfrak{R})^{-1/3} - 1]^{-1},$$

see formulae (1) and (2) in section 2.5.1. As was shown there, as the reduction progresses, a transition from a rate-determining phase-boundary

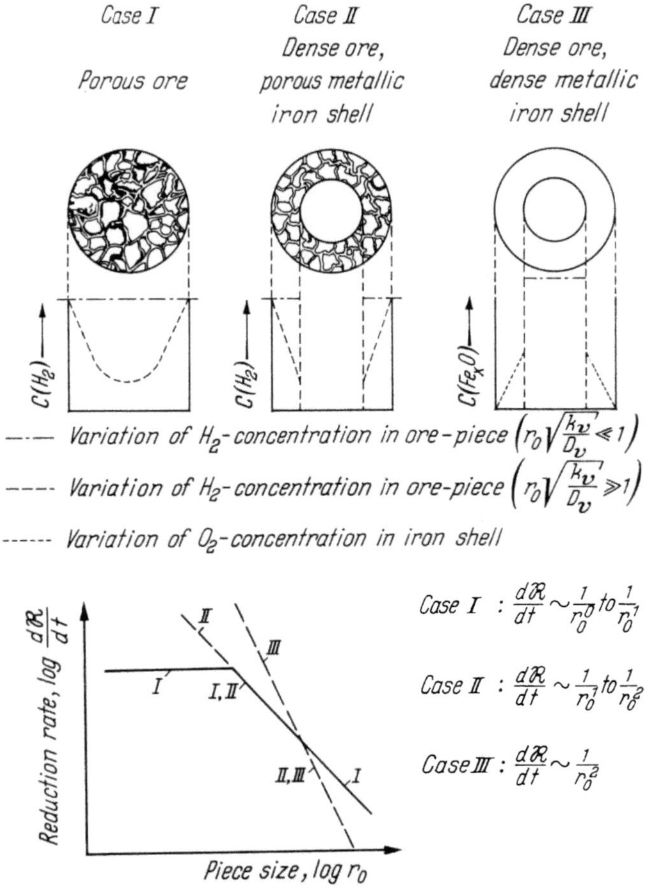

Fig. 97. Ore structure and formation of metal (diagrammatic)

reaction to a rate-determining diffusion process can occur, so that the total variation of the reduction rate with degree of reduction corresponds to a combination of both formulae. Under certain assumptions the resulting curve is almost rectilinear, as shown in Fig. 82, section 2.5.1, and therefore corresponds formally to a first-order relation

$$\dot{\mathfrak{R}} \sim (1 - \mathfrak{R}).$$

In Fig. 98 these three laws are again plotted in the same diagram. The relation

$$\dot{\mathfrak{R}} \sim (1 - \mathfrak{R})$$

has frequently been confirmed experimentally during the reduction of ores[241–243]). It can be obtained from the formula (5) of section 2.4 with

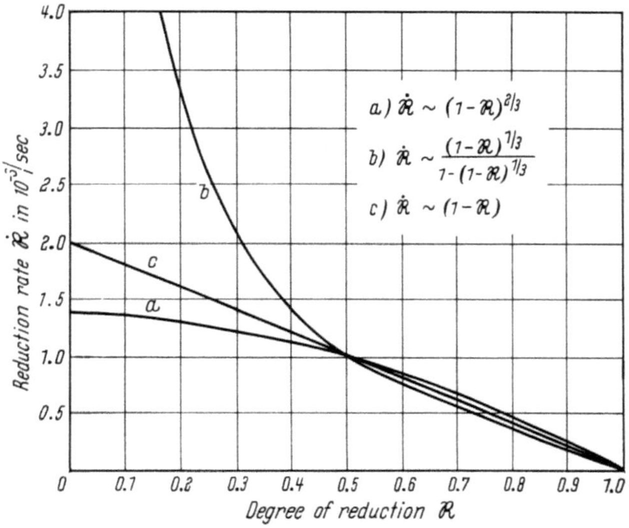

Fig. 98. Effect of the degree of reduction on the rate of reduction

the assumption, that the reacting surface F is proportional to the amount of unreduced oxide $n(t)$ or $(1 - \mathfrak{R})\, n_0$ [241, 251a]).

2.6. Observations on the Formation and Growth of Nuclei

Due to the removal of oxygen from the surface of the oxide a local variation of the Me/O ratio occurs. Because of the limited phase width of the oxides only a more or less limited amount of the liberated metal can be dissolved in the oxide. After a certain supersaturation has been reached, the *metal* or a *lower oxide* must separate out on the surface of the oxide. The first separation takes place in the form of individual *nuclei*. The number and distribution of the nuclei depends in the first place on the surface structure of the oxide, if heterogeneous, nucleation on lattice defects, surface irregularities and inclusions occurs. Further, the free energies of the interfaces and the supersaturation affects the nucleation, as discussed in section 1.2.6. If the nuclei are to *grow*, there must be transport of the metal to the nuclei from the points where oxygen removal occurs. This takes place by diffusion of metal ions and electrons through the

a)

b)

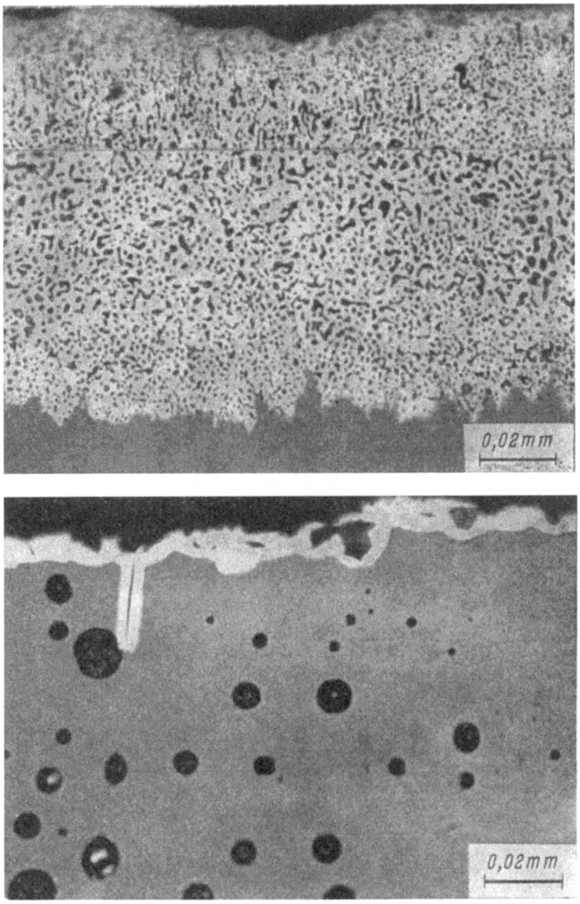

Fig. **99**. Spongy iron (a) and non-porous iron films (b) on reduced wüstite[248])

oxide lattice or along the surface. It is the rate of oxygen removal on the one hand and of metal diffusion on the other hand which determine whether in the regions between the originally formed nuclei the supersaturation of metal will be built up which is needed for additional nuclei to form and so increase the density of distribution of the nuclei. The density of distribution, the type and mode of growth of the nuclei, all these determine whether the nuclei grow together to form a covering layer of the new phase which will separate the underlying oxide from the gas phase or whether the new phase will grow as a sponge on the oxide through the pores of which the gas will have direct access to the original phase[245]).

a)
b)
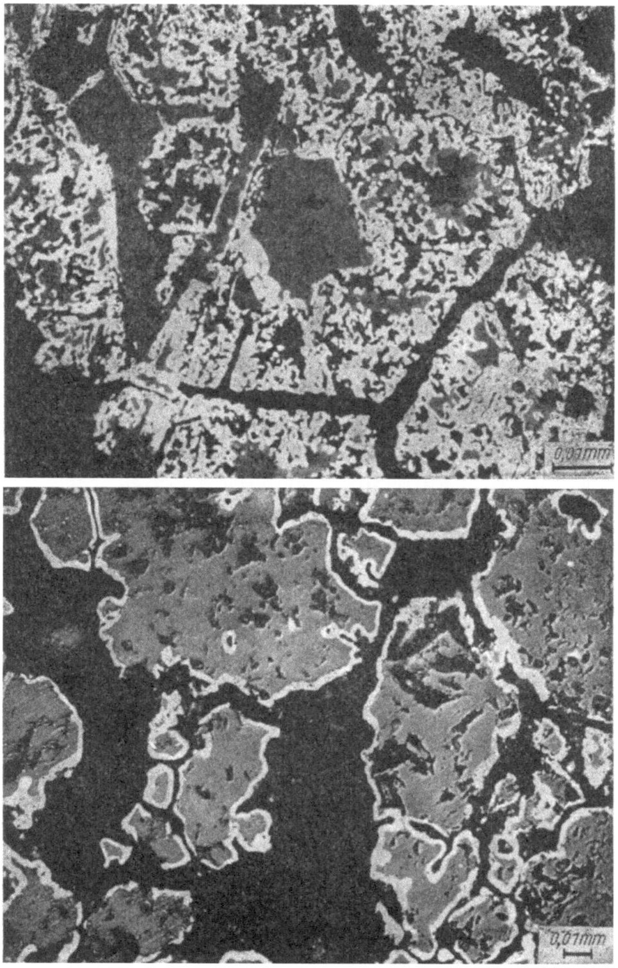

Fig. 100. Structure of metallic iron
a) With highly reducible hematite ore (750 : 1, original 850 : 1)
b) With low-reducibility magnetite ore (about 400 : 1, original 450 : 1)

In reduction down to metal the formation of *spongy iron* has usually been observed; but in some cases of reduction by hydrogen[246–249]) the formation of dense, non-porous *layers of iron* has been observed. The simultaneous appearance of the iron in both these forms on different sides of oxide laminae produced by oxidation of iron sheet has also been demonstrated, making it probable that impurities affect the formation of the nuclei. Figure 99 shows the different forms of development of the metal phase on synthetic oxide specimens; Figs. 100a and 100b those in the reduction of natural ores.

Many investigations have been carried out to determine the influence of nucleus formation on the progress of reduction. The occurrence of an incubation period at the beginning of an experiment is considered as a proof of this influence; a sharp rise in the reduction rate being shown at the end of this period. At temperatures of 200 to 330°C this effect occurs in the reduction of Fe_3O_4 to iron, but not in the reduction of Fe_2O_3 to Fe_3O_4 [250a]. In the reduction of Fe_3O_4 to wüstite no inhibition due to nucleus formation occurs, presumably because of the similarity of the lattices and the consequently reduced boundary-surface tension. Also in the reduction of wüstite to iron at temperatures of 800 to 1050°C the reduction rate increases at the beginning of the experiment[250b]; however, other authors[248] interpret this as due to the demonstrated increase in surface area.

In the reduction of porous ores needle-shaped metal crystals are often observed which grow inwards into the oxide grains. Since the removal of oxygen can take place only at the oxide/gas phase-boundary, diffusion of iron through the lattice or along the phase boundaries is necessary from the oxide surface to the points of growth of the metal crystals. As the driving force for the diffusion process, an iron activity, $a(\mathrm{Fe}) > 1$ must be assumed at the surface of the oxide. The separation of the metal in the oxide produces tensions which can lead to the breaking up of the grains and hence to the formation of new surfaces. Also at the pore surfaces of the ore the formation and growth of nuclei of the metal phase produce changes of volume and consequently changes in the porous structure[245].

A supersaturation of the oxide by iron as a prerequisite for the initiation of formation of nuclei of the metal phase affects the kinetics of the reduction since the activities of the metal and the oxygen in an oxide of composition Fe_nO_m are connected by the relation

$$a^n(\mathrm{Fe})\, a^m(\mathrm{O}) = \mathrm{const.}$$

An increase in the activity of the metal therefore produces a decrease in the activity of the oxygen in the oxide phase. In accordance with formula (40) of section 1.2.5 this change in oxygen activity affects the rate of the phase-boundary reaction.

In an investigation[251a] to determine what effect inhibition of the formation of nuclei has on the reduction of wüstite, a powder of this oxide was prepared by the reduction of hematite powder at 700 to 900°C. The amount of the lattice defects, determined radiographically, which were produced in this process depends on the duration of the reduction treatment. The incubation time in the subsequent reduction of the wüstite to iron, as may be seen from Fig. 101, clearly shows the effect of the apparent lattice distortion, which influences the inhibition of formation of nuclei and consequently the progress of the reaction. This interpretation can be confirmed by microscopic observation of the progress of reduction, which sets in and proceeds at different rates in individual grains of the powder.

In the reduction of ilmenite, too, it can be shown[251b]) that preliminary heat treatment reduces the number of nuclei and hence also the dispersion

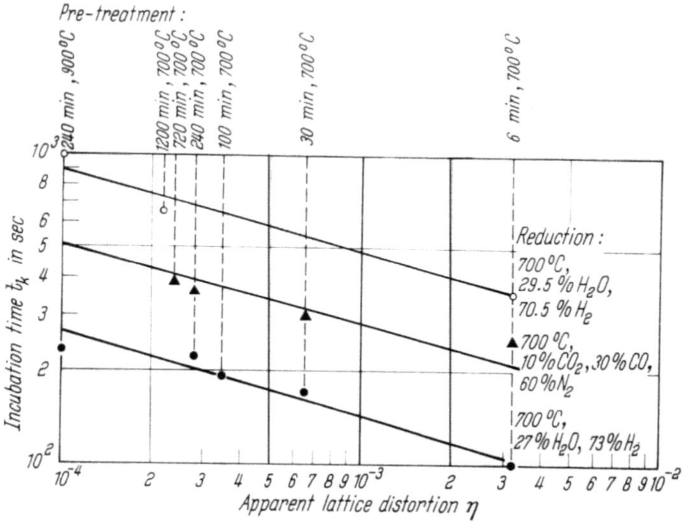

Fig. 101. Incubation time t_k in the reduction of wüstite to iron as a function of the apparent lattice distortion η. Crystallisation treatment of the wüstite as indicated on top of the figure. Pre-treatment gas composition: 0.48 atm CO_2; 0.52 atm CO

of the metal phase formed, and increases the size of the metal particles. In cold-pressed Fe_2O_3-material which was reduced at 520°C the same effect was observed, but at the same time the reduction rate was decreased by a factor of 1.5, due to repair of the lattice defects[251c]). However, at higher reduction temperatures, this effect no longer occurred, evidently because of the rapid repair of the lattice defects[251d]). CECH[251e]) carried out similar tests with NiO; he regarded the lattice defects as the active centres of the phase-boundary reaction. Lattice defects produced by irradiation also affect subsequent reduction[251f]).

In the reduction of highly porous hematite briquettes at about 350°C SCHAEFER[252]) found the variation of the rate of reduction $d(O/Fe)/dt$ with the degree of reduction O/Fe to be as shown in Fig. 102. The rate of reduction has a minimum at O/Fe ≈ 1.33. It was shown radiographically that before the minimum the briquettes contain no metallic iron, and after the minimum no hematite. The minimum therefore occurs if the specimen is completely reduced to magnetite. SCHAEFER explains the decrease in the reduction rate and the absence of iron before the minimum by the hypothesis that, in the briquettes, there is a comparatively high partial

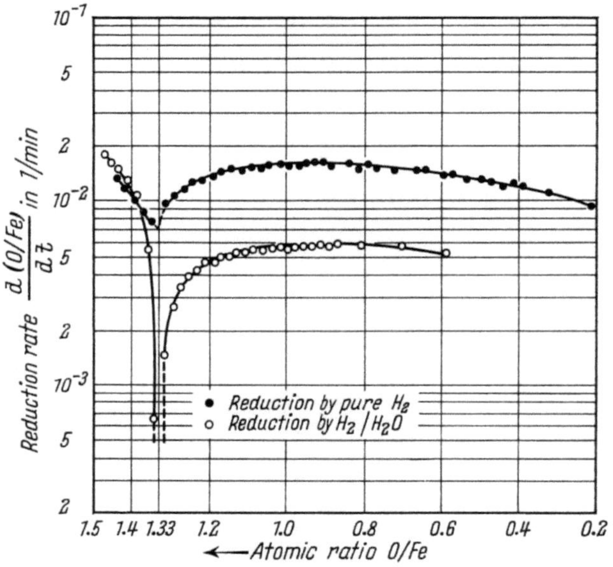

Fig. 102. Effect of water vapour on reduction rate for hematite, according to MORAWIETZ and SCHAEFER[252])

For reduction by H_2/H_2O, $p(H_2O)/p(H_2)$ = 33.6% of the Fe_3O_4/Fe equilibrium value

pressure of water vapour in the gas or that there is a high oxygen activity so long as the hematite is still being reduced, and that with this high oxygen activity the magnetite does not achieve the supersaturation in iron that would be necessary for iron nuclei to form. Only after the complete decomposition of the original hematite does the water-vapour content of the gas in the pores fall so far that the iron activity of the magnetite becomes sufficient for the formation of nuclei. The slow increase in the reduction rate is to be attributed to the slow growth in the number of nuclei.

In agreement with this interpretation, the minimum of the reduction curve at O/Fe = 4/3 becomes more pronounced if water vapour is added to the reducing gas. If the ratio of water vapour to hydrogen reaches about 80% of the equilibrium value for magnetite and iron, the reduction stops completely at O/Fe = 4/3. Evidently in this case the supersaturation of the magnetite by iron which is necessary for the formation of nuclei is no longer achieved. From this value it can be estimated that the iron activity for formation of nuclei to begin must exceed the value $a(\mathrm{Fe}) = 1.2$.

It is not excluded that the pronounced effect of the water-vapour content of the gas phase on the low-temperature reduction of magnetite, which McKewan[253]) and Hedden et al.[250]) have ascertained, may also have to be attributed to the effect of the oxygen activity on the formation and growth of the nuclei of the metal phase; see formula (50) of section 1.2.5.

From reduction experiments with superficially oxidized sheets of iron Gellner and Richardson[254]) inferred the occurrence of supersaturation before separation of the metal occurs. They observed that, at temperatures of reduction up to 700°C, the metal phase was formed at the oxide/gas phase-boundary. In contrast, at temperatures of 900°C and over, a very thin sheet of metal forms on the oxide surface at the beginning of the experiment, and subsequently the separation of metal takes place at the boundary surface between the oxide and the sheet of iron. The same sequence for metal separation also occurred even if, before reduction began, the sheet of iron was completely transformed into oxide by oxidation in an H_2O/H_2 mixture so that a metal nucleus was no longer present. If, however, after the consumption of the iron sheet the oxide specimen was heated for some time, then separation of the metal took place uniformly on both surfaces of the specimen. Similar observations were later made by Kohl[248]), but in his reduction experiments on iron sheets which had been oxidized throughout, the metal separated out predominantly on that face of the oxide which had been in contact with the gas during the oxidation. With Kohl, too, the difference in the form of separation-out of the metal was removed by heat treatment. Kohl attributed these phenomena to the formation of a covering layer of metal on one side of the oxide and the formation of spongy metal on the other side; the proven segregation of the impurities in the iron during the transformation into the oxide would determine the form in which the metal separated out during the reduction. Gellner and Richardson's observations cannot be explained by a supersaturation and by inhibition of the formation of nuclei, since the iron separated out on both boundary surfaces of the oxide and the differences in amount were independent of the presence of a metal nucleus.

Direct observation of the formation of the metal phase is possible by carrying out reduction experiments using a hot-stage microscope (see section 2.8). Figure 103 shows a series of photographs of the surface of wüstite during its reduction at 820°C in a gas containing 44% H_2, 2.3% H_2O, and the remainder N_2, according to an investigation by Pluschkell and Yoshikoshi[254a]). Figure (a) shows the structure of the single-

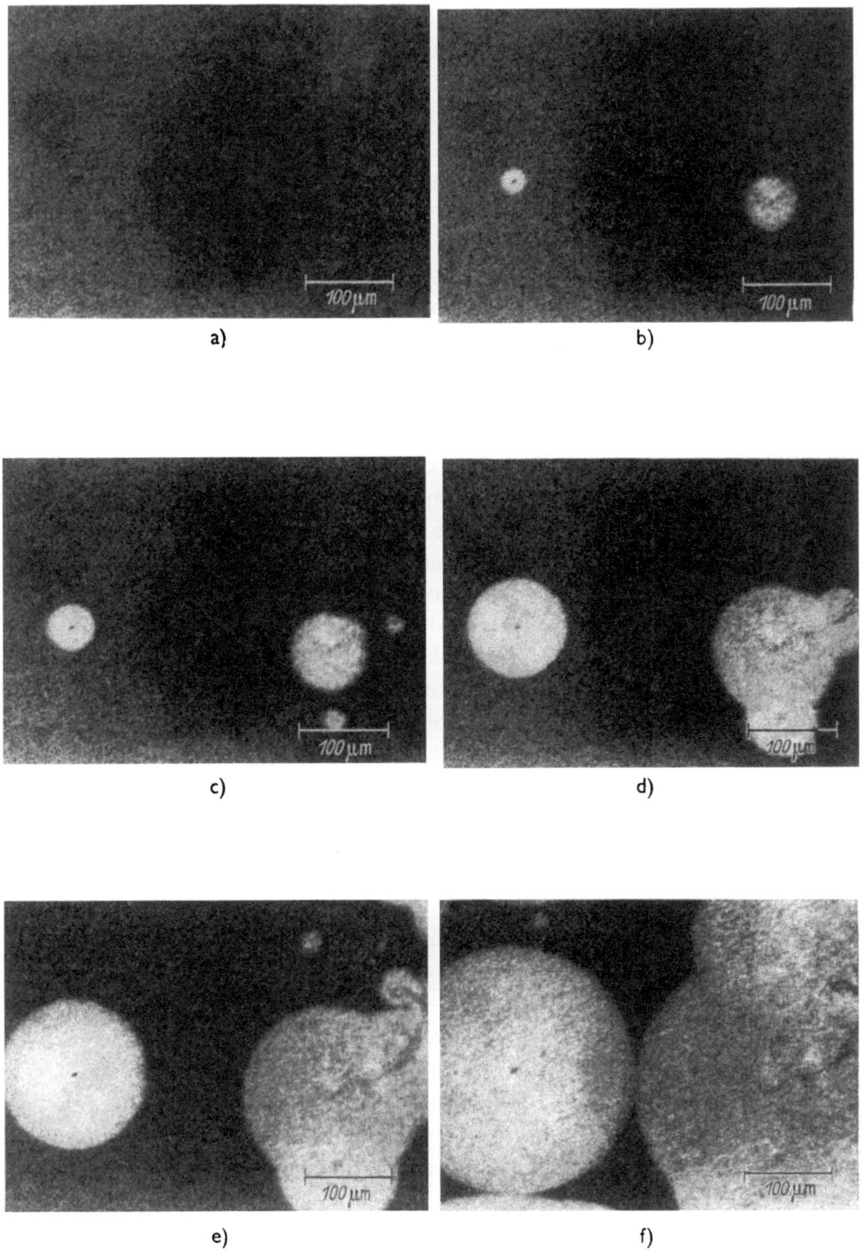

Fig. 103. Reduction of wüstite in the hot-stage microscope at 820°C. Reducing gas: 44% H_2, 2.3% H_2O, remainder N_2. Time from the start of reduction:
a) 15 sec b) 27 sec c) 30 sec d) 35 sec e) 42 sec f) 50 sec

crystalline wüstite specimen with (100)-surface orientation after 15 sec of reduction. 12 seconds later two circular iron nuclei have been formed (b). The other pictures show that the growth of their radius is linear with time. One can conclude that the removal of oxygen at the three-phase boundary iron/wüstite/gas is the rate-determining step.

GRABKE[251]) carried out measurements on the exchange of oxygen on iron and iron oxides during the progress of reduction of magnetite to wüstite and of wüstite to iron, see section 1.2.5. Because of the differing exchange activities of the individual oxides and iron, and by comparison with the values which were obtained on smooth, compact metal and oxide specimens, GRABKE's results lead to the following conclusions regarding the mechanism of the formation of the reduction products: The wüstite grows as a non-porous layer on the Fe_3O_4 during the reduction of magnetite by CO. In the reduction of wüstite, at first isolated nuclei of iron appear, and these spread over the oxide surface until finally they come into contact with one another and grow into a dense, non-porous layer of iron.

From measurements of the variation of the electrical conductivity of wüstite specimens during reduction at 900°C by CO/CO_2 mixtures STOTZ[255]) was able to determine the supersaturation of the oxide by iron which must be exceeded for the formation of metal nuclei to begin. He established that this critical value of the iron activity does not exceed $a(\mathrm{Fe}) = 1.015$. The supersaturation of the oxide phase for formation of nuclei has therefore practically no influence on the kinetics of the phase-boundary reaction in this case, provided that the oxygen activity of the gas phase is not very close to the equilibrium value.

By measurements of the e.m.f. of the solid-electrolyte cell

$$\mathrm{Pt} \mid \mathrm{Ag} \mid \mathrm{AgI} \mid \mathrm{Ag_2S} \mid \mathrm{Pt}$$

having AgI as the solid electrolyte, SCHMALZRIED and WAGNER[256]) were able to determine the silver-activity a^* needed for the formation of nuclei in the reduction of the silver sulphide of the cell to silver

$$\mathrm{Ag_2S} + \mathrm{H_2} \rightarrow \mathrm{H_2S} + 2\,\mathrm{Ag}.$$

They found that, at 400°C,

$$1.017 < a^* < 1.034.$$

In the steady state of the reaction after the onset of formation of nuclei, a supersaturation of the metal persisted, corresponding to a mean value of $a(\mathrm{Ag}) = 1.005$. The excess over the saturation value of the activity is necessary for the diffusion of the silver from the position where the sulphur is liberated to the next silver nucleus.

STOTZ's experiments at 900°C gave very slight supersaturations as prerequisite for the formation of nuclei, while SCHAEFER's measurements at 350°C showed, in contrast, considerable supersaturations as being necessary. It is in accordance with the theory to assume that, in general, the critical supersaturation for formation of nuclei will decrease with increasing temperature. The effect of the formation of nuclei on the rate of reduction will therefore also decrease with increasing temperature. This holds all the more since the coefficient of solid diffusion increases with increasing temperature, more rapidly, in fact, than the rate of removal of oxygen does. The activation energy for the diffusion of iron in wüstite amounts, namely, to 29 kcal, that of the phase-boundary reaction to 12 to 20 kcal, and the reduction-rate increment with temperature for industrial ores is only 2 to 12 kcal. The different dependence on temperature for these processes has the effect that, for equal density of nuclei, a smaller supersaturation between the nuclei is needed at the higher temperatures for the transport of the liberated metal. The metal activity at the oxide surface affects the oxygen activity of the oxide, and this in turn affects the rate of removal of oxygen, i. e., the rate of reduction.

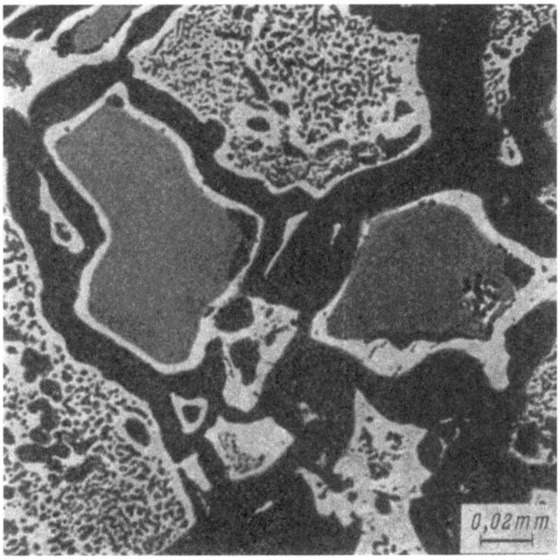

Fig. 104. Reduced magnetite material, in a reduction experiment, according to WIBERG[188])

Altogether, it can be assumed from the results of investigations now available that the influence of the formation and growth of nuclei on the structure of the products formed in the reaction and on the porous structure of the ore is of particular significance for the progress of the reduction. The formation of covering metal layers, for instance, greatly limits the reduction rate because of the low permeability $D(O)\,c(O)$ of iron for oxygen. Figure 104 shows a photo-micrograph communicated by WIBERG[188]), in which the very important difference in the degrees of reduction for grains with spongy iron and with covering iron layers arise in the same reduction experiment. The effect of the supersaturation which is necessary for the formation of nuclei should make itself felt on the progress of the reduction above all at low temperatures and when the composition of the gas differs only slightly from the equilibrium value. Such an approximation of the gas composition to the equilibrium value occurs especially in the pores of the ore.

2.7. The Reduction Kinetics of Mixed Oxides, Oxide Compounds, and Sinters

The kinetics of the reduction of crystals of mixed oxides and oxide compounds is influenced particularly by two basic phenomena:

a) The activity of the iron oxides is reduced by the formation of mixed crystals and compounds. Such formations also change the equilibrium composition of the gas phase, as explained in section 1.1.4. The ratio CO_2/CO or H_2O/H_2 must, however, fall below the equilibrium value if reduction is to start. Moreover, the change in the equilibrium conditions worsens the degree of chemical utilization of the reducing gas which can be achieved. Finally, the change in the equilibrium concentrations, as may be seen from the formulae (5) of section 2.3 and (1) of section 2.4, directly affects the rate of gaseous diffusion in the boundary layer and in the pores of the reduction products and also the rate of the chemical reaction at the phase boundary.

b) In mixed crystals and oxide compounds in which the iron oxide is the more noble component – and this is normally the case in the ores used in the manufacture of iron – when oxygen is eliminated during the reduction process, iron or a lower oxide of iron separates out. A crystal enriched in the second component or the pure second component is the product of further reduction. This decomposition necessitates diffusion in the solid state. The length of the diffusion path depends on the size of the grains or crystals in the original material and on the distribution of the new phase in the host crystals, i.e., on the distance between the nuclei, which, in turn, is determined by the critical supersaturation, the frequency of formation of nuclei, the number of foreign nucleation sites and lattice defects, and the ratio of the rates of reaction and diffusion.

Both these phenomena should affect the rate of reduction of mixed crystals and oxide compounds unfavourably. That this effect does not occur without exception may be due to the fact that the rate constant for the phase-boundary reaction can have a higher value for the mixed crystals or the compounds than it has for the pure oxide. But the main reason will mostly be the influence of the formation and decomposition of the compound or mixed crystal on the porosity of the ore.

In the reduction of mixed crystals of wüstite and manganese(II)-oxide, the alteration in the equilibrium concentration of the reducing gas can be determined from the variation in the activity of wüstite in the mixed crystal, see Fig. 16, section 1.1.3, since the relation

$$K/K^* = a(\mathrm{FeO})$$

holds, where K is the CO_2/CO or H_2O/H_2 ratio for the equilibrium of the mixed crystal with the corresponding metal phase (practically pure iron), and K^* is the corresponding ratio for pure wüstite.

Investigation of the process of reduction of mixed crystals of FeO and MnO gave the following result[257]). Elimination of oxygen at the oxide surface is followed by the separation of crystals of iron, this taking place in the interior of the oxide. As reduction proceeds, the separation front migrates into the interior of the mixed oxide crystals, as may be seen from Fig. 105. This phenomenon can be explained by the fact that owing to the initial separating-out by the iron, a surface region of the mixed

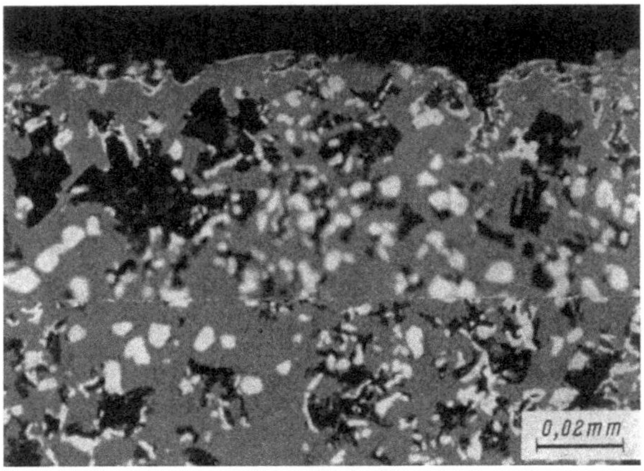

Fig. 105. Iron separation in reduced mixed crystals of FeO-MnO with $N(\mathrm{FeO}) = 0.4$. Reduction temperature: 800°C. Degree of reduction 47% relative to the FeO content

crystal becomes impoverished in iron. When more oxygen is eliminated the liberated metal – largely manganese, since the surface of the mixed crystal has become enriched in manganese – diffuses via the vacancies of the metal ions and the electron deficiencies into the interior of the oxide and cements with the iron in the interior regions which are still rich in iron.

The process, which might be expected next, of inward diffusion of the manganese and simultaneous outward diffusion of the iron via the oxide lattice, which would lead to the formation of a metal phase on the oxide surface, does not take place, although there is a gradient of the iron concentration in the oxide lattice. The reason is that this gradient in the concentration does not result in any corresponding gradient of the chemical potential μ of the iron in the mixed crystal of oxides, and the driving force for diffusion – as for any chemical reaction – arises from differences of chemical potential. For the diffusion flow j_i for component i, we have in general (see section 1.2)

$$j_i = -B_i\, c_i \frac{d\mu_i}{dx},$$

where

B = mobility,
c = concentration (e.g., mole cm^{-3}),
x = a position coordinate.

Putting

$$D_i = B_i\, RT,$$

we get

$$j_i = -\frac{D_i}{V_M} N_i \frac{d\ \ln a_i}{dx}, \tag{1}$$

where

N = mole fraction,
V_M = molar volume,
a = activity.

A simple transformation gives for the flow of iron and manganese

$$\left.\begin{aligned} j(\text{Fe}) &= -D(\text{Fe})\frac{N(\text{FeO})}{V_M}\frac{d\ \ln a(\text{Fe})}{dN(\text{FeO})}\frac{dN(\text{FeO})}{dx},\\ j(\text{Mn}) &= +D(\text{Mn})\frac{N(\text{MnO})}{V_M}\frac{d\ \ln a(\text{Mn})}{dN(\text{MnO})}\frac{dN(\text{FeO})}{dx}.\end{aligned}\right\} \tag{2}$$

The diffusion flow of iron and manganese takes place during the reduction in a region in which the metal has already separated from the mixed oxides, and therefore it can be assumed that there is equilibrium between the oxide phase and the metal phase. If the metal and oxide represent ideal mixtures, then

$$\begin{aligned} a(\text{Fe}) &= K(\text{Fe})\, a(\text{FeO})/\sqrt{p(O_2)} = K(\text{Fe})\, N(\text{FeO})/\sqrt{p(O_2)},\\ a(\text{Mn}) &= K(\text{Mn})\, a(\text{MnO})/\sqrt{p(O_2)} = K(\text{Mn})\,[1 - N(\text{FeO})]/\sqrt{p(O_2)}.\end{aligned} \tag{3}$$

Hence it follows that

$$\sqrt{p(O_2)} = K(\text{Fe})\, N(\text{FeO}) + K(\text{Mn})\,[1 - N(\text{FeO})],$$

and therefore, putting $\varkappa = K(\mathrm{Mn})/K(\mathrm{Fe})$, substitution in (3) gives

$$a(\mathrm{Fe}) = \frac{N(\mathrm{FeO})}{N(\mathrm{FeO}) + \varkappa[1 - N(\mathrm{FeO})]},$$

$$a(\mathrm{Mn}) = \frac{1 - N(\mathrm{FeO})}{1 - N(\mathrm{FeO}) + N(\mathrm{FeO})/\varkappa}.$$

Substituting these expressions in (2), we get

$$\begin{aligned} j(\mathrm{Fe}) &= -\frac{D(\mathrm{Fe})}{V_M} \frac{1}{1 + N(\mathrm{FeO})\,(1-\varkappa)/\varkappa} \frac{dN(\mathrm{FeO})}{dx}, \\ j(\mathrm{Mn}) &= +\frac{D(\mathrm{Mn})}{V_M} \frac{1}{N(\mathrm{FeO}) + \varkappa[1 - N(\mathrm{FeO})]} \frac{dN(\mathrm{FeO})}{dx}. \end{aligned} \tag{4}$$

If, as in the case of mixed crystals of wüstite and manganese(II)-oxide,

$$\varkappa \approx 10^{-5},$$

and if

$$1 > N(\mathrm{FeO}) > 10^{-3},$$

then to a close approximation

$$j(\mathrm{Fe}) \ll j(\mathrm{Mn}) \quad \text{if} \quad D(\mathrm{Fe}) \approx D(\mathrm{Mn})$$

and

$$j(\mathrm{Mn}) \approx \frac{D(\mathrm{Mn})}{V_M} \frac{d \ln N(\mathrm{FeO})}{dx}. \tag{5}$$

This result is in excellent agreement with the observed fact that separation of metal takes place only in the interior of the oxide, since the iron cannot diffuse to the surface of the oxide.

The above derivation is exact only if the oxide is an electron conductor, since only in that case an overall equilibrium can be assumed between the individual mobile metal ions and the electrons or holes of the metal phase. A derivation of the transport equations which starts from the fact of the separate diffusion of the metal ions and electrons[257]) also leads to Eq. (5).

The separation of metal in the interior of oxide specimens has been observed in the reduction of mixed crystals of FeO and MnO and also in reduction experiments on FeO-MgO and NiO-MgO. In order to prove that this phenomenon is not perhaps explained by penetration of gas via the pores of the sintered material, molten manganese was also used as the "reducing agent"[257]). Figure 106 shows that in such an experiment the reaction

$$(\mathrm{Mn})_{\mathrm{metal}} + (\mathrm{Fe})_{\mathrm{oxide}} \rightleftharpoons (\mathrm{Mn})_{\mathrm{oxide}} + (\mathrm{Fe})_{\mathrm{metal}} \tag{6}$$

also leads to the separation of the iron in the interior of the oxide, since the manganese can certainly diffuse into the oxide, but the iron cannot diffuse out of the oxide to the metal phase.

The above derivation is generally applicable, and not merely to mixed crystals of wüstite and manganese(II)-oxide. It has been shown, for example,

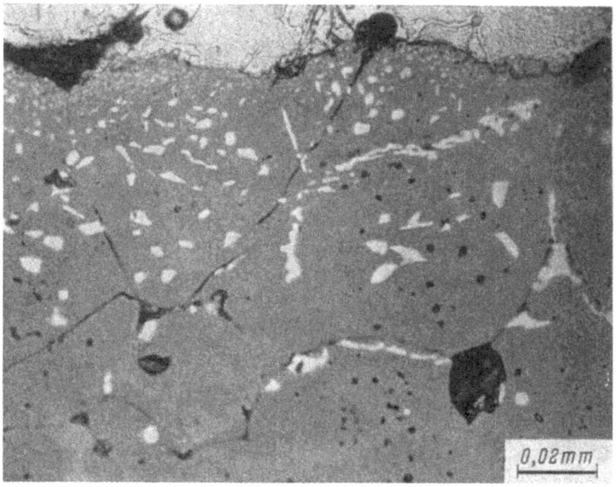

Fig. 106. Cementation of iron in solid mixed crystals of FeO-MnO by molten manganese. $N(\mathrm{FeO}) = 0.4$

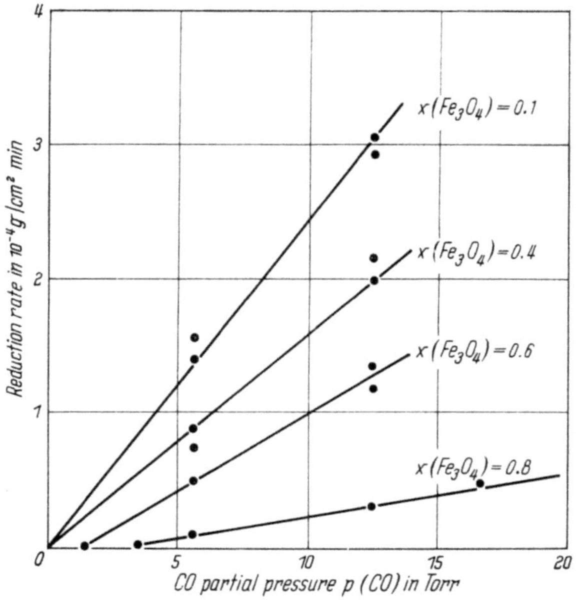

Fig. 107. Dependence of reduction rate on the partial pressure of carbon monoxide and on the chemical composition, for Fe_3O_4-Mn_3O_4 mixed oxides. Reduction temperature: 900°C; p (CO) = const = 50 Torr

that in the reduction of mixed crystals of Mn_3O_4 and Fe_3O_4 to a FeO-MnO mixed phase, the more noble oxide, here the MnO, crystallizes out in the interior of the original phase[258]).

No precise assertions can be made about the absolute value of the reduction rate or about its relative value compared with pure oxides on the basis of the researches carried out, because of the unknown surface area of the sintered material. However, ULRICH[258]) has shown that the reduction rates for the mixed-crystal series of magnetite and hausmannite show about the same variation as the oxygen equilibrium-pressures or the oxygen activities in the mixed oxide, Fig. 107. According to this rule, the

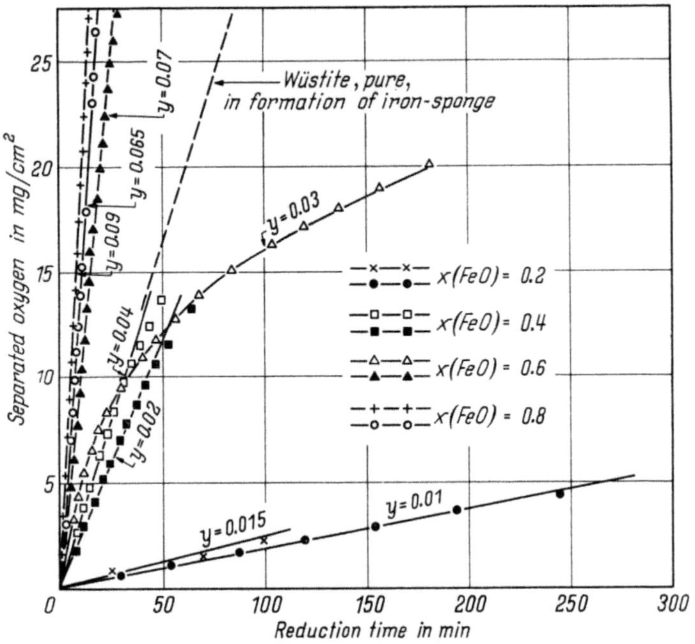

Fig. 108. Progress of reduction of mixed oxides of iron and manganese of various compositions and with various initial concentrations of vacancies, y. The reduction took place at 800°C in a mixture of H_2O and H_2 of about 680 Torr total pressure and with 2.55% water vapour

reduction rate of mixed crystals of FeO-MnO increases with the wüstite content, see Fig. 108. The curve drawn for pure wüstite was obtained from compact wüstite specimens prepared by oxidation, and is therefore only partially comparable with the results of measurements on sintered material of mixed oxides.

The reduction behaviour of oxide compounds with iron oxides has, it is true, often been studied, because of its importance in connection with sinters and pellets, but no fundamental clarification has yet been achieved of the sub-processes which occur. Most of the investigations in which the reduction rate of such phases have been measured suffer, moreover, from the fact that the porosity and surface area of the specimens under in-

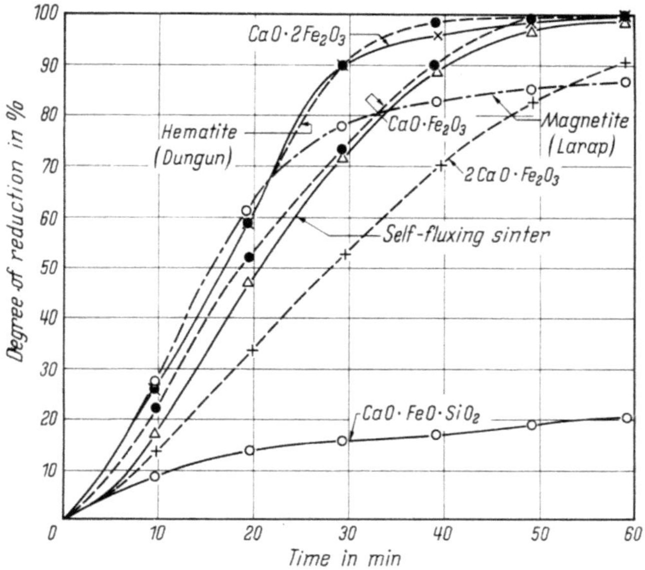

Fig. 109. Reduction of minerals containing iron oxides by H_2 at 900°C [259]

vestigation have not been determined or not sufficiently precisely. The possibility of making general assertions is therefore very limited. Some results of measurements with various kinds of minerals are given in Fig. 109 and Table 8 [260].

Table 8. *Degree of reduction of sinter minerals after reduction for 40 minutes in pure CO at 850°C*

	%
Fe_2O_3	49.4
$CaO \cdot Fe_2O_3$	49.2
$2\,CaO \cdot Fe_2O_3$	25.5
$CaO \cdot 2\,Fe_2O_3$	58.4
$3\,CaO \cdot FeO \cdot 7\,Fe_2O_3$ . . .	59.6
$CaO \cdot Al_2O_3 \cdot 2\,Fe_2O_3$ . . .	57.3
$4\,CaO \cdot Al_2O_3 \cdot Fe_2O_3$ . . .	23.4

According to these results, the calcium ferrites rich in iron oxide and also the compound $CaO \cdot Al_2O_3 \cdot 2\,Fe_2O_3$ are as reducible as, or more highly reducible than, hematitic ores, but dicalcium ferrite $2\,CaO \cdot Fe_2O_3$ and the silicate $CaO \cdot FeO \cdot SiO_2$ are less easily reducible. According to RUECKL's investigations[261]) the iron-bearing silicates become perceptibly reduced in CO only at temperatures above 1200°C. RUECKL concluded from an analysis of the kinetics of the progress of the reduction that the phase-boundary reaction is the rate-determining sub-process for the reduction of calcium ferrite.

Comparison shows, however, that the values obtained by RUECKL for ferrites in CO are greater by a factor of nearly 10^2 than the rate constants for the reduction of magnetite in CO, as found by ULRICH[258]). This difference may be due to a porosity of the ferrites.

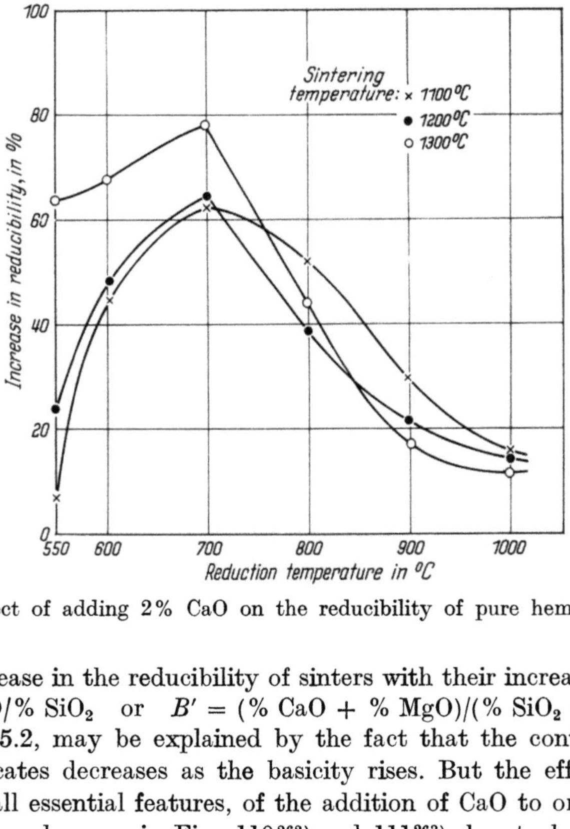

Fig. 110. Effect of adding 2% CaO on the reducibility of pure hematite sinter[262])

The increase in the reducibility of sinters with their increasing basicity $B = \%\,CaO/\%\,SiO_2$ or $B' = (\%\,CaO + \%\,MgO)/(\%\,SiO_2 + \%\,Al_2O_3)$, see section 5.2, may be explained by the fact that the content of iron-bearing silicates decreases as the basicity rises. But the effect, running parallel in all essential features, of the addition of CaO to ores free from silica, as may be seen in Figs. 110[262]) and 111[263]), has to be interpreted in some other way, and this raises the question whether there may not be other factors acting in the case of SiO_2-bearing sinters. SETH and ROSS[262]) explain the good reducibility of sinters containing calcium ferrites by the following reaction mechanism. First, all the higher iron oxides and ferrites rich in iron oxides are reduced until only dicalcium ferrite and wüstite

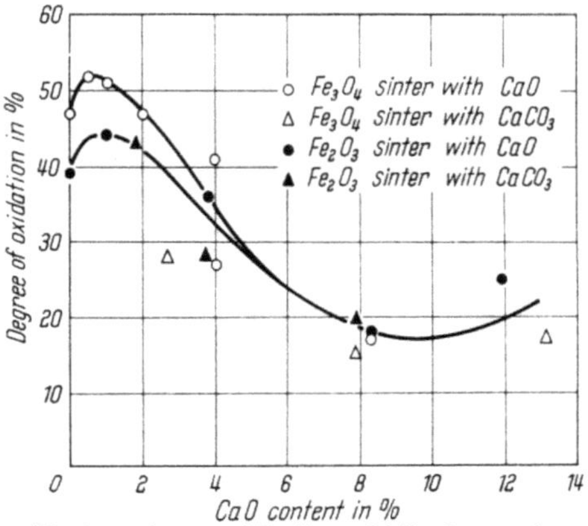

Fig. 111. Degree of oxidation of sinter ($Fe_2O_3 = 100$) after undergoing a non-isothermal reduction treatment in CO/N_2 mixtures

still remain. The gas then attacks predominantly the dicalcium ferrite:

$$2\,CaO \cdot Fe_2O_3 + 3\,H_2 \rightarrow 2\,CaO + 2\,Fe + 3\,H_2O\,. \tag{7}$$

The liberated lime immediately reacts with the wüstite in accordance with

$$2\,CaO + 3\,FeO \rightarrow 2\,CaO \cdot Fe_2O_3 + Fe \tag{8}$$

since lime and wüstite cannot co-exist. This mechanism is particularly probable if we assume that the calcium ferrite formed can be reduced

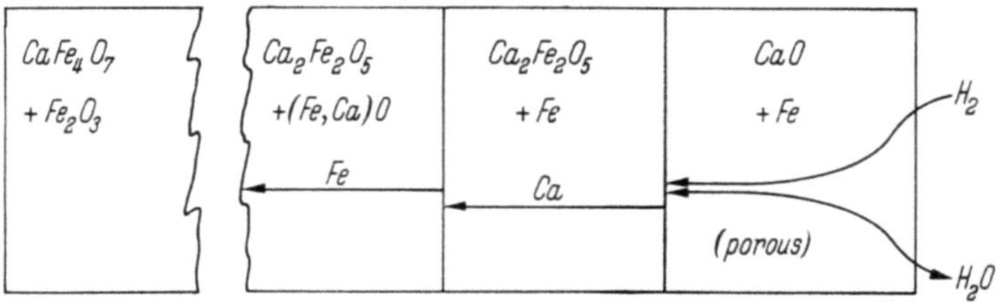

Fig. 112. Diagrammatic course of the reduction of lime-bearing sinter

immediately, that is, practically as soon as it appears, so that it is attacked by the gas when it is in a particularly reactable state. The photo-micrographs published by RUECKL[261]) show that, in the reduction of specimens of $2\,CaO \cdot Fe_2O_3$ and FeO, the wüstite is first consumed; the iron which is formed separates out in the ferrite. This finding may be regarded as proof of the mechanism suggested by SETH and ROSS, since according to the above formulae the wüstite is consumed in the reaction with the CaO, but the ferrite is re-formed after its reduction. On the other hand, the photomicrographs show that wüstite is not present at the oxide-gas phase-boundary, and the formation of dicalcium ferrite in accordance with (8) therefore takes place in the interior of the specimen. A diffusion process must therefore intervene between the reactions (7) and (8). It is improbable that the oxygen has a high mobility. Relying on the findings in the reduction of mixed oxides, we may assume that the reaction proceeds in the following way, which is illustrated in Fig. 112. At the surface of the oxide,

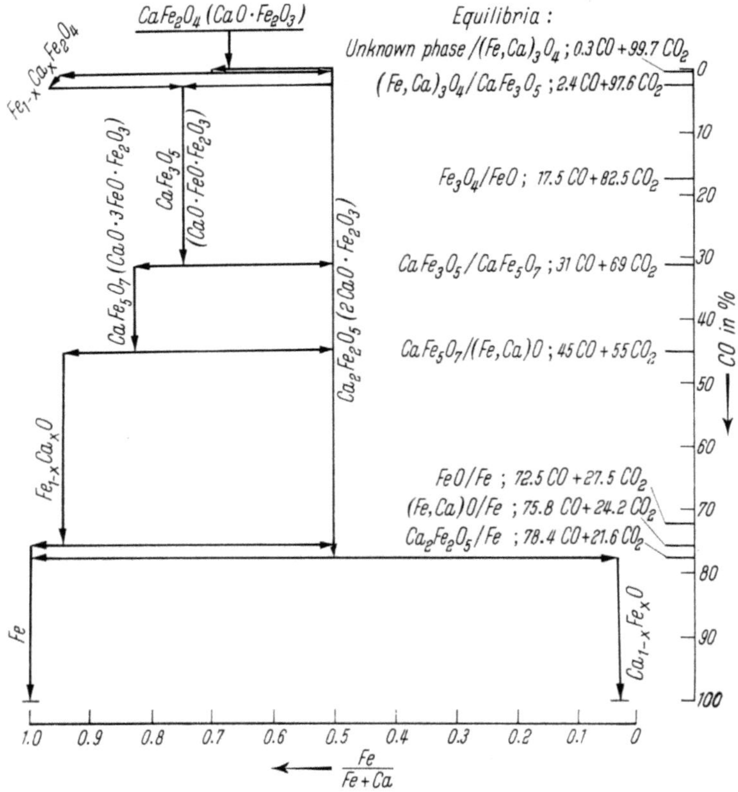

Fig. 112a. Phase equilibria and course of the reaction in the reduction of mono-calcium-ferrite at 1000°C by CO, according to EDSTRÖM[263a])

dicalcum ferrite is reduced. The liberated iron separates out in the oxide phase, and the calcium diffuses in the form of metal ions and electrons via the vacancies into the interior and reacts with the wüstite, and again iron either separates out or diffuses away to the Fe_3O_4.

Diffusion of iron in the layer of dicalcium ferrite and iron is not possible, since the activity of the metal in the layer is constant. Diffusion of oxygen is therefore the only alternative to the diffusion of calcium.

The picture presented is in qualitative agreement with EDSTRÖM's very careful measurements[263a]), in which the equilibrium pressure of the oxygen and the sequence of phases which appear in the reduction of $CaO \cdot 2\,Fe_2O_3$ were determined. These are shown in Figs. 112a and 112b. In making this comparison it should be remembered that Fig. 112 represents the circumstances for a single crystal free from pores up to the outermost layer, but EDSTRÖM's measurements were carried out on porous sinters. LIHL

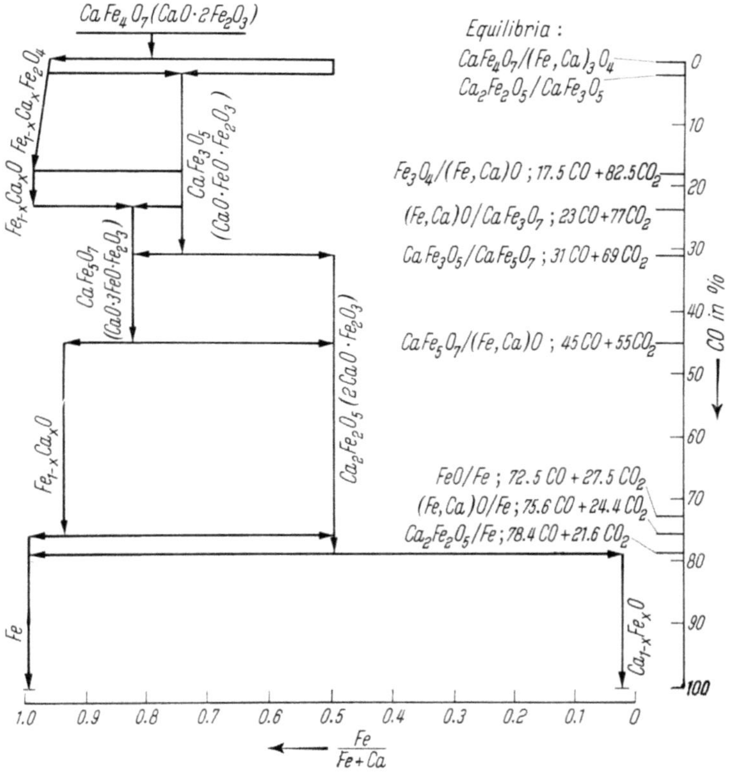

Fig. 112b. Phase equilibria and course of the reaction in the reduction of calcium diferrite at 1000°C by CO, according to EDSTRÖM[263a])

and NEMEC[263b]) have carried out a phase analysis of the reduction products for magnesium ferrite and dicalcium ferrite. These authors give a model which explains the mechanism of the phase changes, the possible diffusion flows and the resulting arrangement of the phases. The following explanation, based on this model, can be given of the transformation of magnesium ferrite into FeO-MgO mixed crystals during the removal of oxygen. At the surface of the spinel transport of cations from the tetra-

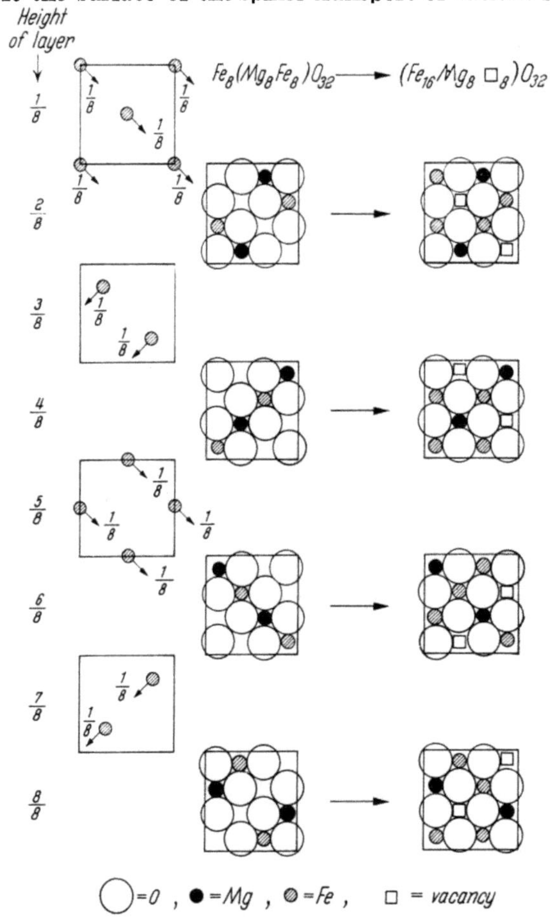

Fig. 112c. Scheme of the transformation of magnesia-ferrite into magnesia-wüstite by a change of position process along (111) in a spinel unit cell

hedral to the octahedral positions takes place, as shown in Fig. 112c. In this way a crystal is formed with spinel structure and the composition of MgO-FeO mixed crystals:

$$Fe^{3+}(Mg^{2+}Fe^{3+})O_4 \rightarrow 4\,(Mg^{2+}_{1/4}Fe^{3+}_{1/2}\square_{1/4})O\,.$$

The reducing gas removes the excess oxygen, and thus magnesio-wüstite forms and the vacancies $\square$ disappear:

$$4\,(Mg^{2+}_{1/4}Fe^{3+}_{1/2}\square_{1/4})O + H_2 \rightarrow 3\,(Mg^{2+}_{1/3}Fe^{2+}_{2/3})O + H_2O\,.$$

It is to be assumed that, as the reaction progresses further, the oxygen removal occurs at the surface of the MgO-FeO mixed crystals. A steady layer thickness for this phase will be reached if 1/4 of the metal ions liberated in this process migrate via the vacancies in this lattice and the corresponding electrons to the phase boundary (Fe, Mg)O/$MgFe_2O_4$ and change the spinel there into magnesio-wüstite:

$$Fe^{3+}(Mg^{2+}Fe^{3+})O_4 \rightarrow 4\,(Mg^{2+}_{1/4}Fe^{3+}_{1/2}\square_{1/4})O\,,$$

$$4\,(Mg^{2+}_{1/4}Fe^{3+}_{1/2}\square)O_{1/2} + \tfrac{1}{4}\,(Fe, Mg) \rightarrow 4\,(Fe, Mg)O\,.$$

Due to the removal of oxygen at the interface (Fe, Mg)O/gas, the activity of the iron — $[a(Fe) \sim a(FeO)/p^{1/2}(O_2)]$ — increases until it becomes possible for metallic iron to separate out.

The reduction of dicalcium ferrite can be explained along similar lines.

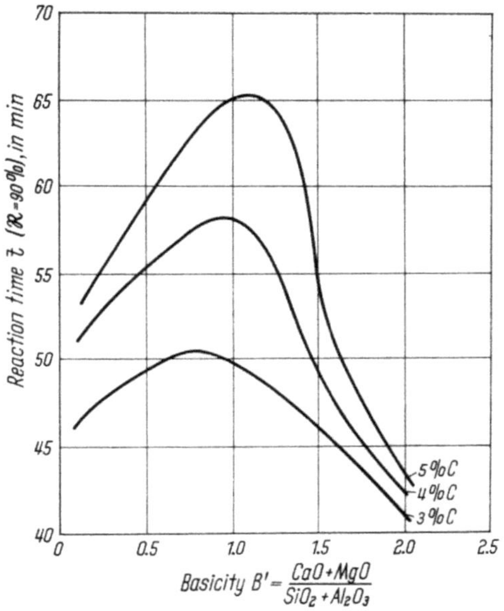

Fig. 113. Reaction time for 90%-complete reduction of sinter in H_2 at 900°C

In most investigations it has been found that, with sinters free from SiO_2, the addition of more than 5 to 10% lime does not improve reducibility but may even decrease it. A similar state of affairs holds for the reduction by hydrogen of sinters containing silica, as may be seen from Fig. 113 [264]).

Seth and Ross also established that in the reduction of iron oxides free from lime a strongly fissured phase boundary between the wüstite and iron appeared, as has also been found in the reduction of wüstite having a high content of vacancies[265]) and in the reduction of wüstite to iron[258]). It was explained in section 2.2 that this increase in surface area arises through the separation of iron ion vacancies and the removal of oxygen at the surface of the oxide. In the reduction of sinters containing lime, on the other hand, SETH and ROSS found that the wüstite layer had a smooth surface. This, too, is understandable if the reaction takes place according to the scheme which we have described, since the wüstite is not in contact with the gas phase and the removal of oxygen takes place at the ferrite surface. The fact that the addition of 2% CaO to the Fe_2O_3 sinter material reduces the thickness of the wüstite and magnetite layers in the reduced specimens is to be attributed to the higher reduction rate, which in the steady state requires a higher diffusion flow of iron for the maintenance of the thicknesses of the layers, and this is possible only by a reduction in the layer thickness in the steady state. We must also take into account the effect which the formation of mixed crystals with the CaO has on the equilibria of the oxide phases – which forms the driving force for the diffusion – and on the diffusion coefficients.

The simple possibility should, however, also be pointed out that the increased reducibility of lime-bearing sinter may be attributable to the increased porosity, a possibility which has been confirmed experimentally many times. Thus STANGERAY and ROSS[265a]) and SUBAT[265b]) concur in finding that a CaO-additive clearly increases the porosity of Fe_2O_3 sinter material and changes the pore spectrum. The micro-pores, particularly, increase in number[265b]). Moreover, during the reduction of these specimens a marked increase of porosity, as compared with pure Fe_2O_3, takes place[265a]).

Data about the reduction behaviour of other combinations of oxides with iron oxides are mostly limited to thermodynamic investigations and considerations. An investigation of the reduction kinetics for ilmenite at temperatures between 500 and 1200°C was carried out by ÖSTBERG[265c]) using hydrogen, carbon monoxide, and solid carbon as the reducing media. The reduction rate was approximately the same as for magnetite. The iron separates out in an extremely finely divided form; its dispersivity can be reduced by preliminary heat treatment or by adding NaCl and Na_2CO_3 to the sinter. The latter method is a fundamental and industrially important effect, which is attributed by the author to a change in the surface energy. At higher temperatures of reduction, titanium sub-oxide is formed.

2.8. Reduction of Slags Containing Iron Oxides

In the lower part of a blast-furnace shaft, temperatures are reached at which the oxidic components of the burden, i.e., the added flux, gangue, coke ash, and unreduced iron oxides, can melt. It is to be assumed that in this region there is no longer any trivalent iron present in the ore. The resulting melt has the composition of a blast-furnace slag very rich in iron, depends on the composition of the burden, and may – as a simplification – be regarded as a FeO-CaO-SiO_2 slag including variable amounts of Al_2O_3.

The melting slag runs as beads over the column of coke, with the rising reducing gases – here still very rich in CO – flowing round it, until finally it joins the slag bath in the hearth, with the hot metal below. In this way a removal of oxygen can take place, due to the contact with the coke and the reducing gas. The iron produced may at first be solid, but later melts, and falls in drops through the slag bath, joining the molten metal below it. The iron itself, due to its carbon content, may exert a reducing effect on the slag.

The processes described produce an additional reduction, the extent of which depends, *inter alia*, on the degree of reduction of the slag being formed, the size of the drops, the track of the drops, and the area of contact and time of contact of the slag with the coke and iron. Moreover, to estimate the progress of reduction, a knowledge of the individual rate constants and the diffusion coefficients of the reacting components in the melt is necessary. A compilation of the values of the diffusion coefficients for Ca, Si, Al, O, and S in CaO-Al_2O_3-SiO_2 slags is given by KOROS and KING[265p]).

KRAINER, BEER, and BRANDL[265d]) investigated the reaction of molten slags containing iron oxides with graphite and coke crucibles suspended from a balance in a tubular furnace. The phase-boundary reaction proved to be rate-determining in this process.

Slags with 70% FeO and a CaO/SiO_2 ratio of 1 : 1 were reduced about 7 times as quickly as fayalite, $2\,FeO \cdot SiO_2$ (70.5% FeO).

Coke reduced the slags investigated just as well as graphite; but the ash content of the coke caused an improvement in the reduction of fayalite by dissolving the SiO_2 which separates out, but impaired the reduction of a FeO-CaO-SiO_2 slag. For a slag with 70% FeO, 14.4% CaO and 15.5% SiO_2 in a coke crucible, the rate constant for the phase-boundary reaction at 1300°C was about $2 \cdot 10^{-4}$ cm/sec, and at 1500°C about 10^{-3} cm/sec. Graphite and coke were poorly wetted by the slags investigated, as measurement of the angle of contact showed. The reason for a somewhat better wetting of the coke is not certain, but could be attributable to its ash content. The iron formed separates from the slag in droplets of diameter 10^{-4} to 10^{-1} cm. The droplets take up carbon so long as they remain in contact with coke or graphite. In the interior of the slag the droplets are decarburized. The decrease of C-content takes place uniformly over the entire cross-section of a drop. The diffusion of the carbon therefore takes place quicker than the phase-boundary reaction. Experiments with the addition of iron saturated with carbon yielded no improvement in the rate of the slag reaction.

Kondakov, Ryzhonkov, and Golenko[265c]) reduced molten FeO in a graphite crucible and established that the reduction rate definitely increases with increasing pressure. It is conjectured that the gas phase participates in the reaction and that the overall reaction is controlled by the Boudouard reaction. The authors base this assumption on the measured activation energy of 38.2 kcal/mole, which, in their view, is similar to the activation energy of the Boudouard reaction. The dependence of the reaction rate on pressure was confirmed by Gel'd, Vlasov, and Serebrennikov[265f]) by reduction experiments on Cr_2O_3 and MnO.

When the radius of the drops and the diffusion coefficients for the slag phase are known, the rate of diffusion in the slag droplets as they roll down can be estimated. If we adopt the usual value of $5 \cdot 10^{-5}$ cm²/sec for the diffusion process in the melt, and use the value given above for the first-order rate constants, we find that, at 1300°C, the diffusion has an appreciable effect on the rate of reaction if the radius of the drops exceeds 0.1 cm. However, due to the small contact area between the coke and the slag droplets[265d]) and the convection in the rolling droplets, the actual value of this limit should be rather higher.

If we assume that the droplets have a diameter of about 0.2 cm, that the contact area is 25% of the surface area of the droplets, and that diffusion controls the reaction, then, if the duration of contact is 3 minutes, the degree of reduction of the iron content in the slag will be 50%. A degree of reduction of this order was found in model experiments[265g]) in which the slag trickled over a column of coke. As shown in Fig. 113a, the slag was admitted as a solid powder via a sample balance into an iron crucible in which it was melted, and it then trickled through a coke charge heated in a second furnace and with nitrogen circulating through it. Figure 113b shows the experimental results obtained by analysis. Slags with 70% FeO and 15% each of CaO and SiO_2, and also a fayalite slag, again containing 70% FeO, were used. The temperature, height of the coke column, and the velocity of the slag (relative to the empty tube) were varied. The rates of flow of the slag chosen here correspond approximately to those in a blast-furnace hearth. The experimental results show that slags which are sufficiently basic are already almost completely reduced after trickling through a coke column only 25 cm high. The same result can therefore be assumed for the progress of this reaction in a blast furnace.

The reduction of iron-oxide slags by iron-carbon melts, as it occurs at the boundary between the metal bath and slag bath in the hearth, has been investigated by various Russian authors[265h]). They state that the rate of reduction is determined by the phase-boundary reaction at temperatures below 1200°C and by diffusion at temperatures above 1260°C. At the metal-slag phase-boundary potential differences are set up which correspond roughly to the equilibrium values calculated thermodynamically.

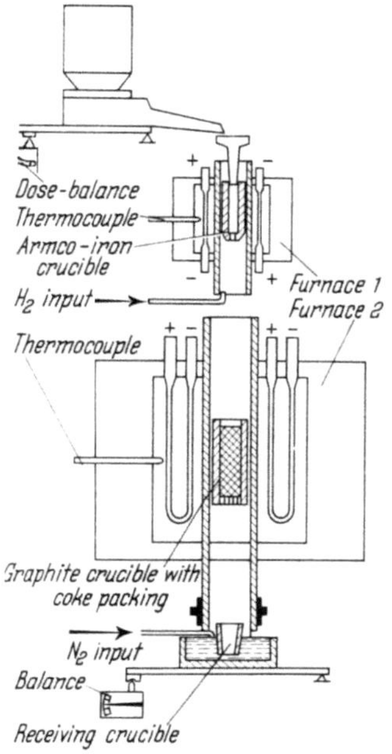

Fig. 113a. Test equipment

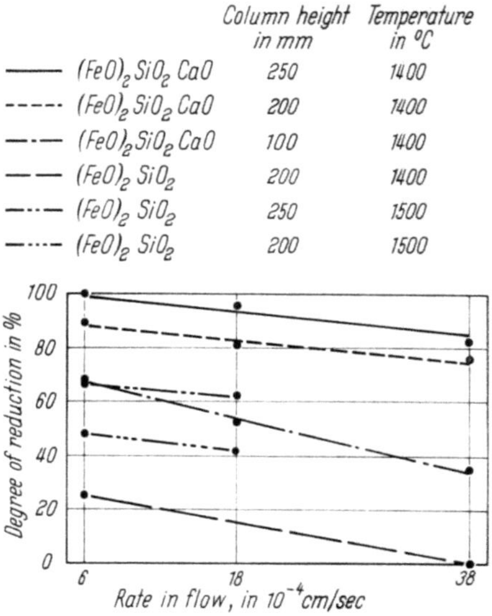

Fig. 113b. Relation between degree of reduction and rate of application of a molten, FeO-bearing slag to a coke column. Grain size: 12 to 15 mm

Dancy[265i]) investigated the reaction between small amounts of molten wüstite and magnetite with large amounts of pig iron containing carbon. He obtained about 40 kcal/mole for the activation energy. In these experiments, the decrease in the reaction rate as the reaction proceeds, which corresponds formally to a first-order reaction, may be attributable to the decrease in the amount of slag and reaction surface.

Philbrook and Kirkbride[265k]) allowed slags with about 5% FeO, 37% SiO_2, 14% Al_2O_3, and 44% CaO to react with carbon-saturated iron in a graphite crucible. They found a second-order reaction with a rate constant of

$$0.02 \frac{\%\,\text{FeO/min}}{(\%\,\text{FeO})^2}$$

at 1460°C.

The reduction of silicon from silicate slags by iron-carbon melts was investigated by Turkdogan et al.[265l]). The slags contained 34 to 64% SiO_2, the remainder being CaO or CaO + BaO. The overall reaction proved to be controlled by the phase-boundary reaction, the rate at 1600°C being about 10^{-3} mole/cm² h. From this it can be calculated that the amount of silicon taken up by the pig iron in dripping through the slag bath (diameter of drops ≥ 0.2 cm, time of fall ≤ 3 sec) and also by exchange between the slag bath and the hot-metal bath between two tappings is insignificant even at such an extremely high temperature.

Ward et al.[265 m)] were able to show that the overall process of reduction of silicon from a slag by carbon-saturated iron takes place as two electrochemically coupled sub-processes:

$$2\,C + 2\,(O^{2-}) \rightarrow CO + 4e^-,$$

$$(Si^{4+}) + 4e^- \rightarrow [Si].$$

To do this they used a cell in which the evolution of CO took place at a carbon rod dipping into the slag, the silicon exchange occurring at the surface of the metal. The current flowing between metal and carbon had the value calculated from Faraday's laws.

The reduction of silicon from solid SiO_2 by saturated iron-carbon melts was studied by Sharma and Ward[265 n)]. Their results can be expressed by

$$\frac{d[\%\ Si]}{dt} = k[\%\ Si]^{-1/2}.$$

The authors conclude from this that the diffusion of oxygen into the metal melt and out again from the melt to the SiO_2 surface is rate-determining. It is not easy to understand, however, what would induce the oxygen to take such a path. The given relation between the reaction rate and the silicon content of the metal would also follow if the reaction between the dissolved carbon and the oxygen in the SiO_2 is rate-determining. Assuming that the silicon has the same activity in the metal melt and at the SiO_2 surface, and also that

$$[a^2(O) \cdot a(Si)] = \text{const.} \quad \text{in } SiO_2,$$

and

$$a(Si) \sim [\%\ Si] \quad \text{in Fe},$$

it follows that

$$\frac{d[\%\ Si]}{dt} = k''\,a(O) = k'/a^{1/2}(Si) = k[\%\ Si]^{-1/2}.$$

The experimental results are therefore not incompatible with the process's being controlled by the phase-boundary reaction.

Turkdogan and Grieveson[265o)] determined the rate of reduction of trivalent to divalent iron in pure molten iron oxide by a CO/CO_2 mixture. At 1550°C the reaction is controlled by the diffusion. If we apply this result to the reduction of divalent iron in blast-furnace slag to metallic iron, we arrive at the proposition that the reduction by gas of the slag droplets rolling over the coke must preponderate over the reduction by the solid carbon, since the area of a droplet which is exposed to the gas is greater than the slag-coke phase-boundary. This conclusion is to be doubted, however, because TURKDOGAN carried out his measurements on melts having a layer thickness of several centimetres, that is, with a long diffusion path. Also, the phase-boundary reaction for the reduction of divalent iron to metallic iron from a slag must certainly take place more slowly, because the corresponding oxygen potential is smaller than that for the reduction of trivalent to divalent iron, while the diffusion coefficient in both cases amounts to about 10^{-5} cm^2/sec. At the present time, therefore, it is not possible to decide whether the reduction of the slag droplets as they trickle down takes place predominantly by reaction with the coke or by reaction with CO.

2.9. Reaction Kinetics of the Gasification of Coal and Coke

In many reduction processes the gas needed as the reducing medium is manufactured by the conversion of coal or coke by oxygen to carbon monoxide either in the reduction plant itself or in a separate reactor connected in series. In the combustion of coal both carbon monoxide and carbon dioxide may be formed:

$$2\,C + O_2 \rightarrow 2\,CO, \tag{1}$$

$$C + O_2 \rightarrow CO_2. \tag{2}$$

The carbon dioxide formed during the reduction process may react again with the carbon, so that carbon monoxide is regenerated. This reaction and the reverse reaction, the separation of carbon from carbon monoxide,

$$CO_2 + C \rightleftharpoons 2\,CO, \tag{3}$$

are known as the *Boudouard reaction*[266]).

In the manufacture of producer gas (or water-gas) steam reacts with carbon at very high temperatures to form carbon monoxide and hydrogen:

$$H_2O + C \rightarrow H_2 + CO. \tag{4}$$

A detailed survey of investigations into the kinetics of these reactions has been given by ILSCHNER[267]) and so we shall not go into details here.

The primary reaction between carbon and oxygen at the surface of the solid at temperatures between 900 and 2000°C leads almost exclusively to the formation of CO[268, 269]). The results of other authors[270]) who give a higher CO_2-content in the reaction product are to be attributed to the after-burning of the CO in the excess oxygen, as may be seen from the graph, in Fig. 114[271]), of the dependence of the partial pressure of the CO in the exhaust gas on the duration of contact between the gas and the hot coal surface. BAETKE[272]) is led to similar conclusions.

As MEYER[273]) established, the reaction order n relative to the oxygen is equal to 1 at temperatures of above 1000—1100°C. This is possible only if the collision yield of the oxygen is independent of the oxygen

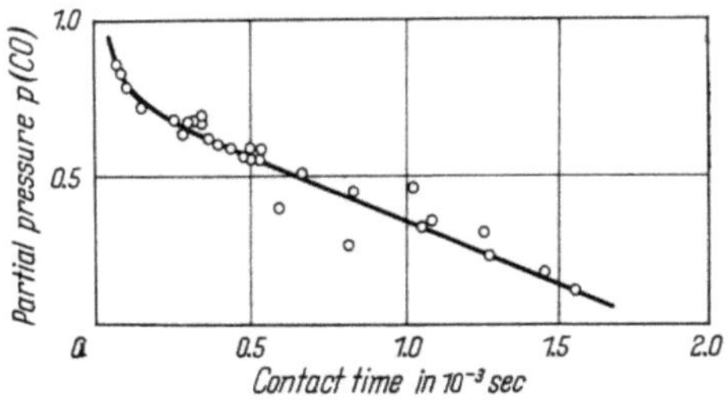

Fig. 114. Partial pressure of CO at the outlet of a coal channel as a function of the contact time, according to TSUKHANOVA[271])

pressure. LETORT[274]) was able to confirm this fact independently. According to other authors[275]) n is said to be equal to $\frac{3}{4}$ or even, for the proper reaction free from the effect of pores, $n = \frac{1}{2}$.

According to WICKE and HEDDEN[276]) the temperature increment of the phase-boundary reaction amounts to 57 ± 3 kcal/mole at temperatures below 1200°C, but is only 36.7 kcal/mole at temperatures above 700°C according to GULBRANSEN and ANDREW[277]). Above 1100 to 1200°C the reaction rate is independent of the temperature[275, 276]).

As already mentioned in section 1.2.5 in discussing the kinetics of the removal of oxygen, the behaviour of the solid partner in the reaction has to be included in considering the order of the reaction. In the oxidation of graphite, in particular, HENNING[278]) has shown that the disintegration of the graphite lattice takes place only where there are *steps and defects in the lattice*. On the other hand, dissolution at active positions and two-dimensional nuclear gaps in the hexagonal basis surface can occur, according to HENNING, by the action of *catalysts* such as metal particles, for instance.

According to DUVAL[269]) the surface centres of the carbon which are capable of reacting are created by the action of the gas, and are annihilated again by surface diffusion of the carbon atoms. Since the gas collision number depends only weakly on temperature, but the surface diffusion depends strongly on temperature, this hypothesis explains the sharp decrease with temperature, above 1100°C, of the collision yield of the gas, i.e., the number of reacting molecules relative to the number of incident molecules. The order of magnitude of the collision-yield is between 10^{-5} and 10^{-3}, and Fig. 115 shows the effect of temperature.

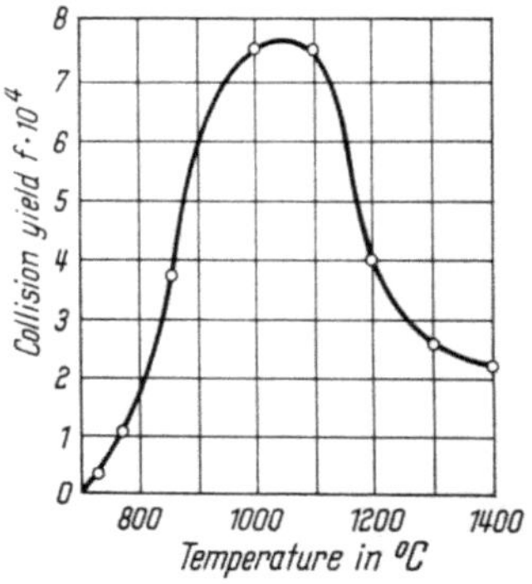

Fig. 115. Collision-yield f of the reaction between O_2 and C as a function of temperature, according to LETORT[274])

Being a heterogeneous reaction between gas and solid, the gasification of coal takes place as a combination of the sub-processes: adsorption or chemisorption of oxygen, surface reaction, and desorption of carbon monoxide. It is usually assumed nowadays that the chemisorption and splitting of the O_2-molecule or the surface reaction is the rate-determining sub-process.

Figure 116 shows a molecular model for the course of the reaction, according to ROSSBERG[279]).

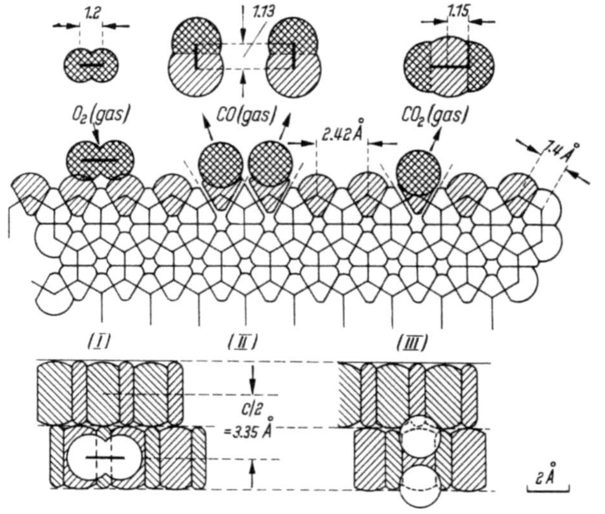

Fig. 116. ROSSBERG's model for representing the combustion process based on a graphite lattice in STUART-BRIEGLEB's spherical representation

Top: model for the gases, O_2, CO, and CO_2, present during combustion

Middle: view of the lattice model

Bottom: elevation

Most authors[280, 281] consider that oxidation of carbon by CO_2 takes place according to the following scheme:

$$CO_2 \rightleftharpoons CO + O_{ads}, \tag{4a}$$

$$O_{ads} + C \rightarrow CO. \tag{4b}$$

The reaction of adsorbed oxygen with the carbon of the lattice and the subsequent desorption of the CO-molecules formed are rate-determining up to about 1200°C, while, on the other hand, the reaction (4a) is to be regarded as an established equilibrium. This mechanism for the reaction has been confirmed by GRABKE's more recent investigations using $^{14}CO_2$ [283], by which the hypothesis[282] that the adsorption or dissociation of the CO_2 is rate-determining has been contradicted.

The rate of oxidation is proportional to the activity of the oxygen which gets into the surface. Under the usual conditions of the reaction this activity is determined by the equilibrium (4a) and is therefore proportional to the partial pressure of the oxygen in the gas phase.

$$v = k' a(O) = k\, p(CO_2)/p(CO).$$

According to HEDDEN[285] the temperature increment of the rate of the total reaction is 85 kcal provided there is no inhibition of gas diffusion. This value is largely independent of the type of coal used.

The sub-reaction (4a), the dissociation of the CO_2, has, according to GRABKE[283], a lower effective activation energy than the sub-reaction (4b), and it will therefore be rate-determining at very high temperatures. Consequently, at temperatures above 1200°C, it has also been found that the overall reaction is of first order relative to the carbon dioxide. The structure of the surface of the solid is found to have a similar effect to that in the oxidation of coal by oxygen. The collision-yield has a maximum value of $2.5 \cdot 10^{-7}$ at about 1280°C [286]. Compared with the maximum value, $7.5 \cdot 10^{-4}$, of the collision-yield in the O_2-C reaction, this value is extremely small; gasification of coal by CO_2 is altogether 10 to 10^3 times slower than gasification by oxygen. The existence of a maximum collision-yield is also to be attributed in this case to the fact that the processes which must occur to create and annihilate the reaction centres on the surface of the solid depend on temperature in different ways[287].

The reaction is severely inhibited[284] by the addition of carbon monoxide to the gas, and by the carbon monoxide formed by the reaction; this might be due to the blocking of the reaction centres by chemi-sorption of the carbon monoxide. It is more probable and consistent with the foregoing representation of the progress of the reaction to interpret this inhibition[280, 281, 283] as being due to the reverse reaction in (4a), which reduces the oxygen activity on the surface of the carbon and thus reduces the reaction rate of the sub-reaction (4b). Consequently hydrogen also inhibits oxidation of carbon by carbon dioxide[289]) because it affects the oxygen activity.

In the reaction of steam with carbon to form carbon monoxide in accordance with formula (4), the stages in the reaction occur similarly to those in gasification of coal by carbon dioxide.

The reaction rate for this reaction and also for the Boudouard reaction is determined by the phase-boundary reaction between the adsorbed oxygen and the surface carbon atoms or by the desorption of the CO according to (4b)[283]). The formation of adsorbed oxygen in accordance with

$$H_2O \rightleftharpoons H_2 + O_{ads} \tag{5}$$

is also to be regarded here as a rapid and established equilibrium. For the same partial pressures of the oxygen, therefore, the same rates of gasification of coal are obtained in CO_2/CO mixtures as in H_2O/H_2 mixtures[283]). The temperature increment for the overall reaction is said to be 65 to 75 kcal/mole[276, 281, 286, 287]). A detailed analysis[281]) of the course of the reaction shows that a uniform dependence of the reaction on temperature is not to be expected.

The reaction is strongly inhibited by hydrogen but not by carbon monoxide[289]). As shown in Fig. 117[286]) the collision-yield has a maximum at about 1400 °C and attains at this temperature nearly the same value as the

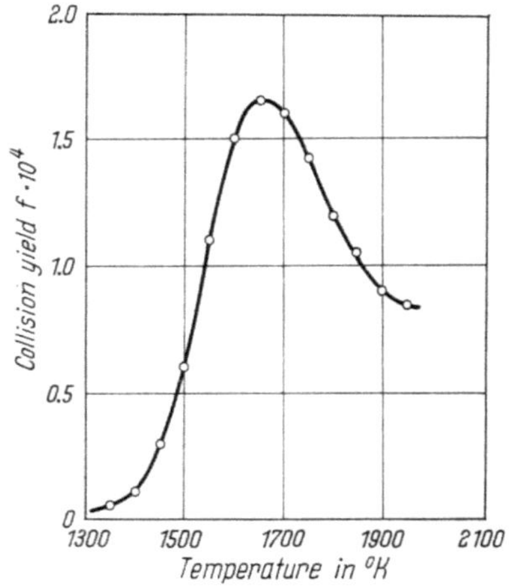

Fig. 117. Collision yield f in the gasification of coal by water vapour with $p(H_2) = 3 \cdot 10^{-3}$ Torr, according to BOULANGIER et al.[286])

collision-yield in the combustion of coal by oxygen at about 1050 °C. The gasification of coal by steam in accordance with (4) is initiated, according to some authors[276, 281]), by the chemi-sorption of the steam in the form of H and OH, both adsorbates being bound, it is said, by a single valency to C-atoms on the boundary.

All reactions in the gasification of coal are greatly accelerated by special activating additives such as metals, oxides, and carbonates. A list of the older references is given by KRÖGER[289]). KOHL and MARINČEK[138a]) have shown that this catalytic effect is also of importance in the reduction of wüstite by solid carbon, cf. section 1.2.8. The processes can also be just as strongly inhibited by certain *inhibitors*. $POCl_3$, $PbCl_3$, $PbBr_3$, and organic halogenides are particularly effective inhibitors. The mechanism of the action of inhibitors was investigated in detail by HEDDEN et al.[285]), and that of the catalysts by LONG and SYKES[290]).

The data so far given relate to the course of the phase-boundary reaction. As in the case of ores, so with coal too this process is linked up with a preliminary and a subsequent gas diffusion process; in metallurgical processes with forced draught of the gas, for example, in a blast furnace operation, the material transport across the boundary layer is of lesser importance, and the gas diffusion in the pores of the reacting solid is of more importance. Both transport processes may be discussed in the same way as for ores, and reference may therefore be made to sections 1.2.2, 2.1, and 2.4. The special features to be taken into account in the gasification of porous coal are presented by WICKE[291]), DAHME and JUNKER[292]), PETERS[293]) and HEDDEN[294]). Just as in the case of porous ores, the progress of the reaction is determined at low temperatures by the phase-boundary reaction, and by porous diffusion at high temperatures, as might be deduced from the difference in the dependence on temperature in the two temperature ranges, see Fig. 118[295]). This diagram should be compared with the corresponding one, Fig. 68, for the reduction of ores.

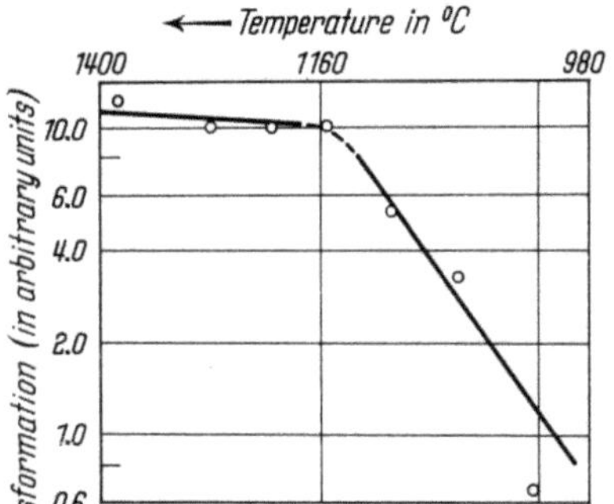

Fig. 118. Transition from the temperature-dependent region to the temperature-independent region in the gasification of coal by steam, according to HUNT et al.[295])

The decomposition of carbon monoxide, with the resulting deposition of carbon on iron ores and iron, i.e., the reverse Boudouard reaction (3), is of great significance in the reduction of ores because of its effect on the temperature distribution and the chan ges in gas flowcaused in a blast furnace. Figure 119 shows the pressure-temperature region in which, according to measurements by BAUKLOH and HENKE[296]) and by SŌMA[297]), decomposition of carbon monoxide takes place at iron oxides or at the iron formed from them by reduction. The more gradual increase of the critical partial pressure of the CO with increasing temperature given by

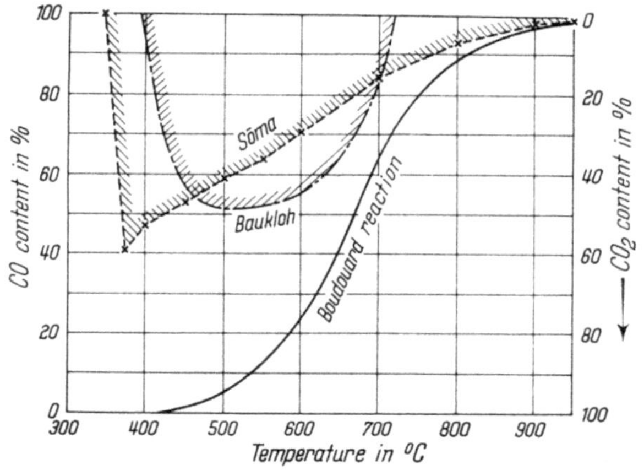

Fig. 119. The region of appreciable decomposition of CO, according to BAUKLOH[296]) and SŌMA[297])

SŌMA corresponds to SCHENCK and MASCHLANKA's data[298]). Figure 120 shows the course of the reaction with time for reduced sinters according to SŌMA's measurements. It can be seen that the reaction sets in only after a clearly marked incubation time, which, in general, is greater than the time the blast-furnace gas remains in the critical-temperature region in the blast furnace.

The decomposition of carbon monoxide is accelerated by hydrogen and the critical-temperature region is extended to higher temperatures[299, 300]).

The decomposition of carbon monoxide in the homogeneous gas phase or at the surface of carbon takes place so slowly that it is of no real importance for the reduction of ores. Of all the components of the blast-furnace burden the iron forming from the ore is the most effective as a nuclear surface or catalyst for carbon deposition, from carbon monoxide, as MEYER[301]) recognized. The lower temperature limit of the carbon deposition region thus depends perhaps on the temperature limit for perceptible reduction, and would then vary from ore to ore.

A similar explanation could be given for the incubation time shown in Fig. 120. This view is supported by measurements[302]) which enable the catalytic activity for the decomposition of CO and the rate of reduction to be compared in the same ores. Similarly a preliminary reduction of the ore increases the activity for CO-decomposition[298]).

As inhibitors of CO-decomposition H_2S and SO_2 are particularly effective, their action being irreversible, i.e., it still continues if the contact

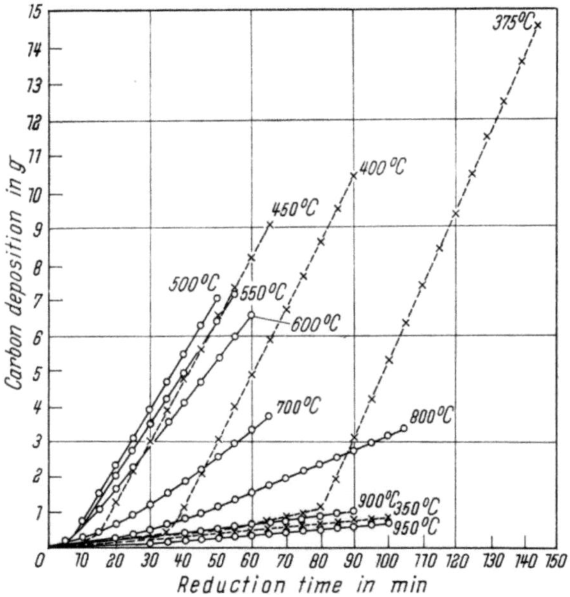

Fig. 120. Carbon deposition during the reduction of self-fluxing sinters, according to SŌMA[297])

50 g sinter, 2.5 to 5 mm; 1 litre CO/min; tube diameter: 41 mm

poison is removed from the gas phase[298]). NH_3, Cl_2, NO_2, and CN inhibit the reaction reversibly[296, 298]).

The deposition of carbon in the pores of the ore exerts an explosive action on the pieces, causing disintegration and a marked decrease in burden permeability[303]). Pressures up to 2000 kg/cm² have been demonstrated and can also be calculated theoretically[304]).

2.10. The Water-Gas Reaction

The different rates of reduction of iron ore by CO and by H_2, and the marked effect which even slight amounts of H_2 contained in a CO/CO_2 mixture have on the reduction rate, were already in 1928 being interpreted by WÜST[304a]) as due to the fact that the hydrogen is the actual reducing component in such gas mixtures. The CO was said to serve mainly to reduce the resultant steam back to hydrogen:

$$H_2 + FeO \rightarrow H_2O + Fe, \tag{1}$$

$$H_2O + CO \rightarrow H_2 + CO_2. \tag{2}$$

The second sub-process of this reaction is known as the *water-gas reaction*. It is well-known that this process needs a catalyst. In a blast furnace all the constituents of the burden come under consideration as catalysts; particularly active is solid iron[304b]), and next the iron oxides.

Later, other authors agreed with this view[304c, d]). Because of the attention which hydrogen-containing additives to the blast or to the reducing medium have received in the last decades, conjectures and investigations on this topic have been published time and again. An estimate of the contribution which the water-gas reaction makes to the reduction has, however, only become possible from the considerations and measurements of recent years[304d, e, f]). These articles agree in presenting the following picture of the reaction with Fe_3O_4, FeO, and Fe as catalysts.

In the catalytic water-gas reaction the oxygen adsorbed on the catalyst plays the role of an intermediate product; it is generated in one of the two stages of the reaction, viz.,

$$H_2O \rightarrow H_2 + O_{ads}, \tag{2a}$$

and is used up again in the second stage

$$CO + O_{ads} \rightarrow CO_2. \tag{2b}$$

The slower, and therefore the rate-determining, stage of the reaction sequence is the transfer of the oxygen from the catalyst to the CO.

Because of the equilibrium set up between the interior of the crystals and the adsorption layer, however, this reaction is identical, for oxides, with the phase-boundary reaction in the reduction by CO [103]); cf. section 1.2.5, and formula (39) in particular. This immediately implies that the water-gas reaction with the oxides cannot take place more quickly than their reduction by CO.

Since the rate constants for the exchange of oxygen between CO_2/CO mixtures and wüstite do not differ appreciably from those for the reaction of such mixtures with magnetite[103]), the simultaneous and adjacent appearance of both these phases does not alter anything in the above circumstances. On the other hand, if the picture of the reaction is correct, then, according to the experimental evidence[304b, g]), metallic iron must be an excellent water-gas catalyst. According to Grabke, at 800°C and with $p(CO_2)/p(CO) = 0.5$, the rate constant for oxygen exchange according to formula (2b) is $2.5 \cdot 10^{-9}$ for wüstite[103]), but is $1.3 \cdot 10^{-7}$ for iron[304h]) (in mole $sec^{-1} \cdot$

cm^{-2} atm^{-1}). Consequently the water-gas reaction can be expected to make an appreciable contribution to the reduction only in those regions of the blast furnace where metallic iron is already present.

The fact – established by probe tests – that the water-gas reaction fails to be set up in the upper part of the shaft, that it "freezes up", is to be attributed more to the lack of iron than to the lower temperatures.

The process of reduction of iron ores in CO/CO_2 mixtures containing H_2 is therefore to be understood, when metallic iron is present, as a reaction sequence: sub-reaction (1), the reduction proper, takes place at the oxides; sub-reaction (2), the regeneration of the hydrogen by the water-gas reaction, takes place at the iron.

The spatial separation of the two sub-reactions requires their connection by a transport process, which must take place as a gaseous diffusion or surface diffusion by one of the participants in the reaction. The optimal conditions occur at the three-phase boundary metal/oxide/gas.

3. Gas Flow and Heat Transfer in Granular Materials

There are numerous possible technological processes for carrying out the reduction of ores by gases containing CO or H_2. All conceivable processes are characterized by the common endeavour to achieve in the smallest possible volume the most rapid and complete chemical conversion and heat transfer possible. A uniform circulation of the reducing gases round the ore pieces offers the best conditions. The following sections deal with gas flow in ore charges, with the region of stability for various types of gas flow, and with heat transfer. According to WICKE and BRÖTZ[305]), in general, the possible states of a granular material through which a fluid is flowing are those shown in Table 9; from left to right in this table, the states shown correspond to increasing fluid velocity or decreasing grain size. Those of particular importance in the reduction of ores are:

a) the static bed, or packed column;
b) the fluidized bed;
c) particulate fluidization (or fluidized dust).

3.1. The Static Bed

3.1.1. Gas Flow in a Static Packed Column

If $\dot{V}$ is the volume of gas flowing per unit time through a static packed column of ore, the gas velocity given by

$$u = \dot{V}/F,$$

where

F = cross-sectional area of the column,

is called the *empty-tube velocity*.

Of course, part of the cross-sectional area F is blocked to the flow by the charge present, and thus the actual velocity of gas flow between the grains is considerably greater than u.

$\dot{V}$, or u, cannot be increased arbitrarily, because, firstly, the pressure drop, and consequently the force of compression, would greatly increase, and secondly, the charge would become unstable, i.e., it would be fluidized, and at very high velocities would even be, at least in part, blown out of

Table 9. *Synopsis of states of a granular material in a fluid stream*

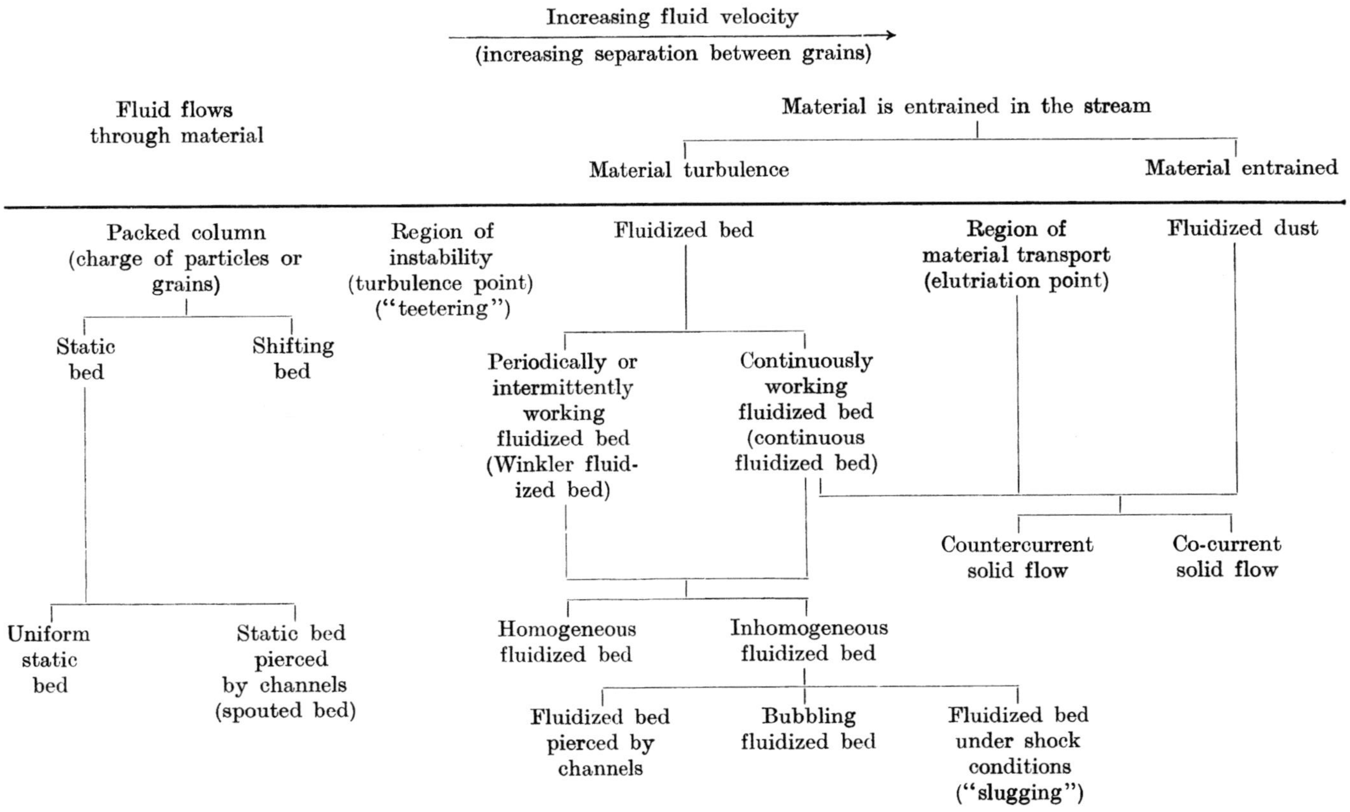

the containing vessel. Not infrequently the maximum rate at which gas can be supplied limits the throughput capacity of this reduction process, and this is therefore discussed in detail later.

The case most easily investigated mathematically is where all the grains of the charge have the same diameter (equi-granular charges). It is true that this case occurs only seldom and is at best only approximately realized in industrial processes; however, it represents a simplification from which it is possible to pass comparatively easily to the multi-granular charges usually encountered industrially.

3.1.1.1. Equi-Granular Charges

At low velocities a laminar state of flow of the gas is set up between the grains; at higher velocities a gradual change takes place into a turbulent regime, in which the particles of the charge still remain at rest. The first systematic work on the resistance to fluid flow through ore charges was published as early as 1929 by FURNAS[306]), but CARMAN[307]) was the first to succeed in finding a mathematical formulation which handles the connection between the flow resistance and the properties of the flowing medium and those of the packed charge through which the flow takes place. This treatment is quite general and deals both with the laminar regime, investigated earlier by KOZENY[308]), and with the industrially more important turbulent regime. This formulation has later been refined[309–312]). The use of dimensionless parameters, which is customary in fluid dynamics, proves useful here too. The resistance parameter ψ is given by

$$\psi = \Delta p\, \varepsilon^3\, d\, \Phi / H\, \varrho\, u^2 (1 - \varepsilon), \qquad (1)$$

and the Reynolds number by

$$N'_{\text{Re}} = \varrho\, u\, d\, \Phi / \eta$$

or in a different form by [309])

$$N_{\text{Re}} = N'_{\text{Re}} / (1 - \varepsilon) = \varrho\, u\, d\, \Phi / \eta (1 - \varepsilon), \qquad (2)$$

where

d = grain diameter (all data are in cgs units),
Δp = pressure drop over the height of the charge,
ε = void ratio (relative volume of grain interstices in the charge),
H = height of charge,
ϱ = density of the flowing medium,
η = viscosity of this fluid,
u = stream velocity, relative to the empty cross-section at the working temperature and pressure, i.e., the empty-tube velocity,
Φ = form factor:
$\Phi = 1.0$ for spheres, $0.6 < \Phi < 1$ for bodies of different geometrical shape.

Plotting ψ against N_{Re} from experimental data gives an approximately universal curve which holds for both gases and liquids as the flowing medium, provided the grain sizes are almost uniform. According to more recent investigations by BRAUER[310]) the curve shown in Fig. 121 is best described by the formula

$$\psi = 160/N_{\mathrm{Re}} + 3.1 N_{\mathrm{Re}}^{-0.1}, \tag{3}$$

which is often called the Carman-Kozeny equation. For small Reynolds numbers ($N_{\mathrm{Re}} < 10$, i.e., for laminar flow) the first term predominates,

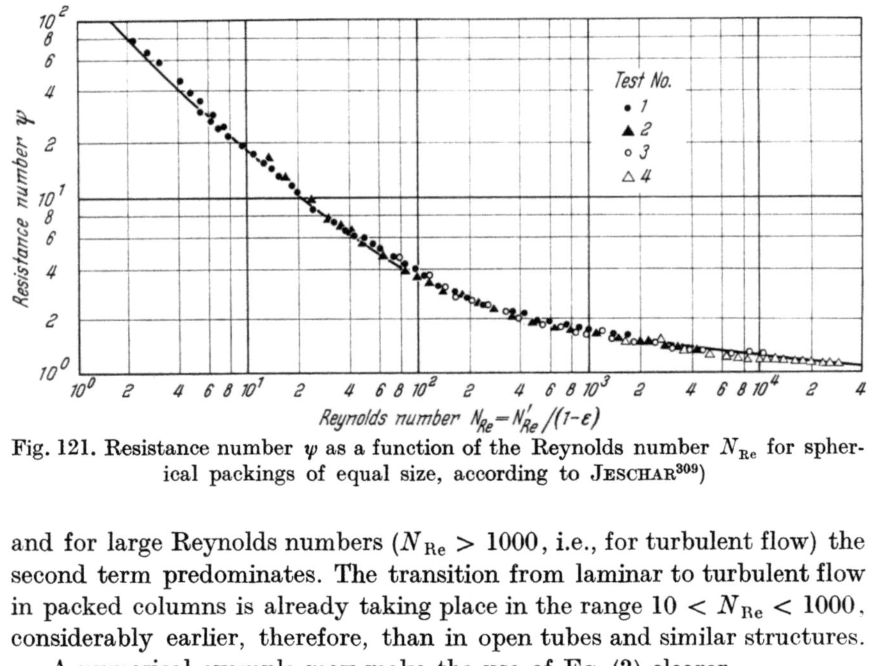

Fig. 121. Resistance number ψ as a function of the Reynolds number N_{Re} for spherical packings of equal size, according to JESCHAR[309])

and for large Reynolds numbers ($N_{\mathrm{Re}} > 1000$, i.e., for turbulent flow) the second term predominates. The transition from laminar to turbulent flow in packed columns is already taking place in the range $10 < N_{\mathrm{Re}} < 1000$, considerably earlier, therefore, than in open tubes and similar structures.

A numerical example may make the use of Eq. (3) clearer.

In a cylindrical container ($D = 100$ cm, $H = 500$ cm) there are spherical ore pellets with $d = 2$ cm, $\Phi = 1$, $\varepsilon = 0.37$. [This is the approximate value for ε when the diameter of the grains is very small compared with the diameter of the container, $d \ll D$; see Fig. 122. We may remark that for spherical packings of grains of equal size in large containers, the voids ratio ε lies between 0.36 and 0.40, corresponding to the tightest and loosest packings. But in spite of this comparatively narrow range of variation of ε, $\psi \sim \varepsilon^3/(1 - \varepsilon)$ is sensitive to changes in ε, and so it is generally necessary to determine ε experimentally.]

To reduce the pellets, hydrogen is passed through them at a mean temperature of 600°C and at normal pressure, at the rate 2500 m³/h at NTP.

Hence the empty-tube velocity is

$$u = \frac{2500 \cdot 10^6}{3600} \frac{873}{273} \frac{1}{\pi \cdot 50^2} \frac{\text{cm}}{\text{sec}} = 283 \frac{\text{cm}}{\text{sec}}.$$

At 600°C the density of H_2 is

$$\varrho = 0.028 \cdot 10^{-3}\ \text{g/cm}^3$$

and its viscosity[313]) is

$$\eta = 1.83 \cdot 10^{-4}\ \text{g cm}^{-1}\ \text{sec}^{-1},$$

whence, by Eq. (2),

$$N_{\text{Re}} = \frac{\varrho\, u\, d\, \Phi}{\eta(1-\varepsilon)} = 138,$$

from Eq. (3)

$$\psi = 3055$$

and from Eq. (1)

$$\Delta p = 21\,400\ \text{g cm}^{-1}\ \text{sec}^{-2} = 218\ \text{mm water gauge.}$$

The pressure drop in the column is therefore comparatively small.

With the numerical values given in this example the state of flow is already partly in the turbulent regime. In the laminar regime Δp increases linearly as the velocity u increases, but approximately quadratically in the turbulent regime [accurately, from Eqs. (1), (2), (3), we have $\Delta p \sim u^{1.9}$ in the turbulent regime].

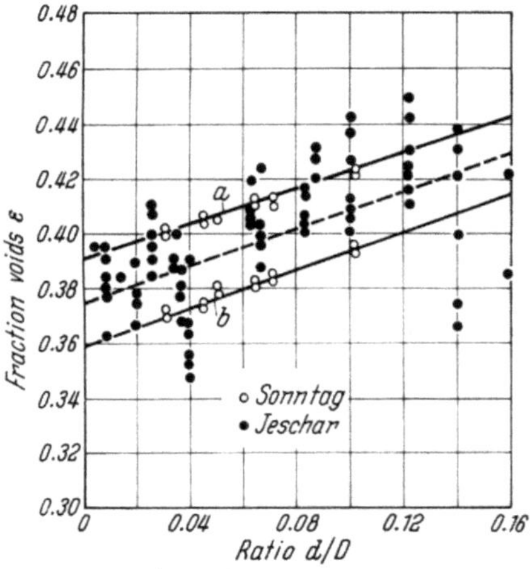

Fig. 122. Mean fraction voids ε as a function of the ratio d/D, according to JESCHAR[309])

a Loosest packing, *b* Tightest packing, according to SONNTAG

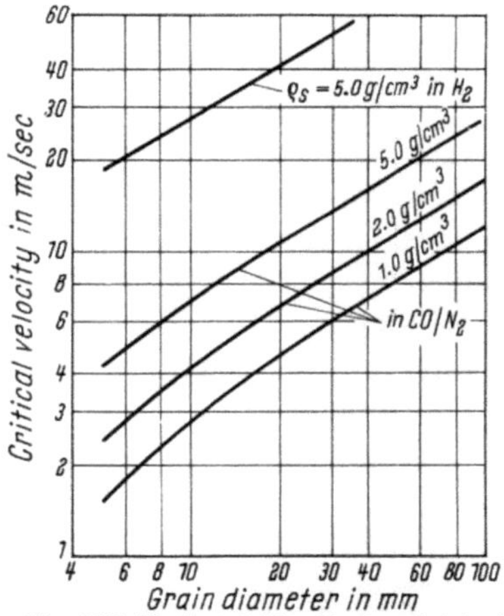

Fig. 123. Bed-stability limits. Critical velocity (relative to empty tube), or turbulence point, as a function of particle diameter

800° C, 1 atm, equivalent spherical packing $\varepsilon = 0.4$

ϱ_s = weight per unit volume of the packed particles

The Stability Limit. As the gas velocity increases still further, the bed of solid particles expands upwards to some extent, the dynamic pressure Δp increases roughly quadratically and finally becomes equal to the weight per unit area of the charge column, i.e.,

$$\Delta p = H(1 - \varepsilon)\,\varrho_s\, g\,, \tag{4}$$

where

ϱ_s = specific weight of the packed particles (volumetric weight),
g = 981 cm/sec², the acceleration of gravity.

At this point the static packed column loses its stability ("teeters") and with any further slight increase in the gas velocity it passes over into the state of a fluidized bed[314]). By combining Eqs. (1) to (4) we obtain, for the *turbulence regime*, the critical velocity u_{cr} at which fluidization sets in:

$$u_{cr} = \frac{0.54\,(\varrho\, g\, \varepsilon^3)^{0.526}}{[(1 - \varepsilon)\,\eta]^{0.0526}}\;\frac{(d\,\Phi)^{0.579}}{\varrho^{0.473}}\,. \tag{5}$$

In Fig. 123 the critical velocity u_{cr} has been plotted for some particular cases of Eq. (5). [It should be noted that the bending of the lines in the lower region of this diagram is due to the laminar terms in Eq. (3) becoming effective; Eq. (5) then needs to be modified.] We see from Fig. 123 that very much higher gas velocities are needed than are customary in most technological processes in order to exceed the stability limit for a packed column and to produce the state of a fluidized bed. [In a blast furnace, the gas velocity has a maximum value of about 4 m/sec (empty-tube velocity) and the Reynolds number is in the range 1000 to 3000 [315]), i.e., the turbulent regime always prevails.] However, this general conclusion is valid only for a uniform flow through a packed column and only when all the grains have the same diameter (equi-granular packings). As soon as there is a deviation from these ideal conditions, essentially less favourable values of the stability limit obtain.

Equation (5) also enables an estimate to be made of the effect of an increase of pressure on the possible throughput of gas. From the relevant quantities in Eq. (5), the gas density varies in accordance with $\varrho \sim p$. Since $u \sim 1/\varrho^{0.473}$, the critical gas velocity and the gas throughput vary as shown in the following table:

p atm	$u_{cr}/u_{cr,0}$	Volume of gas per unit time $\dot{V}/\dot{V}_0$
1	1	1
2	0.72	1.44
4	0.52	2.08

In this table $u_{cr,0}$ and $\dot{V}_0$ are the critical velocity and gas throughput per unit time (in m^3/time) at NTP for $p = 1$ atm. We see from this small table that, as the pressure is increased, the maximum gas velocity which can be used falls, so that, e.g., when the pressure is doubled, the gas throughput increases only by 44%.

3.1.1.2. Multi-Granular Charges

FURNAS[306]) drew attention to the fact that packings in which several different grain diameters simultaneously occur cannot be characterized by the mean value of the grain diameter.

As Fig. 124 shows, in such mixtures, the small grains have a tendency to concentrate in the gaps between the larger grains and thus to hinder the passage of the gas.

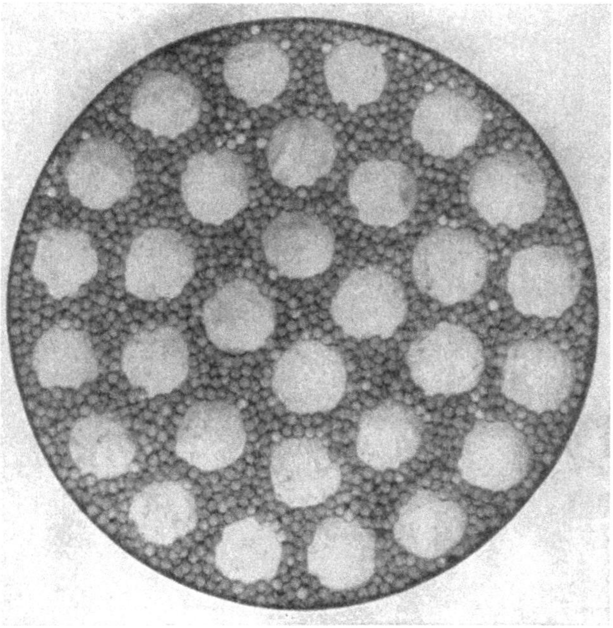

Fig. 124. Representation of a multi-granular packing of spheres with $d_1 = 80$ mm and $d_2 = 10$ mm, according to JESCHAR[309])

Figure 125 shows diagrammatically that ε_m, the voids fraction of the mixture, and consequently also the free cross-sectional area available for the gas flow, decrease all the more, the greater the difference between diameters in a mixture of two grain sizes. The minimum for ε and hence also for permeability occurs at about 60 to 70% large-grain and 40 to

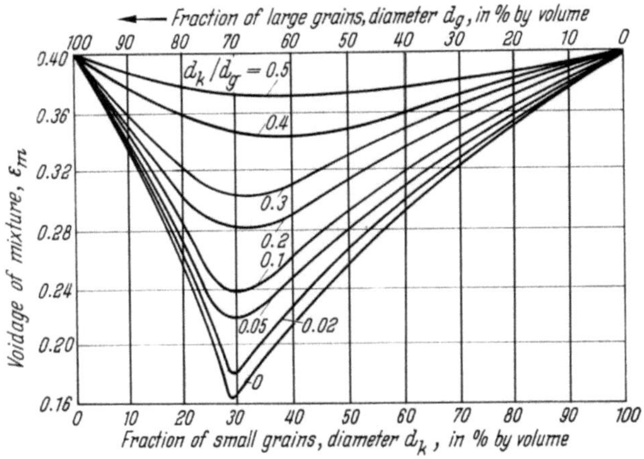

Fig. 125. Voidage in mixture of broken solids of two particle sizes, in relation to the composition, according to FURNAS[306]) and JESCHAR[309])

30% small-grain in the mixture. At this value the permeability is greatly reduced.

Measurements of the flow resistance of spherical packings of uniform diameter and of mixed diameters were carried out by JESCHAR[309]) using

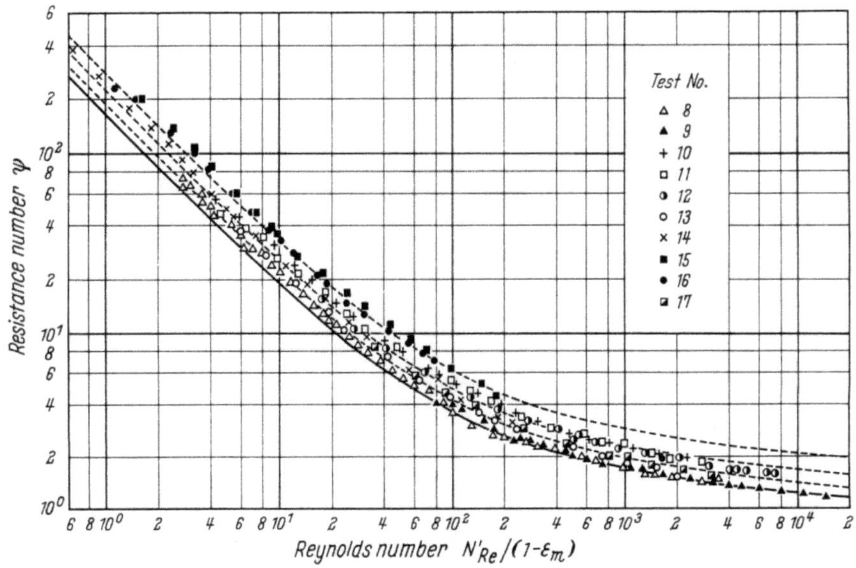

Fig. 126. Resistance number ψ as a function of the Reynolds number N_{Re} for mixed spherical packings of several diameters, according to JESCHAR[309])

the apparatus shown in Fig. 60. The results for packings of uniform size are shown in Fig. 121, and for mixtures of different sizes in Table 10 and Fig. 126.

Table 10. *Spherical packings of mixed diameters*

Test No.	Nominal diameter of spheres mm	Mean diameter of spheres mm	Volume ratio V_n/V	Equivalent diameter of spheres d mm	Height of packing H mm	Voids fraction of the mixture ε_m
8	10	9.9	0.50	13.2	753	0.362
	20	19.9	0.50			
9	30	33.6	0.51	46.9	1000	0.372
	80	80.4	0.49			
10	10	9.9	0.36	22.6	1000	0.263
	80	80.4	0.64			
11	10	9.9	0.26	28.2	993	0.236
	80	80.4	0.74			
12	20	19.9	0.31	41.5	1000	0.312
	80	80.4	0.69			
13	10	9.9	0.65	14.3	1000	0.319
	80	80.4	0.35			
14	2	2.1	0.65	3.2	493	0.313
	30	33.6	0.35			
15	2	2.1	0.28	6.6	500	0.202
	30	33.6	0.72			
16	2	2.1	0.31	6.5	500	0.195
	80	80.4	0.69			
17	10	9.9	0.24	20.9	1000	0.303
	20	19.9	0.26			
	30	33.6	0.24			
	80	80.4	0.26			

The lowest resistance parameters are obtained, as might be expected, with mixtures in which the difference between the diameters is small (tests 8 and 9). As the difference between the diameters increases, the resistance parameters increase considerably (tests 14 to 16). Here it should be borne in mind that the pressure drop in packings having a fine component increases even more rapidly than the resistance parameter, since ε is small. From the measured values JESCHAR was able to deduce that, for multi-granular fillings, the following law holds for the resistance[309]:

$$\psi = \left(\frac{160}{N_{\mathrm{Re}}} + \frac{3.1}{(N_{\mathrm{Re}})^{0.1}}\right)\left(\frac{\varepsilon_k}{\varepsilon_m}\right)^{0.75}, \tag{7}$$

where

ε_k = voids fraction for the component with the smallest grain diameter,
ε_m = voids fraction for the mixture.

Corresponding measurements with charges of non-spherical shape have been communicated by POLTHIER[318]). His results show that the laws which hold for spherical packings can be extended to packings of particles of irregular shape. In the expressions for the parameters ψ and N_{Re} the quantity $d \cdot \Phi$ is to be replaced by an equivalent grain diameter given by

$$d\,\Phi = \left[\sum_{i=1}^{n} (c_i/d_i\,\Phi_i)\right]^{-1},$$

where

d_i = diameter of sphere equal in volume to an irregular particle of type i,

$$\Phi_i = \frac{\text{surface area of sphere equal in volume to an irregular particle of type } i}{\text{surface area of an irregular particle of type } i},$$

c_i = volume component of the irregular particles of type i,

so that

$$\sum_{i=1}^{n} c_i = 1.$$

From Eq. (7) we have calculated the effect of the fine component ($d_k = 2$ mm) on the gas permeability of spherical packings with $d_g = 20$ mm and 40 mm. As may be seen from Fig. 127, with even the smallest amount

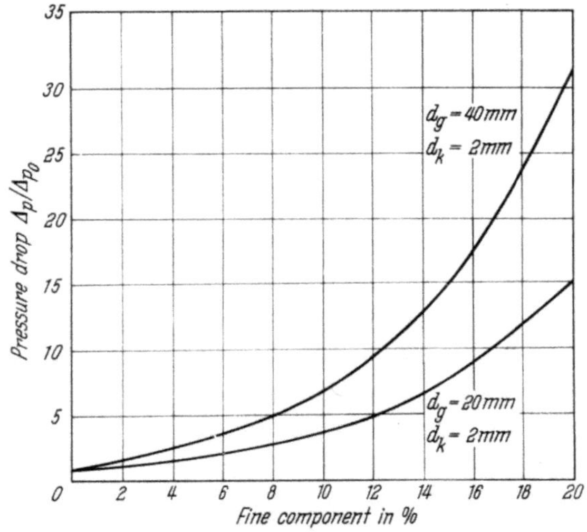

Fig. 127. Pressure drop in packings of spheres of two different sizes, calculated according to JESCHAR[309])

Diameter of furnace shaft: 7.0 m. Blowing-rate: 90000 m³/h at NTP. Temperature $\Theta = 1000$°C. Density and viscosity of medium flowing at temperature Θ: $\varrho = 0.265 \cdot 10^{-3}$ g/cm³, $\nu = 1.792$ cm²/sec. $\varepsilon_k = \varepsilon_g = 0.38$

of the small spheres, there is a marked increase in the necessary pressure difference Δp. With a 10% fine component, the pressure drop in the packing has to increase 3.5 to 4 times to produce the same gas throughput. Actually, during charging of industrial furnaces the fine components are hardly ever uniformly distributed, but become concentrated in particular regions. There are juxtaposed regions with very little fine material, which have a high gas permeability (they form flow channels), and regions rich in fine material, with a greatly reduced permeability (in some circumstances, reduced to a tenth). The gas flow through the charge becomes non-uniform.

This consideration illustrates very clearly the need to keep shaft-furnace charges as free from fine components as possible.

The Stability Limit. For equi-granular charges the stability limit is given by the beginning of turbulence, the formation of a fluidized bed, as shown in Fig. 123. For multi-granular charges, which have regions of different gas permeability, even at gas velocities whose mean value over the cross-section is low, local disturbances may arise through a phenomenon which has been called a *spouted bed*[315a]). It can be investigated using apparatus like that shown in Fig. 128.

In one example of an experiment using the apparatus shown in Fig. 128 the charge consists of grains of equal size. The gas inlet is concentrated over a small part of the cross-section of the packed column, and in the conical connecting section the gas flow can distribute itself over the whole cross-section. As the gas throughput increases, the pressure drop at first increases (Fig. 129), until, at the point *b* a flow channel begins to form through which, in preference, the gas flows; when the gas throughput is further increased, the channel becomes more pronounced and the pressure drop becomes smaller. Finally, at the point *e*, the originally static filling is in a state of intense, circulatory motion, as Fig. 128 indicates. However, the theory of this process is incomplete. For packed columns of comparatively small dimensions (container diameter up to 300 mm) Mathur and Gishler[316, 316a]) give the following relation from which the minimum empty-tube velocity u for the spouted-bed state can be calculated:

$$u = \frac{d}{D}\left(\frac{D_0}{D}\right)^{0.33}\left(\frac{2g\,H(\varrho_s - \varrho)}{\varrho}\right)^{0.5}, \tag{8}$$

where

D = diameter of container,
D_0 = diameter of the gas inlet tube,

and the other symbols have their earlier meanings.

Although there is so far no experimental proof, it is nevertheless conceivable that phenomena of this kind occur in blast furnaces and may even determine the limit of their operational capability. For, if flow channels have been formed because of a non-homogeneous distribution of the fine component of the burden, it must be assumed that when a certain limiting velocity is exceeded (and this may be far below that for the formation of a fluidized bed), the spouted-bed phenomenon will appear in the

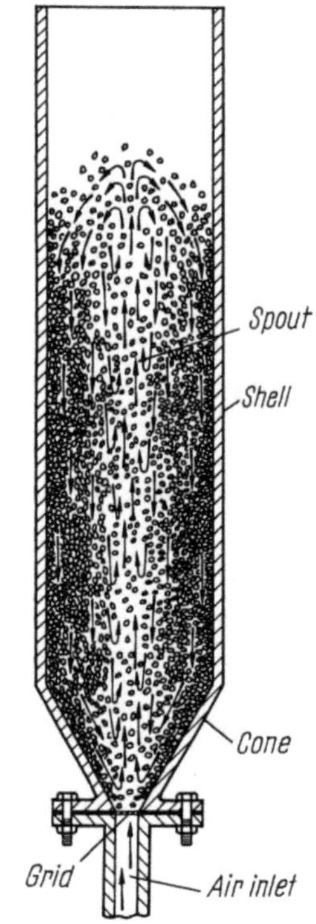

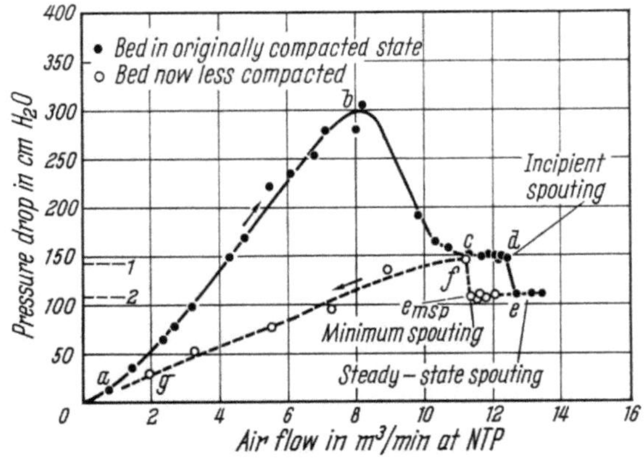

Fig. 129. Typical pressure-drop flow diagram for a spouted bed, according to LEVA[315a]). Note that stable spouting occurs over flow range $e - e_{msp}$. Data observed for 610 mm diameter column carrying 1790 mm of wheat, 45° cone angle, 102 mm orifice for air

1 Theoretical pressure drop, calculated on basis of 0.89/cm^3 bulk density, *2* Observed operating spouting-pressure drop

Fig. 128. Schematic diagram of a spouted bed, according to LEVA[315a])

flow channels. Once such regions become extensive, the behaviour of the burden must become irregular, i.e., the region of stability has been transgressed.

3.1.2. Heat Transfer in Static Beds

The heat transfer from flowing gases to static charges and *vice versa* is comparatively good, since the gas flow is split into many small channels in the interstices of the charge and there is already turbulence even at small values of the Reynolds number. In principle, the highest heat-transfer coefficients are to be expected if the gas and charge move in opposite directions and if the permeability of the charge to the gas flow is the same in all volume elements. The second condition is an idealization, since actually, in all technological processes, a non-uniform flow through the material occurs because of the separation of the differently sized fractions during charging, and often also because of the changes in the grains during chemical decomposition. In all technological processes in shaft furnaces, travelling grates, or similar plant, where static charges are being worked, it is therefore of the greatest importance to ensure uniform permeability and thereby optimal heat transfer and material transport.

The theoretical treatment of heat transfer in charges has already been given, in essentials, in section 2.2.

For the conditions which hold in blast furnaces, for instance, calculation shows that the heat-transfer coefficients are very large. HANSEN[315] has estimated the Reynolds numbers over the height of the blast furnace; in the upper part of the stack $N_{\mathrm{Re}} \approx 1000$, which implies, according to section 2.2, that $N_{\mathrm{Nu}} \approx 40$.

Taking

$$\lambda = 1.6 \cdot 10^{-4} \text{ cal/cm sec °C},$$

$$\varepsilon = 0.4,$$

$$d = 3 \text{ cm (spheres)},$$

we find

$$\alpha = (N_{\mathrm{Nu}}\, \lambda/d)\,(1 - \varepsilon)/\varepsilon = 3.2 \cdot 10^{-3} \text{ cal/cm}^2 \text{ sec °C}.$$

For the reduction to unit volume of the charge we need the total geometrical surface area F of all the grains in the volume V_R:

$$F/V_R = 6(1 - \varepsilon)/d = 1.2 \text{ cm}^{-1}.$$

We get for the heat-transfer coefficient per unit volume of the charge:

$$\alpha_V = \alpha\, F/V_R = 3.84 \cdot 10^{-3} \text{ cal/cm}^3 \text{ sec °C} = 13800 \text{ kcal/m}^3 \text{ h °C}.$$

The temperature gradient in the grains is, in general, comparatively small (cf. sections 2.2 and 4.6).

3.1.2.1. Effect on the Heat Transfer of Longitudinal and Transverse Intermixing of the Gas Flow in Packed Beds

Because of the heat transfer from the flowing gas to the charge material, the gas is cooled along its flow path. Hence a longitudinal and transverse mixing is set up which tends to equalize the temperature in the flowing gas and which reduces the heat transfer from the value given by the simplified model. The longitudinal and transverse mixing can be treated as a heat-transfer process superimposed on the gas flow.

In the steady state the temperature of the gas flowing uniformly through a cylindrical charge reactor is governed by the equation

$$\alpha \frac{f}{c\,\varrho\,\varepsilon}(\Theta - \Theta_a) = -\frac{u}{\varepsilon}\frac{d\Theta}{dx} + M_r\left(\frac{d^2\Theta}{dr^2} + \frac{1}{r}\frac{d\Theta}{dr}\right) + M_a\frac{d^2\Theta}{dx^2},$$

where

M_r, M_a = effective thermal diffusivities in the radial and axial directions respectively,
α = heat-transfer coefficient,
f = surface area of charge material relative to the reactor volume,
c = specific heat of the gas,
ϱ = density of the gas,
u = empty-tube velocity,
Θ = temperature of the gas,
Θ_a = temperature of the charge material,
x, r = axial and radial coordinates.

The left-hand member of the equation expresses the amount of heat transferred from the gas to the material; on the right-hand side, the first term is the flow term, the second the transverse mixing, and the third the longitudinal mixing.

The characteristic quantities for the longitudinal and transverse mixing are the *Peclet numbers*

$$N_{\mathrm{Pe},r} = u\, d_h/\varepsilon\, M_r \quad \text{and} \quad N_{\mathrm{Pe},a} = u\, d_h/\varepsilon\, M_a .$$

FROMENT has recently published a comprehensive survey of measurements of the Peclet numbers, from which it appears that the Peclet numbers are already constant at comparatively low Reynolds numbers, and in fact

$$N_{\mathrm{Pe},r} \approx 10, \quad N_{\mathrm{Pe},a} \approx 2.$$

Longitudinal and transverse mixing are of considerable importance for the theoretical description of the heat-transfer process because of the comparatively large heat-transfer coefficients. Thus, in the illustrative example on p. 206, on taking the important longitudinal mixing into account, the possible heat transfer is reduced by 20%.

3.1.2.2. The Heating of a Charge of Spheres in a Countercurrent of Gas, Taking into Account the Heat Conduction in the Spheres

To elucidate the heating-up process in countercurrent reactors it is essential to know the state of the gas flow (the Reynolds and Peclet numbers) and the heat transfer possible for these values (the Nusselt number) from the gas to the charge material.

However, a knowledge of the state of the gas alone is insufficient to determine how the material in the countercurrent heats up, since the actual transfer of heat from the gas to the material is coupled with the heat transfer process into the interior of the grains.

For the coupled process of heat transfer and heat transport under countercurrent conditions JESCHAR[318 a)] has formulated the system of equations and described its exact solution in the case of the heating-up of a flat plate. The theory has been extended by WARTMANN and MERTES[318 b]) to the important case of the heating-up of a batch of spheres in counter-current conditions, taking into account the heat conduction in the spheres, the heat transfer from gas to spheres, and the heat balance. The original

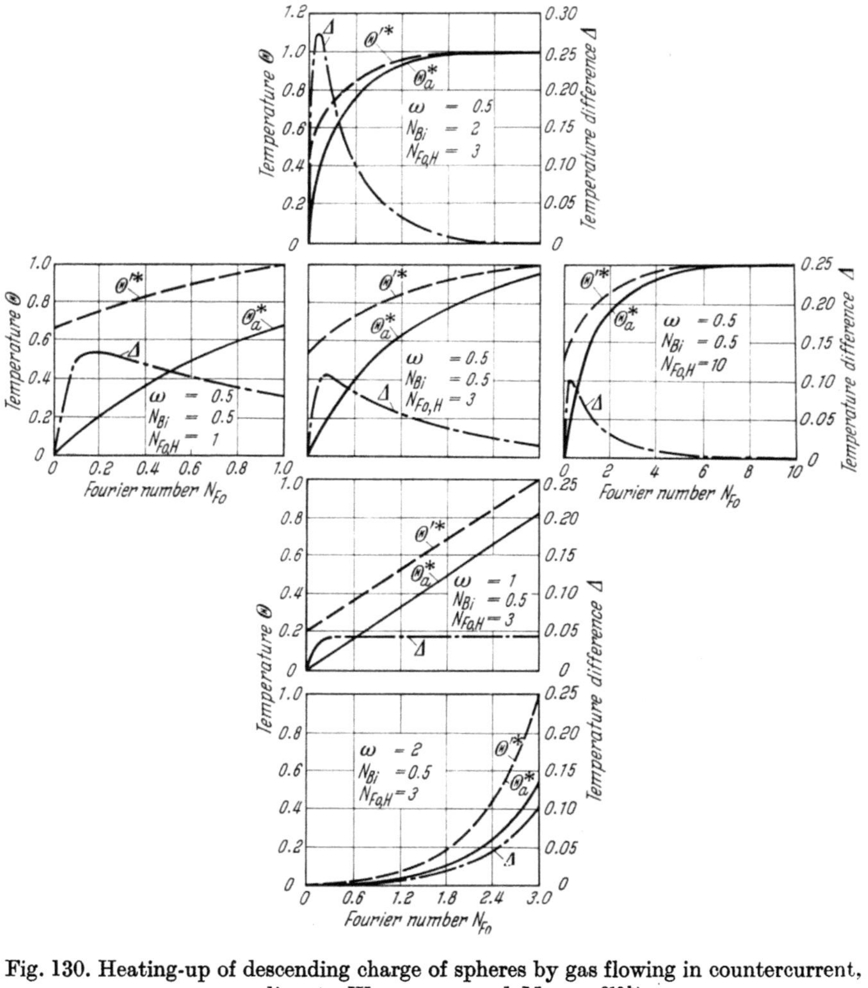

Fig. 130. Heating-up of descending charge of spheres by gas flowing in countercurrent, according to WARTMANN and MERTES[318 b])

papers[318 a, b]) should be consulted for the details of the calculations and the simplifications necessary to ensure a solution in closed form.

In Fig. 130 some particular solutions are represented graphically. For convenience various dimensionless parameters are used:

the "Fourier number", N_{Fo}, is a dimensionless group characterizing the duration for which the spheres are in the shaft, or, the shaft depth:

$$N_{\mathrm{Fo}} = \frac{h}{w} \frac{4\lambda}{d^2 c \varrho},$$

where

h = shaft depth,
w = descent speed of the spheres,
λ = coefficient of thermal conductivity of the spheres,
c = specific heat of the spheres,
ϱ = density of the spheres;

the dimensionless gas temperature, Θ'^*:

$$\Theta'^* = \frac{\Theta' - \Theta_0'}{\Theta_0 - \Theta_0'};$$

the dimensionless surface temperature of the spheres, Θ_a^*:

$$\Theta_a^* = \frac{\Theta_a - \Theta'}{\Theta_0 - \Theta_0'};$$

and the dimensionless temperature difference Δ between the centre and the surface of the spheres:

$$\Delta = \frac{\Theta_a - \Theta_z}{\Theta_0 - \Theta_0'},$$

where

Θ' = temperature of the gas,
Θ_0' = temperature of the gas at inlet,
Θ_0 = temperature of the spheres at inlet,
Θ_a = surface temperature of the spheres,
Θ_z = temperature at the centre of the spheres.

Θ'^*, Θ_a^*, and Δ are plotted as functions of N_{Fo}.

Figure 130 shows how the variation in the gas and spheres' temperatures depends on the properties of the spheres and of the gas and also on the conditions of the process.

The properties of the spheres can also be described by the Biot number N_{Bi} instead of the Fourier number:

$$N_{\mathrm{Bi}} = \frac{\alpha}{\lambda} \frac{d}{2},$$

where

α = heat-transfer coefficient,
λ = coefficient of thermal conductivity,
d = diameter of spheres.

N_{Bi} specifies the ratio of the heat-transfer coefficient to the coefficient of thermal conductivity; a large Biot number thus implies that the corresponding sphere can be heated throughout only with difficulty, as, e.g., in the case of coke. The conditions of the process are described by either the Fourier number (time in reactor, descent speed) or by the ratio of water equivalents ω:

$$\omega = \frac{c\,B}{c'\,B'},$$

where

c, c' = specific heat of the spheres and the gas respectively,
B, B' = throughput of spheres and of gas per unit time respectively.

Accordingly, a low water-equivalent ratio implies that a large amount of heat from the gas is expended in heating up the spheres. High exhaust-gas temperatures are then to be expected.

Figure 130 shows clearly the effects of the individual parameters. For instance, if the Biot number is large, calculation shows there is a big temperature difference between the surface and the centre of the spheres. As the water-equivalent ratio increases, the exhaust-gas temperature decreases. For higher descent speeds, or – with a given shaft height – for shorter retention times in the shaft, i.e., for smaller values of $N_{\mathrm{Fo,\,H}}$, an isothermal zone no longer occurs; compare the diagram for $N_{\mathrm{Fo,\,H}} = 3$ with that for $N_{\mathrm{Fo,\,H}} = 10$. In such a case, the whole volume of the reactor is utilized for the heating-up process.

The theory has been extended by JESCHAR, WARTMANN, and MERTES[318c,d] to cases where, not just one sort of sphere, but batches of spheres which may differ both in size and in their material properties are heated under countercurrent conditions in the shaft. In principle, therefore, the process of heating up the burden in a blast furnace can be described mathematically. For details the original articles should be consulted.

In Fig. 131 the solution is plotted for a particular case. The cooling of the gas and the heating-up of the charge was considered in the first 8 m shaft depth of the reactor. The shaft was assumed to be cylindrical. In this numerical example, the charge descent rate, w = 6.4 m/h, the gas throughput = 8300 m³/m² h at NTP, the specific heat of the gas = 0.375 kcal/m³ °C at NTP, and the heat transfer coefficient α = 110 kcal/m² h °C.

Curve *1* shows the variation in the gas temperature if the reactor is filled only with coke. The data for the coke are:

Coke throughput	3744	kg/m² h
Radius of pieces	0.03	m
Density	1300	kg/m³
Specific weight of charge	585	kg/m³
Specific heat	0.3	kcal/kg °C
Coefficient of thermal conductivity	0.5	kcal/m h °C

Curve *2* shows the variation in the gas temperature if the same reactor, using the same gas, the same gas throughput, and the same charge descent rate, is now filled with a burden having the following properties:

Burden throughput	14400	$kg/m^2 h$
Radius of pieces	0.015	m
Density	5000	kg/m^3
Specific weight of charge	2250	kg/m^3
Specific heat	0.2	kcal/kg °C
Coefficient of thermal conductivity	1.4	kcal/m h °C

Comparison of curves *1* and *2* shows clearly that the utilization of heat is considerably worse when heating up the charge of coke only, than it is when heating up the charge of ore only; in the latter case the heat of the gas is utilized almost completely (quasi-adiabatically). In particular, the comparison of the two curves shows that, with the charge of ore only, the whole volume of the reactor is utilized almost uniformly for transferring the heat from the gas.

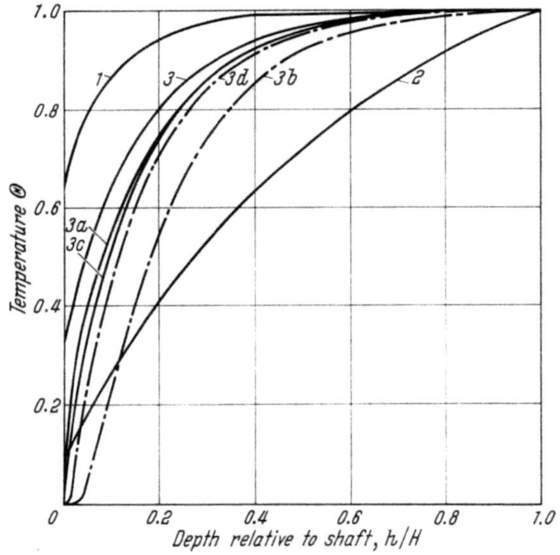

Fig. 131. Heating-up of descending ore and coke charges by gas flowing in counter-current, according to JESCHAR et al.[318c])

1 Gas temperature if only coke in reactor
2 Gas temperature if only ore in reactor
3 Gas temperature if ore and coke in reactor; amount of coke added: 322 kg/ton burden
3a Surface temperature of a piece of coke in reactor
3b Temperature at centre of a piece of coke in reactor
3c Surface temperature of a piece of ore in reactor
3d Temperature at centre of a piece of ore in reactor

Case	Material	N_{Bi}	$N_{Fo, H}$	ω	$\Sigma\,\omega$
1	Coke	6.60	1.78	0.36	0.36
2	Ore	1.18	7.78	0.925	0.925
3a, b	Coke	6.60	1.78	0.22	0.68
3c, d	Ore	1.18	7.78	0.46	

Curves *3* and *3a–d* show the temperature variation in the same reactor when the charge is a homogeneous mixture of ore and coke – roughly the same as in a blast furnace. It is again assumed for the calculation that the same gas, with the same throughput, and the same charge descent rate are used as in the first two cases. The coke and the ore are also assumed to have the same material properties. Only the throughput of ore and coke has been altered. The mixing ratio in this case is taken to be 322 kg coke to 1000 kg ore. Curve *3* shows the variation in the gas temperature for this charge under the chosen conditions. Naturally, curve *3* lies between the two extreme curves, *1* for the charge of coke only, and *2* for the charge of ore only. Curves *3a* and *3b* show the heating-up of the coke, *3a* giving the surface temperature, *3b* the temperature at the centre of a lump of coke. Because of the low thermal conductivity, there is a considerable difference in the temperatures of the surface and core of the coke; the coke warms through only slowly. In contrast, the ore mixture (curves *3c* and *3d*) warms through with little difficulty; the corresponding temperature difference is small. The temperature conditions in the reactor will naturally change if the properties of the charge or of the gas, or the throughput rates, are changed.

Comparison of the calculations by Jeschar, Wartmann, and Mertes[318b, c] with measurements in blast furnaces[559] show that, in principle, the conditions in the blast furnace can be found correctly by calculation. Depending on the mode of operation of the blast furnace, the variation of the gas temperature in the stack is more or less gradual.

3.2. The Fluidized Bed and the Transition to Particulate Fluidization

A *fluidized bed* is formed from a settled bed of solid particles through which an upward stream of gas is passed, and it is formed, in fact, when the gas flow rate exceeds a critical lower limit which depends on the properties of the solid charge and of the gas. At first, the bed expands upwards to some extent; thereafter the solid particles acquire a circulatory motion similar to that in a boiling liquid (Fig. 132). The upper surface of the fluidized bed remains horizontal; the angle of contact is about 90°.

Figure **133a** shows an example of such a fluidized bed. The solid particles in a fluidized bed are in a state of vigorous motion, and so the right conditions are provided for good transference of heat and high rates of reaction between the gaseous and solid phases. Fig. 133b shows the crater-like formations produced when gas bubbles break through the upper surface.

The name "fluidized bed" has been adopted because of the similarity in appearance to a bubbling liquid.

The fluidized-bed principle was first employed on full engineering scale by WINKLER[321]) in the well-known generator for gasification of coal, which is now named after him. Other instances are in the roasting of iron pyrites

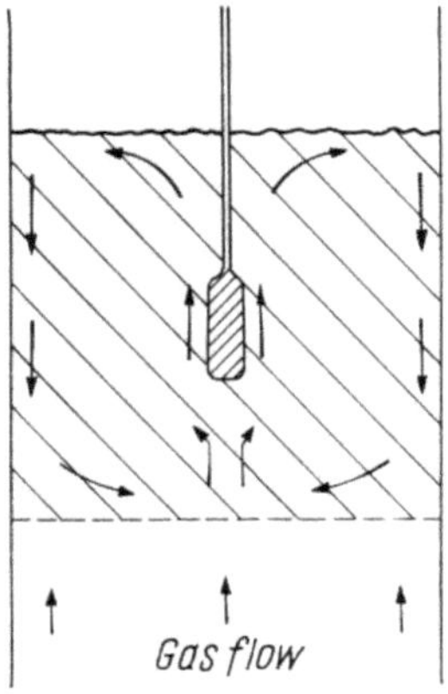

Fig. 132. Diagram of particle flow as observed in a fluidized bed

(Fig. 134) and catalytic gas reactions, in which the catalyst works as a fluidized bed (Fig. 135). The plant for cracking mineral oil, shown in Fig. 135, consists essentially of two fluidized-bed reaction chambers, in

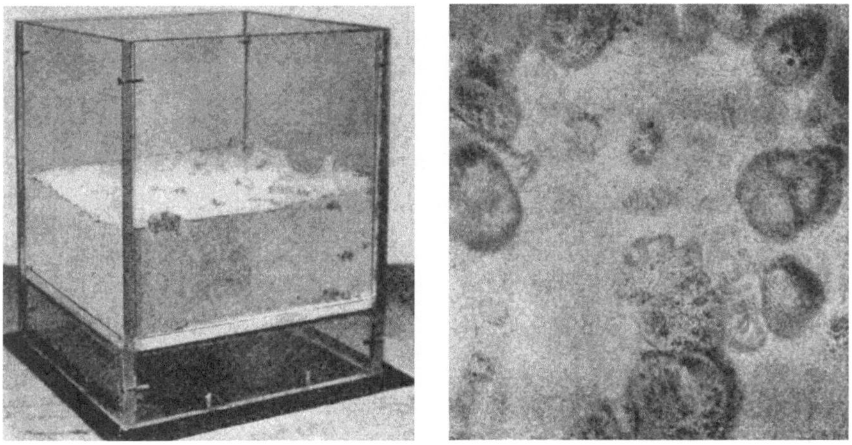

a) b)

Fig. 133. Bubbling fluidized bed in a perspex apparatus, according to BRÖTZ[320])
a) View of the apparatus b) Top view of the fluidized bed

the first of which the mineral-oil cracking process takes place, while in the second the catalyst is freed from the cracking residues by a reaction with hot air. Plants of this type may have an output of 2000 tons a day

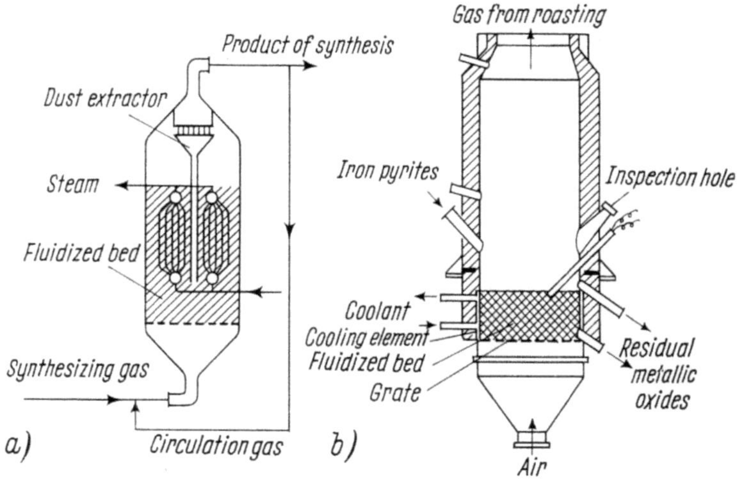

Fig. 134. Fluidized-bed processes for exothermic reactions, according to NONNENMACHER[331])

a) Fischer-Tropsch synthesis b) Roasting of iron pyrites

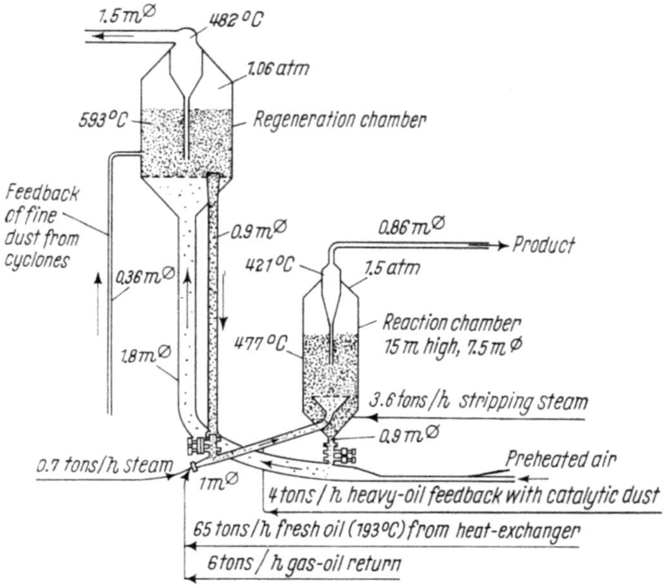

Fig. 135. Catalytic cracking plant based on the fluidized-bed process with a capacity of about 2000 tons/day, according to NONNENMACHER[331])

and more. According to PERRY[322]) the total capacity of the existing fluidized-bed cracking plants amounts to 220000 tons of crude oil a day.

There are many comprehensive publications dealing with fluidized-bed technology, from which we may select[322–326, 330]).

In fluidized beds, solid materials can be treated whose grain size lies in the range 1 μm to 64 mm, but the best size is between 10 μm and 0.25 mm. Too fine a material has a tendency to agglomerate, too coarse results in non-homogeneous fluidization, shock effects, etc. Mixtures of different grain sizes can be handled — within certain restrictions. The empty-tube velocities of the gas in fluidized-bed reactors lie generally within the range 15 to 300 cm/sec, and the height of the fluidized bed may be between 30 cm and 15 m [322]).

3.2.1. The Domains for Existence of Fluidized Beds and of Particulate Fluidization

The existence domain and properties of fluidized beds have been investigated by many workers. Figure 136 shows one example of the apparatus, taken from a paper by WICKE et al.[319]). The gas flow was measured at the orifice plate *2* and the pressure drop in the bed on the manometer *3* (left).

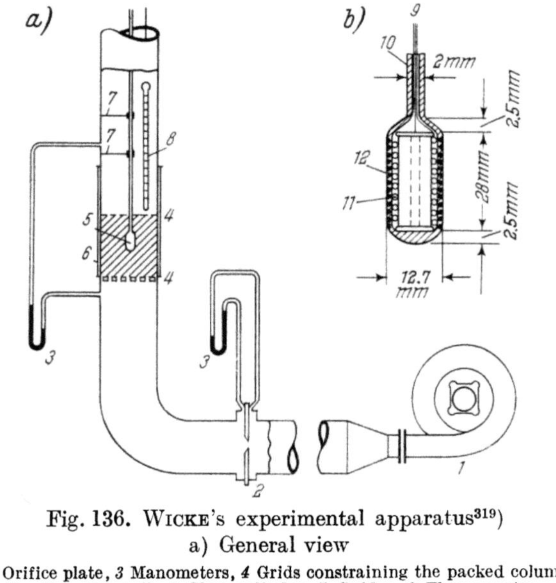

Fig. 136. WICKE's experimental apparatus[319])
a) General view
1 Pump, *2* Orifice plate, *3* Manometers, *4* Grids constraining the packed column, *5* Probe (heating element), *6* Glass cylinder, *7* Guides, *8* Thermometer
b) Heating element (in section)
9 Leads, *10* Nickel-silver tube, *11* Heater winding, *12* Thermometer winding

Further, the heat transfer from the electrically heated cartridge (Fig. 136b) to the bed was measured. As the gas velocity u (relative to the empty cross-section at the working temperature) was gradually increased, the pressure drop in the bed increased in accordance with the Carman-Kozeny equation, in the region corresponding to a static bed (Fig. 137).

At the critical velocity u_L the pressure drop Δp_L of the gas in the packed column (i.e., the total pressure drop reduced by the pressure drop at the base filter on which the charge rests; the loss due to the filter alone has to be determined in separate experiments without the charge), this

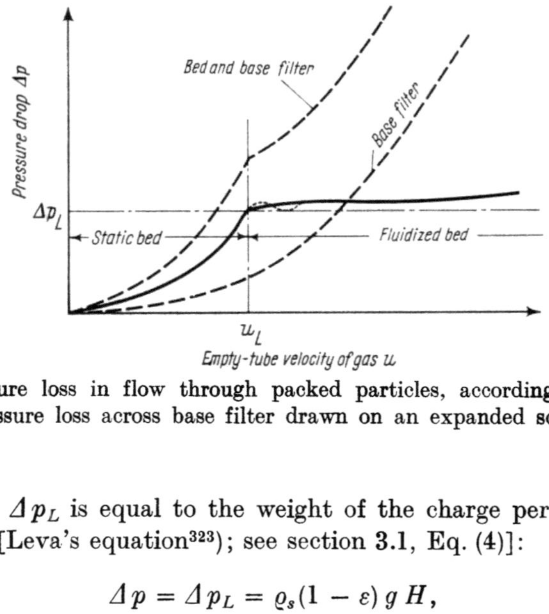

Fig. 137. **Pressure loss in flow through packed particles, according to** WICKE[319]) (pressure loss across base filter drawn on an expanded scale)

pressure drop Δp_L is equal to the weight of the charge per unit area of cross-section [Leva's equation[323]); see section 3.1, Eq. (4)]:

$$\Delta p = \Delta p_L = \varrho_s (1 - \varepsilon)\, g\, H, \qquad (1)$$

where

ϱ_s = specific weight of the particles,
g = gravitational constant,
H = height of charge,
ε = voids ratio.

This equation holds for fluidized beds of solids in gases if the density of the medium $\varrho_f \ll \varrho_s$.

The domain of existence of a fluidized bed begins once u_L is exceeded; the pressure drop Δp increases only insignificantly. The stream velocity u_L at which $\Delta p = \Delta p_L$, i.e., at which the transition to the turbulent regime takes place, can be calculated from the Carman-Kozeny equation (see section 3.1).

With the grain sizes and velocities used in fluidized beds, we are almost always in the laminar regime, and consequently the first term of the Carman-Kozeny equation suffices for the formulation*.

From section 3.1 and Eq. (1) above, we have

$$\psi = \frac{160}{N_{\mathrm{Re}}} = \frac{\Delta p\, \varepsilon^3\, d\, \Phi}{H\, \varrho\, u^2 (1-\varepsilon)} = \frac{160 \eta (1-\varepsilon)}{\varrho\, u\, d\, \Phi}. \tag{2}$$

For $\Delta p = \Delta p_L$, then, we have

$$u = u_L = \frac{\varrho_s\, g\, d^2\, \Phi^2}{160 \eta} \frac{\varepsilon^3}{1-\varepsilon}. \tag{3}$$

Equation (3) is valid for uni-granular fillings. The equation also remains applicable when there are small variations of d. In multi-granular fillings with widely differing grain diameters, conditions become more complicated; see [323]).

The velocity in fluidized beds cannot be increased arbitrarily. If the stream velocity u exceeds the downward velocity of the material particles, these will be elutriated with the gas flow. This velocity therefore is the upper bound for the domain of existence of a fluidized bed and marks the beginning of "particulate fluidization". By venturi-like extensions at the top of the reaction tube the elutriation of the particles of the charge can be repressed[324]). This upper limit u_F can be calculated using Stokes's

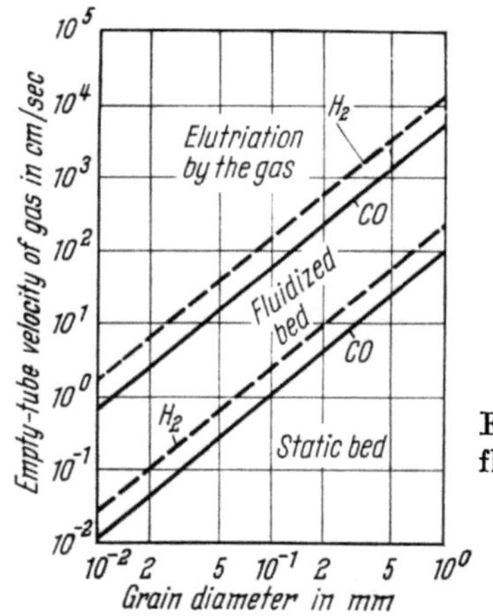

Fig. 138. Range of existence of a fluidized bed for powdered ore with $\varrho_s = 5$ g/cm³ (700°C) [327])

* In Eq. (5) of section 3.1 it was the *second* term of the Carman-Kozeny equation which was used to determine the stability limit for charges of coarse materials, since there is almost always turbulence with predominantly large grains and with high velocities. Figure 123 in section 3.1 showed the gradual superimposition of the "laminar" term, which is determinative for fluidized beds consisting of fine-grained materials.

law[325]): The terminal velocity for spherical particles in a medium of viscosity η is, according to Stokes's law, given by

$$3\pi\, d\, \eta\, u_F = m\, g = \tfrac{1}{6}\pi\, d^3\, \varrho_s\, g,$$

or

$$u_F = \frac{\varrho_s\, g\, d^2}{18\eta}. \tag{4}$$

Figure 138 shows the calculated domain of existence of the fluidized bed in a numerical example for Fe_2O_3, $\varrho_s = 5\ \mathrm{g/cm^3}$, $T = 700°\mathrm{C}$, in H_2 and CO. The lower and upper critical velocities, u_L and u_F, are separated by a factor of exactly 10^2 and increase with the square of the diameter of the particles. The effect of the lower viscosity of H_2 compared with that of CO is clearly seen. If there are particles of various diameters d in a fluidized bed, the smallest value of d should be substituted in Eq. (4) in order to determine when elutriation begins.

3.2.2. Special Features in the Industrial Application of Fluidized-Bed Reactors

Because of the circulatory motion of the solid particles in a fluidized bed it is impossible, in principle, to direct the gas and the solid material in counterflow and thus to obtain the optimal yield or maximal degree of transformation. This difficulty can be overcome by arranging several fluidized beds one above the other, as in Fig. 139. The mathematical treatment then becomes similar to that for a rectifying-column, and a calculation has been carried out by MEISSNER and SCHORA[328]) on the assumption that equilibrium is attained in each section. Further theoretical and practical investigations into the possibility of bringing the control of fluidized-bed reactions into line with the countercurrent principle can be found in papers by VAN DEEMSTER[328a]), BLOCK[328b]), and LAGO et al.[328c]).

If a fluidized bed is of unsuitable dimensions (e.g., has a large height H and a small diameter D, especially if $H/D > 5$), then instability in the fluidized bed is often observed; this is connected with excessive values of Δp (Fig. 140). Shock conditions as in delayed boiling, and the formation of piston-like gas bubbles ("slugging"), can occur when the dimensions have been chosen unsuitably, and they render the control of the process difficult (Fig. 141).

Heat transfer takes place very rapidly in fluidized beds, as may clearly be seen from Fig. 142. Mixing of the materials in fluidized beds also takes place very rapidly. If two layers, one on top of the other, of materials which do not differ too greatly in density, grain size and shape are brought into turbulence, very rapid mixing takes place. The higher the gas velocity

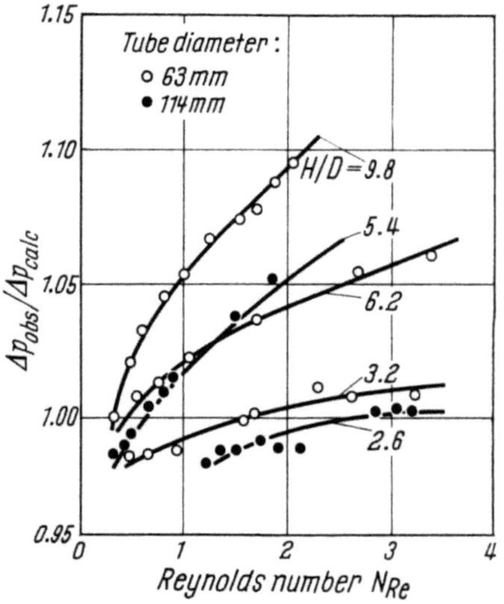

Fig. 140. Deviation of the measured pressure drop Δp_{obs} from the calculated pressure drop Δp_{calc} in a fluidized bed formed by passing air through a column of glass beads, for various ratios of height H to diameter D of the column, according to BRÖTZ[320])

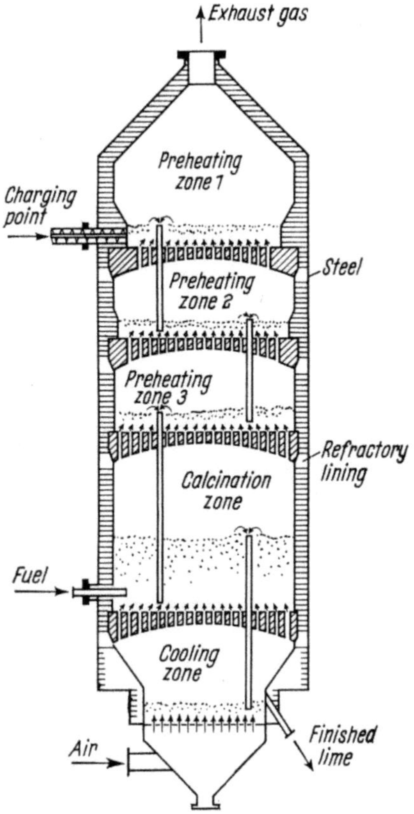

Fig. 139. Multi-stage fluidized-bed reactor for calcining limestone, according to PERRY[322])

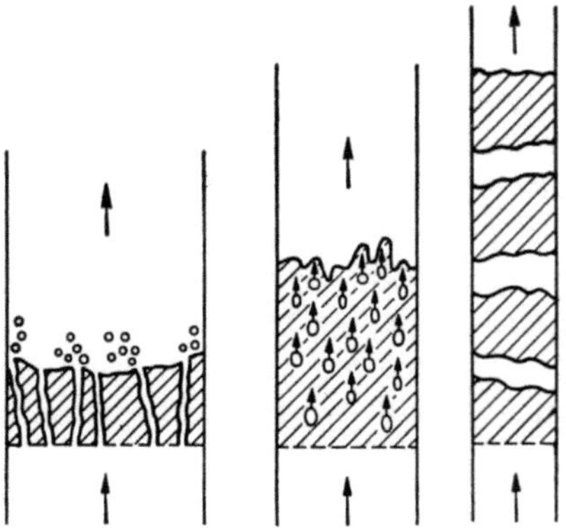

Fig. 141. Non-homogeneous flow in solid-gas systems, according to TRAWINSKI[329]). Left: channel formation. Centre: bubble formation. Right: shock conditions ("slugging")

and the larger the grain diameter, the faster the mixing. The mixing can be characterized by a mixing coefficient (Fig. 143), defined similarly to a coefficient of diffusion[325]).

Grains of different specific weight can be separated in fluidized beds, as in flotation techniques.

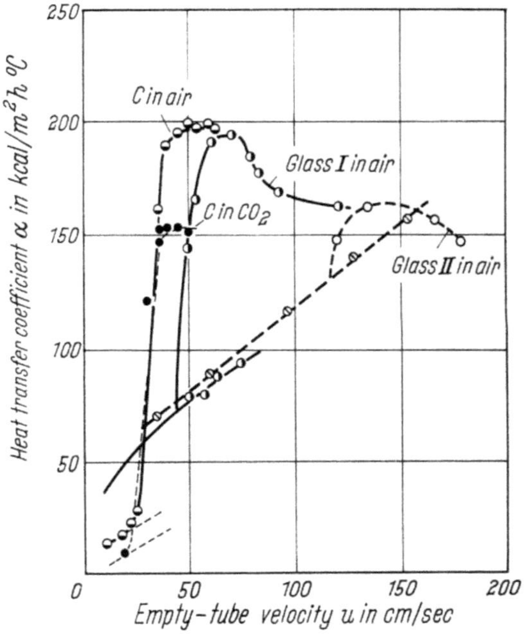

Fig. 142. Heat transfer coefficient α as a function of the gas velocity u for glass beads and active coal grains in air and CO_2, according to WICKE[319])

Glass I: Uniform glass spheres with smooth surface, 0.9 mm; *Glass II:* Similar glass spheres, 3 mm; *C* Active coal grins, 0.31 mm

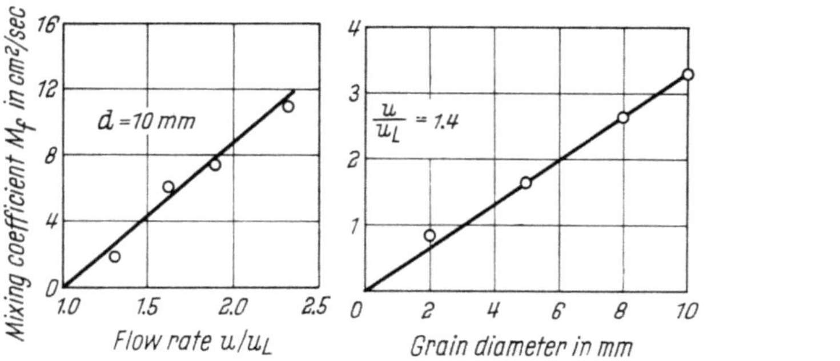

Fig. 143. The mixing coefficient M_f for granular material in relation to the flow rate and the grain diameter, according to BRÖTZ, as cited by GRASSMANN[325])

u_L = velocity at the turbulence point. M_f is defined like a coefficient of diffusion, and was measured by determining the degree of mixing of differently coloured glass spheres (by counting) after subjection to turbulent flow for various times

The usually undesired removal of solid material in the exhaust gas is generally quite high in fluidized beds, since fine abraded dust is produced because of the turbulent motion, and, in accordance with Eq. (5), this dust is entrained even at low gas velocities.

4. Techniques for the Reduction of Iron Ores, Apart from the Blast Furnace

4.1. Statement of the Problem

The reduction of iron ores to the metal is the basis of steel manufacture. The reduction is carried out mainly in blast furnaces, which produce up to 7000 tons Fe a day in the form of molten pig iron, at a very moderate cost for the conversion. The blast furnace may be regarded as a piece of equipment which has almost reached perfection. It is suitable for handling almost all naturally occurring iron ores, and its product, the molten pig iron, can be used directly without further treatment in the existing processes for steel manufacture. World production of pig iron is now, as ever, sharply increasing (Fig. 144). One drawback of the blast-furnace process,

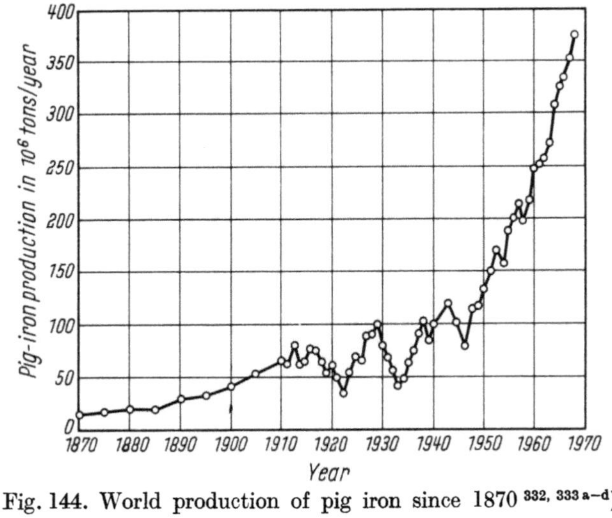

Fig. 144. World production of pig iron since 1870 [332, 333a–d])

however, is that the coke which is used as the heat-*supplying* and reducing medium must be of "metallurgical" quality, and this coke is expensive and can be manufactured only from special kinds of coal. It is to be regarded as a lucky accident that these kinds of coal are available in the traditional iron-making regions, especially in Europe and the United States.

Many countries with a rapidly developing iron industry have, however, no supplies of coking coal. Still, other valuable fuels such as natural gas, petroleum, and coal which is not easily cokable, are often available in large quantities at economic prices. In these countries the need arises to smelt the frequently available indigenous iron ores using indigenous fuels and hence to provide a favourable basis for national iron-making industries.

The problem of reducing iron ores to the metal using fuels other than coke is therefore a very real one. Despite all endeavours, however, it has not been possible so far to find a process which approximately matches the blast furnace in technical perfection.

Table 11 shows, according to the available information, the present capacity for the production of iron by *direct reduction processes*, i.e., apart from the blast furnace.

Table 11. *Global capacity of operational reduction processes, apart from blast furnaces*[334–343])
Rounded figures as at 1967; figures in brackets denote plant under construction. Total capacity about 7.7 million tons a year.

Process	Product	Capacity (1000 tons/year)
Fluidized-bed process		
H-iron	Sponge iron	55
FIOR	Sponge iron	100
Novalfer	Sponge iron	18
Retort processes		
HyL	Sponge iron	400
Höganäs	Sponge iron	115
Rotary vessel processes		
SL-RN	Sponge iron	60
SL-RN	Sponge iron	(150)
SL-RN	Pre-reduced ore	(360)
Krupp-Renn process*	Nodules	3000
Basset	Molten pig iron	75
Dored	Molten pig iron	80
Travelling-grate processes		
DLM	Pre-reduced ore	(180)
Shaft furnace processes		
Wiberg process	Sponge iron	85
PUROFER	Sponge iron	165
Echeverria	Sponge iron	20
Phelps-Dodge	Sponge iron	15
Rotary vessels with electric furnace		
Elkem	Molten pig iron	1000
Electro-reduction		
Tysland Hole furnace	Molten pig iron	1600
Sorel	Molten pig iron	250

* Some furnaces can also be operated in the Krupp sponge-iron process.

Compared with the world output by blast furnaces of more than 350 million tons of pig iron a year, the production capacity of direct reduction processes appears at present decidedly modest (about 2% of the total production of pig iron); a considerable part of the capacity mentioned (e.g., in the Krupp-Renn process) is no longer operating.

In the following sections our object is to explain, from the point of view of physical chemistry and particularly of reaction kinetics, how the many and very different groups of processes for direct reduction may be expected to perform. The efficiency of the reduction processes will therefore be judged primarily according to the extent to which it has been possible, in any technological solution of the problem, to apply the basic laws of thermodynamics, of reaction kinetics, and hydrodynamics to achieve the highest performance and degree of utilization of the materials, taking into account the properties of the ores and fuels used.

4.1.1. Available Raw Materials

4.1.1.1. Ores

There is so much well-presented information, e.g., [343]), on the amounts and properties of the iron ores available that it is unnecessary to give details here.

4.1.1.2. Fuels

Figure 145 gives a synopsis of the properties of solid fuels. With increasing degree of carbonization, i.e., from wood to coke, the percentage of volatile components and the reactivity of solid fuels decrease. The values shown apply to the material when free from moisture and ash (d.a.f. = dry, ash free). In any particular case, the percentage moisture and ash have to be taken into account. More detailed information may be found in the literature[344]).

Table 12 gives the characteristics of the most commonly used fuel oils, and Table 13 those of gases obtained from solid fuels.

Table 14 shows the composition of some samples of natural gas. However, the amounts of the individual components even in gas from a single field can vary so widely that the values shown must be regarded only as a guide. The H_2S component – of which there is sometimes quite a high percentage – must be removed during the production of natural gas, since otherwise, if condensation of moisture occurs, the corrosion in the pipes will be unacceptably severe.

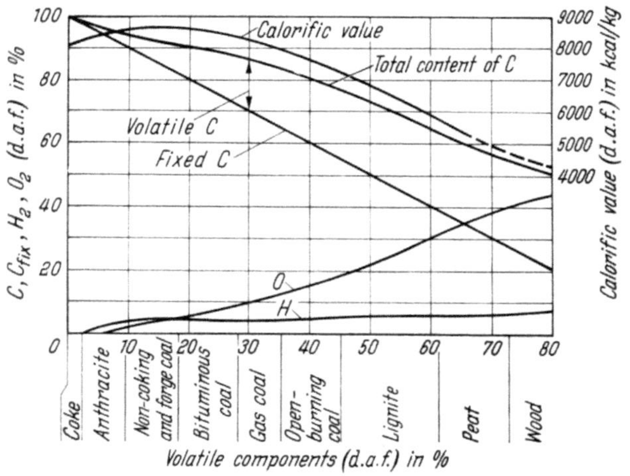

Fig. 145. Composition in terms of elements for solid fuels (dry and ash free) [344])

Table 12. *Physical and chemical constants for fuel oils*[344])

Property		Constants for			
		fuel oil L *	fuel oil M **	fuel oil M ***	fuel oil S ****
Density at 15°C	kg/l	0.86—0.88	up to 1.10	1.10—1.18	0.95—0.97
Pensky-Martens flash-point. .	°C	80—90	>85	>85	>100
Solidifying-point	°C	>−20	—	—	−12 to +20
Calorific value (H_u)	kcal/kg	10000	9000	9000	9500
C-content . . .	%	≈84	≈90	≈90	≈85
H_2-content. . .	%	≈12	≈ 6	≈ 6	≈11
Sulphur	%	up to 0.8	up to 1.0	up to 1.0	up to 3.5
Asphalt	%	up to 0.1	—	—	up to 6.0
Water	%	up to 0.2	up to 1.0	up to 0.5	up to 0.3
Ash.	%	up to 0.05	up to 0.05	up to 0.2	up to 0.1

* Light mineral oil.
** Coal-tar oils (distillate without pitch additive: Type IV).
*** Coal-tar oil with up to 40% pitch additive (special smelting fuel oil).
**** Heavy mineral oil (earlier known as mazut or Bunker C oil).

Table 13. *Characteristic values for various kinds of gas (mean values)*[345]

		Coke-oven gas	City gas	Producer gas from	
				open-burning coal	brown coal briquettes
Composition:					
CO_2	%	1.5—2.4	2.0	3—4	3.5—4.5
heavy hydrocarbons	%	1.8—2.5	2.0	0.2	0.2
O_2	%	0.3—0.6	0.5	—	—
CO	%	4.7—6.7	5.5	28—30	30—32
H_2	%	52—62	56.0	12—14	12—14
CH_4	%	24—27	24.0	1.8—2.5	2.5
N_2	%	5—12	10.0	49—51	49—50
Tar content	g/m³ at NTP	—	—	10—25	up to 22.5
Calorific value (H_u)	kcal/m³ at NTP	4000—4200	4070	1300—1400*	1300—1500*
Density	kg/m³ at NTP	0.4—0.5	0.485	1.25	1.2
Moisture content	g/m³ at NTP	10—25	1—2	30—60	100—120
Theoretical amount of air needed	m³/m³ at NTP	4.0—4.3	4.1	1.4—1.6	1.5
Theoretical volume of exhaust gas, moist	m³/m³ at NTP	4.6—5.0	4.8	2.3—2.5	2.4

* Including sensible heat and tar (crude gas) 1600 and 1650 kcal/m³ at NTP respectively.

Table 14. *Examples of composition of natural gas* (% by volume)[338, 346]

	CH_4	C_2H_6	C_3H_8	Higher hydrocarbons	CO_2	N_2	H_2S
Bentheim (Germany)	89.6	1.0	0.5	—	2.8	5.5	0.6
Lacq (France)	70.0	2.9	—	2.1	10.0	0.5	14.5
Hassi R'Mel (Sahara)	81.3	6.8	2.3	4.3	0.5	4.8	—
Venezuela	80.7	5.5	2.3	1.1	10.3	—	—
Texas	58.7	16.5	9.9	8.5	—	—	6.4
Canada	88.1	5.6	1.8	0.8	—	3.6	—

4.1.2. Classification of Operational Reduction Processes

The variety of methods which have been proposed is bewilderingly large, but most of them have not got beyond the laboratory stage. Indeed, some of the proposals contravene the basic laws of dynamics and fluid reaction kinetics. The possible and proposed methods for reduction of iron ores will be critically examined below, considering essentially the following aspects:

a) The conversion possible per unit time and in unit volume or unit cross-section of the reaction space. This quantity determines the size of plant necessary and hence also the investment and maintenance costs.

b) The chemical and thermal utilization of the fuel or reducing gas employed. This determines the fuel costs per ton Fe.

c) The different kinds of energy source which can be employed, e.g., coal, heavy fuel oil, light oil, natural gas, electrical energy. This determines whether a particular process will be usable with the locally available energy sources.

d) The different grades of ore which can be used, e.g., rich or lean in gangue, lumps, fines, pellets, sinter, briquettes.

The flow and heat transfer in various kinds of apparatus for the decomposition of solids by gases have been discussed in section 3. This leads to a classification of the possible processes for ore reduction. WENZEL[347, 347a]) has given a more detailed classification.

The following processes or types of plant for ore reduction, other than the blast furnace, will be dealt with in later sections of this chapter:

fine particle dispersed cloud and shower furnaces,
fluidized beds,
retorts,
rotary vessels,
shaft furnaces,
electro-reduction.

4.1.3. Limiting Values for the Required Amounts of Reductant and Heat

Figure 146 shows the distribution of ballast material, volatile components, slag-forming components, amount of oxygen to be removed, and amount of reduced metal for a selection of iron ores, whose chemical composition is shown in Table 15.

The minimum heat requirement for reduction and heating-up is given by the known enthalpy values. Figure 147 shows the theoretical heat requirement, according to MINTROP's investigations[348]), for the reduction and melting as a function of the Fe-content of the ore, i.e., as a function of the proportion of non-participating material. The actual heat expenditure in the whole process must be considerably greater than the theoretical amount, since in no case are the reduction gases and fuels which are employed fully utilized.

The reduction gases may be utilized chemically either maximally up to equilibrium or approximately so (with the numerical values given in chapter 1). If the reduction is carried out isothermally and in one stage down to the metal, then a lower limit for the amount of gas needed is

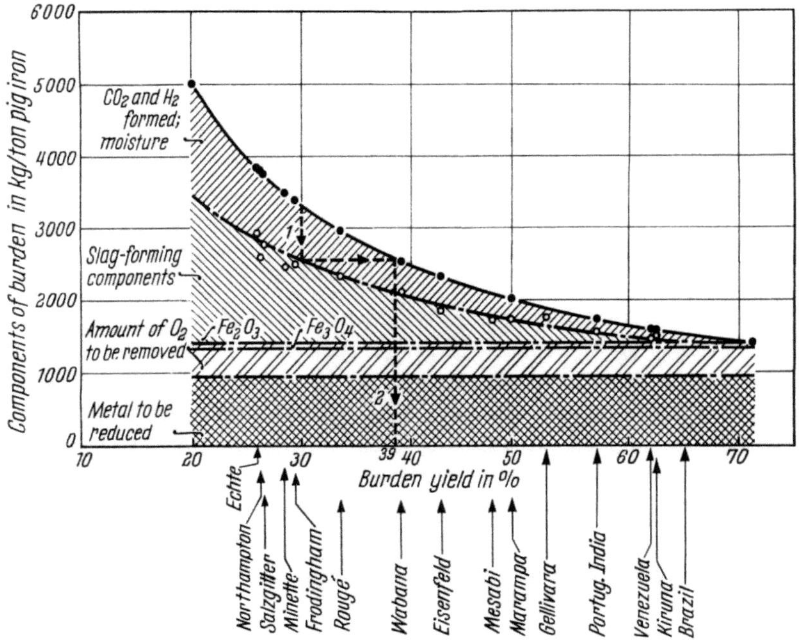

Fig. 146. Structure of ore burdens of various origins, according to MINTROP[348])

The burden yield is defined as the amount of molten pig iron produced, divided by the amount of burden charged (ores plus additions such as limestone less amount lost as dust)

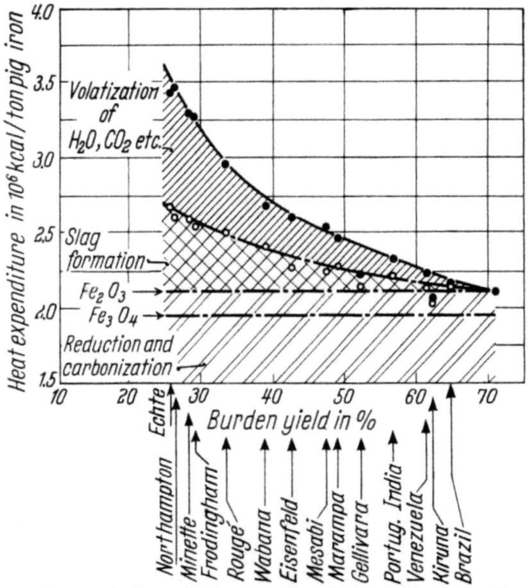

Fig. 147. Heat requirement in relation to burden yield, according to MINTROP[348])

Analysis of pig iron: 4 % C, 1 % Si, 1 % Mn, or: 3.6 % C, 0.2 % Si, 1.8 % P

Table 15. *Chemical composition of various ores*[348a])

Name of ore	Moisture %	Fe %	Mn %	P %	S %	SiO_2 %	Al_2O_3 %	CaO %	MgO %
Brazil	1.0	68.2	0.06	0.024	0.015	0.53	0.83	0.35	0.10
Venezuela . . .	7.8	59.6	0.04	0.11	0.03	0.81	1.55	0.28	0.10
Kiruna D. . . .	1.2	58.2	0.11	1.69	0.02	4.31	1.08	6.32	1.28
Portug. India . .	6.1	54.2	0.41	0.053	0.04	2.25	7.65	0.4	0.13
Wabana	1.52	49.3	0.17	0.92	0.037	13.10	5.88	3.35	0.71
Northampton . .	15.2	32.5	0.24	0.6	0.1	14.70	6.10	2.7	0.4
Frodingham . .	10.7	22.7	0.96	0.31	0.16	8.1	5.1	18.2	1.0
Rougé	4.2	43.5	0.03	0.49	0.04	16.04	6.68	0.38	0.13
Mesabi	13.17	51.4	0.62	0.052	—	6.06	1.51	0.09	0.08
Gellivara	1.3	56.6	0.06	0.85	0.09	9.24	2.38	3.55	1.78
Marampa	6.8	50.8	1.51	0.034	0.06	4.47	9.00	0.37	0.10
Eisenfeld	1.7	41.8	0.20	0.11	0.09	8.23	2.02	9.54	4.94
Minette (Lux.). .	9.9	23.5	0.18	0.51	0.05	10.44	4.47	19.79	0.81
Salzgitter. . . .	13.3	32.9	0.15	0.47	0.20	14.43	6.28	3.94	1.59
Echte	6.4	23.6	0.12	0.49	0.10	11.87	8.33	17.69	1.97

given by the equilibrium

$$Fe - FeO/H_2 - H_2O$$

or

$$Fe - FeO/CO - CO_2 .$$

The corresponding values are plotted in Fig. 148 – as the upper curve in each case.

A mixture of H_2O/H_2 or of CO_2/CO which is in equilibrium with FeO/Fe and is thus unable to reduce any more of the FeO still has a strong reducing action on the higher oxides of the iron. If we think of the reduction reactor as separated into two stages, so that first the fresh reducing gas reduces Fe_3O_4 or FeO to Fe, and then the exhaust gas from this first stage is made to act at the same temperature on Fe_2O_3 in a second stage, then the minimum gas throughput required is smaller (Fig. 148, lower curves).

A reduction reactor operating isothermally is an abstraction. In general the perceptible heat of the reducing gases covers the heat requirement for the heating-up and reduction of the burden provided the reduction is endothermic. By applying the countercurrent principle we endeavour to utilize this amount of heat to the fullest extent, we try to achieve the lowest possible exhaust-gas temperatures, i.e., to obtain a sort of adiabatic method of operation as opposed to isothermal reduction.

Figure 149 shows that a considerable gas throughput is necessary to cover the heat requirement for the reduction and heating-up of pure Fe_2O_3. Since the specific heats of the gases CO, H_2, and N_2 have approximately the same value, it is not necessary to distinguish between them

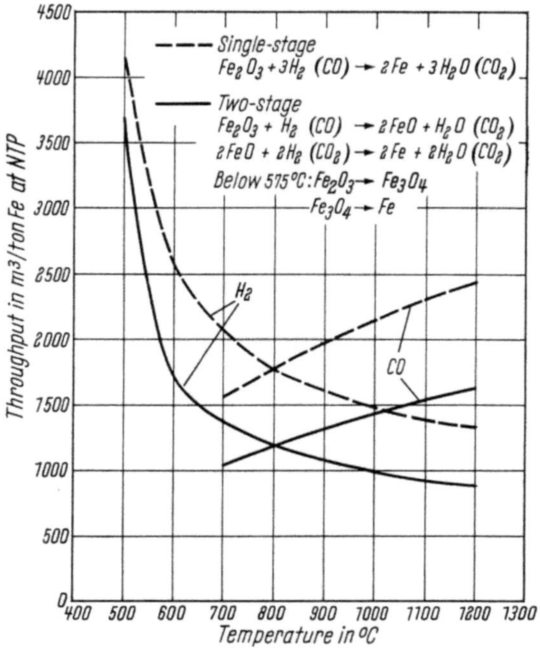

Fig. 148. Minimum consumption of H_2 or CO in the isothermal reduction of Fe_2O_3 to Fe (gas utilized up to equilibrium)

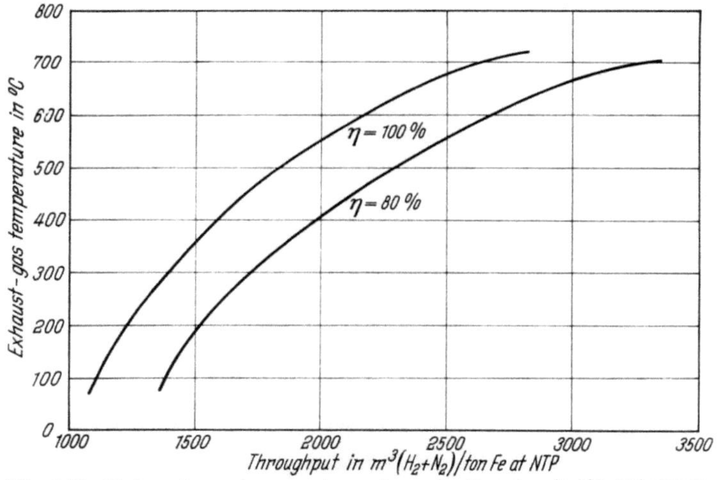

Fig. 149. Exhaust-gas temperatures in reduction by H_2/N_2 mixtures.

$$Fe_2O_3 + 3\,H_2 \rightarrow 2\,Fe + 3\,H_2O$$

H_2-input temperature: 1100°C, Fe-output temperature: 1000°C
Heat requirement covered entirely by sensible heat of the H_2/N_2 mixture
η = heat given to the burden: (sensible heat of gas mixture at 1100°C − sensible heat of exhaust gas at exhaust temperature)

as regards their heat yield. It may clearly be seen from Fig. 149 how considerably the necessary gas throughput is reduced if it is possible to depress the exhaust-gas temperature to a low value. It can be deduced at once from Fig. 148 that when a reducing gas rich in H_2 is used, a high degree of chemical utilization or a low expenditure of H_2 can be achieved by carrying out the reduction at the highest possible temperature; in the two-stage reduction at 1000°C, 992 m^3 H_2 per ton Fe at NTP suffice in the equilibrium case. To cover the requirement in sensible heat, however, this amount of gas is far from sufficient (Fig. 149).

The supply of sensible heat can, however, be considerably increased by the admixture of an inert gas, such as N_2, with the hydrogen. These considerations are particularly important for reduction in shaft furnaces with counterflow of the ore and gas, and will be dealt with in more detail later (section 4.6).

4.1.4. The Products and their Further Treatment

Provided that the melting point of Fe is exceeded, a product similar to the pig iron from blast furnaces is also obtained from these processes (e.g., in electro-reduction, in some rotary furnace processes, and in jet smelting). In some cases it is possible to produce a sort of semi-steel, something between pig iron and steel. In general, however, these smelt products require refining treatment in the usual steel-making plants.

In most direct reduction processes the working temperature is below 1100°C, and consequently the exterior form of the lumps of ore does not greatly alter. The metal appears in a solid sponge-like form and is called *sponge iron*, *pre-reduced pellets*, etc.

If the temperature is raised to about 1200°C, the sponge iron agglomerates into pasty lumps which are called nodules or luppen and which represent a transitional stage between the solid and the molten phases.

An essential disadvantage of sponge iron is its great susceptibility to re-oxidation as a result of its large internal surface area, arising from the reduction. By compressing the iron sponge powder or lumps in briquette presses, the liability to re-oxidation can be reduced.

Table 16 shows the results of weathering tests on sponge iron of various manufacture; similar results have been obtained in other investigations (Fig. 150[350, 350a]). We see from the given values that the decline in the content of metallic iron, i.e., the re-oxidation or rusting, can be greatly reduced by compression during the briquetting, but cannot be prevented. Sponge-iron briquettes can be made much more resistant to exposure by coating them with an asphalt-base rust inhibitor or by protecting them

Table 16. *Examples of weathering tests on sponge iron (stored in the open air)*[349]

Test sample	Exposure time (days)	Total amount of Fe(%)	Amount of metallic Fe (%)	Metallic Fe / Total Fe
Sponge iron pellets, without additional compression	0	95.04	90.06	0.948
	16	90.09	65.04	0.722
	30	84.64	53.84	0.636
Sponge iron subsequently strongly compressed into 5-kg compacts:				
Specimen I	1	84.7	76.6	0.904
	32—37	83.2	68.2	0.819
Specimen II	0	91.7	81.6	0.890
	90	91.3	78.4	0.859

with plastic sheeting, Fig. 150a[350b]). The asphaltic rust inhibitor with a hydrocarbon basis (essentially of heavy oil) was added to the briquettes at the rate 1% by weight. The plastic sheeting was about 0.1 mm thick[350c]).

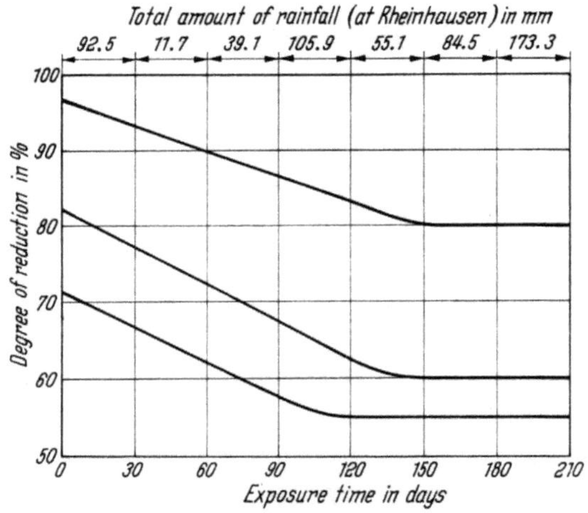

Fig. 150. Decrease in degree of reduction with exposure time for Carol-Lake pellets in the open air without covering[350a])

If it is desired to save the comparatively high cost of briquetting and the subsequent back-oxidization which is still inevitable without further protection of the briquettes, it is expedient to carry out immediately the further processing of the sponge iron at the place of its manufacture.

In the experimental blast furnace of the U.S. Bureau of Mines[363, 363b]) and in a full-scale test with a furnace of 5.5 m hearth diameter[363a]), pellets which had been largely reduced were charged, melted to pig iron, and the effect on the performance and specific coke requirement was assessed. With a 90% degree of reduction a 50% improvement in performance of the test furnace was found, with a reduction of about 250 kg/ton in the specific coke requirement (to a minimum value of 316 kg/ton pig iron). The chemical utilization of the CO in the blast furnace was sharply reduced. In a full-scale test with 30% reduced pellets (90% metallization), with and without the addition of natural gas, the improvement in performance and the saving in coke were not so favourable: the specific furnace performance increased from 47.7 to 58.6 tons pig iron per day per m² of hearth area

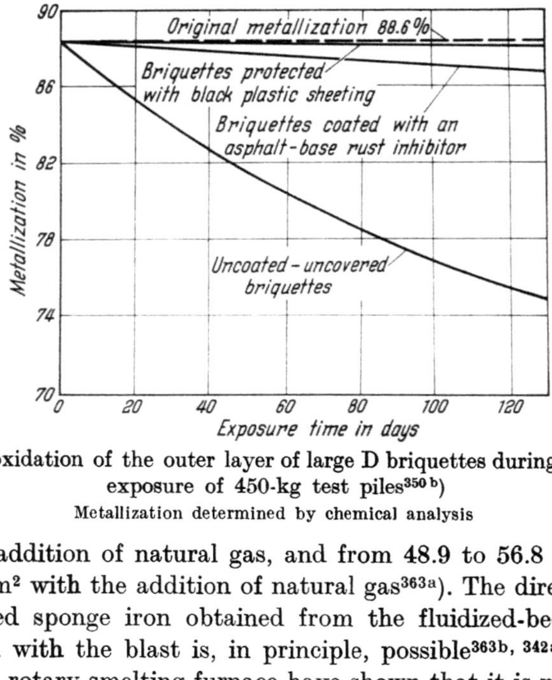

Fig. 150a. Re-oxidation of the outer layer of large D briquettes during environmental exposure of 450-kg test piles[350b])
Metallization determined by chemical analysis

without the addition of natural gas, and from 48.9 to 56.8 tons pig iron per day per m² with the addition of natural gas[363a]). The direct utilization of fine-grained sponge iron obtained from the fluidized-bed process by blowing it in with the blast is, in principle, possible[363b, 342a]).

Tests in a rotary smelting furnace have shown that it is possible, using sponge iron alone as the metallic charge, to convert it into steel. In melting down pre-reduced pellets of 30 mm diameter in an experimental 3-ton rotary furnace, a smelting performance of 0.46 tons/m² h was achieved, the heat requirement being $2.13 \cdot 10^6$ kcal/ton; the maximum depth of the bath was 200 mm[351]). In the Bouchet rotary smelting furnace, which operates with a bath depth of only 90 mm and a furnace capacity of about 1.7 tons, the heat requirement for melting iron powder is $1.25 \cdot 10^6$ kcal/ton, and the smelting performance 1.4 tons/m² h[351a]).

In the electric-arc furnace, in which, up to now, charging is usually discontinuous, 70% or more of the metallic charge can be in the form of sponge iron[352, 353, 355, 355a, b]). Rapid slag production and removal of the mostly acid gangue is important, since otherwise the basic furnace-lining is rapidly eroded. The amount of electrical energy needed is very dependent on the degree of reduction of the sponge iron used (Fig. 151).

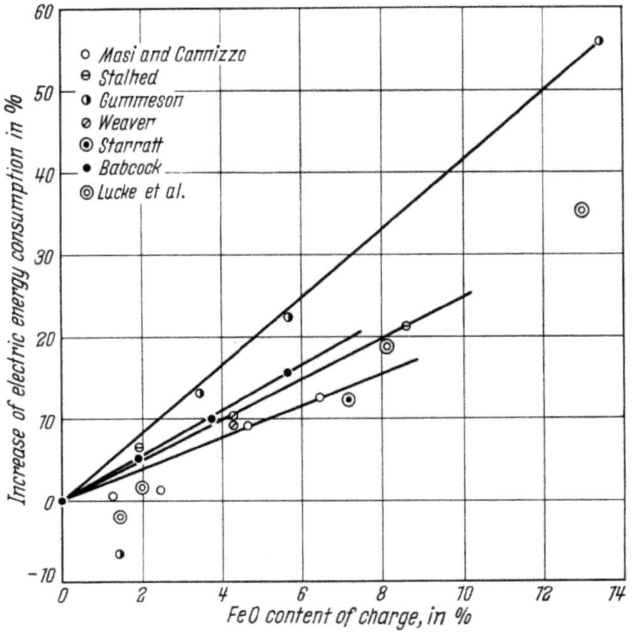

Fig. 151. Percentage of iron(II)-oxide in charge and increase of electric energy consumption for the smelting sponge iron in electric steel furnaces, according to LUCKE et al.[339])

It is only very recently that use has been made of the advantages afforded by sponge iron as the charge material. In tests carried out with continuous feeding of the sponge iron through the furnace roof at a rate such that the supply of sponge iron did not exceed the smelting capacity available, two different techniques have proved successful[352a–c]). Fig. 151a shows a method of operation which has been employed in Canada with a 25-ton furnace and a 70-ton furnace[352b]). The sponge iron is led towards each electrode and falls between the electrode and the wall on to the bath in the neighbourhood of the point of arcing. The temperature of the bath is regulated by the rate of feed so that a tapping can be made shortly after the last addition of sponge iron. Through the elimination of the interruptions for after-charging, the smelting performance rose by 20% when the input was 100% sponge iron. The pellets used contained:

Total Fe	91.80 %	S	0.028 %
Metallic Fe	90.20 %	P	— %
Residual oxygen . .	0.60 %	C	0.008 %

The refining period could be entirely omitted, and so the overall furnace performance rose by 45%. But in spite of the elimination of the refining period, when the proportion of sponge iron is high, the total energy require-

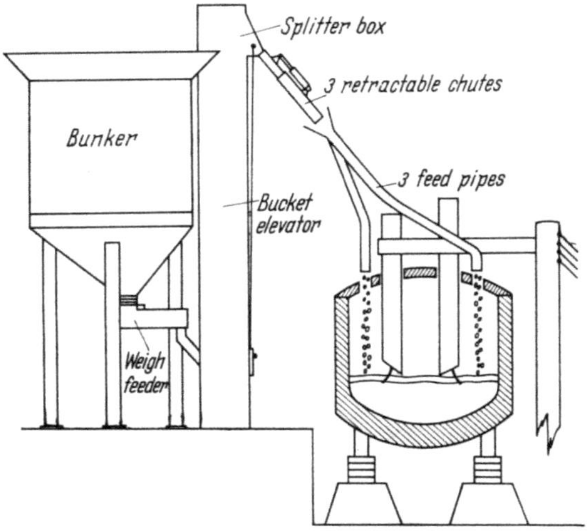

Fig. 151a. Continuous charging apparatus, according to SIBAKIN et al.[352b)]

ment is sometimes higher than the value for an input of 100% scrap iron (see Fig. 151b).

Another process for continuous iron-smelting from sponge iron has been developed by IRSID[352c)]. The feed-in takes place centrally between the electrodes. During the whole smelting period a cone of sponge iron of constant weight is left on the bath between the electrodes, and only as much is added as can be instantaneously smelted. The bath temperature is only slightly above the liquidus temperature, and so after smelting further heating up to the tapping temperature is required. Due to the partial covering of the bath by sponge iron and the low temperature of the bath, the losses due to radiation are kept small. With an input of 20% scrap and 80% sponge iron the smelting performance in a 3-MVA-6-ton furnace was increased by 60% over that for ordinary scrap. The electrical energy required for smelting amounted to 545 kWh/ton compared with 585 kWh/ton for scrap.

The sponge iron used had the following analysis:

Total Fe	93.0 %	C	0.16 %
Metallic Fe	91.9 %	P	0.010 %
Residual oxygen	0.38 %	S	0.015 %
Gangue	5.4 %		

The filling cone between the electrodes, apart from forming a partial cover for the bath, serves the purpose of directing the added sponge iron towards the three arcs, since the melting down takes place most effectively directly in the arcs. The bigger the furnace and therefore the greater the distance between the electrodes, the smaller will be the fraction of the sponge iron running out of the filling cone which will be led directly into

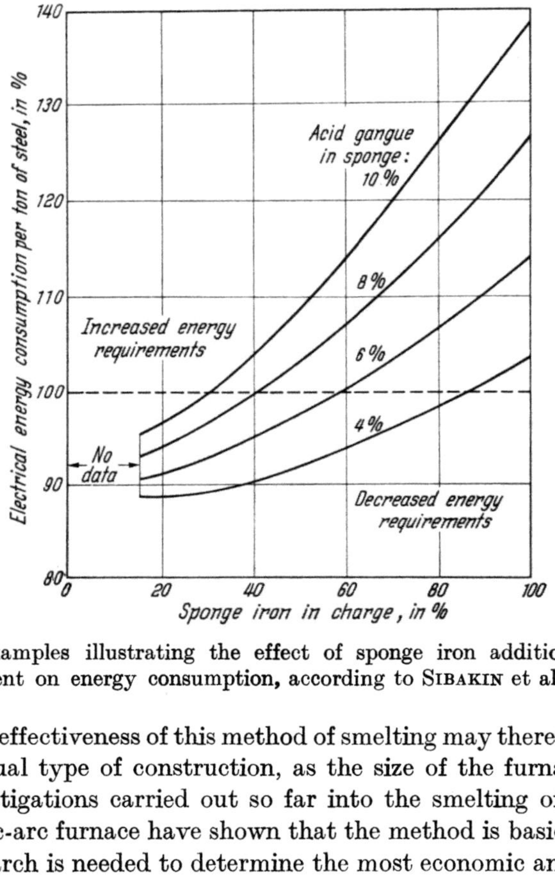

Fig. 151 b. Examples illustrating the effect of sponge iron addition and gangue content on energy consumption, according to SIBAKIN et al.[352b]

the arcs. The effectiveness of this method of smelting may therefore decrease, given the usual type of construction, as the size of the furnace increases.

The investigations carried out so far into the smelting of sponge iron in the electric-arc furnace have shown that the method is basically feasible. Further research is needed to determine the most economic and technically efficient method of operation.

In oxygen converter processes for steel-making, sponge iron can be used instead of, and to the same extent as, scrap iron as a cooling medium. Tests have been carried out in LD-vessels of 3 tons[354], 5 tons[354a], and 35 tons, in a 40-ton LDAC-vessel[359], and in a Kaldo furnace[362]. Various results[354, 354b, 359] have been found as regards the cooling effect, depending on the nature of the material charged and the charging method. The equivalence ratio compared with scrap was found to be between 0.83 and 1.4 kg of sponge iron per kg of scrap iron replaced; the larger additions were possible with continuous charging[354b]). The chemical composition, size analysis, and bulk density of the sponge iron charged are shown in Table 16a.

Table 16a. *Composition and properties of the sponge iron used as the cooling medium in the oxygen converter*

	RN-briquettes[350, 359])	Bomi-Hill sponge iron[354])	Itabira sponge iron[354b])
Chemical composition (%)			
Total Fe	90–95	82.2 –86.0	94.0
Metallic Fe	80–89	69.7 –77.2	87.5
Metallization	89–94	84.8 –89.8	93.0
Gangue	max. 3	11.2 –12.1	≈ 8.2
C	≈ 0.2	0.23– 0.40	0.20
P	max. 0.03	0.12– 0.15	0.04
S	max. 0.03	0.08– 0.11	0.05
Size analysis (%)	dia. 95 mm 40–75 mm high	75% < 1 mm	>10 mm, 16.2 5–10 mm, 41.8 3– 5 mm, 17.0 1– 3 mm, 16.0 < 1 mm, 9.0
Bulk density (tons/m^3)	3.5	≈ 2*	1.7

* Estimated.

Tests in the 5-ton LD-converter with additional heating through a heating/refining lance and with continuous feeding of sponge iron through the nozzle were successful, but with 30% highly reduced pellets in the input, additional heat amounting to 560000 kcal per ton of the total input was required[354a]).

The susceptibility of sponge iron to re-oxidation militates against its use in the open hearth furnace; nevertheless, in small furnaces small proportions of iron sponge have occasionally been used[350, 355]).

Many direct-reduction processes yield sponge iron in dust form, and attempts have also been made to handle the fine-grained material directly in electro-steel furnaces[358, 342a]) and to use it as a cooling medium in oxygen converters[342a]).

Attempts to transform sponge iron into rolled products without smelting were successful, in principle[360, 361]), but the difficulty of producing concentrates of high purity is an obstacle to using the method on a large scale.

Table 16b. *Characteristics of the processing of ore lumps in direct-reduction plants with a daily output of 500 tons Fe*[357c])

Locations: A_1, A_2 = Rhine-Ruhr; B = Rotterdam

Characteristic		Krupp sponge-iron	SL-RN	HyL	Purofer
Ore, 67.9% Fe					
amount delivered . . .	kg/ton Fe	1880	1957	2056	2060
amount charged . . .	kg/ton Fe	1500	1507	1496	1498
size range*	mm	3.5—27	3—27	10—40	10—40
screened fine ore . . .	kg/ton Fe	380	450	560	562
Flue dust loss and					
Fe-loss, amount . . .	kg/ton Fe	60	77	32	32
Fe-content	%	30	30	50	50
Rough dolomite, 0—3 mm	kg/ton Fe	60	50	—	—
Ash	kg/ton Fe	100	100	—	—
Returned coal	kg/ton Fe	300	500	—	—
Fuel					
A_1 Natural gas, $H_u = 8000$	m^3/ton Fe at NTP	—	—	708**	415
Browncoal, 0—12 mm, $H_u = 5100$	kg/ton Fe	373	670	—	—
Coke fines, 0—3 mm, $H_u = 6200$	kg/ton Fe	235	—	—	—
A_2 Natural gas	m^3/ton Fe at NTP	—	—	708**	415
Imported coal, $H_u = 7000$	kg/ton Fe	480	489	—	—
B Natural gas	m^3/ton Fe at NTP	56	—	708**	415
Imported coal, $H_u = 7000$	kg/ton Fe	416	489	—	—
Electrical energy	kWh/ton Fe	35	47	10.5	93
Water					
circulating water . . .	tons/ton Fe	—	20	90	60
added water	tons/ton Fe	2.2	3.0	9.0	5.0
Catalyst	kg/ton Fe	—	—	0.068	—
Personnel					
operators' posts*** . .	—	1 + 4	1 + 3	1 + 8	1 + 3
staff	—	2	2	2	2
Iron sponge, 93% Fe, $\Re = 95\%$					
sulphur content . . .	%	0.030	0.030	0.020	0.010
carbon content . . .	%	0.1—0.2	0.1	<1.80	<0.5
Excess steam	kg/ton Fe	—	—	64	—
Plant floor space	m^2	7500	10000	7000	6000
Size of reduction chamber	m	4.2∅ × 68	4.0∅ × 60		3.1∅ × 13
Total heat requirement .	Gcal/ton Fe****	3.36	3.42	5.66	3.32

* Screening free from smaller grain sizes.

** Degree of reduction 85%.

*** On 35 standstill days only two posts occupied.

**** G = 10^9.

Investigations have been made into the economics of using sponge iron as a charge in the blast furnace[356, 357, 357a,b]) in electric-arc furnaces[357c]), and in the LD-converter. The difficulty lies in the variability of prices from place to place and from time to time when an attempt is made to introduce suitable parameters.

Evaluation is particularly difficult when pre-reduced material is used as the input in a blast furnace, since the blast-furnace process itself must be regarded as being still in a state of evolution and as having not yet reached the limit of its possible performance (see section 5.5). In the investigations carried out to date[363, 363a,b]), the increased performance and economy in coke which have been achieved using pre-reduced burdens are strictly valid only for the particular operational conditions at the time of the investigation. The increase in throughput, found in an operational test, from 47.7 to 58.6 tons of pig iron per day per m^2 of hearth area when using 30% of highly reduced pellets in the burden[363a]) loses its significance when we compare it with the output of 56.3 tons pig iron per day per m^2 which can now be achieved without additional fuel consumption or pre-reduction[363c]).

The increase of 42.5% in output, when using continuous charging, which was found in the evaluation of sponge iron as a substitute for scrap in the electric-arc furnace[357c]), also needs to be checked by further experiments on the large scale, especially considering the recent UHP technique (ultra high power).

The characteristics for the processing of screened lump ore are given in Table 16b for 2 locations and 4 different direct-reduction processes. Further details of this comparison, e.g., the benefits of green (unburned) pellets, are compiled in[357c]).

4.2. Reduction in Dispersed-Cloud Apparatus and Shower Furnaces

4.2.1. Limits of the Region of Applicability

4.2.1.1. Maximum Rates of Reduction

The reduction rate is particularly high with very small particles since they offer a particularly large surface area for the reaction between the gas and the ore. Figure 152 contains, as a guide to the upper limit for the reaction rate, the values measured by KNACKE[364]) and also by BENARD et al.[365]) for the phase-boundary reaction

$$FeO + H_2 = Fe + H_2O$$

for wüstite samples of definite surface area.

The measured reduction rates v are given by the following equation:

$$v = 131\, n_{H_2} \exp(-15200/RT) \quad (\text{mole O/cm}^2\ \text{sec}),$$

where

n_{H_2} = H_2-concentration (mole/cm²),
R = gas constant (cal/mole °K),
T = absolute temperature (°K).

However, the rates calculated from this equation will vary depending on the previous treatment of the wüstite, and, in particular, on the amount of lattice disorder.

Ezz and Wild[366]) reduced very fine particles (50 to 150 µm) of ores of various origin, whilst suspended as a dispersed cloud in H_2, and found

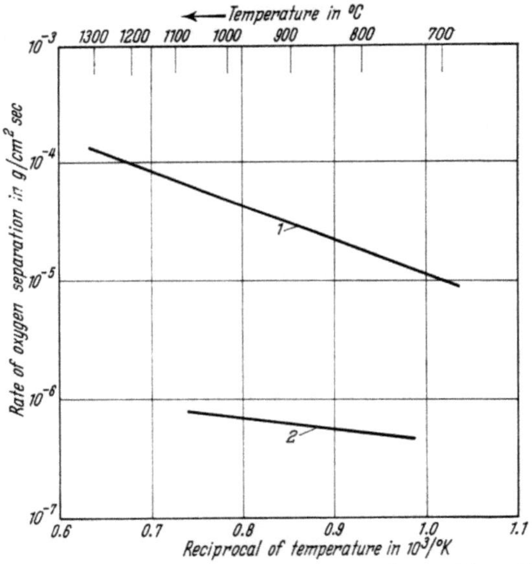

Fig. 152. Rate of removal of oxygen for very small particles. Comparison with the phase-boundary reaction in pure hydrogen

1 Phase-boundary reaction: $131 \cdot \exp[-15200/RT] \cdot n_{H_2}$ (mole O/cm² sec); n_{H_2} = H_2-concentration in the gas phase, in mole/cm³; according to Knacke[364])
2 Observed values for fine ore particles (d = 50 to 150 µm); material: Rif ore; 95% Fe_2O_3; interior surface area (as supplied) about $1.5 \cdot 10^4$ cm²/g; according to Ezz and Wild[366])

that in the temperature range 700 to 1100°C the time for an 80% reduction lay between 20 and 30 sec, and that this time was, in fact, almost independent of the kind of ore and of the grain size (see Figs. 153 and 154). If the measured rates of the reaction are related to the internal surface area (although, certainly, the latter can only be roughly estimated for the state of the ore as supplied), then the lower line in Fig. 152 is obtained,

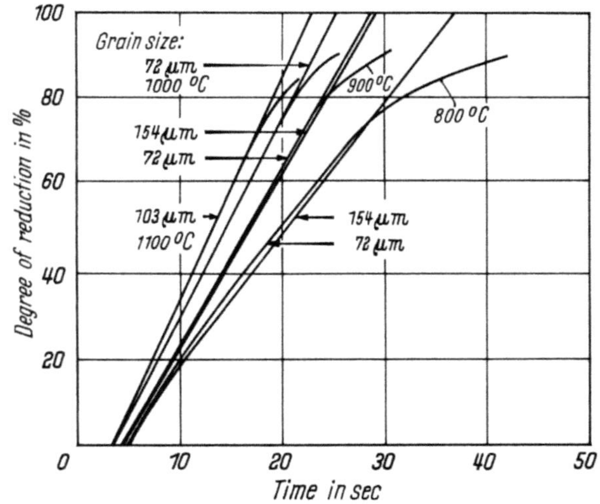

Fig. 153. Degree of reduction as a function of time (very fine grains, at various temperatures in H_2), according to Ezz and WILD[366])

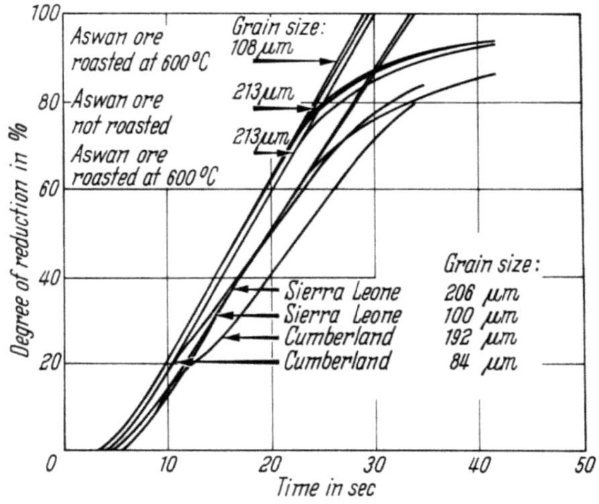

Fig. 154. Degree of reduction as a function of time (three different ores, 900° C in H_2), according to Ezz and WILD[366])

with values less by a factor of 10^{-1} to 10^{-2} than those for the unaffected phase-boundary reaction rate. The fact that in Ezz and WILD's experiments the gas phase becomes leaner in H_2 through enrichment with H_2O explains only part of this discrepancy. More important must be the fact that diffusion in the reacting solid has a retarding influence, since the particle diameters are still always large compared with the mean distance between the Fe-nuclei. An appreciable acceleration of the reduction is again to be expected only if the grain diameter is less than about 1 μm.

Since in full-scale industrial plants for ore reduction a particle size of about 50 μm must, in general, represent the lowest feasible limit, we shall, in the following considerations, take the minimum time needed for complete reduction to be 20 sec, the least reduction time observed by Ezz and Wild.

4.2.1.2. The Most Favourable Operating Conditions

The maximum conversion per unit volume of the reaction vessel for a fine-dust-cloud process can be deduced from the following considerations.

Suppose that in a reaction chamber, assumed to be cylindrical and represented in Fig. 155, heated H_2 as the reducing gas comes from above,

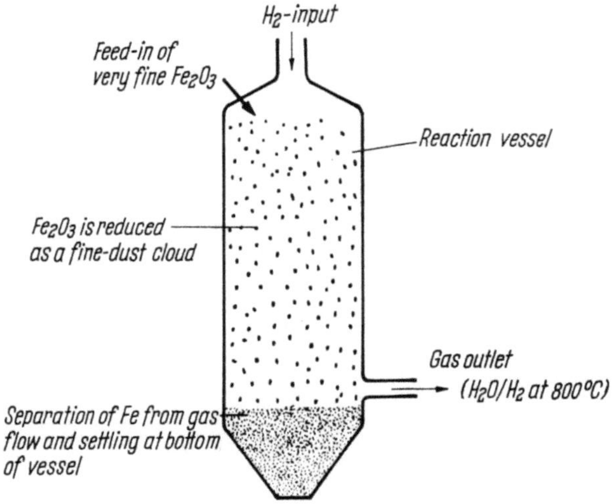

Fig. 155. Reduction in a fine-dust cloud (diagrammatic)

forms a dust cloud with the fine ore, and then leaves the chamber at the lower end, a special arrangement ensuring that the reduced ore is separated from the gas flow. The relative velocities of the ore particles, which are assumed to be very small, may be neglected compared with the gas flow.

4.2.1.2.1. Consumption of Reducing Gas for the Chemical Reaction. If the ore and gas are moving in the same direction, the maximum possible gas utilization is determined by the equilibrium Fe-FeO/H_2-H_2O or Fe-FeO/CO-CO_2. The minimum gas throughput needed is therefore given by the upper curves in Fig. 148, section 4.1. Thus, at a reaction temperature of 800°C, for instance, a minimum gas throughput of 1770 m^3 H_2(CO) per ton Fe at NTP is required.

4.2.1.2.2. Gas Requirement for Heat Transfer. If – as may probably usually be assumed – only the reducing gas is available as the carrier for the heat needed for the heating-up and reduction, then a comparatively high gas throughput will be needed, because the co-current principle is being used, to supply the heat requirement.

Under the conditions of 4.2.1.2.1, for example, the hydrogen on entering must have a temperature of about 1500°C in order that, with a throughput of 1770 m^3 H_2/ton Fe at NTP, after covering the heat requirement for the reduction and heating of the Fe_2O_3 there should still be an exhaust-gas temperature of 800°C, equal to the reaction temperature for this case. If the preheat temperature is allowed to go lower, then a longer reduction time will be needed because the exhaust-gas temperature also falls, and further, if H_2 is used, the maximum attainable chemical utilization of the gas will be worse, as shown in Fig. 148 of section 4.1. From these considerations, therefore, a high requirement of heating and reducing gas is to be expected.

4.2.1.2.3. Maximum Loading of the Reduction Vessel. It appears to be possible to put arbitrarily large amounts of gas through the reaction vessel per unit time without endangering the result of the reaction. This is one of the advantages of processes operating with dispersed clouds as compared with fluidized beds or stationary charges.

Nevertheless there is a limit to the throughput capacity of the reaction vessel: for if the gas velocity is substantially higher than the rate of fall of the ore particles in the reducing gas, then the relative velocity of the solid in the gas is negligible. This happens with gas velocities above 2 m/sec for particles of grain diameter below 80 μm. The gas must then remain in the reaction vessel for at least the period needed for complete reduction.

Figure 156 shows, for a reduction temperature of 800°C and normal pressure, the attainable throughput performance per unit volume of reaction vessel as a function of the time needed for the reduction. In comparison with blast-furnace performance (up to 2.5 tons Fe per day per m^3 of working volume) even the most favourable performance figures conceivable (0.5 tons Fe per day per m^3 at a reduction time of 20 sec) seem very low. Increases are, however, possible at higher pressures, and in fact the performance is proportional to the working pressure of the gas in the reaction space.

The calculation which leads to Fig. 156 may be explained by means of an example. Suppose the reduction time is $\tau = 25$ sec, the volume of the reactor $V_R = 1\ m^3$, the gas requirement $1770\ m^3\ H_2$/ton Fe at NTP (approximating to the equilibrium requirement), and the mean temperature in the reaction vessel 1150°C. Then the volume of gas to be put through is

$$1770 \cdot \frac{1423}{273} = 9230\ m^3/\text{ton Fe}.$$

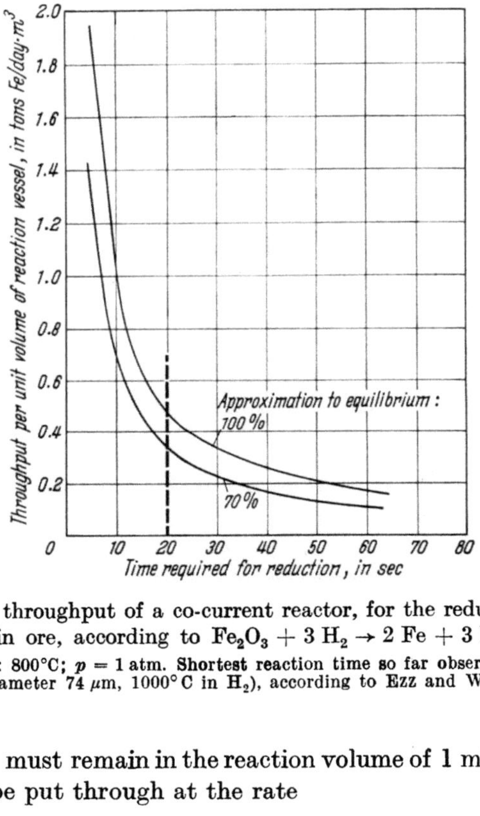

Fig. 156. Maximal throughput of a co-current reactor, for the reduction of very fine grain ore, according to $Fe_2O_3 + 3\ H_2 \rightarrow 2\ Fe + 3\ H_2O$

Reaction temperature: 800°C; $p = 1$ atm. Shortest reaction time so far observed is 20 sec (Rif ore, grain diameter 74 μm, 1000° C in H_2), according to Ezz and Wild[366])

The ore and gas must remain in the reaction volume of $1\ m^3$ for $\tau = 25$ sec. So the gas can be put through at the rate

$$\frac{1\ m^3}{25\ sec} = \frac{1\ m^3}{25\ sec}\ \frac{3600\ sec}{1\ h} = 144\ m^3/h.$$

The rate of output of iron is therefore

$$144/9320 = 0.0155\ \text{ton Fe/h}\ m^3 = 0.374\ \text{ton Fe/day} \cdot m^3.$$

If the reduction times were reduced to less than 5 sec, then the performance of large, modern blast furnaces would be equalled. According to Ezz and Wild[366]), however, reduction times of this order are conceivable only if the requirement for a high degree of utilization of the gas is dropped.

The conditions are similar in *falling curtains of fine particles.* In this case, if the velocity of the rising gas flow is small, then the countercurrent principle roughly holds. However, the rate of descent of the grains must be greater than the upward velocity of the gas (see section 3.2, Fig. 138).

4.2.2. Techniques for Carrying out the Reduction of Ores in the Form of Suspended Clouds and Falling Curtains of Fine Particles

CAVANAGH[367]) developed a process called "jet smelting" (Fig. 157) which approximates quite closely to the theoretical model shown in Fig. 155.

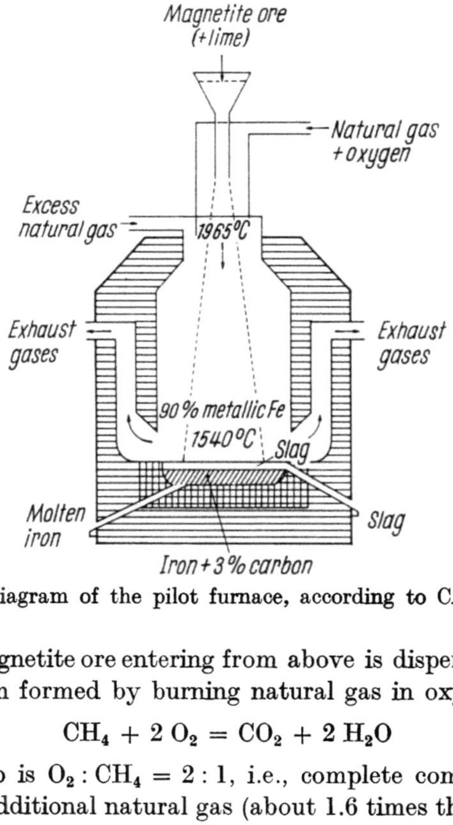

Fig. 157. Diagram of the pilot furnace, according to CAVANAGH[367])

Finely ground magnetite ore entering from above is dispersed in a strongly heated gas stream formed by burning natural gas in oxygen:

$$CH_4 + 2\,O_2 = CO_2 + 2\,H_2O$$

(the volume ratio is $O_2 : CH_4 = 2 : 1$, i.e., complete combustion). By the introduction of additional natural gas (about 1.6 times the combustion gas requirement) in an annular space, a reducing atmosphere is formed. The heat requirement for reduction, heating-up and melting of the ore particles must be covered by the sensible heat content of the gas. The separation of the reduced material from the gas stream takes place in the lower part of the apparatus, the molten iron collecting in a hearth and being tapped from time to time. A separate tap hole is provided for the slag formed. The pilot furnace demonstrated the practicability of the principle, but continuous operation data have not yet been published. The heat and gas requirements were given[367]) as 1510 m³ natural gas/ton pig iron at NTP and an additional 1170 m³ O_2/ton pig iron at NTP, i.e., about $14.6 \cdot 10^6$ kcal/ton pig iron. This extraordinarily high energy requirement is the result of using the co-current principle, which also means working with an exhaust gas

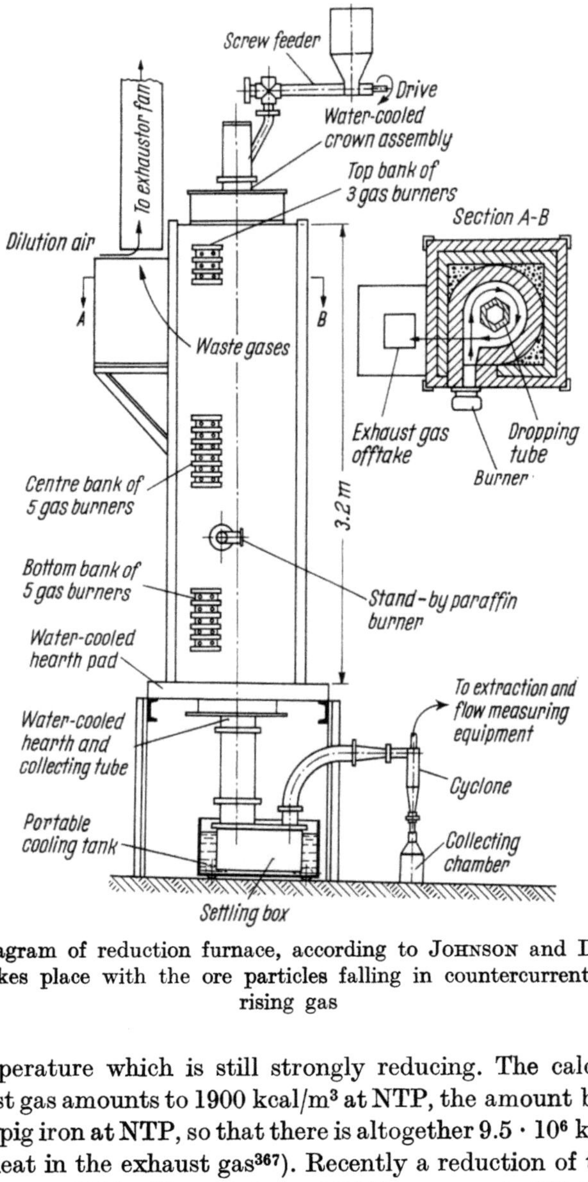

Fig. 158. Diagram of reduction furnace, according to Johnson and Davison[368]). Reduction takes place with the ore particles falling in countercurrent against the rising gas

of high temperature which is still strongly reducing. The calorific value of the exhaust gas amounts to 1900 kcal/m³ at NTP, the amount being about 5000 m³/ton pig iron at NTP, so that there is altogether $9.5 \cdot 10^6$ kcal/ton pig iron latent heat in the exhaust gas[367]). Recently a reduction of the natural gas requirement to 1000 m³/ton pig iron at NTP was reported[369]). In the jet-smelting process, carbon is added to the pig iron in the hearth. If the ore particles are not completely reduced during their flight, the residual reduction can take place by means of this carbon in the bath. Thus in this process natural gas and carbon are to some extent interchangeable, but the overall fuel requirement must be taken into account when comparing with competing processes.

In the BISRA experiments[368]) with the "flame smelting process" the size of the ore grains and the gas velocity were adjusted to one another so that a countercurrent partial reduction was possible, as mentioned in

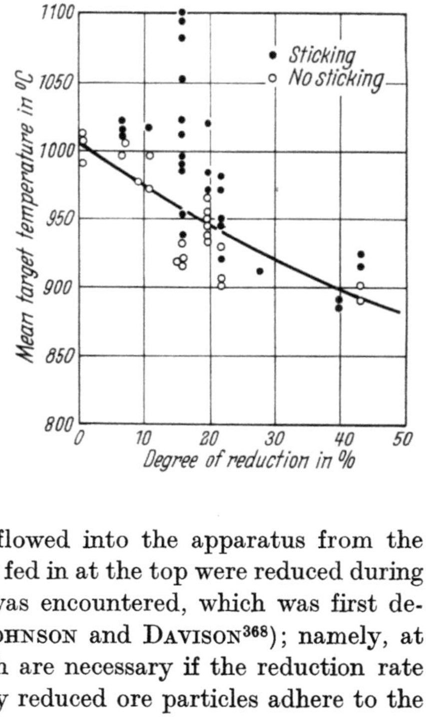

Fig. 159. Effect of temperature and degree of reduction on the adhesion of ore to the refractory lining, according to JOHNSON and DAVISON[368])

Table 9. As shown in Fig. 158, gas flowed into the apparatus from the bottom upwards, and the ore particles fed in at the top were reduced during their fall. Here a further difficulty was encountered, which was first described in this type of process by JOHNSON and DAVISON[368]); namely, at the temperatures above 900°C, which are necessary if the reduction rate is to be sufficiently high, the partially reduced ore particles adhere to the walls of the chamber. The further the reduction has proceeded, the lower the temperatures at which adhesion takes place (Fig. 159).

In the VÖEST process[369a]) the ore is reduced and melted as a "falling curtain" in a narrow shaft using a gas mixture of 75% CO and 25% H_2 at a temperature of 1550 to 1650°C. The gas is generated from wet coal and oxygen in a furnace chamber heated by electric arc. Depending on the operating-temperature, the product is pig iron or steel with 0.1 to over 2% carbon. Lime is added with the coal for the removal of sulphur. In the pilot plant producing 12 tons per day, 865 kg of (wet) coal, 730 m^3 O_2 at NTP, and 110 kWh were needed per ton of molten steel.

Because of the stated fundamental limitations and technical difficulties—a small throughput or a necessarily high pressure in order to obtain an acceptable performance, unsatisfactory chemical and thermal utilization of the reducing gas, the sintering (adhesion) of the ore particles—caution is needed in assessing the prospects of reduction processes using dispersed clouds of falling curtains of ore particles. Similar processes which use cyclone burners and smelting chambers[370]) will have to overcome similar difficulties.

4.3. Reduction in Fluidized Beds

4.3.1. Possible Charge Material and Limits of Operation

The possibilities and limits for iron reduction in the fluidized bed can already largely be stated from the earlier chapters on the theory and from laboratory-scale tests.

As shown in section 3.2, fluidized beds are limited to a range of gas-flow velocities bounded above by the elutriation velocity (Stokes's terminal velocity) and bounded below by the fluidization point. This domain of existence for fluidized beds is depicted in Fig. 138, section 3.2, for a temperature of 700°C and either H_2 or CO (N_2 behaves almost the same as CO as regards fluidized-bed characteristics). The specified ranges of the flow velocity are valid if all the particles have the same diameter. For mixtures of several grain sizes, such as always occur in industrial applications, the upper limiting velocity is the elutriation velocity for the *smallest* particles, since it is, of course, these which are first elutriated.

Since the reduction of fine particles by heated gas takes place extraordinarily rapidly, the chemical reaction is not, in general, rate-determining. With these high rates of transformation the reduction gas comes into equilibrium after a short contact time and can effect no further reduction. The amount of gas which can be fed into the fluidized bed per unit time therefore has a marked effect; the upper limiting velocity as given in Fig. 138 thus determines the maximum throughput of the reactor. If we start from these upper limiting velocities, dependent on the grain size, and assume the most favourable case to be when the reducing gas is chemically utilized according to equilibrium values, then the performance limits shown in Fig. 160 are obtained, for 700°C and for either H_2 or CO. We see that a performance similar to that of the blast furnace can be achieved only if the ore used has no appreciable component of grain size below 0.2 mm. But as the grain size goes up substantially, the previous assumption of a rapid approximation to equilibrium between gas and ore no longer holds. However, by increasing the working pressure in the fluidized bed the performance can be considerably improved. Compared with the performance shown in Fig. 160 (calculated for atmospheric pressure), the

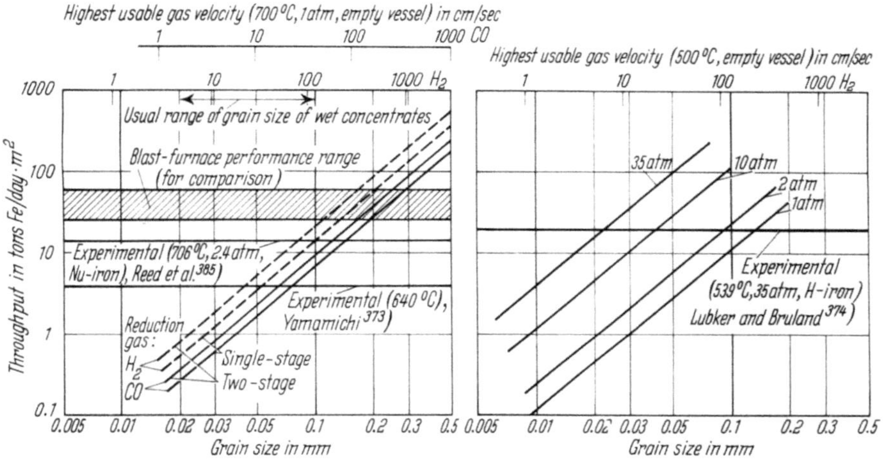

Fig. 160/161. Maximum throughput of a fluidized-bed reactor per m² cross-section as a function of grain size (gas utilized down to equilibrium)

Fig. 160. 700°C, 1 atm Fig. 161. H_2, 500°C, at various pressures

attainable performance increases approximately proportionally to the pressure [see Eq. (3) in section 3.2], because the available amount of gas is increased without increasing the flow velocity (Fig. 161).

Sub-division of the reduction into two or more fluidized beds arranged one above the other (Fig. 162) has a beneficial effect; not only the degree of chemical utilization of the gas (see section 4.1) but also the throughput performance (Fig. 160) is increased.

Figures 160 and 161 show the throughput performance per square metre of the reactor base area for fluidized-bed processes, some data being for operational reactors and some for pilot reactors. Here too the disparity from the blast furnace may be seen; this is the result of the small grain sizes used, 0.04 to 0.25 mm in the experiments of YAMAMICHI[373]), and 0.06 to 0.83 mm in the H-iron process[374]). The use of a coarser fine ore, from which the finest component has been removed, suggests itself, but will considerably increase costs and thus limit the employment of fluidized-bed processes; or, as was done in a Russian investigation, the fine ore may be made into pellets of 2 to 3 mm diameter[375]). But this step, too, is expensive and is contrary to the original idea of the fluidized bed, which was to avoid agglomeration.

There are many references in the literature to the difficulties arising in the reduction of ore in the fluidized bed owing to the ore grains sintering together. This endangers the stability of the fluidized bed and the feasibility

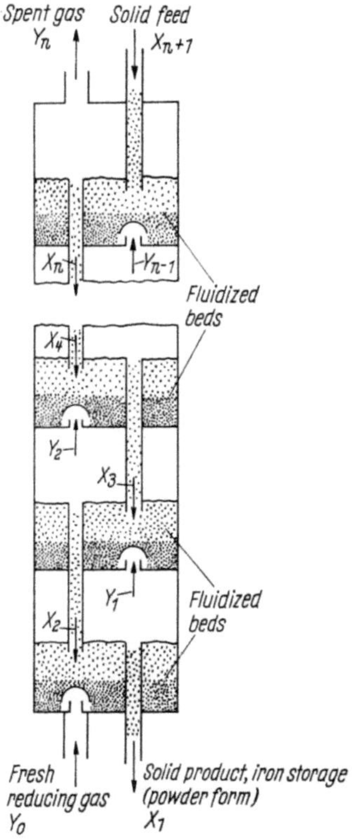

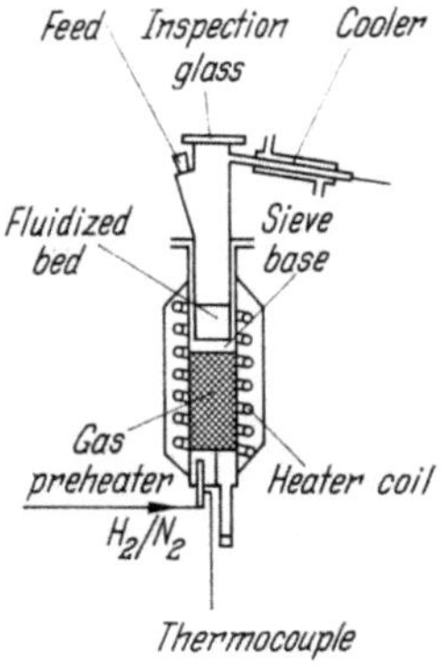

Fig. 163. Test rig for reduction of ore in a fluidized bed, according to WINZER[376])

Fig. 162. Diagram of multi-stage fluidized-bed reactor for ore reduction, according to MEISSNER and SCHORA[372])

Column containing n fluid beds or stages. Gases rise through this column, countercurrent to the descending solids. Representation of downcomer pipes and bubble caps on each stage is schematic

of the process. This phenomenon has therefore been examined in a special test rig[371, 376]).

Figure 163 shows a laboratory reactor working on H_2-gas. The behaviour of the fluidized bed could be observed through an inspection glass at the upper end of the reaction chamber. It was found that, especially at high operating temperatures, after some time the layer began to sinter together (Fig. 164). This sintering was at first seen in a small region and spread within a few seconds over the whole bed. The charge then stayed motionless on the porous base, and the reducing gas penetrated through a few channels without producing any worthwhile reduction. Figure 165 shows that this sintering action takes place after only a few minutes at temperatures above 700°C, that is, a long time before reduction is complete. This holds for the hematite used, with grain sizes in the range 0.06 to 0.1 mm. With larger grain size and ore of lesser purity this sintering process is mitigated but not completely repressed. Systematic measurements of these effects are not yet available, so far as we know.

However, the temperature must be regarded as the main factor which determines when sintering will take place; processes which are to be adopted industrially must therefore work either at low temperatures (below 500°C)

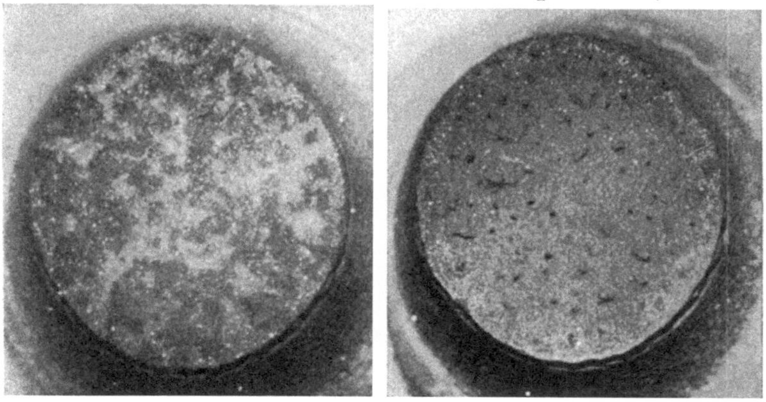

a) b)

Fig. 164. Fluidized beds of ore during reduction[376])
a) Before sintering occurs b) After sintering occurs

or else with the ore remaining for the shortest possible time in the fluidized bed, and consequently with a low degree of reduction.

If the reducing gas contains CO, then at low temperatures the Boudouard reaction becomes of importance in this respect that, from 600°C downwards, the separation of carbon from the CO on to the ore would disturb the reduction. When the reducing gas is generated from natural gas or oil, however, it is always primarily CO which is formed (see section 4.8), and for low-temperature reduction processes this CO must be completely removed from the reducing gas before it is used. Repression of C-separation by means of anti-catalysts, such as traces of H_2S, is conceivable but does not appear to have been tried yet in fluidized-bed processes[377]).

In a larger test rig YAMAMICHI[373]) has tested a two-stage fluidized-bed process (Fig. 166). The ore fed in had grain sizes mainly between 0.04 and 0.25 mm (Table 17). The gas velocity was around 35.6 cm/sec and therefore

Table 17

Mesh size . . . . mm	+ 0.25	+ 0.15	+ 0.10	+ 0.074	+ 0.044	− 0.44
Ore I . . . . . %	0	4.5	4.0	19.3	56.0	16.2
Ore II . . . . %	25.0	18.0	18.0	12.0	20.0	7.0

higher than the admissible maximum velocity for the fine component of the ore. As might be expected, this had the result that about 30% of the feed was elutriated by the gas and was collected in a cyclone (Fig. 167;

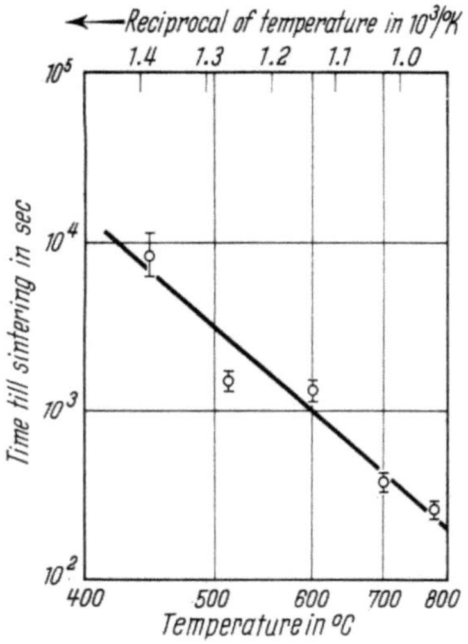

Fig. 165. Time till sintering of hematite ore (grain size 0.06 to 0.1 mm) reduced by hydrogen, according to WINZER[376])

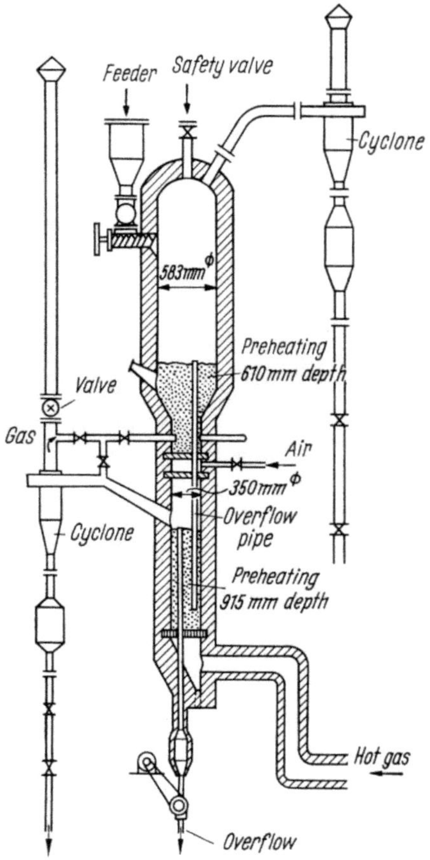

Fig. 166. Fluidizing reduction furnace for iron ore, according to YAMAMICHI[373])

and see Table 19). Since this elutriated fraction showed only a slight degree of reduction, in an industrial plant it would have to be fed back into the reactor, and would probably circulate several times. YAMAMICHI was able to eliminate the sintering of the charge by reducing the operating temperature from about 700° to about 600°C, in agreement with the results shown in Fig. 165.

SCHENCK et al.[378]) were able to carry out the reduction in a fluidized bed at 1000°C, the fine ore being suspended in a fluidized bed of fine coke. Here the coke reinforced the reducing action of the gas, so that quite high rates for ore throughput were obtained. At 900°C it was possible to achieve a high degree of reduction (Fig. 168).

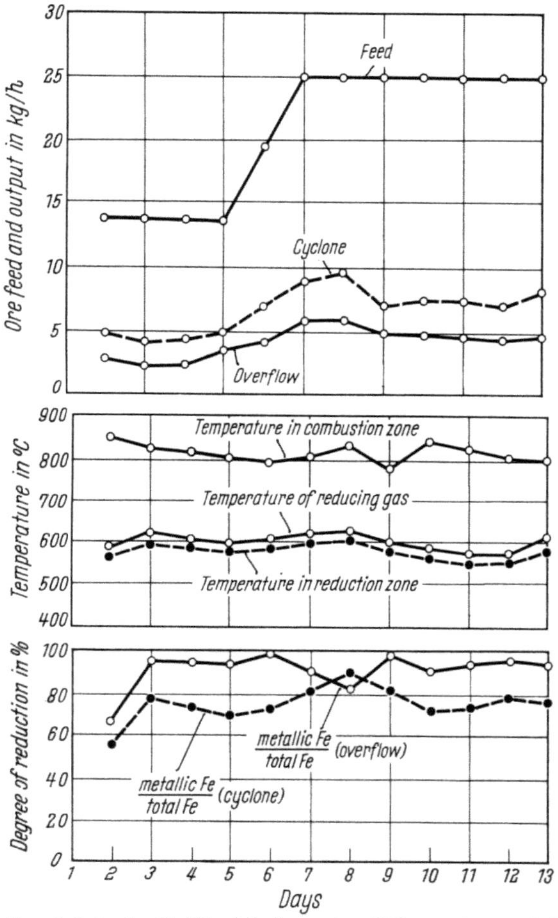

Fig. 167. Operational data for fluidized-bed reactor (350 mm internal diameter), with gas mixed with 4% air in the preheating zone, according to YAMAMICHI[373])

LANGSTON and STEPHENS[379]) made deliberate use of the tendency of fine ore to sinter by first setting up a fluidized bed of ore with comparatively coarse grading, 0.30 to 0.84 mm and then continuously adding fine ore (100% below 0.04 mm) during the reduction. This very fine ore sintered, during reduction, on to the coarse ore particles, thus increasing their diameter (Fig. 169).

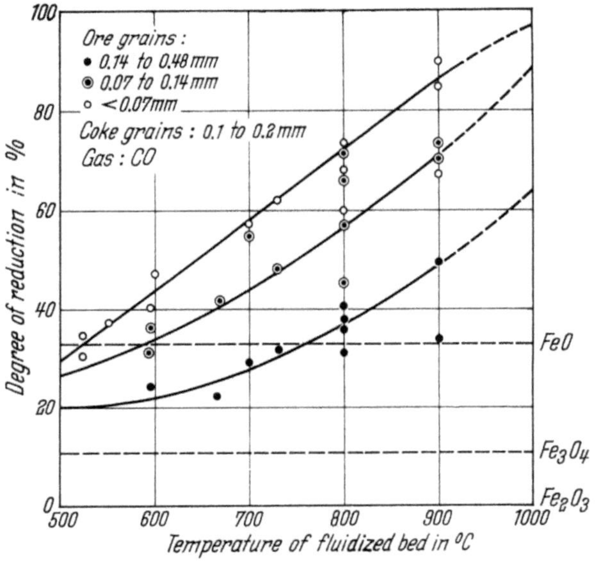

Fig. 168. Effect of temperature of fluidized bed on the degree of reduction obtained with ores of various grain sizes, using CO as the fluidizing gas, according to SCHENCK et al.[378])

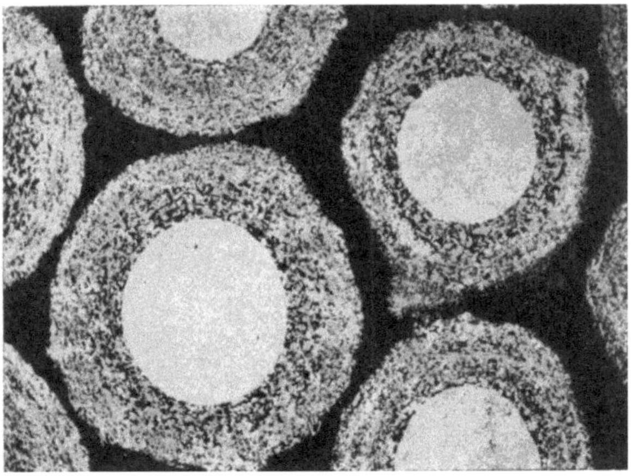

Fig. 169. Reduced ore particles from the self-agglomerating fluidized bed; reduction at 820°C (cross-section, 50 : 1); according to LANGSTON and STEPHENS[379])

Fluidized-bed reduction tests (at 762°C in H_2) with different kinds of ore[380]) showed that ores containing a lot of magnetite are much more difficult to reduce (Table 18).

Table 18. *Fluidized-bed reduction of various fine ores*

Grain size: 0.15 to 0.84 mm Temperature: 762°C Gas flow velocity: 18.3 cm/sec

Name of ore	Time required for 90% reduction					
	Percentage by weight (dry)				Magnetic	$t_{90\%}$
	Fe	O_2	SiO_2	Al_2O_3	%	min
Venezuela	66.5	28.3	0.9	2.4	0.0	40
Labrador	60.8	25.6	7.9	1.7	0.0	42
Mexico	64.0	28.1	1.6	1.1	32.3	45
Siderite	40.5	14.1	10.9	4.2	0.0	26
Marcona	62.8	26.3	6.7	—	19.8	74
Rusk Crude	38.8	14.6	15.5	6.5	0.0	30
Rusk Conc.	42.2	15.4	16.3	7.2	0.0	34
Brazil	66.4	28.7	0.6	0.7	1.6	52
Chile	64.7	26.4	5.4	1.1	77.8	>140
Liberia	65.1	26.7	3.5	1.2	72.8	>140

Feinmann[394]) has established that, in agreement with theory, in reduction by H_2 (at 600 to 700°C) in the range 0.21 to 1.19 mm, grains of all sizes are reduced at the same rate. Venezuela ore of high porosity was used.

4.3.2. Review of Processes

Many proposals for ore reduction in fluidized beds have been made. Some have been extensively developed industrially. A review of these methods up to 1962 is contained in [381]). The results obtained in these processes are largely in accordance with the above theory and with laboratory tests (Table 19). Essentially there are three groups of working conditions:

1. Low temperature, pure H_2, high pressure (H-iron process)[374, 382, 383]).
2. Medium temperature, pure H_2, medium pressure (e.g., Nu-iron process)[384–387]).
3. High temperature, H_2/CO mixture, slight positive pressure (i.e., above atmospheric)[373, 388, 389]).

It has been possible to develop the H-iron process[383]) up to units of 100 tons Fe per day output (Fig. 170). Concentrates are reduced in the fluidized-bed reactor with pure H_2 at about 500°C and 35 atm quite thoroughly (more than 90%). The reducing medium is obtained from natural gas, coke-oven gas, or fuel oil by the following reaction stages:
cracking with excess oxygen

$$CH_4 + \tfrac{1}{2} O_2 = CO + 2 H_2 (+ H_2O + CO_2),$$

and conversion of the CO by excess steam

$$CO + H_2O = CO_2 + H_2 (+ H_2O),$$

Table 19. *Characteristics for fluidized beds used in the reduction of ores*

	Unit	YAMAMICHI	H-iron Small	H-iron Medium	Nu-iron	ERL	Blast-furnace for comparison
1. *Reactor dimensions*							
Internal diameter of the reactor (lowest bed)	m	0.350	0.85	1.65	0.419	0.30	9 (hearth diameter)
Height of fluidized bed	m	0.915	9.0*	29*	5.3	9.0*	
Cross-section of reactor	m^2	0.096	0.57	2.14	0.138	0.071	63.6 (hearth cross-section)
Number of fluidized beds		2	3	3	3	2	
2. *Throughput*							
Iron	$\frac{\text{tons Fe}}{\text{day}}$	0.36	10	50	2.0	1.0	3000
Iron relative to reactor cross-section	$\frac{\text{tons Fe}}{\text{day} \cdot m^2}$	3.8	17.5	23.4	14.5	11.0	55
Degree of reduction	%	≈ 80	98	98	90	85	100
3. *Ore charge*							
Chief constituent		Fe_2O_3	Fe_3O_4		Fe_2O_3	Fe_2O_3	
Grain size	mm	7% < 0.04 27% < 0.06 39% < 0.10 75% < 0.25	0.040—0.833	0.06—0.833	0—2	0—3**	
Rejected fraction	%	30	No data		No data	30—100	1—5
4. *Reducing gas*							
Input temperature	°C	640	540	540	706	930	
Composition		Coke-oven gas	H_2	H_2	H_2	$H_2/CO/N_2$***	
Pressure	atm	≈ 1	36	36	≈ 2.4	1—2	
Linear flow-velocity at input†	cm/sec	35.6	33.5	21.5	45.7	90	250—400
Chemical utilization in one circuit through reactor	%	≈ 18	≈ 5	≈ 5	≈ 24	≈ 35	≈ 45
5. *Estimated fuel consumption for large plant*	$\frac{\text{kcal}}{\text{ton Fe}}$	Coking-gas $3.8 \cdot 10^6$††	Natural gas $4.8 \cdot 10^6$ or heavy oil $4.9 \cdot 10^6$†††		Fuel oil $3.51 \cdot 10^6$	Natural gas $3.77 \cdot 10^6$†††	Coke $4.0 \cdot 10^6$

* Total height of plant.
** Least possible amount of fine-grained component to be used.
*** From decomposition of CH_4 with air.
† Relative to empty cross-section and at working temperature and pressure.
†† $H_u = 3800$ kcal/m³ at NTP. ††† $H_u = 8580$ kcal/m³ at NTP.

followed by removal of the H_2O and CO_2 in a hydraulic scrubber. There remains pure hydrogen, which is brought up to the reactor input temperature of about 540°C by passage through a heat exchanger and a gas heater. The CO component, which would in itself have been usable for the

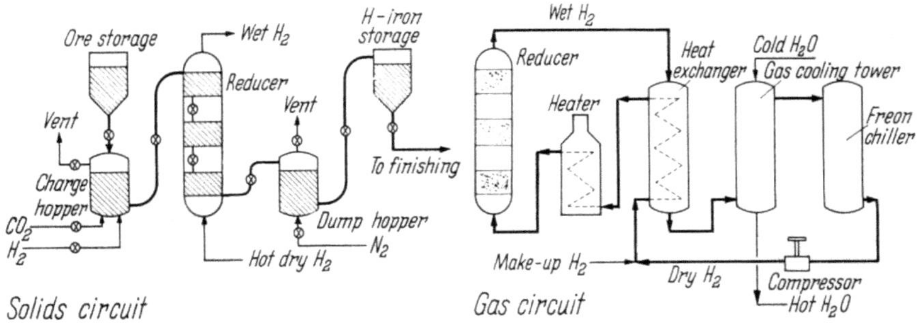

Fig. 170. Diagram of the H-iron process, according to LUBKER and BRULAND[374])

reduction, must be replaced as above by H_2, since otherwise the Boudouard reaction would take place in the ore at 500°C and would thus disturb the reduction through the separation of carbon. The reactor works with three superimposed fluidized beds, two for the reduction, one for preheating, so that countercurrent action between gas and ore is approximated. Nevertheless, because of the low temperature, the reduction proceeds slowly (Fig. 171), and the chemical utilization of the H_2 is quite low because of the unfavourable equilibrium position and the poor approximation to the equilibrium condition (Fig. 172). The purified spent gas

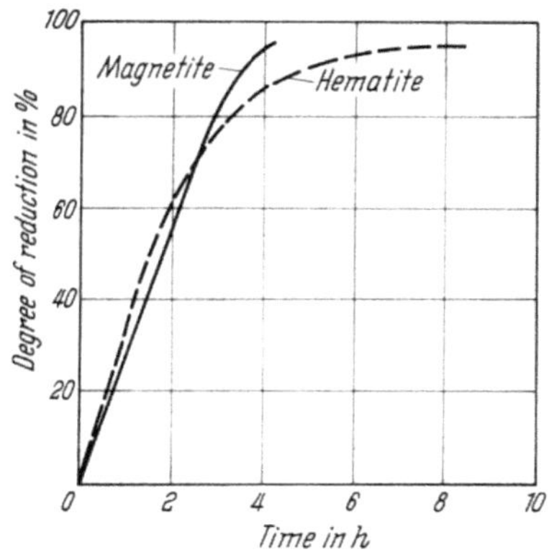

Fig. 171. Rate of reduction in the H-iron process, according to LUBKER and BRULAND[374])

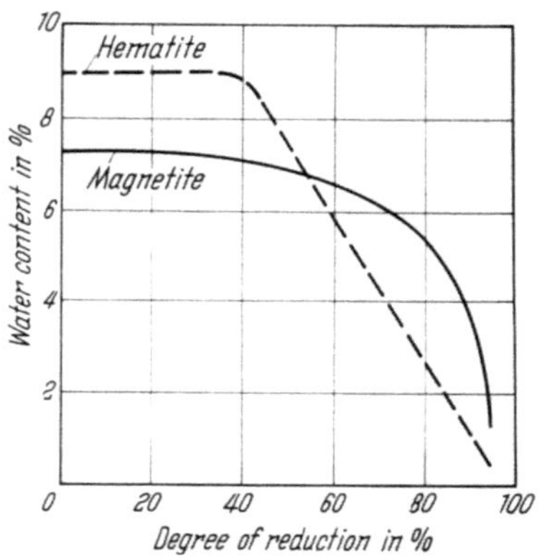

Fig. 172. Transformation of hydrogen to water in the H-iron process, according to LUBKER and BRULAND[374])

from the reactor (about 94% of the total quantity) is therefore re-circulated, and the H_2O formed by the reduction process is removed during each cycle in the hydraulic scrubber.

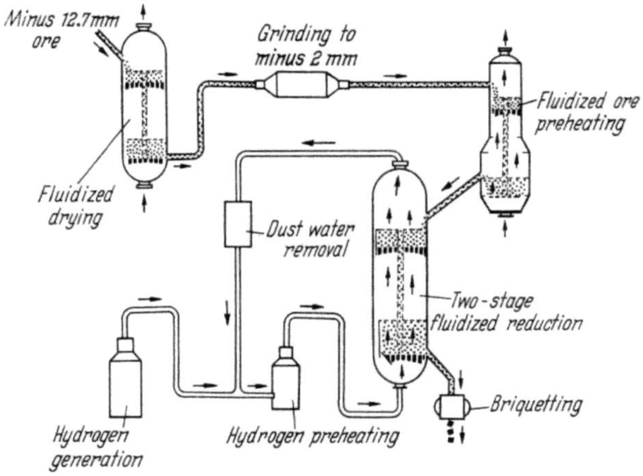

Fig. 173. Diagram of the Nu-iron process, according to Reed et al.[385])

The Nu-iron process (Fig. 173) differs from the H-iron process essentially in using a higher reaction temperature (600 to 760°C) and a lower gas pressure. The gas utilization is improved, however at the cost of some

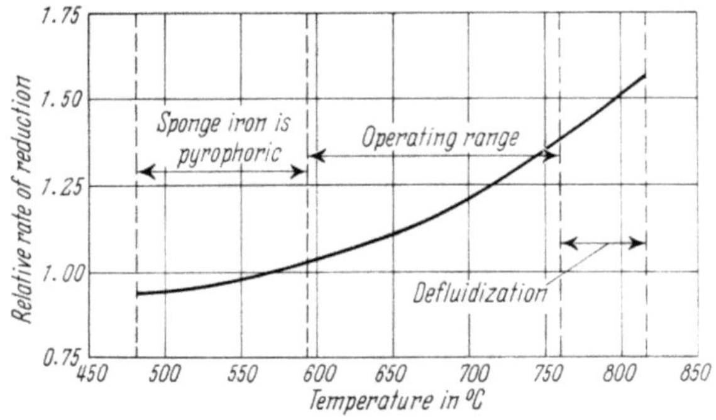

Fig. 174. Effect of temperature upon rate of reduction in Nu-iron process, according to Reed et al.[385])

uncertainty of control of the fluidized bed, since the sinter point is approached more closely (Fig. 174). Even with small bed heights, the gas utilization is close to the equilibrium value (Fig. 175). The unused H_2 is re-circulated after removal of the H_2O. There is in existence, so far, a semi-industrial plant with a daily output of 2 tons Fe; a big plant is under construction.

The Esso-Research-Little (ERL) process works without gas feedback at even higher temperatures. Consequently the CO which is formed in the reducing gas by cracking natural gas or oil can remain, and thus the pre-

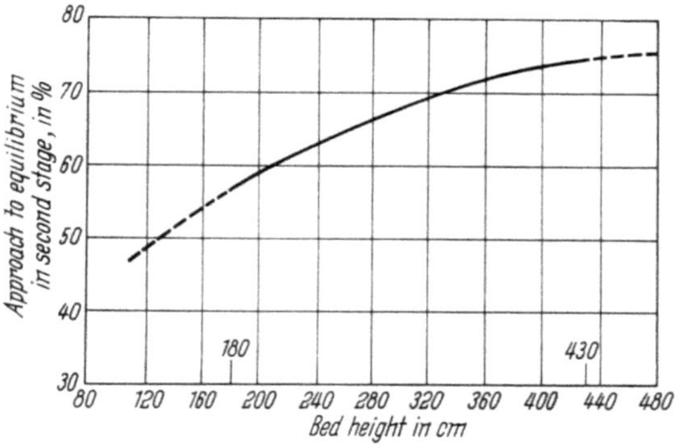

Fig. 175. Effect of bed height on the approach to equilibrium in Nu-iron process, according to REED et al.[385])

paration of the gas is considerably simplified. However, the fluidized bed can then be kept stable only by breaking off at about 85% reduction.

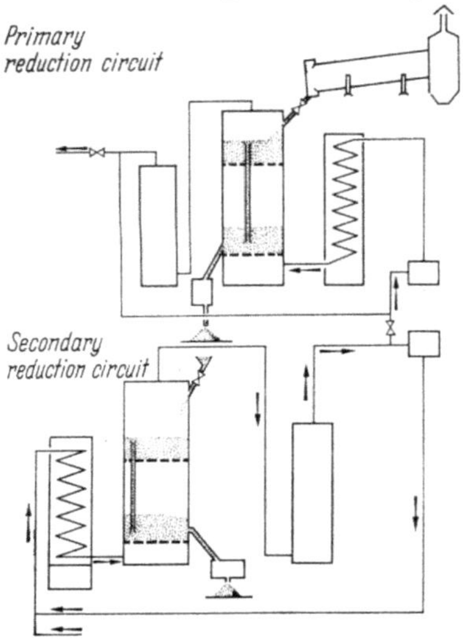

Fig. 176. Diagram of two-stage reduction process[389b])

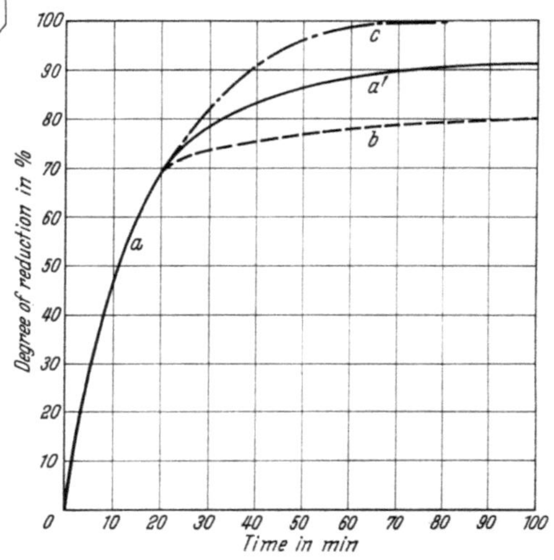

Fig. 177. Reduction of a hematite with 69% iron in two stages

a Start of reduction at 700°C, *a'* Continuation of reduction at 700°C without interruption, *b* Continuation of reduction at 550°C without interruption, *c* Continuation of reduction at 550°C after complete cooling and reheating

Similar to the ERL-process is the Fluid Iron Ore Reduction (FIOR) process[389a]), developed by Esso, in which up to 97% reduction can be achieved. At the end of 1965 a test plant with 300 tons per day output had been put into operation, but no detailed information about the process has yet been published.

The Novalfer-Onia-process[389b]) operates with two stages, each having two fluidized beds (Fig. 176). Figure 177 shows for Mauretanian concentrate with 69% Fe a phenomenon which is observed in the reduction of many ores by H_2. At a temperature of 700°C and when reduction is about 75% complete, the rate of oxygen removal decreases very sharply. After complete cooling, followed by reheating to about 550°C, the further reduction to more than 90% takes place without difficulty (curve *c* in Fig. 177). In the two-stage reduction the energy requirement is reduced in about the same ratio as the amount of hydrogen circulated (Table 19a). At present there is a pilot plant in operation with an output of 60 tons/day.

Table 19a. *Characteristics of the 12-ton/day Novalfer plant, in single-stage and two-stage reduction*[389d])

		Process		
		Single-stage	Two-stage	
Degree of reduction	%	91	90	
Ore preheated to	°C	800	800	
Temperature in fluidized beds:			1st stage:	2nd stage:
1st fluidized bed	°C	640	705	440
2nd fluidized bed	°C	570	680	600
Ore	kg/ton Fe	1510	1515	
Hydrogen	m^3/ton Fe at NTP	922	702	
Natural gas	m^3/ton Fe at NTP	222	189	
Oxygen	m^3/ton Fe at NTP	44	16	
Electrical energy	kWh/ton Fe	355	178	
Hydrogen circulated[389d])	m^3/h ton Fe at NTP	4700	2500	

In the reduction of dense hematite (Sierra Leone) by H_2 and H_2/H_2O mixtures, Rist et al.[389c]) have established the following special features of the kinetics. Reduction by a gas mixture which reduces to the metal produces a porous, sponge-iron layer round a dense oxidic core (Fig. 178a). If the ore then reacts with a gas mixture of composition corresponding roughly to that for the Fe/FeO equilibrium, then porous wüstite forms uniformly over the whole volume (Fig. 178b). On further reduction at about 570°C, after the wüstite has been in the fluidized bed for a certain time, the greater part of each grain is uniformly reduced (to about 96% degree of reduction), while the remainder is practically non-metallic. This phenomenon may be explained by assuming an incubation period for the iron nucleation and that finally the phase-boundary reaction is rate-determining.

In the CO-C-iron process[390, 391]), as in the Stelling process[392]), the Boudouard reaction is intentionally promoted, a fluidized bed of ore being

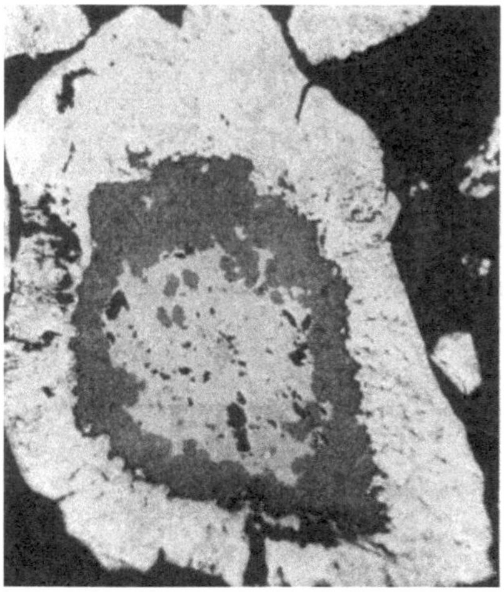

Fig. 178a. "Frontal" reduction of a particle of hematite in iron; iron (on the outside), magnetite-wüstite and hematite (on the inside) (reflected natural light, about 170 :1, original 250 : 1)

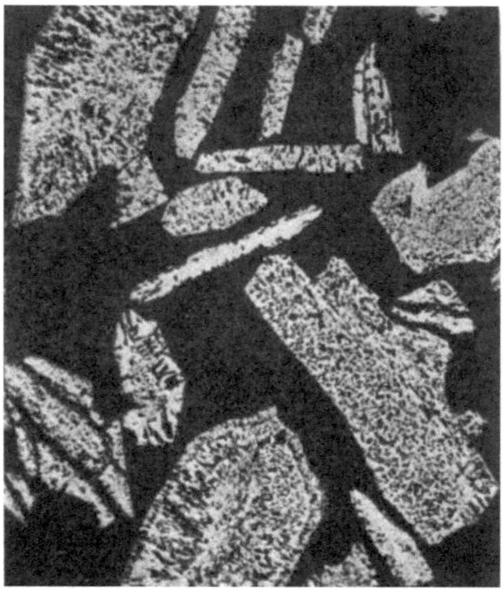

Fig. 178b. Porous wüstite obtained by controlled reduction of hematite (reflected natural light, about 85 :1, original 125: 1)

operated with CO-rich gas at about 500°C. Carbon precipitates on the ore particles in a controlled amount sufficient to enable subsequent reduction according to

$$Fe_2O_3 + 3\ C = 2\ Fe + 3\ CO\ (+\ CO_2)$$

in a second stage working at a temperature above 900°C. Because of the high temperature of operation this second stage takes place not in a fluidized bed but in a static bed; the process cannot therefore be regarded as a proper fluidized-bed process. The problem is to bring in sufficient heat to cover the needs of the strongly endothermic reaction.

The second stage could be carried out in batteries, similar to coke ovens. However, the amount of heat to be supplied for the reaction, per ton of product, is considerably greater than in coking.

4.3.3. Energy Requirement for Fluidized-Bed Processes and Comparison with the Blast Furnace

The energy requirement is composed of

a) the energy required for the cracking and reforming of the fuel and the manufacture of the reducing gas, and

b) the energy required for the reduction.

As to a), the cracking and reforming of the fuel (e.g., oil, or natural gas) and preparation of the reducing gas are accomplished in very different ways, depending on the type of fuel and process chosen, and will be discussed in section 4.8 in connection with methods of decomposition.

As to b), in the reactor, gas and heat energy are consumed in:

b 1) *Reduction*. The stoichiometric requirement is

602 m^3 H_2 or CO per ton Fe at NTP when Fe_2O_3 is the input, and
535 m^3 H_2 or CO per ton Fe at NTP when Fe_3O_4 is the input.

The actual gas requirement is determined by the degree of chemical utilization of the gas; it is this which gives the volume of gas to be supplied per ton Fe. When H_2 is used as the reducing gas (H-iron process, Nu-iron process), the H_2O is dried out from the gas and the remaining H_2 is recirculated for the further reduction. Consequently the actual H_2-consumption is then little greater than the stoichiometric quantity needed to remove the oxygen.

b 2) *Heating the Charge*. Either (as in the case of a H_2/CO mixture) part of the gas is burnt with air in the reactor and is used in an initial fluidized bed to heat the ore as far as possible above the operating temperature in order to accelerate the subsequent reduction, or, when H_2 is the reducing medium, additional fuel is used to heat the plant. Because of the heat expenditure in preparing the gas and also because the spent gas leaves the fluidized bed at a high temperature, the total heat expenditure is greater than the theoretical requirement, as Table 19 shows, although the systems are designed to utilize the heat in the exhaust gas, as may be seen from Figs. 170, 173, and 176.

As regards their performance per square metre of reactor cross-section, fluidized beds fall below the blast furnace, since, as Table 18 shows, it is expedient to work only with low gas velocities. Even the considerable increase in working pressure (in the H-iron process) cannot quite bridge the considerable performance gap up to the blast-furnace level.

Improvement of the output per m^2 of reactor cross-section by still further increasing the pressure would seem to be quite possible. On the other hand, increasing the height of the bed has little effect, since the gas throughput per m^2 of the reactor cross-section may not be increased. Agglomeration of the ore or its classification into narrow ranges, of about

3 mm, of grain size might also lead to considerable improvement in performance, by raising the admissible flow velocity. However, the reaction rate would then decrease, and hence also the degree of chemical utilization of the gas and the performance would decrease.

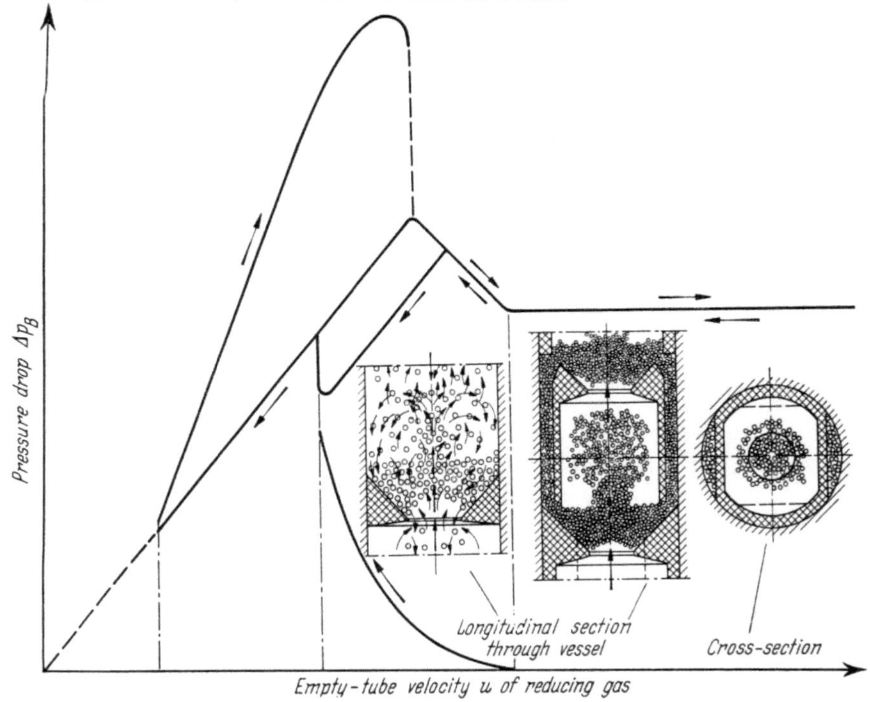

Fig. 179. Reduction of small pellets in the spouted bed (diagrammatic), according to VAVILOV[393])

In this case the spouted-bed technique can be used in ore reduction, as the example in Fig. 179 shows.

Nevertheless, such methods do not appreciably improve the heat balance.

4.4. Reduction in Retorts

Ore lumps or agglomerates are reduced in movable or fixed containers – retorts – to sponge iron by gas containing H_2 or CO. Retort processes work discontinuously in the following operational cycle:

charging – reduction – emptying – charging, etc.

An approximation to the counterflow principle is possible if two or more retorts are operated in series, the gas flow being arranged so that the gas first encounters ore which has already been largely reduced and then the gas is used in the preliminary reduction of fresh, unreduced ore.

4.4.1. Fundamentals

Performance data for ore-reduction retorts can be predetermined to a close approximation from laboratory experiments. Figure 180 exhibits the basis of the method. The reducing gas – in this case, H_2 – enters the retort from below and is replaced by the reduction product, H_2O, as it rises. It will be assumed in the following treatment that the charge has the same temperature at all points in the vessel (isothermal retort), a condition which is easily satisfied in laboratory experiments and approximately satisfied under actual working conditions. The reducing-power of the gas gradually grows less as it becomes enriched in H_2O and the reduction of the ore in successive layers becomes slower and slower. The upper layers lag behind in their degree of reduction.

One difficulty in the following discussion arises from the fact that the equilibrium values for the reduction reaction are completely different for the various iron oxides; the value will depend on which particular iron oxide is being reduced. Figure 181 shows (in accordance with section 1) how the equilibrium stages for the Fe-O-H system at 600 and 900°C depend upon the degree of reduction. A gas mixture which, in the lower part of the retort, has completed the reaction in the FeO-Fe region, i.e., has attained the equilibrium composition, is still able to perform a reduction in the upper part of the retort where the degree of reduction is less. Unfortunately, the correspondence between the equilibrium concentration and the degree of reduction is not one-to-one. Because the reduction is usually a topochemical process (see section 2), metallic Fe has often already formed on the outside even of a region where the degree of reduction is lowest, i.e., where there are still large amounts of Fe_2O_3 and Fe_3O_4 for instance; the gas reacts with FeO and the decomposition of the higher oxides takes place by solid diffusion (see section 2). Effectively, when the degree of reduction is low, the equilibrium concentration of the gas is less than would correspond to the mean degree of reduction, particularly when the rate of reduction is high. As a practical approximation therefore, the procedure indicated in Fig. 181 will be adopted: viz., for a degree of reduction in the range $\mathfrak{R} = 0$ to 33.33%*, i.e., up to the FeO-stage, it will be assumed

* In the equations of Table 20 and the following calculations, the degree of reduction is given simply as a numerical ratio and not as a percentage. The omission of the % sign simplifies the formulae. However, the degree of reduction in the text and in the diagrams is given as a percentage, obtained simply by multiplying the numerical ratio

$$\frac{\text{amount of oxygen removed from the ore}}{\text{original amount of oxygen in the ore}}$$

by 100.

It is also assumed, as is usually the case, that the Fe-content of the ore is in the form of Fe_2O_3. For lower stages of oxidation in the iron ore, e.g., Fe_3O_4, the equations must be modified.

that the equilibrium concentration decreases linearly with increasing degree of reduction. Hence we obtain the equilibrium concentrations for H_2/H_2O mixtures as functions of the degree of reduction, as indicated in Table 20.

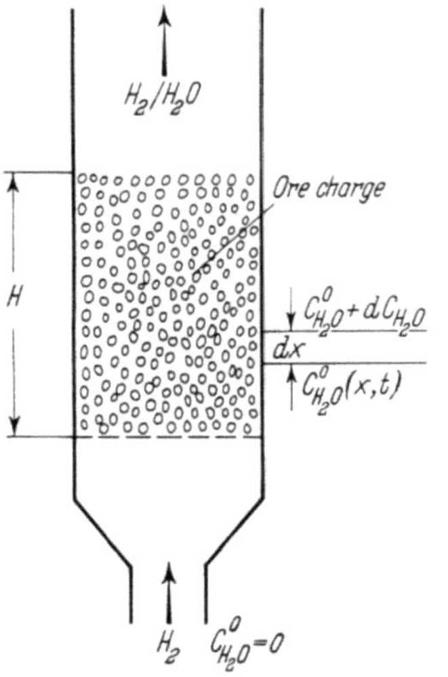

Fig. 180. Isothermal reduction of a static ore charge in a retort (non-steady reduction), schematic

$C^0_{H_2O}$ = concentration of H_2 in the gas phase (by volume)

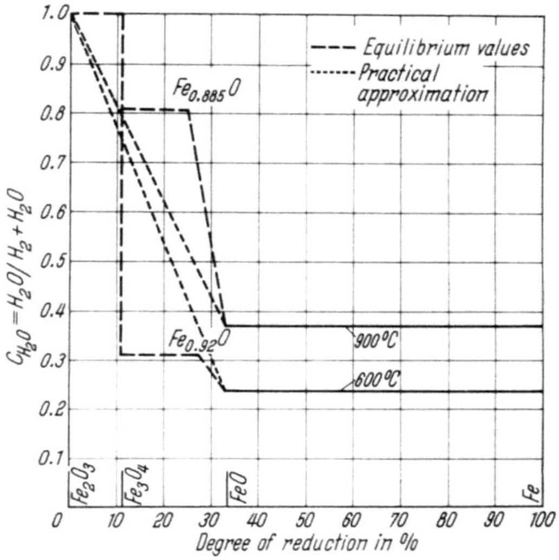

Fig. 181. Equilibrium between Fe-oxides and H_2/H_2O mixtures at 900°C and 600°C

Table 20. *Practical approximate values for the equilibrium of the reduction of ore by* H_2 *as functions of the degree of reduction*

$$Fe_2O_3 + 3H_2 = 2Fe + 3H_2O$$

Temperature °C	Maximum conversion of the H_2 $C^{eq}_{H_2O} = \frac{(\%\ H_2O)}{(\%\ H_2) + (\%\ H_2O)}$ for $\mathfrak{R} \leqq 0.333$	for $\mathfrak{R} \geqq 0.333$
900	$1.00-1.892 \cdot \mathfrak{R}$	0.370
800	$1.00-1.981 \cdot \mathfrak{R}$	0.340
700	$1.00-2.130 \cdot \mathfrak{R}$	0.290
600	$1.00-2.298 \cdot \mathfrak{R}$	0.235
	for $\mathfrak{R} \leqq 0.111$	for $\mathfrak{R} \geqq 0.111$
500	$1.00-7.70 \cdot \mathfrak{R}$	0.145
400	$1.00-8.02 \cdot \mathfrak{R}$	0.110

The rates of reduction must be measured as functions of the temperature, gas composition, and degree of reduction for the input material in each case. Measurements of this kind have been carried out[395]) for one example (Swedish Fe_2O_3-pellets containing 1.5% gangue, 31 mm in diameter, and with 19% pores). In the experiment, at various temperatures T and with various amounts, $C^0_{H_2O}$, of H_2O in the hydrogen, the progress of isothermal removal of oxygen was observed as a function of time, i.e., $\mathfrak{R} = \mathfrak{R}(t)$. By constructing the tangents to the curves so obtained at the points corresponding to various degrees of reduction, the rates of reduction were obtained for the several experimental conditions, i.e., we obtained

$$d\mathfrak{R}/dt = f(\mathfrak{R}, T, C^0_{H_2O}).$$

The distance from equilibrium, $C^0_{H_2O} - C^{eq}_{H_2O}$, enters linearly[395, 396]), so the following simplification can be made:

$$v = \frac{d\mathfrak{R}}{dt} \frac{1}{C^{eq}_{H_2O} - C^0_{H_2O}}, \tag{1}$$

where v is the reduced rate of reduction.

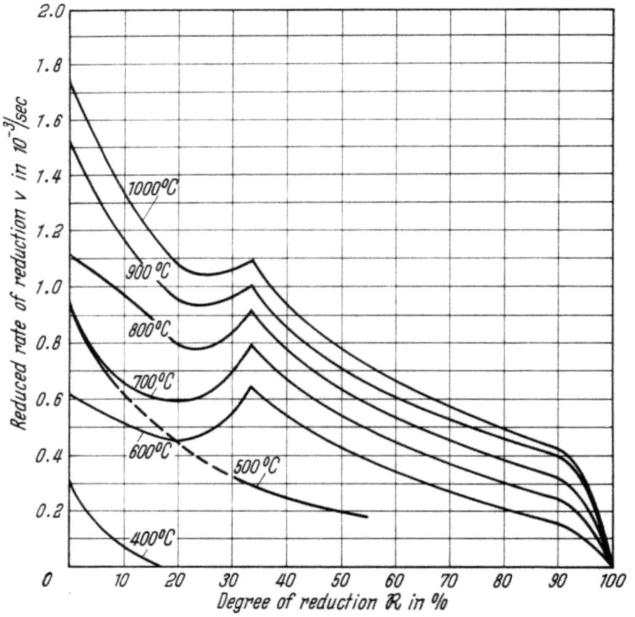

Fig. 182. Rate of reduction in pure H_2. Dependence on temperature and degree of reduction $\mathfrak{R}$

Fe_2O_3-pellets, 31 mm diameter, 19% pores, $p = 1$ atm, $d\mathfrak{R}/dt = v \cdot \Delta C$ in sec^{-1}, $\Delta C = C^{eq}_{H_2O} - C^0_{H_2O}$ (volume ratio), $C^{eq}_{H_2O}$ = maximal H_2-content in equilibrium with the Fe-oxides or Fe present, $C^0_{H_2O}$ = actual H_2O-content of the gas

The values of v calculated in this way from experiments have been plotted in Fig. 182 as functions of the temperature and $\mathfrak{R}$. From the values of v in Fig. 182 the rate of reduction $d\mathfrak{R}/dt$ for any chosen conditions can be found as follows:

$$\frac{d\mathfrak{R}}{dt} = v(C_{H_2O}^{eq} - C_{H_2O}^{0}) \quad (\text{sec}^{-1}), \tag{2}$$

where $C_{H_2O}^{eq}$, which, dimensionally, is a ratio of volumes, is to be taken from Table 20, and $C_{H_2O}^{0}$ is the concentration of the H_2O in the free gas space, again a ratio of volumes.

For a horizontal section, of small thickness dx, across the reactor, shown in Fig. 180, the mole rate of reduction, $\dot{n}_O$, in mole O/sec, can be expressed as[396])

$$d\dot{n}_O = F\, dx\, M_{Fe} \cdot 2.69 \cdot 10^{-2} \frac{d\mathfrak{R}}{dt} \quad (\text{mole O/sec}), \tag{3}$$

where

F = area of cross-section of reactor (cm^2),
M_{Fe} = Fe-content in the charge (g Fe/cm^3),
$2.69 \cdot 10^{-2}$ = O-content of the ore (mole O/g Fe).

Combined with (1), Eq. (3) gives

$$d\dot{n}_O = F\, dx\, M_{Fe} \cdot 2.69 \cdot 10^{-2}\, v(\mathfrak{R})\, [C_{H_2O}^{eq}(\mathfrak{R}) - C_{H_2O}^{0}], \tag{4}$$

in which we indicate that both v and $C_{H_2O}^{eq}$ depend on $\mathfrak{R}$ by writing $v(\mathfrak{R})$ and $C_{H_2O}^{eq}(\mathfrak{R})$.

The H_2O-content in the rising H_2-gas changes, due to reaction, in accordance with

$$d\, C_{H_2O}^{0} = \frac{d\dot{n}_O}{\dot{V}_{H_2}}, \tag{5}$$

where

$\dot{V}_{H_2}$ = throughput of H_2 (mole/sec).

Combining (4) and (5) we get

$$\frac{dC_{H_2O}^{0}}{dx} = \frac{F}{\dot{V}_{H_2}} M_{Fe} \cdot 2.69 \cdot 10^{-2}\, v(\mathfrak{R})\, [C_{H_2O}^{eq}(\mathfrak{R}) - C_{H_2O}^{0}]. \tag{6}$$

Integrating (6) and using the boundary condition that $C_{H_2O}^{0} = 0$ when $x = 0$ and the fact that $\mathfrak{R}$ is independent of x, we find

$$C_{H_2O}^{eq}(\mathfrak{R}) - C_{H_2O}^{0} = C_{H_2O}^{eq}(\mathfrak{R})\, e^{-\alpha x}, \tag{7}$$

where, for brevity, we have put

$$\alpha = \frac{F}{\dot{V}_{H_2}} M_{Fe} \cdot 2.69 \cdot 10^{-2}\, v(\mathfrak{R}). \tag{8}$$

Substituting (7) into (4) and integrating with respect to x over the whole height of the reactor, i.e., from $x = 0$ to $x = H$, we get

$$\dot{n}_O = \dot{V}_{H_2} C_{H_2O}^{eq}(\mathfrak{R}) \cdot (1 - e^{-\alpha H}) \quad (\text{mole O/sec}). \tag{9}$$

Relative to the volume of the charge,

$$F\,H = V_R \quad (\text{cm}^3),$$

the rate of removal of oxygen is

$$\frac{\dot{n}_O}{V_R} = \frac{\dot{V}_{H_2}}{V_R}\, C^{eq}_{H_2O}\,(\mathfrak{R})\,(1 - e^{-\alpha H}) \quad (\text{mole O/cm}^3\ \text{sec}). \tag{10}$$

The exponent of the exponential is the "Damköhler number"

$$\alpha H = \frac{F\,H}{\dot{V}_{H_2}}\, M_{Fe} \cdot 2.69 \cdot 10^{-2}\, v(\mathfrak{R})$$
$$= \frac{V_R}{\dot{V}_{H_2}}\, M_{Fe} \cdot 2.69 \cdot 10^{-2}\, v(\mathfrak{R}). \tag{11}$$

The quotient of the rate $\dot{V}_{H_2}$ at which the gas is passed through and the volume $V_R = F \cdot H$ of the charge appears in Eq. (10) as a coefficient and also in the exponential; apart from a coefficient of proportionality, the ratio of these two quantities gives the time for which the gas is in the charge. For, at a pressure of $p = 1$ atm and a working temperature T (in °K), $\dot{V}_{H_2}$ becomes

$$\dot{V}_{H_2} \cdot 22400\, T/T_0 = 82.1\, T\, \dot{V}_{H_2} \quad (\text{cm}^3/\text{sec}), \quad \text{since} \quad T_0 = 273\,°\text{K}.$$

The time for which the H_2 remains in the charge (reduced for the working temperature T, for 1 atm, and for the empty chamber) is thus

$$\tau_{H_2} = \frac{V_R}{82.1\, T\, \dot{V}_{H_2}} \quad (\text{sec}). \tag{12}$$

It follows from Eqs. (10) and (12) that, with the same ore input, same working temperature, and same composition of the reducing gas, reducing retorts have the same output per unit volume whatever their size may be, provided that the retention time of the gas in the charge is kept the same. Consequently, the results of laboratory tests can be applied directly to full-scale plant.

This consequence is valid not only for the case when $\mathfrak{R}$ is independent of x, but also quite generally, since in any retort the total volume can be divided into elements in each of which $\mathfrak{R}$ is independent of x.

From (10) we can draw conclusions for the limiting cases

$$\text{(a)}\ \alpha H \ll 1 \quad \text{and} \quad \text{(b)}\ \alpha H \gg 1.$$

a) If $\alpha H \ll 1$, i.e., if the time τ_{H_2} is very short, then the output per unit volume is

$$\frac{\dot{n}_O}{V_R} = 2.69 \cdot 10^{-2}\, M_{Fe}\, v(\mathfrak{R})\, C^{eq}_{H_2O}\,(\mathfrak{R}), \tag{13}$$

i.e., the reduction proceeds undisturbed by any enrichment in H_2O, the performance is maximal, but the degree of utilization of the gas is approximately zero.

b) If $\alpha H \gg 1$, then

$$\frac{\dot{n}_O}{V_R} = \frac{\dot{V}_{H_2}}{V_R} C^{eq}_{H_2O}(\mathfrak{R}), \tag{14}$$

i.e., the gas entering is utilized down to equilibrium; the performance is low, but the gas utilization is maximal.

The technological and economic optimum lies between these two limiting cases.

In principle, the formulation as in Eq. (10) is also applicable if the degree of reduction is different in different parts of the chamber, i.e., if $\mathfrak{R} = \mathfrak{R}(x, t)$ depends on position and time, as usually happens. The integration can be carried out by adopting for $v(\mathfrak{R})$ the relation shown in Fig. 182 and for $C^{eq}_{H_2O}$ the values shown in Table 20, but, of course, the result cannot be expressed in closed form and so is omitted here. The numerical calculation can be carried out without further difficulty, and this is explained in the next section.

4.4.2. Numerical Calculation for Reduction in a Retort[403])

For convenience of presentation the following example has been deliberately simplified and will show only the principle. In cases of practical concern the calculation must be *ab initio* taking the properties of the ore in question into account.

Let the retort, with $F = 1\ m^2$ and $H = 1$ m, be divided into 5 zones, one above the other, as in Fig. 183, and suppose the charged ore has the reduction rates corresponding to Fig. 182. We take the working temperature as 900°C, and let

$$M_{Fe} = 1.61\ \text{g Fe/cm}^3.$$

Then the amount of Fe per zone is (since $\Delta x = 20$ cm, $F = 10^4\ cm^2$)

$$\Delta x \cdot F \cdot M_{Fe} = 1.61 \cdot 10^4 \cdot 20 = 3.22 \cdot 10^5\ \text{g Fe/zone},$$

the corresponding amount of oxygen is $0.43 \cdot 3.22 \cdot 10^5 = 1.383 \cdot 10^5$ g O/zone, and therefore the stoichiometric H_2-requirement is 194 m^3 H_2 at NTP.

The time τ_{H_2} will be varied, and first we take

$$\tau_{H_2} = 1.0\ \text{sec, i.e., } H_2\text{-amount} = 1\ m^3/\text{sec} = 0.232\ m^3/\text{sec at NTP}.$$

For the beginning of the reduction we consider a time-phase 1 of duration 0.5 min = 30 sec.

In all zones at the start let $\mathfrak{R} = 0$, i.e., as in Table 20, $C^{eq}_{H_2O} = 1$; moreover, for zone 1, $C^0_{H_2O} = 0$ and therefore, from Fig. 182, the rate of reduction is

$$d\mathfrak{R}/dt = 1.48 \cdot 10^{-3}\ \text{sec}^{-1} \quad (\Delta C_{H_2O} = 1).$$

So in 30 sec the degree of reduction increases by

$$\Delta \mathfrak{R} = 30 \cdot 0.148 \cdot 10^{-2} = 0.0444 .$$

The amount of steam formed is therefore

$$\Delta H_2O = 0.0444 \cdot 194 = 8.61 \text{ m}^3 \text{ at NTP},$$

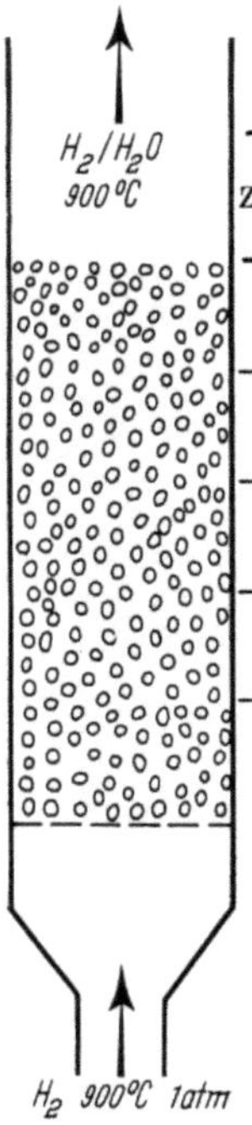

Zone	H_2O-content at end of the zone	Initial degree of reduction	Rate of reduction $d\mathfrak{R}/dt$	Increase in degree of reduction
	$C^{nE}_{H_2O}$			
5	$C^{n4}_{H_2O}$	$\mathfrak{R}_{n5}$	$v(\mathfrak{R}_{n5};\ 900°C)\,(C^{eq}_{H_2O} - C^{n5}_{H_2O})$	$\left(\frac{d\mathfrak{R}}{dt}\right)_{n5} \cdot \Delta t_n$
4	$C^{n3}_{H_2O}$	$\mathfrak{R}_{n4}$	$v(\mathfrak{R}_{n4};\ 900°C)\,(C^{eq}_{H_2O} - C^{n4}_{H_2O})$	$\left(\frac{d\mathfrak{R}}{dt}\right)_{n4} \cdot \Delta t_n$
3	$C^{n2}_{H_2O}$	$\mathfrak{R}_{n3}$	$v(\mathfrak{R}_{n3};\ 900°C)\,(C^{eq}_{H_2O} - C^{n3}_{H_2O})$	$\left(\frac{d\mathfrak{R}}{dt}\right)_{n3} \cdot \Delta t_n$
2	$C^{n1}_{H_2O}$	$\mathfrak{R}_{n2}$	$v(\mathfrak{R}_{n2};\ 900°C)\,(C^{eq}_{H_2O} - C^{n2}_{H_2O})$	$\left(\frac{d\mathfrak{R}}{dt}\right)_{n2} \cdot \Delta t_n$
1	$C^{n0}_{H_2O}$	$\mathfrak{R}_{n1}$	$v(\mathfrak{R}_{n1};\ 900°C)\,(C^{eq}_{H_2O} - C^{n1}_{H_2O})$	$\left(\frac{d\mathfrak{R}}{dt}\right)_{n1} \cdot \Delta t_n$

n = phase, duration Δ in sec, 1st index = time phase, 2nd index = zone

Fig. 183. Isothermal reduction of a static bed in the retort. Simplified system of calculation

and the gas phase is enriched in H_2O by

$$\Delta C^0_{H_2O} = \frac{8.61}{30 \cdot 0.232} = 1.24$$

(remember this is a volume ratio), i.e., by more than the given H_2.

This result can only mean here that the rate of reduction is sufficient to convert the whole of the H_2 entering to H_2O and accordingly to produce a degree of reduction

$$\Delta \mathfrak{R} = \frac{30 \cdot 0.232}{194} = 0.036 = 3.6\% .$$

No reaction can take place in zone 2 since the H_2-gas has already been converted into H_2O in zone 1.

The calculation is continued using gradually increasing time phases (2.5, 6, 10 min). We give the numerical calculation for the 4th phase as a further example.

Duration: 10 min = 600 sec. Amount of H_2 entering: 139 m^3 at NTP.

Zone 1: $C^0_{H_2O} = 0 \quad \mathfrak{R} = 0.421 \quad v = 0.85 \cdot 10^{-3}\ sec^{-1}$

$C^{eq}_{H_2O} = 0.37$

$d\mathfrak{R}/dt = 0.85 \cdot 10^{-3} \cdot 0.37 = 0.0314 \cdot 10^{-2}\ sec^{-1}$

$\Delta \mathfrak{R} = 600\, d\mathfrak{R}/dt = 0.189$

$\Delta H_2O = 0.189 \cdot 194 = 36.6\ m^3$ at NTP

$\Delta C^0_{H_2O} = 36.6/139 = 0.264$

Zone 2: $C^0_{H_2O} = 0.264 \quad \mathfrak{R} = 0.224 \quad v = 0.93 \cdot 10^{-3}\ sec^{-1}$

$C^{eq}_{H_2O} = 0.576$

$d\mathfrak{R}/dt = 0.0290 \cdot 10^{-2}\ sec^{-1}$

$\Delta \mathfrak{R} = 600\, d\mathfrak{R}/dt = 0.174$

$\Delta H_2O = 0.174 \cdot 194 = 33.8\ m^3$ at NTP

$\Delta C^0_{H_2O} = 33.8/139 = 0.242,$ etc.

The error in this method of calculation arises from the fact that, e.g., in the first stage we assumed that $\mathfrak{R} = 0$ and $C^0_{H_2O} = 0$ throughout the whole time interval and throughout the whole space of zone 1, i.e., that these initial values will remain unchanged. Actually, the enrichment in H_2O and the increase in the degree of reduction both exert an inhibiting effect on the reaction. To obtain a more exact result, both the length of the zones and the duration of the time phases must be greatly reduced. The amount of calculation increases multiplicatively, but electronic computers can be used.

Figure 184 shows the collected results of the calculation, the time change of the reduction in each of the 5 zones being plotted. The lag of the degree of reduction in the upper part of the retort is clearly visible, and also the very high degree of chemical utilization of the hydrogen initially, $C^E_{H_2O}$. Figure 185 shows the corresponding, calculated results with 50 times the throughput of H_2, i.e., $\tau_{H_2} = 0.1$ sec. Because of the larger throughput of gas, the H_2O-enrichment of the reducing gas is considerably less, and the chemical utilization becomes worse. Consequently the reduction takes place at about the same rate in all the zones.

Both diagrams suffer from the defect that the calculated course of the reduction appears to be more rapid than is actually the case; this is due to comparatively crude sub-division into volume and time intervals. With modern computing machinery, finer sub-divisions can be used, and then the calculated results fit the actual circumstances more closely.

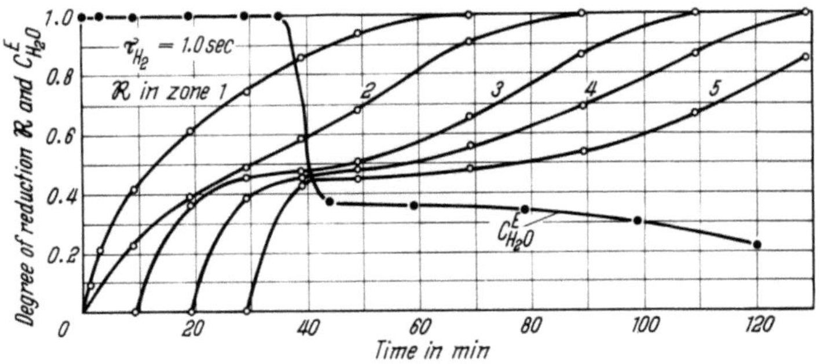

Fig. 184

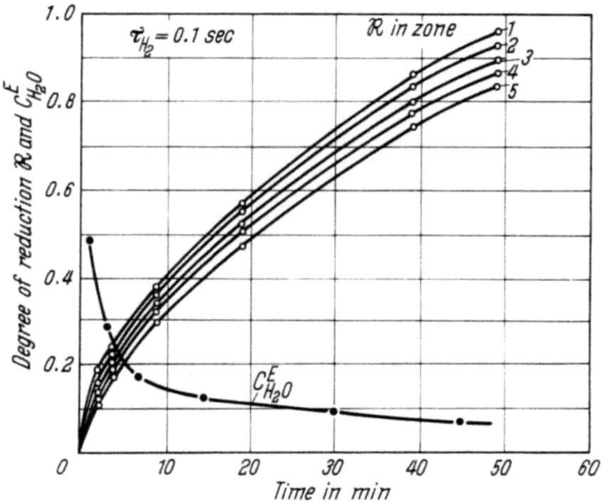

Fig. 185

Fig. 184/185. Progress of reduction in zones of an isothermal retort for $\tau_{H_2} = 1.0$ sec and 0.1 sec

Fe_2O_3-pellets, 19% pores, 31 mm diameter. H_2 900°C, 1 atm
τ_{H_2} = retention time of H_2 in the charge (900°C, empty vessel), $\Re$ = degree of reduction = (amount of oxygen removed):(initial amount of oxygen in charge), $C^E_{H_2O}$ = concentration of H_2O in exhaust gas

4.4.3. Experimental Investigation

Experimental investigation by laboratory tests is even simpler than the numerical investigation. As already mentioned, Eqs. (10) to (12) imply that, if laboratory experiments are carried out using the same ore with the same grain size and under the same conditions for reduction as in industrial working (as regards temperature, composition of the gas, and its retention time), then the results obtained will be independent of the reactor volume.

Using the small retort (internal diameter 140 mm) represented in Fig. 186b, Swedish Fe_2O_3-pellets of 31 mm diameter and 19% pores were reduced in H_2 at 900°C. The retention time of the gas was varied by changing the height of the charge and the rate of gas throughput. The reduction was followed by observing the loss in weight. Reduction was terminated

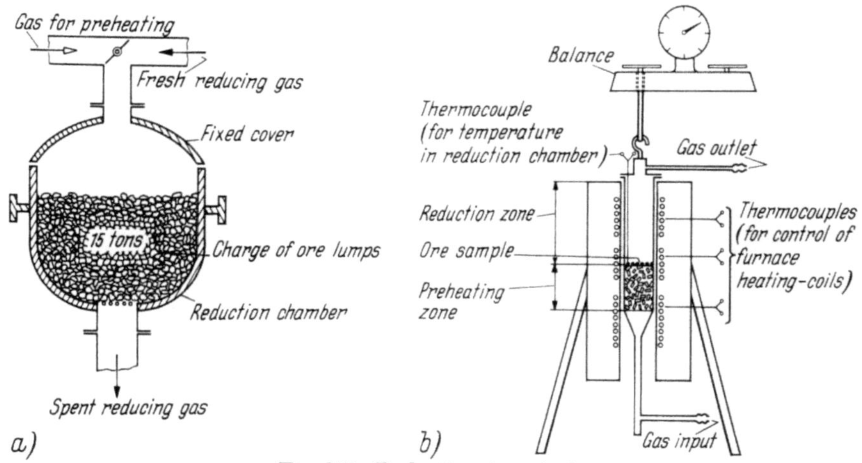

Fig. 186. Reduction in retorts

a) HyL-process, operational plant b) Test apparatus

when 90% complete. The gas requirement was determined from the measured rate of flow, and the performance per unit volume for the retort was calculated from

$$\frac{L}{V_R} = \frac{M}{V_R t(90\%)} \left(\frac{\text{tons Fe}}{\text{day} \cdot \text{m}^3}\right),$$

where

M = amount of iron charged (as Fe_2O_3) (tons Fe),
V_R = volume of the charge (m^3),
$t(90\%)$ = time in days needed for 90% reduction.

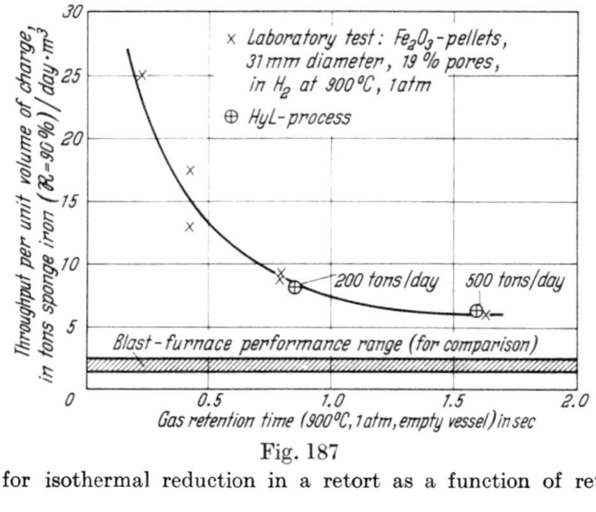

Fig. 187

Throughput for isothermal reduction in a retort as a function of retention time

Figure 187 shows the performance per unit volume for the test retort as a function of the gas retention time in the charge τ_{H_2}. We see that, in agreement with the calculated results (Figs. 184 and 185), as the time τ_{H_2} increases, the performance decreases, and indeed, as we should expect,

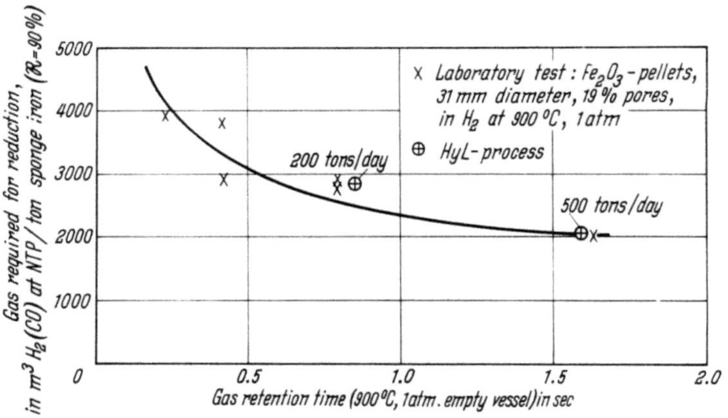

Fig. 188. Reducing gas required in isothermal reduction in a retort as a function of retention time

appreciably less rapidly than the reciprocal $1/\tau$. The measured performance reaches values of 15 tons Fe per day per m^3 and more, i.e., the output per unit volume is much greater than that for a blast furnace. This is due to the high reducing efficiency of hydrogen.

Figure 188 shows the corresponding graph of the amount of gas needed (m^3 H_2/ton Fe at NTP) as a function of τ_{H_2}.

4.4.4. Retort Reduction on the Commercial Scale

Ore lumps and agglomerates were reduced by gas in retorts comparatively early[397, 398]), but only since 1958 has there been in continuous operation a large installation working with the HyL-process[399, 400]). The retorts in the first plant were similar to those sketched in Fig. 186a. With 5 retorts, holding 13 to 15 tons of ore each, this plant produces 200 tons of sponge iron a day. In 1960 it was followed by a plant producing 500 tons of sponge iron a day[401, 401a]), and in 1967 a further 500-ton/day plant went into operation[401b, c]).

The reducing gas is obtained by reforming natural gas with steam according to the reaction

$$CH_4 + H_2O = CO + 3\,H_2.$$

(See section 4.8 for further details.)

The reformed gas for the reduction has the approximate composition:

$$H_2 : 74\%, \quad CO : 13\%, \quad CO_2 : 8\%, \quad CH_4 : 5\%.$$

The steam content is not specified, but at the outlet gas temperature of 30°C from the coolers[401c,d] and at the outlet pressure of 4.5 atm[401e] it may be about 1%. The temperature of the gas on input into the retort is about 1040°C, the exhaust-gas temperature about 870°C[400]. The assumed isothermal mode of operation is therefore only roughly satisfied. The heating of the reducing gas to above 1000°C is effected by recuperative preheating at 700 to 800°C, followed by partial combustion with air, also preheated to 700 to 800°C[401c], the latter process impairing its reducing capability somewhat.

As Fig. 189 shows, the reducing gas first flows through two retorts, connected in parallel, in which final reduction takes place, and it is then

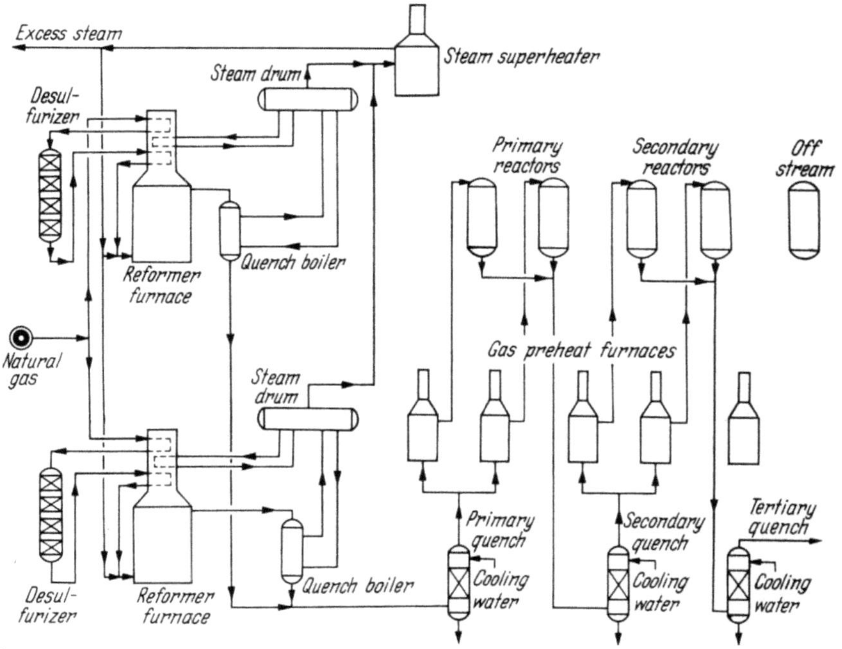

Fig. 189. Flow diagram for the HyL-process, 200-ton/day plant, according to CELADA et al.[399]

cooled, partially freed from H_2O, reheated to 1040°C and led through two more retorts in parallel, in which a pre-reduction of the ore takes place. Meanwhile, one retort is being either charged or emptied*.

* A similar retort process with a simplified gas preparation has been developed in Rumania and operated in a pilot plant with a daily output of 1 ton sponge iron[404].

Figure 190 shows schematically the cycle of operations schedule for the 5 retorts. In the 200-ton/day plant the sponge iron was removed from the retorts by tilting after the end of reduction. The 500-ton/day plant operates with four fixed retorts each holding 100 tons of ore[401, 401a–d].

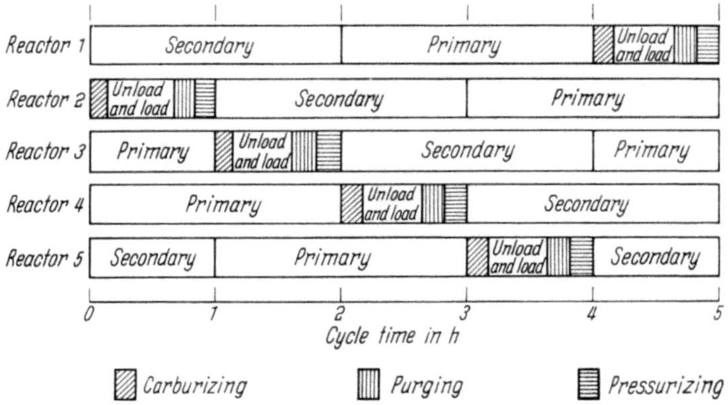

Fig. 190. Diagram of operating cycle for five retorts in a 200-ton/day HyL-plant, according to CELADA et al.[399]

When reduction is finished, the sponge iron is completely cooled down by cold reducing gas, and is emptied on to a conveyor belt by means of a scraper-arm which bores from below into the partially agglomerated charge[401b]. The 200-ton/day plant works with 3 gas reformers, the first 500-ton/day plant with 5, while the new 500-ton/day installation has only one gas reformer[401c].

As regards the ore classification, various data are available. According to[401d] the fractions 3 to 12 mm and 12 to 36 mm were charged separately, in the 200-ton/day plant, while in the 500-ton/day plant the fraction was 12 to 36 mm. In the literature the grain-size ranges 6 to 25 mm[400] and 12.5 to 50 mm[401c] are also given. The ore and sponge iron show the following analysis[401d]:

	Ore %	Sponge iron %
Fe_{total}	65.29	86.66
Residual oxygen	—	3.32
Metallization	—	86.6
S	0.047	0.02
P	0.380	0.34
C	—	1.80
Gangue	6.76	7.86

The sponge iron has a bulk weight of 1.6 tons/m³. The natural-gas consumption per ton sponge iron is 750 m³ at NTP for the 200-ton/day plant[400], and for the 500-ton/day plant has been given as 480 m^3 [401a]), 560 to 580 m^3 [401f]), 580 m^3 [401g]), and 700 m^3 [401c]) at NTP. The last value is the mean total consumption per ton sponge iron for the production period August 1966 to January 1967. However, the value, 700 m^3/ton Fe at NTP, is not identical with the amount of gas required for the reduction, since the top gas from the process cannot cover the total heat requirement and therefore has to be enriched with natural gas[401c]). In [401g]) the gas requirement for reduction is put at 580 m^3/ton Fe at NTP.

From the published data[399–401,401d,f,g]) the kinetic characteristics shown in Table 21a can be calculated.

In this table, the charge volume per retort is evaluated from the daily output, the number of times the retorts are emptied per day, the bulk weight, and the Fe-content of the sponge iron.

Table 21a. *Characteristics for the HyL-process*

		Plant 1	Plant 2
Production L	tons sponge iron/day	200	500
Number of retorts under reducing gas, n		4	2
Ore charge per retort	tons	13	100
Charge volume per retort, V	m^3	6	45
Charge volume reduced, $V_R = n\,V$	m^3	24	90
Performance per unit volume of charge, L/V_R	tons sponge iron/day · m^3	8.33	5.55
Gas requirement	m^3 natural gas/ton Fe at NTP	750	580
$H_2 + CO$ requirement	m^3 $(H_2 + CO)$/ton Fe at NTP	2640	2040
Effective gas throughput, $\dot{v}$	$m^3(H_2 + CO)$/sec (at 900°C, 1 atm, empty vessel)	26.3	50.8
Gas retention time $V_R/\dot{v}$	sec	0.913	1.77

Table 21b. *Degree of reduction in different zones of a large, industrial retort (HyL-process) after 4 hours' reduction*
(Gas remains 0.913 sec in the charge; 900°C, 1 atm, empty vessel)

	Zone 1 0–305 mm	Zone 2 305–610 mm	Zone 3 610–915 mm	Zone 4 915–1220 mm	Zone 5 1220–1525 mm
Metallic Fe %	90.0	86.8	79.4	73.0	64.0
Total Fe %	93.4	93.0	90.8	89.8	87.4
$\frac{\text{Metallic Fe}}{\text{Total Fe}}$	0.963	0.933	0.874	0.813	0.732
C %	1.3	1.36	1.64	1.62	1.34
S %	0.06	0.02	0.04	0.03	0.02

The values given in Table 21a for the HyL-process have been plotted in Figs. 187 and 188. It may be seen that, in spite of the unavoidable differences between the methods of operation in laboratory experiments and industrial plants, there is surprisingly good agreement. The characteristics of the Rumanian process also fit in well with these curves[405]). Also compare Table 21b with Fig. 184.

In the full-scale industrial plants the reducing gas is not pure hydrogen as it is in the laboratory experiment; also, part of the H_2O arising as a reduction product is removed between the two stages of the reduction. The reducibility of the ore charge is probably better than that of the Fe_2O_3-pellets used in the laboratory test. Nevertheless, the calculations and laboratory tests conducted in the way described can be used to forecast, at least within an order of magnitude, the performance data for full-scale plant, and hence to avoid a big part of the laborious and expensive operational trials on the latter.

4.5. Reduction in Rotary Kilns

4.5.1. Possible Methods of Working

Ores of various kinds can, by the addition of solid reducing agents (fine-grained coal, coke), be reduced to the metal in rotating, cylindrical containers (Fig. 191). Because the reaction

$$Fe_2O_3 + 3\,C = 2\,Fe + 3\,CO$$

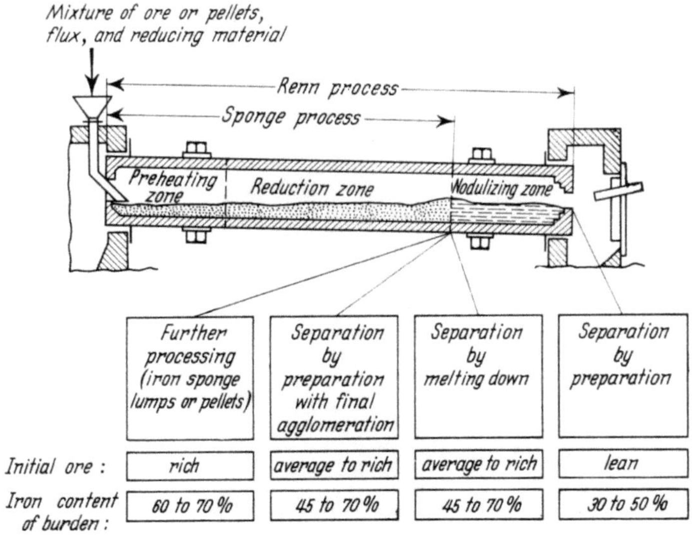

Fig. 191

Possibilities for iron production in rotary kilns, according to Lucke et al.[406])

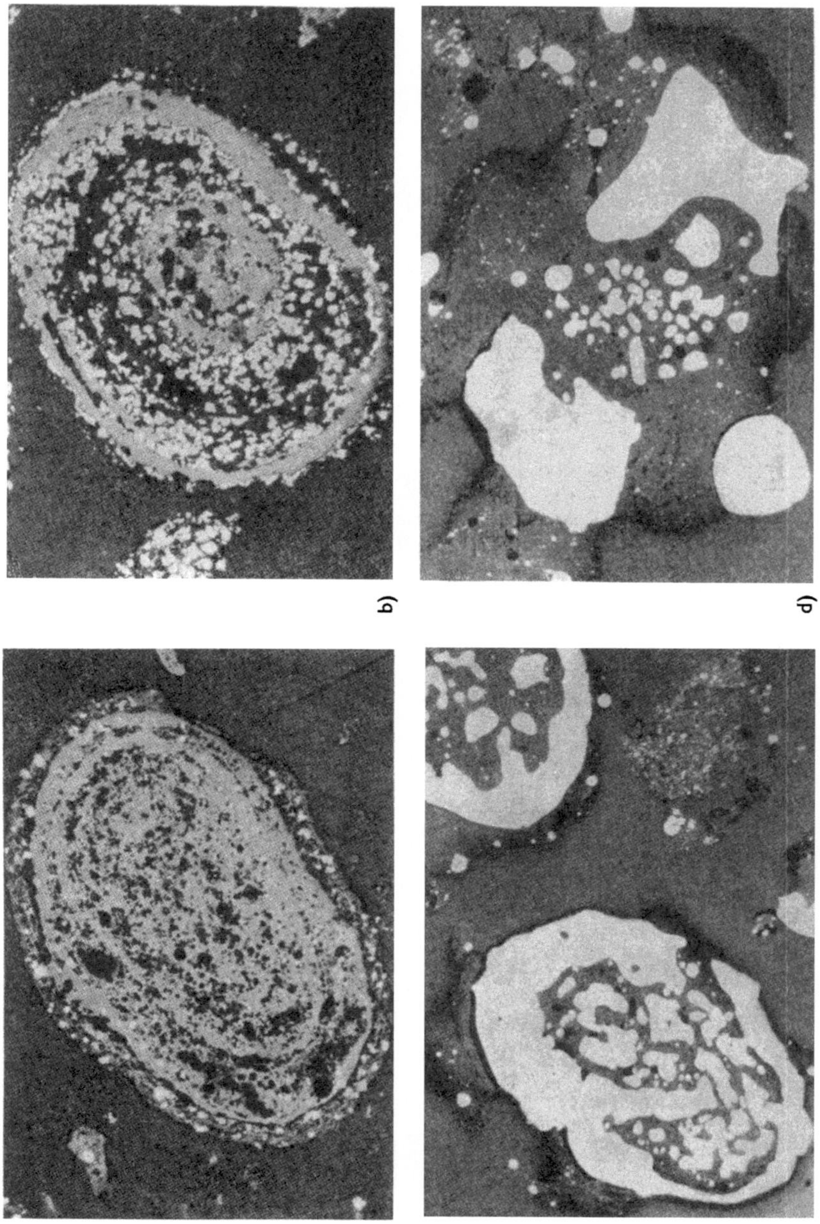

a) b) c) d)

Fig. 192. Reduction in lean ore, according to WAHLSTER[407]) at
a) 1000°C; b) 1050°C; c) 1200°C; d) 1230°C

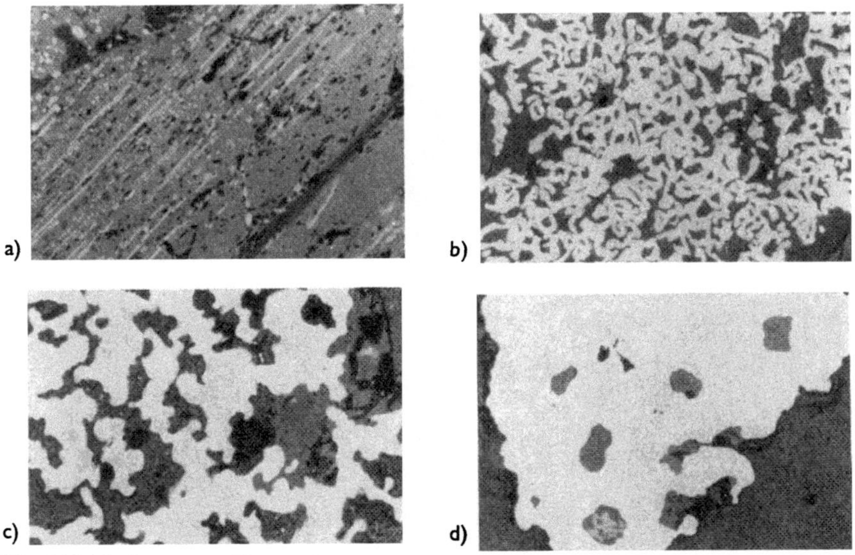

Fig. 193. Reduction in Kiruna-D ore (heating-up test), according to WAHLSTER[407]) at a) 1000°C; b) 1100°C; c) 1200°C; d) 1230°C

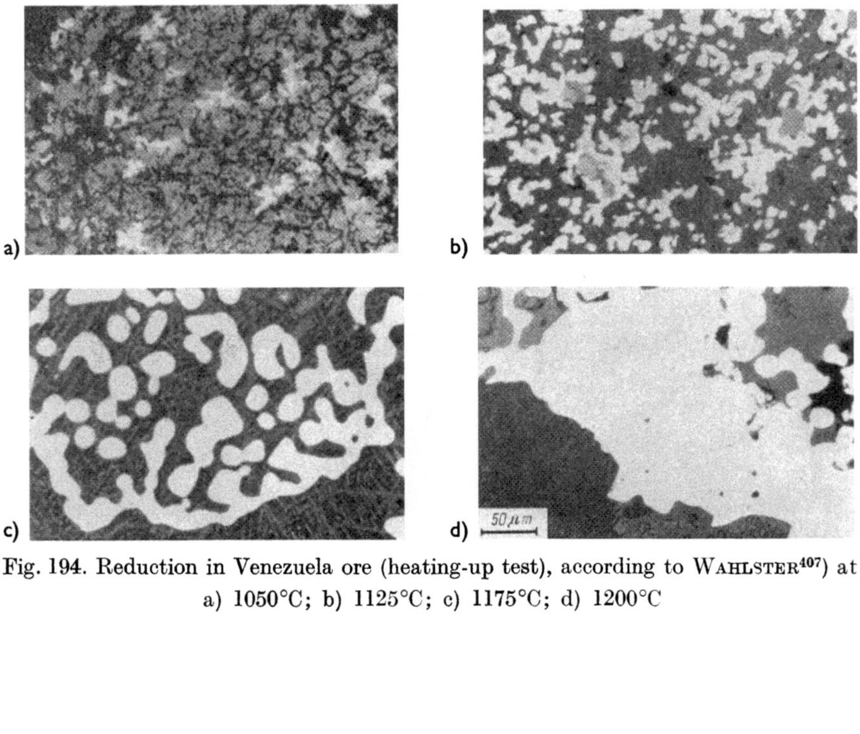

Fig. 194. Reduction in Venezuela ore (heating-up test), according to WAHLSTER[407]) at a) 1050°C; b) 1125°C; c) 1175°C; d) 1200°C

is strongly endothermic, heat must be provided, and in rotary kilns this is usually accomplished by burning natural gas, coke-oven gas, oil or coal dust above the charge. The part of the rotary kiln into which the charge first passes acts as a "preheating zone". As soon as the charge has been sufficiently heated up, the reduction of the iron oxide begins, in the "reduction zone". At low working temperatures, up to about 1100°C, the ore, while keeping its outer form the same, will be reduced to sponge iron. Further, if the temperature is raised to about 1250°C, the surface area of the material is reduced and it forms *nodules*, in the "nodulizing zone"; and at still higher temperatures, fusion occurs. According to the results of laboratory investigations, the transition from sponge iron to nodulizing is progressive, as shown in Figs. 192, 193, and 194.

As HEILIGENSTÄDT has proved[408]), the heat transfer is accelerated by the rotary motion principally at the lower temperatures. At higher temperatures radiation becomes the main mode of heat transfer, and the effect of the rotary motion is diminished. In the reduction zone two very different gas atmospheres exist: above the charge there is the oxidizing exhaust gas from the burner-heaters, and inside the charge a strongly reducing atmosphere of CO. The CO arises once the reactivity of the reducing agent has become sufficiently great, with increasing temperature, for the decomposition of the CO_2 arising from the furnace atmosphere and the ore reduction, according to the Boudouard reaction

$$CO_2 + C = 2\,CO\,.$$

Figure 195 shows the two gas atmospheres in proximity, one above the other.

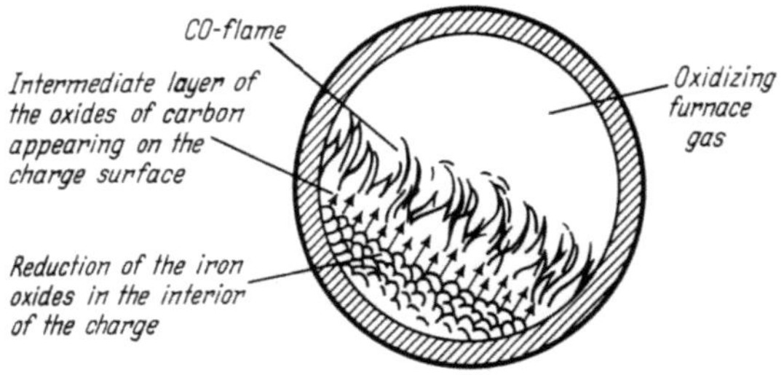

Fig. 195. Processes in the reduction zone, according to JOHANNSEN[409])

4.5.2. The Reaction Kinetics

Reduction takes place essentially in two sub-processes:

a) formation of the reducing gas:

$$CO_2 + C = 2\,CO\,,$$

b) removal of the oxygen from the ore:

$$Fe_nO_m + m\,CO = n\,Fe + m\,CO_2.$$

Thus the Boudouard reaction a)—which is not always desirable in the shaft of a blast furnace because it uses heat—is absolutely essential in rotary furnaces, because by it the reducing gas is formed (or is regenerated). The faster the reaction a) proceeds, the smaller will be the CO_2-content and the faster will be the reduction. To ensure a sufficient transformation the solid fuel must have a high reactivity.

For the kinetics of the Boudouard reaction the results of HEDDEN[411]) are used:

We define the rate constant k_{eff} by the equation

$$v_C = \frac{d n_C}{dt} = k_{eff}[n^0_{CO_2} - n^{eq'}_{CO_2}], \tag{1}$$

where

$n^0_{CO_2}$ = CO_2-concentration in the gas phase (mol/cm³),
$n^{eq}_{CO_2}$ = CO_2-concentration in the equilibrium with C and CO (the Boudouard reaction),
v_C = molar transformation of charge per cm³ per sec.

Then the experimental results give the curves for k_{eff} shown in Fig. 196. We see that there are three regions:

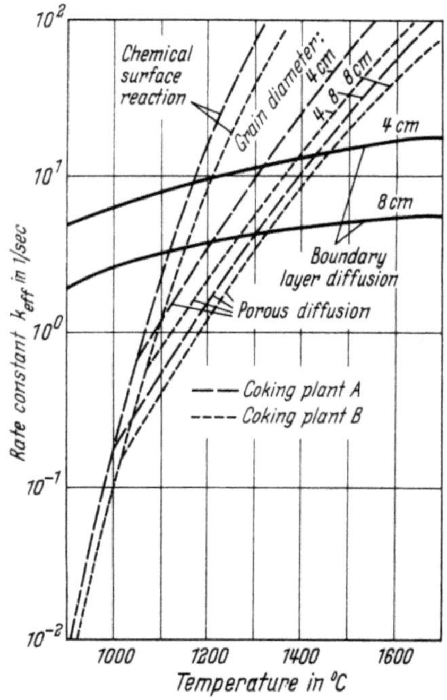

Fig. 196. Effective rate constants for the Boudouard reaction for coke from coking plants A and B, according to HEDDEN[412a])

at low temperatures up to 1000—1100°C the temperature dependence is at its most pronounced, and the chemical reaction is rate-determining;

at higher temperatures porous diffusion predominates;

and finally, diffusion in the boundary layer of the gas flow is the governing process.

Further, by definition[411]),

$$k_{eff} = k_M M_C \eta, \tag{2}$$

where

M_C = amount of carbon per cm³ of charge,
η = Thiele's utilization-of-pores factor (section 2.5).

For the region in which the phase-boundary reaction is rate-determining k_M satisfies the relation[411])

$$k_M = H_C \exp[-86000/RT] \qquad (\text{cm}^3 \text{sec}^{-1} \text{g}^{-1})$$

and if $\eta = 1$

$$k_{eff} = M_C H_C \exp[-86000/RT] \qquad (\text{sec}^{-1}), \tag{3}$$

where

H_C = reactivity factor for the solid fuel.

Depending on the reactivity and bulk weight of the fuel, $M_C H_C$ is between $0.4 \cdot 10^{13}$ and $0.4 \cdot 10^{15}$ sec^{-1}. The activation energy has approximately the same value, 86000 cal/mole, for all the fuels examined[411]). So far as working with fine coal or coke (< 8 mm) is concerned, even up to 1200°C the phase-boundary reaction can be taken to be rate-determining, i.e., $\eta = 1$ in practically the whole of the domain which is of interest.

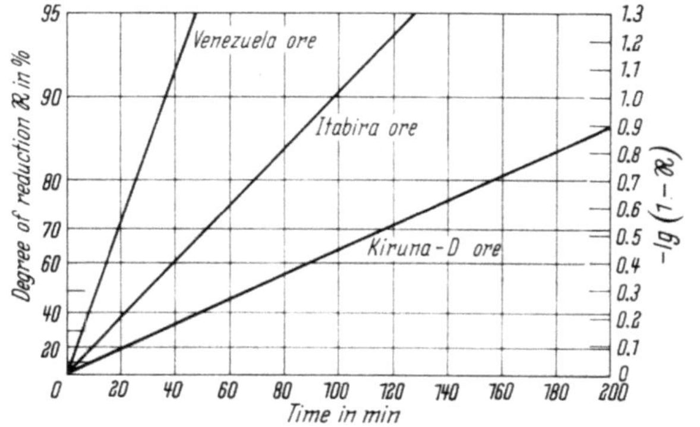

Fig. 197. Degree of reduction in CO as a function of time at 900°C[435])

Ore	Grain diameter mm	Fe %	FeO %	Fe_2O_3 %	$(d\Re/dt)_{40\%}$ %/min
Venezuela	≈ 5	65.26	0	93.32	3.75
Itabira	≈ 13	68.8	0	98.38	1.40
Kiruna-D	≈ 20	58.94	24.51	57.11	0.625

The reduction rates for the iron ores which come into consideration are also known. Figure 197 shows some examples of possible reduction processes at 900°C and 1 atm in pure CO. Making some simplification, the reduction curve can be represented by the following equation[414]):

$$\frac{d\mathfrak{R}}{dt} = k(1 - \mathfrak{R}) = 1.667 \left[\frac{d\mathfrak{R}}{dt}\right]_{40\,\%} (1 - \mathfrak{R}) \quad (\mathrm{sec}^{-1}), \tag{4}$$

or, after integration,

$$-\log(1 - \mathfrak{R}) = 0.724 \left[\frac{d\mathfrak{R}}{dt}\right]_{40\,\%} t,$$

where

$\mathfrak{R}$ = degree of reduction = $\dfrac{\text{amount of oxygen removed}}{\text{initial amount of oxygen in charge}}$,

t = time (sec),

$\left[\dfrac{d\mathfrak{R}}{dt}\right]_{40\,\%}$ = rate of reduction for $\mathfrak{R} = 0.4 = 40\%$ (sec^{-1}).

The rates of reduction for the three ores shown in Fig. 197 were measured in a gas mixture of 40% CO and 60% N_2 (cf. section 5.3) and extrapolated to 100% CO by multiplying by 2.5. As in the previous section, the degree of reduction is expressed as the fraction of the oxygen which has been liberated, i.e., $\mathfrak{R}$ goes from 0 to 1 (equivalent to 0 to 100%) and the time is in seconds. The numerical value can be expressed in several ways; e.g.,

$$\left[\frac{d\mathfrak{R}}{dt}\right]_{40\,\%} = 0.625\,\%/\mathrm{min} = 0.625 \cdot 10^{-2}\ \mathrm{min}^{-1} = 1.042 \cdot 10^{-4}\ \mathrm{sec}^{-1}.$$

To understand events in the rotary furnace, we embark on the following computation[435]).

If we assume that the reduction is governed essentially by the gaseous diffusion in the pores of the sponge iron which is formed, and this is almost always justified for the ores of average grain size which are considered, then Eq. (4) can be extended as follows:

$$\frac{d\mathfrak{R}}{dt} = k(1 - \mathfrak{R}) \left(\frac{T}{T_0}\right)^2 \frac{n_{CO_2}^{eq} - n_{CO_2}^{0}}{n_{CO_2}^{eq}} \quad (\mathrm{sec}^{-1}), \tag{5}$$

where

T_0 = temperature at which the progress of the reduction was measured (°K),

T = the actual temperature,

$n_{CO_2}^{eq}$ = CO_2-concentration in the gas phase in the equilibrium with Fe-FeO/CO or with FeO-Fe_3O_4/CO (mole/cm^3),

$n_{CO_2}^{0}$ = actual CO_2-concentration.

If a charge contains M_{Fe} (in g Fe/cm^3) in the form of the ore put in, then the amount of oxygen removed per unit time and unit volume is given by

$$v_O = 2.69 \cdot 10^{-2} M_{Fe} \frac{d\Re}{dt} \quad (\text{mole O/cm}^3\,\text{sec}). \tag{6}$$

The ore contained 0.43 g O/g Fe (hematite) = $2.69 \cdot 10^{-2}$ mole O/g Fe.

From (4), (5), (6) we find

$$v_O = 4.48 \cdot 10^{-2} M_{Fe} \left[\frac{d\Re}{dt}\right]_{40\,\%} (1 - \Re) \left(\frac{T}{1173}\right)^2 \frac{n_{CO_2}^{eq} - n_{CO_2}^0}{n_{CO_2}^{eq}}$$
$$(\text{mole O/cm}^3\,\text{sec}), \tag{7}$$

or introducing the abbreviation

$$H_{Fe} = 4.48 \cdot 10^{-2} M_{Fe} \left[\frac{d\Re}{dt}\right]_{40\,\%}$$

we have

$$v_O = H_{Fe} (1 - \Re) \left(\frac{T}{1173}\right)^2 \frac{n_{CO_2}^{eq} - n_{CO_2}^0}{n_{CO_2}^{eq}} \tag{8}$$

if, as in the present case (Fig. 197), the measurement of the course of reduction is made at 900°C = 1173°K. The factor H_{Fe} characterizes the reducibility of the ore and can be taken to be between the limits

$$H_{Fe} = 0.2 \cdot 10^{-5} \text{ to } 2 \cdot 10^{-5} \quad (\text{mole O/cm}^3\,\text{sec}).$$

As an abbreviation we introduce a "rate constant" for the reduction

$$k_{Fe} = H_{Fe} (1 - \Re) \left(\frac{T}{1173}\right)^2 \frac{1}{n_{CO_2}^{eq}} \quad (\text{sec}^{-1}). \tag{9}$$

Then we have

$$v_O = k_{Fe} (n_{CO_2}^{eq} - n_{CO_2}^0) \quad (\text{mole/cm}^3\,\text{sec}). \tag{10}$$

As regards the dependence of the equilibrium concentration, $n_{CO_2}^{eq}$, on the degree of reduction of the ore, we proceed as in section 4.4, i.e., as the degree of reduction increases, $n_{CO_2}^{eq}$ will be supposed to decrease linearly until, at $\Re = 0.333$, the FeO-Fe equilibrium holds.

In the steady state, at every point in the charge a CO_2-concentration must be set up which remains constant in time and which is determined in the following way from the rates of the two basic reactions.

The charge in a disc of the vessel, of thickness dl and cross-sectional area F, normal to the axis, has a volume

$$dV = F\,dl.$$

Suppose that, because of the rotation of the furnace, there is no difference, either positionally or with time, in the charge contained within this disc, apart from the topochemical process of the reduction.

The molar rates for, on the one hand, the gasification of the carbon:

$$v_C = k_C\,(n^0_{CO_2} - n^{eq'}_{CO_2}) \quad \text{(mole C/cm}^3\text{ sec)}, \tag{11}$$

where, according to HEDDEN (see above),

$k_C = k_{eff}$ = rate constant for the C-gasification,

and, on the other hand, for the reduction of the ore:

$$v_O = k_{Fe}\,(n^{eq}_{CO_2} - n^0_{CO_2}) \quad \text{(mole O/cm}^3\text{ sec)},$$

must therefore be equal

$$v_O = v_C. \tag{12}$$

Substitution from (10) and (11) into (12) gives the concentration, $n^0_{CO_2}$, of the CO_2 in the charge as a function of the rate constants for the two reactions, viz.,

$$n^0_{CO_2} = \frac{n^{eq}_{CO_2} - k_C\,n^{eq'}_{CO_2}/k_{Fe}}{1 + k_C/k_{Fe}}. \tag{13}$$

We see immediately that,

$$\text{if} \quad k_C \gg k_{Fe}, \quad \text{then} \quad n^0_{CO_2} = n^{eq'}_{CO_2},$$

and

$$\text{if} \quad k_C \ll k_{Fe}, \quad \text{then} \quad n^0_{CO_2} = n^{eq}_{CO_2}.$$

The rate of reduction becomes

$$v_O = k_{Fe}\,n^{eq}_{CO_2}\left(1 - \frac{1 - k_C\,n^{eq'}_{CO_2}/k_{Fe}\,n^{eq}_{CO_2}}{1 + k_C/k_{Fe}}\right). \tag{14}$$

Thus the rate of reduction depends on the rate constants of both reactions and there are the two limiting cases

α) reactivity of fuel low compared with that of the ore, i.e., $k_C \ll k_{Fe}$, and then

$$v_O \approx 0,$$

β) reactivity of fuel high compared with that of the ore, i.e., $k_C \gg k_{Fe}$, and then

$$v_O = v_{O\,max} = k_{Fe}\,(n^{eq}_{CO_2} - n^{eq'}_{CO_2}).$$

This last expression therefore gives the maximum reduction rate at the prevailing temperature.

Now, at temperatures above 900°C we have

$$n^{eq'}_{CO_2} \ll n^{eq}_{CO_2}$$

and then

$$v_{O\,max} = k_{Fe}\,n^{eq}_{CO_2},$$

and (14) becomes

$$v_O = k_{Fe}\,n^{eq}_{CO_2}\left(1 - \frac{1}{1 + k_C/k_{Fe}}\right).$$

From the last two equations we get

$$\frac{v_O}{v_{O\,max}} = \frac{k_C/k_{Fe}}{1 + k_C/k_{Fe}}. \tag{15}$$

This relation between the ratio of the actual to the maximum reduction rate and the ratio of the rate constants is plotted in Fig. 198.

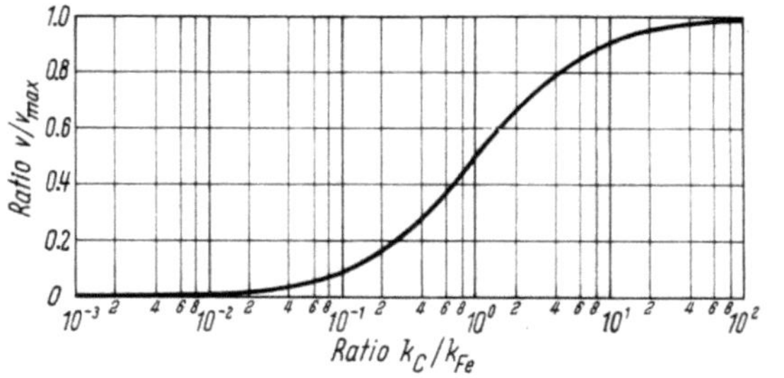

Fig. 198. Ratio of the actual rate of reduction to the maximum possible rate (i.e., in pure CO) as a function of the ratio of the rate constants[435])
v_{max} = rate of reduction in pure CO and $T \geqq 900°C$

From Fig. 198 and from measurements which can be carried out in the laboratory with simple equipment, the best match between ore and fuel can easily be found. Figure 199 shows the possible limiting values for

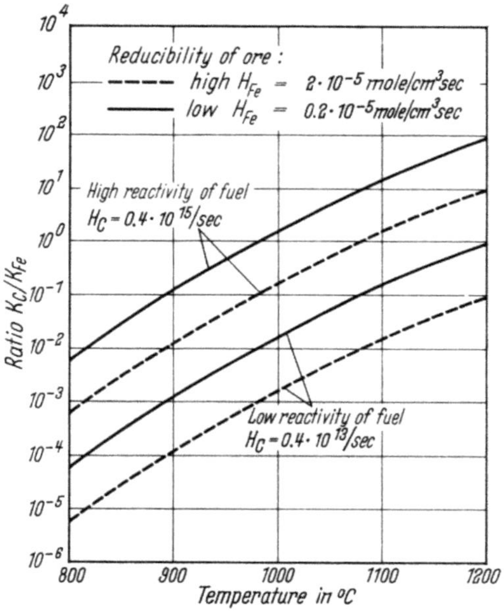

Fig. 199. Ratio of rate constants for ore reduction K_{Fe} (after 50% separation of oxygen) and K_C for C-gasification with CO_2[435])

the ratio k_C/k_{Fe} as a function of temperature, for different values of the reactivities, H_{Fe} and H_C, of the ore and fuel. We can go a step further and calculate the amounts that can be converted, as follows.

Equation (14) enables the rate of removal of oxygen per unit of time and per unit of volume of the charge to be obtained as a function of the

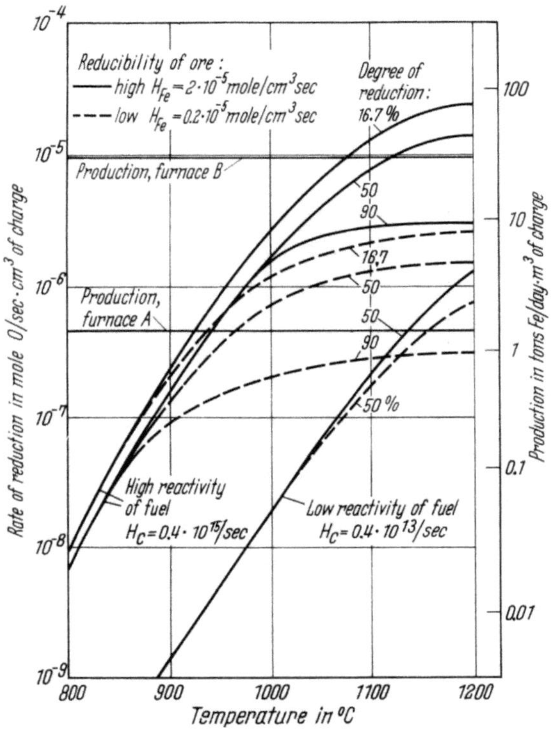

Fig. 200. Rate of reduction in a charge of ore (10 to 20 mm) and fine fuel (0 to 8 mm)[435]
A Krupp largescale test, 110 m rotary furnace, $T = 800$ to 1050 °C; *B* Stora rotary furnace, 2.13 m rotary kiln, $T \approx 1300$ °C

reactivities of the ore and fuel and also in dependence on the temperature and degree of reduction. These rates are plotted in Fig. 200. The values can easily be converted into the Fe-output (tons Fe per day per m³ of charge) as shown by the ordinate scale on the right. The strong influence which the reactivity of the fuel has can be seen; this leads, among other results, to the whole process of ore reduction being unusually strongly dependent on the temperature. The diagram also shows in what region the reactivities of the ore and fuel must lie if the desired throughput is to be achieved.

4.5.3. Comparison with Experiment

A large number of informative experiments have been carried out in a rotary furnace, 14 m long, in which the temperature distribution was controlled by feeding in air for combustion through shell burners; the arrangement is sketched in Fig. 201.

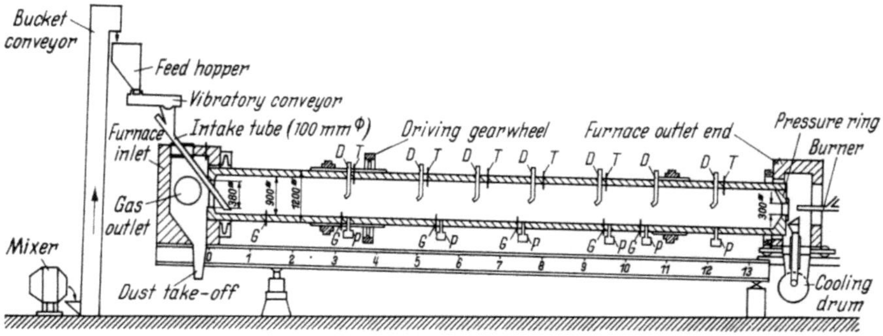

Fig. 201. Rotary kiln plant for reduction tests (14 m long, 1.2 m diameter), according to LUCKE et al.[410])

T Temperature measuring, *D* Nozzle (air supply), *G* Openings for gas sampling, *P* Openings for material sampling

It was found that the theory correctly predicts the essential characteristics of the reduction process. First, Fig. 202 shows the composition of the gas atmosphere existing within the charge, and it may be seen that, as we should expect, the ratio $CO_2/(CO + CO_2)$ lies between the equilib-

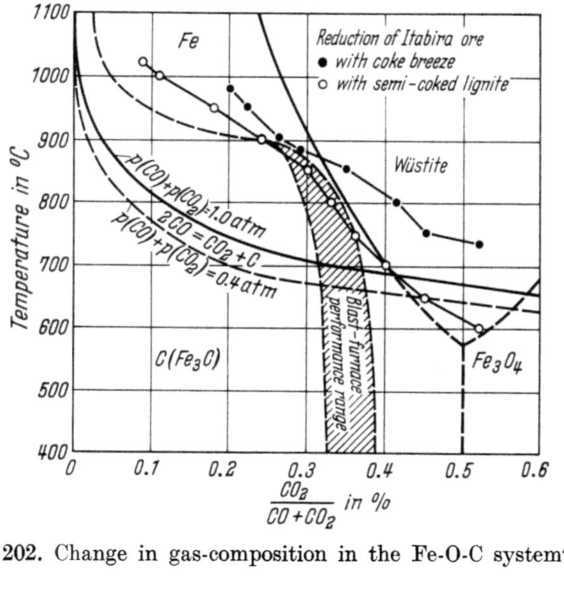

Fig. 202. Change in gas-composition in the Fe-O-C system[410])

rium curve for the Boudouard reaction and that for the reduction of an iron oxide, namely the reduction of wüstite to the metal at the higher temperatures, and the reduction of magnetite to wüstite at the lower temperatures.

Again, Fig. 203 shows that the assumption that the reaction in a lump of ore proceeds topochemically is correct.

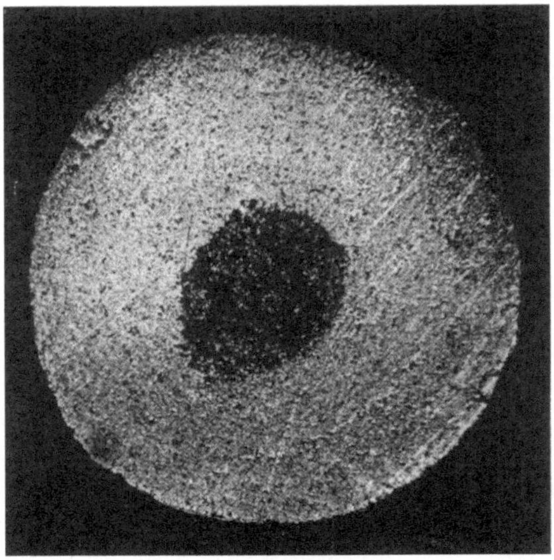

Fig. 203. Cross-section of partially reduced pellet from the SL-RN-process. showing the topochemical progress of reduction, according to SIBAKIN[416])

From Figs. 204 and 205 we see that, as expected from Eq. (14) and Fig. 200, the reactivity of the ore and the reactivity of the fuel both clearly affect the progress of the reaction, and we also observe the marked dependence on the temperature.

Figure 206 shows the variation in the temperatures of the gas and the charge along the length of the furnace. It can clearly be seen that the whole length is marked off into a *preheating zone* (up to a charge temperature of 800°C, with very little reduction) and into a *reduction zone* (where there is little increase in temperature). Reduction to appreciable extent begins above 800°C in a temperature region in which, according to Fig. 200, appreciable reactions are to be expected with fuels having the reactivity of those actually used.

In view of the satisfactory agreement between the theoretical calculations and the actual course of the reduction, we may legitimately draw some further conclusions.

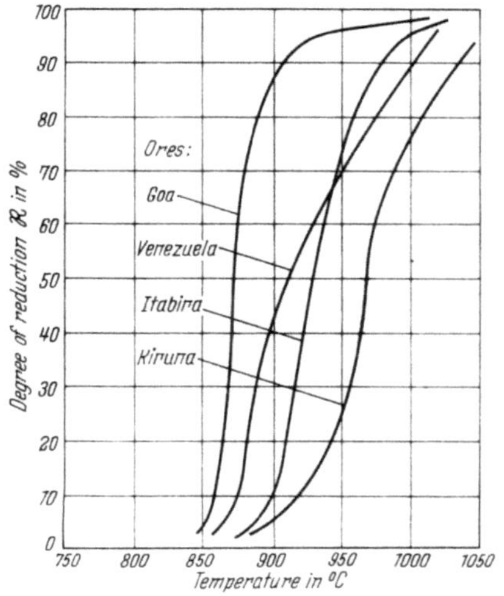

Fig. 204. Dependence of the degree of reduction on the temperature, various ores, grain size 5 to 30 mm (charge: 100 kg ore/h), fuel: coke breeze in a rotary kiln (14 m long), according to LUCKE et al.[406])

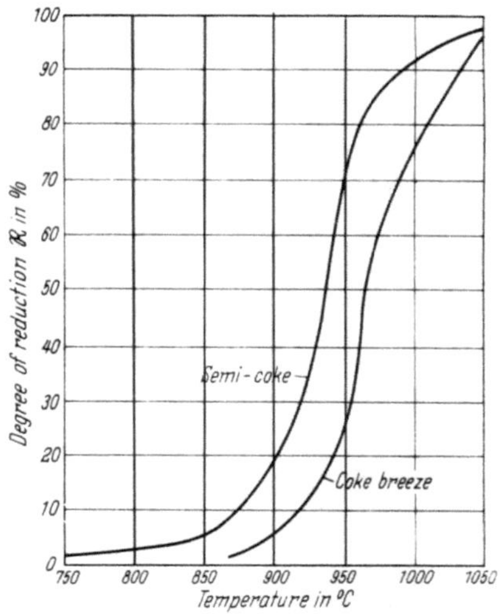

Fig. 205. Dependence of degree of reduction on the temperature in the reduction of Kiruna-C_2 ore, of grain size 5 to 30 mm, when using different reducing agents[407])

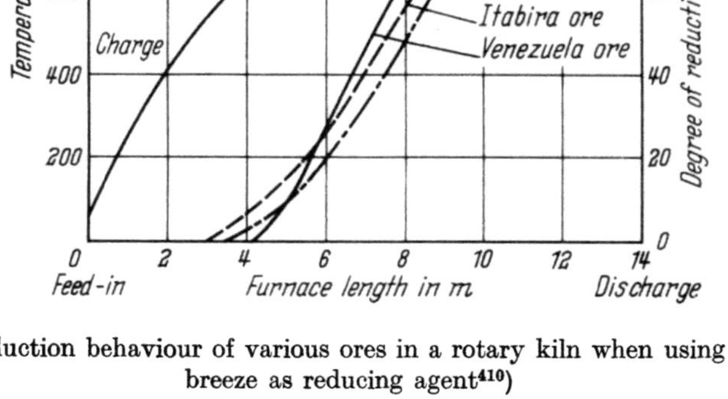

Fig. 206. Reduction behaviour of various ores in a rotary kiln when using 60% coke breeze as reducing agent[410])

4.5.4. Heat Transfer

A comprehensive explanation of the heat transfer in rotary furnaces has been given by HEILIGENSTÄDT[408]), and so we shall not enter into a detailed discussion here. In the range up to about 800°C the heat transfer is comparatively slow, but at higher temperatures, i.e., in the reduction zone, it is more rapid because of the increasing radiation of the furnace gas. Taking the balance across a section for the heat required and the heat supplied, and also estimating the heat-transfer numbers, we find that an appreciable throughput performance becomes possible only when the spent-

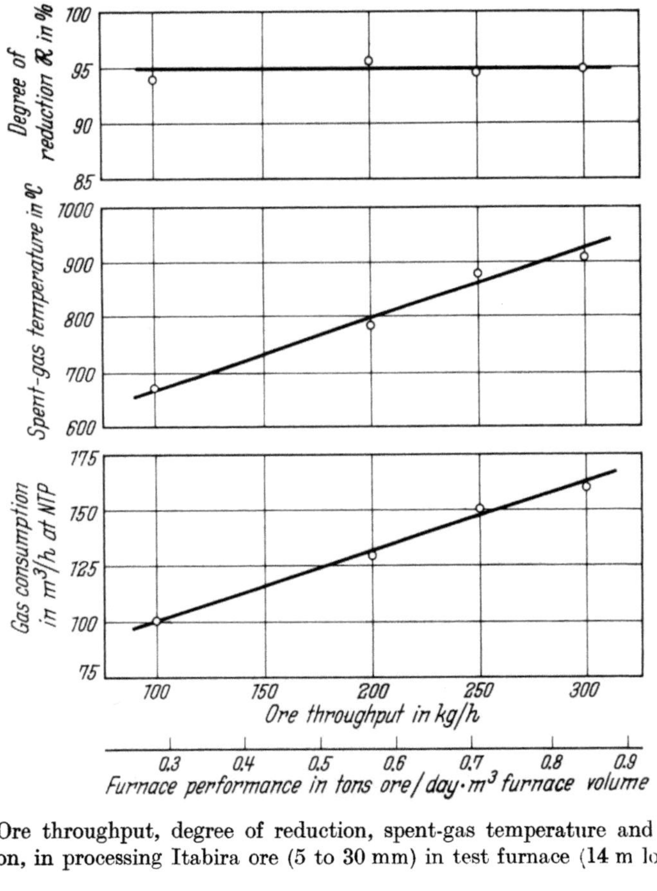

Fig. 207. Ore throughput, degree of reduction, spent-gas temperature and gas consumption, in processing Itabira ore (5 to 30 mm) in test furnace (14 m long)[410])

gas temperature is above 500°C[408]). If the heat supplied per unit time in the gas phase is increased, then the heat transfer increases as the spent-gas temperature increases. The preheating zone contracts in favour of the reduction zone, and the plant output increases, in accordance with experiment (Fig. 207). To preheat what may be taken as a typical charge for obtaining sponge iron (1430 kg Fe_2O_3, 500 kg C, 100 kg gangue, 100 kg moisture, all per ton Fe) to a temperature of 800 °C, a quantity of heat $Q = 480 \cdot 10^3$ kcal/ton Fe is needed. To transfer this amount of heat to the charge, the latter must have a surface area F_v in the preheating zone, given by

$$F_v = L\,Q/(24q), \tag{16}$$

where

L = required throughput performance,

or

$$f_v = F_v/L = Q/(24q), \tag{17}$$

where

q = heat transfer (kcal/m²h),
f_v = necessary surface area of the charge for heat transfer in the preheating zone to give the desired performance.

Figure 208 shows the relation between the necessary surface area of the charge and the mean heat transfer in the preheating zone. According to HEILIGENSTÄDT's calculations[408]), the lower limit for the specified heat

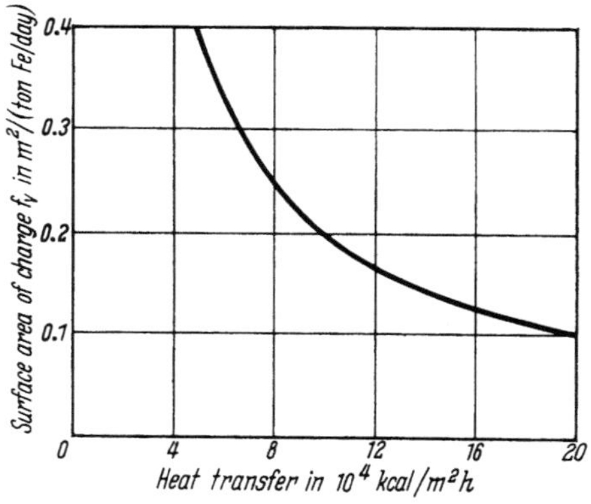

Fig. 208. Preheating zone of rotary kiln (0 to 800°C charge temperature). Necessary charge surface area f_v for the desired throughput as a function of the mean heat transfer[435])
Charge per ton Fe: 1430 kg Fe_2O_3, 500 kg C, 100 kg SiO_2, 100 kg moisture

transfer, $q = 40000$ kcal/m²h can be associated with a spent-gas temperature of about 500°C, and the upper limit, $q = 200000$ kcal/m²h, with spent-gas temperatures of about 1000°C.

There is also a considerable heat requirement in the reduction zone. In this zone, however, the heat transfer is essentially more rapid due to the increasing radiation, and is therefore no longer the factor which determines the throughput. This deduction can be justified by the fact that, above 800°C, the temperatures of the gas and charge are getting close together (Fig. 206).

4.5.5. Pre-Determination of the Output

Hitherto the procedure has mostly been that the ore/fuel mixture to be tested has been made to react in a small rotary furnace and the results have been determined as regards the performance, fuel consumption, and degree of reduction achieved, in relation to the chosen working conditions. Such tests let us see whether the chosen charge materials are basically suitable for the process, whether scabbing (deposits) occur, whether the charge materials fail to mix[429]), etc.

For prediction of the throughput and the optimum dimensions of a full-scale plant the following procedure can be applied:

4.5.5.1. Determination of the Size of the Preheating Zone

As shown in section 4.5.4, the heat transfer is the determining process and it depends on the size of the surface area, F_v, of charge which is exposed to the gas flames. To determine the necessary area, the heat Q (in kcal/ton Fe) needed for heating the charge to 800°C must first be determined. Depending on the desired throughput performance L (in tons Fe/day) and the heat transfer to be adopted according to Fig. 208 (allowing for the heat losses in the spent gas), the necessary surface area of charge in the preheating zone is given by Eq. (16).

4.5.5.2. Determination of the Size of the Reduction Zone and of the Whole Plant

From the reactivities of the ore and fuel, $\left[\frac{d\,\Re}{d\,t}\right]_{40\%}$ or H_{Fe} and H_C, which are determined from laboratory tests, the possible throughput per m³ of the charge volume, L/V_R (in tons Fe/day m³) is obtained from Eqs. (7), (9), (14) and Fig. 200. In the reduction zone, therefore, a charge volume V_R must be provided, corresponding to the desired performance L (in tons Fe/day) and to the measured reactivities, which is given by

$$V_R = \frac{L}{L/V_R} \quad (\mathrm{m}^3). \tag{19}$$

If the other circumstances are kept constant, then, the performance of the rotary furnace in the reduction of iron ores depends essentially on the *surface area of the charge* in the *preheating zone* and on the *volume of the charge* in the *reduction zone*. In the preheating zone, therefore, a shallow charge is desirable, but as deep as possible a charge in the reduction zone. In this connection it may be mentioned that BARRET[431]) as early as 1939 manufactured sponge iron in a rotary furnace whose diameter at the outlet end was considerably greater than in the upper portion; however, this scheme has not been adopted again since that time.

The volume of the empty furnace is not a convenient reference quantity, or, at least, cannot be used to deduce the performance of a large furnace from that of a similar small furnace, because the ratios

$$\text{furnace volume} \quad : \quad \begin{matrix}\text{charge volume}\\ \text{(``reduction zone'')}\end{matrix} \quad : \quad \begin{matrix}\text{charge surface}\\ \text{(``preheating zone'')}\end{matrix}$$

vary considerably with the size of furnace.

4.5.6. Reduction in Rotary Furnaces Carried out on the Commercial Scale

The large number of proposals which have been made in the literature and some of which are commercially operative cannot be fully described here, chiefly because some of the methods differ little in principle from one another. A publication by the High Authority of the European Community for Steel and Coal, Luxembourg[417]), provides a good survey. We shall here select from each group that process which can, by virtue of its operational procedure, be compared with the theoretical calculations.

4.5.6.1. Production of Sponge Iron[406, 407, 410, 416, 418–420])

The working temperature in the vessel is kept so low (usually below 1050°C in the charge) that no agglomeration takes place and the ore is reduced whilst keeping its external shape the same as when it was charged. The plants carrying the trade names R-N[420]) (Republic Steel, National Lead), Krupp sponge-iron[406, 407, 410, 410a,b]), S-L[416, 419]) (Stelco-Lurgi), Kalling-Avesta[417]), are of this type. It should be mentioned here that the SL-process and the RN-process have recently been amalgamated into one complex, known as the SL-RN-process[419a]).

In all cases the ore charge is mixed with a higher proportion of solid reducing material than would stoichiometrically be necessary, in order to ensure that a reducing atmosphere is maintained within the charge. Gas-rich coal cannot be added to the charge mixture because the volatile components would be driven out within the first few metres and would thus be lost to the process both as heating and as reducing material. Moreover, heavy hydrocarbons and tar would lead to difficulties in the purification of the spent gas because of condensation. If the coal contains 25 to 30% or more of its calorific value in volatile components, then it is introduced in the discharge region of the furnace and carbonized there. The gas from this low-temperature carbonization is used for the heating, and the coke discharged is added to the input mixture.

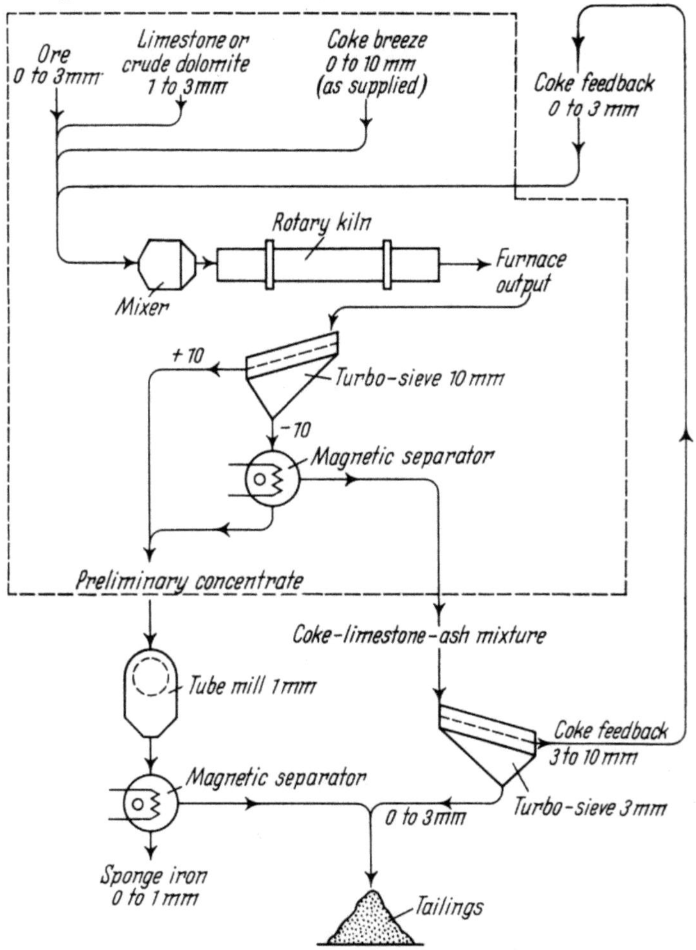

Fig. 209. **Flow** sheet for an industrial plant (Krupp sponge iron), according to Lucke et al.[406])

An interesting variation has been suggested by Pavlovic[432]) and has been realized on a small-scale, namely that, instead of a solid reducing material such as fine coal or coke breeze, oil is used, being injected into the highly heated ore mixture. The oil is thus decomposed into carbon and reducing gases which serve in the reduction and heating of the furnace.

A method which is based on the same principle and which has been tried out in an 8-m pilot furnace is designed to make use of the volatile components in gas-rich coal for the reduction[419b]). By means of high-pressure air the coal is injected from the discharge end far into the interior of the furnace. A uniform distribution over the surface of the charge is achieved by regulating the velocity of the HP-air, the inclination of the blast nozzles, and the grain size of the coal. Through the rotation of the furnace the coal is mixed with the charge and is gasified in its interior. The evolving hydrocarbons are decomposed and the resulting carbon remains almost entirely within the ore.

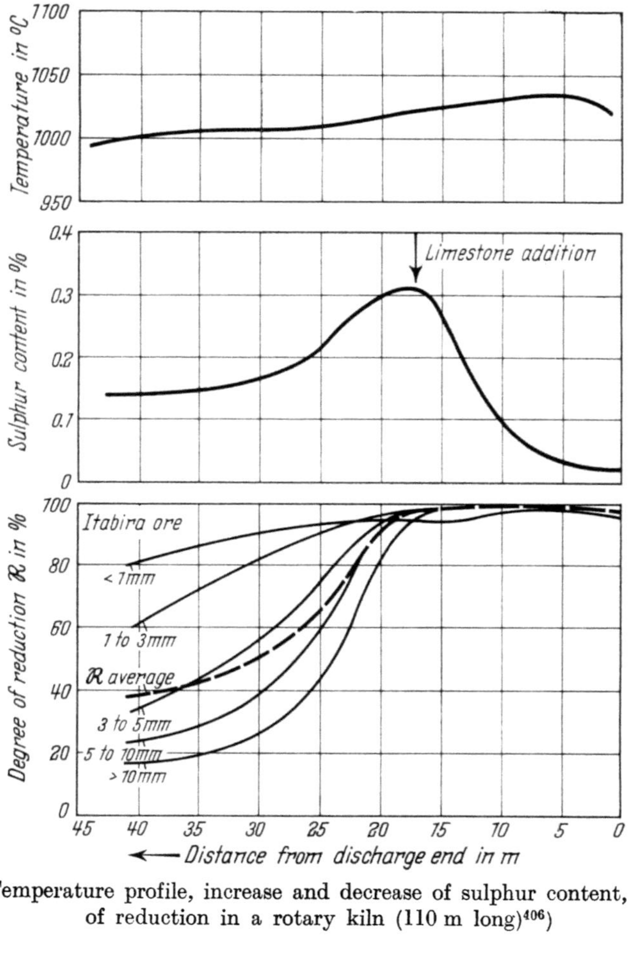

Fig. 210. Temperature profile, increase and decrease of sulphur content, and degree of reduction in a rotary kiln (110 m long)[406])

To remove the sulphur from the fuel and hence to prevent sulphuration of the sponge iron, basic materials such as lime, limestone, and dolomite are added[418]) (see Fig. 209).

The sponge iron is separated magnetically from the excess coke and desulphurizing material, as shown, e.g., in Fig. 209. In the RN-process the charge was of lean ore and final grinding of the magnetic component and separation from the gangue was necessary. When the RN-process was combined with the SL-process to form the SL-RN-process, the use of lean ore was abandoned in favour of rich ores or concentrates. Preliminary preparation of the ore is in most cases more economical than the reduction and subsequent enrichment of lean ores[419a]).

Screened ore lumps, or pellets with 60% Fe or higher, are used as charge material.

The course of the reaction corresponds to the kinetics calculations and to small-scale tests (cf. Fig. 210 with Figs. 206 and 200). Table 22, derived from information in [406]) and [416]) gives the most important data for large-scale trials of the Krupp sponge-iron process (600 tons/day throughput of ore) and the SL-RN-process (120 tons/day throughput of ore).

An essential difference between the two processes is that, in the Krupp-process, control of the heat supply over the length of the furnace is achieved

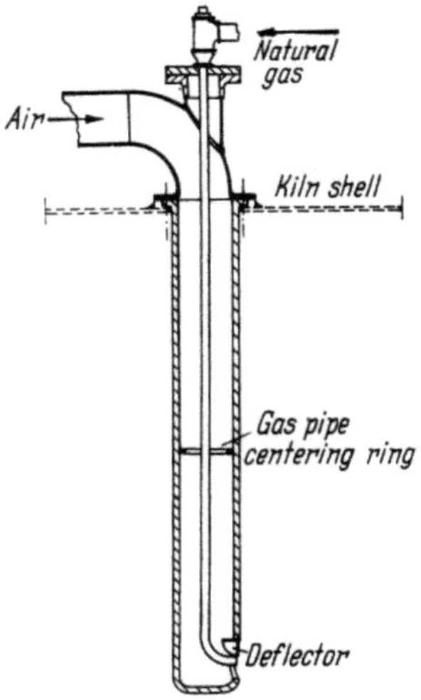

Fig. 211. Shell burner for the rotary furnace (SL-RN-process), according to SIBAKIN[416])

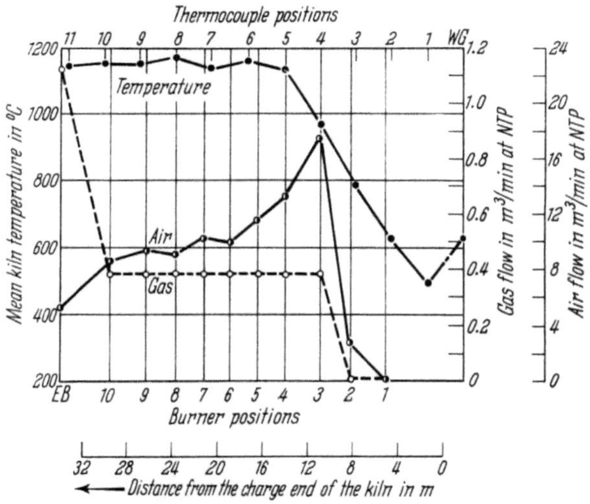

Fig. 212. Typical temperature profile in operation of pilot plant. Gas and air distribution also shown, according to SIBAKIN[416])

by the air supply to the whole furnace zone (Fig. 201), while in the SL-RN-process, additionally, fuel gas is supplied through the "shell burners" (Fig. 211) spaced along the furnace length. In this way the furnace space temperature can be closely controlled (Fig. 212). According to the previous theoretical considerations this makes higher throughputs possible.

Table 22. *Characteristics of the reduction process in rotary furnaces*

		Krupp sponge-iron		SL-RN	
		per ton sponge iron	per ton metallic iron	per ton sponge iron	per ton metallic iron
Ore					
Amount	kg	1480	1680	1368	1578
Fe-content	%	68.4		66.5	
Moisture content	%	3		unknown	
Anthracite					
Amount: charged	kg	592	671	541	624
returned	kg	149	168	207	239
gasified	kg	443	503	334	385
C_{fix}-content	%	84.5		80.9	
Amount of C gasified	kg		426		311
Additives					
Limestone	kg	86	98	—	—
Dolomite	kg	—	—	40	46
Heating gas		Coke-oven gas		Natural gas	
H_u	kcal/m³ at NTP	4100		≈ 9120	
Amount	m³ at NTP	155	176	88	102
Air needed	m³ at NTP	≈ 2600	≈ 3000	1960	2260
Spent gas					
Amount	m³ at NTP	≈ 3400	≈ 3900	2560	2960
Temperature	°C	650		≈ 650	
$CO + H_2$	%	2.5		≈ 0.5	
Net heat consumption					
From gas	10^6 kcal	0.635	0.720	0.800	0.930
From anthracite	10^6 kcal	2.995	3.400	2.140	2.460
Total	10^6 kcal	3.630	4.120	2.940	3.390
Contribution from returned anthracite	10^6 kcal	1.000	1.130	1.330	1.540
Sponge iron					
Total Fe-content	%	91.8		91.9	
Metallic Fe-content	%	88.2		86.7	

Further, in the SL-RN-furnace three damming rings are built in to enable a large proportion of the space to be filled, so that there is a considerable amount of material within the furnace volume. The performances of the plants are compared in Table 23[406, 416]).

The effect of the higher and more uniform furnace temperature in the SL-RN-process and also of the higher filling-ratio, V_R, can clearly be seen.

If we assume that about two-thirds of the volume of the charge corresponds to the reduction zone, then the performance per m³ of the reduction zone is (0.68 to 1.64). $\frac{3}{2}$ = 1.02 to 2.46 tons Fe/day · m³.

Table 23. *Comparison of the Krupp sponge-iron process and the SL-RN-process*

		Krupp sponge-iron (a)	Krupp sponge-iron (b)	SL-RN
Furnace				
Length, l	m	14.5	110	35.1
Internal diameter, D . .	m	1.2	4.6	2.29
Furnace volume, V_0 . .	m³	9.23	1330	145
Charge volume, V_R† . .	m³	≈ 2	≈ 300	≈ 44
Performance				
Ore throughput	tons/day	2.16	600	120
Metallic Fe, L	tons/day	1.35	365	72
Specific performance				
per m³ furnace volume, L/V_0	tons Fe/day · m³	0.146	0.274	0.497
per m² area, $L/l\,D$	tons Fe/day · m²	0.078	0.72	0.90
per m³ charge volume, L/V_R	tons Fe/day · m³	≈ 0.68	≈ 1.22	≈ 1.64

† Unfortunately, the volume V_R of the charge can be estimated only crudely from the constructional data given, such as the angle of inclination etc. These numbers are therefore uncertain, but from the point of view of the preceding theoretical considerations it would seem important for the future to determine them accurately.

According to Fig. 200 this corresponds quite accurately to the reduction temperature adopted of about 900°C for ores and fuels of average to good reactivity.

It would be possible to increase the size of the reduction zone at the expense of the preheating zone by heating the charge up to the reaction temperature outside the rotary furnace, by using the heat of the spent gas in a shaft furnace[410a]), in a travelling grate, a rotating hearth furnace, or in some similar way.

Further improvement in performance could be expected from an increase in the reactivity of the ore and the fuel and also by increasing the filling-ratio of the furnace. The result of such measures can be predicted on the basis of laboratory tests and the method of calculation derived above.

By suitable injection, as already described, of gas-rich coal and its gasification at the site of the reaction it was possible in a laboratory test in an 8-m furnace to obtain a specific performance of as much as 2 tons Fe/day · m³ [419b]). With a filling-ratio of about 30% this corresponds to a performance of 6.7 tons Fe per day per m³ of charge volume, and, with the achieved temperature of the charge of 1100°C and a 98% reduction, this agrees well with the theoretically determined values for fuels of high reactivity as given in Fig. 200.

Charge of Green Pellets. Very recently, green (unburned) pellets have been successfully reduced in rotary furnaces[419b,c]). The disintegration which frequently occurs with ore lumps and burnt pellets during reduction did not occur. The bulk stresses[435a]) occurring due to the phase changes during the reduction are absorbed in the relatively loose structure of the green pellets. The yield consisted of about 95%, in pellet form, of the Fe in the charge, the remaining Fe being in the form of abraded dust[419b]) (Table 23a).

Table 23a. *Iron distribution in sponge iron products*[419b])

Size fraction mm	Lump ore 6—15 %	Lump ore 6—15 %	Burned pellets 10—12 %	Green pellets 10—12 %
>10	41.7	72.7	54.3	89.1
1—10	49.0	24.0	44.0	8.2
<1	6.4	2.1	0.7	0.8
Σ	97.1	98.8	99.0	98.1

The compressive strength and abrasive strength*, [419c]) increase rapidly with the degree of reduction (Fig. 212a). For rotary furnaces the fraction < 1 mm is particularly harmful because of its tendency to separate and

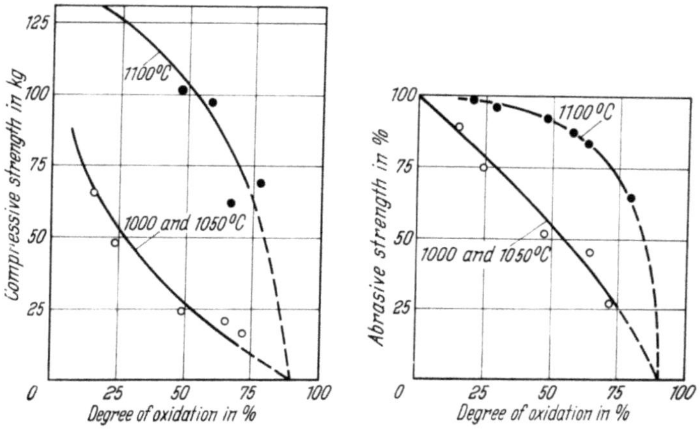

Fig. 212a. Reduction of green unfired pellets in a kiln at various temperatures, according to DAHL [419c])

to form scab deposits. The abrasive strength and compressive strength of burnt pellets attain a minimum at the FeO-stage, having about the same value as for green pellets[419c]). At higher degrees of reduction the green

* The abrasive strength was determined as follows: 5000 revolutions at about 2500 rev/h of 150 cm^3 of pellets in a drum of diameter 200 mm, 100 mm long, with 4 ribs round the circumference. The fraction with size > 1 mm is taken to be the abrasive strength. The stress is about 3.5 times as great as in the usual Micum test.

pellets are superior to burnt pellets. There is a connection between the compressive strength, the plasticity, and the abrasive strength. A sufficient compressive strength can be achieved[419b]) by drying and the use of binding agents before the pellets are fed into the rotary furnace, without the need for preliminary reduction.

4.5.6.2. Production of Molten Pig Iron

In certain circumstances it is possible by raising the furnace temperature at the output end to melt the ore charge after it has been reduced (Basset and Stürzelberg processes)[417]). The performance then falls off, because an extra process has to take place in the same vessel.

High performance can, however, be achieved by increasing the working temperature throughout the whole charge volume to about 1300°C as in the Stora process[421]) (see Fig. 213) — but note that since 1965, when a plant

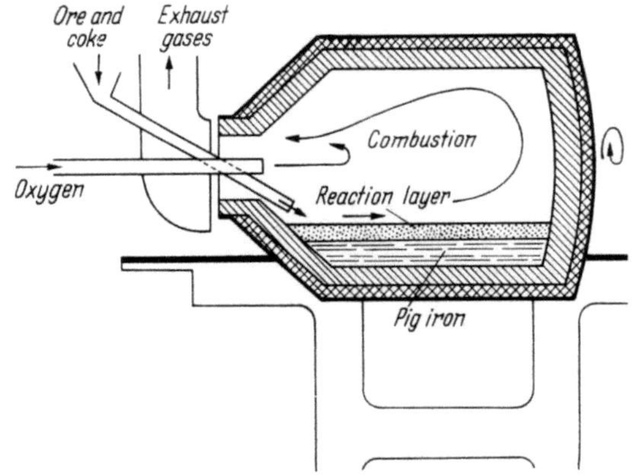

Fig. 213. Schematic diagram of the Stora rotary iron-making furnace, according to BENGTSSON et al.[421])

with an output of 80000 tons a year was put into operation, this is now known as the Dored process[421a]) — and also in the process described in[433]). The iron is produced directly in liquid form. A performance of 10.8 tons pig iron a day is achieved with a very small charge volume. Relative to the volume of the solid, reacting charge, floating on the pig iron, the performance reaches an order of magnitude of 40 tons of pig iron per m^3 of the charge volume[430]), and agrees in order of magnitude with the theory, see Fig. 200. (Note, however, that this agreement is only qualitative, since the mathematical formulae are valid only up to about 1200°C.) The weakness of rotary-furnace processes — the poor heat transfer as compared with shaft furnaces — is nevertheless particularly noticeable here. The spent gas leaves the furnace at a temperature of at least 1500 °C; this necessitates the use of oxygen instead of air for the combustion in order to reduce the volume of the spent gas and to keep the loss of its sensible heat within acceptable bounds, and may possibly require the introduction of a second rotary furnace for preheating the charge (Fig. 214). It would seem to be not impossible that, in this way, we could get from iron to steel in a single

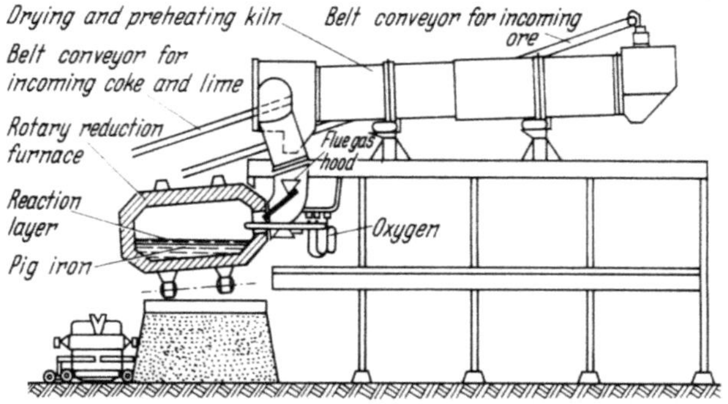

Fig. 214. Planned arrangement for the Stora rotary reducing furnace with a drying and preheating kiln, according to BENGTSSON et al.[421])

vessel[433]). The heat consumption, however, is given as 17 to 35 · 10^6 kcal per ton of product, and is thus very high.

The Bouchet process[433a]) for reducing and smelting fine ores with coal uses a rotary smelting furnace with an oxygen burner and an additional supply of oxygen for the combustion of the CO formed during the reduction. The weight of the charge is about 1.7 tons, the depth of the bath only 90 mm at its maximum. About 4.5 · 10^6 kcal of heat and 385 m^3 at NTP of oxygen must be supplied per ton of steel. The smelting performance is about 0,55 ton/m^2 h. The process is also suitable for smelting iron powder. The smelting performance then rises to 1.4 ton/m^2 h, the energy consumption falls to 1.25 · 10^6 kcal, and the oxygen consumption is 185 m^3/ton of steel at NTP.

4.5.6.3. Production of Iron Nodules

The Krupp-Renn process[409, 422, 423]) for the production of iron balls or nodules stands midway between the above two processes. The working temperature at the output end is about 1200 °C in the charge, and the metallic Fe coagulates into nodules and can therefore be easily separated magnetically from the gangue. There is nothing new as regards the reduction process. The process is used exclusively with very lean, and mainly acid, ores ($CaO/SiO_2 \leq 0.2$). Because of the large proportion of non-reacting ballast, the heat transfer is largely the determining factor[408]).

4.5.6.4. Pre-Reduction

If the reduction is incomplete, the performance of the rotary furnace can be increased and the load on the subsequent smelting plant can be reduced; of course, the degree of reduction then varies markedly with the

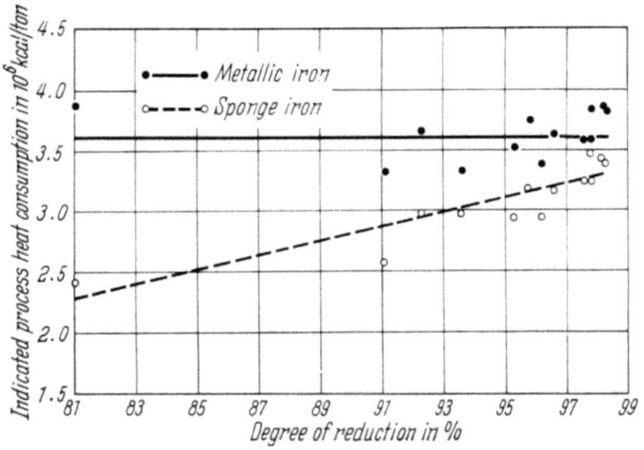

Fig. 215. Relationships between indicated heat consumption and degree of reduction (SL-RN-process), according to SIBAKIN[416])

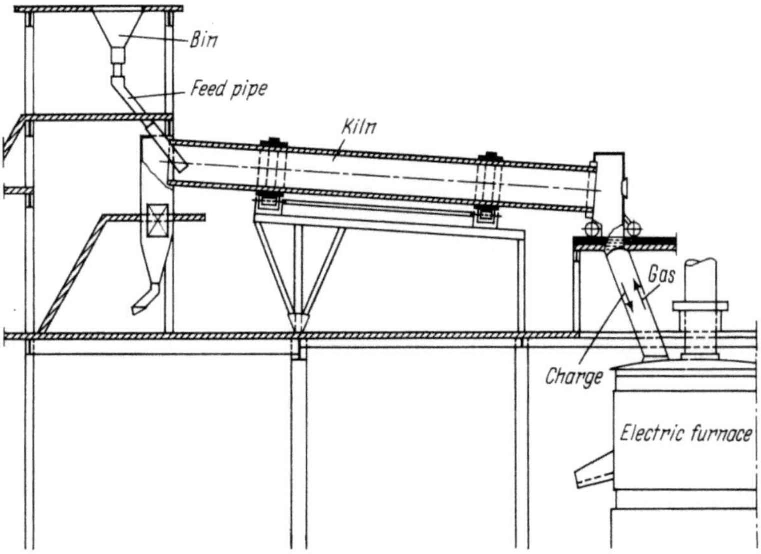

Fig. 216. Pilot plant comprising a rotary kiln and an electric furnace; Strategic-Udy process[428])

charge grain size (Fig. 210). A uniform degree of reduction can be achieved with a charge of green pellets because there is no disintegration (Table 23a). The fuel consumption in the rotary furnace relative to metallic iron is not necessarily reduced by this (see Fig. 215). Rotary furnaces for pre-reduction are used preferably in connection with electro-reduction and smelting plants (Fig. 216), e.g., electric low-shaft furnaces[424, 428]), that is, to save

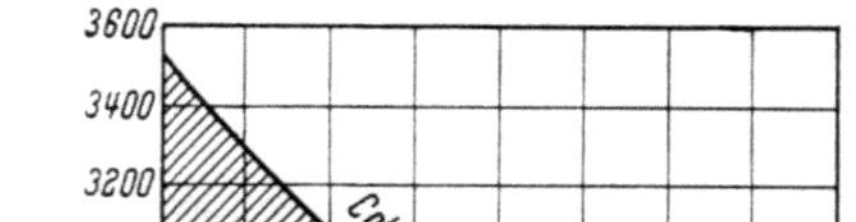

Fig. 217. Effect of iron content and preliminary treatment of ore on the electrical energy required for pig iron smelting in an electric reduction furnace[428])

electrical energy (Fig. 217) (Strategic Udy, Orcarb, Elektrokemisk processes, etc.). This combination of processes will be discussed further in connection with electro-reduction (section 4.7).

4.6. Reduction in Shaft Furnaces (Without Melting)

4.6.1. Theoretical Basis

4.6.1.1. Restrictions Imposed by Fluid Dynamics on the Throughput

Application of the counterflow principle requires that the charge shall sink uniformly as charging is maintained, i.e., in each volume element there must be no appreciable stagnation or upward movement of the charge. The rate of gas flow must therefore be kept to values below the turbulence point. The formation of a *spouted bed* is also to be avoided. The conditions and limits for a satisfactory, even distribution of the gas flow were described in section 3.1. With adequate reduction rates, the theoretical maximum conversion per m² of the reactor cross-section is found from:

(i) that amount of gas which can just be put through without disturbing the stability of the packed column, and

(ii) the maximum reducing power of the gas, which is fixed by the equilibrium relations.

For the reduction of ore charges at 900°C and at $p = 1$ atm the relation shown in Fig. 218 is obtained between the usable rate of gas flow and the throughput performance per m² of the shaft cross-section; separate curves are given for gases with 40% and 100% reducing component (H_2 and CO) and for a degree of chemical utilization

$$\eta = \frac{CO_2 + H_2O}{CO + H_2 + CO_2 + H_2O}$$

of either 0.3 or 0.5.

Figure 218 shows that, for gas velocities greater than 5 m/sec, throughput performances are obtained which are greater than those in a blast furnace. According to Fig. 123, gas velocities up to 40 m/sec are possible when using H_2-rich gas and ore lumps greater than 20 mm. What is important now is that the exchange of material and heat between gas and

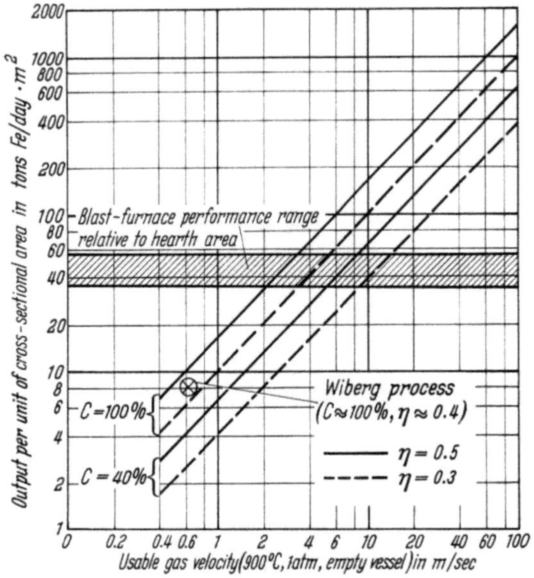

Fig. 218. Performance of a shaft furnace in the reduction of Fe_2O_3 with a $H_2/CO/N_2$ mixture in relation to: u, the applicable velocity; C, the concentration in the reducing gas of $CO + H_2$; and η, the degree of chemical utilization of the $CO + H_2$

charge shall take place rapidly enough in order to achieve an acceptable degree of utilization of the gas at these high velocities. Thus, kinetics and heat transfer limit the throughput.

4.6.1.2. Kinetics of Reduction in Counterflow

The counterflow principle certainly brings about the best throughput performances with, at the same time, good chemical and thermal utilization of the reducing gas, but it also makes the mathematical treatment very difficult.

Here too the gas retention time in the charge determines the attainable reduction performance. It is true that, by increasing the height of the shaft and, if necessary, by introducing an inert gas as a heat carrier, it would be possible, even with the highest gas velocities, to make the retention time have the desired value, and thus to approximate to equilibrium between gas and ore. But here, of course, we come up against constructional limitations.

The problem, from the point of view of reaction kinetics, is: to calculate the relation between, on the one hand, the throughput time of gas and ore, and, on the other hand, the total conversion, i.e., the degree of reduction of the ore and the chemical and thermal utilization of the gas; and in finding this relation we are to take into account variations in all the parameters which affect the reduction, such as type of ore, size

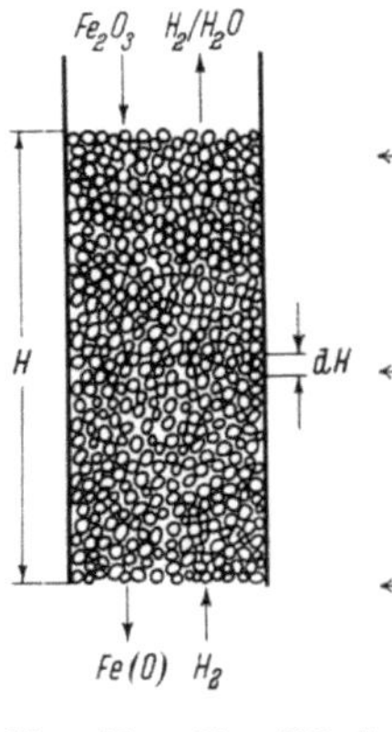

Charge		Gas		
Temperature	Degree of reduction	Temperature	H_2O-content	N_2-content
20°C	0	θ_g	$C^E_{H_2O}$	0 or 50%
T	$\mathfrak{R}$	θ	$C^0_{H_2O}$	0 or 50%
T_e	$\mathfrak{R}_e$	θ_a	0	0 or 50%

Fig. 219. Simplified scheme for a countercurrent reactor for the reduction of ore

of ore lumps, working temperature, type of gas, etc. From the preceding considerations from the point of view of fluid mechanics, gas mixtures rich in hydrogen will be of particular interest.

By virtue of the relation which we are about to calculate, then, we can specify, for a given desired performance per unit area of cross-section of the shaft, the height of the shaft for which a prescribed degree of utilization of the gas and a prescribed output of Fe will be attained.

The principle of the calculation will be explained below in terms of the following model:

A cylindrical reduction shaft, of cross-sectional area F and height H, (Fig. 219) is charged from above with Fe_2O_3-pellets, which, after reduction has been carried out, are drawn off from below. The reducing gas is to consist essentially of hydrogen flowing upwards from below, in counterflow against the ore charge. Several stages of approximation to the actual conditions can be made, as shown in the following treatment.

4.6.1.3. First Approximation (a Purely Reaction-Kinetic Approximation)[437]

4.6.1.3.1. Basis of the Method of Calculation. A calculation in closed form for the reduction process in a countercurrent reactor is unfortunately not possible, because, as may be seen from the example in Fig. 182, the rate of reduction depends in too complicated a manner on the degree of reduction and on the temperature. A numerical computation can be carried through, but it is very laborious. Because of the fundamental importance of the relations, we give in greater detail an example of the simplified numerical calculation to explain the method.

The reactor, assumed cylindrical as in Fig. 219, will be operated as follows.

Reducing gas: input temperature $\Theta_a = 1000°C$

composition – either (a) 100% H_2,

or (b) 50% H_2, 50% N_2.

The reduced material (sponge iron), which is removed from the bottom of the cylinder, is at first supposed to have an exit temperature T_e and a degree of reduction $\mathfrak{R}_e$.

The amount of gas V_{gas} for reduction per kg Fe-throughput, measured in m^3/kg Fe at NTP, will now be determined by trial. This gas enters the reducing column at the bottom and is gradually cooled by the heat transferred from it for heating up and reducing the material. The shaft will be divided into zones whose length corresponds to a fixed, chosen decrease in the temperature of the gas. As an initial simplification, it will be assumed that the temperature difference between the gas Θ and the charge T, $(\Theta - T)$, is negligible in the zones where reduction is taking place. (Actually, there is a temperature difference, although it is very small; in section 4.6.1.4 its size will be calculated and its effect allowed for.) The zones are strongly coupled one with the next, since the reducing gas which leaves one zone is the input gas for the next zone, and a similar circumstance holds for the solid charge in the opposite direction. The numerical treatment is continued over all the zones. When this is done, it may happen that the reducing power of the gas is exhausted comparatively early, while in some zone the calculated H_2O-content attains or exceeds the equilibrium value. This would mean that the initial conditions had been wrongly chosen and that the selected degree of reduction could not be achieved; either a lower final degree of reduction $\mathfrak{R}_e$, or a larger amount of gas per kg Fe (V_{gas}), must be taken as the initial value.

As a result we get the time required for the reduction in each zone and hence their lengths, and then, by addition, the necessary total retention time for the charge in the shaft and the height of the latter. From this, since the bulk weight of the charge is known, we get the throughput performance per m³ of the charge volume.

4.6.1.3.2. Properties of the Charge Materials. Let the charge consist of Fe_2O_3-pellets with the properties already laid down in section 4.4: e.g.,

$$\text{diameter } d = 3.1 \text{ cm},$$

$$\text{bulk weight } M_{Fe_2O_3} = 2.3 \text{ g/cm}^3$$

or

$$M_{Fe} = 1.61 \text{ g Fe/cm}^3 \text{ or tons Fe/m}^3.$$

For these pellets the rates of reduction in hydrogen have already been depicted in Fig. 182 of section 4.4 as functions of the degree of reduction, temperature, and gas composition. As explained, the gas composition operates via the factor

$$\Delta C = C^{eq}_{H_2O} - C^{0}_{H_2O},$$

where

$C^{eq}_{H_2O}$ = equilibrium concentration of H_2O, to be inserted taking into account its dependence on the degree of reduction (Table 20),
$C^{0}_{H_2O}$ = H_2O-concentration in the free gas space,

both these quantities being taken as volume ratios.

The further use of Fig. 182 has already been explained in section 4.4.

The heat content of the Fe_2O_3 and the H_2 will also be required. According to[438, 439]), on average, 28.6 kcal/kg Fe will be needed for heating the charge for each 100°C. This is 85% of the theoretical value for Fe_2O_3, and therefore on average rather too high, since there is mainly Fe_2O_3 in the upper part of the shaft but mainly Fe in the lower part. However, the error introduced by disregarding this fact balances the heat loss of the reactor which also has not been considered.

For the complete reduction

$$Fe_2O_3 + 3\,H_2 = 2\,Fe + 3\,H_2O\,,$$

the amount of heat required is[439])

$$\Delta H = 21.8\ \text{kcal/mole} = 195\ \text{kcal/kg Fe}\,.$$

An increase of 1% in the degree of reduction therefore requires, on average, a heat input of 1.95 kcal/kg Fe. The heat content of hydrogen is correspondingly inserted[438]). The change in the heat content of the gas mixture as it ascends the shaft due to enrichment in H_2O is not at first taken into consideration, in order to be able to see clearly the state of affairs from the point of view of the reaction kinetics. In the second approximation these three factors which have been neglected will be taken fully into account.

4.6.1.3.3. Numerical Example. The following further assumptions are adopted in this calculation:

Θ_a = 1000°C, gas input temperature, at the bottom,
T_e = 1000°C, material output temperature, at the bottom,
$\mathfrak{R}_e$ = 99.5%, degree of reduction at output.

The thickness ΔL of each zone is to be chosen so that the gas temperature Θ falls in each zone by 100°C (or 50 or 25°C). All the conditions are now fixed. For the kinetic balance we find:

First trial:

Gas throughput chosen to be: $V_{\text{gas}} = 2.5\ \text{m}^3\ H_2$/kg Fe at NTP (100% H_2 in the gas entering).

Zone 1

$$\Theta = 1000°\text{C} \to 900°\text{C}, \quad \Delta\Theta = 100°\text{C}, \quad \bar{\Theta} = 950°\text{C},$$
$$T = 900°\text{C} \to 1000°\text{C}, \quad \Delta T = 100°\text{C}, \quad \bar{T} = 950°\text{C}.$$

Heat supply:

$V_{\text{gas}}\,\Delta\Theta\,c_p = 2.5 \cdot 100 \cdot 0.330$	= 82.5 kcal/kg Fe
Requirement for $\Delta T = 100°\text{C}$	= 28.6 kcal/kg Fe
Amount of heat remaining for reduction	= 53.9 kcal/kg Fe

Possible increase in degree of reduction (100% reduction corresponds to 195 kcal/kg Fe):

$$\Delta\mathfrak{R} = 53.9/195 = 0.277 \quad (\text{from } 71.8 \text{ to } 99.5\%).$$

The H_2O-content of the hydrogen is thereby increased as follows:

$$C^0_{H_2O} = \frac{0.277 \cdot 0.430}{16}\,\frac{22.4}{2.5} = 0.0668\,.$$

On average the following conditions prevail in zone 1:

$$\bar{\mathfrak{R}} = 85.7\%\,; \quad \bar{T} = 950°\text{C}; \quad C^0_{H_2O} = 0.0668/2 = 0.0334\,.$$

Table 24

a) 2.5 m^3/kg Fe at NTP; 100% H_2; 1000°C; $\mathfrak{R}_e = 99.5\%$

1	2	3	4	5	6a	6b	6c	7	8	9a	9b	9c	10	11	12	13
					Heat					Mean data						
Zo-ne	Θ input	$\Delta\Theta$	T input	ΔT	supplied from gas	required for ΔT	remain-ing for re-duction	$\Delta\mathfrak{R}_j' = \frac{(6c)}{195}$	$\Delta C^0_{H_2O}$	$\overline{\mathfrak{R}}$	$\overline{T}$	$\overline{C^0_{H_2O}}$	ϑ	$C^{eq}_{H_2O}$	$\frac{d\mathfrak{R}}{d\tau} = 10\cdot(11-9c)$	$\tau = \frac{\Delta\mathfrak{R}}{\frac{d\mathfrak{R}}{d\tau}}$
No.	°C	°C	°C	°C	kcal/kg Fe	kcal/kg Fe	kcal/kg Fe		Volume ratio		°C	Volume ratio	sec^{-1}	Volume ratio	sec^{-1}	sec
1	1000	−100	900	+100	82.5	28.6	53.9	0.277	0.0668	0.857	950	0.0334	$0.44\cdot10^{-3}$	0.385	$15.5\cdot10^{-5}$	$1.79\cdot10^3$
2	900	−100	800	+100	81.5	28.6	52.9	0.271	0.0652	0.583	850	0.0994	$0.60\cdot10^{-3}$	0.355	$15.3\cdot10^{-5}$	$1.77\cdot10^3$
3	800	−100	700	+100	80.5	28.6	51.9	0.266	0.0640	0.314	750	0.1640	$0.80\cdot10^{-3}$	0.354	$15.2\cdot10^{-5}$	$1.75\cdot10^3$
4	700	−100	600	+100	79.5	28.6	35.3	0.181	0.0436	0.090	650	0.2178	$0.62\cdot10^{-3}$	0.801	$36.2\cdot10^{-5}$	$0.50\cdot10^3$
5	600	−170	20	+580	133.9	133.9	—	—	—	—	—	—	—	—	—	$2.10\cdot10^3$
								0.995	0.2396							$\sim 7.91\cdot10^3$

$L/V_R = 17.6$ tons Fe/day m^3 $\tau_{gas} = 0.49$ sec $\tau_{charge} = 2.19$ h

b) 2.0 m^3/kg Fe at NTP; 100% H_2; 1000°C; $\mathfrak{R}_e = 99.5\%$

1	2	3	4	5	6a	6b	6c	7	8	9a	9b	9c	10	11	12	13
1	1000	−100	900	+100	66.0	28.6	37.4	0.192	0.0578	0.899	950	0.0289		0.385		
2	900	−100	800	+100	65.2	28.6	36.6	0.188	0.0566	0.709	850	0.0861		0.355		
3	800	−100	700	+100	64.4	28.6	35.8	0.184	0.0553	0.523	750	0.1420		0.315		
4	700	−100	600	+100	63.6	28.6	35.0	0.180	0.0542	0.341	650	0.1968		0.262		
5	600	−100	500	+100	63.2	28.6	34.6	0.178	0.0536	0.162	550	0.2507*		0.200*		
								0.922	0.2775							

* Reduction stops according to calculation because $C^{eq}_{H_2O} < C^0_{H_2O}$.

From Fig. 182 in section 4.4 and Table 20 we then find for the rate of reduction

$$\frac{d\Re}{dt} = 0.44 \cdot 10^{-3}\,(0.385 - 0.0334) = 15.5 \cdot 10^{-5}\,\text{sec}^{-1}.$$

The necessary retention time for the charge comes out to

$$\tau_1 = \frac{\Delta\Re}{d\Re/d\tau} = \frac{0.277}{15.5 \cdot 10^{-5}} = 1.79 \cdot 10^3\,\text{sec}.$$

This calculation is repaeted in the same way for the successive zones (see Table 24, a).

Summary: First trial (Details in Table 24, a)

	Θ °C	T °C	$C^0_{H_2O}$	$\Re$ %	τ_{charge} sec
Zone 1 beginning	1000	1000	0	99.5	1790
2 beginning	900	900	0.0668	71.8	1770
3 beginning	800	800	0.1320	44.7	1750
4 beginning	700	700	0.1960	18.1	500
5 beginning	600	600	0.2396	0	2100*
5 end	430	20	0.2396	0	
					7910 ≈ 2.19 h

* This value was taken from heating tests on ore samples[448]).

The total retention time for the charge in the reactor must therefore amount to $\tau = 2.19$ h in order to ensure heating up and complete reduction under the given conditions.

The variation of the most important quantities in relation to the retention time, i.e., over the shaft height, is shown in Figs. 220 and 221.

The performance per unit volume of the reactor is then calculated as follows:

$$\frac{L}{V_R} = \frac{M_{Fe}}{\tau} \cdot 24 = \frac{1.61}{2.19} \cdot 24 = 17.6\,\frac{\text{tons Fe}}{\text{m}^3\,\text{day}}.$$

and

$$\Theta_g = 430\,°\text{C (temperature of gas issuing at the top)},$$

$$\eta_{H_2} = 23.96\%\ \text{(chemical utilization of } H_2).$$

Second trial:

$V_{gas} = 2.0\,\text{m}^3\,H_2/\text{kg Fe at NTP}$, $\Theta_a = 1000\,°\text{C}$ (100% H_2 in the reducing gas).

Result: No complete reduction possible (details of calculation in Table 24,b).

Further evaluations can be found in [437]). The results are shown in the condensed Table 25 and in Fig. 222.

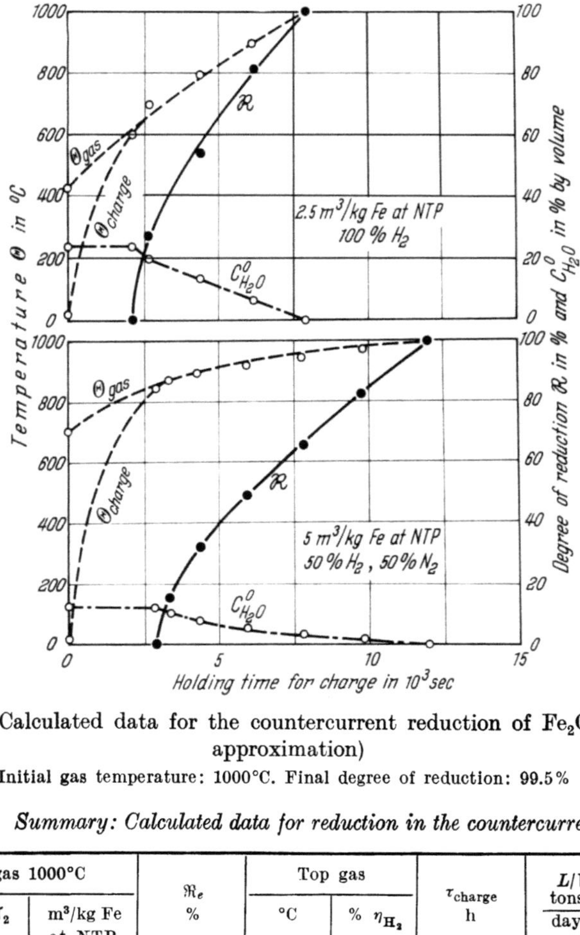

Fig. 220/221. Calculated data for the countercurrent reduction of Fe_2O_3-pellets (1st approximation)

Initial gas temperature: 1000°C. Final degree of reduction: 99.5%

Table 25. *Summary: Calculated data for reduction in the countercurrent reactor*

Reducing gas 1000°C			$\Re_e$	Top gas		τ_{charge}	L/V_R	τ_{gas}
% H_2	% N_2	m³/kg Fe at NTP	%	°C	% η_{H_2}	h	tons Fe / day m³	sec
100	—	up to 2.0	incompl.	—	—	—	—	—
100	—	2.1	99.5	320	28.5	3.60	10.7	0.95
100	—	2.3	99.5	380	26.0	2.31	16.7	0.56
100	—	2.5	99.5	430	24.0	2.19	17.6	0.49
100	—	3.0	99.5	530	20.0	2.00	19.3	0.37
100	—	4.0	99.5	645	15.0	1.93	20.0	0.27
100	—	5.0	99.5	720	13.0	1.89	20.5	0.21
100	—	7.0	99.5	820	8.6	1.77	21.3	0.14
50	50	up to 2.2	incompl.	—	—	—	—	—
50	50	2.3	99.5	380	52.0	47.20	0.8	11.40
50	50	2.5	99.5	430	47.9	11.00	3.5	2.44
50	50	3.0	99.5	530	40.0	4.46	8.7	0.83
50	50	4.0	99.5	645	29.9	3.72	10.4	0.41
50	50	5.0	99.5	720	24.0	3.33	11.5	0.37
50	50	7.0	99.5	820	17.1	3.06	11.9	0.26

The retention times for the gas in the charge, τ_{gas}, shown in Tables 24 and 25, are derived from the retention time, τ, for the charge in the following way:

$$\tau_{gas} = \frac{273\,\tau}{V_{gas}\,M_{Fe}\,1000\,T},$$

where

V_{gas} = amount of gas put in (m^3/kg Fe at NTP),
T = mean temperature of the gas (°K),
M_{Fe} = amount of Fe per unit volume of the charge (tons/m^3),
τ_{gas}, τ = retention times for gas and charge, in the same units for both (h or sec).

The mean working temperature was chosen to be T = 1100°K (827°C).

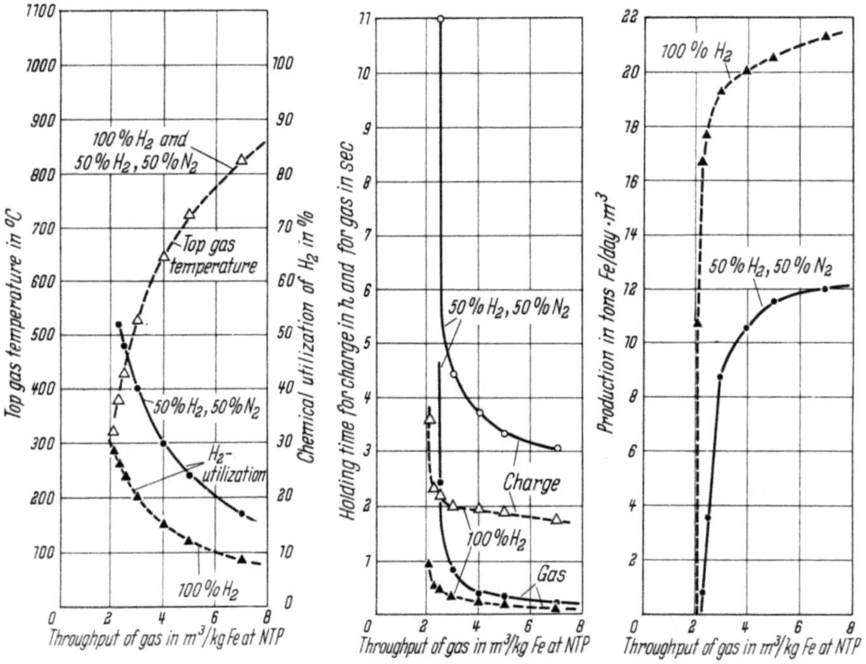

Fig. 222. Calculated data for the countercurrent reduction of Fe_2O_3-pellets (1st approximation)

Shaft cross-section: 1 m^2. Initial gas temperature: 1000°C. Final degree of reduction: 99.5%

4.6.1.3.4. Limitation of the Results. Carrying out all the evaluations showed how, as the amount of gas was decreased, the degree of reduction $\Re = 0$ is reached at lower and lower temperatures, until finally the degree of reduction $\Re = 0$ can no longer be attained. Since the rising gas (see Figs. 220 and 221) begins perceptibly to reduce from about T = 500°C, the correct input is precisely that for which $\Re = 0$ is attained at about 500°C; that is, in the case of pure hydrogen, an amount of gas equal to

2.1 m^3/kg Fe at NTP, and, in the case of a 50% H_2/50% N_2 mixture, a gas amount of about 2.3 m^3/kg Fe at NTP. In cases where $\mathfrak{R} = 0$ is attained at temperatures of more than about 500°C, the actual course of the reduction is such that $\mathfrak{R} = 99.5\%$ is already achieved at lower temperatures.

Figure 220 shows the results for pure H_2 when the gas amounts have been chosen approximately correctly; Fig. 221 shows the results for a 50% H_2 mixture, for which $\mathfrak{R} = 0$ is already attained at $T = 800°C$. In this case the actual course of the reduction is not correctly reproduced (earlier and slower increase of $\mathfrak{R}$, early achievement of $\mathfrak{R} = 99.5\%$).

4.6.1.4. Second Approximation (Setting up a Mathematical Model with the Heat-Transfer Process Taken into Account)

The procedure which we have called the first approximation clearly shows that a calculation based on reaction kinetics can be carried through and how this is to be done. Even though it was in some respects crude, it still yielded justifiable results, and in particular one characteristic feature of this kind of calculation was made clear: namely, the calculation can be carried through to a final result only if it is started with suitable initial values. Mathematically speaking, we are dealing with a boundary-value problem; for at one boundary (the lower end of the reactor) the temperature and composition of the gas are *prescribed*. The temperature and composition (i.e., degree of reduction) of the charge at the lower end of the reactor are the *required results*. In contrast, at the upper end of the reactor, the temperature and composition (i.e., the H_2O-content) of the gas are the unknown quantities.

So far as the reaction kinetics are concerned, even when using a more elaborate mathematical model, the procedure cannot be fundamentally different from that in the first approximation, even though many refinements may be introduced; from what has been said in section 2.2 and 3.1 it follows that the heat transfer from the gas to the charge can be calculated quite independently of the assumptions made in the first approximation.

4.6.1.4.1. The Temperature Difference between the Core and Outer Surface of the Pellets. Since there is a possibly large difference in temperature between the core and outer surface of the pellets, it is not permissible when calculating the heat transfer to start from the mean temperature of a pellet; rather, it is the surface temperature which must be used. Consequently the difference between the temperature of the surface and the core must be known, or between the surface temperature and the mean temperature over the cross-section.

The temperature difference without regard to reduction. Let

T_k = temperature of core,
T_0 = temperature of outer surface,
T = mean temperature of a pellet,
$\dot{T}_0$ = rate of change of surface temperature, i.e., the rate of heating (°C/h).

Then in almost all cases arising in practice, i.e., so long as $\dot{T}_0$ does not exceed 1000°C/h during the heating up and the diameter of the pellets (spheres) is not greater than about 8 cm, the temperature difference is given by

$$T_0 - T_k = \frac{1}{240} \frac{c_p \varrho}{\lambda} \dot{T}_0 \, d^2 \tag{1a}$$

and

$$T_0 - T = \frac{1}{600} \frac{c_p \varrho}{\lambda} \dot{T}_0 \, d^2, \tag{1b}$$

where

c_p = specific heat (kcal/°C kg)
ϱ = density (t/m³)
λ = coefficient of thermal conductivity (kcal/m h °C)
d = diameter (cm)
} of the pellets at the temperature T.

This result is obtained from the equation of heat conduction in a sphere. It is also seen that this temperature difference is set up in so short a time that we may neglect this interval.

The temperature difference taking reduction into account. During the process of reduction heat is liberated inside the pellet if CO is the reducing gas, or is absorbed if H_2 is used. So the equation of heat conduction must be extended to include the corresponding terms, if it is to be used to determine the temperature difference.

Let

$\hat{\varrho}$ = density of the material (of the pellets) in the unreduced state,
$\hat{g}$ = proportion of oxygen by weight in the unreduced state (t/m³),
ΔH_{H_2} = heat of reaction in the reduction by H_2 (kcal/kg O),
ΔH_{CO} = heat of reaction in the reduction by CO (kcal/kg O),
$\dot{\mathfrak{R}}_{H_2}$ = rate of change of degree of reduction in the reduction by H_2 (h^{-1}),
$\dot{\mathfrak{R}}_{CO}$ = rate of change of degree of reduction in the reduction by CO (h^{-1}).

Then the difference between the mean and the outer surface temperature is given by

$$T_0 - T = \frac{1}{600} \frac{1}{\lambda} [c_p \varrho \dot{T}_0 + \hat{\varrho} \hat{g} (\Delta H_{H_2} \dot{\mathfrak{R}}_{H_2} + \Delta H_{CO} \dot{\mathfrak{R}}_{CO})] \, d^2. \tag{2}$$

In this formula it must be noticed that c_p, ϱ, and λ now depend not only on the mean temperature but also on the degree of reduction $\mathfrak{R}$ of the material, because the reduced Fe differs considerably from the unreduced material (Fe_2O_3) as regards these properties. As regards the thermal conductivity, the porosity, and its increase as the reduction proceeds, is to be taken into account.

Numerical example for reduction by H_2. For the pellets used, of diameter $d = 3$ cm, and for a rate of heating equal to 250°C/h, the difference between the temperatures of core and surface is found to be:

at the start of heating-up	$(T \approx 0°C)$	$T_0 - T_k = 7°C$,
at the start of the reduction	$(T \approx 400°C)$	$T_0 - T_k = 15°C$,
towards the end of the reduction and heating-up	$(T \approx 800°C)$	$T_0 - T_k = 2°C$.

If the rate of heating is only 100°C/h, these differences are reduced to 3, 6, 1°C, and can be neglected in the subsequent calculation.

In reduction with CO, there is again a sharp increase in this temperature difference at the start of the reduction.

4.6.1.4.2. Reduction of a Charge (of Uniform Grains or Spheres) in a Countercurrent Process; Setting up the Differential Equations. In a countercurrent reduction of a charge (pellets) the material exchange and the heat exchange at each instant have to be considered. So the relations have to be set up for the instantaneous balance of materials taking into account the rate of reaction, and for the instantaneous heat balance and for the instantaneous heat transfer, taking into account the reaction kinetics. The first approximation left the latter out of account and considered temperature ranges instead of the instantaneous state of affairs.

The following mathematical treatment is comparatively complicated and requires the introduction of extra notation, which differs to some extent from the symbols previously used; the latter have, however, been retained, to facilitate the link-up with the previous work. Since this treatment may only be of interest in special cases, the reader may prefer to turn to the results, given in section 4.6.1.4.4.

Material exchange. In agreement with Fig. 219, let

C_{H_2}, C_{H_2O}, C_{N_2} = volume proportions of H_2, H_2O, N_2 in the gas flow through the charge,
$C_{H_2O,a}$ = volume proportion of H_2 in the gas at input (no H_2O, the remainder N_2),
w = amount of gas flowing in (m^3/m^2 h at NTP),
B = rate of throughput of pellets (t/m^2 h).

Then, from the material balance,

$$C_{H_2O} = \frac{M}{\mu_0} \hat{g} \frac{B}{w} (\Re_e - \Re), \tag{3}$$

where

$\mu_0 = 16$ kg/kmole = atomic weight of O,
$M = 22.4\ m^3$/kmole at NTP = molar volume,
$\Re_e$ = degree of reduction of pellets on discharge from the reactor,
$\hat{g}$ = content of removable oxygen (kg O/ton pellets),

and further

$$C_{H_2} = C_{H_2O,a} - C_{H_2O}, \quad C_{N_2} = 1 - C_{H_2,a} \quad \text{(constant)}.$$

The rate of reduction (see above) is given by

$$\dot{\Re} = v\,\Delta c = v(C^{eq}_{H_2O} - C^{0}_{H_2O}) \quad \text{with } v \text{ in } h^{-1}.$$

Putting

$$\frac{C^{eq}_{H_2O}}{C_{H_2O} + C_{H_2}} = K$$

(dependent on the temperature T of the material and on the degree of reduction $\Re$, cf. Table 20) we obtain

$$\dot{\Re} = v(K\,C_{H_2,a} - C^{0}_{H_2O}). \tag{4}$$

Instantaneous heat balance. The first approximation sets up an approximate heat balance for a definite range of temperature of the charge or the gas in each case. If the heat balance is determined more accurately and if the temperature range is then made smaller and smaller, the differential equation for the heat exchange is finally obtained.

Let

c_{pi} = specific heat of the component i (kcal/°C m³ at NTP or kcal/°C kg, per m³ at NTP for gases and per kg for solids),

$c'_{pi} = dc_{pi}/dt$ = its derivative with respect to the temperature t.

All the c_{pi} are empirically determined functions of t or T, as was the case for v and K, and their values are to be taken from [438–440]).

Further, for brevity, we write (μ = molar weight):

$$\Delta = \frac{1}{3\mu_0}[2\mu_{Fe}\,c_{p\,Fe} - \mu_{Fe_2O_3}\,c_{p\,Fe_2O_3}]\,T \qquad \text{(kcal/kg O)}, \tag{5}$$

$$\delta = 10^{-3}\frac{M}{\mu_0}[c_{p\,H_2O} - c_{p\,H_2}]\,t \qquad \text{(kcal/kg O)}, \tag{6}$$

$$I = \frac{1}{3\mu_0}[2\mu_{Fe}\,c_{p\,Fe}\,\Re + \mu_{Fe_2O_3}\,c_{p\,Fe_2O_3}\,(1 - \Re)]\,T \quad \text{(kcal/kg O)}, \tag{7}$$

$$i = 10^{-3}\frac{w}{\bar{g}B}[\hat{c}_{p\,H_2}\,c_{p\,N_2} + (1 - \hat{c}_{p\,H_2})\,c_{p\,N_2}]\,t \qquad \text{(kcal/kg O)}, \tag{8}$$

$$w = \Delta H_{H_2} + \Delta + \delta. \tag{9}$$

Finally,

$$\frac{dI}{dT} = I' = \frac{1}{3\mu_0}[2\mu_{Fe}\,(c_{p\,Fe} + c'_{p\,Fe}\,T)\,\Re + \\ + \mu_{Fe_2O_3}\,(c_{p\,Fe_2O_3} + c'_{p\,Fe_2O_3}\,T)\,(1 - \Re)], \tag{10}$$

$$\frac{di}{dt} = i' = 10^{-3}\frac{w}{\bar{g}B}[\hat{c}_{p\,H_2}\,(c_{p\,H_2} + c'_{p\,H_2}\,t) + \\ + (1 - \hat{c}_{p\,H_2})\,(c_{p\,N_2} + c'_{p\,N_2}\,t)], \tag{11}$$

$$\frac{d\delta}{dt} = \delta' = 10^{-3}\frac{M}{\mu_0}[(c_{p\,H_2O} + c'_{p\,H_2O}\,t) - (c_{p\,H_2} + c'_{p\,H_2}\,t)]. \tag{12}$$

Using these quantities, the differential equation for the heat exchange can be written

$$\dot{t}[i' + \delta'(\mathfrak{R}_e - \mathfrak{R})] = \dot{T} I' + \dot{\mathfrak{R}}[\Delta H_{H_2} + 2(\Delta + \delta)] + \frac{2\sqrt{\pi/{}^\circ L}}{10^3 \hat{g}\, \varrho_s} Q_v, \quad (13)$$

where

$$\dot{t} = \frac{dt}{d\tau} \quad \text{and} \quad \dot{T} = \frac{dT}{d\tau},$$

τ being the time (in hours) for which the charge has already been in the reactor at the position and the instant considered,

and

Q_v = heat loss through the wall of the reactor (kcal/m² h),
$^\circ L$ = cross-sectional area of the reactor (m²),
ϱ_s = bulk density of the pellets (see above) (t/m³).

Comparison of Eq. (13) with the calculation in the first approximation shows that the following improvements have been made:

1. The term Q_v gives the heat loss through the walls.

2. The terms δ and Δ take into account the difference in the specific heats of H_2 and H_2O on the one hand and of Fe_2O_3 and Fe on the other hand, and also the change in the evolution of heat in H_2-reduction when $t > 18°C$ and $T > 18°C$.

3. In I the difference in heat content of the already reduced Fe compared with the unreduced Fe_2O_3 is taken into account.

4. The heat content of the gas is now calculated taking into account the existing composition of the gas.

The instantaneous heat transfer. Writing

$$A = 10^3\, g\, \varrho_s = 690 \text{ kg O/m}^3,$$

we can put (13) into the form

$$\frac{Q}{A} = \dot{t}[i' + \delta'(\mathfrak{R}_e - \mathfrak{R})] - \dot{\mathfrak{R}}\,\delta - \frac{2\sqrt{\pi/{}^\circ L}}{A} Q_v = \dot{T} I' + \dot{\mathfrak{R}}\,(\Delta H_{H_2} + \Delta), \quad (14)$$

and then Q is the amount of heat exchanged, from the gas to the charge, measured in kcal/m³ h.

We now have to set up the relation governing the heat transfer for this amount of heat.

Let

$\eta_{H_2}, \eta_{H_2O}, \eta_{N_2}$ = dynamic viscosities (kg sec/m²),
$\lambda_{H_2}, \lambda_{H_2O}, \lambda_{N_2}$ = thermal conductivities (dependent on t) (kcal/m h °C),
$g = 9.81$ m/sec² = gravitational acceleration,
α = heat transfer coefficient (gas/charge) (kcal/m² h °C),
ε = voids fraction for the charge,
w^* = actual gas velocity (m/h),

$\tilde{w}$ = gas velocity reduced to normal temperature and pressure (m/h),
w = gas velocity reduced to gas at normal temperature and pressure and relative to empty volume (see above) (m/h or m³/m² h at NTP),
$\varrho_{H_2}, \varrho_{H_2O}, \varrho_{N_2}$ = densities of H_2, H_2O, N_2 (kg/m³),
$\varrho_{0H_2}, \varrho_{0H_2O}, \varrho_{0N_2}$ = densities of H_2, H_2O, N_2 at normal temperature and pressure (kg/m³ at NTP).

Further, let

$\bar{\eta}_g, \bar{\lambda}_g, \bar{\varrho}_g, \bar{\varrho}_{0g}, \bar{c}_g$ = corresponding quantities for the gas mixture,
N_{Re}, N_{Nu}, N_{Pr} = Reynolds, Nusselt, and Prandtl numbers.

In accordance with section 2.2 and [174]), the heat transfer is governed by the following relations:

$$N'_{Nu}(N_{Pr,0}/N_{Pr})^{1/3} = 2\varepsilon/(1-\varepsilon) + N'_{Re}\beta + (N'_{Re})^{1/2}, \tag{15}$$

where

$N'_{Nu} = N_{Nu}\varepsilon/(1-\varepsilon)$,
$N'_{Re} = N_{Re}/(1-\varepsilon)$,
$N_{Pr,0} = 0.7$,
$\beta = 0.005$.

Hence

$$\alpha = 100\frac{1-\varepsilon}{\varepsilon}\frac{\bar{\lambda}_g}{d}(N'_{Re})^{1/2}\{(N_{Pr,0}/N_{Pr})^{1/3}[1+\beta(N'_{Re})^{1/2} + (N'_{Re})^{-1/2}\,2\varepsilon/(1-\varepsilon)]\} \tag{16}$$

and here the quantity in the braces { } changes very little.

The amount of heat transferred is given by

$$Q = 600\frac{1-\varepsilon}{d}\alpha(t-T_0). \tag{17}$$

Substitution of (16) into (17) gives the differential equation governing the heat transfer. Simultaneous solution of (14) and (17) with $\dot{\mathfrak{R}}$ replaced by (4) gives the variation of temperature and of degree of reduction in the reactor, and therefore, in particular, T_e and t_g and, most important, $\mathfrak{R}_e$. The range of integration here is from $\tau = 0$ to $\tau = \tau_e$, where τ_e is the charge retention time in the reactor, measured in hours.

4.6.1.4.3. Solution of the Differential Equations. The pair of Eqs. (14) and (17) cannot be integrated to obtain a solution in closed form. Consequently, as already mentioned, we have to proceed as in the first approximation by integration steps, and the rule is: the shorter the intervals, the more accurate the result.

Single integration interval. On replacing $\mathfrak{R}$ in (14) and (17) by its expression (4), only the derivatives $\dot{t}$ and $\dot{T}$ occur. If we sub-divide the reactor into n zones of length $\Delta L = L_e/n$, we can put

$$\dot{t} = \frac{B}{\sigma} \frac{\Delta t}{L_e/n} = \frac{n}{\sigma} \frac{B}{L_e} \Delta t = C \Delta t, \tag{18}$$

$$\dot{T} = \frac{B}{\sigma} \frac{\Delta T}{L_e/n} = C \Delta T, \tag{19}$$

where

$$C = B\, n/\sigma\, L_e.$$

If we denote the zone boundaries by $j = 0, 1, \ldots, n$ and the temperatures at these boundaries by t_j and T_j, then (14) and (17) imply

$$\begin{aligned} Q_j &= A\,C\,(t_j - t_{j-1}) \left[\overline{i'_j} + \overline{\delta'_j}(\Re'_j - \overline{\Re'_j})\right] A\, \overline{\dot{\Re}}_j\, \overline{\delta}_j - 2\, Q_v \sqrt{(\pi/{}^\circ L)} \\ &= A\,C\,(T_j - T_{j-1})\, \overline{J'_j} + A\, \overline{\dot{\Re}}_j (w_j + \overline{\Delta}_j), \end{aligned} \tag{20}$$

$$Q_j = 600 \frac{1 - \varepsilon}{\alpha} \bar{\alpha}_j (t_j - \overline{T}_{0,j}), \tag{21}$$

where

$$A\,C = n \frac{10^3 \hat{g}\, B}{L_e} \left(\frac{\text{kg O}}{\text{m}^3\,\text{h}}\right),$$

and the bars above the various quantities denote their mean values between the boundaries j and $(j - 1)$. Since, however, on starting such an integration step, only the values on the boundary $(j - 1)$ are known, these have at first to be used instead of the mean values. Having found t_j and T_j the mean values can be calculated, and, repeating the integration, new values of t_j and T_j are obtained. Iteration is continued until t_j and T_j have been found to the desired accuracy. The j-th integration interval is then completed, and the next interval can be treated in the same way.

Initial and end conditions. The integration must start at $L = 0$ and be continued until $L = L_e$, i.e., the values are first calculated for zone 1, then for zone 2, and so on up to zone n.

Apart from the quantities put into the model as *system parameters*, such as L_e, etc., the quantities

B = throughput of pellets (tons/m² h),
w = amount of gas (m³/m² h at NTP),
C_{H_2} = proportion of H_2 in the gas,

can be arbitrarily prescribed. Still more important is the quantity

t_a = temperature of the gas at input (°C);

the initial temperature of the charge may be taken generally as 20°C.

Thus for the first integration step we have given

$$T_0 = 20,$$

and for the last integration step

$$t_n = t_a.$$

But in order to be able to begin on the first integration step we still need to know the value of t_0 (which is the same as t_g) and also $\mathfrak{R}_e$. As already pointed out in connection with the first approximation, here too we shall arrive at the end of the last integration step at the correct values of t_a and $\mathfrak{R}_e$ only if the integration has been started with a suitable value of t_g and the corresponding assumed result $\mathfrak{R}_e$. Since neither value is known initially, t_g and $\mathfrak{R}_e$ can be found only by repeatedly carrying out the whole integration with various values of t_g and $\mathfrak{R}_e$ chosen by trial, i.e., by iteration. How this iteration can be carried out most economically will not be discussed here. When the process has been successfully completed, then, we shall have t_g and $\mathfrak{R}_e$ and also the variation of all the other quantities of interest over the height of the reactor.

4.6.1.4.4. Results from the Mathematical Model. Finally we shall present some further results given by a calculation of the type just described. As a particular case, we consider a reactor with a cross-section of 1 m^2 and a height of 3 m.

Figure 223 shows the variation of the degree of reduction, the H_2O-concentration in the gas, and the gas temperature, over the height of the reactor for three different rates of throughput of pellets and different rates of supply of H_2. We see that in the case 1000 kg Fe/h and 1200 m^3 H_2/ton Fe at NTP, complete reduction can just still be attained, while in the other two cases the pellets are already reduced at half the reactor height or at an even higher level.

Figure 224 shows the variation in the temperatures of gas and charge in the cases mentioned. It is noteworthy that the temperature of the pellets very soon approaches within 1°C of the gas temperature. This slight difference suffices to maintain the heat transfer still necessary.

The three-dimensional view shown in Fig. 225 illustrates how, as the gas supply is diminished and the throughput of pellets is increased, a boundary is encountered beyond which complete reduction is no longer possible. The optimal performance, i.e., the biggest utilization of the reactor volume, is attained precisely on this boundary. In Fig. 225 the results for the particular reactor are reduced to unit volume of the reaction space, keeping the retention times for ore and gas the same. [It is not taken into account here that, e.g., when the shaft height is doubled keeping the retention time of the ore and of the gas in the charge the same, the heat transfer is improved because the gas velocity is doubled. Neglect of this fact is,

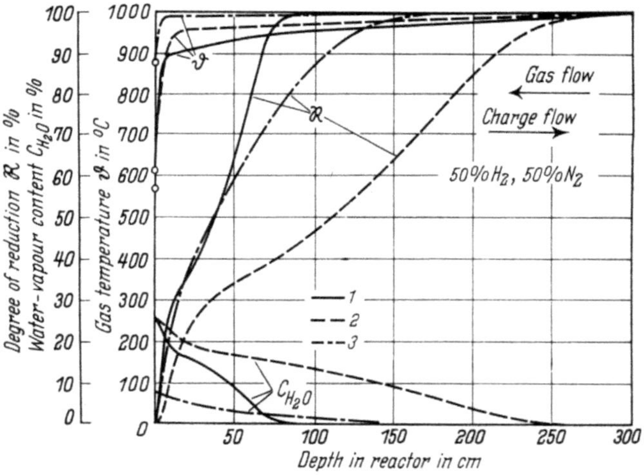

Fig. 223. Variation of degree of reduction, gas composition, and gas temperature with depth, in a countercurrent reactor (2nd approximation)[437])

Reactor depth: 3 m. Reactor cross-section: 1 m^2. Initial gas temperature: 1000°C. Gas composition 50% H_2, 50% N_2

Curve	Production kg Fe/h	Reducing gas m^3 H_2/ton Fe at NTP	Retention time for charge sec
1	250	1200	72000
2	1000	1200	18000
3	1000	4000	18000

however, unimportant, since in any case the heat transfer takes place very quickly (see Fig. 224).]

Figures 226 and 227 show this boundary curve in the reduction by a H_2/N_2 mixture (repetition of Fig. 225) and by pure H_2, and also show the corresponding degree of utilization of the gas and the resulting gas

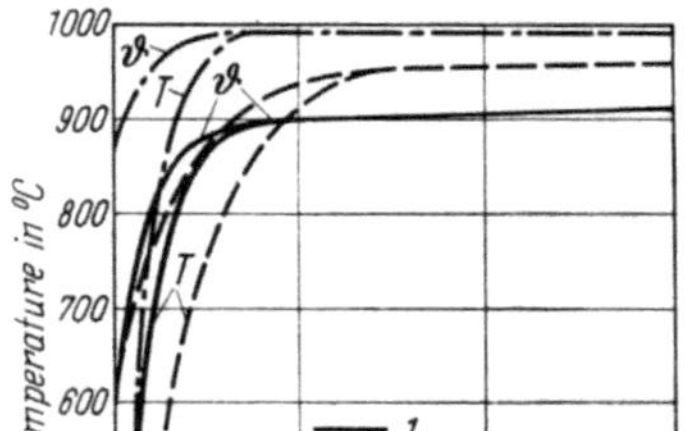

Fig. 224. Approach of the charge temperature to the gas temperature in a countercurrent reactor (2nd approximation)[437])

Reactor, initial gas temperature, and gas composition as in Fig. 223
ϑ = gas temperature, T = mean temperature of pellets. Values for production and reducing gas as in Fig. 223

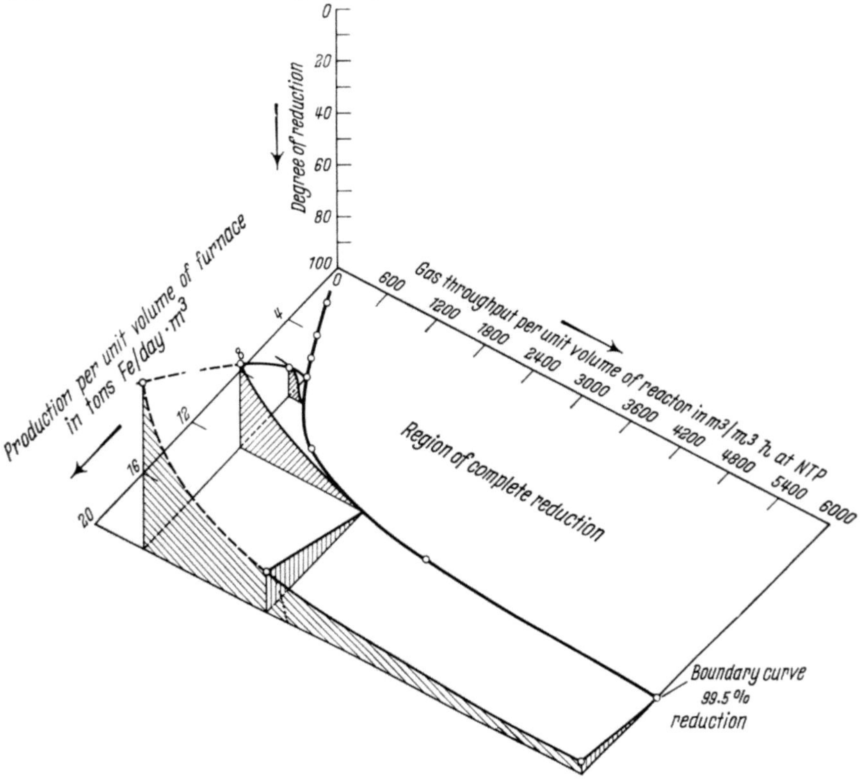

Fig. 225. Relation between degree of reduction, production of sponge iron, and throughput of reducing gas, in the reduction of hematite pellets in a countercurrent reactor (2nd approximation)[437])

Reactor, initial gas temperature, and gas composition as in Fig. 223

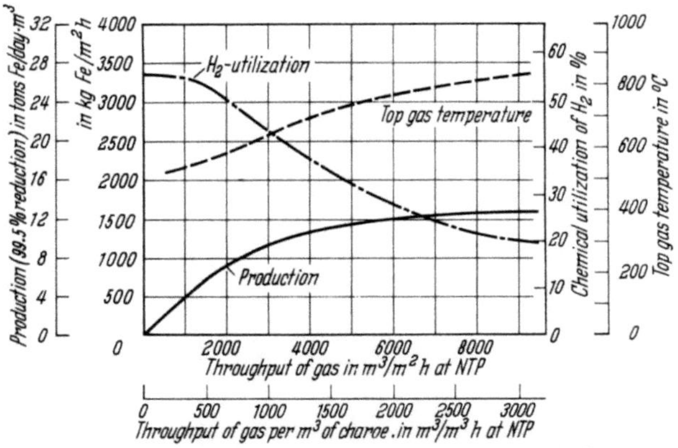

Fig. 226. Reduction of hematite pellets in a countercurrent reactor (2nd approximation)[137])

Reactor, initial gas temperature, and gas composition as in Fig. 223.

temperature at outlet. It can be seen that increasing the gas supply and therefore the gas velocity has scarcely any effect after a certain value has been reached.

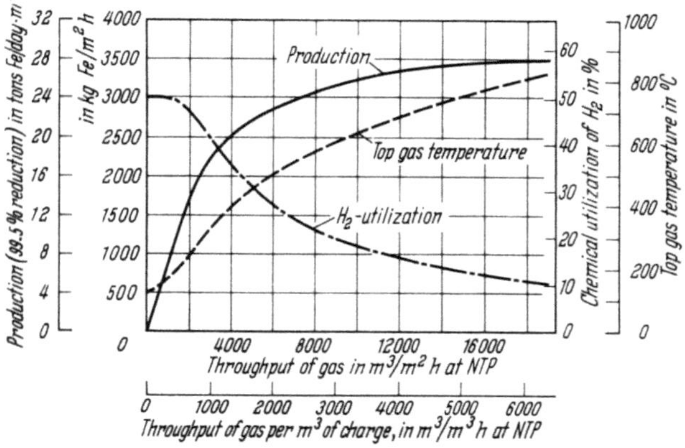

Fig. 227. Reduction of hematite pellets in a countercurrent reactor (2nd approximation)[437])

Reactor and initial gas temperature as in Fig. 223. Gas composition: 100% H_2

4.6.2. Countercurrent Reduction in Shaft Furnaces on the Commercial Scale

Figures 228 and 229 show examples, one operating on full-scale, the other on half commercial scale. In the Wiberg process (Fig. 228) a reducing gas of chiefly CO is manufactured by the gasification of coke with H_2O and CO_2 using electrical energy, and after desulphurization in a dolomite tower the gas is fed at about 950°C into the reducing shaft where countercurrent reduction of the ore takes place. In the upper part of the shaft pre-reduction to FeO takes place and the charge is heated to about 900°C by burning part of the furnace gas with air. A characteristic feature of the process is that in the main reduction zone, (FeO → Fe), the process is almost isothermal, due to the preheating of the ore and the composition of the reducing gas (when the ratio of CO to H_2 is about 3 to 1 the reducing reaction takes place with almost no evolution of heat, because the reduction by CO is weakly exothermic and by H_2 is strongly endothermic). The Finsider shaft furnace reducing process works on similar lines[449]).

In the Wiberg process a hot-gas blower conveys most (60 to 75%) of the reducing gas from the main reduction zone back to the gas generator, where it is regenerated into reducing gas according to the equations

$$H_2O + C = CO + H_2 \quad \text{and} \quad CO_2 + C = 2\,CO.$$

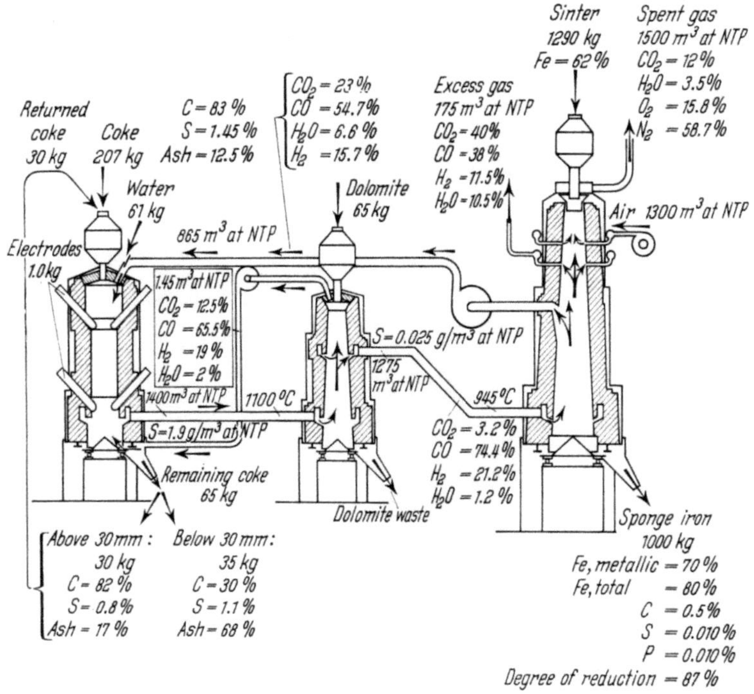

Fig. 228. Materials balance for 1 ton of sponge iron in the Wiberg-Söderfors process, according to STÅLHED[442])

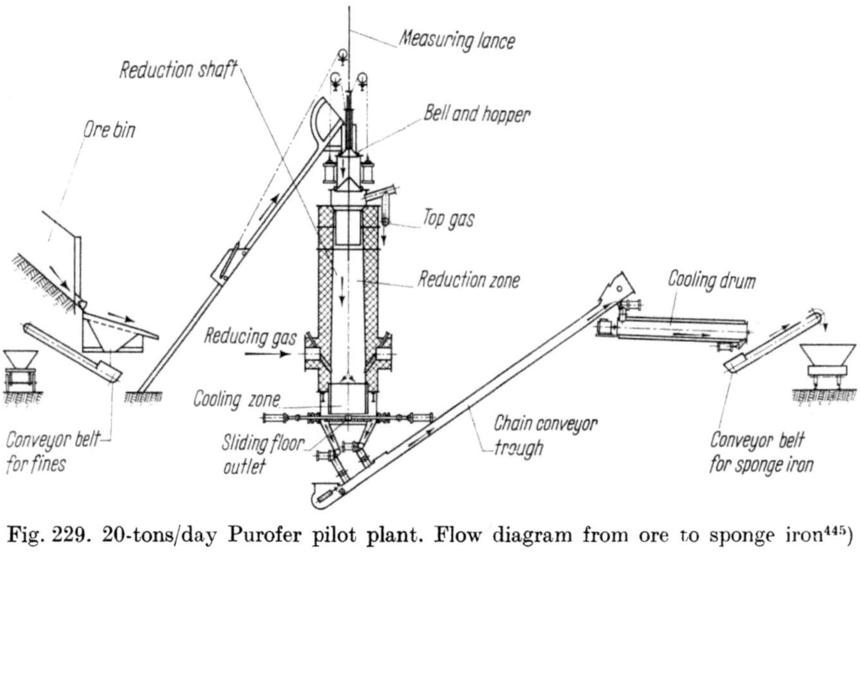

Fig. 229. 20-tons/day Purofer pilot plant. Flow diagram from ore to sponge iron[445])

The pressure generated at the blower determines the gas throughput per unit time for the whole plant and thus determines the output as well. Because of the difficult operating conditions (the dust contained in the gas, the temperature around 900°C) the increase of pressure which can be achieved by the blower is comparatively small. The performance of the Wiberg plant amounts to about 0.67 tons of sponge iron per day per m^3 of the charge volume. Table 26 gives other data for the performance of the Wiberg furnace. The degree of chemical utilization is very high and is mainly around $\eta_{CO, H_2} = 38.7\%$. The gas velocity, relative to the empty cross-section, of 0.61 m/sec can only yield a low performance per m^2 of shaft cross-section; the performance achieved is in good agreement with the performance to be expected (Fig. 218).

In the Purofer process developed by Hüttenwerk Oberhausen AG[445–447], a H_2-rich reducing gas is generated from CH_4-rich gases (see section 4.8), and this gas is used to reduce the charge in a counterflow process. Because of the high H_2-content, short reduction times can be achieved (Fig. 230).

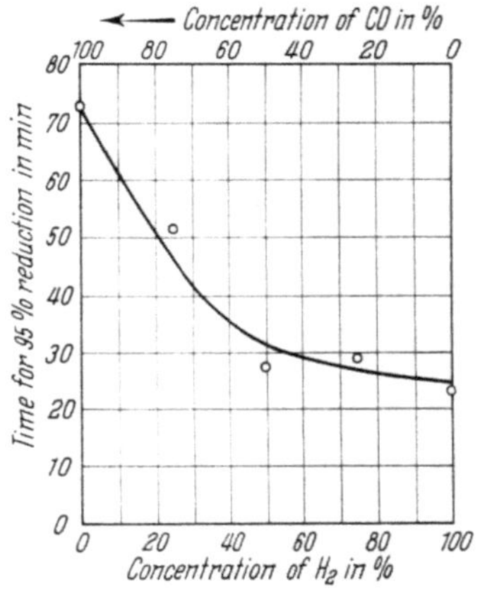

Fig. 230. Time for 95% reduction as a function of the H_2-content and CO-content of the gas[445])

Input: 400 g iron(III)-oxide pellets. Depth of bed: 50 mm. Inner diameter of vessel: 70 mm. Temperature: 900°C. Flow: 40 litres/min at NTP

The plant resembles the upper part of a blast furnace without the bosh and the smelting zone. Figure 231 gives the flow diagram for the whole plant.

The operational data obtained when using pellets, of diameter $d = 3.1$ cm, of Fe_2O_3 are given in Table 26. The analysis and rate of reduction of the pellets are the same as given in section 4.6.1.2 and used in the theoretical calculation. The output per unit volume of charge amounts to

$$L/V_R = 4.95 \text{ tons Fe/day} \cdot \text{m}^3$$

and the degree of chemical utilization of the gas is $\eta_{CO, H_2} = 40.0\%$.

The relation between the throughput performance and the degree of chemical utilization of the reducing gas as given by the mathematical model using both the first and second approximations is shown in Fig. 232.

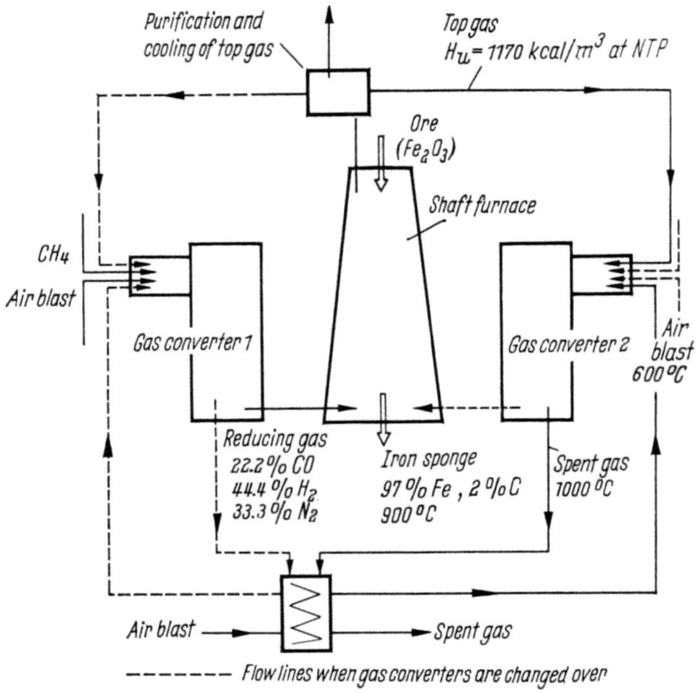

Fig. 231. Flow diagram for a Purofer plant

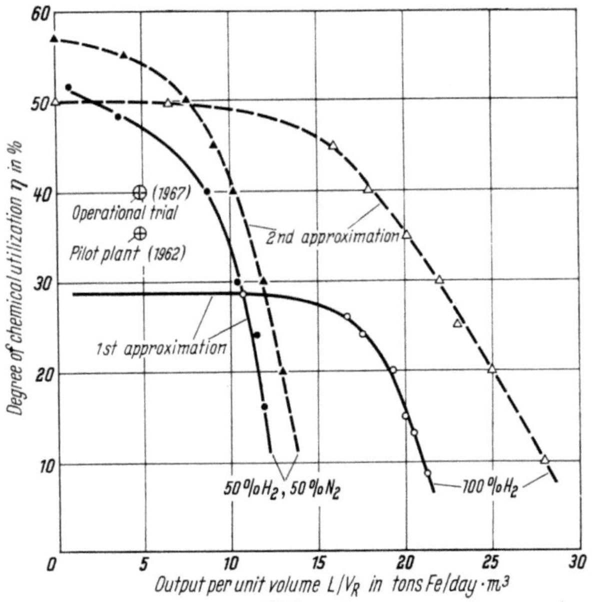

Fig. 232. Output per unit volume, L/V_R, for a countercurrent reactor and degree of chemical utilization of gas in the reduction of Fe_2O_3-pellets by hydrogen[447])

Initial gas temperature: 1000°C. Final degree of reduction: 99.5%

Table 26. *Data of the shaft-furnace reduction process*

		Wiberg[441]	Purofer[446]	
1. *Plant Dimensions*				
Height of charge . .	m	15	3.7	2.65*
Inner diameter of shaft at bottom .	mm	3000	700	1300
Effective volume (Discharge level to charging line) . .	m^3	≈ 85	1.2	3.28**
2. *Reducing gas*				
Degree of chemical utilization. . . .	%	38.7***	35.4	40.0
Amount†	m^3/ton at NTP	1400†	1665†	1460†
Composition:				
CO	%	74.4	26.7	19.2
CO_2	%	3.2	0.8	1.0
H_2	%	21.1	58.1	38.9
H_2O	%	1.2	< 25 g/m^3 at NTP	1.7
CH_4	%	—	1.8	0.7
N_2	%	—	11.8	38.9
Input temperature .	°C	950–1000	1000	1000
Input velocity (relative to empty cross-section) . .	cm/sec	61	177	232
3. *Sponge iron*				
Output	tons/day	57.0	5.8††	16.3
$\frac{\text{Output}}{\text{Effective volume}}$.	$\frac{\text{tons}}{\text{day m}^3}$	0.67	4.83	4.95
Output per unit area of reactor cross-section at bottom	$\frac{\text{tons}}{\text{day m}^2}$	8.06	15.1	11.5
Composition:				
Total Fe	%	86.0	95.3	87.5
Metallic Fe . . .	%	77.0	90.5	79.9
$\frac{\text{Metallic Fe}}{\text{Total Fe}}$ . . .	%	≈ 90	≈ 95	91.5
P	%	0.007	0.015	0.034
S	%	0.007	0.012	0.005
C	%	0.7	1.5	0.87
Gangue	%	11.7	1.7	8.3

* 3 m depth to the charging line.
** Gas-inlet level to the charging line.
*** In the main reduction zone.
† Relative to gas free from N_2.
†† Output restricted by inadequate gas supply.

Table 26 (continuation)

		Wiberg[441]		Purofer[446]			
4. *Specific Consumption*		per ton sponge iron	per ton metallic iron	per ton sponge iron	per ton metallic iron	per ton sponge iron	per ton metallic iron
Coke	kg/ton	115	149	0	0	0	0
Heavy oil	kg/ton	50	65	0	0	0	0
Electrical heating energy	kWh/ton	900	1170	0	0	0	0
Natural gas (CH_4) .	m^3/ton at NTP	0	0	400†††	442†††	355†††	388†††
Oxygen	m^3/ton at NTP	0	0	40	44.2	0	0
Total heat required. of which	kcal/ton	$2.06 \cdot 10^6$	$2.63 \cdot 10^6$	$3.43 \cdot 10^6$	$3.79 \cdot 10^6$	$3.04 \cdot 10^6$	$3.32 \cdot 10^6$
Proportion supplied by electricity is .	%	37.6		0	0	0	0

††† Calculated for large plants from the operational data for the pilot plant (30% higher there).

The relations calculated by electronic computer using the second approximation are to be preferred because they are based on a more refined argument. The experimental results lie between the first and second approximations. The incomplete agreement may be attributable to the presence of CO_2 and H_2O components in the reducing gas and to the high heat loss through the walls of the small pilot shaft (700 or 1300 mm). It was already apparent from the first-approximation calculation that if the temperature of the gas in the upper part of the shaft falls too soon, then the rate of reduction, and hence also the throughput, is greatly reduced. Moreover, the gas supply in the first test plants was inadequate. It follows from these facts that in larger plants, having an adequate gas supply and smaller heat losses, still higher throughput performance per unit volume of charge and presumably a closer approximation to the theoretical value can be expected; and this is confirmed by the operational results with the second installation of 500 tons/day capacity. The rough approximation which is now already seen between the calculated and observed values shows that the theoretical formulation given above is suitable for predicting the performance of countercurrent reduction equipment*.

* The Echeverria process[450]) also uses a shaft furnace for the production of sponge iron. In contrast to the processes previously described the lump ore charge is reduced with solid fuel at 1050°C. The heat needed for heating up and reducing the charge is provided by heating the shaft from without. With high output, difficulties arise in the transfer of the heat and consequently a high specific heat consumption is to be expected. A pilot plant at Legazpia in Spain produces about 20000 tons of iron sponge a year, that is, about 60 tons per day.

Further, it shows that, using reducing gases rich in hydrogen, production of iron sponge in shaft furnaces is possible and that the specific performances are higher than in the blast furnace. This is true whether the performance is calculated relative to the volume or the cross-sectional area of the working vessel.

At the same time the degree of chemical and thermal utilization achieved is of the same order of magnitude as in the blast furnace. This should mean that reduction processes in shaft furnaces have a decisive advantage over other direct reduction processes for producing sponge iron.

Because there is no coke to provide a structure to support the load of the charge column, the softening of the ore which begins in the zone of the highest temperatures has for a long time been regarded as seriously endangering the flow of gas through the charge. This problem can, however, be solved by correctly adjusting the input temperature of the gas to the temperature at which the charge softens*.

4.7. Electro-Reduction

Electro-reduction processes may be regarded as those processes, for the production of molten pig iron, in which the bulk of the heat required for the reduction, heating up and melting of the charge is provided by electrical energy. For the conversion

$$Fe_2O_3 + 3\,C = 2\,Fe + 3\,CO \tag{1}$$

the theoretical requirement is $1.047 \cdot 10^6$ kcal/ton Fe and 322.2 kg C/ton Fe, to which is to be added the heat needed for heating up and melting. If it is possible to convert the CO-gas, produced in the reduction, into CO_2 by means of the ore, then the requirement in energy and carbon is markedly decreased (Fig. 233). However, the amount of gas formed is small. In complete conversion according to Eq. (1), 602 m^3 CO/tons Fe at NTP are produced. If this gas is assumed to have, when it is formed, about the same temperature as the molten pig iron (1350°C), then its sensible heat is sufficient to preheat a charge free from gangue up to about 500°C. Actually, this amount of heating-up is never achieved, because of the inactive material in the charge and because of the heat losses. At temperatures below 500°C,

* In the Phelps-Dodge process[415a]) sponge iron is produced for copper cementation from the iron-rich slag (65% Fe) resulting from the blowing of copper ore, using a low shaft furnace. The granulated slag is preheated to 1100°C in a rotary cylindrical furnace before entering the shaft furnace. It is reduced by a reducing gas manufactured continuously from natural gas and air (see section 4.8). In the 15000 tons per year plant the energy consumption per ton Fe is $6.7 \cdot 10^6$ kcal and 209 kWh, and the degree of chemical utilization of the reducing gas is 22 to 23%.

however, the transformation

$$Fe_2O_3 + 3\,CO \rightarrow 2\,Fe + 3\,CO_2$$

and even the preliminary reduction

$$Fe_2O_3 + CO = 2\,FeO + CO_2$$

are still very slow. An appreciable "indirect" reduction, as in a blast furnace, is not therefore to be expected. Since, almost always and every-

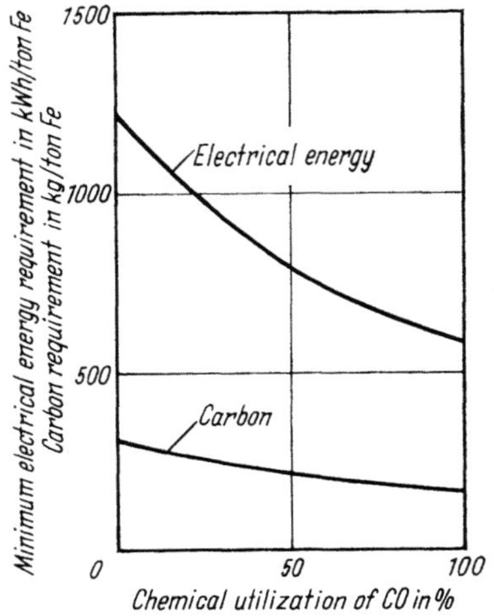

Fig. 233. Electro-reduction of iron ores. Minimum requirement in electrical energy and carbon for reduction according to $Fe_2O_3 + \frac{3}{1+\eta_{CO}} C = 2\,Fe + 3\frac{1-\eta_{CO}}{1+\eta_{CO}} \cdot CO + \frac{3\,\eta_{CO}}{1+\eta_{CO}} CO_2$ in relation to degree of chemical utilization of CO, $\eta_{CO} = \frac{CO_2}{CO + CO_2}$

where, electrical energy is the most expensive form of energy, the attempt is nevertheless made to utilize as fully as possible the CO-gas produced, and thus to save electrical energy.

Most of the processes so far tried out fall into the following categories:

a) *Electro-reduction bed* (molten pig iron and molten slag) *with a lump-charge column above it* (steady bed), moving in counterflow to the gas produced (electric low-shaft furnace, Tysland Hole furnace)[452, 453]).

b) *Electro-reduction bed* (molten pig iron and molten slag) *with a fine-grain charge column above it,* in which the CO-gas produces a fluidized bed [Lubatti process[456]), electro-fluidized bed[457])].

c) As in a) but preceded by a reduction plant, usually a *rotary kiln*, for pre-reduction and preheating of the charge [Elkem process, Strategic-Udy process, Orcarb process, DLM process[454, 459, 461, 463a, 458])].

4.7.1. Electro-Reduction Bed with Lump-Charge Column above it

Figure 234 shows diagrammatically an electric low-shaft furnace. The ore is distributed as uniformly as possible over the cross-section of the furnace from several hoppers above it. Three Söderberg electrodes are submerged in the charge down to the slag layer. Under the action of the heat generated from the electrical energy, the fuel (mainly coke of high reactivity, or coal) reacts with the oxygen in the ore, and the resulting iron and slag melt, are tapped off from time to time. Blast-furnace coke

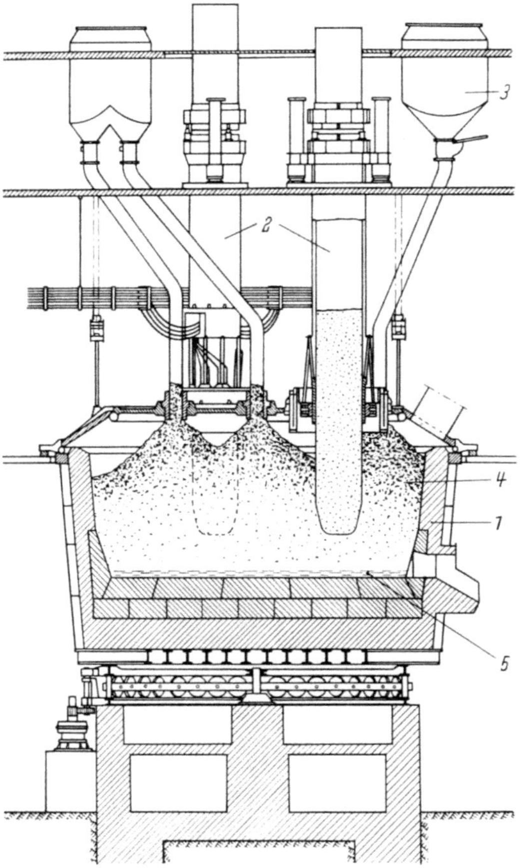

Fig. 234. Electric low-shaft furnace, according to MARINČEK[452])
1 Reduction and smelting vessel, *2* Electrodes, *3* Feed hopper for ore and coke, *4* Charge, *5* Smelt

of high strength and low reactivity is not required; this is one advantage of the process. On the other hand it is very necessary that the charge be of well crushed and screened lumps.

Tables 27 and 28 show typical material and heat balances for the process. Because the electrical energy can be fed directly to the main zone of reaction, the heat is well utilized.

The production performance of the electric low-shaft furnace is, nevertheless, remarkably poor.

The largest plant in the world, known at the present time, in Matanzas, Venezuela, with furnaces of 11.4 m internal diameter, has, according to WIEDER[453]), an output of 200 tons pig iron per day per furnace. The height of the charge is about 3 m, and the electrodes are 1.5 m in diameter; the electric load of one furnace is 20000 kW, the applied voltage being between 90 and 240 volts. The ore for the charge is screened at 10 mm, and the fine component is sintered before use. The following operating characteristics – which agree approximately with Tables 27 and 28 – are achieved[453]):

Coke consumption		380 to 420 kg/ton pig iron
Basic fluxes		about 300 kg/ton
Electrical energy		2300 to 2600 kWh/ton
Electrode consumption		13 to 18 kg/ton
Amount of slag		400 kg/ton
Pig-iron analysis:	C	4.1%
	Si	0.8%
	Mn	0.5%
	P	0.13%
	S	0.045%

The performance per unit area of the reactor cross-section (diameter 11.4 m) is

$$L/F = 200/102 = 1.96 \text{ tons pig iron/day} \cdot \text{m}^2,$$

which is very low when compared with that for a blast furnace

$$L/F = 30 \text{ to } 50 \text{ tons pig iron/day} \cdot \text{m}^2 \text{ hearth area.}$$

Relative to the usable charge volume V_R (about 300 m³), the performance figure is

$$L/V_R = 0.667 \text{ tons pig iron/day} \cdot \text{m}^3$$

compared with $L/V_R = 1.5$ to 2.5 tons pig iron/day · m³ for the blast furnace.

The unsatisfactory performance of the electric low-shaft furnace is due essentially to the restricted possibility of supply and the strong concentration of the electrical energy and to the fact that only a small amount of gas is produced and this is insufficient to produce any appreciable preheating and pre-reduction of the charge. Any further increase in the diameter of the vessel with the object of improving the performance is unlikely, because of the increased complexity of the arrangements for ensuring a uniform distribution of the charge over the furnace cross-section.

The method is therefore restricted to special cases. Nevertheless at the present time there are some 50 electric low-shaft furnaces in existence, with a total power of about 650 MW[454]).

Some of these special cases are very interesting, metallurgically speaking. This is because the amount of the reducing medium can be controlled

Table 27. *Material balance for a steel or pig-iron producing electro-metallurgical furnace*

a) Composition of the components of the charge in % (reduced to the dry material)

Components of charge	Fe	Fe_2O_3	FeO	MnO_2	P_2O_5	CaO	MgO	SiO_2	Al_2O_3	C fix.	Volat. comp.	CO_2	S	Moisture
Sinter	48.6	69.5	—	0.8	0.4	7.7	0.5	10.8	8.6	—	—	1.5	0.2	1.2
Coke	0.9	(various) 0.4	—	—	—	0.7	0.2	4.5	3.3	(86.9) 87.3	1.8	—	0.9	4.8
Electrode mass	0.5	—	—	—	—	0.2	0.1	3.3	1.8	77.8	16.1	—	0.6	—
Limestone	0.6	0.8	—	—	—	(55.1) 55.4	—	0.2	0.2	—	—	43.4	—	0.9

b) Weight of charge in kg/ton pig iron

Raw material	Moist	Dry	Fe	Fe_2O_3	FeO	MnO_2	P_2O_5	CaO	MgO	SiO_2	Al_2O_3	C fix.	Volat. comp.	CO_2	S	Moisture
Sinter . .	1969.9	1946.5	946	1352.8	—	15.6	7.8	149.9	9.7	210.2	167.4	—	—	29.2	3.9	23.4
Coke . .	388.8	371.0	3.3	(various) 1.5	—	—	—	2.6	0.7	16.7	12.2	322.4	6.7	—	3.3	17.8
Electrode mass .	20.0	20.0	0.1	—	—	—	—	—	—	0.1	—	15.6	3.2	—	0.1	—
Limestone	235.7	233.6	1.4	1.9	—	—	—	129.4	—	0.5	0.5	—	—	101.4	—	2.1
Charge . .	2614.4	2571.1	950.8	(various) 1.5	—	15.6	7.8	281.9	10.4	227.5	180.1	338.0	9.9	130.6	7.3	43.3
In the reduction product			946		—	12.7	6.9	—	—	10.7	—	338.0	—	—	0.5	—
In the slag	—	707.5	(4.8)	(various) 1.5	6.2	2.9	0.9	281.9	10.4	216.8	180.1	—	—	—	6.8	—
Slag (overall weight percentage)			—	0.2	0.9	0.4	0.1	39.8	1.5	30.6	25.5	—	—	—	1.0	—

Table 28. *Total power consumption of the electric low-shaft furnace in kWh per ton pig iron*[452]).

	kcal	kWh	%
A. *Pure pig-iron component of charge*			
Heat of reduction	972455	1131	43.6
Heating-up and melting	287000	334	12.9
Top-gas heat	72931	85	3.3
	1332386	1550	59.8
B. *Slag and other components of charge*			
Moisture	33908	39	1.5
Water of hydration	—	—	—
Carbon dioxide	138775	161	6.2
Slag	277340	323	12.5
	450023	523	20.2
C. *Complete charge*	1782409	2073	80.0
D. *Furnace losses* (20%)	445602	518	20.0
E. *Total heat consumption or power consumption*	2228011	2591	100.0

without reference to the amount of heat which it is desired to generate. In this way, for example, it has been possible to work with iron ores containing titanium; pig iron almost free from titanium (0.02%) and a slag rich in TiO_2 (up to 72% TiO_2) are obtained (the slag is used in the manufacture of pigments)[455]).

In Japan, too, work has been done on the basis of manufacturing pig iron from ferruginous sand containing titanium by means of electro-reduction furnaces[465]).

Iron alloys can also be produced in electric low-shaft furnaces and electric furnaces of the type discussed; see [464]) for a detailed survey.

4.7.2. Electro-Reduction with a Fine-Grained Charge Column Operating as a Fluidized Bed

The desire to utilize fine-grained ores and fine coal led to these materials being used directly on an electro-reduction bed. The Lubatti process (Fig. 235) in principle works like the electric low-shaft furnace. Three or even six electrodes dip into the molten slag and produce there the heat needed to cover the energy requirement for heating up, reducing, and melting the charge. Above the slag bath there is a column, 20 to 50 cm high, of fine-grained ore and fuel, through which the CO-gas produced during the reduction flows and forms a fluidized bed (Fig. 236). The furnace is open at the top. The molten pig iron collects at the bottom and is tapped off from time to time. The electrical power consumption is of the same order as in electric low-shaft furnaces, about 2500 kWh/ton pig iron[456]). With small outputs, the process is used less for the manufacture of pig iron than for the regeneration of waste containing noble metals, for instance.

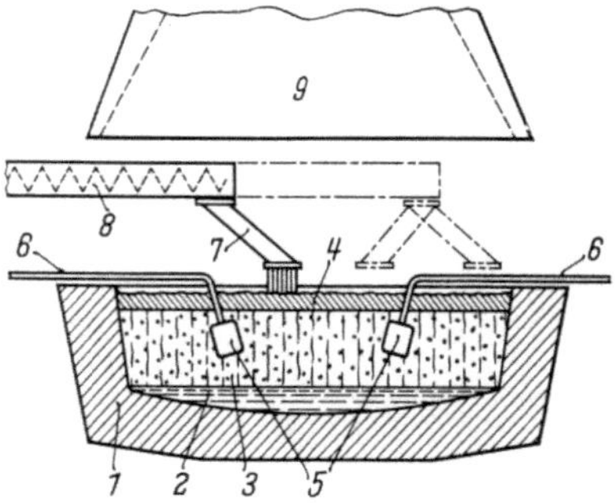

Fig. 235. Lubatti furnace, according to PECORARI[456])

1 Lining, *2* Liquid metal, *3* Molten slag, *4* Floating burden layer, *5* Submerged head-shaped electrodes, *6* Water-cooled electrode arms, *7* Rotatable distributor, *8* Swivelling conveyor screw, *9* Gas flue hood

In the electro-fluidized-bed reduction process described in [457]) there is again a fluidized bed of fine ore and fine-grained reducing material above the melt (Fig. 237). The electrodes do not dip into the slag, but heat is

Fig. 236. 3500-kVA Lubatti furnace. Appearance of the surface of the charge with the furnace in full operation, according to PECORARI[456])

produced from the current carried between them by the coal particles of the fluidized bed. In order to obtain a thorough reaction between the rising gas and the charge, the height of the charge column has been considerably increased. But there are still the restrictions imposed by the fluidized-bed process on the usable gas velocity and hence on the performance (cf. sections 3.2 and 4.3) and by the difficulties of effecting a countercurrent motion of gas and ore. The device is operated with gas velocities of 15 to 25 cm/sec (relative to the empty vessel), depending upon the grain size of the coal[458]). Outputs are reported of about 4 tons pig iron per day per m^2 of the vessel's cross-section[466]), i.e., twice as much as

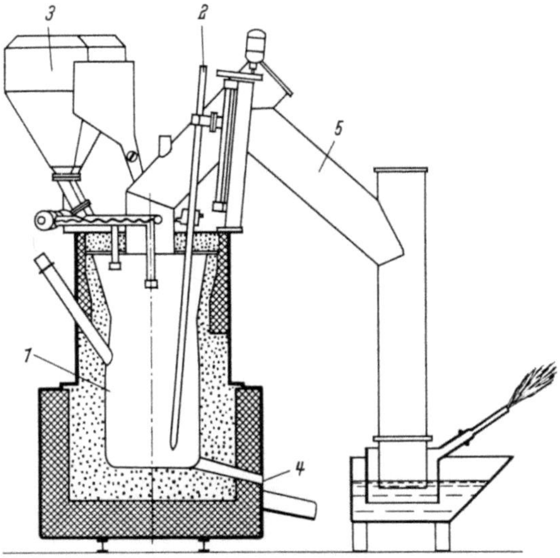

Fig. 237. 200-kW electric fluidized-bed furnace, Institut für Eisenhüttenwesen, Aachen, according to WENZEL[457])

1 Reduction and smelting vessel, *2* Electrodes, *3* Feed hopper, *4* Tap hole, *5* Gas flue

in the electric low-shaft furnace, but still less than in the blast furnace. The energy consumption is 2500 kWh/ton pig iron and about 400 kg coke breeze are used per ton pig iron.

This method is also being applied for the production of iron alloys and of CaC_2.

4.7.3. Electro-Reduction Using a Pre-Reduced Charge

By pre-reduction and preheating of the charge, the metallurgical task of the electric furnace is lightened, and, because of the lower specific consumption, a higher production rate can be achieved with the same electrical power input. The combination of a rotary kiln and an electric furnace has become customary in this respect[459–461]). The spent gas from the electric furnace is used in the pre-connected rotary kiln. It is therefore no longer necessary to pass the CO-gas through the charge in the electric furnace. The charge in the electric furnace can be differently and more favourably distributed, and fine ore can be used directly. The electrodes dip only a short distance into the charge (Fig. 238)[461]). The charge is distributed near the walls of the vessel because of the need to keep the electrically conducting, pre-reduced material of the charge as far as possible from the electrodes, since otherwise the electrical efficiency would be sharply decreased. A more detailed account of the course of the reduction in rotary kilns can be found in section 4.5. Figure 216 has already

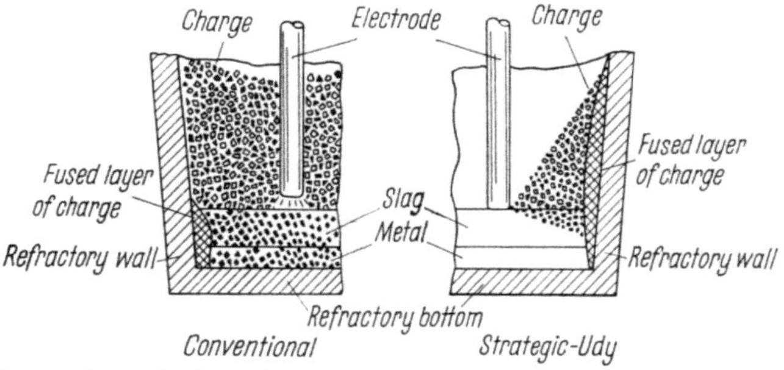

Fig. 238. Comparison of electrode environment in the conventional electric smelting process and the Strategic-Udy process, according to UDY[461]) and COLIN[454])

shown how the rotary kiln and electric furnace are connected, and Fig. 217 showed the connection between pre-reduction and preheating and the consumption of electrical energy for ores of different Fe-content.

In a semi-industrial plant[461]) it was possible by carrying out a 55% pre-reduction in a rotary kiln to reduce the electrical energy consumed to 1080 kWh per ton pig iron. In addition, 430 kg coal per ton pig iron was used, and in the 1000-kVA furnace it was still possible to produce 15

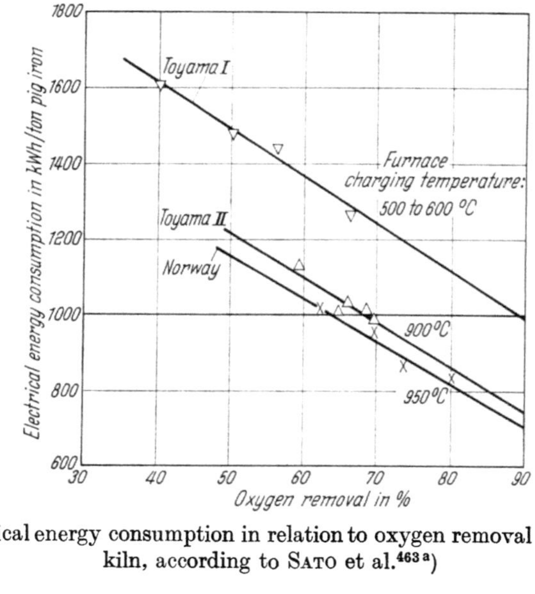

Fig. 239. Electrical energy consumption in relation to oxygen removal in pre-reduction kiln, according to SATO et al.[463a])

to 20 tons pig iron per day[460]), about twice as much as without pre-reduction. The marked dependence of the power consumption on the degree of pre-reduction and the temperature of the sponge iron charge is shown in Fig. 239[463a]). The applicability of the process to large-scale operation has been tested in a big installation in Venezuela with a rotary kiln 107 m long and 3.95 m in internal diameter. Coal was fed in to 4 regions of the vessel and the air for combustion was blown in through 11 blowers in the casing. The charge consisted of fine ore without agglomeration and a native coal with about 44% volatile constituents.

Iron sand containing titanium can also be worked in rotary kilns[463a]).

In the Dwight-Lloyd-McWane (DLM) process ore/coal-pellets are pre-reduced to 55 to 60% on a travelling grate instead of in a rotary kiln and are then charged hot to the electric furnace and melted to pig iron[458]); 850 kg coal and about 1000 kWh per ton melt are needed.

4.7.4. Possible Special Applications of Electro-Reduction

Wimmer[462]) uses electrical energy for the production of a FeSi-alloy rich in aluminium (53% Si, 26% Al, 16% Fe), with which, in a second stage, iron ore is reduced to molten iron or steel by an alumino-thermal or silico-thermal process. De Sy[463]) introduces the energy needed for reduction into the system by induction.

As regards the electro-reduction processes described above and also the many similar processes which have been suggested, application on a large scale is likely only under very special, and up to now rarely occurring, local conditions. If, however, with the further development of nuclear reactors for generating electricity, electrical power should become markedly cheaper, the situation will presumably change in favour of electro-reduction.

4.8. Preparation of Fuels for the Reduction

Only in very rare cases can primary energy sources such as raw coal, petroleum, or natural gas be used without previous treatment in the reduction of ores. Reducing media are mostly obtained by special preliminary processes such as coking, gasification, and chemical or thermal cracking. A brief survey of such processes is given below, but, since comprehensive publications about coking are already available, e.g., [467, 468]). coke will be only briefly discussed.

4.8.1. Solid Fuels as Reducing Agents

Some processes are capable of producing the reducing gases, CO and H_2, from uncoked coal within the reducing plant itself. This occurs, for instance, in the reduction processes taking place in rotary kilns, in which fine coal can be used. The blast furnace, too, can use a small proportion of ordinary commercial fine coal (section 5.4).

The main reducing material and fuel for the blast furnace, however, and hence for iron manufacture in general, is coke, the improved product of certain kinds of coal. Reference may be made to the extensive literature, e.g., [467, 468]), for information about coking processes and the effect of the quality of the coke.

The requirements for a good blast-furnace coke are*:

a) a high content of carbon and a low content of ash[470a]) and sulphur;

b) a narrow range of grain sizes, since the gas flow through the charge largely depends on this[470b]). According to American investigations[470c,d]), if the range of lump size of the ore is also small, the performance of the blast-furnace increases as the proportion of coke of size 20 to 50 mm increases. If the range of size of the ore is wider, this increase in performance is very small or there may even be a loss in production. In parallel with this, a lower consumption of coke is observed. German ironworks[470e,f]) obtained similar results on changing from ordinary blast-furnace coke to classified coke of 40 to 60 or 80 mm. In high slag volume practice, the coke should not be chosen too small because of the danger of choking (see section 5.5.4).

c) high strength at low bulk density;

d) low reactivity with CO_2 at temperatures below 1100°C, in order that the heat-consuming reaction

$$CO_2 + C = 2\,CO$$

shall be so far as possible repressed in the blast-furnace shaft.

The reactivity of coke has been measured in various ways[471–474, 474a, 496, 743]). Figure 240 shows one example. Here two different kinds of coke were made to react with CO_2; the lower the ordinate, the higher the re-

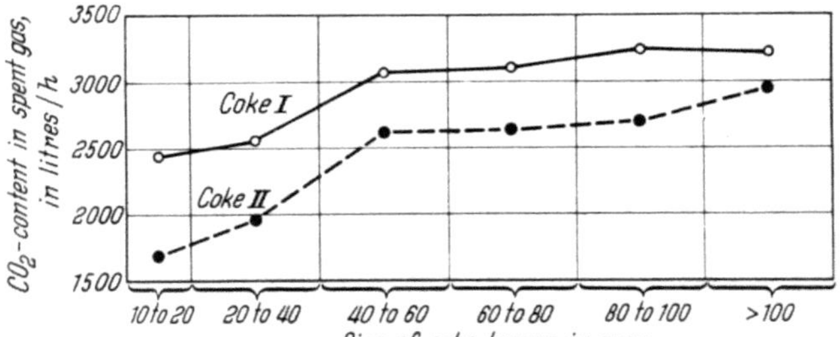

Fig. 240. CO_2-content in spent gas as a function of the coke size, in an investigation of the reactivity of two kinds of coke, according to PÜCKOFF[474])

* See [469,470]) regarding limiting values and methods of testing.

activity. According to this, the grain size has not too great an effect, because, as already explained in section 4.5, below 1100°C the phase-boundary reaction is rate-determining. In order to see clearly the effect of the coke reactivity on blast-furnace performance, operational tests were carried out using coke of varying reactivity with CO_2, the different reactivities being obtained by using different kinds of coke and different times and temperatures in the coking-oven. It was found that a limited significance can be ascribed to this quantity[475]) (Fig. 241). In another

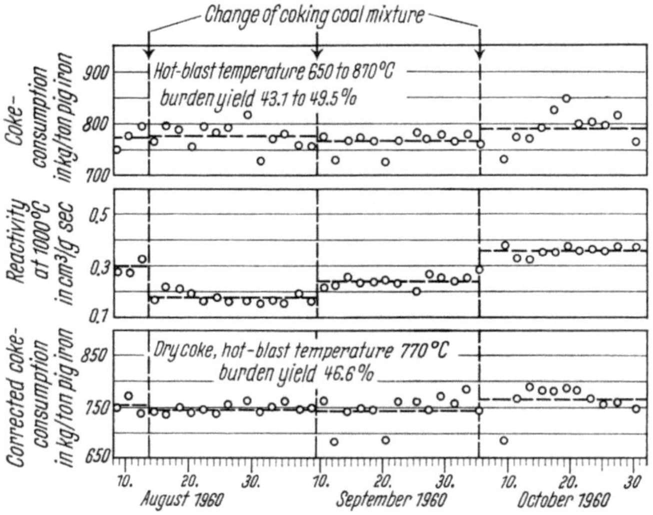

Fig. 241. Effect of change of coking coal mixture on the reactivity of the coke and the coke consumption in the blast furnace, according to KAHLHÖFER et al.[475])

investigation[471]) doubling the reactivity of the coke at 1100°C resulted in a 50 to 70 kg/ton pig iron increase in the coke consumption in a blast furnace. Further American investigations[496]) have shown that, other conditions remaining the same, the reactivity of coke with CO_2 depends strongly on the ash content, and with 5% ash is only about half the value of that with 10% ash. Further[475a]), by blending petroleum coke with the coking mixture the reactivity can be perceptibly reduced.

4.8.2. Gaseous Fuels and Reducing Materials

4.8.2.1. Statement of the Problem

"Gas-reduction" processes, that is, essentially, the retort process, the fluidized-bed process, and the shaft-furnace processes without smelting, need as the reducing medium either H_2, or CO, or a mixture of both, possibly with some inert component, such as N_2. For this purpose solid, liquid, or gaseous fuels are interchangeable, provided, of course, that by some suitable reaction the required reducing gases can be obtained from them. Figure 242 shows the stages of the process which are, in principle, necessary.

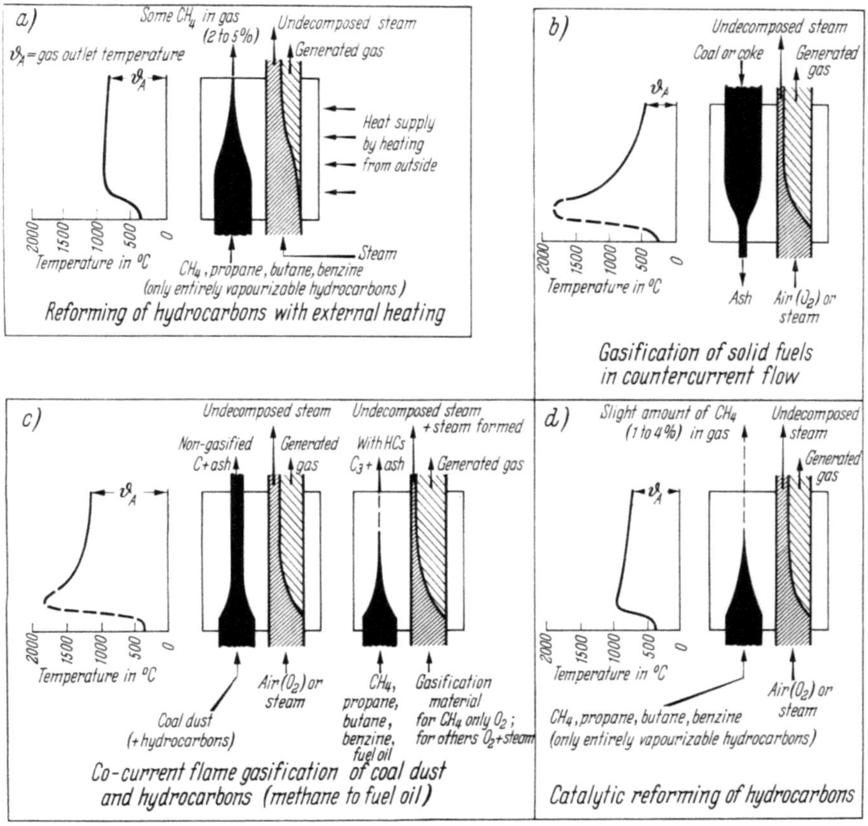

Fig. 242. Diagrammatic representation of processes for producing reducing-gases, according to VON GRATKOWSKI[498])
b) to d) Autothermal processes

There is a multiplicity of reactions and methods which can be applied (see, e.g., [478, 495]), for a list of them); the cost of such processes depends very decidedly on the following factors:

a) the level of the working temperature and pressure (which are determined by the position of the equilibrium and by the reaction rates);

b) the amount of heat to be supplied or removed. The most suitable reactions are to be regarded as those in which the enthalpy of reaction is just sufficient to make the reducing gas be formed with the desired composition and to raise the temperature to that required for the next stage. These are called "autothermal processes".

For the gasification of coal to CO or even to CO/H_2 mixtures many processes are known which have been well tested industrially and on which von Gratkowski has reported in detail[477, 478]). At the present stage of development it should, in general, be economic to use solid fuels without pre-treatment in rotary-kiln reduction processes. Consequently we shall deal below mainly with the transformation of natural gas and fuel oils into reducing gases.

Different reduction processes require the manufacture of different gases; the requirements are:

either (a) H_2, containing the smallest possible amount of CO, CO_2, H_2O, and undecomposed hydrocarbons,

or (b) CO, containing the smallest possible amount of CO_2, H_2O, and undecomposed hydrocarbons,

or (c) an H_2/CO mixture, containing the smallest possible amount of CO_2, H_2O, and undecomposed hydrocarbons.

4.8.2.2. Equilibria

Essentially it is possible to achieve the required transformations by the reactions described by the following equations:

(a) $CH_4 = C + 2\,H_2$ (thermal cracking), $\Delta H = 798$ kcal/m³ CH_4 at NTP.
(b) $CH_4 + H_2O = CO + 3\,H_2$ (reforming), $\Delta H = 2198$ kcal/m³ CH_4 at NTP.
(c) $CH_4 + CO_2 = 2\,CO + 2\,H_2$ (reforming), $\Delta H = 2637$ kcal/m³ CH_4 at NTP.
(d) $CH_4 + \frac{1}{2}\,O_2 = CO + 2\,H_2$, $\Delta H = -380$ kcal/m³ CH_4 at NTP.
(e) $CO + H_2O = CO_2 + H_2$ (conversion), $\Delta H = -439$ kcal/m³ CO at NTP.

For fuel oil the reactions are similar, e.g.,

(f) $(CH_2) + \frac{1}{2}\,O_2 = CO + H_2$, $\Delta H = -1466$ kcal/kg oil.

The fact should be noted that the exothermic heat of cracking is greater for the fuel oil than for CH_4 and the ratio H_2/CO is smaller [cf. reactions (d) and (f)].

The reaction (d) takes place in several stages: first the CH_4 is partly burnt to CO_2 and H_2O which then react with excess CH_4 according to equations (b) and (c). The equilibrium positions for these reactions are known (Figs. 243 to 246).

It can be seen from these diagrams that, at normal pressure and temperatures above 900°C, the equilibrium is almost entire on the right-hand side in all three of these reactions.

In commercial-scale conversions the formation of large amounts of soot is always a disturbing factor and is therefore suppressed as much as possible.

For this reason reaction (a) is out of the question for industrial-scale production of reducing gases. Reaction (a) was carried out earlier in the

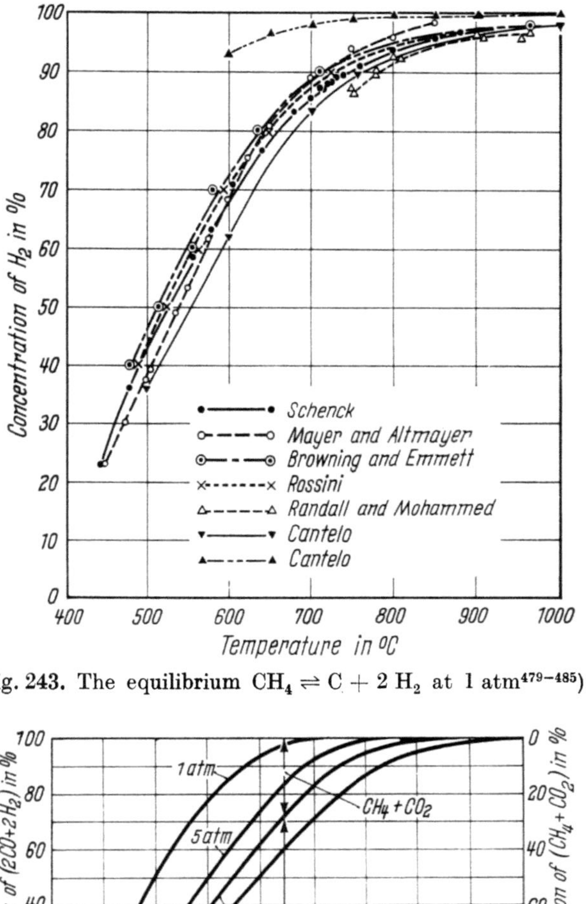

Fig. 243. The equilibrium $CH_4 \rightleftharpoons C + 2\,H_2$ at 1 atm[479–485])

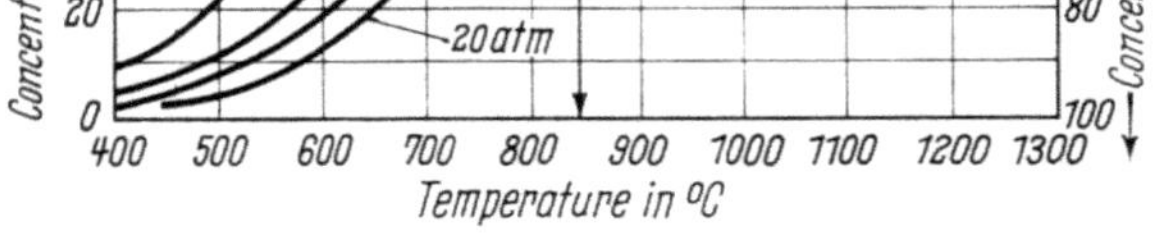

Fig. 244. The dependence of the equilibrium $CH_4 + CO_2 = 2\,CO + 2\,H_2$ on temperature, according to HILLER[486])

U.S.A. to generate hydrogen and obtain carbon black in Cowper hot-blast stoves at 1200°C. By present-day standards, the results were unsatisfactory (disturbing formation of soot and also acetylene, naphthalene, and tarry

Calculated composition of the reformed gas with $CH_4:H_2O=1:1$

°C	400	500	600	700	800	900	1000
CO_2	4.2	7.0	6.0	2.4	0.7	0.2	0.0
H_2O	37.0	24.9	13.2	5.4	2.0	0.7	0.4
CH_4	41.2	32.0	19.2	7.8	2.5	0.9	0.2
CO	0.2	2.0	9.4	18.7	23.0	24.4	24.8
H_2	17.4	34.7	52.2	65.7	71.8	73.8	74.8

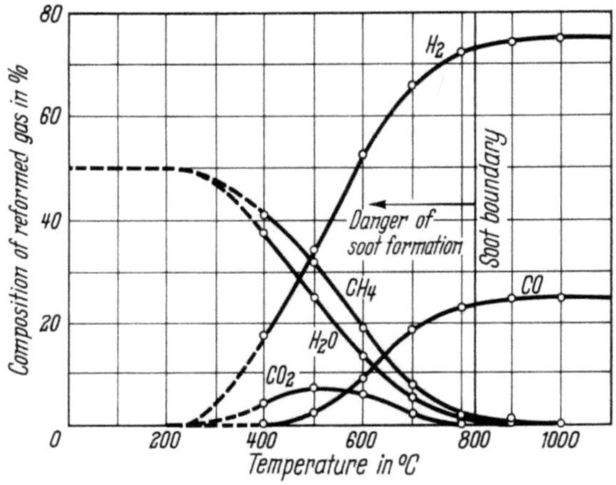

Fig. 245. Equilibrium relation between the composition of the reformed gas with $CH_4:H_2O = 1:1$ and the temperature, according to PETERS and KAPPELMACHER[487])

Calculated composition of the reformed gas with $CH_4:CO_2=1:1$

°C	400	500	600	700	800	900	1000
CO_2	44.9	33.2	16.2	5.0	1.3	0.3	0.1
H_2O	1.5	3.8	4.2	2.8	1.1	0.5	0.2
CH_4	46.5	36.9	20.8	7.8	2.5	0.9	0.2
CO	5.1	16.8	33.8	45.0	48.7	49.6	49.9
H_2	2.0	9.3	24.6	39.4	46.4	48.7	49.6

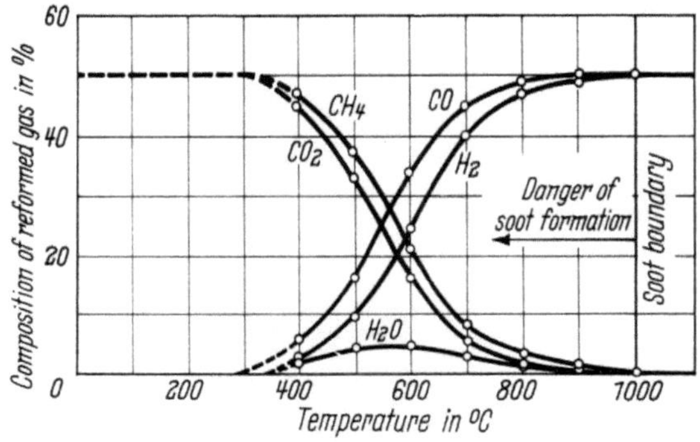

Fig. 246. Equilibrium relation between the composition of the reformed gas with $CH_4 : CO_2 = 1 : 1$ and the temperature, according to PETERS and KAPPELMACHER[487])

substances in the gas). After development work by the Ruhrgas A.G. and Universal Oil Products, it is now possible to split CH_4 at about 900°C, using Ni catalysts, into $2\,H_2 + C$, so that the carbon is left adhering to the catalyst. In a second stage the catalyst is regenerated and reheated to the cracking temperature, by burning the carbon to carbon monoxide. The catalyst thus also serves as an extra, solid heat-carrier. The catalyst circuit is between two fluidized-bed reactors or between two reactors with static bed plus pneumatic delivery from the regenerator to the cracking reactor.

In the reactions (b), (c), and (d), however, soot is very often produced in considerable amounts. The reaction conditions for preventing soot formation in reaction (d) have been elucidated by the researches of MONTGOMERY, WEINBERGER, and HOFFMANN[488] (Fig. 247). It is mainly a question of using the proper O_2/CH_4 ratio.

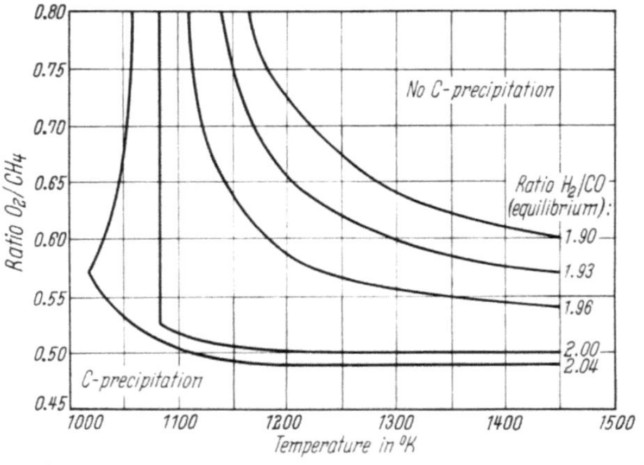

Fig. 247. Ratio O_2/CH_4 as a function of temperature for various H_2/CO-ratios in the reformed gas at a total pressure of 760 Torr, according to MONTGOMERY et al.[488])

From the thermodynamics, therefore, the conditions for the production of reducing gas of the desired composition, by the reactions (b), (c), and (d) are known. These conditions are:

the establishment of stoichiometric mixtures for the reaction, i.e.,

$$CH_4/CO_2 = 1 \quad \text{or} \quad CH_4/H_2O = 1 \quad \text{or} \quad CH_4/O_2 = 2;$$

a temperature of at least 900°C;

and use of a suitable catalyst.

For the water-gas reaction, temperatures of at most 450°C are adequate (Fig. 248). With new kinds of catalysts, such as have been developed by BASF, for instance, the conversion reaction can be carried out at a temperature as low as 200°C when using very pure natural gas or light petrol.

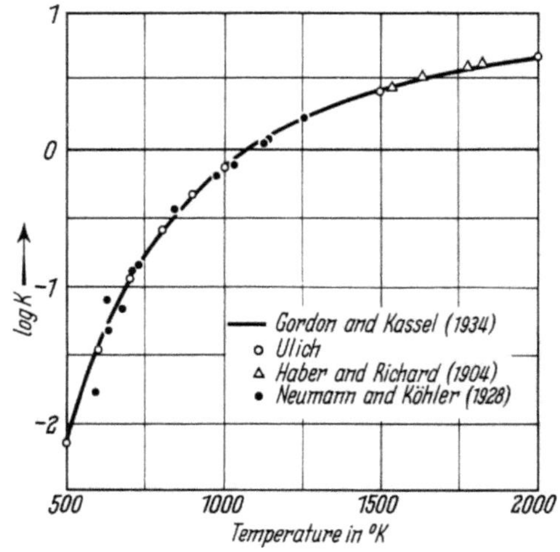

Fig. 248. Water-gas equilibrium $CO_2 + H_2 \rightleftharpoons CO + H_2O$, according to ULICH and JOST[489])

$$K = \frac{CO \cdot H_2O}{CO_2 \cdot H_2}$$

4.8.2.3. Reaction Rates. Industrial Operation

The main features of the reactions for splitting CH_4 by H_2O, CO_2, or O_2 were described by FISCHER and TROPSCH[490]) back in 1928. With nickel catalysts the reactions proceed at 800 to 900°C. In the meantime the chemical industry has developed numerous large plants in which gases containing CH_4 (such as natural gas, coke-oven gas, and residual gases derived from these) are split into hydrogen by CO_2, H_2O, or O_2. The manufacture of hydrogen for synthesis purposes according to the following reactions, for instance, has become of great importance in the ammonia industry:

1. "Reforming" process according to reaction (b), i.e., splitting with Ni-catalysts at $< 900°C$ with a large excess of steam.
2. Conversion of CO according to reaction (e) at 200 to 450°C.
3. Washing out the CO_2 with suitable washing solutions (e.g., ethenolamine, K_2CO_3-solutions, or even water), condensation of the excess steam under pressure.

Even in the optimum temperature range, the conversion reaction takes place only incompletely, with an economically acceptable excess of steam, because of the position of the equilibrium point. If pure hydrogen is required, the residual CO-content (0.5 to 4%) must be washed out using either an ammoniacal Cu(I)-salt solution under high pressure or liquid N_2. After washing with liquid nitrogen, however, the hydrogen contains a considerable N_2-component.

The hydrogen used in the H-iron process or in the Nu-iron process is obtained by one of these well-tried methods.

The unsatisfactory feature of these processes is that stages 2 and 3 of the reaction seem to be unnecessary, since a CO/H_2 mixture, such as may be produced in stage 1 alone, is well-adapted to the reduction of iron oxides at a temperature of 900°C. A considerable simplification, with a resultant reduction in costs, of the reforming processes for natural gas and oil would be achieved if the CO/H_2 mixture could be generated "in one heat" and fed into the reducing plant without loss of temperature, e.g., without washing processes.

The requirement of the cracking process is, then, the following: to generate a H_2/CO mixture with little residual CH_4 and as little CO_2 and H_2O as possible, since these constituents limit the reducing effect on iron oxides. Defining the *degree of oxidation* of the gas as the ratio

$$(CO_2 + H_2O)/(CO + H_2 + CO_2 + H_2O),$$

we may say that the maximum degree of oxidation tolerable is 0.05 or 5%.

There are several possible solutions, as follows.

4.8.2.3.1. Endothermic Cracking Process. In accordance with the reaction

$$CH_4 + H_2O = CO + 3\,H_2,$$

natural gas, after desulphurization (because sulphur poisons the catalyst), is mixed with an excess of steam and passed through an externally heated system of tubes in which, at a temperature of about 900°C, the desired reaction takes place through contact with the Ni-catalyst. The amounts of heat to be transferred are considerable (the reaction enthalpy is 2198 kcal/m³ CH_4 at NTP, and in addition about 1200 kcal/m³ CH_4 at NTP is needed for heating up the gas), so that a large heat exchange area is necessary (i.e., a tubular furnace). The resulting mixture, at about 900°C, contains a high proportion of excess steam, so that cooling, removal of water, and reheating may be necessary [see section 4.4 and [491])]. The plant erected by Hojalata y Lamina at Monterry, Mexico is one which employs this process[491]).

4.8.2.3.2. "Autothermal" Cracking Process[495, 497]). The reaction

$$CH_4 + \tfrac{1}{2}\,O_2 = CO + 2\,H_2, \qquad -380\ \text{kcal/m}^3\ CH_4\ \text{at NTP}$$

is weakly exothermic. If, additionally, the reacting gases are preheated, it appears that, in principle, an *autothermal* process is possible, i.e., a cracking process in which no further heat need be supplied for the reforming reaction to proceed. Figure 249 presents the heat balance for several conceivable cases:

Preheating of the CH_4: nil.

Preheating of the cracking medium: up to 1200°C.

Composition of the cracking medium: 25% O_2, 50% O_2, 100% O_2 } remainder N_2.

Mixing ratio:

a) $O_2/CH_4 = 0.5$, i.e., the stoichiometric ratio in order that no CO_2 or H_2O be formed;

b) $O_2/CH_4 = 0.6$, i.e., 20% excess of O_2 above the stoichiometric ratio for the cracking reaction, thus giving a theoretical degree of oxidation of the gas, when there is complete conversion of CH_4, of

$$(CO_2 + H_2O)/(CO + H_2 + CO_2 + H_2O) = 0.067 = 6.7\%.$$

For Fig. 249, the heat gain – per unit amount of the split gas – which arises due to the preheating and the heat of the reaction was calculated. The reaction can be conducted *autothermally* only if this

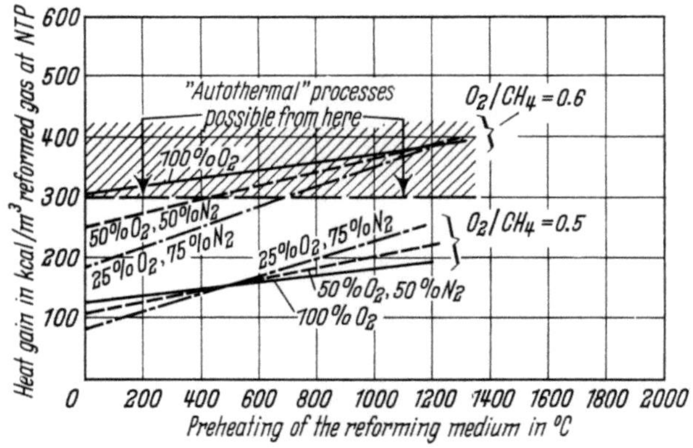

g. 249. Heat gain from the reforming reaction $CH_4 + \frac{1}{2} O_2 = CO + 2 H_2$ as a function of the preheating and O_2-content of the reforming medium

heat gain amounts to at least 300 kcal/m³ of the cracked gas at NTP, i.e., if the heat gain is sufficient to heat the gas mixture produced up to 900 to 1000°C. If the heat gain is less than this, then the amount of heat lacking must either be introduced from outside the reaction, or the reaction will take place otherwise than to produce $CO + H_2$ as desired. It can be seen from Fig. 249 that the reaction cannot be conducted autothermally without a considerable excess of O_2. If there is an excess of O_2, an unwelcome deterioration in the quality of the cracked gas must be reckoned on. A 20% excess of O_2, which is barely sufficient to produce the amount of heat needed, means (see Fig. 249, upper part) that 6.7% of the result ing CO/H_2 mixture has already been oxidized to $CO_2 + H_2O$. Since, e.g., at 820°C only about 35% of a CO/H_2 mixture can be utilized for the reaction

$$FeO + H_2(CO) = Fe + H_2O(CO_2),$$

the reducing power of the gas mixture is already reduced by

$$6.7/35 = 19.2\% .$$

It seems therefore that the known autothermal processes in which natural gas or fuel oil is decomposed in the flame, and for which the construction of the plant is simple, are still not the optimum for the generation of reducing-gas mixtures. Tables 29 to 31 show the corresponding balances for the Koppers-Totzek, Texaco, and Shell autothermal processes. The H_2O-contents of the generated cracked gases are not given, but they must be quite considerable because of the proportion of O_2 used. The CO_2-contents are already inadmissibly high.

Table 29. *Cracking of heavy fuel oil in the Koppers-Totzek generator* (*Low-pressure cracking*) [VON GRATKOWSKI[477]].

Composition of fuel %		Composition of gases %	
Water	0.2	CO_2	10.2
Ash	0.2	CO	42.8
C	87.2	H_2	45.2
H	11.0	CH_4	trace
O	0.5	N_2	1.6
N	0.5	H_2S	0.2
S	0.6		

Calorific value, H_u: 9825 kcal/kg

		Consumption and production per ton oil	per m³ $(CO + H_2)$ at NTP
Consumption:			
O_2	m³ at NTP	872	0.33
Steam	tons	1	
Electrical energy	kWh	98	
Production:			
Gas	m³ at NTP	2943	
$CO + H_2O$	m³ at NTP	2590	
Spent steam (15 atm, 280°C)	tons	1.75	

Table 30. *Cracking by the Texaco process* [von Gratkowski[477]].

a) Cracking of an American natural gas

Composition of the natural gas %		Composition of the cracked gas %	
CO_2	0.9	CO_2	2.2
CH_4	81.9	CO	38.0
C_2H_6	7.0	H_2	59.5
C_3H_8	6.2	CH_4	0.1
C_4H_{10}	2.4	N_2	0.2
N_2	1.6		

		Consumption and production	
		per m³ natural gas at NTP	per m³ (CO + H_2) at NTP
Consumption:			
Natural gas	m³ at NTP	1	0.329
O_2	m³ at NTP	0.826	0.272
Production:			
Cracked gas	m³ at NTP	3.12	1.025
CO + H_2	m³ at NTP	3.04	1
Possible spent steam	kg	≈ 1.7	≈ 0.57

b) Cracking of fuel oil

Composition of the fuel %		Composition of the cracked gas %	
Ash	0.06	CO_2	3.65
C	85.59	CO	47.97
H	11.38	H_2	47.45
N	0.72	CH_4	0.26
S	1.96	N_2	0.22
O	0.35	H_2S + COS	0.46

Weight: 0.977 kg/litre at 23°C

		Consumption and production	
		per ton oil	per m³ (CO + H_2) at NTP
Consumption:			
Oil		1 ton	0.351 kg
Steam		0.38 ton	0.1335 kg
O_2	m³ at NTP	719	0.2523
Production:			
Cracked gas	m³ at NTP	2980	1.046
CO + H_2	m³ at NTP	2850	1

The Texaco process does not provide a waste-heat boiler. Data on the soot content of the gas are not available.

Table 31. *Cracking of higher hydrocarbons by the Shell process* [VON GRATKOWSKI[477]]

	Fuel		
	Heavy fuel oil	Light petrol	Propane
Density . . kg/litre	0.97	0.668	
H_u . . . . kcal/kg	9840	boiling begins 37°C, ends 119°C	
Composition	%	%	mole-%
	Ash . . 0.07	C 84	C_2 . . . 0.2
	C . . . 85.4	H 16	C_3H_6 . . 28.2
	H . . . 11.0		C_3H_8 . . 69.3
	N . . . 0.2		C_4 . . . 1.1
	S 2.7		$N_2 + CO_2$ 1.2
	Remainder 0.63		
Gas composition:			
CO_2 . . . %	4.0	4.0	3.2
CO . . . %	46.3	41.3	39.0
H_2 . . . %	48.2	53.9	56.8
CH_4 . . . %	0.3	0.5	0.7
N_2 . . . %	0.3	0.3	0.3
$H_2S + COS$ %	0.9	—	—
Consumption per kg fuel:			
O_2 . . . m³ at NTP	0.75	0.77	0.79
Steam . kg	0.4	0.4	0.4
Consumption per m³ $(CO + H_2)$ at NTP:			
O_2 . . . m³ at NTP	0.267	0.245	0.242
Production per kg fuel:			
Crude gas m³ at NTP	3.0	3.32	3.40
$CO + H_2$ m³ at NTP	2.83	3.16	3.26
High-pressure exhaust steam kg	2.45	2.45	2.45

If such a process is adopted for the generation of reducing-gas mixtures, it is inevitably necessary to undertake extensive after-treatment (at least cooling and washing out of the CO_2 and condensation of the H_2O).

Using different apparatus it was possible in conjunction with the Hüttenwerk Oberhausen AG's Purofer process to solve the problem of generating a reducing-gas mixture lean in CO_2 and H_2O without the need for subsequent washing treatment[492]). Figure 250 shows the apparatus used. In a cyclone burner CH_4 with an oxygen-containing gas (air or recyclea gas from the reduction furnace) in precisely the stoichiometric proportion $[(\frac{1}{2}O_2 + H_2O + CO_2)/CH_4 = 1]$ is converted in a flame reaction; the resulting mixture passes finally through a preheated catalyst layer where, with some heat consumption, the conversion is carried through, in the desired direction, until complete. Table 32 shows the composition and temperature of the resulting gas mixture, which, in this case, can be used without restriction and without further treatment

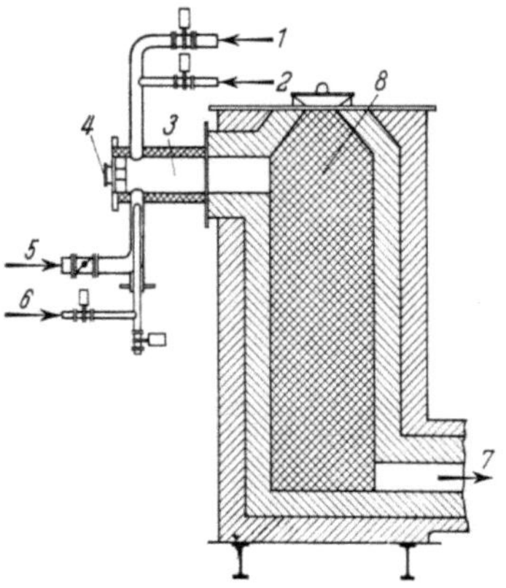

Fig. 250. Diagram of a HOAG gas converter[492])

1 Top gas from reduction shaft-furnace, *2* Natural gas, *3* Cyclone burner, *4* Observation hole, *5* Air blast, *6* Steam, *7* Reducing gas, carbon monoxide and hydrogen, 1000 to 1100°C, *8* Catalyst

for the reduction of iron ores. Two such gas converters are operated in alternate cycles, like the blast preheaters for a blast furnace. One gas converter is being heated by the furnace gas from the ore reduction, while the other is generating the required reducing gases. The function of the

Table 32

Gas	Composition %						Calorific value H_u kcal/m³ at NTP
	CO_2	C_nH_m	CO	H_2	CH_4	N_2	
Residual gas	0.9	1.9	11.4	9.8	54.1	21.0	5545
Natural gas	0.6	0.0	0.0	0.8	97.8	0.8	8406

Cracked gas from	Composition %				
	CO_2	CO	H_2	CH_4	N_2
Residual gas + O_2 .	0.8	26.7	58.1	1.8	11.8
Natural gas + O_2 .	1.7	25.4	63.1	4.7	4.8

The steam content is less than 25 g/m³ at NTP. The temperature of the cracked gas is between 1050 and 1100°C.

gas-reforming furnace is thus twofold: it serves as a regenerative heat accumulator, and as a catalyser for the gas-reforming reaction.

The Phelps Dodge Corporation (Arizona) also manufacture a directly usable reducing gas for operating their sponge-iron plant[498a, b]). In this process, natural gas, preheated to 400°C, is converted with air preheated to 500°C ($\lambda = 0.28$) using a nickel catalyst. The reducing gas has a temperature of 980°C and a degree of oxidation of 6 to 7%.

In construction and method of operation all reforming processes are much simpler when the initial material is natural gas or other gases containing CH_4 than when starting from oil. The oil has to be gasified by additional subsidiary processes, and the tendency to form soot during these processes is particularly high.

5. The Blast Furnace Process

5.1. An Approach to Physico-Chemical Concepts

As a vessel for reducing and smelting iron ores, the blast furnace has already demonstrated the creative genius and originality of its concept, if only by the fact that the principle of the process has scarcely required any modification over a period of one hundred years. As proof of this, Fig. 251 presents four profiles in simplified diagrammatic form, illustrat-

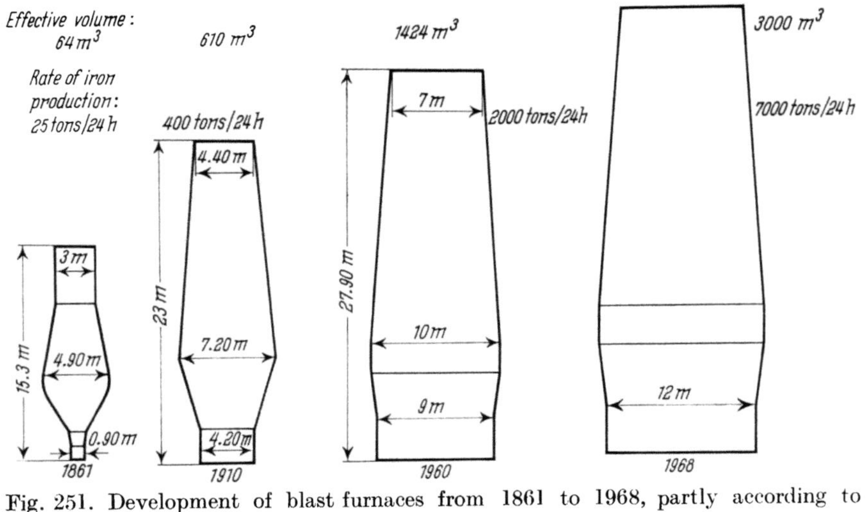

Fig. 251. Development of blast furnaces from 1861 to 1968, partly according to SPRINGORUM[499])

ing the development of blast furnace lines throughout a period of almost a century. It is not intended, at this point, to discuss details of blast furnace construction since comprehensive work on this subject is available[517, 518]).

Definite improvements have, however, been made in size, output and efficiency and particularly rapid progress has been made since 1950. Figures 252 and 253 indicate the marked increase in output per furnace and the reduction in coke rate at ironworks in the German Federal Republic and the U.S.A. American blast furnace outputs developed from a higher initial level and therefore exhibit a lower percentage increase as compared with

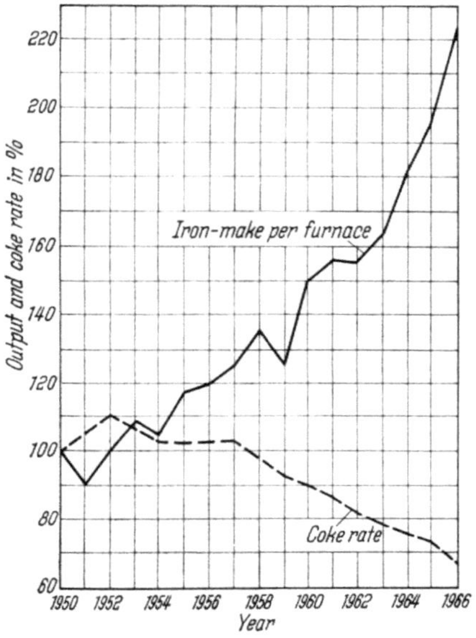

Fig. 252. Development of blast furnace performance in the Federal Republic of Germany (1950 = 100)[499a])

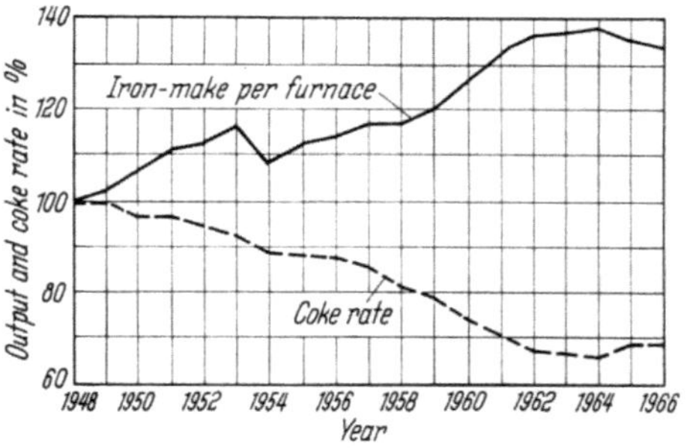

Fig. 253. Development of blast furnace performance in the United States of America (1948 = 100)[526a])

German furnaces. Blast furnaces in Japan have now the lowest average coke rates of the entire ironmaking industry.

These developments have been accompanied by comprehensive research into the physical chemistry of the blast furnace process.

Initially, the thermodynamic aspects of the blast furnace were investigated[503–505]). Subsequently, the principal features of burden preparation were properly recognised and presented in the classic work of WAGNER, HOLSCHUH, and BARTH[501]) and of JOSEPH et al.[502]).

The essentially thermodynamic investigations were continued more recently by WESEMANN[506]), GÖRGEN[507]), and SCHÜRMANN[508]).

It was shown that a combination of the heat and materials balances offers numerous possibilities of double checking, thus providing a more detailed understanding of blast furnace operating data. Such calculations have become an indispensable aid in assessing and exploiting plant trials.

In line with, and supplementing, this work, a thorough investigation has been made of the thermodynamic functions, especially the equilibria and chemical reaction heats occurring in the blast furnace. It can now be affirmed that, for the most part, they have been clarified with acceptable accuracy (see section 1 for further details).

The heat required for converting ores into molten pig iron (hot metal) can now be stated with good precision [see section 4.1 and references [507, 509]]. Compared with the actual fuel consumption, the extent to which the heat is utilised is far below 100% and varies a good deal as Fig. 254 shows.

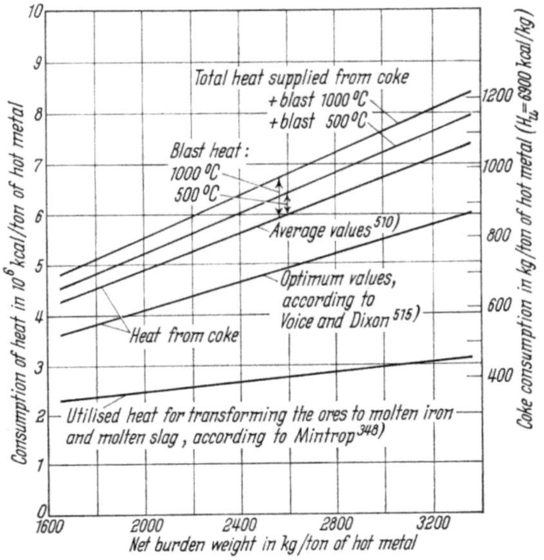

Fig. 254. Quantities of heat supplied and utilised in the blast furnace in the production of pig iron with varying net burden weight

The coke rate curve with normal burden preparation stems from a mathematical/statistical treatment of the operational data available in 1961 [510]), the optimum values being reported in reference [515]) (see section 5.2 for further details). The extent to which the heat input is utilised varies from 35 to 60%, see Table 33. Heat losses are mainly due to:

a) cooling and radiation;

b) sensible heat content of the top gas;

c) the CO and H_2 component which remains unused in the blast furnace and recurs in the top gas.

The losses from a) are associated with blast furnace design and are of the order of 2 to 5% of the total heat input (compare Table 33). The sensible heat content of the top gas (item b) corresponding to the prevailing temperatures of 100 to 400°C in blast furnace throats when furnaces are operated under normal conditions likewise amounts to only a few percent of the total heat input. Item c) is the most important factor, namely the fact that a considerable part of the CO is not consumed in reducing the ores, thus liberating heat, but leaves the furnace unused.

Table 33. *Approximative data for blast furnace heat balances*[520])

	%
Heat gains:	
Coke heat	90 to 95
Blast heat	5 to 10
Heat expenses:	
Useful heat	30 to 45
Wall losses	2 to 5
Waste gas heat	50 to 65
Separate items:	
Reduction Fe	15 to 25
Reduction of elements accompanying Fe	1.5 to 3.5
Calorific value of C in iron	3 to 6
Heat content of liquid iron	2.5 to 4.0
Heat content of liquid slag	3 to 5.5
Preparation of the charge	2 to 5.5
Heat of formation of the slag	0.8 to 1.5
Sensible heat of the gas	3 to 6
Calorific value of the gas	45 to 60

The chemical utilisation of the CO in the blast furnace can be represented by the expression

$$\eta_{CO} = \frac{CO_2}{CO + CO_2}$$

in which CO_2 and CO are the respective proportions by volume in the top gas. CO_2 from the decomposition of limestone must be incorporated (com-

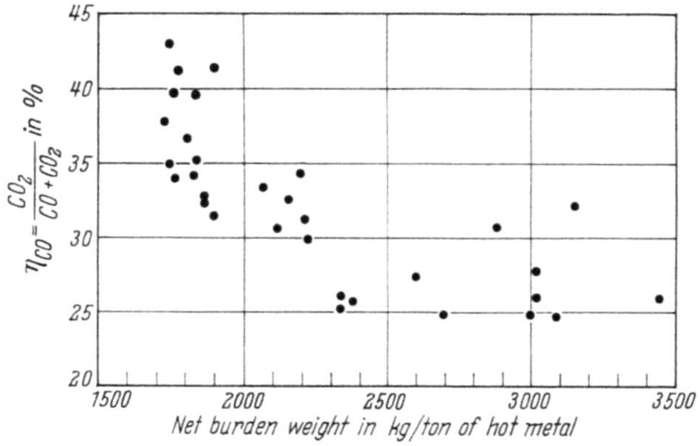

Fig. 255. Chemical utilisation of the top gas η_{CO} in relation to the net burden weight, plotted from data in reference[510])

pare section 5.2). Figure 255 calculated from data in reference[510]) shows that η_{CO} varies from 25 to 45% and is, in fact, generally at its peak level

Table 34. *Record performances for blast furnaces*[512a–c, 526f, 597, 746]

		Plant/furnace							
		Mannes-mann AG Duisburg-Huckingen [746]	Hüttenwerk Oberhausen AG, Furnace A [597]	Bethlehem Steel Corp. Sparrows Point [746]	Chere-povets [512a]	Chiba No. 5 [526f]	Nippon Kokan K.K., Fuku-yama No. 1 [512c]	Sumitomo Metals, Wakayama No. 4 [512c]	Nippon Kokan K.K., Fuku-yama No. 2 [512b]
Hearth diameter . .	m	7.0	8.5	8.76	9.75	10.00	9.80	11.00	11.20
Working volume . .	m^3	821	1449	1430	2000	1874	1735	2187	2255
Ores	kg/ton of iron	255	704	11	3	756	—	—	522
Sinter	kg/ton of iron	—	972	397	1645	848	1145*	920*	1055
Pellets	kg/ton of iron	1320	—	969	—	30	—	220*	—
Net burden.	kg/ton of iron	1897	1648	1570	1648	1627	1635*	1645*	1563
Coke rate (dry) . . .	kg/ton of iron	578	557	500	434	503	475	483	473 raw
Oil	kg/ton of iron	—	—	44	—	36	45	28.5	27
Natural gas	m^3/ton of iron at NTP	—	—	—	110	—	—	—	—
Iron make	tons/24 h	2091	2893	3177	4240	3883	4612	4924	6154
Utilisation of furnace volume (BF coefficient) .	tons/(m^3 · 24h)	2.547	2.00	2.222	2.12	2.07	2.658	2.251	2.73
Output per unit of hearth area. . . .	tons/(m^2 · 24 h)	54.3	51.0	52.7	56.5	49.5	61.2	51.8	62.5
Slag make	kg/ton of iron	386	345	235	361	308	285.5	294	263
Top gas pressure . .	atm gauge	—	—	0.70	1.77	0.78	0.58	1.03	0.95

* Estimated data calculated from slag quantity.

with a rich burden, i.e., with a low net burden weight. In any case, a considerable proportion of the heat generated by the fuel is lost in the top gas as the heat of combustion of CO and H_2.

The economics of blast furnace operation depend closely on the chemical utilisation η_{CO}.

When replacement fuels are used, hydrogen utilisation

$$\eta_{H_2} = \frac{H_2O}{H_2 + H_2O}$$

is also important. It is, however, closely associated with η_{CO}[662] (Fig. 343).

The world output record of 6060 metric tons (6680 American short tons or 5965 English long tons) of pig iron daily (as a monthly average) is held by a Japanese blast furnace plant[512a]. The blast furnace

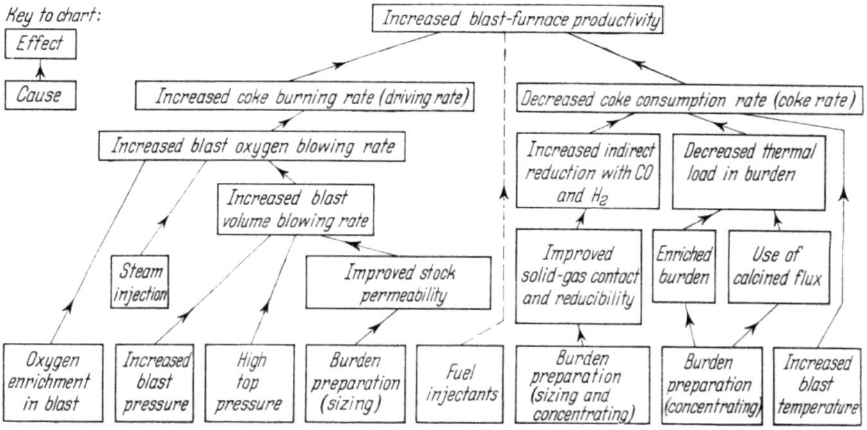

Fig. 256. Factors affecting blast furnace productivity, according to VOICE and DIXON[515])

hearth diameter is 11.20 m, the hearth area 98.50 m² and the production rate 62.5 metric tons of iron per m² of hearth area per day. Record values for other blast furnaces are presented in Table 34, the output also being referred to the working or effective volume and the hearth area of the blast furnaces to facilitate comparison. Opinions differ widely as to which burden materials, method of burden preparation and ironmaking practice most favour optimum performance. This divergence of opinion is not surprising when one considers the multiplicity of measures available to the ironmaker (see Fig. 256).

There is still no evidence that the records mentioned are the limits of possible achievement in the blast furnace.

The following important problems awaiting theoretical consideration clearly emerge from what has been discussed:

a) determination of the limits of blast furnace performance as regards output and fuel consumption;

b) prediction of optimum ore preparation and blast furnace practice, particularly as regards lump ore crushing and screening and agglomeration of ore fines;

c) calculation of the operational results, particularly output and coke rate, to be expected for given burdening and operational conditions;

d) partial replacement of coke by cheaper fuels; forecasting of operational data which might be anticipated from additions of gaseous, liquid and solid fuel of all kinds;

These problems must be dealt with in relation to numerous subsidiary conditions, particularly burden permeability, heat transfer, burden stability and many others.

A complete and reliable prediction of blast furnace operational data is not yet possible.

The still prevalent thermodynamic approach is useful only for a subsequent interpretation of operational results and for checking data for accuracy and completeness.

Reaction velocities and, consequently, output and coke consumption cannot be forecast on the basis of materials and enthalpy values (heat balances), i.e., on the basis of thermodynamics alone.

Since 1950, extensive laboratory and plant measurements of the rate of chemical reactions in the blast furnace have been made and these can be evaluated on the basis of reaction kinetics (see section 2). The hitherto available material, however, still provides a picture which is incomplete in some important aspects. The processes involved in the softening and melting of the burden are still being investigated and can only be slightly touched upon. In accordance with the aim of this monograph, the next section concentrates on the application of reaction kinetics, thermodynamics and the theory of fluid flow and heat transfer.

Approximate forecasts and ore valuations have until now been successful only by applying either general reaction kinetics or operational experience using mathematical statistics, see, for instance, references [510, 516]).

An account will now be given of results which have been achieved by applying first principles in blast furnace practice and by associating them with practical experience. At the same time, those areas in which it is logical and even essential to carry out further calculations and laboratory and plant trials will be indicated.

5.2. Preparation of the Burden

5.2.1. Principal Features of Process Technology

As early as 1932, WAGNER, HOLSCHUH and BARTH[521]) pointed out that the particle size range of ores before charging in the blast furnace must be closely defined within a carefully selected range if optimum output and fuel consumption are to be achieved. Similar ideas at about the same time in the U.S.A. produced noteworthy results[521a]). Fine grained material below about 5 mm is injurious because, when charging, this forms local accumulations and regular obstructions in certain zones of the stock column. In section 3.1, attention has already been drawn to the fact that even relatively small proportions of fine grained material in the stock column are sufficient to cause a considerable loss of permeability. Fragments larger than about 60 mm are just as harmful as fine grained material because they greatly retard total heat transfer and reduction. Closer definition of the particle size range by burden preparation which WAGNER et al. referred to as "physical burdening" has been adopted at the great majority of blast furnace plants in the last decade, see, for instance, references [523–526, 526a]).

Efforts toward developing burden preparation have previously been concentrated on the ore side. For some years, however, the preparation of blast furnace coke and particularly the need for a closer definition of its particle size distribution have also become increasingly important[526b]).

There are essentially four distinct types of installations for burden preparation. Their range of application depends on the nature of the ores*.

5.2.1.1. Ore-Blending Installations

Ores vary quite considerably in chemical composition and particle size, this being adversely reflected in plant performance and in the uniformity of the product during subsequent processing.

These irregularities can be reduced to a tolerable level in stacking and reclaiming or ore bedding plants. Thus, for instance, it is possible to reduce the standard deviation in a run-of-mine ore containing 9.0% SiO_2 from 1.94 to 0.52%[526c]).

A stacking and reclaiming or ore bedding plant consists of at least two mixing beds or piles, one of which is laid down longitudinally and the other reclaimed transversely[526d]) (Fig. 257).

---

* W. LUYKEN has reviewed the older literature on burden preparation, particularly as regards the installations used[522]).

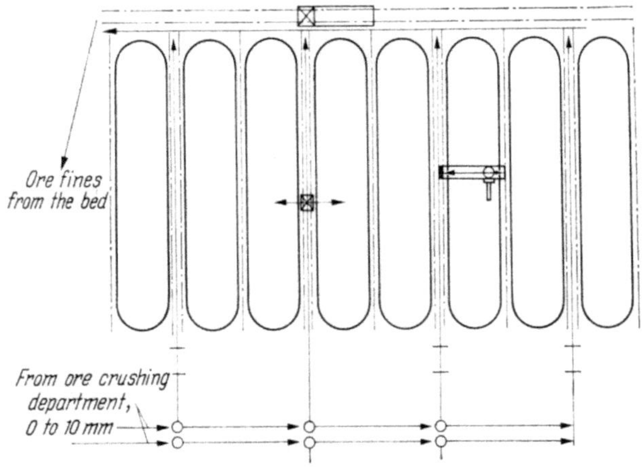

Fig. 257. Bedding plant with eight beds, according to FRIEDETZKY[526 d])

Fig. 258. Aerial photograph of an ore preparation plant[523])

A bed or pile can contain 25000 to 80000 metric tons of ore. Ore bedding plants have proved to be useful both for blending ores of greatly varying composition[526c]) and for blending ore mixtures[526e]). They are used both for run-of-mine ores and for graded lump ore and fine ores or mixtures[526f]).

Because of the considerable amount of floor space required, it will often be difficult to install ore bedding plant in existing iron and steel works. The latest development in the field of proportioning belt weighers enables ores to be withdrawn from storage bins with a precision of $\pm 0.5\%$. Depending on the homogeneity of the individual components of the mix, it should be possible to achieve the same standard deviations with proportioning belt weighers as are attained with bedding installations.

5.2.1.2. Ore Crushing and Screening Plant

The run-of-mine ore is reduced to the requisite lump size in crushers, the fines being subsequently separated by screening. Figure 258 presents an aerial view of the ore crushing and screening installations of a West German iron and steel plant[523]), the site of which, between ore discharging (water and rail transport) and sinter plant, may be con-

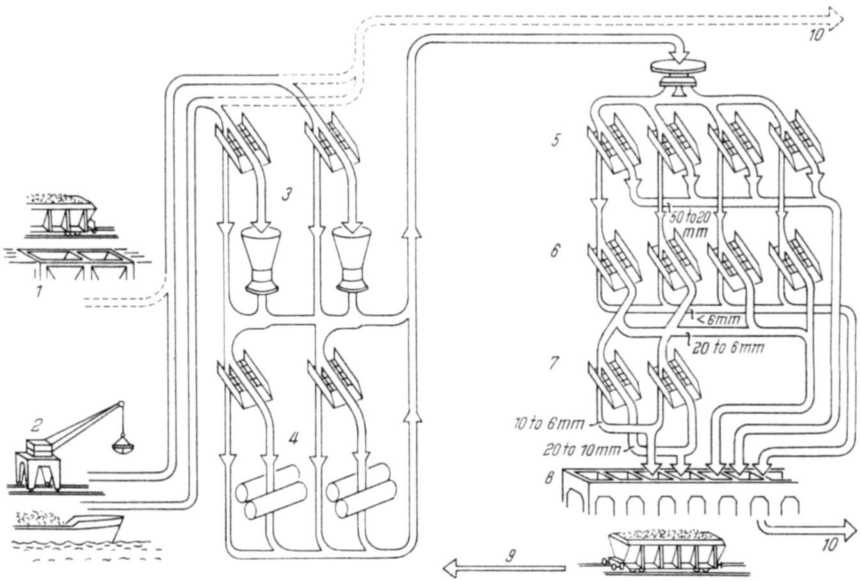

Fig. 259. Flow diagram of an ore crushing and screening plant[523])

1 Ground storage bins, *2* Cranes, *3* Primary crushing, from 300 to 50 mm, *4* Secondary crushing, from 50 to 20 mm, *5* Primary screening, 20 mm square-mesh screen, *6* Secondary screening, 6 mm harp-type screen, *7* Tertiary screening, 10 mm square-mesh screen, *8* Fourteen ore hoppers (470 m^3), *9* To the blast furnaces, *10* To the sinter plant sidered typical of this kind of installation. Figure 259 presents a schematic layout of the flow of materials. On discharging, the ore passes through two stages of crushing, cone and roll crushers which can be relieved by a scalping operation, with, finally, four parallel lines of screens. It will suffice to point out that the size range originally varying from 0 to 200 mm is reduced to the three following fractions:

a) 0 to 6 mm fines (for agglomeration);

b) 6 to 20 mm size (for the blast furnace), and

c) approximately 20 to 60 mm lump ore (for the blast furnace).

Figure 260 shows some examples of actual size distribution of fractions b) and c). Many ore crushing and screening plants produce only a single lump ore fraction[526]). The advantages and disadvantages of these two methods will be examined in more detail below.

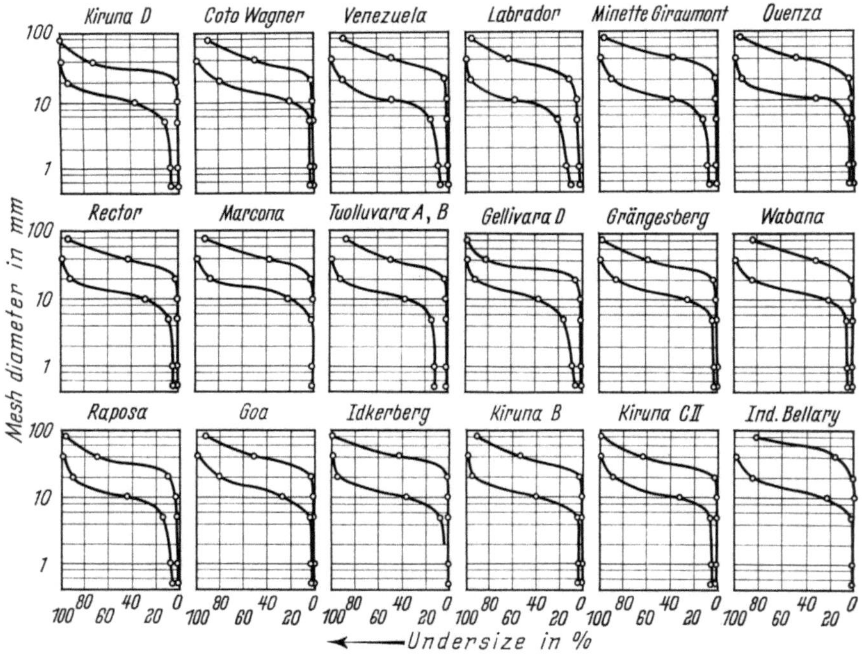

Fig. 260. Screen analyses of lump ores after passing through ore crushing and screening plants[523])

Screening out to 20 to 60 mm (upper curves) and 6 to 20 mm (lower curves)

5.2.1.3. Sinter Plants

Ore fines occur when ores are crushed and screened and sometimes even in the mine or in the dressing of low grade ores. Provided that the proportion of particles of less than 0.1 mm in the ore fines is not too high, the classic sinter plant (Fig. 261) is still eminently suitable and extensively used for agglomeration. Almost all present day sinter plants operate on the down-draft principle, see, for instance, references [523, 527–529], although in some cases rotary kilns are used[530] as in the manufacture of cement clinker. In down-draft sinter plants, the ore fines are mixed in trommel mixers with about 4 to 10% fuel addition, mostly coke breeze,

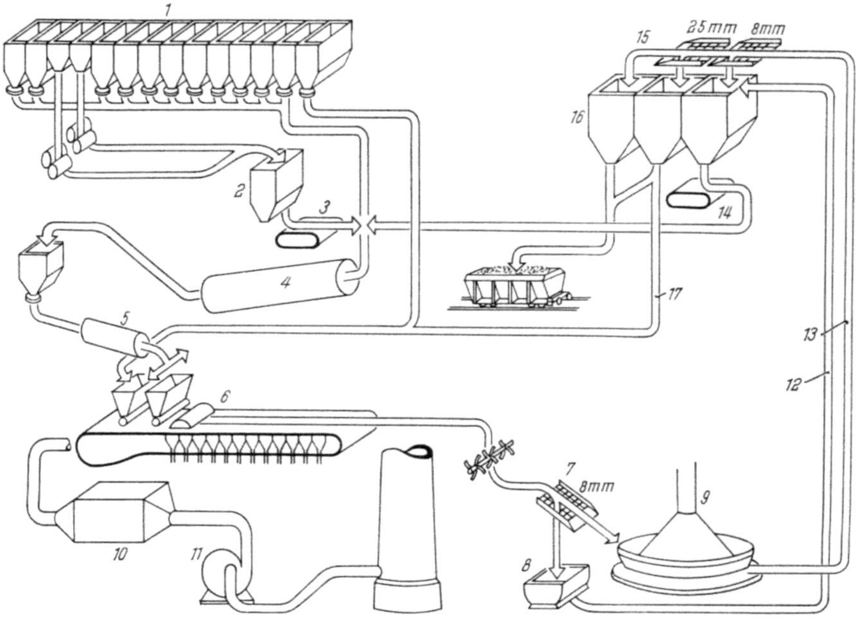

Fig. 261. Flow diagram of the continuous sintering process[523]

1 Two bins for coke breeze, eleven for ore fines, one for hearth layer, *2* Coke breeze minus 3 mm, *3* Regulating belt feeder for coke breeze, *4* Pre-mixing drum, *5* Secondary mixing drum, *6* Sinter machine: 2.5 m wide, 37.5 m long, 93.75 m^2 suction area, *7* Hot screening, *8* Return fines cooler, *9* Sinter cooler: (255000 m^3 of cooling air per hour, 180 mm wg, *10* Dust content after the two-stage electrofilters below 0.100 g/m^3, *11* Fan: 150°C, 1100 mm wg, 550000 m^3/h, *12* Cooled return fines, *13* Cooled sinter, *14* Regulating belt feeder for return fines, *15* Cold screening, *16* Finished sinter, *17* Hearth layer

crumbled and sintered on travelling grates after ignition of the coke breeze by means of gas or oil burners. Air is sucked through the mixture, in which temperatures of 1200 to 1400°C are attained. Bed depth may be as much as 35 cm. The available induced draft is limited to about 1100 mm wg. Sintering iscomplete when the fuel in the mixture has been burnt.

The results of sintering are influenced by numerous plant and operational factors, the latter being associated with the nature of the feed. Thus, for instance, the rate of sintering, and consequently plant performance, depends on the uniform and rapid supply of air for the combustion of the particles of coke breeze in the bed as Voice (Fig. 262) and Rausch and Cappel[540] point out. Here again, therefore, high and uniform

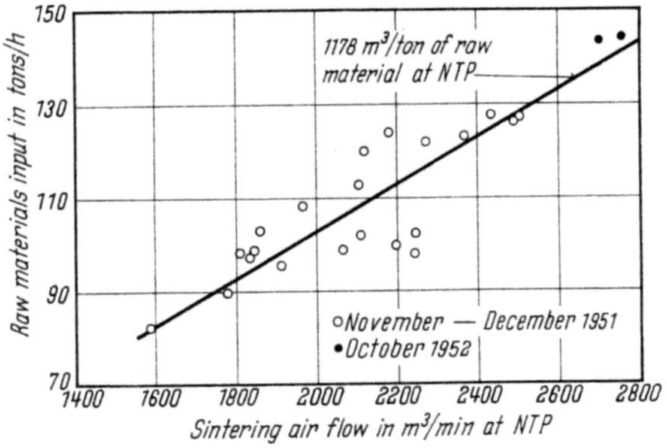

Fig. 262. Effect of sintering air flow on the plant material input rate, according to VOICE et al.[531])

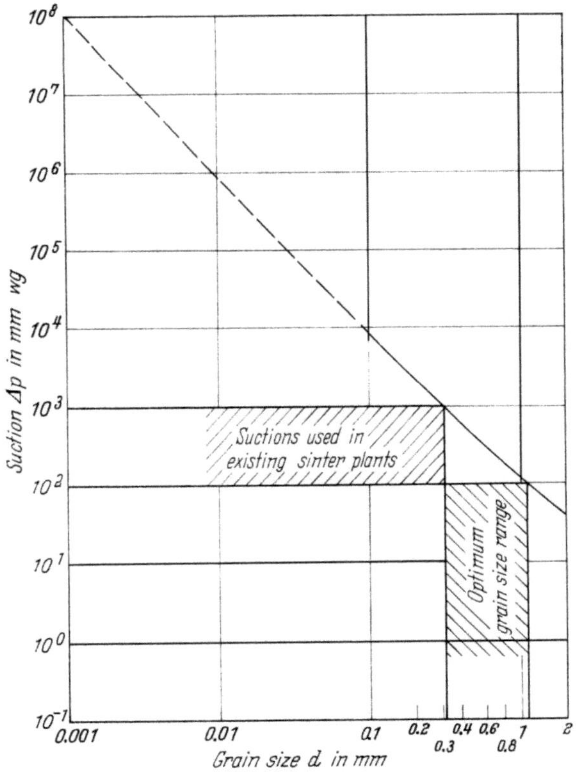

Fig. 263. Suction required in sintering in relation to grain size of the sinter mix[533])

permeability of the feed is a decisive factor. Figure 263, calculated by the Carman-Kozeny equation (for the example in Fig. 264), shows that the feed must not contain marked proportions of the − 0.3 mm fraction

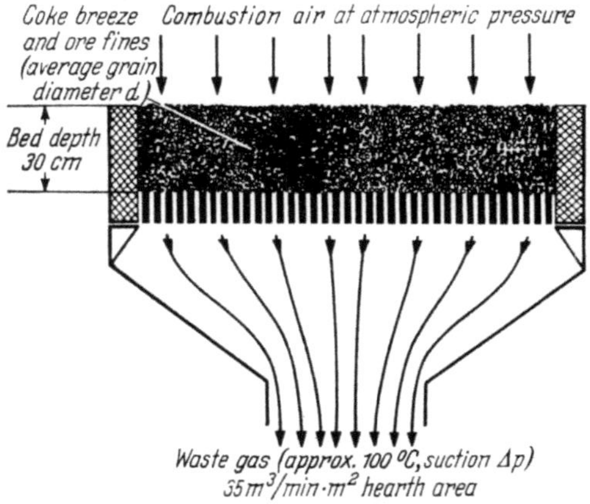

Fig. 264. Diagram of down-draft sintering of iron ores[533])

if high permeability, good heat transfer[540a]) and consequently high sinter output are to be ensured. Figure 265 confirms that better sinter output is achieved with higher feed permeability.

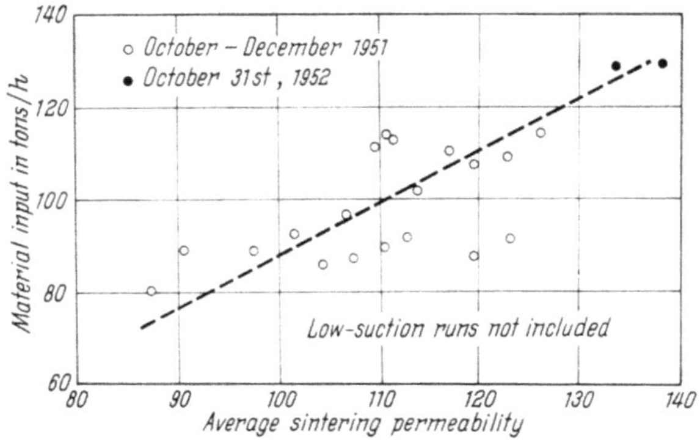

Fig. 265. Influence of permeability on material input rate[531])

Since nearly all types of fine ores contain considerable amounts of − 0.3 mm, they must be finely crumbled between mixing and feeding to the sinter strand, using the surface tension of the water. In modern sinter plants, this task is performed by separate granulating or rotating drums or, in some cases, disks. This process greatly reduces sintering time, thus increasing output as has been frequently demon-

Table 35. *Analysis of the raw materials examined with the optimum*

Type of ore	Optimum content					Return fines <10mm %
	H_2O %	C %	Lime* (CaO) %	Dolomite* %	Dust %	
Millscale	6.0	3.0	—	—	—	18.0
	6.0	3.0	3.0	—	—	19.3
	6.0	3.0	—	6.0	—	18.0
	6.0	3.0	—	—	6.0	17.0
Swedish concentrate C	8.0	4.0	—	—	—	13.4
	8.0	4.0	4.0	—	—	—
	8.0	4.0	—	5.0	—	23.4
	8.0	4.0	—	—	7.5	15.7
Allerheiligen ore	14.0	4.0	—	—	—	—
	14.0	4.0	5.0	—	—	25.5
	14.0	4.0	—	4.0	—	20.4
	14.0	4.0	—	—	15.0	20.0
Echte fines	13.0	5.0	—	—	—	28.3
	13.0	5.0	2.0	—	—	—
	13.0	5.0	—	4.0	—	25.7
	13.0	5.0	—	—	10.0	23.0
Wet concentrate	13.0	3.5	—	—	—	22.2
	13.0	3.5	7.0	—	—	—
	13.0	3.5	—	4.0	—	23.5
	13.0	3.5	—	—	25.0	22.6
Geislingen ore (fine)	14.0	6.0	—	—	—	21.4
	14.0	6.0	2.0	—	—	22.9
	14.0	6.0	—	2.0	—	18.0
	14.0	6.0	—	—	15.0	19.0
Blast furnace flue dust	16.0	8.7	—	—	—	25.6
Blast furnace flue dust with purple copper ore	17.0	5.0	—	—	—	19.4
Spanish pyrites residue	21.0	5.0	—	—	—	—
	21.0	5.0	6.0	—	—	35.1
	21.0	5.0	—	6.0	—	31.1
	21.0	5.0	—	—	15.0	31.6
Pyrites residue from Bochum-Riemke	25.0	6.0	—	—	—	—
	25.0	6.0	5.0	—	—	—
	25.0	6.0	—	5.0	—	—
Purple copper ore	18.0	5.0	—	—	—	35.5
	18.0	5.0	2.0	—	—	48.4
	18.0	5.0	—	4.0	—	41.7
	18.0	5.0	—	—	15.0	38.1

* The lime processed in the trials contained 66.5 % CaO.

values determined with the addition of water, fuel, lime, dolomite and dust[538])

Chemical analysis of raw material								
FeO %	Fe_2O_3 %	Σ Fe %	Mn %	P %	SiO_2 %	Al_2O_3 %	CaO %	MgO %
49.56	39.99	67.91	0.90	0.60	2.14	1.09	0.15	0.23
24.32	65.14	64.40	0.24	1.28	2.36	1.38	2.85	0.34
1.30	49.10	35.48	0.42	0.60	21.82	8.47	6.05	1.26
14.65	18.03	23.97	0.28	0.38	15.93	11.50	16.73	3.33
4.59	49.88	38.48	0.36	0.54	15.32	6.83	4.69	2.03
2.40	46.82	34.64	0.58	0.40	21.68	7.35	6.95	1.52
26.30	19.57	34.09	0.94	1.02	15.94	5.37	11.85	1.39
6.30	74.90	57.31	0.17	0.16	6.52	3.05	0.61	0.39
3.50	59.55	44.28	0.29	0.024	16.56	6.12	2.56	1.17
1.25	85.92	61.12	0.09	0.056	6.18	2.44	0.65	0.28

Table 35

Type of ore	Optimum content					Return fines <10mm %
	H_2O %	C %	Lime* (CaO) %	Dolomite* %	Dust %	
Yugoslavian pyrites residue .	26.0	5.0	—	—	—	—
	26.0	5.0	10.0	—	—	—
	26.0	5.0	—	7.5	—	—
	26.0	5.0	—	—	10.0	—
Portuguese pyrites residue . .	15.0	6.0	—	—	—	—
	15.0	6.0	3.0	—	—	—
	15.0	6.0	—	6.0	—	—
	15.0	6.0	—	—	5.0	—
Return fines, 10 mm	8.4	2.9	—	—	—	17.3
	7.0	2.6	—	—	—	16.5

* The lime processed in the trials contained 66.5% CaO.

strated[533–536]. Maximum particle size of the sinter mix should not exceed about 5 mm or heat transfer and agglomeration will be inhibited[532a].

In older sinter plants, use has also been made of existing trommel mixers which have been enlarged to such an extent that the desired crumbling effect is achieved at this particular stage[537]. Crumbling can, of course, be successfully achieved only if the optimum water content has been imparted to the ore mixture. KOSMIDER, BERTRAM and SCHENCK[538] have determined these optimum moisture values and the optimum fuel addition experimentally for different ores. Table 35 shows that these values differ widely from ore to ore. The effects of particle shape, size and wettability are to be noted in this respect.

If the moisture contents of the raw sinter mix are too high, condensation of the water vapour can occur toward the bottom of the sinter bed and this may well increase the moisture in the sinter mix, thus reducing permeability and, in an extreme case, clogging up the sinter grate. The lower the humidity of the waste gas and the higher the temperature of the ore mixture, the less is the condensation of water vapour (Fig. 266). Several investigators[538b, c, d] consequently recommend that the hot return fines arising under operational conditions should be added to increase the ore mix temperature, thus avoiding water vapour condensation. When hot return fines have been added, an increase of up to 30% in output has been observed[538c, f].

When the mix is fed to the sinter strand, appropriate measures must be taken to prevent size segregation of the feed across the width of the machine[538a]. When size segregation occurs in a vertical plane, the fine

(continuation)

Chemical analysis of raw material								
FeO %	Fe_2O_3 %	Σ Fe %	Mn %	P %	SiO_2 %	Al_2O_3 %	CaO %	MgO %
2.05	76.65	55.10	0.10	0.032	12.81	3.70	0.35	0.2
1.11	82.80	58.82	0.12	0.02	3.24	2.65	0.71	0.13
9.15	61.16	49.86	0.33	0.36	12.36	3.45	5.60	1.13
7.23	56.23	44.92	0.27	0.36	14.18	2.81	8.74	1.17

grain components of the feed which are, at the same time, richer in fuel segregate in the top part of the sinter bed. This has a favourable thermal effect on the sintering process and consequently a deliberate

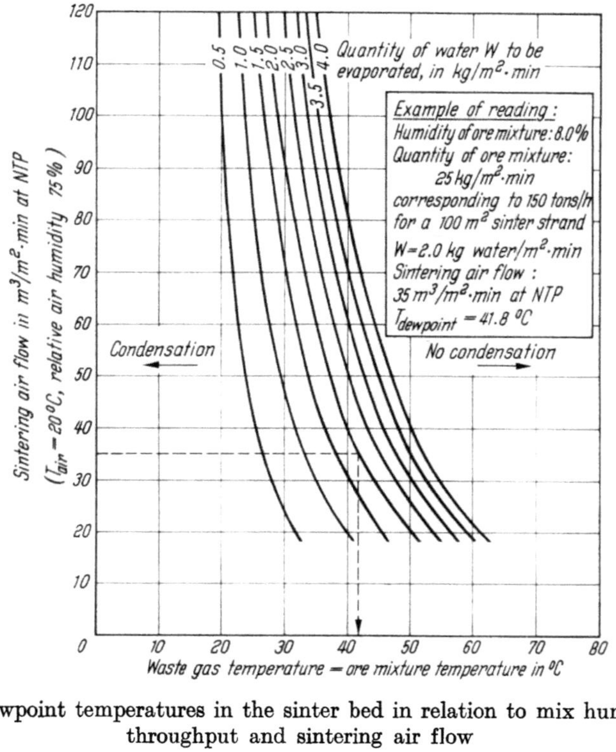

Fig. 266. Dewpoint temperatures in the sinter bed in relation to mix humidity, mix throughput and sintering air flow

attempt is frequently made to achieve vertical size segregation by, for instance, inserting sheet metal deflectors.

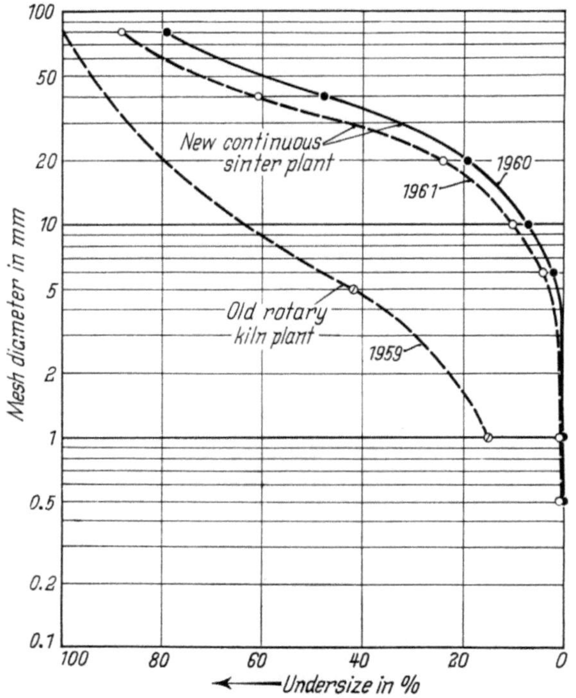

Fig. 267. **Screen analysis of sinter**[523])

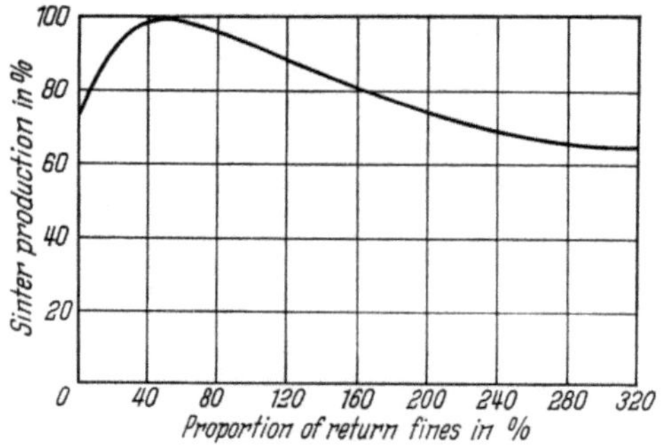

Fig. 268. Sinter **production in relation to proportion** of return fines, according to RAUSCH and CAPPEL[540])

Sinter grates are either connected in an endless rotating belt [Dwight Lloyd plants], or are in the form of single mobile pans[539]). Trials have recently been carried out in an attempt to depart from the principle of the horizontal and to carry out the sintering process in a vertical strand[541]).

In all modern sinter plants, great importance is attached to a perfect separation of the residual fine particles from the finished sinter. A comparison of the particle size curves for sinter produced on a modern strand sinter plant with those for the product of an obsolete rotary kiln without any screening equipment shows how necessary this is (Fig. 267). The separated fines are returned to the feed in the form of return fines. Within certain limits, depending on the type of ore blending, return fines improve the permeability of the feed and increase sinter output[540]) (Fig. 268).

5.2.1.4. Pelletising Plants

Considerable amounts of fines (0 to 0.1 mm) are being produced to an increasing extent in the wet preparation of low grade ores, e.g., taconites in the Great Lakes region of the U.S.A.[542, 543, 543a]). Quite a long time ago, SENGFELDER recognised that fundamentally different processes would have to be used for the agglomeration of very fine ore[544]). The crumbling effect familiar from the classic down-draft sinter plants was perfected to form the basis of a new process. By adding water and small amounts of binder constituent (e.g., special clays such as bentonite, NaCl, $CaCl_2$ or $FeSO_4 \cdot 7\ H_2O$), pellets of 10 to 30 mm diameter are formed from concentrates in tilted, slowly rotating disk pelletisers or drums (Fig. 269). These pellets are bonded together by the surface tension of the water as was shown by TIGERSCHIÖLD[545]) (Fig. 270). The wet balls, known as green pellets, are subsequently indurated or hardened at temperatures above 1000°C in shaft furnaces, rotary kilns and on travelling grates. On travelling grates, part of the heat is supplied by the combustion of fine solid fuels incorporated into the outer zone of the green pellets during their manufacture and having a similar effect as in the classic sintering process. Figures 271 and 272 present some examples of installations in diagrammatic form. On being heated up, the pellets pass through a critical temperature range when the water binder constituent has already evaporated but strengthening by means of recrystallisation and sintering has still not set in. Here, some slight degree of cohesion is ensured by the previously mentioned addition of binder constituents such as NaCl, etc. These bind the ore particles together to form a sort of mortar-like crust. In shaft furnaces, this phase is more difficult to control than is the case on travelling grates. The finished product is known as pellets and a considerable market has been developed for these in recent years[547, 547a]).

While the total capacity of existing pelletising installations in the U.S.A. and Canada[547b]) was only one million tons per annum in 1956, it

Fig. 269. Section of a disc pelletiser in operation, according to MEYER[546])

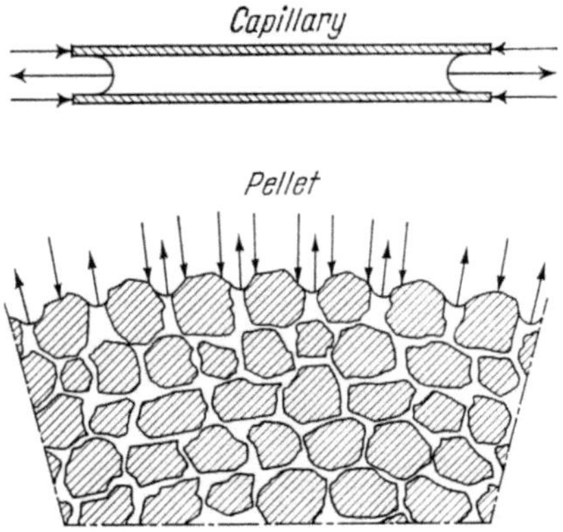

Fig. 270. Compression and tensile stress in a capillary tube and in the pore system of a pellet, according to TIGERSCHIÖLD[545])

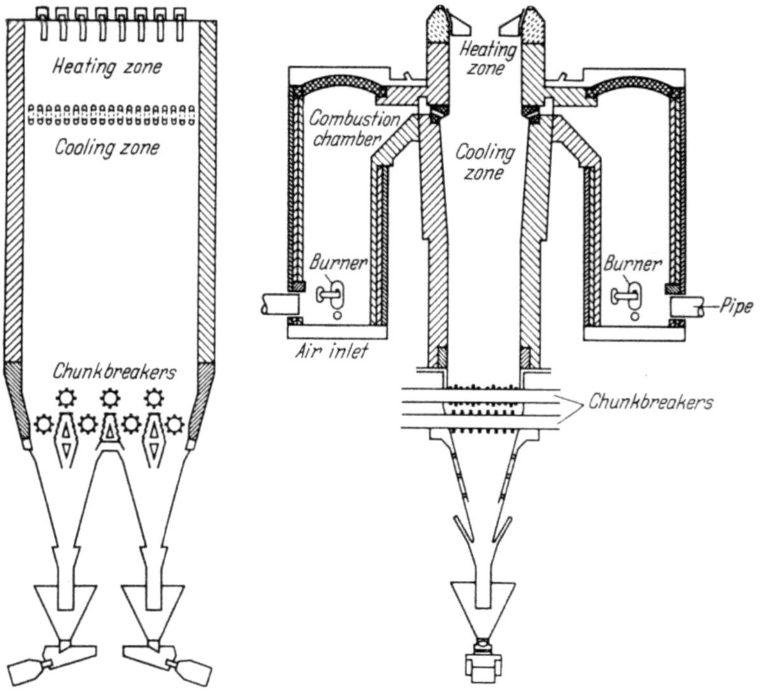

Fig. 271. Shaft furnace belonging to the Erie Mining Co. used for firing pellets, according to MEYER[546])

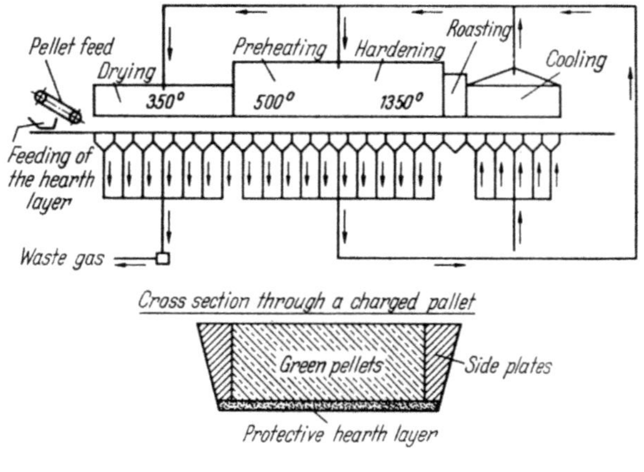

Fig. 272. Pelletising machine with several firing zones, according to MEYER[546])

had risen to 52.6 million tons per annum by 1966[606a]). Including planned installations and plants under construction, the total annual capacity in the U.S.A. and Canada now amounts to about 70 million tons.

5.2.1.5. Other Agglomeration Processes

Various attempts have been made to compact fine ores and blast furnace flue dust by pressing at room temperature, i.e., by briquetting to produce a suitable charge for the blast furnace[533, 548, 548a]). Figure 273

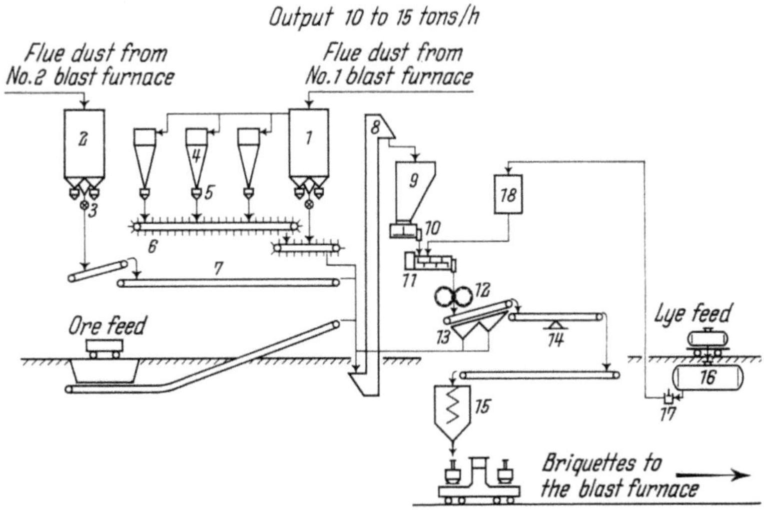

Fig. 273. Flow sheet of a briquetting plant, according to Wolf and Wysocki[548])

1 Dust catcher, furnace No. 1, *2* Dust catcher, furnace No. 2, *3* Bucket wheel charging valve, *4* Cyclone, *5* Conical charging valve, *6* Chain conveyor, *7* Belt conveyer, *8* Bucket elevator, *9* Feed hopper, *10* Table feeder, *11* Mixer, *12* Briquette press, *13* Wire tissue belt, *14* Belt weigher, *15* Briquette hopper, *16* Sulphite-spent liquor storage tank, *17* Sulphite-spent liquor pump, *18* Sulphite-spent liquor feed tank

presents an example of such an installation. It has been shown that this type of agglomerate is not capable of withstanding the severe compression and abrasion in large blast furnaces. Quantitative proof of this was provided by means of radioactive isotope techniques[533]) (see also section 5.3).

Recent attempts have been made to improve briquettes by heating them up to 1000°C during the briquetting process and possibly using a partially reduced charge [hot briquetting[550, 550a])]. Final results are still awaited particularly as regards the high tool wear which is expected.

5.2.2. Application of Reaction Kinetics to Burden Preparation and Operational Results

5.2.2.1. Uniform Permeability of the Stock Column

The blast furnace may be regarded as a materials and heat exchanger operating on the principle of counterflow between the ascending gas and the descending charge which is initially solid, later pasty and finally liquid. Efficient furnace operation, i.e., a high rate of throughput with low fuel consumption, can be achieved only if the ascending gases leave the furnace

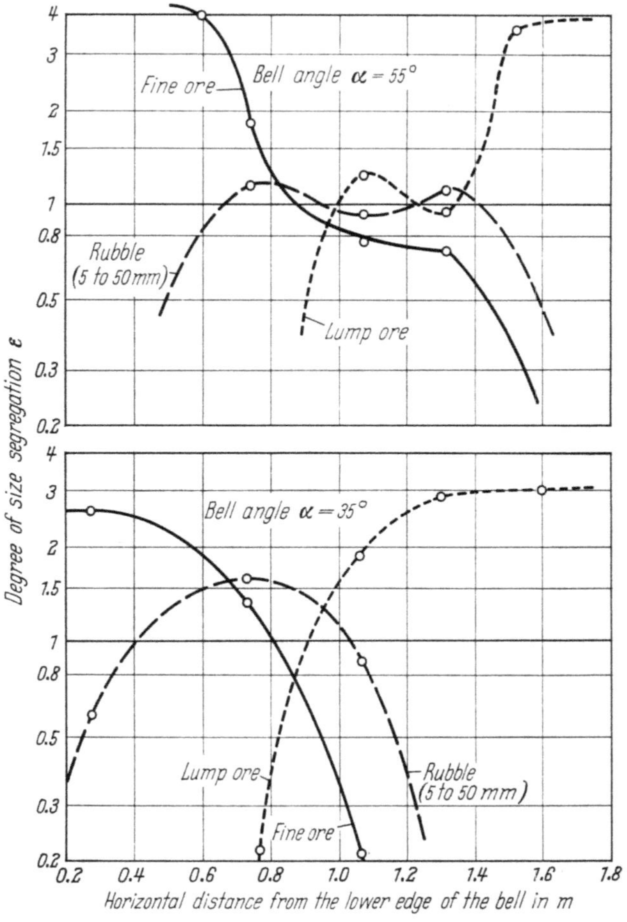

Fig. 274. Size segregation of Gellivara ore with various bell angles, according to TOUSSAINT et al.[552])

after being utilised to the fullest possible extent, both thermally and chemically, i.e., at the lowest possible temperature and with the highest possible CO_2 content. High utilisation efficiencies are, however, possible only if the gas flows uniformly around the lumps of ore throughout the cross section of the charge, i.e., the permeability across the stock column must be approximately constant. This can be achieved only if the ores

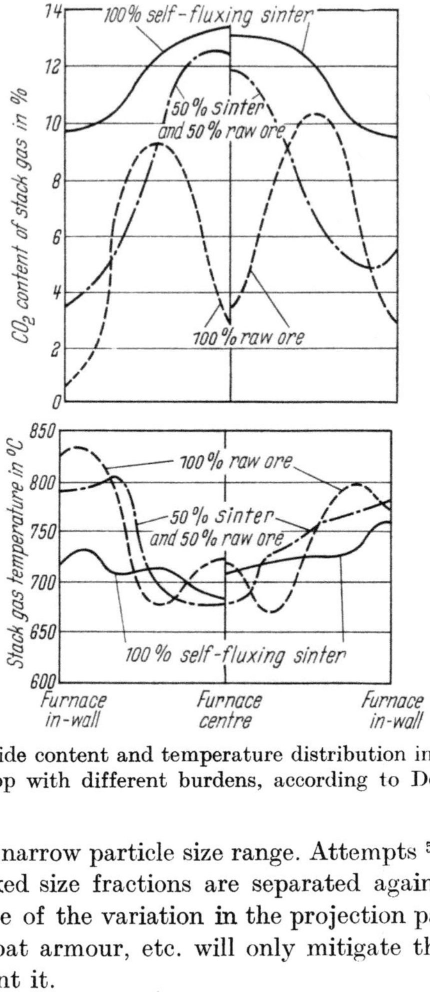

Fig. 275. Carbon dioxide content and temperature distribution in the stack gas about 6 m below the top with different burdens, according to DOI and KASAI[553])

exhibit a relatively narrow particle size range. Attempts [551, 552]) have shown that even well mixed size fractions are separated again when sliding off the big bell because of the variation in the projection parabola (Fig. 274).

The use of throat armour, etc. will only mitigate this state of affairs but will not prevent it.

Only by charging closely graded lump ores, well screened sinter or pellets, is uniform permeability attained. Evidence of this is presented in Fig. 275 by means of data obtained from permeability measurements by DOI and KASAI[553]) and similarly obtained by other investigators[554–557]). Temperature and concentration distribution of the stack gas certainly become more uniform when changing over from a raw ore burden to a sinter burden with efficient screening. These horizontal probe measurements were made at depths of about 6 m below the top. It would be interesting to repeat

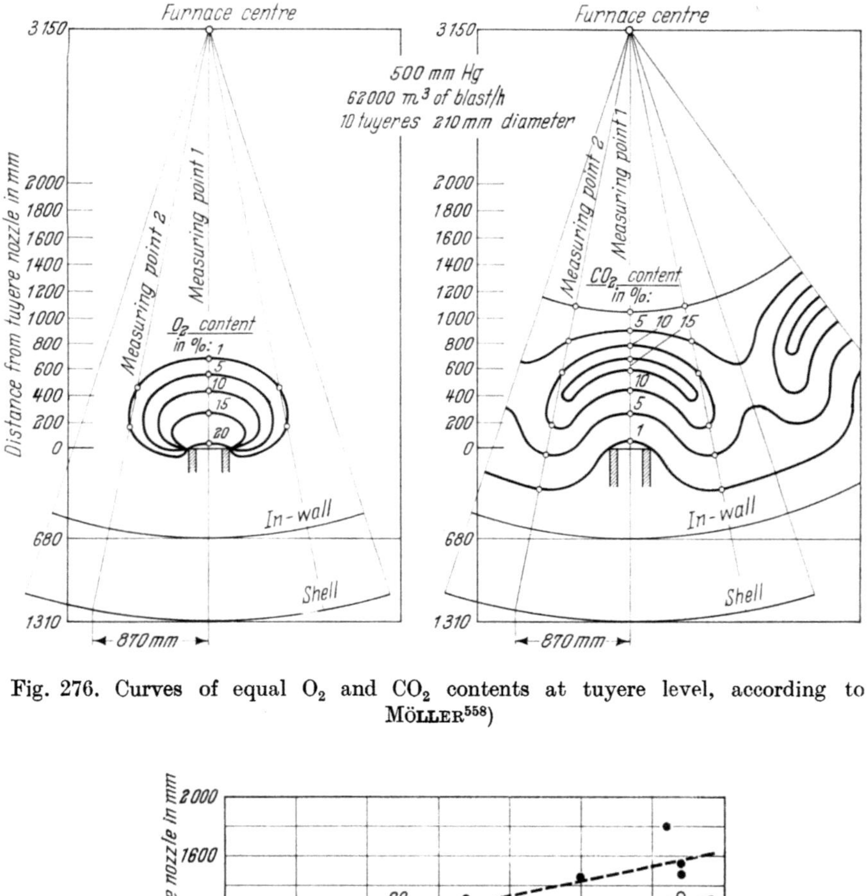

Fig. 276. Curves of equal O_2 and CO_2 contents at tuyere level, according to MÖLLER[558])

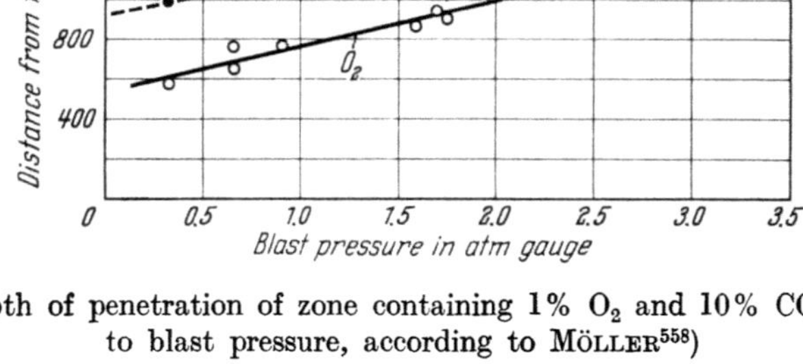

Fig. 277. Depth of penetration of zone containing 1% O_2 and 10% CO_2 in relation to blast pressure, according to MÖLLER[558])

such measurements at lower levels in order to ascertain in which zone the gas flow, initially concentrated in a narrow annulus in the region of the tuyeres, expands over the entire cross section of the furnace. There is no doubt that the greater the depth at which this transition occurs, i.e., the smaller the region of poor permeability (the "dead man"), the more efficient will be the utilisation of the stack gas. As numerous probes have shown, the effective hearth area (sometimes referred to as the active tuyere zone) is about 1 m wide (Fig. 276). With a 9 m diameter hearth, the highly permeable proportion of the cross section at tuyere level is thus only about 40%. By increasing the blast pressure, this zone can be slightly extended (Fig. 277).

5.2.2.2. Treatment of Lump Ores

First of all, a narrow size range must be attained when lump ores are crushed and graded. The degree of crushing can be varied to a great extent and this affects the reducibility of the ores. Reducibility improves with decreasing lump size but crushing costs and the consequent proportion of fines are increased.

In order to determine the optimum conditions, it is first necessary to be clear about the progress of the reduction reaction in the blast furnace.

The conversion of the descending ores is reflected in the variation in the composition and temperature of the stack gas. Temperature and composition of the stack gas are determined by means of vertical probes[559–561]) (Fig. 278) which descend with the burden, see Fig. 279. From top to bottom, there is initially a zone of steeply rising temperature (up to about 800°C). In this zone, there is scarcely any change in gas composition. Here the burden is heated to reduction temperature. The zone of indirect reduction follows. This is characterised by almost constant temperature and a considerable change in the CO_2 content of the gas. Below this, in the zone of direct reduction from about 1100°C, the CO_2 contents are low in spite of rapid reduction due to the simultaneously occurring Boudouard reaction.

Hupfer and Weidenmüller[560]) carried out similar consecutive measurements in the same blast furnace with two very different burdens:

a) a readily reducible lump ore burden and

b) a sinter burden of medium reducibility (63.3% sinter, 19% graded lump ore).

The expression $CO_2/(CO + CO_2)$ was formulated from the CO and CO_2 contents measured at each point and plotted in the equilibrium diagram against the temperature measured at the same point (Fig. 280).

With burden a), considerably lower temperatures were found for incipient reduction than with burden b), being discernible in the drop in

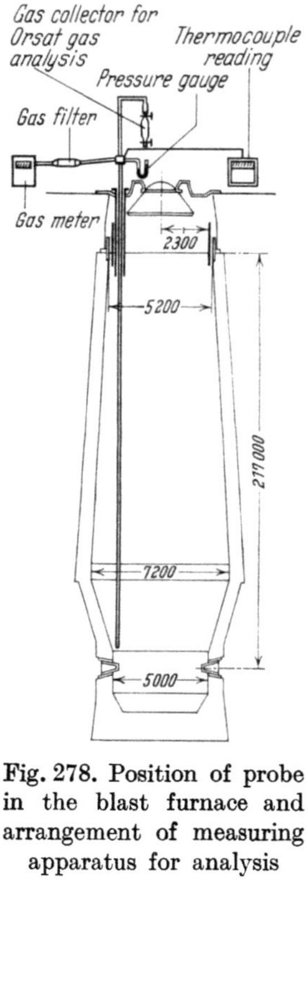

Fig. 278. Position of probe in the blast furnace and arrangement of measuring apparatus for analysis

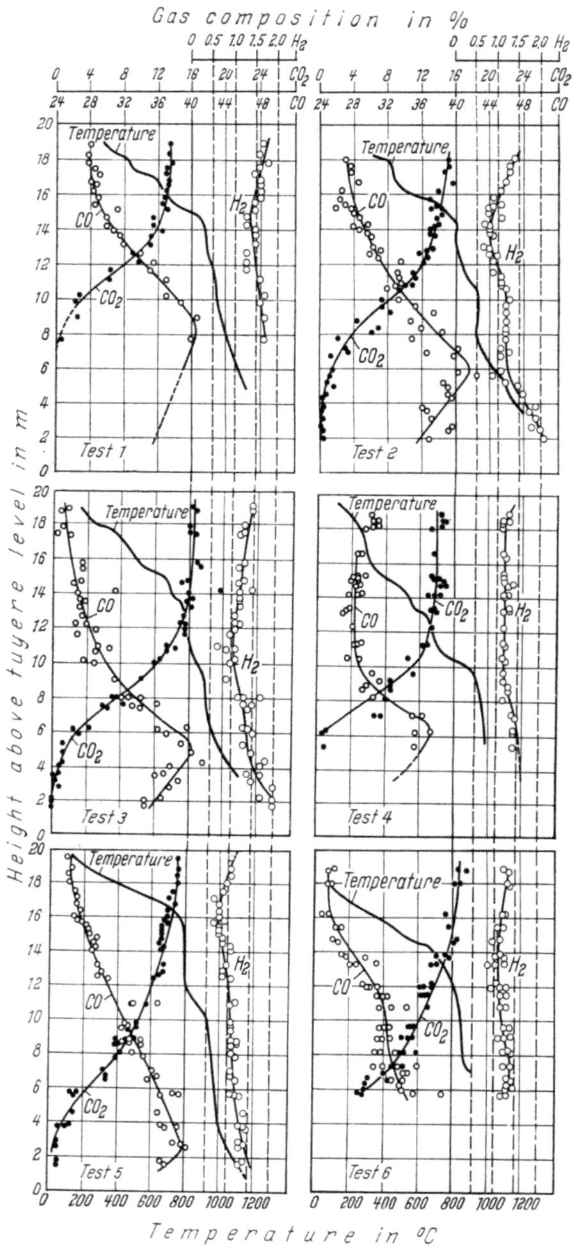

Fig. 279. Temperature profile and gas composition in the blast furnace, according to SCHÜRMANN et al.[559])

the $CO_2/(CO + CO_2)$ values above 400 and 800°C respectively. The trials were supplemented by direct determination of the progress of reduction in a lump of ore. A basket was attached to the probe (Fig. 281). Samples of ore were lowered in this basket and reacted with the ascending gas. Subsequent analysis showed that in the indirect reduction zone, i.e., where the changes in the CO_2 content of the stack gas would be measured, greatly varying degrees of reduction were attained with the same ores, depending

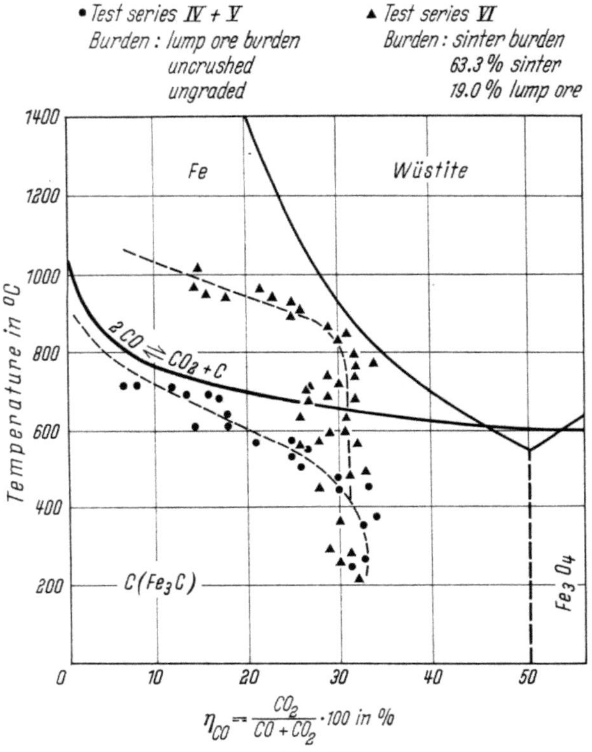

Fig. 280
Variation in gas composition in the Fe-O-C diagram, according to HUPFER and WEIDENMÜLLER[560])

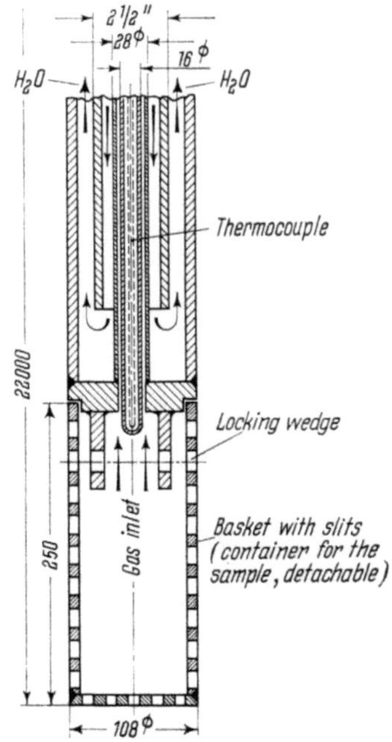

Fig. 281
Water-cooled probe, according to HUPFER and WEIDENMÜLLER[560])

on the average reducibility of the burden and on furnace practice. Figures 282a and b show the dissociated proportion of oxygen $\mathfrak{R}_{CO}$ at 8 m and 14 m respectively above tuyere level.

Because the reduction in each lump of ore in the blast furnace affects, and is affected by, that in every other lump of ore in the furnace, the results of such probe measurements are never valid for other than the particular case in question and are not generally applicable. It may, of course, be generally concluded that the rate at which ores are reduced should be as high as possible at temperatures below 1000 to 1100°C, i.e., in the zone of indirect reduction.

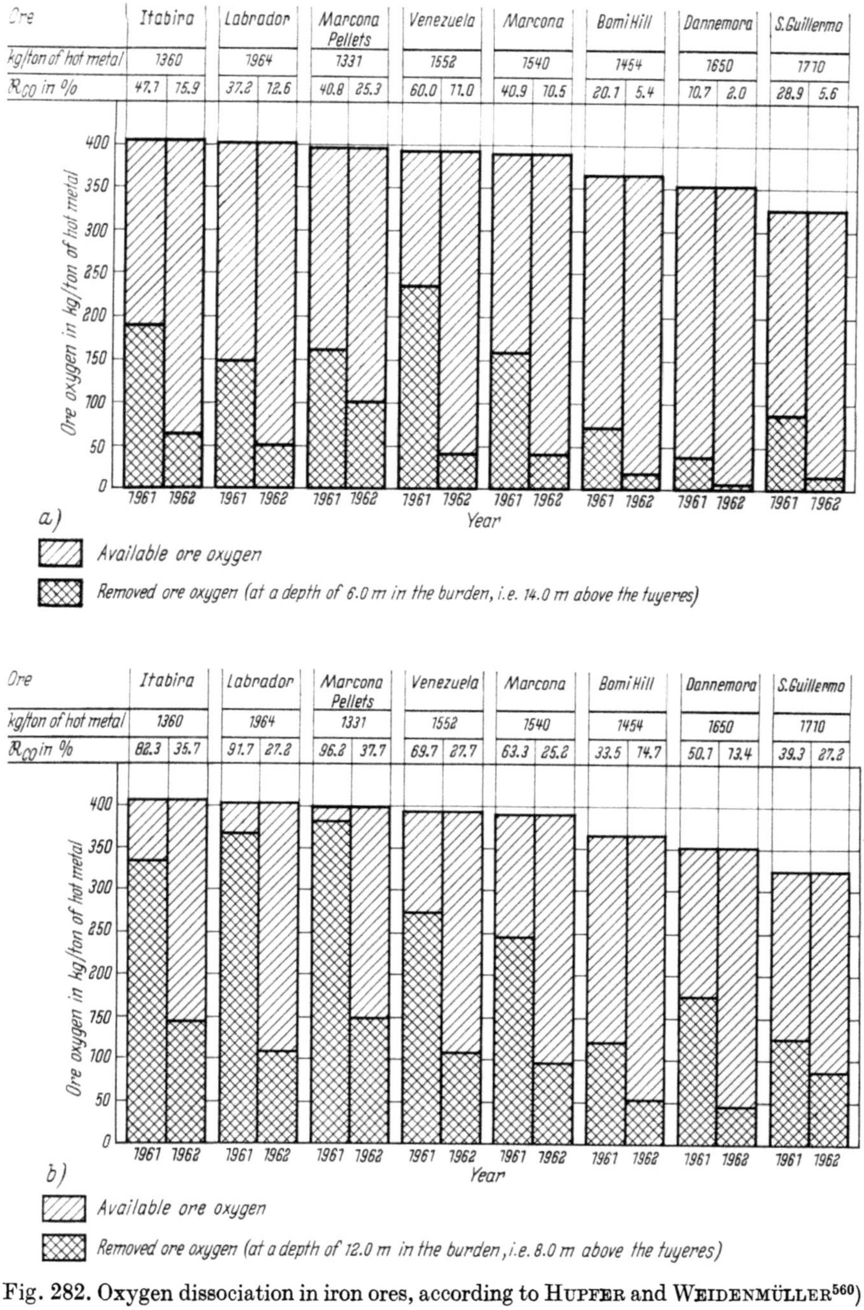

Fig. 282. Oxygen dissociation in iron ores, according to HUPFER and WEIDENMÜLLER[560])
Series of trials 1961: lump ore burden, uncrushed and ungraded
Series of trials 1962: sinter burden; 63.3% sinter, 19.0% lump ore, graded

With lump ores, the rate of reduction can be influenced by way of lump size control as Fig. 283 shows. Reduction occurred in accordance with Stahleisen-Prüfblatt 1770-64[562a]).

The validity of these experimental results for blast furnace operation was tested as described in the following section[562, 605]).

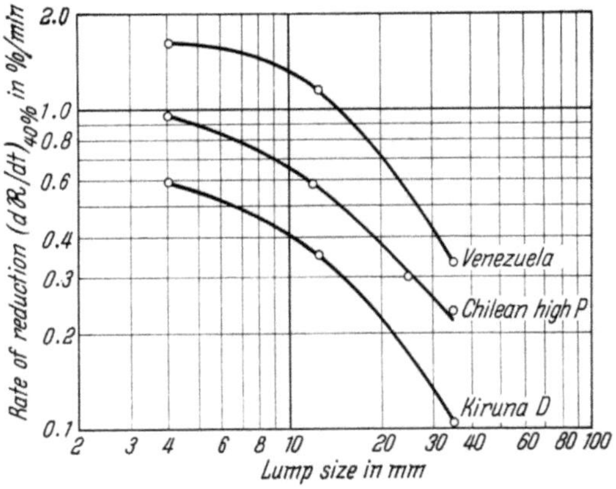

Fig. 283. **Relationship** between lump size and rate of reduction in ores, tested **according** to Stahleisen-Prüfblatt 1770-64 [639])

5.2.2.2.1. Fixing the Upper Limit of Ore Lump Size. The first trial of this nature[562]) took place in a blast furnace with a 6 m hearth diameter. This furnace was charged for a period of eight weeks with an almost constant burden with about 35% consisting of Kiruna ore which is not easily reducible. For the first period of four weeks, this ore was charged in the 6 to 20 mm size and for the second four week period, in the 20 to 60 mm size. Figure 283 shows that the rate of reduction was markedly affected. Table 36 presents the most important data for both periods of the trials. There was a marked drop in the coke rate for a given level of pig iron production when the more readily reducible ore was used. Figure 284 shows that with the small size of Kiruna ore, the coke rate is of the same order as would be expected on the evidence of statistical evaluation for a well graded burden with the same proportion of inert materials, i.e. of the same net burden weight (see the straight line in Fig. 284). These "expected values" will be dealt with in detail in what follows. The heat balance for both periods of the trials (Fig. 285) proves that the saving in fuel was essentially achieved by improved chemical utilisation of the ascending gas.

Similar trials were carried out at Huckingen[605]). A blast furnace with a 6.2 m hearth diameter was given an approximately constant basic

Table 36. *Operating trial with variously graded Kiruna ore*[562])

Particle size of Kiruna ore Period of trial		20 to 60 mm 3rd to 27th April, 1959	6 to 20 mm 3rd to 30th May, 1959
Iron production	tons/24 h	696	693
Coke consumption	kg/ton of iron	876	760
Net burden	kg/ton of iron	2142	2084
Net burden yield	%	46.68	47.95
Flue dust make	kg/ton of iron	73	94
Iron from:			
Kiruna ore	%	36.9	33.1
Wabana (ungraded)	%	3.4	3.7
Sinter	%	19.2	20.5
German ores ($<$100 mm)	%	13.6	11.6
Venezuela ore (6 to 60 mm)	%	4.1	10.6
Metallic charge	%	11.1	7.6
Slags	%	6.3	5.7
Other material	%	5.4	7.2
		100	100
Proportion of $<$1 mm in burden	%	7.4	8.7
Hearth load	kg of coke/(m² h)	898	776
Blast temperature	°C	703	649
Blast pressure	cm Hg	60.3	61.8
Top gas temperature	°C	229	185
Top gas calorific value H_u	kcal/m³ at NTP	1046	951
Composition:			
CO_2	%	7.5	10.7
CO	%	33.3	30.0
H_2	%	1.6	1.8
Iron temperature	°C	1352	1351
Chemical composition of iron:			
C	%	3.71	3.80
Si	%	0.34	0.34
Mn	%	0.82	0.77
P	%	1.83	1.72
S	%	0.059	0.059
Slag volume	kg/ton of iron	639	608
Slag composition:			
CaO	%	43.33	43.08
SiO_2	%	32.39	32.33
FeO	%	0.9	1.07
MnO	%	1.16	1.21
S	%	1.02	1.04
MgO	%	5.10	4.54
P_2O_5	%	0.61	0.41
$Al_2O_3 + TiO_2$	%	11.53	10.75
Basicity CaO/SiO_2		1.34	1.33
Coke (constant) composition:			
C	%		87.84
Ash	%		9.02
Moisture	%		2.11
S	%		0.76

charge with about 38 to 45% of the Fe burden in the form of screened lump ores of varying reducibility. Here again there was a distinct drop in the coke rate when burden components difficult to reduce were replaced by more reducible ones (Fig. 284).

Figure 283 indicates the reducibility of types of ore having a lump size range corresponding to sections *a 1* to *c 2* of the trials, e.g., for Kiruna D

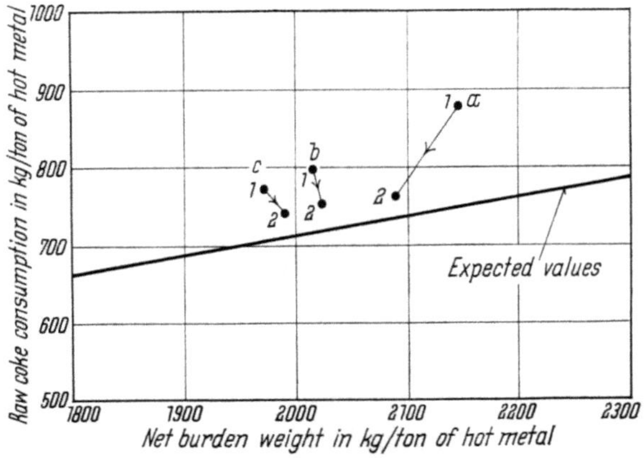

Fig. 284. Influence of reducibility on coke rate (cf. Fig. 283)

Test	Burden Fe contribution from individual components in %	Type	Size range in mm	Works
a 1	36.9	Kiruna D	20 to 60	Oberhausen
a 2	33.7	Kiruna D	6 to 20	
b 1	38.0	Kiruna D	8 to 40	Huckingen
b 2	37.5	Venezuela	12 to 70	
c 1	42.0	Chile	15 to 70	Huckingen
c 2	45.0	Chile	8 to 50	

ore 6 to 20 mm with a mean, therefore, of 13 mm particle size, $(d\mathfrak{R}/dt)_{40\%}$ = 0.36%/min (according to Stahleisen-Prüfblatt 1770-64, 40% CO, 60% N_2, 900°C).

These values exhibit a clear correlation with the coke rate (Fig. 286).

Based on the measured reduction behaviour of the ores, the upper limit of lump size should be fixed at 20 to 60 mm depending on reducibility so that a reducibility $(d\mathfrak{R}/dt)$ = 0.3%/min can be attained or exceeded. This applies to burdens of the type used in sections *a 1* to *c 2* of the trials.

5.2.2.2.2. Fixing the Lower Limit of Ore Lump Size. As production trials have shown, the blast furnace can deal with particle sizes down to 3 mm[564]. To obtain a distinct size grading with this particle size requires special measures, however. The most commonly used dry vibrating screens etc. generally fail even when screening out about 5 mm size, the lump

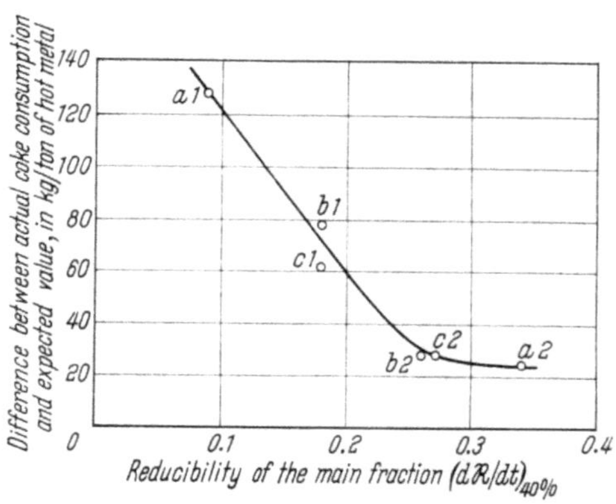

Fig. 286. Relationship between reducibility and coke consumption, according to Stahleisen-Prüfblatt 1770-64 [639])
The main fraction was between 35 to 45% of the burden Fe. The remainder was about constant with average reducibility
Trial periods *a 1* to *c 2* as in Fig. 284

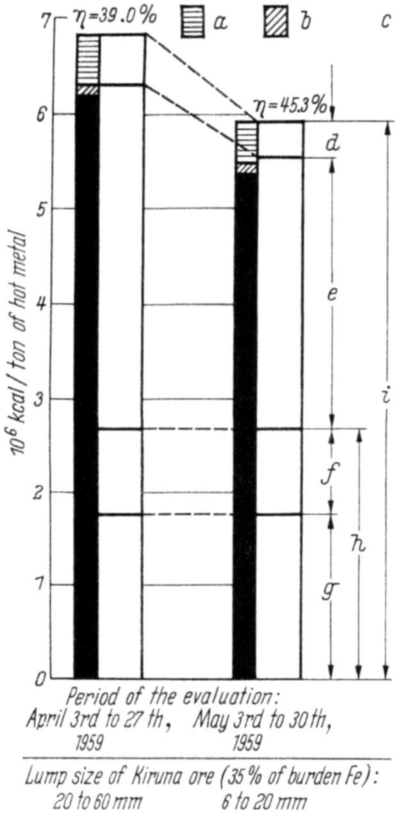

Fig. 285. Heat balance of a blast furnace with a 6 m hearth diameter when charging graded Kiruna ore [523])
a blast heat, *b* heat of solution, *c* coke heat, *d* losses, *e* calorific value, top gas and flue dust, *f* smelting and stack work, *g* reduction and carburisation, *h* effective heat, *i* total heat

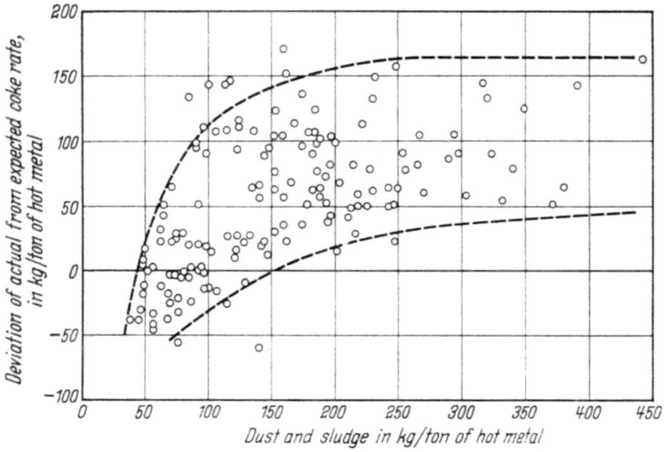

Fig. 287. Deviation of coke rate from the expected value in relation to dust and sludge loss[523])

ore fraction then still containing considerable proportions of off-size, i.e. below 5 mm. Small particle size adversely affects the uniformity of permeability, thus increasing the coke rate and lowering output. Proof of this is offered by Fig. 287 in which the specific flue dust make is used as an indicator of the amount of fines in the burden. Table 37 shows the general deterioration in furnace operating results even with relatively small proportions of fines in the burden [compare also references [565, 566, 604])]. It is, therefore, most important to fix the lower screen size limit so that complete separation of the fine particle size from the burden will be ensured.

5.2.2.2.3. Method of Operating Ore Crushing and Screening Plants. When lump ores are crushed in normal operational cone and roll crushers, a good deal of extra fines below 5 mm is incurred. These fines must be screened out and subsequently agglomerated by sintering. This gives rise to considerable costs. There is, therefore, every incentive to keep the amount of fines arising to a minimum. Operational trials[567]) on a full scale ore crushing and screening plant handling several types of ore showed the considerable extent to which the proportion of fines increases, the smaller the ore is crushed (Figs. 288 and 289).

This means that ores should not be crushed to sizes smaller than what are required to achieve uniform permeability and an adequate rate of reduction. This indicates a size range for the most readily reducible ores of between about 7 and 60 mm and, for the ores difficult to reduce, a range of about 7 to 20 mm. Ores having a particle size range of 7 to 60 mm tend, however, to segregate during handling and when being charged into the blast furnace with all the disadvantageous consequences for furnace operation. It then becomes advisable to provide an additional size grading to subdivide the lump ore into two fractions, e.g., 7 to 30 and 30 to 60 mm. These fractions must be charged separately into the blast furnace. Such practice can frequently not be adopted because the number of ore types which have to be conveyed and stocked separately is thereby greatly increased. In such cases, it is better to crush the lump ores to a greater extent than is necessary even when the ores themselves exhibit good reducibility, to charge the lump ore fraction in the now more closely defined size range without separating it, so that uniform permeability is ensured and to accommodate the higher sinter costs. Both methods of operation are used in practice. The first is used particularly when only relatively few types of ore are smelted simultaneously. Details can be obtained from the extensive literature on the subject [see summaries [524, 562, 568]) and individual reports [569–572])]. It is clear that when the burden is cleaned up and only a few types of ore are treated, considerable technical and economic advantages may be expected.

Table 37. *Operating data before and after screening out of pellets in Bethlehem*[565])

		Before screening	After screening
Month		May 1962	August 1963
Blast furnace		B	D
Hearth diameter	m	8.76	8.76
Monthly output	tons	83575	91292
Output	tons/day	2696	2945
Specific furnace output	tons/(m^2 day)	44.7	48.8
Coke consumption	kg/ton of iron	586	552
Net burden yield	%	61.5	61.4
Composition of burden:			
Grace pellets	kg/ton of iron	997	762
Own sinter	kg/ton of iron	345	648
Screened Erie ore	kg/ton of iron	77	—
Bought sinter	kg/ton of iron	—	11
Open hearth slag	kg/ton of iron	113	68
Scrap	kg/ton of iron	86	64
Limestone	kg/ton of iron	87	102
Gravel	kg/ton of iron	—	12
Gross burden	kg/ton of iron	1705	1667
Blast furnace flue dust (including filtration sludge)	kg/ton of iron	78	39
Slag make	kg/ton of iron	268	274
Blast volume	m^3/h at NTP	189000	200000
Hot blast temperature	°C	724	777
Top gas temperature	°C	238	232
Top pressure	atm gauge	0.18	0.20
CO/CO_2 ratio		1.50	1.47

Size grading of Grace pellets

Particle size	"As received" (unscreened) % by weight	After screening % by weight
over 15 mm	12 to 15	13
10 to 15 mm	70 to 75	80
6 to 10 mm	5 to 8	5
under 6 mm	7 to 8	1 to 2

Size grading of sinter

Particle size	May 1962 % by weight	August 1963 % by weight
over 25 mm	20	25
6 to 25 mm	70	68
under 6 mm	10	7

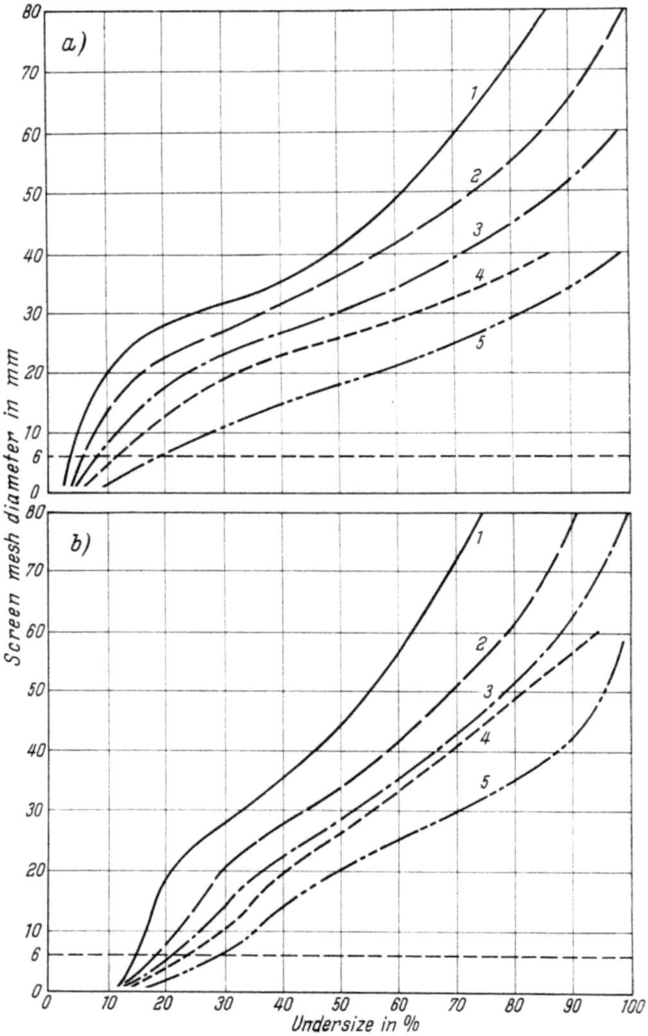

Fig. 288. Variation in the screening data when crushing ore, according to HERR[567])

a) Venezuela ore

1 "As received" condition, *2* After cone crusher, *3* to *5* After cone crusher and roll crusher, width of gap 42.5 mm (*3*), 32.5 mm (*4*), 22.5 mm (*5*)

b) Kiruna ore

1 "As received" condition, *2* After cone crusher, *3* to *5* After cone crusher and roll crusher, width of gap 35 mm (*3*), 25 mm (*4*), 15 mm (*5*)

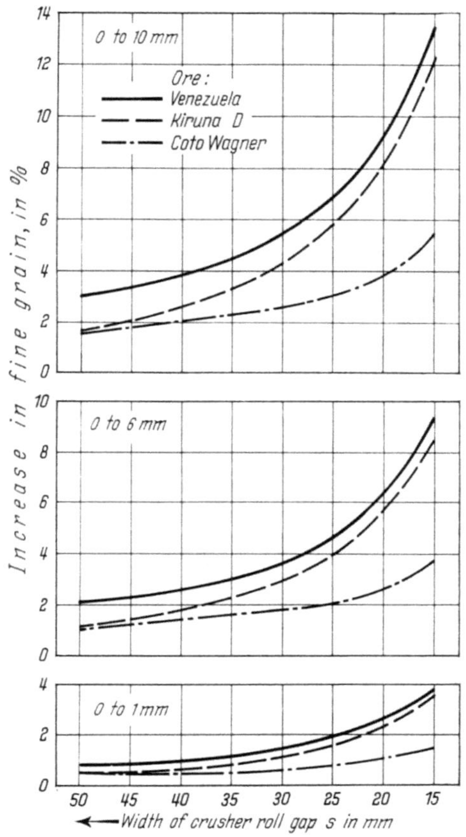

Fig. 289. Resulting proportion of fines when crushing ores with the same initial grain size range in relation to the width of gap *s*, according to Herr[567])

5.2.2.3. Treatment of Fine Ores

In the sintering process, in addition to agglomeration, it is possible to change the ore mineralogically so as to produce phases most acceptable to the blast furnace. These conditions are essentially controlled by additions to the sinter mix, by the proportion of return fines and by way of the fuel content which governs the temperature and the CO and CO_2 contents of the gas phase within the mix during the combustion process. Figure 290 indicates that oxidising, reducing and again oxidising zones are formed one above the other in the bed during sintering. The extent of these zones depends on the extent of the fuel addition and has a decisive effect on sinter quality.

5.2.2.3.1. Nurturing the Most Suitable Phases for Reduction during the Sintering Process. Other things being equal[575, 576]), the rate of reduction

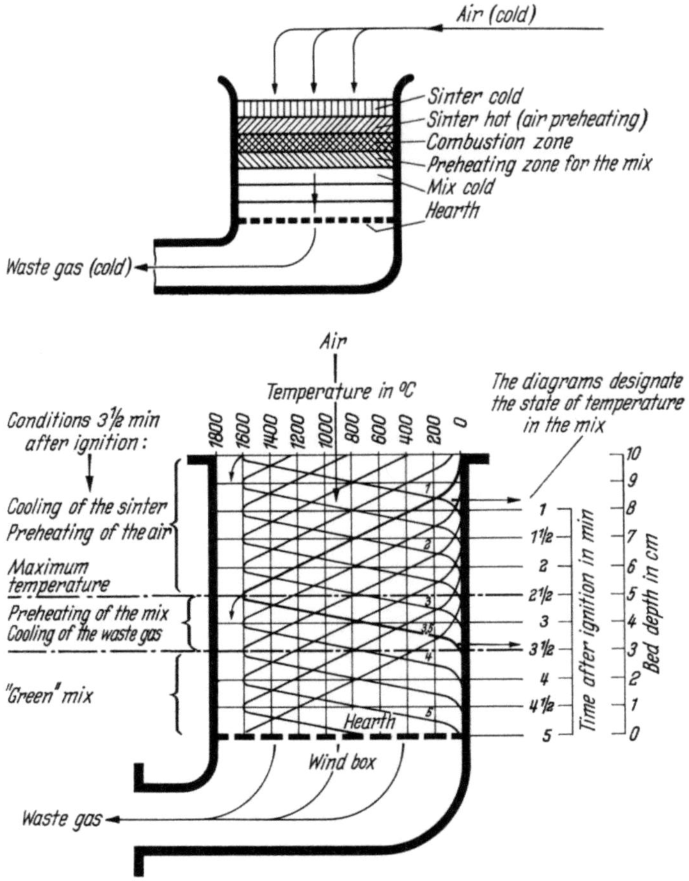

Fig. 290. Diagram of temperature conditions in sections of the sinter bed (a single batch) during down-draft sintering, according to Wendeborn[573])

is determined by equations of the following type:

$$v = k\,F\,(C^{eq}_{CO_2} - C^{0}_{CO_2}),$$

in which

F = gas-solid interface available for reaction,
k = velocity constant depending on temperature, degree of reduction, etc.,
$C^{0}_{CO_2}$ = CO_2 content of the gas phase,
$C^{eq}_{CO_2}$ = maximum possible CO_2 concentration at equilibrium.

The factors k, F and $C^{eq}_{CO_2}$ can be influenced by the sintering process.

It is expedient to produce phases compatible with the highest possible $C^{eq}_{CO_2}$-values. Investigations by Schenck et al.[577, 578]) show that these values are very different for the most important phases, namely for pure iron oxides and for compounds of iron oxides with CaO and SiO_2 (see section 1.1.4, Fig. 21).

Starting from pure iron oxides, the highest $C^{eq}_{CO_2}$ values are to be found in the important wüstite range. The combination of FeO and SiO_2 is the least favourable.

It further follows from the above equation that the reaction interface F, i.e. the surface of the sinter particle should be as large as possible. With low fuel additions and consequently low sinter temperatures, higher porosity and thus high F values can be attained as VOR DEM ESCHE and STEINHAUER[579]) have indicated. A limit is fixed because the sinter must have sufficient strength to withstand handling and transportation and the stresses in the blast furnace.

Finally, the reaction rate constant k can be influenced. According to recent investigations, k has a particularly high value when the FeO arising as an intermediate product of reduction exhibits a high degree of disorder. This usually occurs when the Fe in the sinter is found in the form of Fe_2O_3 instead of Fe_3O_4. Moreover Fe_2O_3 does not combine so strongly with the components of the gangue (Al_2O_3, SiO_2, MgO) during sintering as does FeO which forms spinel with Al_2O_3, fayalite with SiO_2 and a solid solution with MgO. Fe_2O_3 combines readily with lime only when it occurs locally in excess of silica and forms, in a smaller amount, compounds or solid solutions with existing spinels such as $CaAl_2O_4$. Moreover, the combination of Fe_2O_3 with CaO merely inhibits reduction with low degrees of reduction. If reduction to FeO is rapid and occurs at low temperature, further reduction of the FeO must be expected to occur more rapidly than the chemical combination of FeO with silica, alumina and magnesia.

As early as 1950, TIGERSCHIÖLD[580]) recognised these relationships empirically when he was seeking high "degrees of oxidation" of sinter, i.e. a high Fe_2O_3 content. It was actually found – apparently paradoxically – that a lower coke consumption was achieved with a high degree of oxidation of sinter, i.e., with a greater supply of oxygen to the blast furnace[580]). Magnetite concentrates with a high Fe content were available as raw material. With small fuel additions during sintering, a complete reaction with atmospheric oxygen

$$2\,Fe_3O_4 + \tfrac{1}{2}\,O_2 = 3\,Fe_2O_3$$

was successfully achieved on the sinter strand. By consistently observing these rules of procedure, Swedish blast furnace operators have, for many years, been able to achieve the lowest coke rates in the world[581, 582]).

The demand for a high degree of oxidation in sinter is synonymous with a low FeO content (Fig. 291). FeO, normally not present in the sinter, is found in wet chemical analysis for bivalent iron, if Fe_3O_4 is present ($Fe_3O_4 = FeO \cdot Fe_2O_3$) or in the presence of fayalite ($2\,FeO \cdot SiO_2$) inter alia. With fine hematite ores, high fuel additions in sintering are associated with a partial reduction to Fe_3O_4. KLEIN[564, 583]) was able to show, with these types of ore, that the best coke rates were attained when the degree of oxidation of the sinter was as high as possible or the FeO content was kept low (Fig. 292). Table 38 presents the most important data on the sinter from the trial periods. In addition to reduc-

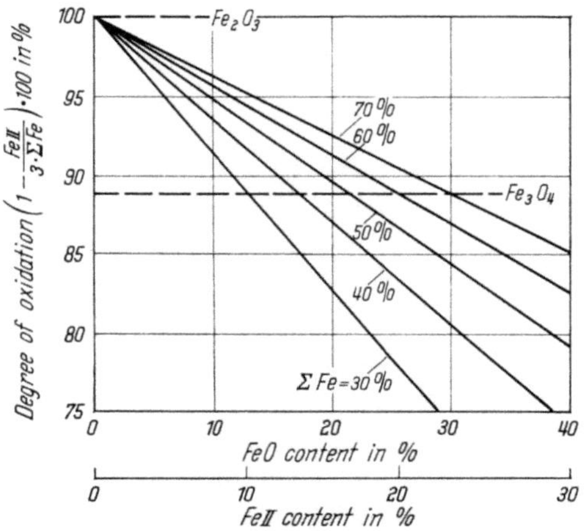

Fig. 291. Degree of oxidation (ratio of the O_2 actually combined with Fe to the maximum O_2 content) in relation to the bivalent Fe (FeII) content and the total Fe content (Σ Fe)

ing the FeO content and thus improving reducibility, the screening technique at the sinter plant was, of course, also improved at the same time so that the effect of improved reducibility is masked. At the time of writing, however, the authors are not aware of any controlled operational trial in which all conditions have been maintained constant and only sinter reducibility has been varied. Such a trial would be of great theoretical and practical importance.

With sinter low in Fe content (under about 45% Fe), further mixed and iron oxide phases, in addition to those already mentioned in section 1.1.4, become increasingly prominent. COHEN[584]), DOI and KASAI[585]), VOR DEM ESCHE and STEINHAUER[579]) and others have determined the composition of these phases[585a, b]).

With these types of sinter, the attainment of adequate strength, even at the cost of reducibility, is decisive as ELLIOTT has convincingly demonstrated on low grade English ores[586]).

5.2.2.3.2. Self-Fluxing Sinter*. For reasons of desulphurisation most blast furnaces produce slags with a basicity range of $p = CaO/SiO_2 = 1.0$ to 1.3. For many years, additions of limestone or dolomite to the burden were, therefore, normal when ores containing SiO_2 formed a large part of the

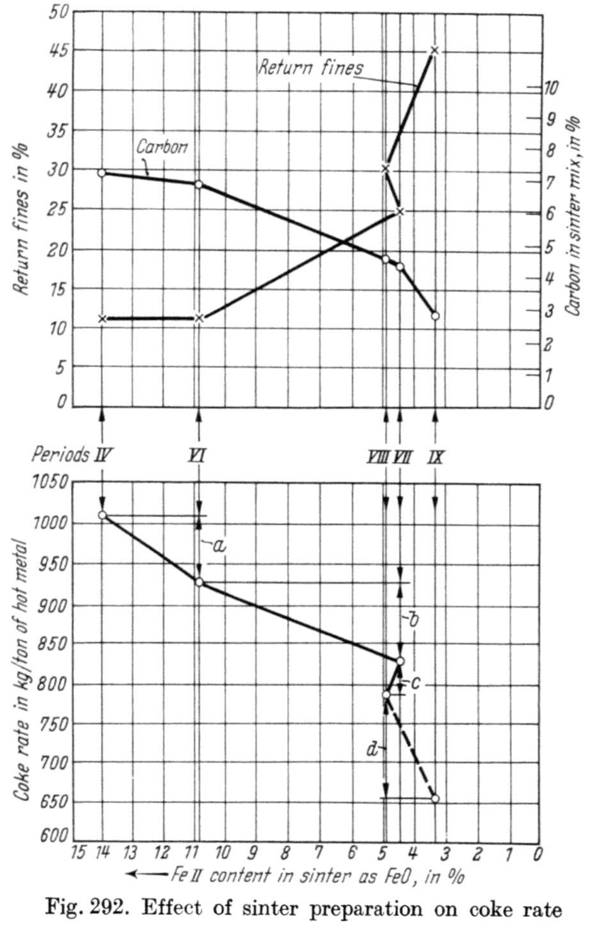

Fig. 292. Effect of sinter preparation on coke rate

a Combined effect of FeO + SiO_2 + screen size, *b* Combined effect of FeO + SiO_2 + lime + rescreening, *c* Effect of rescreening, *d* Combined effect of FeO + experimental rescreening

charge. Gradually, however, it was recognised that considerable advantages ensued if the requisite amounts of CaO or MgO were already incorporated in the sinter so that the increase in basicity could be effected at an earlier stage on the sinter strand. Apart from promoting the slag-forming react-

* Compare detailed survey in reference [525]).

Table 38. *Results of sinter trials[564]). Details of periods IV to IX as shown in Fig. 292**

Period	Sinter mixture %			Sinter analysis %				Production data		
	Coke	Flue dust	Return fines	Fe as Fe_2O_3	Fe as FeO	CaO	SiO_2	Sinter charged	Prod. per furnace per day tons	Coke rate kg/ton
IV. Minus 9.5 mm ore sintered	4.5	7.7	±11	46.0	14.0	0.2	11.5	39.4	458	1010
VI. Minus 3.2 mm ore sintered	3.7	8.1	±11	51.4	10.9	0.2	8.5	40.5	495	927
VII. Minus 3.2 mm ore sintered plus limestone	3.6	3.2	±24.5	54.0	4.5	5.4	7.7	37.9	518	830
VIII. Minus 3.2 mm ore sintered plus limestone plus rescreening . . .	4.0	2.9	±29.7	52.6	4.9	5.5	6.6	38.6	540	790
IX. Minus 3.2 mm ore sintered plus limestone plus experimental rescreening	1.9	2.8	±46.0	55.6	3.4	5.7	7.9	75.0	620	658

* Ore fractions: minus 50.8 mm to 19 mm, minus 19 mm to 3.2 mm.

ion, there are also possibilities of considerable improvements in the reduction reaction.

On the one hand, reducibility is thus increased (Fig. 293 and section 2.7) and furthermore the disturbing effect of the decomposition of the limestone in the blast furnace is avoided. Kinetic calculations and a mathematical/statistical evaluation of numerous operational data[525]) were carried out with the following results:

The decomposition of the limestone

$$CaCO_3 = CaO + CO_2$$

with the high thermal requirement of $\Delta H = 42.4$ kcal/mole $= 965$ kcal/kg CO_2 absorbs considerable quantities of heat from the blast furnace in a sensitive region. At 892°C, $p_{CO_2} = 760$ mm Hg[588]) and above this temperature, the decomposition reaction proceeds with great rapidity at normal pressure. Wuhrer and Radermacher[589]) pointed out that the decomposition of the limestone lumps takes place from the outer to the inner zones and that the heat transfer within the lumps determines the overall reaction. Consequently, considerable superheating of the surface is required and the decomposition times t_d increase as the square of the lump diameter d:

$$t_d \approx d^2.$$

Thus, it follows that limestone is essentially decomposed in the blast furnace only at gas temperatures of between 1000 and 1100°C whereby the stack gas in this range of temperature is relieved of considerable

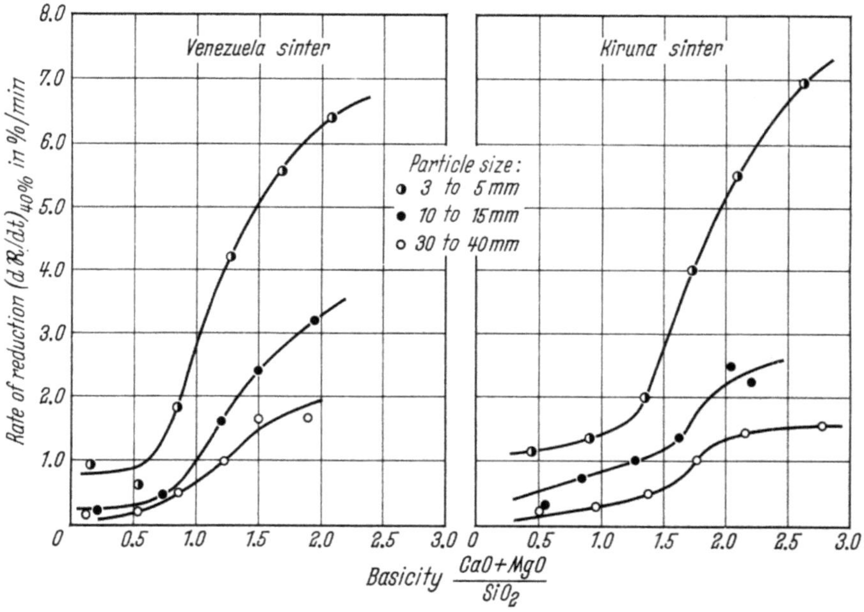

Fig. 293. Relationship between basicity and reducibility for Venezuela sinter and Kiruna sinter, according to MEYER[590])

For comparison, it should be noted that the rate of oxygen removal for a degree of reduction of 40% for Venezuela ore of 10 to 15 mm particle size was about 1.5%/min and for Kiruna ore of 10 to 15 mm particle size, about 0.8%/min

amounts of heat (about 1300 kcal/kg CO_2) (Fig. 294). When the CaO is incorporated in the sinter, thus advancing the decomposition and slag-forming reaction to the sinter strand, the following changes in the thermal requirements of the burden take place:

Heat of dissociation and additional heating calories saved:		1300 kcal/kg CO_2
Heat of slag formation lost:		− 450 kcal/kg CO_2
Heat recovered from CO_2:		− 220 kcal/kg CO_2
	total min	630 kcal/kg CO_2
	total max	850 kcal/kg CO_2

With an effective evolution of heat of 2000 kcal/kg of fuel coke, there is, therefore, a coke consumption of 0.32 to 0.43 kg per kg of CO_2. To this is added a further increment because of kinetic relationships.

During the decomposition of the limestone, the liberated CO_2 (1000 to 1100°C) increases the $C^0_{CO_2}$ in the ascending stack gas, thus retarding reduction*.

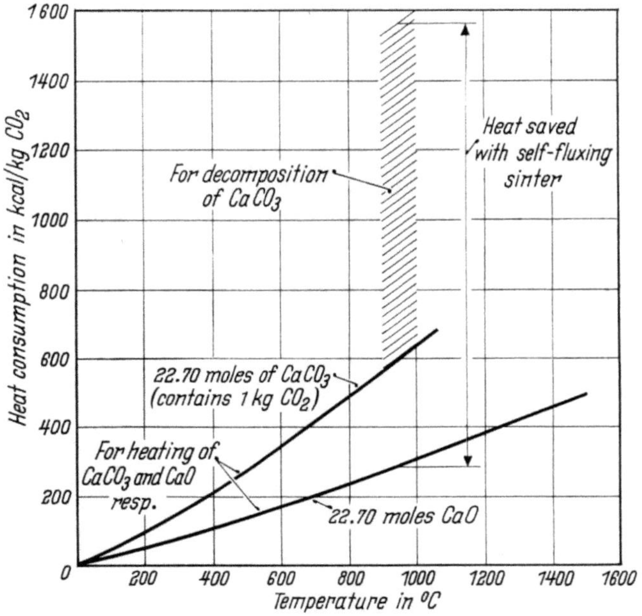

Fig. 294. Heat consumption for decomposition of limestone in the blast furnace [525])

Many systematic operational trials with gradual replacement of the limestone by CaO incorporated in the sinter showed[586, 591, 592]) that the chemical utilisation of the top gas, i.e. the relationship

$$\eta_{CO} = \frac{CO_2}{CO + CO_2}$$

remains constant with varying additions of limestone. A statistical evaluation of the CO_2 contents of numerous blast furnaces operated partly with and partly without limestone, leads to the same conclusion[525]).

It follows therefore that for every mole of CO_2 which arises in the decomposition of limestone, the desired gas-ore reaction is also suppressed

* There is a partial decomposition of the CO_2 in the limestone by coke in accordance with the reaction $CO_2 + C = 2\,CO$ as POKHVISNEV[585e]) found. This does not change the calculation used here to determine coke consumption.

by one mole. On this basis, the deterioration of indirect reduction and the resulting increase in coke consumption can be calculated:

1 kg of CO_2 produces 0.509 m^3 of CO_2 at NTP,

1% release of the total ore oxygen (380 kg) = 3.8 kg O corresponds to the conversion of 5.35 m^3 of CO to CO_2 at NTP.

About 0.36 kg of fuel coke are used to liberate 1 kg of CO_2 and this provides an additional 0.57 m^3 of CO per kg of CO_2. The remaining calculation data and the result are presented in Fig. 295. Thus, it will be seen

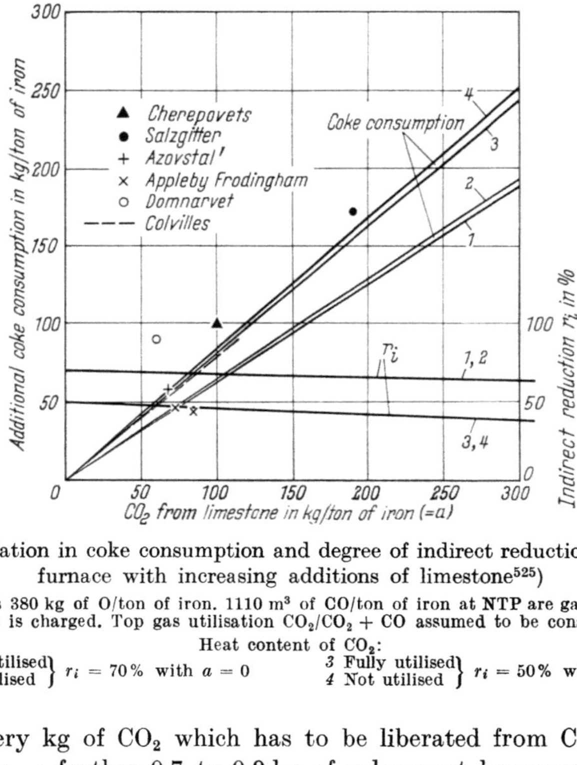

Fig. 295. Variation in coke consumption and degree of indirect reduction in the blast furnace with increasing additions of limestone[525])

Burden contains 380 kg of O/ton of iron. 1110 m^3 of CO/ton of iron at NTP are gasified if no limestone is charged. Top gas utilisation $CO_2/CO_2 + CO$ assumed to be constant

Heat content of CO_2:

1 Fully utilised, 2 Not utilised } $r_i = 70\%$ with $a = 0$; 3 Fully utilised, 4 Not utilised } $r_i = 50\%$ with $a = 0$

that for every kg of CO_2 which has to be liberated from $CaCO_3$ in the blast furnace, a further 0.7 to 0.9 kg of coke must be consumed*. Thus the CO_2 in the limestone may be regarded as by far the most expensive ballast from a fuel point of view.

5.2.2.3.3. Operational Results. A comparison of systematic practical tests with calculations shows general agreement [Fig. 295 and reference [593])].

* With smaller initial coke additions, i.e. a richer burden than that which corresponds to the calculation given as an example, the calculated saving in coke per kg of CO_2 is somewhat reduced as SCHÜRMANN[587]) found on the basis of complementary calculations.

The amount of fuel used for decomposing the limestone in the sinter plant is only a fraction of the coke saved in the blast furnace. About 50 to 70 kg of coke breeze is required per ton of limestone charged[525]). It is, therefore, generally very economical to decompose the limestone in the sinter strand itself and to charge the blast furnace with self-fluxing sinter. As regards slag-forming reactions, too, there are advantages which can, however, not be dealt with in detail here. Prevention of the frequently encountered degradation of self-fluxing sinter during stocking is, on the other hand, important. Attention is drawn to the specialist literature for an account of the steps which must be taken[525]).

When selecting sinter basicity, it must be remembered that sinters with a lime/silica ratio of between 1.3 and 1.8 pass through a minimum level of mechanical stability[545a]) (Fig. 295a). With more highly basic sinters, in addition to a greatly improved mechanical strength, higher outputs and greater reducibility can be attained. A prerequisite for processing highly basic sinter is the simultaneous addition of acid sinter or possibly ores low in lime content.

Tables 39a and b present data on blast furnaces operating with high sinter burdens with a greatly varying ore basis. If the coke rate is plotted against the net burden weight and separately for self-fluxing sinter burden and

Table 39a. *Operating results using*

Blast furnace plant	Blast furnace No.	Sinter in burden %	Net burden kg/ton of iron	Blast temperature °C	Pig iron production t/24 h	Coke (raw) kg/ton of iron	Slag kg/ton of iron	Slag $\frac{CaO}{SiO_2}$	Blast furnace Hearth diameter m	Blast furnace Working volume m^3
Appleby Frodingham 1953 . .	9/10	95	2700	624	655	880	1283	1.09	7.62	1131
Appleby Frodingham 1957 . .	Queen Victoria	98	2513	683	1250	770	1250	1.20	8.7	1256
Dillingen	2	96	2224	675	495	721	892	1.21	5.0	485
Domnarvet . . .		100	1780	895	438	609	430	1.30	4.5	221
Kokura	1	94	1753	836	555	570	439	1.16	5.6	470
Magnitogorsk .	B	85	1670	880	2344	633	471	1.09	8.0	1371
Salzgitter . . .		73	2602	813	722	814	1123	0.68	6.3	860
Seraing	7	94	2163	848	444	687	820	1.21	5.0	497
Steel Co. of Canada, 7th to 27th March, 1958	B	100	1764	690	952	624	480	0.95	5.49	587

for burdens with limestone additions (also completely screened and/or sintered), two scatter bands are obtained (Fig. 296). The regression lines for these data are calculated as follows:

a) self-fluxing sinter burden

$$K = 175 + 0.247\ M \quad (\%\ \text{explained} = 92\%),$$

b) limestone burdens

$$K = 200 + 0.257\ M \quad (\%\ \text{explained} = 82\%).$$

Here K indicates coke consumption in kg/ton of pig iron and M, the net burden weight in kg per ton of iron.

The % explained is relatively very high, the residual scatter correspondingly small. The regression line for self-fluxing sinter runs about 45 kg of coke per ton of iron below the other curve corresponding to the case where CO_2 is added combined with limestone at the average rate of 113 kg/ton of iron. When changing over from self-fluxing sinter burden with $M = 2000$ kg/ton (Fig. 296 contains an example of a reading) to a charge which is not self-fluxing with suitable limestone additions (113 kg CO_2/ton of iron), the coke rate is raised by a total of 77 kg/ton, i.e. about 0.7 kg of coke per kg of CO_2 and here again there is satisfactory agreement with the calculated figures.

burden free from limestone[525])

Hearth area (m^2)	Coke throughput (tons/m^3/24 h)	Coke throughput (tons/m^2/24 h)	Top gas: Volume m^3/ton of iron at NTP	Top gas: Calorific value kcal/m^3 at NTP	Top gas: CO %	Top gas: CO_2 %	Special features
45.6	0.51	12.6	3510	896	~28.8	~ 9.5	Additional tuyeres in bosh
59.4	0.77	16.2					Regular checking practice
19.6	0.74	18.2	2720	864	27.3	11.7	
15.9	1.21	16.8	2090	783	25.0	17.5	
24.6	0.673	12.8	1990	720	22.8	17.2	
50.2	1.08	29.6				~15.5	19.69 g H_2O/m^3 of blast at NTP 0.96 atm gauge top pressure
31.2	0.68	18.8	2930	935	29.5	11.0	
19.6	0.61	15.6	2660	821	26.1	12.6	
23.7	1.01	25.1	2420	767	24.6	16.0	Slag contained 13.3% MgO

Blast furnace plant	Blast furnace No.	Sinter in burden %	Net burden kg/ton of iron	CO_2 from limestone kg/ton of iron	Blast temperature °C	Pig iron production tons/24h	Coke (raw) kg/ton of iron	Slag kg/ton of iron	Slag CaO/SiO_2
German Federal Republic									
HOAG	A	27	2120	89	735	1524	725	573	1.32
HOAG	8	64	1830	84	820	486	644	402	1.32
Hoesch	6	70	1769	2	855	1526	619	400	1.29
Mannesmann	3	~50	2336	143	840	830	789	840	
Mannesmann	3	~50	2341	139	840	814	805	831	
Mannesmann	3	~50	2376	140	785	821	811	807	
Phoenix-Ruhrort	1	60	1751	11	804	852	606	365	1.22
Salzgitter	4	69	3016	194	706	514	1045	279	1.09
Thyssen	10	50	1901	37	864	1720	662	1505	1.28
Völklingen		57	2669	116		420	775		
Great Britain									
Appleby Frodingham		0	3570				1190		
Appleby Frodingham		35	3090	~170	438	508	1130	1195	1.18
Appleby Frodingham		42	3020	~142	491	525	1100	1195	1.13
Appleby Frodingham		63	3000	~ 97	562	647	980	1280	1.11*
Colvilles Clyde		44	1843	79	699	580	663	341	1.47
Colvilles Clyde		50	1731	42	677	656	621	350	1.48
Colvilles Ravenscraig		47	1842	55	759	1042	655	421	1.31
Dorman Long		98	1812	94		575	689	~470	
Dorman Long		2	2211	141		538	824	~550	CaO/SiO_2
Steel Co. Wales Margam	3	63	2200	~ 75	612	970	738	558	1.34
Minette ore smelting									
Belval		65	2890		794	457	988		1.22
Chiers	4	51	2890		725	378	875		
Knutange Fontoy		63	2670		~750	400	825		1.45
Mont St. Martin		24	3446	~250	710	322	997		1.35
Mont St. Martin		32	3146	~205	641	407	884		1.34
Mont St. Martin		44	2878	~152	664	451	811		1.32
Sweden									
Domnarvet		95	1800	60		157	700	~400	
Soviet Union									
Azovstal		79	2135	102	834	1430	822	765	1.13
Cherepovets	2	100	~2055	~ 70	950		657	815	1.15
Magnitogorsk	C		1872	36	848		680	513	1.06
U.S.A.									
Algoma	5 1954	61	~2050	138	529	879	671	521	
Edgar Thomson		21	2071	126	507	1442	812	477	1.28
Fontana	1	42	~1900		682	1205	679		
Gary	12	82	1863	61	660	1160	684	474	1.04

* $\frac{CaO + MgO}{SiO_2}$

Blast furnace: Hearth diameter m	Blast furnace: Working volume m^3	Hearth area m^2	Coke throughput tons/m^3/24 h	Coke throughput tons/m^2/24 h	Top gas: Volume m^3/ton of iron at NTP	Top gas: Calorific value kcal/m^3 at NTP	Top gas: CO %	Top gas: CO_2 %	Special features
8.5	1450	56.7	0.76	19.5	2680	919	28.9	12.8	
5.6	673	24.6	0.47	12.7	2450	830	26.2	13.6	
8.0	1359	50.2	0.65	18.8		904	28.6	14.7	O_2/H_2O addition
6.1	664	29.2	0.99	22.4	3010	977	30.9	10.9	
6.1	664	29.2	0.99	22.4	3071	985	31.1	10.4	
6.1	664	29.2	1.00	22.8	3118	976	30.6	10.6	
5.5	561	23.7	0.92	21.8	2160	823	25.8	13.8	
6.3	860	31.2	0.62	17.2	3900	914	29.5	11.3	
9.0	1424	63.6	0.80	17.9	2410	901	28.5	13.1	
5.8	557	26.4	0.58	12.3					
7.62	1131	45.6							
7.62	1131	45.6	0.51	12.6	4830	901	~29.1	~ 9.5	
7.62	1131	45.6	0.51	12.7	4610	910	~29.0	~10.2	
7.62	1131	45.6	0.56	13.9	4030	919	~29.7	~ 9.8	
5.57		24.4		15.8		~825	~26.5	~14.4	
5.57		24.4		16.7		~792	~25.4	~15.5	
7.85	1265	48.4	0.54	14.1		~770	~24.7	~16.2	
5.49	603	23.7	0.66	16.7	2600	~826	26.5	15.2	
5.49	603	23.7	0.74	18.7	3300	~886	28.5	12.9	
7.86		48.4		14.8		~870	27.8	14.6	Checks every 20 min
5.5		23.7		19.1		1001	32.0	7.65	
5.0		19.6		16.9					
5.0		19.6		16.8		830			
4.8	623	18.1	0.51	17.7		975	29.7	10.4	
4.8	623	18.1	0.58	19.9		915	27.8	13.2	
4.8	623	18.1	0.59	20.2	~3140	914	27.9	12.4	
3.1	140	7.5	0.78	14.7					
8.0	1300	50.2		23.4	~3010	~1000			High top pressure furnace
			1.00					14.8	
					2290	887	27.9	13.4	
7.62		45.6		12.9					
8.4	1260	55.4	0.93	21.1	3330	820	26.3	13.2	
7.77	1300	47.5	0.63	17.3		~740	~23.8	~16.8	Ore <25 mm
7.6	951	45.3	0.83	17.5	~2695	845	26.1	12.7	

The lines plotted in Fig. 296 can be used, because of the low residual scatter, to predict approximate coke rates for a given burden. These figures are, therefore, also known as "expected values".

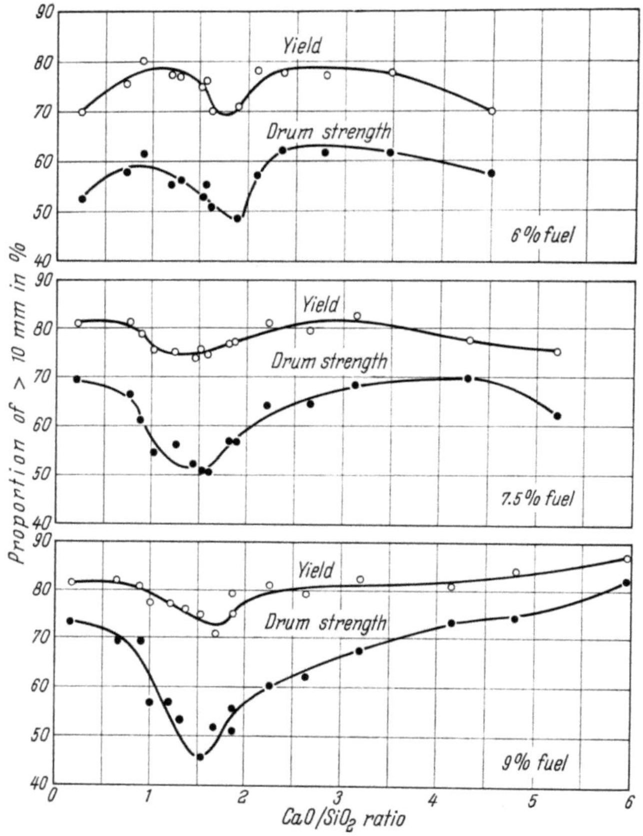

Fig. 295a. Yield of >10 mm and proportion of >10 mm after the drum test in relation to the CaO/SiO_2 ratio, according to WINZER and SCHMITZ[545a])

Figure 297 shows how the data for the entire pig iron production of a plant gradually approach these expected values when all the stipulated measures are logically applied. Each point corresponds to a monthly production of about 100 000 tons of basic Bessemer iron. During the time interval designated as period 1, there was no crushing and screening plant and the sinter screening installations were inadequate. During period 2, ore grading was commenced. The ores were crushed in accordance with their reducibility, the fines were screened out and the remaining lump ore was fed as far as possible in two fractions. Period 3 is characterised by a large

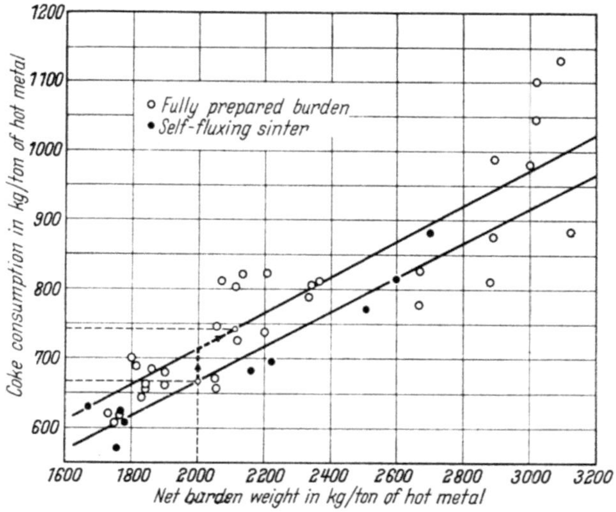

Fig. 296. Coke consumption in relation to net burden weight[525])

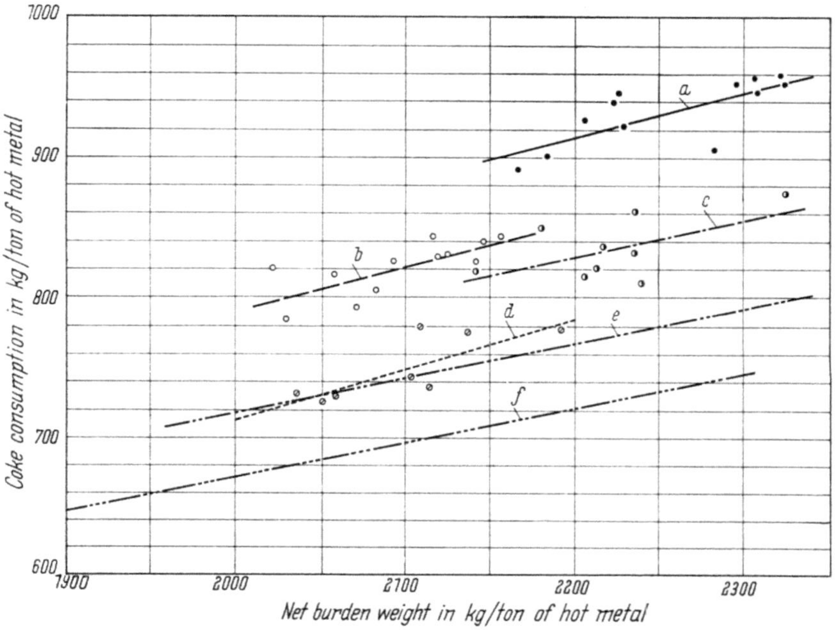

Fig. 297. Coke consumption and net burden weight for basic Bessemer pig iron with various types of burden preparation[523])

a 1st period, *b* 2nd period, *c* 3rd period, *d* 4th period, *e* Expected values with fully graded and sintered burden, *f* Expected values when self-fluxing sinter is charged blast furnace, the more favourable lines of which facilitated lower coke rates. Finally, in period 4, sinter with markedly improved size grading was available. In line with these developments, the coke rate is displaced parallel to the initial curves, till finally the "expected values" are reached.

5.2.2.4. Treatment of Very Fine Ores

5.2.2.4.1. Processes Occurring in the Pellet during Firing. Table 40 presents a typical screen analysis of an "as received" ore and Tables 41a to c, the properties of several kinds of finished pellets. When the green pellets

Table 40. *Example of screen analysis of fine ore suitable on the one hand for sintering and on the other hand for pelletising*[546])

Oversize with DIN screen	Fine ore suitable for sintering	pelletising
mm	% of size fraction	
10.0	4.1	—
8.0	2.3	—
3.0	11.7	—
1.0	7.0	—
0.5	27.1	—
0.3	18.2	—
0.15	20.3	0.4
0.09	2.1	11.6
0.075	4.1	—
0.06	3.1	11.3
under 0.06	—	76.7

are fired, care must be taken to ensure that the sintering atmosphere contains sufficient free oxygen to allow the oxidation of Fe_3O_4 to Fe_2O_3 to be completed or, in the case of an addition of Fe_2O_3, to ensure that no reduction takes place. Because of the varying crystal systems of the Fe_3O_4 and Fe_2O_3 phases, complete recrystallisation takes place during oxidation and this has a strengthening effect. By means of metallographic examination, Cooke et al.[595a, b]) were able to ascribe the strengthening mechanism during firing to well pronounced recrystallisation processes. Figure 298 shows that Fe_3O_4 particles coalesce even at 800°C, being transformed into Fe_2O_3 so that with appropriate treatment (induration at 1200 to 1300°C with grain coarsening) high strengths can be achieved. On the other hand, with Fe_2O_3 (Fig. 299) a strengthening recrystallisation will begin only at very high temperature (above 1300°C). Firing times are therefore longer with Fe_2O_3 ores[547]). Fig. 300 indicates the possible strengthening mechanisms.

Table 41 a. *Chemical composition of various types of pellets*[547])

Pellets	Fe	FeO	Mn	P	SiO_2	CaO	MgO	Al_2O_3	S	H_2O
	Proportion %									
Shaft furnace pellets										
Erie	62.13	0.38	0.37	0.025	9.33	0.91		0.19	0.007	3.39
Grace	56.60		0.07	0.005	2.80	0.50	1.80	0.70		
Hilton	64.80				2.85	0.25	2.61	0.29	0.01	1.30
Malmberget	68.50	1.03	0.05	0.04	0.70	0.50	0.25	0.40	0.04	
Marmora	63.71	0.73	0.09	0.005	4.14	1.80	0.97	0.85	0.012	0.75
Oskarshamn	60.85		0.03	0.08	7.16	2.20	0.78	1.85	0.05	
Grate kiln pellets										
Humboldt	62.18		0.21	0.05	8.48	0.66		0.43	0.009	2.63
Republic Mine										
Acid pellets	63.80	0.75	0.39	0.07	7.40	0.29	0.81	0.50	0.005	1.00
Pellets low in SiO_2	65.00	0.75	0.07	0.39	4.80	0.29	0.81	0.50	0.005	1.00
Self-fluxing pellets	59.94	0.31	0.038	0.036	4.40	4.31	0.47	0.46	0.037	3.33
Straight grate pellets										
Silver Bay	63.00	2.58	0.18	0.06	7.82	1.00	0.60	0.20	0.04	3.28
Ungava	65.40	1.81	0.25	0.01	4.53	0.56	0.33	0.50	0.02	

Table 41 b. *Grain size distribution of various types of pellets*[547])

Pellets	Size grading mm												
	>30	>20	>15	>12	>10	>8	>6	>5	>3	>2	>1	>0.5	>0.2
	Proportion %												
Shaft furnace pellets													
Oskarshamn	18.5				73.5		74.7		84.7		91.0	92.6	94.3
Malmberget	5.0	41.7			78.7			84.3					
Marmora		5.10		42.0	69.9		82.4		88.5		91.6		
Erie		2.0		11.0	56.6			87.5	93.5		96.0		
Hilton		0.6		10.6		83.9	93.7			97.3			97.7
Grace		2.2	10.7	37.7		82.4	93.6		97.1				
Grate kiln pellets													
Republic Mine													
Acid pellets			17.0	68	90.0	90.0	98.0				98.5		
Self-fluxing pellets	2.7			78.8		95.0	97.0			98.2			98.3
Humboldt				53.3	92.7			97.8	98.5		98.9		
Straight grate pellets													
Silver Bay					85.35	88.19		91.14		94.59	95.17		

Table 41 c. *Physical properties of various types of pellets*[547])

Pellets	Bulk density	Density	Porosity	Average crushing strength	Smelting behaviour		
					Softening point	Melting point	Flow point
	g/cm³	g/cm³	%	kg	°C	°C	°C
Shaft furnace pellets							
Oskarshamn .	3.10	4.70	26.00	180			
Malmberget .	3.94	5.14	23.40	756	1560	1615	1660
Marmora . . .	3.60	4.89	25.80	335	1500	1545	1640
Straight grate pellets							
Silver Bay . .	3.21	4.72	31.90	146	1323	1525	1635
Ungava . . .	3.30	4.88	32.70	140	1465	1540	1575

In pellets made of magnetite concentrates, the oxidation zone migrates slowly into the interior of the piece whereby in air, as EDSTRÖM[595]) pointed out, the diffusion of oxygen through the already oxidised layer has a rate controlling effect. Calcium hydroxide and bentonite have proved to be particularly suitable bonding agents which, even in small (about 0.5%)

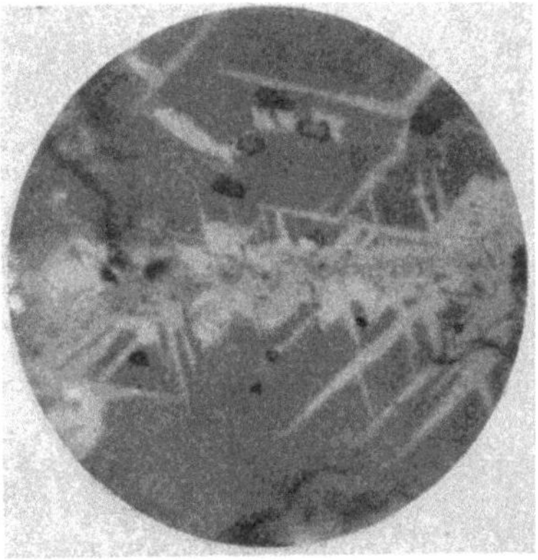

Fig. 298. Hardening in magnetite concentrate pellets: line of cementation between two magnetite blocks (heated to 800°C in air for 2 hours followed by furnace cooling) (1250 : 1, original 2000 : 1); according to COOKE and STOWASSER[595b])

additions, greatly reduce the requisite firing temperatures by means of chemical reaction[594]). With the purest concentrates, OTTOW believes it to be expedient to carry out the firing process in a neutral atmosphere so that the finished pellets consist mainly of Fe_3O_4. In this case, although reducibility is certainly adversely affected, reduction strength is good[595d]).

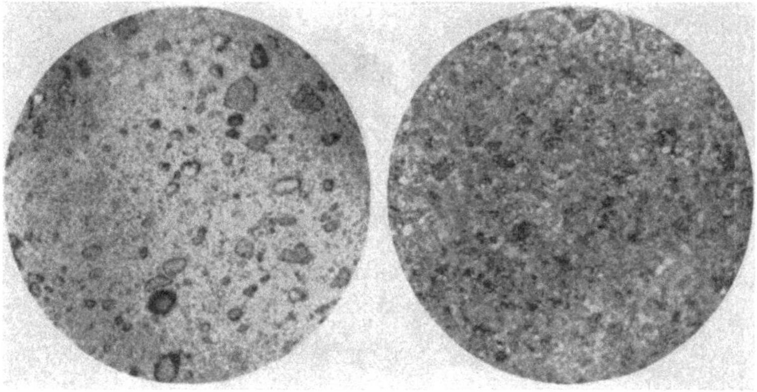

a)

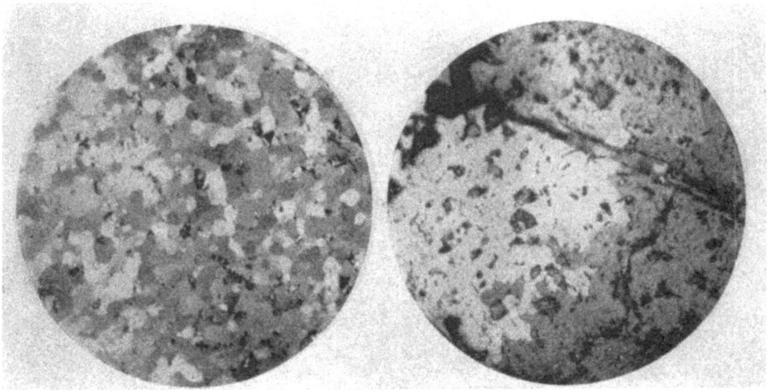

b)

Fig. 299. Hardening in hematite concentrate pellets: recrystallization with rising temperature, analytical grade ferric oxide compacted and heated for 30 min (crossed nicols, 125 : 1, original 200 : 1); according to COOKE and STOWASSER[595b])

a) Left: 1100°C, average grain size 6 μm. — Right: 1200°C, average grain size 7 μm
b) Left: 1300°C, average grain size 20 μm. — Right: 1400°C, average grain size 400 μm

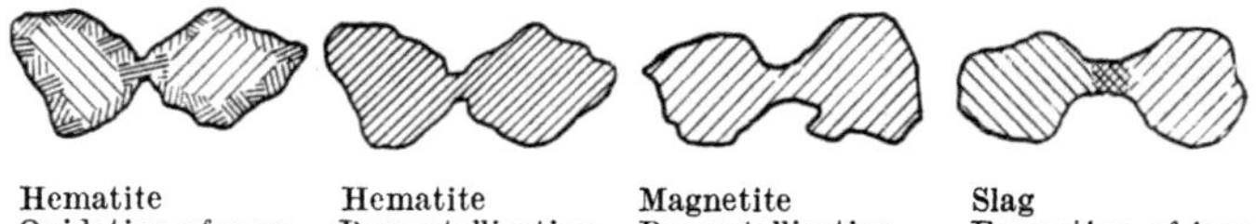

Composition of bond:	Hematite	Hematite	Magnetite	Slag
Mode of formation:	Oxidation of magnetite	Recrystallization of hematite	Recrystallization and grain growth of magnetite	Formation of iron silicates
Atmosphere:	Oxidizing	Oxidizing	Neutral	Neutral or slightly oxidizing
Temperature of formation:	>200 °C	>1100 °C	>900 °C	>1000 °C

Fig. 300. Different bonds between single grains in a pellet made from magnetite concentrate, according to TIGERSCHIÖLD[545])

Concentrates containing 3 to 8% SiO_2 can be made into self-fluxing pellets by means of $CaCO_3$ additions. This, however, means that the $CaCO_3$ must be ground and this is a very costly process [see, for instance, references [595b, 597a]]. Difficulties are also encountered with sticking during the indurating operation[607]).

5.2.2.4.2. Behaviour of Pellets during Reduction. The reduction rate of pellets has frequently been investigated [e.g., references [596, 601]]. Because they are usually very porous, pellets are generally reduced relatively quickly. Figure 301 shows the cumulative frequency distribution of the

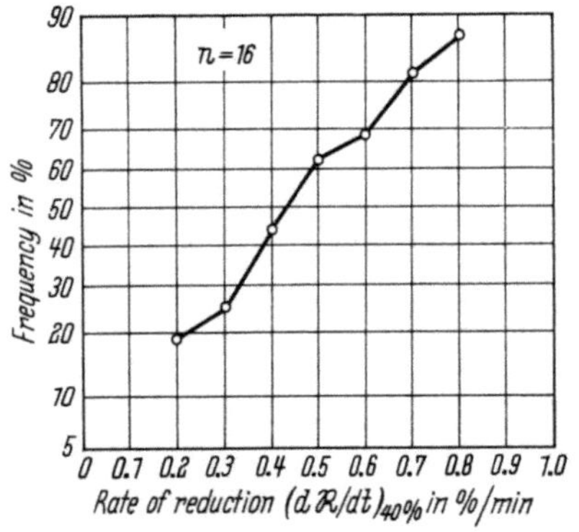

Fig. 301. Frequency distribution of the reducibility of ores and agglomerates (particle size 10 to 15 mm), tested according to Stahleisen-Prüfblatt 1770-64[597])

reducibility (measured as described in section 5.3) for a typical selection of ores and agglomerates of 10 to 15 mm particle size. For a given range of diameter, pellets will be in the top third of the measured reducibility range (Silver Bay pellets at 0.68%/min).

In the reduction of commercial hematite pellets under laboratory conditions, an unusual increase in volume[595c]) (of up to 340%) and a pronounced tendency to degradation were observed in some cases[595f, g]). With the increase in the apparent volume of the pellets ("swelling"), there is a reduction in the mechanical strength which causes an undesirable degradation of the pellets with consequent stoppages and a drop in output[595h, i]) when pellets susceptible to swelling are reduced on a full production scale. The greater the tendency of the pellets to swell, the greater their proportion in the burden and the higher the specific blast furnace productivity, the greater is the drop in output[595j]).

Swelling and the associated tendency to degradation can not yet be explained quantitatively. Investigations of the conditions under which swelling occurs revealed a number of mechanisms which partly explain these processes. In addition, some possibilities of reducing this susceptibility to degradation are known.

Reports in the literature agree that swelling is encountered only in the reduction of hematite pellets. In those few cases where an increase in volume during reduction is denied, magnetite pellets are involved.

Pellets have been observed to swell during reduction with CO, H_2, $CO/CO_2/N_2$ mixtures and H_2/H_2O mixtures. On the other hand, other investigators[595k, l]) find no swelling when pellets are reduced with H_2.

Swelling and degradation are phenomena which are encountered only in pellets with a low gangue content[595d]).

According to EDSTRÖM[595m]), the swelling of ores is caused by the transformation of the hexagonal hematite into cubic magnetite and the resulting lattice disturbances. Lattice disturbances cause pore formation (the Kirkendall effect) whereby there is a considerable increase in the apparent volume of the samples during the transformation from hematite to magnetite. WAHLSTER[595i]) holds that the conditions under which the pellets are manufactured and fired are responsible for swelling because, with ores in the natural condition and of similar composition and with similar porosity characteristics, no such marked increase in the apparent volume has been observed.

According to OTTOW[595d]), the degradation of Fe-rich hematite pellets is caused by the combined effect of two factors, namely the change in volume during reduction (Fig. 301a) and the anisotropic nature of the rate of reduction during the reduction of hematite to magnetite (Fig. 301b).

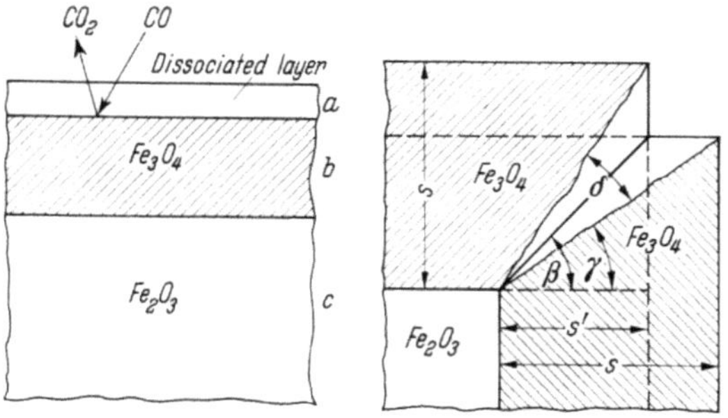

Fig. 301a
Change in volume during reduction, according to OTTOW[595d])

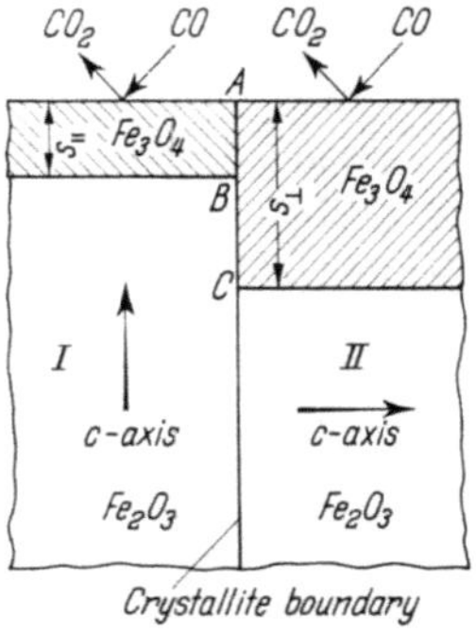

Fig. 301b
Diagram of anisotropy of the rate of reduction in a single grain of hematite, according to OTTOW[595d])

The former mechanism causes a lattice disturbance in both the first stages of reduction. Consequently, even in pellets which consist of magnetite after firing, a loss of strength is to be expected during reduction. OTTOW thus demonstrates experimentally that even magnetite pellets suffer a loss of strength during reduction – even though only to a slight extent.

WATANABE[595k]) ascribes the susceptibility of pellets to swelling and degradation to the fact that the oxidation of magnetite to hematite at relatively low temperatures occurs in such a manner that the (0001)-planes of the hematite which is produced grow on the (111)-planes of the magnetite (epitaxy).

WATANABE observed that pellets made from coarsely crystalline magnetite material by oxidising roasting at a relatively low temperature exhibited a coarse grained structure which still revealed the characteristic grain shape of the raw material used in their manufacture. On examining the reduction behaviour of such pellets in the hot stage microscope, he noticed, even during the heating up period in a neutral atmosphere, lines within the hematite grains representing the lines of intersection of (0001)-planes with the surface of the sample. The reduction of such pellets has a tendency to occur along these lines of intersection. During the transformation from hematite to magnetite, cracks are formed there and these favour very rapid reduction of the hematite, reduce the mechanical stability of the pellets and finally cause degradation.

WATANABE also observed, during the transition from wüstite to metallic iron, a growth of thin, fibrous iron needles from the surface to the interior of the pellet. These needles increase the apparent volume of the pellets. Nothing is yet known of the reasons for their occurrence.

Swelling and degradation of pellets during reduction can be reduced by the following methods:

1. raising the roasting temperature;
2. addition of suitable materials to the pellet mix;
3. manufacture of magnetite pellets;
4. stabilisation of the cubic γ-modification of the Fe_2O_3 (maghemite);
5. enlargement of the Fe_2O_3 crystallites.

Possibilities 4 and 5 have not yet been developed to such an extent that their use on a practical scale appears to be promising.

5.2.2.4.3. Operational Results with Pellets in the Blast Furnace; a Comparison with Sinter. The rounded shape and the almost constant diameter of pellets are useful prerequisites for uniform permeability in the blast furnace. For this reason, and because of the generally good reducibility, it was thought that pellets must be far superior to all other blast furnace additions. Initially, this assumption appeared to be confirmed by operational results which showed that the gradually increased pellet addition had raised the output of a large American blast furnace by about 50% and reduced the coke rate from an initial 900 kg/ton of iron to 650 kg/ton of iron[598]). The initial values were of course very unsatisfactory. According to the results mentioned in the previous section, extremely uniform permeability can also be attained with well graded lump ore and with sinter, the same reduction rates can be attained as with pellets and when self-fluxing sinter is used, limestone additions in the furnace are not required as is the case with a pellet burden. It is, therefore, not surprising that the

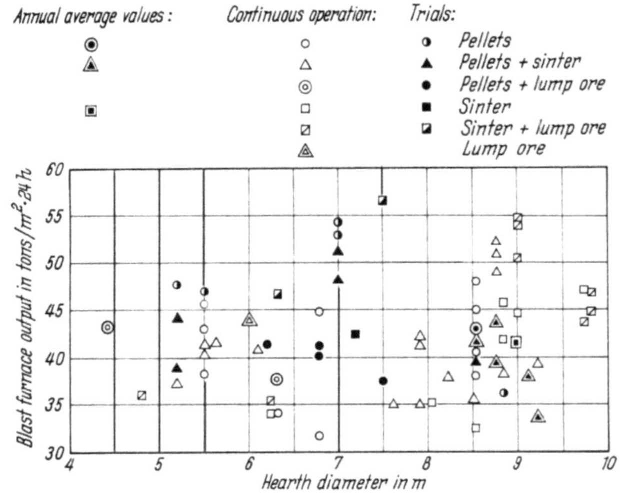

Fig. 302. Output of blast furnaces operating with sinter, pellet and lump ore burdens, according to WINZER[603a])

Table 42. *Operating data for Japanese blast furnaces with various amounts of pellets in the burden*[600])

		Blast furnace No.		
		1	2	3
Hearth diameter	m	7.5	8.8	8.8
Iron make	tons/24 h	1540	2450	2210
Specific furnace output	tons/(m^2·24 h)	34.8	34.2	36.2
Coke consumption	kg/ton of iron	586	543	494
Oil addition	kg/ton of iron	0	33.4	44.7
Mixed ore	kg/ton of iron	592	552	458
Goa ore	kg/ton of iron	63	48	91
Sinter	kg/ton of iron	165	512	524
Pellets	kg/ton of iron	728	449	445
Blast volume	m^3/h at NTP	97000	128000	133000
Blast temperature	°C	944	801	941
CO in top gas	%	24.8	25.7	23.0
CO_2 in top gas	%	17.1	17.3	17.6
Slag volume	kg/ton of iron	391	323	335

thorough examination of all available blast furnace operational data from the literature shows that as regards the attainable blast furnace operating results, there is no definite distinction between sinter and pellets[599–603, 603a, b]) (Fig. 302). These findings can, for instance, be supported by the results obtained in Japanese blast furnaces (Table 42) and by the results of large scale trials carried out at two German blast furnace plants[601, 602]). The furnaces were driven as hard as possible. Referred to the same hearth area or to the same working volume, the outputs were about the same. The coke rate to be sure was higher in the second trial (lump ore/sinter burden). The reason for this, certainly well founded, was the fact that a considerable proportion of the burden was made up of Kiruna ore which is difficult to reduce and which was not of optimum grading. Thus, the superiority or otherwise of pellet burdens has not yet been firmly established.

5.3. Testing Ores

The testing of iron ores and agglomerates is closely connected with ore assessment, i.e., determination of the adequate market price. For iron works not having their own ore supplies, as, for instance, in Germany and Japan, optimum selection of ore on the basis of a correct ore assessment is a matter of life and death. Great attention has, therefore, always been paid to ore testing.

The following concepts are concerned with the scientific and technical basis of ore testing for the blast furnace. Economic aspects are not consciously touched upon directly in the present volume.

The value of an iron ore or agglomerate is essentially determined by the cost of preparing it for smelting and by the coke consumption in the blast furnace, by the throughput which can be attained and by the proportion of charge which leaves the blast furnace in the form of flue dust.

The fuel requirement is determined first by the proportion of ballast (inert material) in the ore as was explained in section 4.1. In addition to this, it is of decisive importance that the reduction reaction should proceed as rapidly as possible, without hindrance and with good chemical utilisation of the rising stack gas. The main prerequisites for this are:

a) uniform permeability of the stock column in the blast furnace, i.e.,

a 1) accurate size grading, particularly as regards separation of the fines;

a 2) good resistance to mechanical abrasion and impact;

a 3) beginning of softening at high temperatures and a short interval between the beginning of softening and complete melting;

b) good reducibility.

The importance of these factors was dealt with in sections 3.1 and 5.2 and in what follows, testing possibilities will be reported.

Lack of space makes it impossible to go into details of numerous earlier works. In spite of intensive work in this field, it is only in recent years that it has been possible – and then only partially – to develop testing methods of value for forecasting and which can be applied in different plants. This partial success has been due mainly to the co-operative work of the competent technical committees of the Verein Deutscher Eisenhüttenleute. The proven methods of testing can be divided into two groups:

a) testing of properties and their consistency, the results of which cannot be translated to blast furnace practice;

b) testing which can be applied directly to blast furnace practice, i.e., which permits quantitative forecasting of the behaviour of ore or sinter in the blast furnace.

Testing methods belonging to the second group are, of course, more valuable for forecasting.

Only those methods of testing which can be directly related to the blast furnace will be considered here and not those methods generally used in the laboratory, e.g., for determining chemical composition, porosity, bulk density and density, etc. Good manuals are available for these, see, for instance, reference [608]).

5.3.1. Grain Size Distribution

There are rules for the control of grain size distribution of ores and sinter which first provide for drying, where required, with subsequent screening on standard screens[608, 608a]). On account of the drying process, this method of testing, is, however, limited to occasional spot checks with relatively small samples. The main problem is concerned with selecting small samples in such a way that they will be representative of the whole.

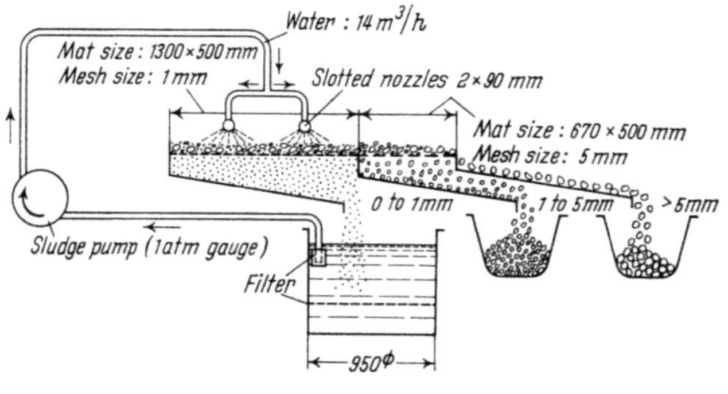

Fig. 303. Wet screening of larger quantities of samples (throughput 3 tons/h)[609])

Larger samples can be tested by way of wet screening. Figure 303 illustrates the type of installations used. The ore is loaded onto the mat without drying and is sprayed off with a sharp jet of water, thus causing the fine fraction, and particularly that portion of the fines which adheres

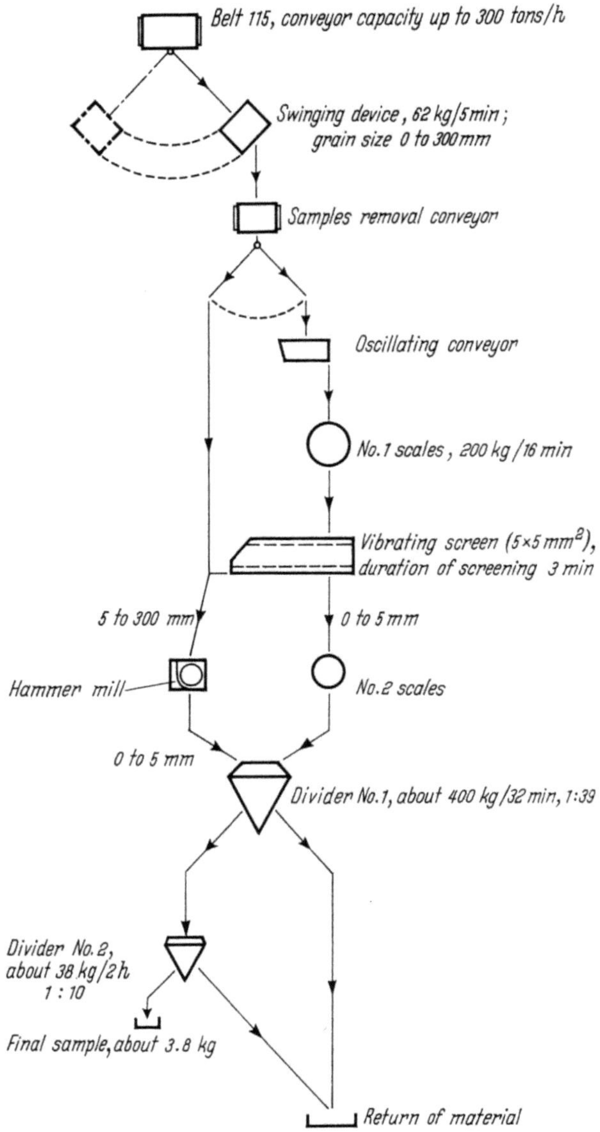

Fig. 304. Automatic device for sampling and testing degradation of sinter, according to Heynert et al.[611])

to the large lumps (known as the "adhering particles"), to be freed from the coarse fractions and be determined. In the case in question, the material was screened first to 1 and then to 5 mm size on an adjacent mat. The fractions larger than 1 mm were weighed and the moisture content determined. Most of the water was first filtered out of the fines smaller than 1 mm and the same procedure as for the larger fractions was then followed so that the weights of the fractions referred to the dry state could then be calculated. Random samples of 1000 kg of screened ore were tested during the preparation of the burden. The fines below 1 mm in size constituted 2.4 to 9.1% of the relevant quantity tested (in the dry state). Most of these values are considerably higher than those resulting from the check screenings in the ore crushing and screening plants. This is mainly to be ascribed to the stress to which the ores are exposed during transportation and handling and also to the adherent fine particles.

With dry agglomerate, e.g., sinter, continuous check screening is possible without a troublesome drying operation. Figure 304 shows an automatic sampling and testing device. The increment to be screened passes over weighing machine 1 and a reciprocating screen. The fines are picked up by weighing device 2. The sample is subsequently crushed and sent for chemical examination. The sinter is subjected to considerable mechanical stressing by the screening process so that not only the amount of the fine fraction present but, to some extent, the susceptibility to disintegration of the sinter is tested.

Trials in which radioactive ore samples graded in the various fraction sizes were charged into the blast furnace showed that a considerable proportion, though not all, of the under 0.5 mm material was discharged. Some of the minus 0.1 to 0.2 mm material did, however, remain in the blast furnace[610]). On the other hand, considerable quantities of the over 0.5 mm fraction were discharged as dust. The blast furnace must, therefore, be regarded as an imperfectly operating air sifter.

5.3.2. Mechanical Stability

5.3.2.1. Mechanical Stability in the Lower Temperature Range*

While being conveyed from the ore screening plant to the blast furnace top, ores are several times subjected to impact and abrasion. In the blast furnace it is mainly a question of abrasion and compression with, at the same time, a rise in temperature in a reducing atmosphere.

* Ambient temperature until the beginning of reduction, i.e., about 400°C.

Table 43. *Chemical composition of the ores investigated*

Ore	Fe	FeO	Fe_2O_3	P	Mn	CaO	SiO_2	H_2O_{comb}	CO_2	Moisture content
				Magnetites						
Bomi Hill	65.25	24.64	65.99	0.14	0.24	0.05	1.70	1.02	0.18	2.0
Dannemora	50.94	23.85	46.26	0.01	2.92	4.46	10.40			1.0
Elba	54.73	24.51	51.09	0.03	0.28	2.90	10.53	1.0	3.83	2.0
Gellivara	62.00	24.30	61.72	0.87	0.09	3.25	5.50	trace	0.1	1.0
Idkerberg	64.80	29.41	60.06	0.71	0.15	2.41	4.88	0.05		1.7
Kiruna D	61.75	26.44	58.99	2.04	0.24	7.10	2.70	0.26	0.47	0.6
May-sur-Orne . . .	43.0	33.67	24.17	0.60	0.45	2.45	15.16	3.86	11.42	1.5
Nyberg	45.55	19.48	43.54	0.02	0.21	5.40	22.65	0.26		0.7
Rector	56.70	23.35	55.19	3.36	0.23	10.4	1.32	0.23	0.33	0.3
Tuollavara	62.60	25.22	61.56	0.048	0.15	0.5	5.8	0.78	0.07	1.0
				Hematites						
Chilean ore	63.82	8.51	81.82	0.45	0.21	1.3	4.11	1.42	0.05	0.7
Costi	58.39	0.65	82.50	0.055	0.74	0.05	2.05	6.90	0.02	2.2
Djerissa	57.10	0.90	80.65	0.02	2.33	3.8	3.13	5.31	3.88	6.8
Franco Belga . . .	56.15	5.0	76.08	0.02	1.33	6.0	12.41	0.32	0.18	0.6
Füsseberg	52.9	0.19	75.60	0.006	9.60	0.2	9.93		0.22	5.2
Indian ore	58.9	2.84	81.08	0.05	0.38	0.05	2.05	4.44	trace	1.3
Itabira	68.22	0.38	97.13	0.04	0.13	0.05	0.51	0.60	0.0	1.0
Labrador	61.54	0.38	87.57	0.037	0.78	0.05	6.2	0.0	3.0	6.0
Marampa	56.9	0.25	81.08	0.039	0.37	0.05	4.4	4.86	0.3	6.8
Marcona	62.0	3.1	85.25	0.055	0.11	0.50	6.0	2.20	0.0	1.2
Malaespera	52.56	6.58	67.87	0.009	1.05	2.07	10.0	0.54	1.29	0.3
Nozay St. Sulpice .	44.55	0.83	62.78	0.71	0.08	0.1	16.67			2.3
Rif I	65.9	8.25	85.08	0.03	0.57	1.3	7.7	2.0	0.4	5.6
Striberg	48.81	6.45	62.65	0.02	0.21	1.0	26.37	10.67	0.0	0.2
Venezuela	63.6	0.52	90.37	0.14	0.16	trace	0.6	4.8	0.07	7.1
Wabana	54.75	8.51	68.85	0.90	0.24	3.1	13.18	2.9	3.61	1.4
				Ores low in Fe content and with a high content of volatile components						
Altenstadt	34.5	2.71	46.33	0.31	0.56	8.7	19.9	6.48	8.57	7.25
Echte	28.3	15.8	22.96	0.46	0.20	18.6	15.3	4.92	15.91	6.7
Kahlenberg	18.9	1.35	25.55	0.32	0.24	28.5	13.07	3.51	25.55	7.0
Klippenflöz	14.1	5.7	13.8	0.22	0.18	34.2	11.31	2.0	26.69	2.7
Minette	32.2	11.35	33.69	0.65	0.30	16.0	8.30	7.84	14.65	11.0
Peine	24.31	3.74	30.61	0.83	2.25	24.7	5.05	4.60	20.64	5.5

Various types of shatter test* have been developed and the resulting data should, therefore, not be generally compared. A test value which is valuable for forecasting will therefore be obtained only if the behaviour of a larger number of ores is determined under constant testing conditions. The ores tested can then be classified as good/medium/poor. An ore with a low shatter strength naturally tends to break down during handling.

Systematic shatter tests were carried out on 32 different ores[609, 612] (the chemical composition is stated in Table 43) in such a manner that

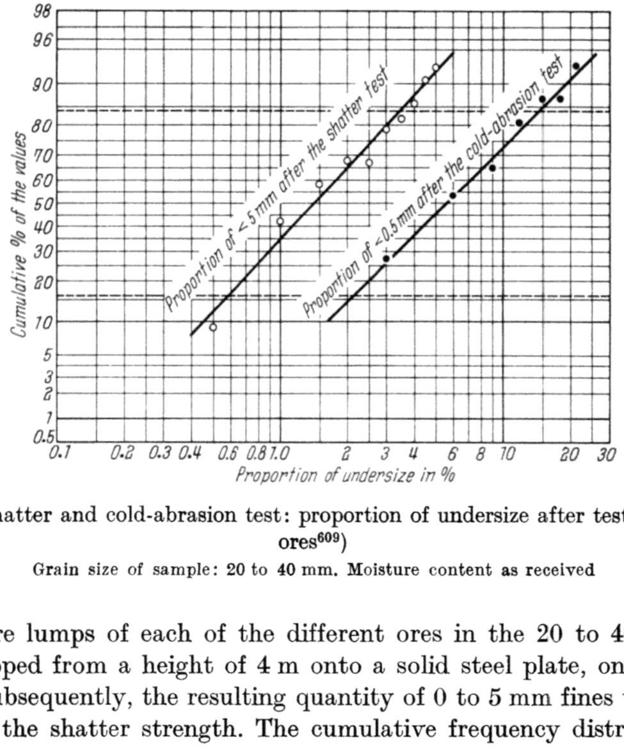

Fig. 305. Shatter and cold-abrasion test: proportion of undersize after testing 32 iron ores[609])

Grain size of sample: 20 to 40 mm. Moisture content as received

3 kg of ore lumps of each of the different ores in the 20 to 40 mm size were dropped from a height of 4 m onto a solid steel plate, one lump at a time. Subsequently, the resulting quantity of 0 to 5 mm fines was determined as the shatter strength. The cumulative frequency distribution of the test data were thus obtained as shown in Fig. 305.

The susceptibility to abrasion is usually tested in trommels [for examples, see references [609–614a])]. A comparable test for abradability of sinter has been described by Send and Weilandt[613]).

* The use of the concept "strength" in connection with the mechanical testing of ores (e.g., shatter strength, abradability), is to be deprecated because the concept of strength always refers to a defined stress which is frequently almost uniaxial and which causes the predetermined change in shape (yield, rupture, etc.). The test stressing of ores, however, is not associated with the production of such defined states of stress.

In Fig. 305, the cumulative frequency distribution of the abrasion data is plotted for the same 32 ores (the charge: 10 kg, grain size: 20 to 40 mm, 4 minutes testing period at 25 rpm, trommel diameter: 1000 mm, other dimensions as in [614]). The criterion is the proportion of 0 to 0.5 mm after trommel treatment. Particularly in supply contracts for pellets, it is becoming more and more the practice to agree limiting values for the mechanical strength ("strength" and "abrasion"). In general, therefore, an attempt is being made to produce an international standard for trommel testing[612a]). A similar apparatus was used to test the abradability at 300°C in an atmosphere of blast furnace gas. The hot abrasion values obtained correlated so closely with the cold abrasion values that no additional information useful for forecasting was hereby obtained[612]) (Fig. 306).

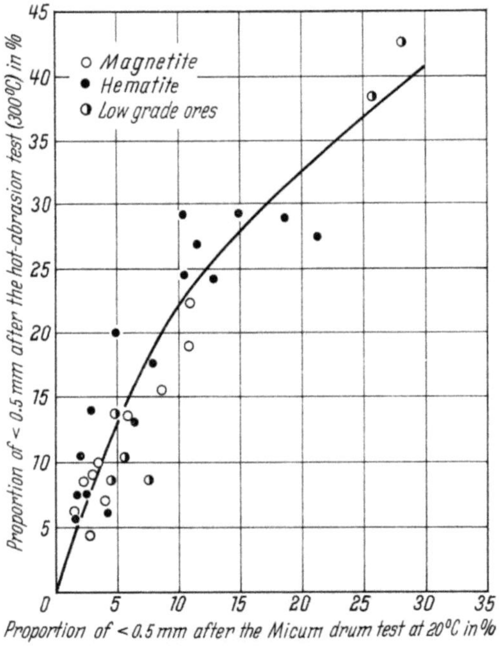

Fig. 306. Comparison of Micum-drum and hot-abrasion test results, according to JANKE[612])

The abrasion figures for the ores could be translated as follows to actual blast furnace operations. After screening and drying, sample lumps of six different ores were bombarded in an atomic reactor with fast neutrons thus becoming activated (by the formation of Fe^{55}, Fe^{59}). When the activated sample lumps were charged into the blast furnace, a predetermined section of the top gas flow was extracted by means of a probe (Figs. 307 and 308) from the top gas main and the amount of dust and its specific radioactivity were examined. It was shown that the proportion of the

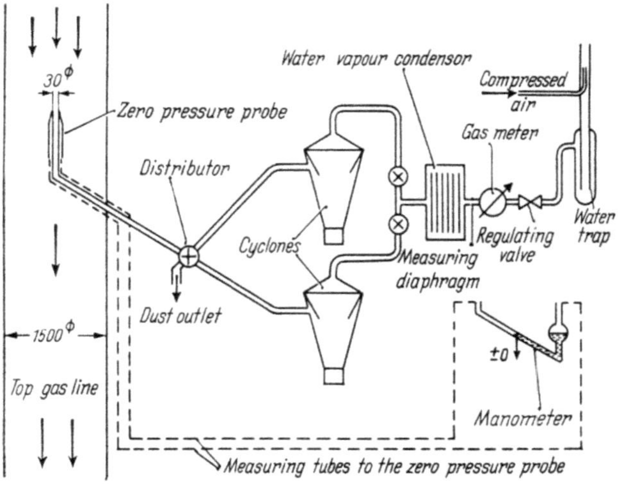

Fig. 307
Equipment for determining the dust content of blast furnace gas[615])

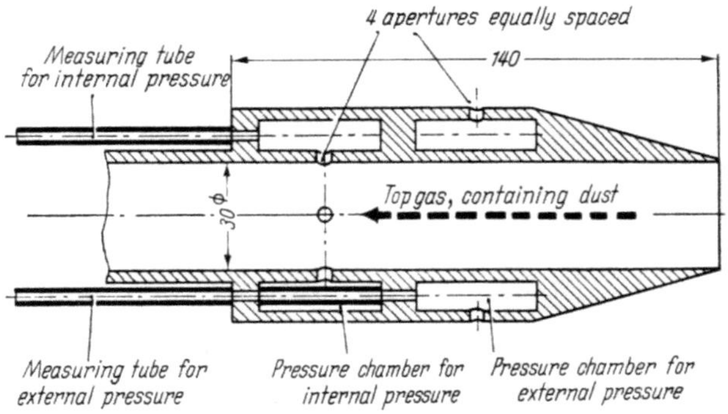

Fig. 308
Zero pressure probe for velocity equalisation in dust sampling[615])

ore discharged varied within wide limits depending on the abradability of the ores. Figure 309 shows two examples (sinter and briquettes made of blast furnace flue dust). For details of the experimental procedure, see the original work[615, 616]).

There was a close relationship between the abrasion figures as determined in the trommel test and the proportion* discharged from the blast furnace (Fig. 310). The test data obtained in the laboratory can therefore be translated to conditions obtaining in the blast furnace.

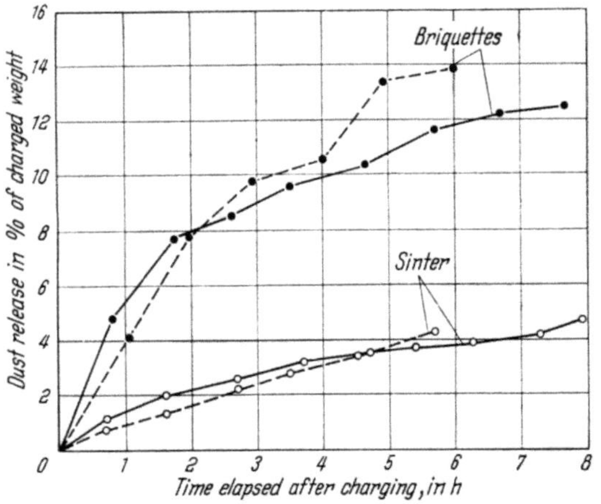

Fig. 309. Total dust released as a function of time after charging in the blast furnace[615])

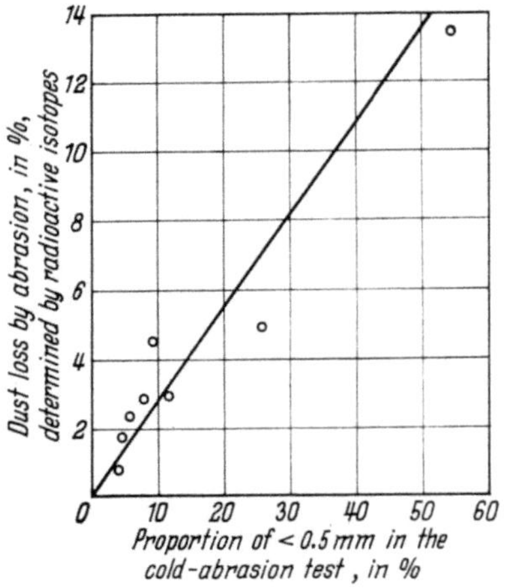

Fig. 310. Dust loss by abrasion of the ore in the blast furnace[609])
Hearth diameter 5.6 m

5.3.2.2. Mechanical Stability during Reduction

Various processes have been used to obtain data on the mechanical properties at elevated temperatures and under reducing conditions:

a) the Linder test[617]);

b) the Gakushin test[617a]);

c) the determination of the compression strength of pellets reduced to various degrees after cooling to the ambient temperature[595i]).

These testing methods are rather costly. The possibility of applying the experimental results directly to blast furnace practice is limited by the fact that, since the reducing gas mixture and the temperature programme are already selected, part of the results has already been anticipated.

---

* In the isotope test, only the discharge resulting from abrasion in the blast furnace was actually involved. Since the ore had been previously dried and well screened, it was impossible for any proportions of fines previously entrained to have been discharged as dust or for any decrepitation to have taken place.

5.3.3. Susceptibility to Decrepitation

Some ores tend, on heating, to burst whereby several large lumps and proportions of powdery fines are produced. This phenomenon was first observed in the minette ore[618, 619]). To effect a closer study of this phenomenon, ore samples were heated in a stream of nitrogen at 400°C in the apparatus illustrated in Fig. 311. In this way, it was intended that the

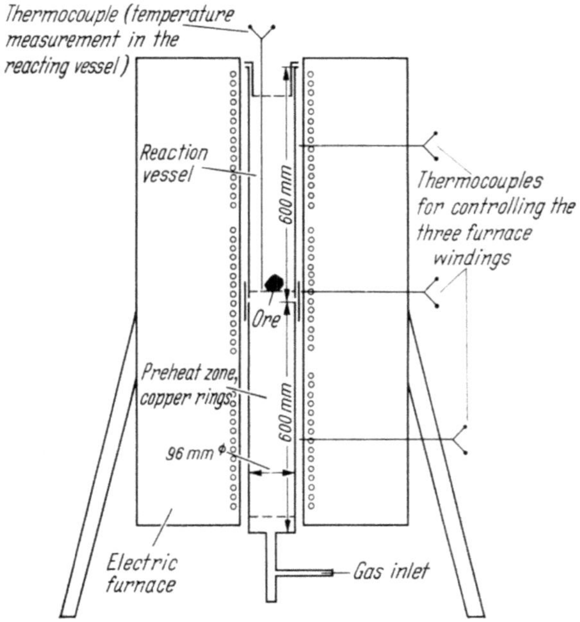

Fig. 311. Test rig for determining decrepitation of iron ores, according to reference [609]) and PEHLKE[620])

heat transfer in the upper regions of the blast furnace should be simulated*. Bursting, also known as "decrepitation", was observed and the resulting proportion of under 0.5 mm was determined by screening out. It was thus shown that with the ores which were examined, decrepitation depends mainly on the moisture content (Fig. 312). Calculations showed that a sudden evaporation of water in the pores – after a superheating – seems to be the cause of bursting[620]). The nucleation of bubbles appears to play a decisive part in this case [see also references [621, 622])].

According to recent experiments carried out by WATANABE[595k]), decrepitation can be explained by two further mechanisms. In particularly

* The temperature which is higher as compared with the top gas approximates to the low rate of flow of nitrogen as regards heat transfer.

soft ores, which, in addition to hematite, contain notable quantities of limonite, decrepitation is initiated by the release of water of crystallisation.

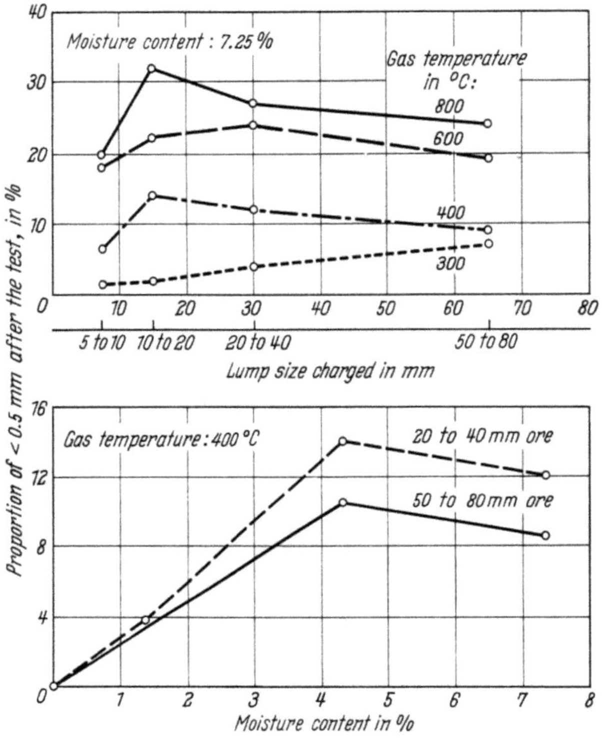

Fig. 312. Effect of lump size and moisture content on the decrepitation of an oolithic ore[609])

Hard, slaty hematite ores which can be split parallel to the (0001)-planes, burst on heating because of the variation in thermal expansion due to anisotropy.

5.3.4. Propensity to Dust Formation

A considerable proportion of the very fine grained material in the charge is discharged from the blast furnace by the rising stack gas or top gas. Very fine grained material arises from three causes:

a) inadequate screening;
b) abrasion;
c) decrepitation.

As regards a), the amount of fines present in the ore can be determined by screening (see section 5.3.1). The proportion of 0 to 0.5 mm is regarded as being susceptible to being discharged and is designated by x_1.

As regards b), the abrasion test in the trommel makes it possible to draw a direct conclusion as regards the quantity which is reduced to dust, x_2 (Fig. 310).

As regards c), during the decrepitation test which is carried out under conditions of heat transfer calculated to simulate those in the blast furnace,

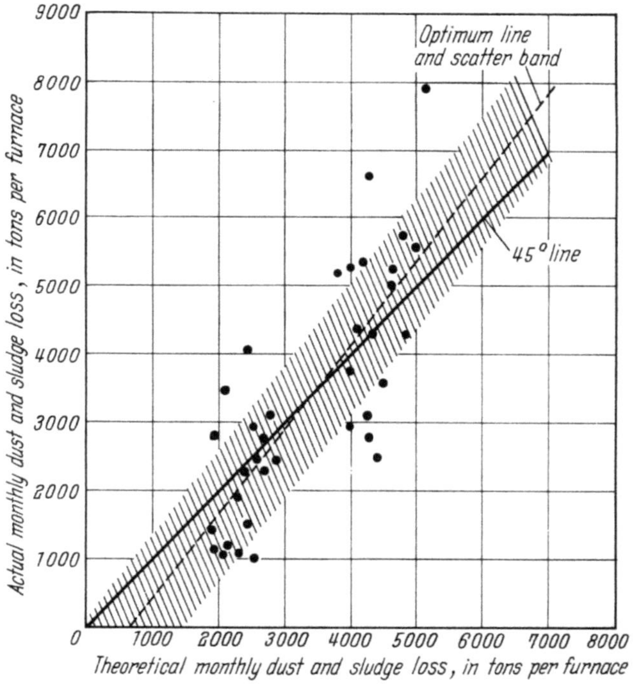

Fig. 313. Comparison of actual and theoretical dust and sludge loss (dry)[609]

Blast furnaces with 6 m hearth diameter

the resulting proportion of 0 to 0.5 mm, as in a), is considered to be susceptible to being discharged and is designated by x_3.

The theoretically expected discharge (dust and solid materials in the sludge) from the blast furnace results therefore as the sum of $x_1 + x_2 + x_3$ for all the burden components charged. Comparison with practice shows, according to Fig. 313, approximate agreement between the theoretical and actual values. The mathematical/statistical evaluation produces the following relationship between the theoretical amount of dust and sludge which ensues $x = x_1 + x_2 + x_3$ and the quantity actually found, y,

$$y = -842 + 1.247(x_1 + x_2 + x_3)$$

with a % explained of 49%.

The dust components x_1, x_2, x_3 are proved separately and as a total statistically ascertained[609]).

It is, therefore, possible to use the dust formation susceptibility figures obtained in the laboratory and theoretically in order to classify ores in a comparative manner and furthermore to forecast the extent of dust formation semi-quantitatively. The higher the proportion of the ores which is reduced to dust, the poorer, generally speaking, is the permeability of the burden in the furnace since a proportion of the fine grained material always remains in the furnace and causes local blockage of the stock column.

5.3.5. Softening

Much experimental work has been carried out on ore softening[623–630]). Two groups of apparatus have been essentially used:

a) furnaces for heating under pressure (Fig. 314). The ore charged is gradually heated up under a compressive stress (usually 2 kg/cm²) and the collapse or "fall point" is observed as in normal refractories testing;

b) hot stage microscopes (Fig. 315). Small specimens are heated on a ceramic stage and the change is observed continuously under a microscope. A lateral pressure can also be exerted.

Initially, attention was paid to the softening behaviour of unreduced ore and temperatures of between 700 and 1350°C were observed for the beginning of softening[625]). At these temperatures, ore in the blast furnace has, however, already been reduced to a considerable extent and has, therefore, undergone essential changes in its mineral make-up and softening behaviour. Experiments were, therefore, started with a systematic variation in the degree of reduction[626–630]).

According to Poos and Decker[626]), maximum softening occurs in the region above $\mathfrak{R} = 50\%$ while, according to Grieve[627]), essential softening sets in even above $\mathfrak{R} = 20\%$, i.e. in the FeO region.

Investigations by Braumann[629]) have shown that it is not only the extent of the degree of reduction which is important but also the manner in which it is attained, as will be seen in Fig. 316 [t_b, t_e are the temperatures corresponding to the beginning and end of softening[630])]. Thus, there is a large number of factors which affect the softening behaviour of an ore. These can be collated because most of these influencing factors take effect by way of porosity which is easy to measure. Figure 317 shows that the higher the porosity, the lower the incipient softening temperature of an

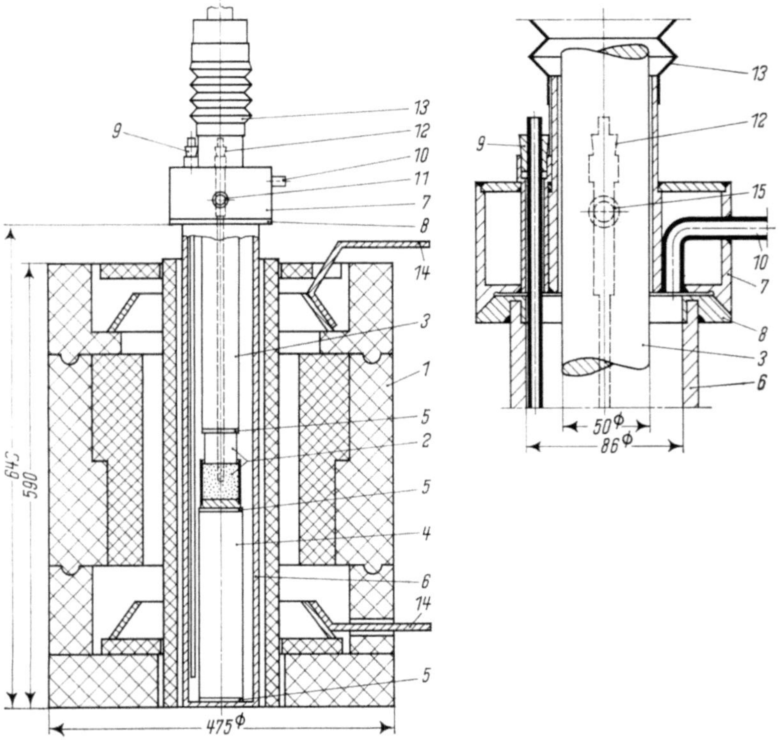

Fig. 314. Experimental furnace for the **softening** test[630])

1 Peacoal furnace (chamotte parts with steel jacket), *2* Sample crucible with seal, *3* Upper pressure piston (90% SiC), *4* Lower pressure piston (90% SiC), *5* Ceramic washers, *6* Steel tube, *7* Water cooling (copper), *8* Bearing shim (copper), *9* Protective gas inlet, *10* Protective gas outlet, *11* Cooling water inlet, *12* Thermocouple sheath (between 2 and 6), *13* Sealing packing ring, *14* Current supply, *15* Cooling water outlet

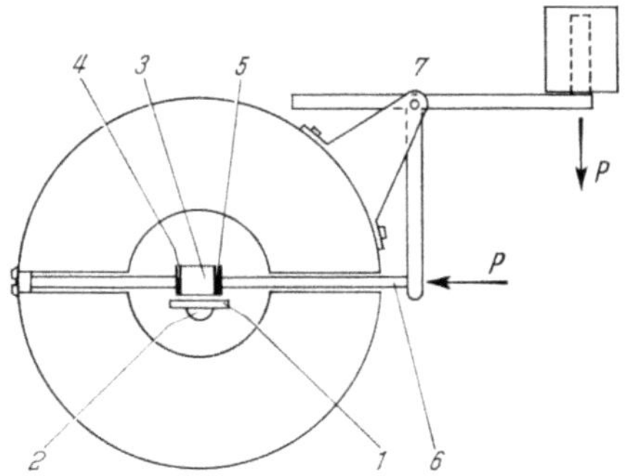

Fig. 315. Device in hot stage microscope for determining softening under load, according to Dickens et al.[625])

1 Crucible for melting specimen, *2* Crucible holder with thermocouple, *3* Specimen, *4* Fixed alumina plate, *5* Moveable alumina plate, *6* Pressure piston, *7* Device for applying load ore, it being immaterial whether the pores were originally there or whether they arose due to reduction conditions. This interrelationship is particularly

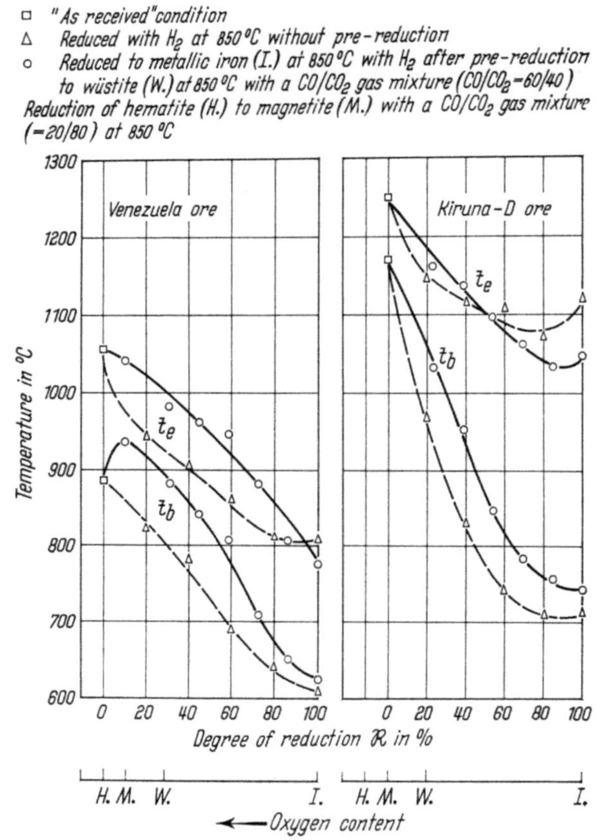

Fig. 316. Effect of pre-reduction on temperatures of equal degree of softening[630])
t_b and t_e = temperatures of the beginning and end of softening (3 mm and 10 mm height reduction respectively)

valid for the end of softening (defined as 10 mm reduction in height of the specimen).

Although no quantitative correlation has yet been established between these test data and actual blast furnace practice, it can be stated qualitatively that a high narrow softening range with degrees of reduction above 50% creates the most favourable prerequisites for trouble-free blast furnace operation. In this sense, ores or agglomerates of maximum porosity cannot be considered to be the optimum in spite of maximum reducibility. When reducibility increases above a certain point, there is no improvement in the coke rate and premature softening occurs because of the relationships presented in Fig. 317, thus interfering, under certain circumstances, with

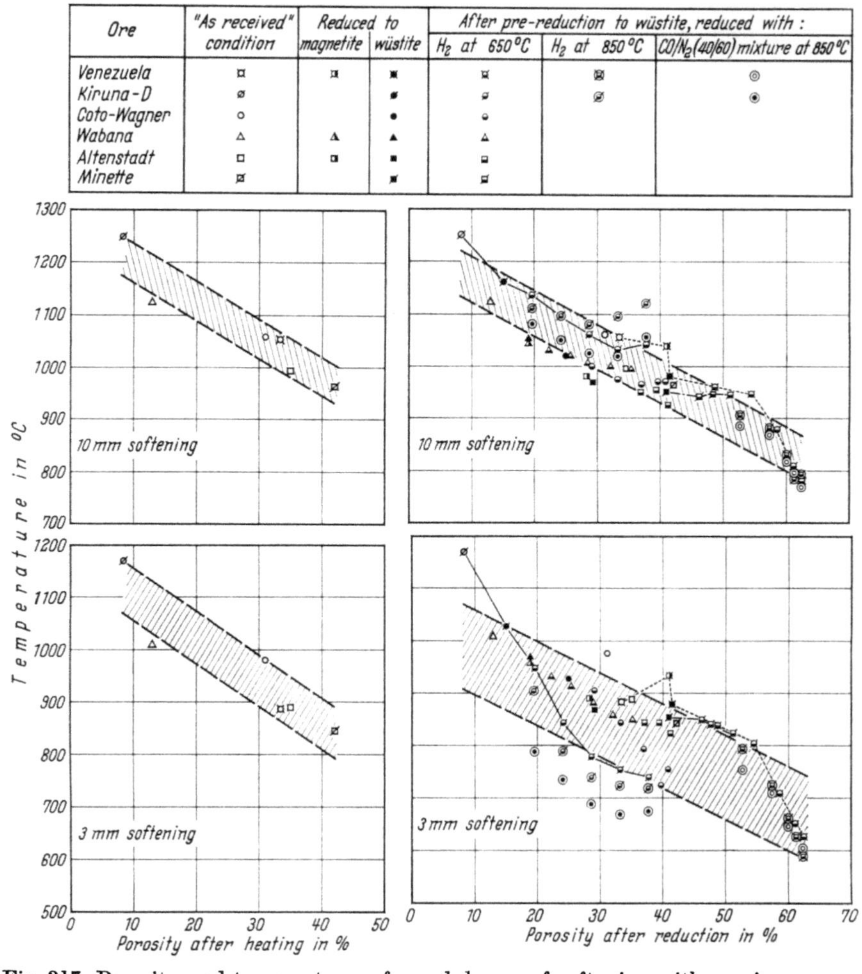

Fig. 317. Porosity and temperatures of equal degree of softening with varying oxygen content[630])

permeability and giving rise to hanging phenomena with all their well known adverse consequences. Thus, it does not appear to be advisable to crush down readily reducible lump ores too far or, for instance, to manufacture pellets with a particularly high porosity, especially as, in every case, there is a very marked increase in porosity with increasing degrees of reduction[630, 631]).

5.3.6. Reducibility

The concept of "reducibility" is generally understood to refer to a coefficient of measure for the rate of reduction.

In the manufacture of agglomerates, it is possible to exert a marked influence on reducibility by suitable control of the process. The reducibility of lump ores can be adjusted by altering the lump size.

Optimum reducibility is shown to be a compromise between various requirements, particularly:

a high rate of reduction,
uniform permeability of the stock column,
softening behaviour,
cost of ore preparation.

The determination of reducibility as an aid to operational management is, therefore, of particular importance. It has been carried out in various investigations using different methods[632–640]).

The purpose of the test, at first, appears to be relatively simple; the ore or agglomerate is subjected to conditions calculated to simulate blast furnace conditions as regards temperature and gas composition and the result of reduction (rate of reduction, degree of reduction after a certain

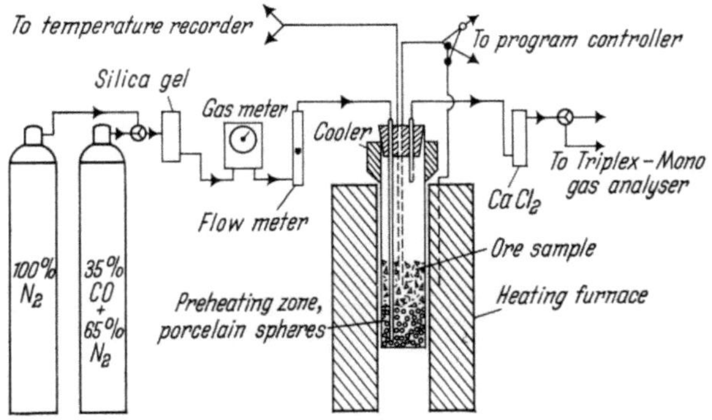

Fig. 318. Test rig for the heating-up test for the reduction of ores[634])

time, etc.) is determined. Many such suggestions have been successfully applied a considerable number of years ago[632, 633]). Further knowledge was thus gained regarding the removal of oxygen from ores.

The heating-up process, which can essentially be traced back to MATHESIUS[632]), but which has now been further refined (Fig. 318), traces the course of the various dissociation reactions with increasing temperature. The ore to be tested (grain size 10 to 15 mm) is heated in a vessel of 50 mm inside diameter and the temperature is allowed to rise at the rate of 4°C/min from 20 to 980°C. At the same time, a reducing gas mixture (35% CO, 65% N_2) flows slowly through the ore. The chemical reactions are contin-

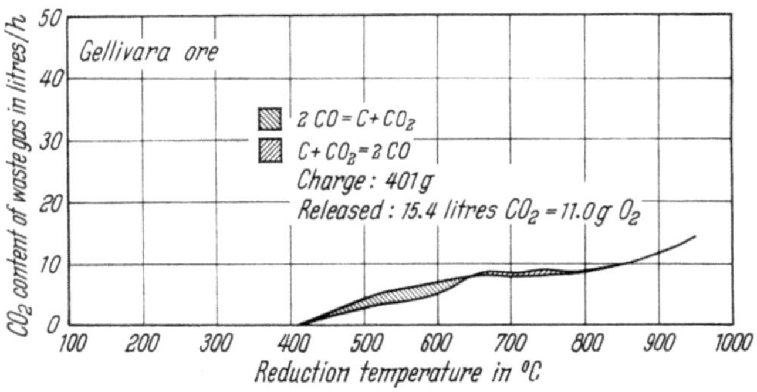

Fig. 319. Reduction test with Gellivara ore, using the heating-up process, according to DICKENS et al.[625])

uously recorded by means of the waste gas analysis (CO_2, CO, H_2). By means of balance calculations, the values obtained can be assigned to the various reactions (reduction, Boudouard reaction, carbonate decomposition, water-gas reaction). Figures 319 and 320 show some examples of meas-

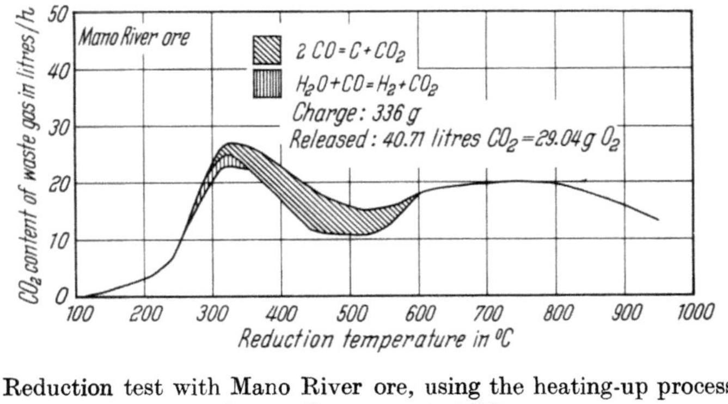

Fig. 320. Reduction test with Mano River ore, using the heating-up process, according to DICKENS et al.[625])

urements. It will be seen that, for instance, the temperature of incipient reduction varies greatly for the ores examined. No clear correlation has yet been established in investigations of this type with blast furnace operating data.

Generally speaking, however, the variation in the scaling of the measured reducibility data has been found to be similar, using numerous testing procedures, as has been shown by Poos and LINDER[645]) who investigated 35 ores and agglomerates by three different reducibility testing methods [heating up process, the Linder process[617]) and an isothermal test method used in Belgium[647])]. In isolated cases, it was possible to demonstrate some similarity between the results of the reduction testing procedures, the degree of indirect reduction in the blast furnace and the coke rate[645, 648, 648a]). The degree of indirect reduction is, of course, not a characteristic which can be generally applied to predict the behaviour of ores in the blast furnace. On the contrary, the probe measurements of HUPFER and WEIDENMÜLLER[560]) have shown that a given ore exhibits quite variable degrees of indirect reduction, depending on how the blast furnace is operated.

Reducibility testing is problematic for two essential reasons:

1. The *practical* difficulty arises from the fact that some of the sets of conditions prevailing in the blast furnace such as, for instance, abrasion, compressive stressing and gas velocity (about 3 m/sec) can be reproduced in the laboratory only at very great expense. As demonstrated by SCHENCK[631]), the pressure of the stock column greatly reduces the reduction rate, particularly with very porous ores.

2. The *fundamental* difficulty results from the additional fact that conditions in the blast furnace, i.e., temperature distribution and gas concentration from top to bottom of the stack and the temperature range in which reduction occurs can by no means be assumed to be constant but are, on the contrary, primarily governed by the reaction rates of the burden components (cf. Figs. 319 and 320 and section 5.2, Figs. 280 and 282). In such cases, any attempt to define reduction behaviour by way of the conditions prevailing in the blast furnace, would be essentially to anticipate the result.

The occurrence of the reactions can be simulated on a smaller scale by means of experimental blast furnaces[633a–c]) and in this way genuine reduction data can be obtained. Because of the complicated and time-consuming handling problems, however, such methods have not proved to be viable in practice for testing larger numbers of types of ore. Many of the results obtained are also difficult to comprehend. The experimental blast furnace of the Bureau of Mines was cooled down as quickly as possible from current operations by the injection of nitrogen and all the intermediate stages of reduction remaining in the furnace were "frozen" and examined analytically zone by zone. It was inexplicably found that Venezuelan ore and sinter exhibit almost the same reduction pattern from top to bottom of the furnace (Fig. 321).

Because of these partly fundamental and partly practical difficulties, the recent tendency in German blast furnace plants has been to abandon the idea of reproducing blast furnace conditions in reducibility testing in favour of the following principles[609, 634–641]).

A testing procedure is demanded which will grade the ores and agglomerates as regards reducibility in accordance with behaviour in the blast furnace. The importance of this test for blast furnace practice must

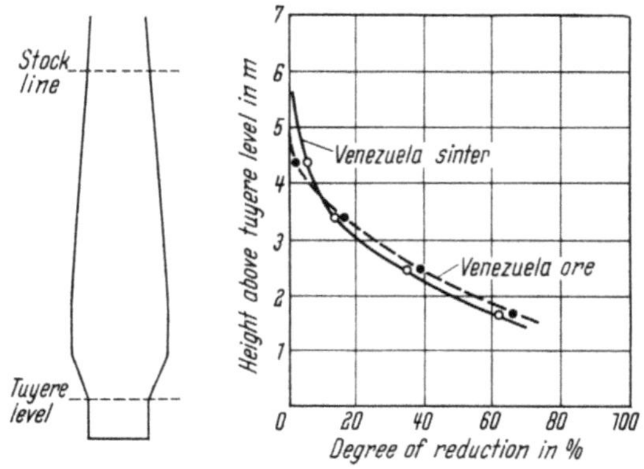

Fig. 321. Reduction of sinter and ore in an experimental blast furnace, according to BOSLEY et al.[633a])

be confirmed by operational trials as has been described, for instance, in section 5.2.

The most important condition for the test is, therefore, that reduction should be carried out so that, in the interest of correct grading (vis-à-vis reaction kinetics), the same partial reaction is rate-determining as with reduction in the blast furnace. This means that the experimental procedure must provide for the following conditions:

a) high gas flow rate so that diffusion and convection processes do not become rate-determining in the free gas space;

b) available temperature range 700 and 1000°C;

c) lump size must be the same as under operational conditions as otherwise if the lump size is altered a change can occur in the reaction mechanism and consequently in the rate-determining step (see section 2.5.3);

d) reduction gas: CO/N_2 mixture, or possibly H_2.

Figure 322 shows an experimental set-up [similar to one already used by BARRETT and WOOD[649])] designed to fulfil the above-mentioned requirements used in most German ironworks in conjunction with standardised testing conditions.

Testing conditions are as follows:

gas temperature: 900°C;

gas composition: 60% N_2, 40% CO;

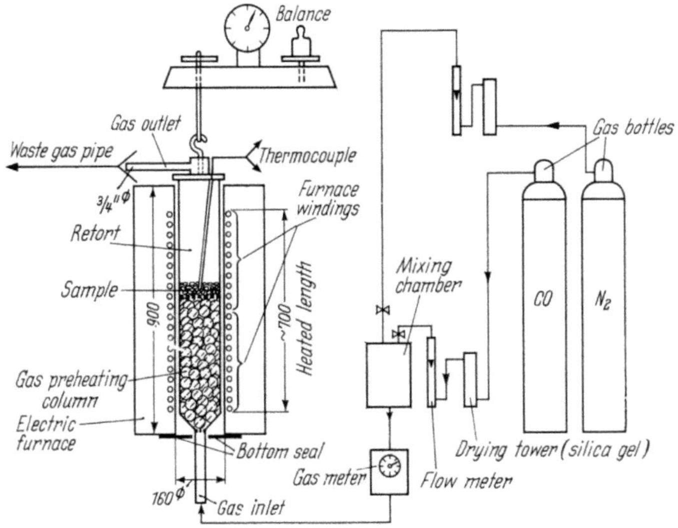

Fig. 322. Test rig for determining reducibility of iron ores and agglomerates at constant temperature, according to Stahleisen-Prüfblatt 1770-64 [639])

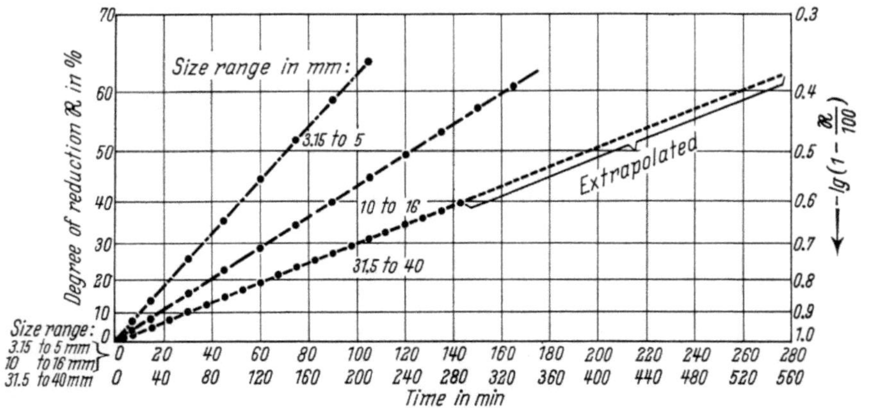

Fig. 323. Relationship between degree of reduction and time (example using Kiruna-D ore), according to Stahleisen-Prüfblatt 1770-64 [639])

Ore type	Reduction test on	Particle size range mm	Chemical composition % Fe	FeO	Fe_2O_3	SiO_2	Oxygen input g	Reduction rate $\left(\frac{d\Re}{dt}\right)_{40\,\%} = \frac{33.6}{t_{60} - t_{30}}$
Kiruna-D	1st Nov., 1963	3.15 to 5	55.43	21.92	54.85	8.0	212.9	33.6:(93.5 − 36.5) = 0.59
		10 to 16	59.20	23.73	58.34	3.55	228.4	33.6:(162 − 63) = 0.34
		31.5 to 40	59.73	24.04	58.81	3.72	229.5	33.6:(520 − 200) = 0.105

gas supply per unit of time: 5 m^3/h at NTP (corresponding to a free cross-sectional velocity of 38.8 cm/sec at 900°C);

inside diameter of vessel: 140 mm.

DICKENS et al.[635]) found that the influence of the flow rate and, consequently, of convection and diffusion within the reaction tube disappeared completely only at higher flow rates (exceeding a free cross-sectional velocity of about 100 cm/sec). The lower flow rate mentioned above may, however, be used without fundamental corruption of the test results.

Weight of sample in tube: 1000 g.

Size grading: three of the following ranges*: 3.15 to 5 mm,
10 to 16 mm,
31.5 to 40 mm,
50 to 63 mm.

The reduction rate at 40% reduction is used as a measure of reducibility. An example of an assessment is shown in Figs. 323 and 324.

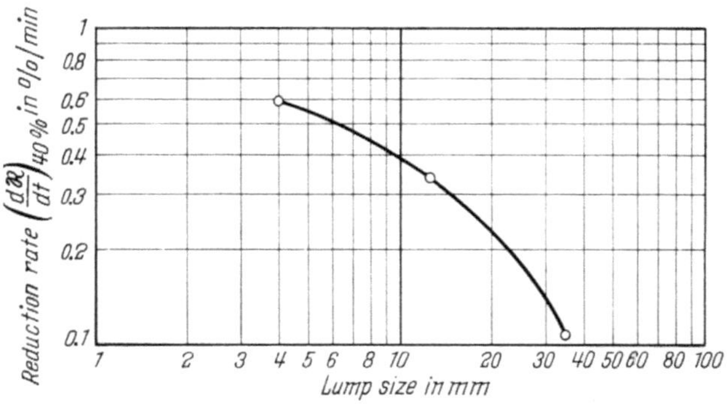

Fig. 324. Relationship between lump size and reduction rate (example using Kiruna-D ore), according to Stahleisen-Prüfblatt 1770-64 [639])

The oxygen removal and, consequently, the degree of reduction are continuously monitored with the balance. The degree of reduction at the end of the test is assessed by chemical analysis of the reduced product for metallic and total iron (Fe_{met} and Fe_{tot}). This analysis can be performed extremely quickly by reason of the magnetic properties of the reduced product[638]).

A mixture of 60% N_2 and 40% CO is used for reduction. Pure CO would cost less and would accelerate testing; it would also be acceptable from the point of view of reaction kinetics. All reduction rates would be accelerated by a factor of 2.5. Unfortunately, pure CO gas decomposes

* The figures for the grain diameter correspond to DIN 4187/88 for perforated plate and wire mesh for test screens (square mesh).

considerably around 500°C due to the Boudouard reaction:

$$2\,CO = C + CO_2$$

with the equilibrium constant

$$K_p = \frac{p_{CO}^2}{p_{CO_2}}$$

and deposits carbon in the preheat zone of the test retort. The equilibrium constant K_p is a quadratic function of p_{CO} and, therefore, of the decom-

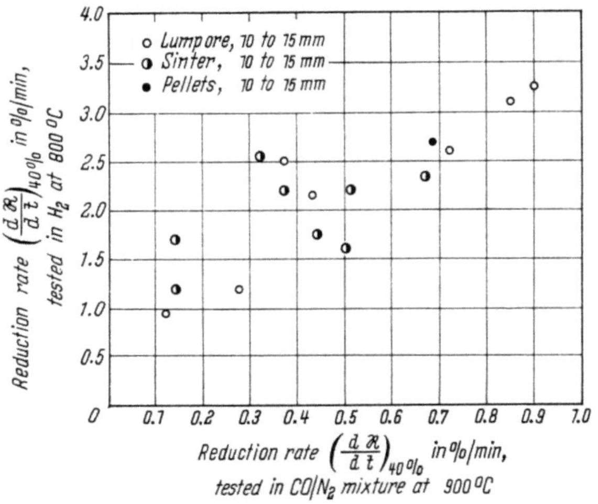

Fig. 325. Isothermal reducibility test. Comparison of reduction rate at 40% reduction in relation to reduction conditions. Test performed with CO/N_2 mixture at 900°C, according to Stahleisen-Prüfblatt 1770-64 [644])

position probability. Dilution with 60% N_2, therefore, considerably inhibits decomposition.

The first reduction tests of this type, concerned with the influence of particle size, used very simple investigational procedures and hydrogen at 800°C[609, 612]). Results were largely the same as those obtained with CO/N_2 mixtures. When a wide range of ores and agglomerates are reduced under both sets of conditions (H_2 at 800°C and 40% CO/60% N_2 at 900°C), there is a clear relationship between the test results (Fig. 325). Indeed, with sinter samples of particularly poor reducibility, reduction with H_2 as compared with CO/N_2 mixture appears to be more rapid than with other samples. This is confirmed by investigations made by KRAINER et al.[643]) and SKODIN[642]) who observed that additions of even small amounts of H_2 to CO-rich reducing gases accelerated reduction more with sinter samples of poor reducibility than with other samples. This indicates a change in the mechanism of diffusion in the micropores[642]). As a reducing gas, H_2 is not, therefore, without its disadvantages as regards kinetics and the CO/N_2 mixture is preferable despite the more complicated handling involved.

Figure 326 shows test results for the reduction of a large number of ores under the standardised conditions of Stahleisen-Prüfblatt 1770-64.

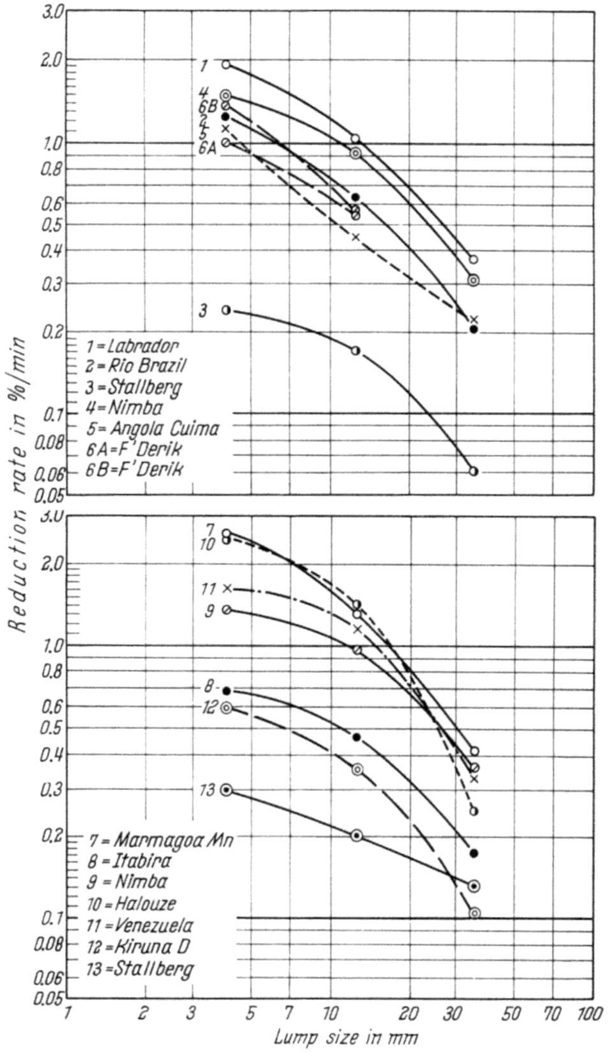

Fig. 326. Reduction rate in relation to lump size of various ores, tested according to Stahleisen-Prüfblatt 1770-64 [646])

The analyses are given in Table 44*. The dependence on particle size is evident and these test data enable a given ore mixture to be controlled to a required reducibility.

Table 44. *Analyses of ores in Fig. 326*

Ore	Fe_{tot}	FeO	Fe_2O_3	Mn	P	SiO_2	Al_2O_3	CaO	MgO
Labrador	60.05	0.0	85.87	0.92	0.13	7.50	0.73	0.25	0.06
Rio Brazil	76.60	0.91	95.65	0.07	0.047	1.11	1.06	0.34	0.16
Stallberg	47.36	25.95	38.95	4.37	0.015	10.00	0.83	6.25	3.96
Nimba	67.02	0.77	94.98	0.16	0.063	0.50	1.68	0.16	0.08
Angola Cuima	64.27	7.55	83.54	0.17	0.055	4.02	1.54	0.25	0.70
F'Derik A	67.57	0.0	96.63	0.04	0.10	1.12	1.00	0.34	0.04
F'Derik B	66.46	0.0	95.04	0.04	0.10	2.30	1.34	0.28	0.04
Marmagoa	46.30	0.0	66.21	7.71	0.040	5.10	7.64	0.34	0.12
Itabira	68.80	0.0	98.38	0.05	0.027	0.45	0.60	0.28	0.08
Nimba	67.69	1.55	95.08	0.22	0.068	0.73	1.06	0.28	0.08
Halouze	49.50	9.11	60.69	0.39	0.84	14.56	6.76	3.47	1.45
Venezuela	59.82	0.0	85.54	0.05	0.09	1.38	0.92	0.15	0.06
Kiruna D	57.89	24.08	56.01	0.05	1.36	4.93	0.99	5.31	1.41
Stallberg	46.13	25.69	37.41	4.36	0.02	9.84	0.72	6.66	4.51

5.4. Injection of Fuels

5.4.1. Statement of the Problem

The idea of partially or completely replacing coke (originally the only fuel used in the blast furnace) by other fuels is relatively old[651–655]. Such replacement fuels might well include gaseous (particularly natural gas and coke oven gas), liquid (light and heavy mineral oils and tar oils) and solid fuels (powdered coal). The relatively late initiation of systematic blast furnace trials with replacement fuel injection[666, 666a, 680, 697, 697a] was soon followed by the adoption of this practice in full-scale iron production. The addition of replacement fuels (also known as "substitute fuels" or "auxiliary fuels") is, of course, a particularly attractive proposition where the heat unit can be provided more cheaply by the replacement fuel than by blast furnace coke. Analysis of different sources of primary energy shows how the proportions have varied as a result of keen competition[656–658b]. Available publications indicate that there is a variation of 1 to 2% in the percentage proportions of energy carriers. This can be ascribed mainly

* Some of the ores in Tables 43 and 44 are of the same type although of slightly differing composition. This is due to the different investigation times (Table 43: 1958/59, Table 44: 1963).

to the use of different calorific values. Figures 327 and 328 show the trend in the Federal Republic of Germany and the United States. Conversion factors for the various fuels are given in Table 45.

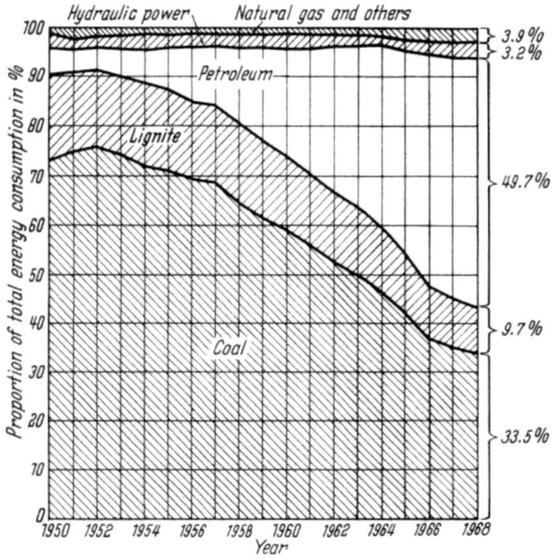

Fig. 327. Primary energy consumption in the Federal Republic of Germany[657, 658b])

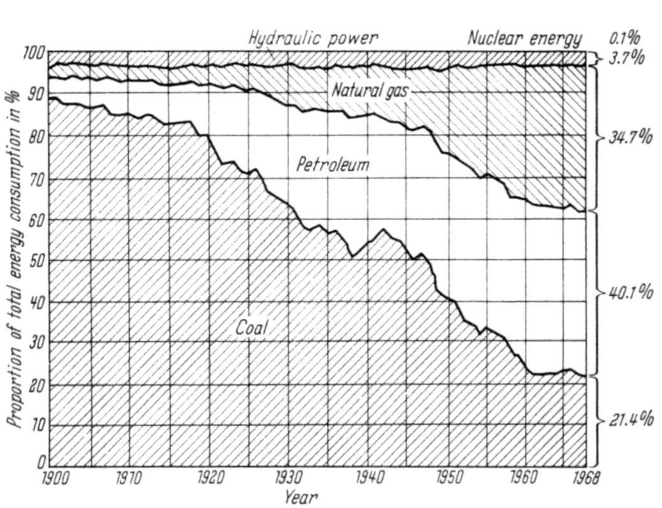

Fig. 328. Primary energy consumption in the U.S.A., according to Laurien and Wedekind[656]) and reference[658b])

Provisional data for 1968

Table 45. *Conversion factors related to the standard calorific unit = minimum calorific value of coal = 7000 kcal/kg*[657])

Coal	kg/kg	1.0
Coal briquettes	kg/kg	1.071
Coke	kg/kg	0.971
Untreated lignite	kg/kg	0.286
Lignite products	kg/kg	0.69
Hard lignite	kg/kg	0.5
Pitch coal	kg/kg	0.71
Coke-oven gas	kg/m³ at NTP	0.57
Natural gas	kg/m³ at NTP	1.29
Natural gas from petroleum deposits . .	kg/m³ at NTP	1.47
Petroleum, mineral oil products	kg/kg	1.43
Electric power	kg/kWh	0.4
Firewood	kg/kg	0.5
Peat for firing	kg/kg	0.43

In considering the replacement of coke by other fuels, it must always be remembered that coke fulfils three primary functions in the blast furnace:

a) it supplies heat;
b) it acts as a reducing agent;
c) it supports the burden.

5.4.1.1. Coke as a Source of Heat

The fuel coke is burned at the tuyeres with preheated blast to form CO. According to the Boudouard equilibrium, combustion cannot proceed beyond the CO stage

$$C + \tfrac{1}{2}\,O_2 = CO;$$

$$\Delta H = -26.416\ \text{kcal/mole} = -2202\ \text{kcal/kg C}.$$

The reaction between the fuel coke and hot blast at the tuyeres produces what is known as tuyere gas. This is composed of about 34.9% CO, 63.5% N_2, and 1.6% H_2 and has a sensible heat content of around 700 kcal/m³ at NTP (this corresponds to a tuyere gas temperature of about 2000°C) depending on blast temperature. As it rises up the furnace, this gas releases most of its heat to the stock column, thus providing the heat required for all the endothermic reactions and the heating and melting of the charge. The gas leaves the top of the furnace at a temperature of between 100 and 400°C.

The quantity of heat available in the furnace is made up of the heat of the reaction C → CO plus the sensible heat of the blast less the heat lost in the top gas and the wall losses.

5.4.1.2. Coke as a Reducing Agent

The carbon monoxide (and, in some cases, the hydrogen) in the tuyere gas act as reducing media:

$$Fe_nO_m + m\,CO = n\,Fe + m\,CO_2\,.$$

As far as this reaction occurs at high temperatures (above about 1100°C), carbon monoxide is formed in addition from the reduction carbon dioxide according to the Boudouard reaction:

$$CO_2 + C = 2\,CO\,.$$

The coke consumed in this way is known as "reduction coke".

5.4.1.3. Coke as a Support for the Burden

Because of its good hot strength, far exceeding that of any other burden component, coke also maintains burden permeability and, consequently,

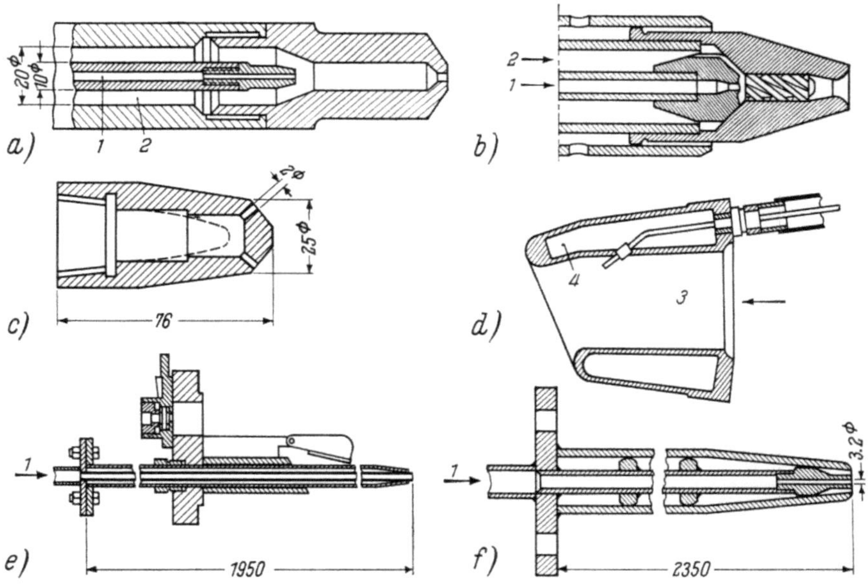

Fig. 329. Systems for injection of fuel oil through blast furnace tuyeres, according to GRIMM[660a])

1 Fuel oil, *2* Atomising air, *3* Tuyere, *4* Cooling jacket

a) Central lance (Low shaft furnace at Liège, Belgium)
b) Central lance (IRSID, France)
c) Central, six-hole lance for fuel oil (Cockerill-Ougrée, Seraing, Belgium)
d) Tuyere atomiser cooled by the tuyere cooling jacket (Steel Company of Wales, Margam, Richard Thomas & Baldwins, Ebbw Vale)
e) Central lance (Richard Thomas & Baldwins, Ebbw Vale)
f) Central lance (Shell-BISRA, England)

smooth running of the furnace, even when all the other components of the burden soften or melt.

The simplest and most reliable method of introducing replacement fuels into the blast furnace is by injection through the tuyeres. Figures 329 and 330 illustrate typical injection systems for heavy fuel oil and powdered

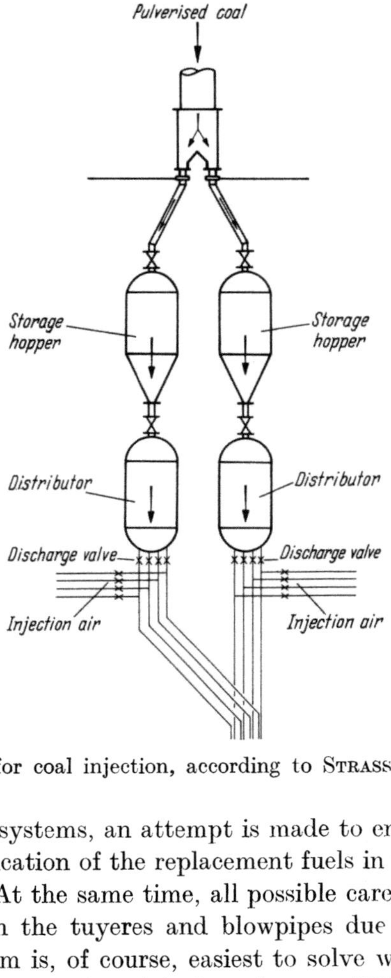

Fig. 330. System for coal injection, according to STRASSBURGER et al.[661a)]

coal. With all these systems, an attempt is made to ensure complete combustion and/or gasification of the replacement fuels in the oxidising tuyere zone of the hearth. At the same time, all possible care is taken to prevent unnecessary wear on the tuyeres and blowpipes due to premature combustion. This problem is, of course, easiest to solve with gaseous replacement fuels, less easy with liquid fuels and most difficult with powdered coals[661a–c)]. An interesting compromise solution is the injection of pumpable suspensions (e.g., 60% pulverised coal with 40% heavy fuel oil)[661b)]. This type of suspension is notably stable (Fig. 331).

About half the blast furnaces in the western world are equipped for injection of replacement fuels.

The replacement fuels are first burnt directly in front of the tuyeres by the excess hot blast to form CO_2 and H_2O and then they are subsequently reduced to CO and H_2 on contact with the excess coke. Altogether, the following reactions occur in the tuyere zone:

$$\text{with natural gas:} \quad CH_4 + \tfrac{1}{2} O_2 = CO + 2\, H_2;$$
$$\text{with oil:} \quad -CH_2- + \tfrac{1}{2} O_2 = CO + H_2.$$

CO and H_2 are also formed as end products with coal in proportions corresponding to its chemical composition.

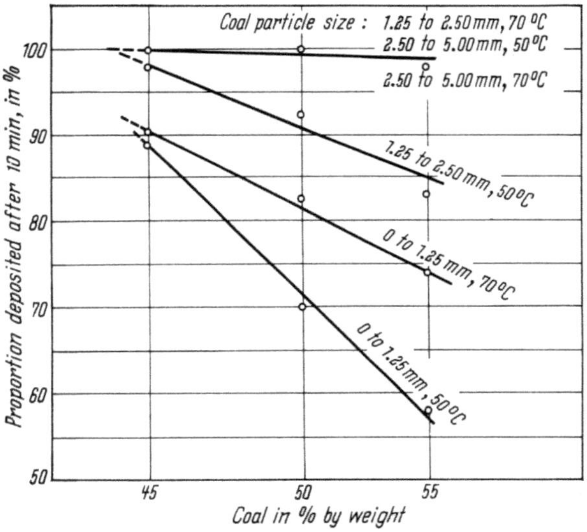

Fig. 331. Sedimentation of pulverised coal of varying particle size in heavy fuel oil after 10 minutes, according to LIMPACH[661c])

Table 46 surveys the properties of the relevant liquid and gaseous replacement fuels. The tuyere gas formed during their gasification contains more hydrogen than the gas formed from coke. Hydrogen is a more powerful reducing agent than CO[664–664b]). An increase in its concentration in the tuyere gas, therefore, initially constitutes an advantage. Even very small H_2 contents greatly accelerate reduction especially with sinter of very poor reducibility[664a, b]). It is clear, however, from Table 46, that the reaction heat from gasification of hydrogenous replacement fuels is less than that obtained with coke, particularly considering the volume of tuyere gas formed and the fact that only the fuel coke, preheated to about 1500°C, undergoes combustion. This preheating is impossible in the case of replacement fuels. According to Fig. 332, the quantity of heat ΔQ ob-

Table 46. *Properties of coke and hydrogen carriers*

		% O_2 in blast	Coke	Heavy fuel oil	Natural gas	Coke-oven gas	Steam
Chemical composition:					**	**	
Carbon	%		87.0	84.0	—	—	—
Ash	%		10.0	0.5	—	—	—
Sulphur	%		1.0	2.5	—	—	—
Hydrogen	%		0.4	12.0	—	57.0	11.11
Oxygen	%		0.6	0.5	—	—	88.89
Nitrogen	%		1.0	0.5	—	6.9	—
Methane	%		—	—	100.0	26.0	—
Carbon monoxide	%		—	—	—	5.7	—
Carbon dioxide	%		—	—	—	2.0	—
Heavy hydrocarbons	%		—	—	—	2.4	—
Calorific value H_u	kcal/kg (m^3 at NTP)		6900 to 7200	9600 to 9800	8580	4200	—
Wind required to transform		21	3.84	3.71	2.38	0.74	—2.96***
into CO and H_2	m^3/kg at NTP	22	3.67	3.54	2.27	0.71	—2.83
	or m^3/m^3 at NTP	23	3.51	3.39	2.17	0.68	—2.70
Tuyere gas from coke or hydrogen carriers:							
Carbon monoxide	m^3/kg at NTP or m^3/m^3 at NTP		1.62	1.57	1.00	0.42	1.24
Hydrogen		21	0.05	1.34	2.02	1.17	1.24
Nitrogen			3.06	2.95	1.88	0.65	—
Tuyere gas total	m^3/kg at NTP	21	(4.67)* 4.73	5.86	4.90	2.24	2.48
	or m^3/m^3 at NTP	22	(4.49) 4.55	5.70	4.79	2.21	2.48
		23	(4.33) 4.39	5.54	4.69	2.18	2.48
Composition of tuyere gas:							
Carbon monoxide	%		34.2	26.8	20.5	18.7	50.0
Hydrogen	%	21	1.1	22.9	41.1	52.3	50.0
Nitrogen	%		64.7	50.3	38.4	29.0	—
Gasification and cracking reaction			$C + \frac{1}{2} O_2 = CO$	$\text{-}CH_2\text{-} + \frac{1}{2} O_2 = CO + H_2$	$CH_4 + \frac{1}{2} O_2 = CO + 2 H_2$	$CH_4 + \frac{1}{2} O_2 = CO + 2 H_2$	$H_2O = H_2 + \frac{1}{2} O_2$
Reaction heat	kcal/kg (m^3 at NTP)		—1912	—1466	—381	—169	+3211

* Without volatile components. ** Percentage by volume, otherwise percentage by weight. *** Blast saved.

tainable in the tuyere zone increases with the C/H ratio. The CO_2 present in natural gas and coke-oven gas considerably reduces the heat of reaction in the cracking and gasification reaction (about 30 kcal/m³ at NTP of natural and/or coke oven gas for each 1% CO_2). With pulverised coals

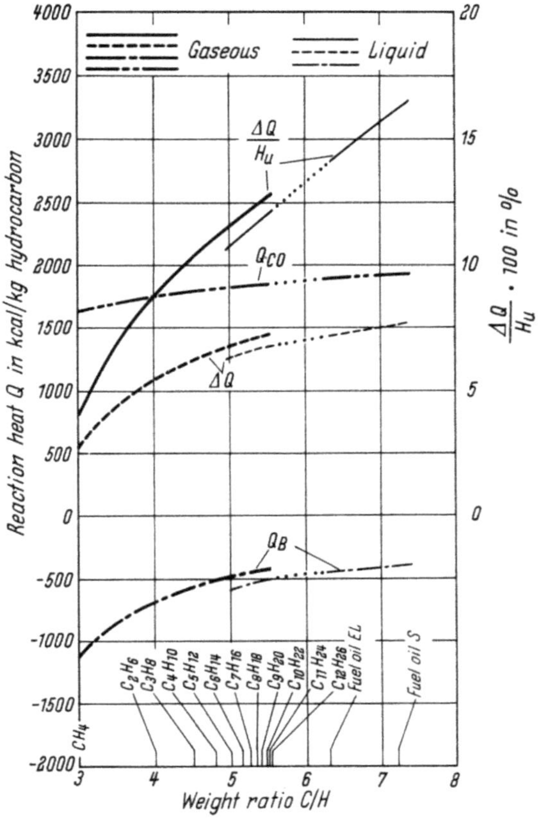

Fig. 332. Reaction heat in thermal cracking and gasification of hydrocarbons. Reference state: 1 atm, 25°C

Q_{CO} = heat of combustion of C → CO, Q_B = heat of formation of the hydrocarbon, $\Delta Q = Q_{CO} - Q_B$, Fuel oil EL = light oil, Fuel oil S = heavy fuel oil

(Table 47a), the heat of the gasification reaction, at lower H_2 contents, is greater than with oil and gas. Table 47b shows the volumes and compositions of tuyere gas obtained with various types of coal. From the thermal viewpoint, therefore, coal injection is preferable to oil or gas on account of its smaller cooling effect, while from the point of view of reaction kinetics, oil and gas have the advantage of accelerating reduction by H_2. The overall effects of these phenomena will be discussed later.

Table 47 a. *Composition and calorific values of solid fuels*

No.	Composition d.a.f*	Anthracite	Coal with 19 to 28% volatile matter	Coal with 28 to 35% volatile matter	Coal with 35 to 40% volatile matter	Lignite
1a	C . . . . %	92.0	88.9	86.6	83.6	68.3
1b	H . . . . %	3.4	4.8	5.3	5.7	5.0
1c	O . . . . %	2.3	3.9	5.9	8.5	25.7
1d	N . . . . %	1.5	1.6	1.5	1.5	0.5
1e	S . . . . %	0.8	0.8	0.7	0.7	0.5
1f	Volatile components %	9.0	22	32	38	55
2	Calorific value, d.a.f.* kcal/kg p.c.**	8390	8420	8280	7980	6037
3	Calorific value of CO kcal/kg p.c.**	5181	5007	4877	4708	3846
4	Calorific value of H_2 kcal/kg p.c.**	975	1376	1519	1634	1433
5	Calorific value of CO and H_2 (3 + 4). . kcal/kg p.c.**	6156	6383	6396	6342	5279
6	Reaction heat, d.a.f.*, devolatilisation + gasification (2 — 5) . kcal/kg p.c.**	2234	2037	1884	1638	758

* d.a.f. = referred to dry and ash-free substance.
** p.c. = pure coke and/or pure coal.
Reaction heat S → SO_2 too small for inclusion.

Table 47 b. *Blast requirement, tuyere gas volume and composition with a blast humidity of* 10 g/m³ *at NTP; ash content of coal:* 6%, *dry*

Coal moisture	Anthracite		Coal with 19 to 28% volatile matter		Coal with 28 to 35% volatile matter		Coal with 35 to 40% volatile matter		Lignite	
in % of raw coal →	3	5	3	5	3	5	3	5	3	5
Blast requirement m^3/kg at NTP (d.a.f.)	3.78	3.72	3.53	3.47	3.44	3.37	3.32	3.16	2.01	1.94
Tuyere gas volume m^3/kg at NTP (d.a.f.)	5.19	5.16	5.08	5.06	5.02	4.99	4.83	4.81	3.49	3.46
Tuyere gas composition										
CO . . . %	33.1	33.3	32.7	32.8	32.2	32.4	32.3	32.5	36.5	36.8
H_2 . . . %	9.0	9.5	12.2	12.8	13.4	14.0	14.8	15.4	17.8	18.8
N_2 . . . %	57.9	57.2	55.1	54.4	54.4	53.6	52.9	52.1	45.7	44.4

Generally, however, with small amounts of replacement fuels and otherwise constant conditions, the thermal efficiency of the furnace is improved by the reducing effect of the H_2. With higher injections of replacement fuel.

however, the reduction in tuyere gas temperature may reduce thermal efficiency.

These conditions soon impose an economic limit on replacement fuel injection. Conditions improve as long as the cooling effect of replacement fuels can be compensated by a corresponding increase in hot blast temperature, thus maintaining constant the theoretical temperature and/or heat content of the tuyere gas. Much higher replacement ratios result in this instance (coke saved per unit fuel injected). A limit is also rapidly imposed in this instance, however, by the heating capacity of the stoves. Bustle main temperatures of 1100°C are about the limit with traditional chamotte-lined stoves, 1200°C, with sillimanite linings and 1350°C, for silica stove fillings[665, 665a]). Maximum hot blast temperatures are often already used under normal driving conditions (i.e., without replacement fuel injection). This makes it impossible to increase blast temperature any further. Certain burden conditions, such as the presence of lump ores with low softening temperatures, allow operation with relatively low hot blast temperatures. Replacement fuel injection, in this case, provides a good opportunity to increase both hot blast temperature and furnace efficiency. The technical and economic optimum will generally lie between the following extremes:

a) injection of replacement fuels with no change in blast temperature;

b) injection of replacement fuels with no change in the tuyere gas heat content (i.e., with increased blast temperature).

The function discussed in section 5.4.1.3 should also be borne in mind. No replacement fuel can fulfil the role of coke as a support for the burden. The minimum coke charge on which the furnace can still operate has not yet been determined. Ischebeck et al.[661]) were able to show that burden permeability decreases appreciably with increasing oil injection, i.e., diminishing coke charge (Fig. 333). A lower coke consumption limit has not been established in reproducible form to date because economic factors have prevented fuel injection from being deliberately pushed to this limit. The lowest coke consumptions were reported by Osaka Iron and Steel Co. 367 kg dry coke/ton iron with a blast temperature of 1078°C, an oil injection rate of 153 kg/ton iron and a slag bulk of 353 kg/ton iron[715]).

The problem is, therefore, to utilise established laws and concepts of thermodynamics and reaction kinetics in conjunction with knowledge acquired in laboratory and plant trials to arrive at precise predictions which can be used as a basis for estimating the probable effect of a given replacement fuel on given blast furnace operating conditions. The following questions, in particular, require explanation:

How do the main operating parameters change (output, coke consumption, top gas volume and composition)?

At what point does replacement fuel injection become economic?

Which replacement fuel is best?

Several attempts have been made to develop a calculation system to answer these questions[662, 663, 667–676a].

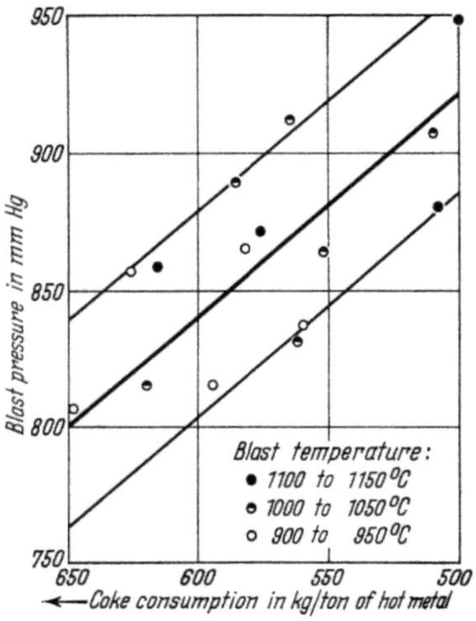

Fig. 333
Rise in blast pressure with decreasing coke consumption, according to ISCHEBECK et al.[661]

5.4.2. Fundamental Considerations in the Prediction of Blast-Furnace Operating Parameters with Fuel Injection

The statements in section 5.1 also apply with fuel injection into the blast furnace. Prediction of the influence of injected fuels, without recourse to operating experience, is only possible if the following are known: all reaction rates, especially for reduction of the ore, gasification of coke by CO_2, heat transfer in the heating and melting of the charge over the entire range of prevailing temperatures, gas compositions and pressures and variations in the burden. Due to lack of adequate data, particularly on the reaction rates, true prediction has not yet been possible.

With the combined application of practical furnace operating data and established thermodynamic and kinetic laws, however, a system of equations can be evolved in which few parameters need to be classed as assumptions only verifiable by industrial trials. Using this method, together with the results of certain specifically-controlled plant trials in which all relevant parameters have been accurately measured, the basic assumptions can be adapted to operating practice and the evolved equation system can be used to predict for many future contingencies with reasonable accuracy.

5.4.3. Calculation Procedure

5.4.3.1. Gasification of Fuels in Front of the Tuyeres

Current knowledge indicates that replacement fuels are most effective when completely gasified by the blast oxygen to CO and H_2 directly in front of the tuyeres. The injection systems must, therefore, be equipped for maximum possible atomisation of the injected fuel in the furnace blast. Combustion should proceed in the oxidising zone immediately in front of the blowpipes. The hot blast and replacement fuel must, therefore, be conveyed separately to the blowpipe and then combined with maximum

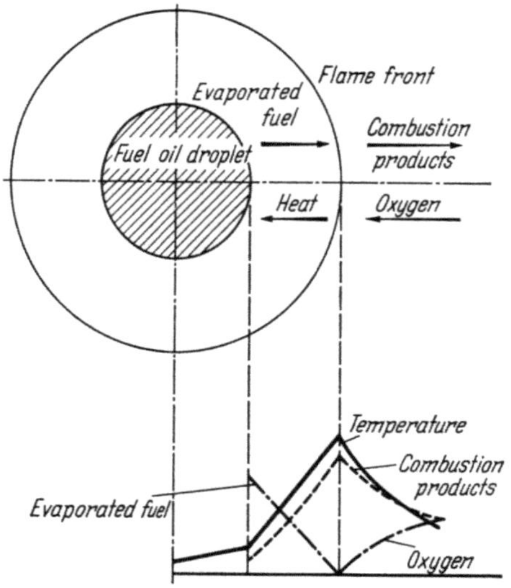

Fig. 334. Model of a burning oil droplet, according to MASDIN and GRIMM[677a])

intensity. Combination of the two within the tuyere is permissible if the hot blast flow rate is higher than the fuel ignition rate. In the case of oil, the gasification and/or combustion process has been explained almost completely. According to GRIMM[677a]) whose ideas largely coincide with those of FRITSCH[677b]), MASDIN and THRING[677c]), the burning oil droplet (Fig. 334) passes through the following stages:

a) heating and evaporation of the fuel;

b) mixing of the fuel vapour with the hot blast;

c) ignition of the fuel vapour and, in certain cases, of the carbon cenospheres;

d) combustion.

With completely evaporable oils, the evaporation rate is the combustion-determining step; the droplet diameter falls according to the equation

$$D_0^2 - D^2 = k(t - t_0)$$

in which

D_0 = initial diameter (at the moment t_0),
D = diameter (at the moment t),
k = evaporation constant (between $1.0 \cdot 10^2$ and $1.5 \cdot 10^2$ cm²/sec for kerosene in air).

This equation shows the necessity of keeping D_0 to a minimum, i.e., of atomising the fuel to the finest possible droplets. If complete combustion

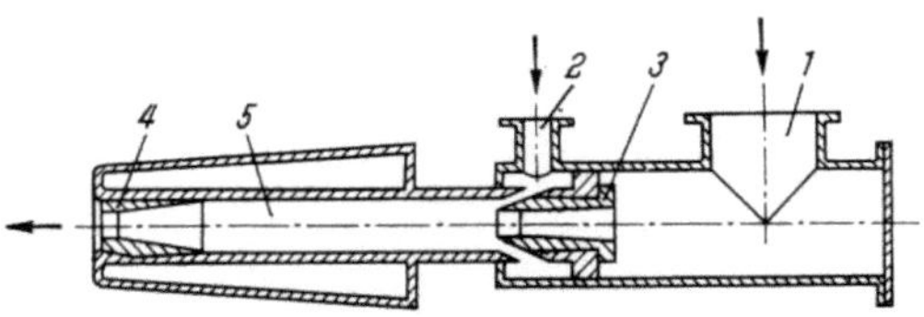

Fig. 335. Tuyere used in a slagging gas producer working on a combination of petrol and coke, according to DUFTSCHMID and MARKERT[679])

1 Entry of gasification medium, 2 Entry of petrol vapour, 3 Mixing device, 4 Tuyere mouth, 5 Mixing zone

and conversion of injected gas or oil to CO and H_2 are not attained in the oxidising zone in front of the tuyeres, the unburnt residues are cracked (particularly in the case of oil) in the furnace stock column, resulting in carbon black deposition in the tuyere and/or hearth gas and eventual fouling of the gas cleaning plant[702]). This is accompanied by decreased fuel utilisation (the heat content of the carbon black is lost to the furnace) and, in the event of heavy sooting, blockages in the stock column.

The conversion of gaseous fuels and petrol in the tuyere zone has been investigated with probes by DUFTSCHMIDT and MARKERT[679]) in the comparable conditions of a slagging gas producer. Figures 335 and 336 show a typical result. Complete conversion to CO and H_2 is achieved shortly after exit from the tuyere mouth. The gaseous oxygen disappears closer to the tuyere mouth than it does with coke alone and the flame shortens. It is, therefore, concluded that heat removal in the tuyere cooling water is increased which means greater heat losses.

Present-day techniques are, however, sufficiently developed to ensure complete gasification and/or conversion of gaseous and liquid fuels to CO and H_2.

With solid fuel injection conditions in front of the tuyere are more difficult to assess. The Russians initially used pulverised coal with a particle size range of 0 to 0.1 mm [680]) to obtain complete gasification to CO and

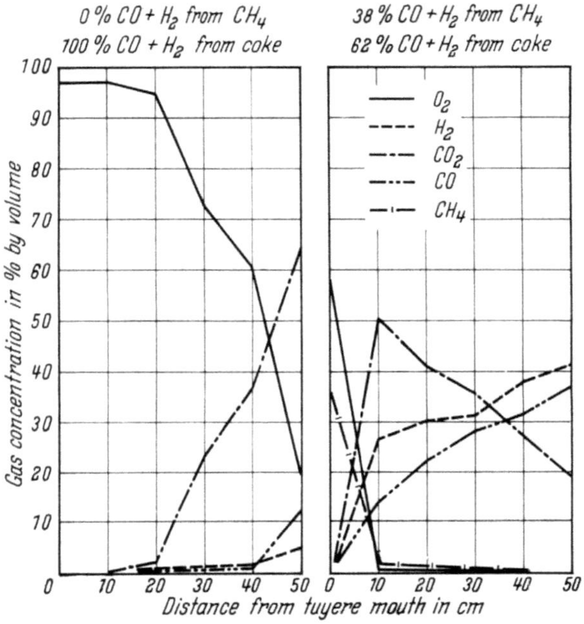

Fig. 336. Progress of gasification in front of the tuyeres, according to DUFTSCHMID and MARKERT[679])

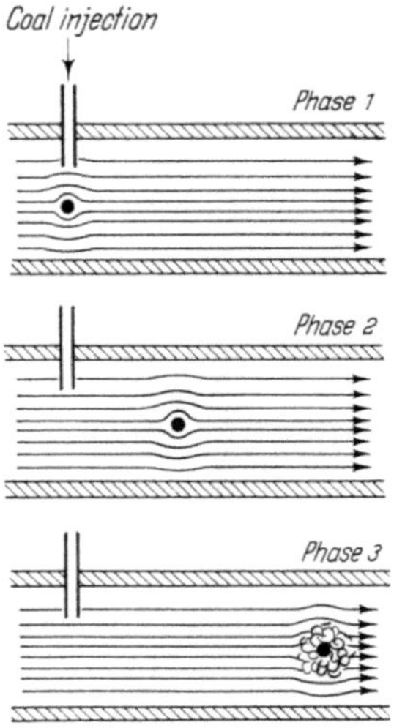

Fig. 337. Processes occurring when a coal particle is injected into the hot blast[663])
Blast velocity $c = 150$ m/sec; $p = 2.0$ atm absolute; $T = 1000°C$

Phase 1: Coal particle enters blast; initial velocity $u = 0$; velocity difference $\Delta u = c - u = 150$ m/sec; substantial heat transfer
Phase 2: Coal particle has accelerated; velocity $0 < u < c$; velocity difference $\Delta u = c - u < 150$ m/sec; reduced heat transfer
Phase 3: Ignition temperature reached, rapid increase in devolatilisation rate and heat transfer

H_2 in the tuyere zone. Pulverisation of coal is an expensive operation, however. Utilisation of fine coal with a particle size of 0 to 3 mm or even 0 to 6 mm is more economic.

Processes within and on the surface of coal particles added to hot blast have recently been investigated[663, 677]).

Figure 337 shows a model rig which served as a basis for assessment[663]). The coal particles are injected at ambient temperature into the hot blast. The blast rate would be of the order of $c = 150$ m/sec (referred to operating conditions, i.e., 2.0 atm absolute and 1000°C). The velocity of the coal particle towards the tuyeres is initially $v = 0$; it gradually accelerates by transferred momentum to $c = 150$ m/sec. Using the known law for resistance of spheres to flow[681]) and the heat transfer equation evolved by Ranz and Marshall[682]),

$$N_{\mathrm{Nu}} = 2 + 0.6 N_{\mathrm{Re}}^{1/2} N_{\mathrm{Pr}}^{1/3}$$

the volume of heat transferred by convection can be calculated. The relevant difference in velocity of the blast and coal particle is represented by Δu in the relation

$$N_{\mathrm{Re}} = \frac{\Delta u\, d}{\nu}.$$

The thermal conductivity of the coal particle is taken as $\lambda = 0.2$ kcal/m h °C[683, 683a]).

The estimated heating per unit time is shown in Fig. 338. It is clear that, within a few thousandths or hundredths of a second, surface temperature increases sufficiently to ignite the gaseous coal.

Figure 339 shows the ignition temperatures in relation to the combustible fraction of coal volatiles. The estimated result is confirmed by experimental findings. Once ignition has occurred, the conditions on which calculation was based no longer apply, heat transfer increases considerably

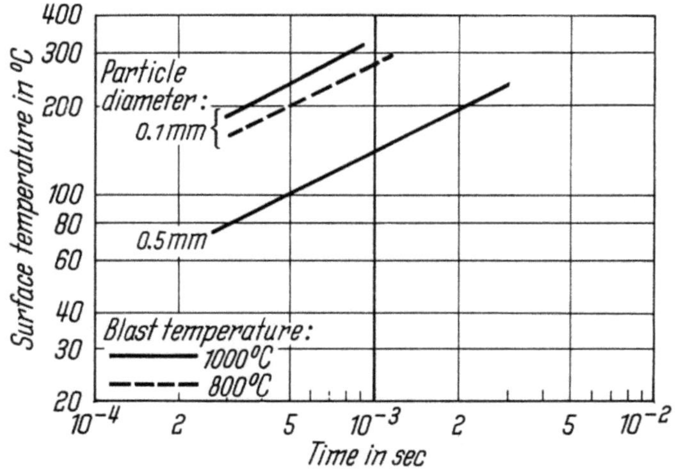

Fig. 338. Surface temperature in relation to time, blast temperature and particle diameter (thermal conductivity 0.2 kcal/m h °C)[663])

and devolatilisation and gasification of the coal particle to CO and H_2 are intensified. PETERS and LEHMANN[677]) separately determined the duration of the individual processes from heating-up to gasification, as shown in

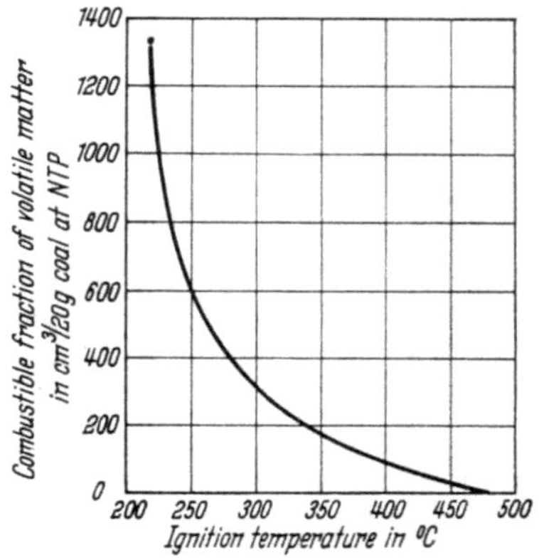

Fig. 339. Relationship between the combustible fraction of volatiles and ignition temperature, according to GUMZ[684])

Fig. 340, and concluded that complete gasification of the coal particles was unlikely in the oxidising zone. The residence times are too short (Fig. 341). Nearly complete devolatilisation can, however, be assumed for

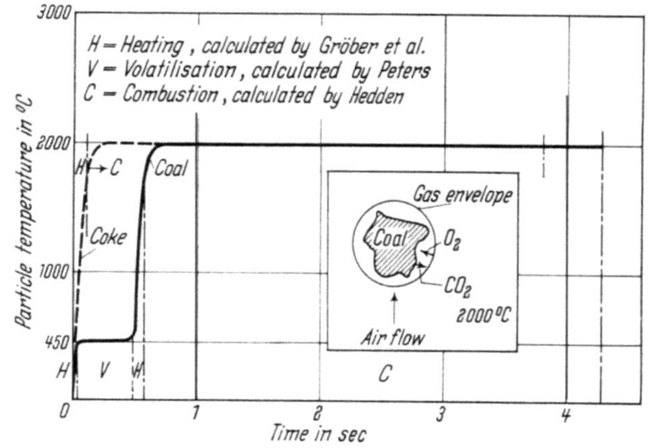

Fig. 340. Progress of combustion of 1-mm coke and coal particles in air at 2000°C and 2 atm absolute, according to PETERS and LEHMANN[677])

particles in the size range above 1 mm. If gasification fails to occur in the tuyere zone, the devolatilised coal particle passes up into the stock column. In adverse conditions, particularly with pulsations in the tuyere blast, the carbon particles become enveloped in slag and, instead of reacting chemically, leave the furnace along with the slag[678]). In favourable conditions, however, the devolatilised coal particles replace some of the reducing coke. The conditions in which either of these alternatives can be expected to predominate have not been fully explained. Only industrial trials can clarify this question. The following calculations are based on favourable

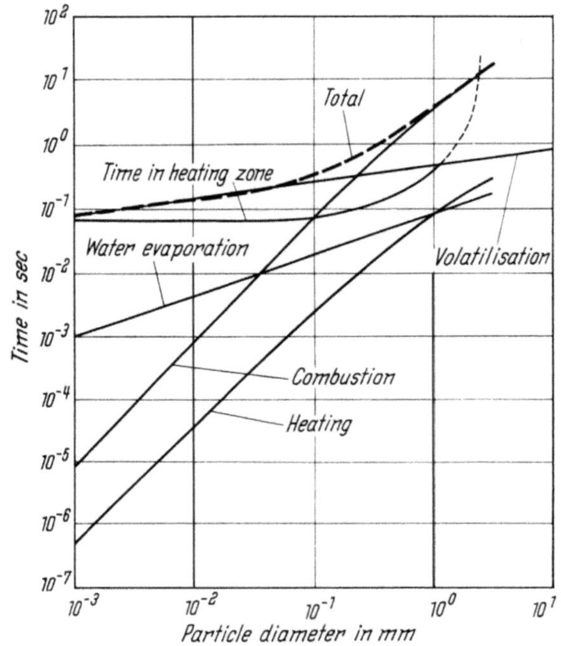

Fig. 341. Dependence of coal combustion on particle size in air at 2000°C and 2 atm absolute, according to PETERS and LEHMANN[677])

conditions. Such conditions have prevailed in most of the industrial trials reported to date.

All subsequent calculation is based on the assumption that the replacement fuels are completely gasified to CO and H_2 in the tuyere zone*.

5.4.3.2. Control of Blast-Furnace Reactions with Maintenance of Constant Tuyere Gas Temperature (i.e., Increasing Blast Temperature)

Some of the fundamental ideas given in section 5.4.1 will now be developed.

The required parameters can be divided into independent and dependent variables, i.e., those which can be freely chosen within certain limits and those which necessarily depend on the initial choice made.

* The calculation is not affected if fine coals enter the stock column in the form of a devolatilised carbon cenosphere instead and replace reducing coke[663]).

The independent variables include:

a) Burden: taken as constant;

b) Heat content of the tuyere gas Q_F (which can be influenced by the injections and the blast temperature): also taken as constant for the first calculation (Q_F = const);

c) Volume of tuyere gas per unit time V_{FZ} (which can be influenced by the blast volume and injections): also taken as constant (V_{FZ} = const);

d) Type and quantity of replacement fuel.

The dependent variables include:

a′) Chemical utilisation of the CO in the stack gas

$$\eta_{\mathrm{CO}} = \frac{\mathrm{CO_2}}{\mathrm{CO} + \mathrm{CO_2}}$$

in which CO_2 and CO are the corresponding volume fractions in the top gas (corrections must be made for the burden CO_2 not combined with CaO);

b′) Chemical utilisation of the H_2 in the stack gas

$$\eta_{\mathrm{H_2}} = \frac{\mathrm{H_2O}}{\mathrm{H_2} + \mathrm{H_2O}}.$$

c′) Utilisation of the sensible heat of the stack gas (measured from the top gas temperature and cooling losses).

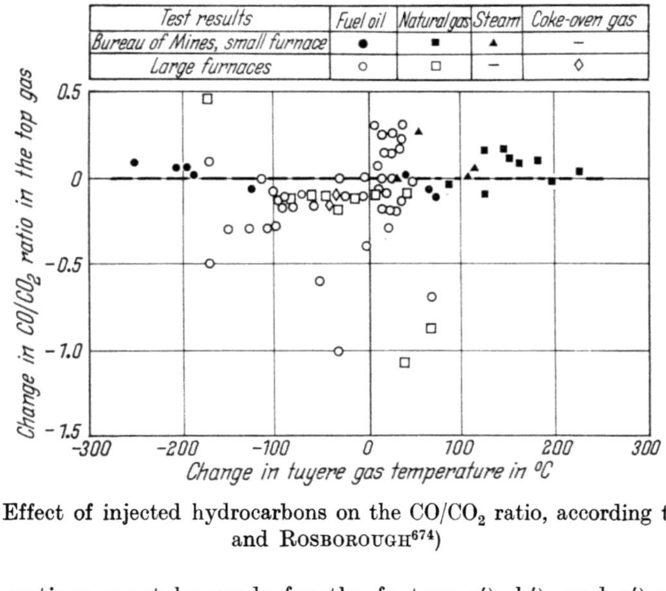

Fig. 342. Effect of injected hydrocarbons on the CO/CO_2 ratio, according to TRENSE and ROSBOROUGH[674])

Assumptions must be made for the factors a′), b′), and c′) which incorporate all aspects of the result (e.g., output, coke consumption, degree of indirect reduction). The correctness of the assumptions must be verified by industrial trials.

Note to a′) *Chemical utilisation of* CO. According to section 5.1, the degree of chemical utilisation η_{CO} generally depends closely on the burden. As industrial trials have shown, η_{CO} does not normally change during replacement fuel injection, if the burden is kept constant (Fig. 342). In the ensuing part of the calculation, η_{CO} is, therefore, initially assumed as constant.

Note to b′) *Chemical utilisation of* H_2. Assessment of industrial trials showed that, with burdens of good reducibility, chemical utilisation of the H_2 is parallel to that of CO.

Regression analysis gave

$$\eta_{H_2} = 0.88\eta_{CO} + 0.10 \quad (\% \text{ explained} = 68\%).$$

This finding was followed up in laboratory trials. The reduction rig used, when working on H_2/CO mixtures, allowed separate determination of the total reduction to H_2 and CO. η_{CO} and η_{H_2} were again found to be parallel (Fig. 343).

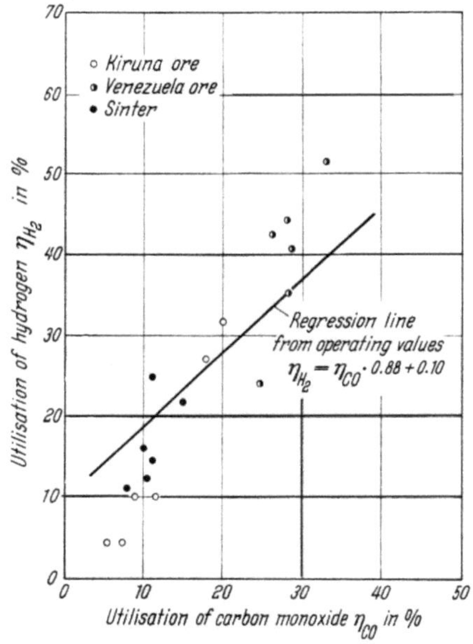

Fig. 343. Utilisation of hydrogen and carbon monoxide during reduction of iron ores in the retort up to a degree of reduction of 60%[662])

Particle size: 10 to 12 mm, carbon monoxide in reducing gas: 35%, flow rate of reducing gas: 40cm/sec

This finding can be attributed to the comparatively rapid establishment of water-gas equilibrium at indirect reduction temperatures[662a,b]) (Fig. 344). The equilibrium constant

$$K_p = \frac{(CO_2) \cdot (H_2)}{(CO) \cdot (H_2O)}$$

is 0.6 at 700°C and 1.6 at 1000°C, i.e., once large proportions of H_2O prevail as a result of reduction, they are partly transformed back to H_2 by the water gas reaction and CO is consumed. According to SKODIN[664a]) and KRAINER[664b]), hydrogen can be expected to have a stronger influence with sinter of poor reducibility.

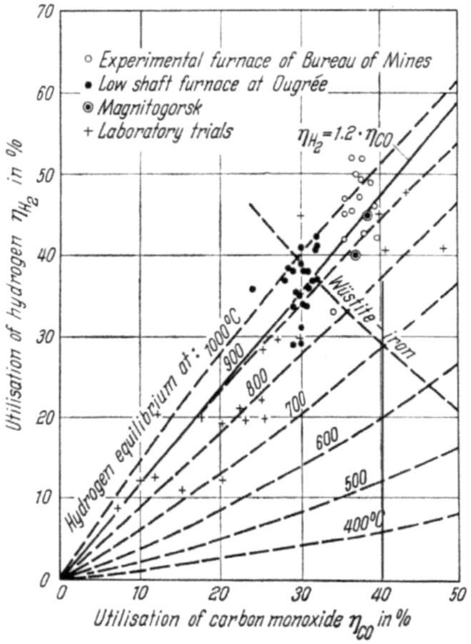

Fig. 344. Utilisation of hydrogen and carbon monoxide in the blast furnace, according to SCHÜRMANN and BÜLTER[662b])

Note to c′) *Utilisation of sensible heat of the stack gas.* During replacement fuel injection, the tuyere gas becomes enriched with H_2. As a result, at high top gas temperatures, a reduction in utilisation can be expected and has actually been observed in some cases though not in others. The possibility of a change in top gas temperature should, therefore, be incorporated in the calculation as a parameter at the outset.

The basic assumptions for the dependent variables are, therefore, as follows:

a′) η_{CO} = burden-dependent but unchanged by fuel injection;

b′) $\eta_{H_2} = 0.88\eta_{CO} + 0.10$;

c′) Utilisation of stack gas sensible heat somewhat improved at high top gas temperatures, otherwise constant.

To keep the tuyere gas heat content Q_F constant during fuel injection, the blast temperature must be increased from ϑ_1 to ϑ.†

† Oxygen addition to the blast at constant blast temperature has a similar effect, as shown by BRANDI et al.[668a]).

For operation using only coke, the tuyere gas sensible heat content is[662])

$$Q_F = \frac{K_{h1}(2432 + 3.84 c_p \vartheta_1)}{K_{h1} \cdot 4.67} \left(\frac{\text{kcal}}{\text{m}^3 \text{ at NTP}}\right).$$

For operation with e.g., fuel oil injection, the sensible heat content is

$$Q_F = \frac{K_h(2432 + 3.84 c_p \vartheta) + H(1466 + 3.71 c_p \vartheta)}{4.67 K_h + 5.86 H}.$$

In these equations,

$\left.\begin{matrix} K_{h1} \\ K_h \end{matrix}\right\}$ = combustion coke required (kg/ton iron) $\left\{\begin{matrix}\text{without} \\ \text{with}\end{matrix}\right\}$ replacement fuel;

2432 kcal/kg = heat content of coke preheated to 1400°C:

heat content from preheating	= 520 kcal
+ enthalpy of combustion of coke to CO	= 1912 kcal
total	= 2432 kcal;

c_p = mean specific heat of hot blast;

$\left.\begin{matrix} \vartheta_1 \\ \vartheta \end{matrix}\right\}$ = hot blast temperature $\left\{\begin{matrix}\text{without} \\ \text{with}\end{matrix}\right\}$ replacement fuel;

H = amount of replacement fuel (e.g. fuel oil) (kg/ton iron);

3.84 and 3.71 = blast required to gasify 1 kg coke and 1 kg fuel oil respectively to $CO + H_2$ (m^3 at NTP);

4.67 and 5.86 = tuyere gas volume/kg coke and kg fuel oil respectively (m^3/kg at NTP);

1466 kcal/kg = enthalpy of combustion of the fuel oil in the tuyere zone (Table 46)

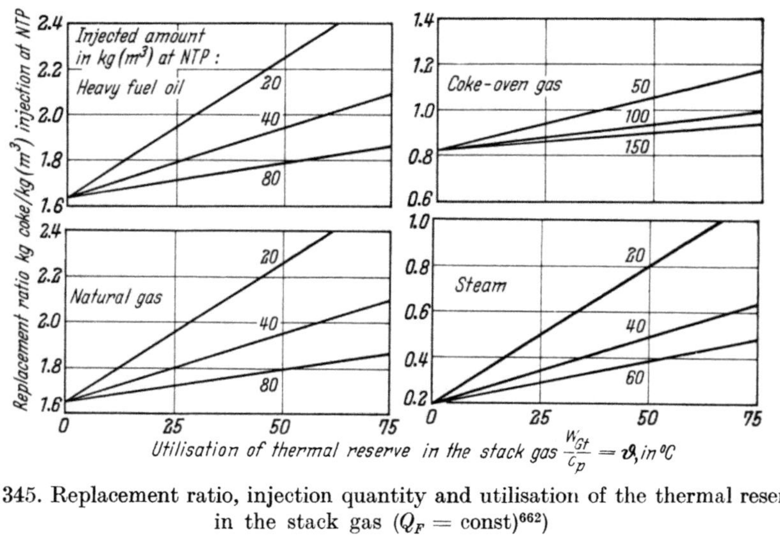

Fig. 345. Replacement ratio, injection quantity and utilisation of the thermal reserve in the stack gas (Q_F = const)[662])

It follows from the two expressions to be equated that the coke saving must be known in order to calculate the blast temperature increase required to maintain Q_F constant. The important furnace parameters are also interdependent and this complicates the calculation. The same conditions apply with injection of gases or pulverised coal.

In performing the calculation, account should be taken of the following changes in blast furnace heat requirements: change in slag volume (less coke ash), possible higher S input, reduction in coke, therefore, less heat required for preheating.

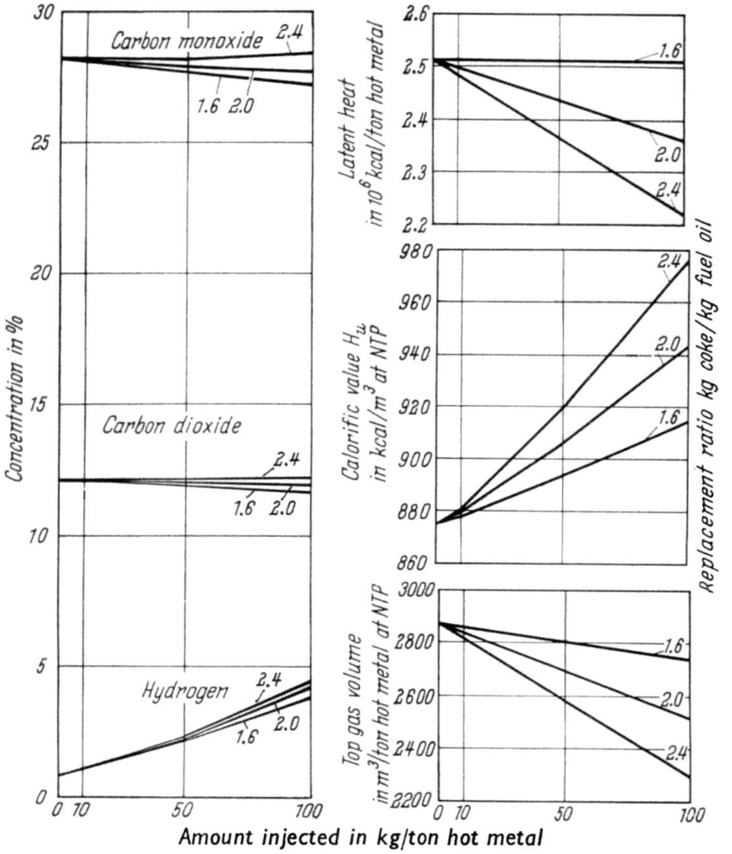

Fig. 346. Characteristics of top gas during heavy fuel oil injection (Q_F = const)[662])

There is consequently a change in the quantity of heat required in the tuyere gas, the volume of tuyere gas and the CO and H_2 input. The degree of indirect reduction also changes. The calculation procedure and result are, however, fixed by the assumptions, therefore, detailed presentation is not required here; this will be found in the original report[662]). Other methods of calculation differ only in the formulation of the basic assumptions which are not always set out with sufficient clarity and provide only a makeshift solution in view of the missing reaction rates. The replacement ratio, i.e., the coke saved per kg replacement fuel is defined, as shown in Fig. 345, by the degree of thermal utilisation of the stack gas, e.g., by the extent to which top gas temperature can be reduced.

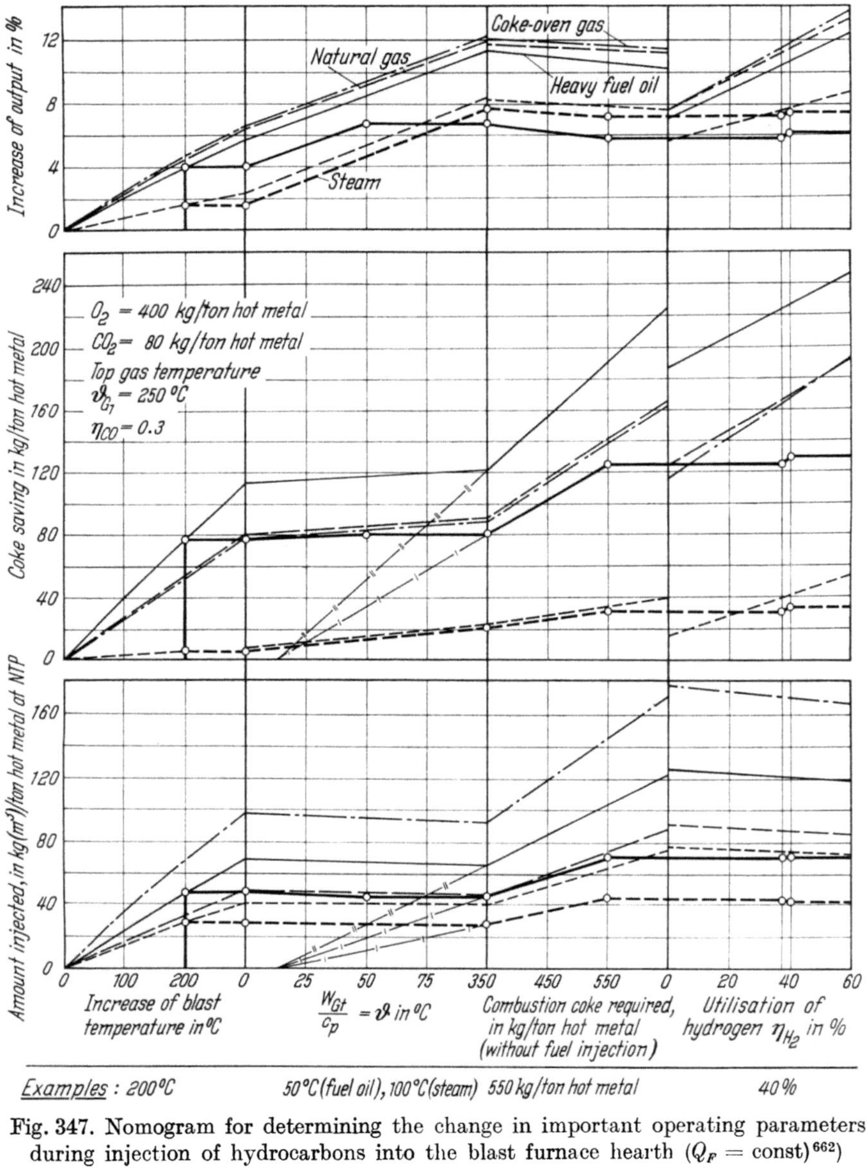

Fig. 347. Nomogram for determining the change in important operating parameters during injection of hydrocarbons into the blast furnace hearth (Q_F = const)[662]

Table 48. *Results of theoretical calculation with constant tuyere gas heat content* (Q_F = const)[662]

Assumed operating conditions		A	B	C
Sinter in burden . .	%	100	50 to 70	below 50
Net burden weight .	kg/ton iron	1800	2200	3000
Coke consumption .	kg/ton iron	620	765	970
Blast temperature .	°C	850	750	600
CO utilisation . . .	%	40	30	25
CO_2 from limestone.	kg/ton iron	—	80	120
Top gas temperature	°C	300	250	150

		A				B				C			
		Fuel oil	Natural gas	Coke-oven gas	Steam	Fuel oil	Natural gas	Coke-oven gas	Steam	Fuel oil	Natural gas	Coke-oven gas	Steam
Blast temperature increase	°C/kg (m^3) injected fuel at NTP	4.3	5.9	3.1	6.9	2.8	4.0	2.0	4.8	2.3	3.3	1.6	4.0
Replacement ratio .	kg coke/kg (m^3) injected fuel at NTP	2.0	2.1	1.1	0.7	1.7	1.8	0.9	0.6	1.62	1.63	0.9	0.4
Increase in iron production . . .	%/kg (m^3) injected fuel at NTP	0.21	0.28	0.16	0.23	0.08	0.12	0.06	0.16	0.04	0.07	0.04	0.07

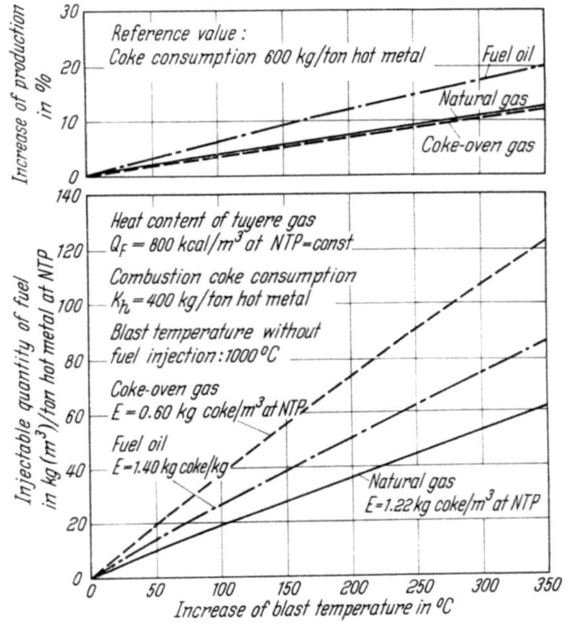

Fig. 348. Increase in furnace iron-make by replacement fuel injection (Q_F = const)

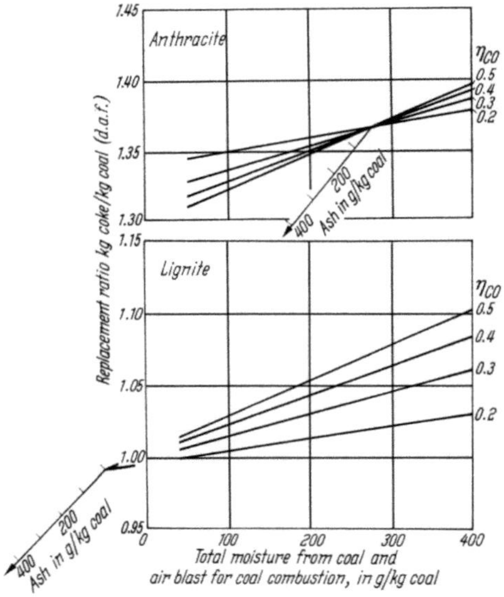

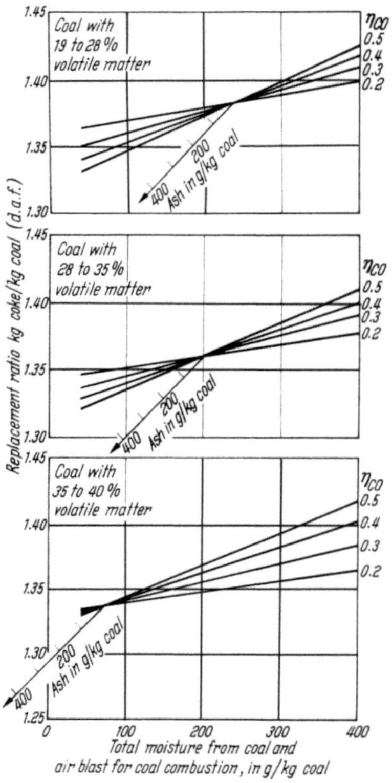

Fig. 349. Replacement ratio for injection of anthracite, lignite, and coals with different volatile contents in relation to coal moisture content and amount of water from the proportion of blast for coal combustion, η_{CO} ($\eta_{H_2} = 0.88\eta_{CO} + 0.1$) with Q_F = const[663])

The result of the calculation, namely, that even steam additions save coke, although, at first sight rather curious, is explained by the simultaneously increased blast temperature. As expected, the H_2 content of the top gas rises whilst the CO content falls, the specific calorific value increases and the top gas volume decreases (Fig. 346). The most important data are assembled in the nomogram in Fig. 347 which can be used for prediction in certain cases (examples of readings are included).

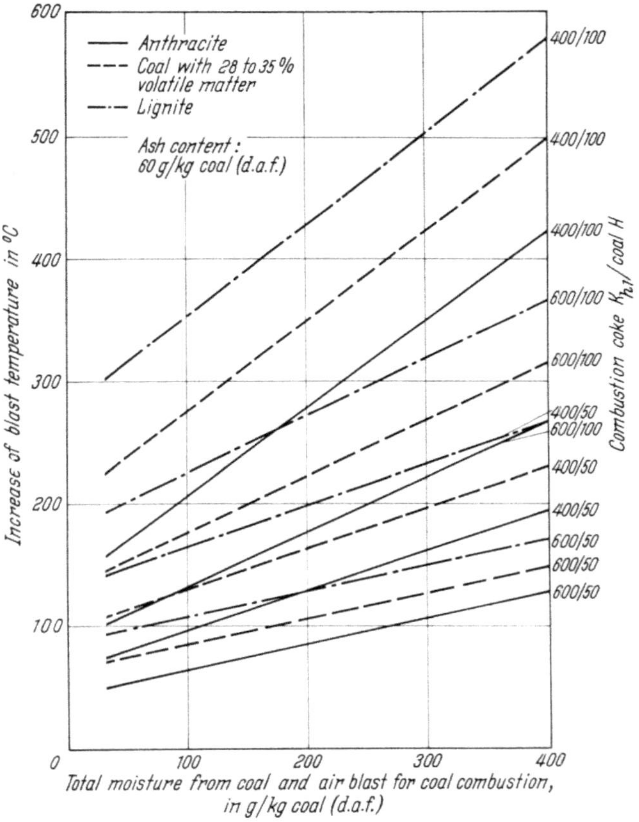

Fig. 350. Blast temperature increase required to keep the theoretical tuyere gas temperature constant in relation to the total moisture from the coal and blast and the ratio of combustion coke K_{h1} to coal H (water-free)[663]

Table 48 shows predictions deduced from the preceding statements for actual cases of rich, medium and lean burdens (screened and/or sintered in each case).

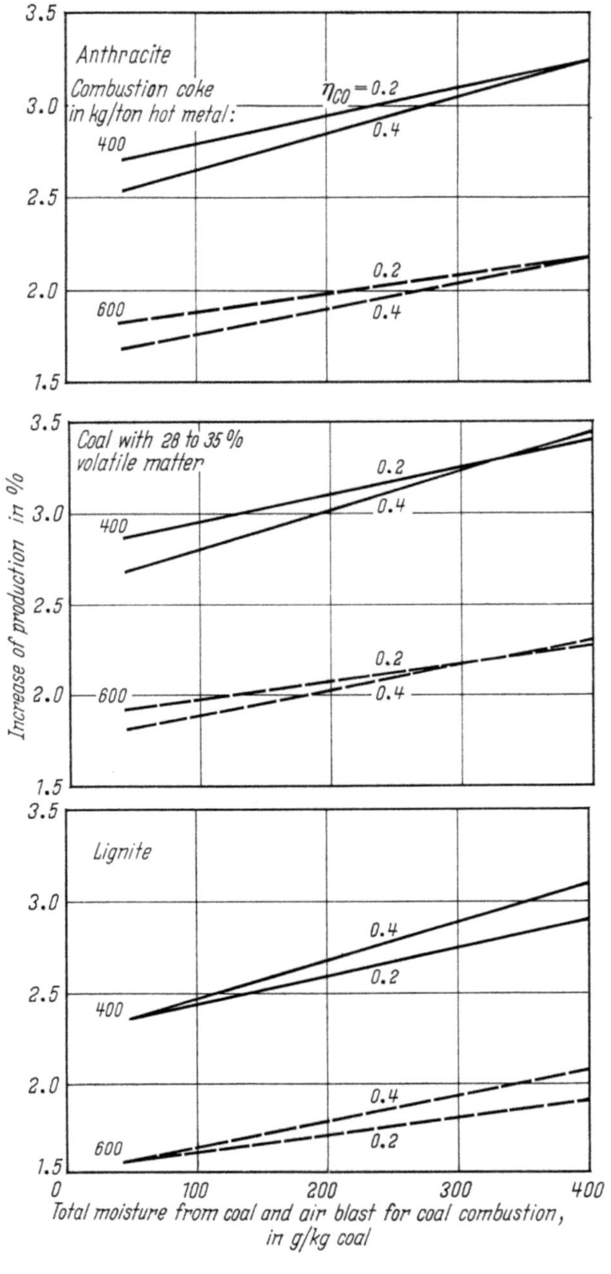

Fig. 351. Improvement in furnace iron-make in relation to total moisture from the coal and the blast involved, η_{CO} and the quantity of combustion coke, coal injection (water-free): 53 kg/ton iron (6% ash) with Q_F = const[663])

With a specified blast temperature increase and Q_F = const, the preceding equations allow estimation of the injectable quantities of replacement fuels and of the improvement in furnace performance (Fig. 348).

Injection of pulverised coals complicates the calculation in that, depending on the pretreatment and/or moisture content of the coal, varying amounts of H_2O enter the furnace. As a result of the reaction

$$C + H_2O = CO + H_2,$$

the H_2O both acts as a powerful coolant in the tuyere zone and further enriches the tuyere gas with H_2.

The effect of using various types of coal can be calculated in the same way as for gas and oil provided there is complete gasification and/or utilisation of the pulverised coal in the tuyere zone[663]). Figure 349 shows the expected replacement ratios (kg coke/kg injected coal) and Fig. 350 shows the blast temperature increase required to keep the tuyere gas heat content constant. Only a small improvement in furnace performance can be expected (Fig. 351).

5.4.3.3. Control of Blast-Furnace Reactions with Blast Temperature Kept Constant (i.e., Decrease in Tuyere Gas Temperature)

If blast temperature cannot be increased during replacement fuel injection, the tuyere gas heat content will diminish. The burden temperature in the tuyere zone can be kept constant by an appropriate increase in the fuel injection, this being reflected by regularity of iron composition, particularly as regards silicon. η_{CO}, the chemical utilisation of the CO, was

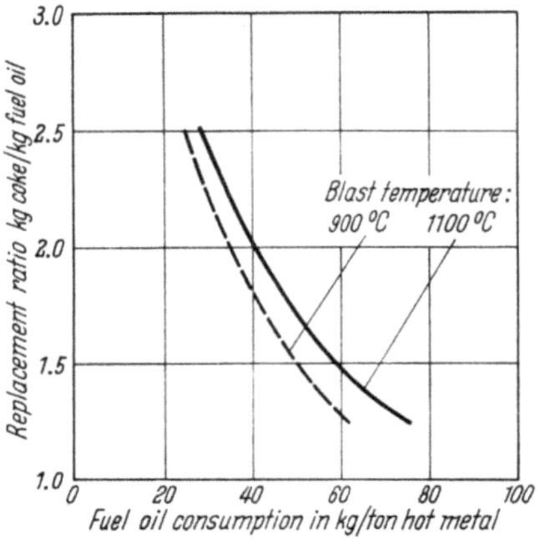

Fig. 352. Replacement ratio with varied oil injections and constant blast temperature[689])

observed by TRENSE and ROSBOROUGH[674]) to remain more or less constant even with varied tuyere gas temperatures. Using assumed values for the thermal utilisation of the stack gas, it is now possible to calculate conditions at constant blast temperature in relation to the injected fuel. Conditions at constant tuyere gas heat content can serve as a point of departure and the calculation can proceed as though the blast temperature were lowered. Information is available on the relationships between blast temperature and furnace operating data [e.g., see reference [676])]. In contrast to conditions with constant tuyere gas heat content, the replacement ratio will decrease with increasing fuel injection and the consequent reduction in tuyere gas heat content (Fig. 352). At an injection rate of 60 kg oil/ton iron, representing a reduction of about 130°C in the theoretical tuyere gas temperature, the theoretical replacement is approximately in proportion with the calorific values:

$$E = \frac{H_u(\text{oil})}{H_u(\text{coke})} = \frac{9800}{7000} = 1.40 \frac{\text{kg coke}}{\text{kg oil}}.$$

Similar ratios are obtained with other replacement fuels (Table 49).

Table 49. *Optimum injectable quantities of coal and mixtures of coal and oil and replacement ratios with blast temperature constant at* 900°C [663])

a	b	c	d	e
Type of fuel	Heat liberated in hearth**	Optimum injectable amount kg/ton iron	Coke saved kg/ton iron	Replacement ratio*** kg coke/kg replacement fuel
Coke	1995			
Fuel oil	1440	46	73.5	1.60
Anthracite*	1792	110	116	1.05
Coal with 19 to 28% volatile matter*	1620	85	90	1.06
Coal with 28 to 35% volatile matter*	1461	81	79	1.06
Coal with 35 to 40% volatile matter*	1268	73	73	1.0
Mixture of 40% fuel oil S and 60% coal with 28 to 35% volatile matter*	1437	62	73	1.18

* Coals containing 3% water and 7% ash (water-free).

** Minus heat required for decomposition of moisture in coal (114 kcal/kg coal) and of moisture in blast (3208 kcal/kg water).

*** $\frac{\text{Column d}}{\text{Column c}}$.

5.4.4. Comparison with Practice

Evaluation of individual plant trial results involves inevitable uncertainties. The fuel replacement ratio E is determined from the data for the reference and trial periods. If M_1 is the iron output (in t) and K_1 is the coke consumption (in t) in the reference period, M_2 and K_2 are the corresponding figures for the trial period and H is the amount of oil injected (in t), the fuel replacement ratio can be calculated as follows:

$$E = \frac{K_1/M_1 - K_2/M_2}{H/M_2}.$$

Unavoidable weighing errors, although relatively small, greatly impair the accuracy of E because the small difference between two large numbers is involved. The relationship shown in Fig. 353 between weighing errors

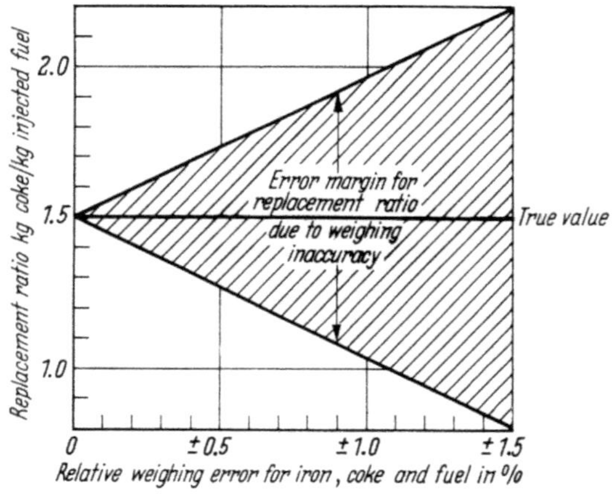

Fig. 353. Replacement ratio error margin[689])

Example: Initial coke consumption: 600 kg/ton iron, quantity of injected fuel: 50 kg/ton iron, quantity of coke replaced: 75 kg/ton iron

and the replacement ratio error margin was calculated as an example using established laws of error calculation. Although electronic scales are of good precision[688]), it is generally necessary in actual blast furnace practice to assume a weighing error of, at least, $\pm 0.5\%$ which means that the range for E in the calculation example given would be 1.25 to 1.75 kg coke/kg oil. The industrially determined replacement ratios should, therefore, be evaluated with these, and often greater, limitations in mind. Frequently, moreover, the burden has not been kept sufficiently regular. A clear indication can be obtained, therefore, only by using statistical analyses.

5.4.4.1. Natural Gas

Russian investigations into the composition of gas in the tuyere zone[696a]) (Figs. 354a and b) show that the correct assumptions have been made and that the combustion zone shortens with natural gas injection, as revealed mainly by the O_2 concentration curve.

It is possible to compare the theoretical predictions with actual operating results of natural gas injection reported in American literature[690–696]),

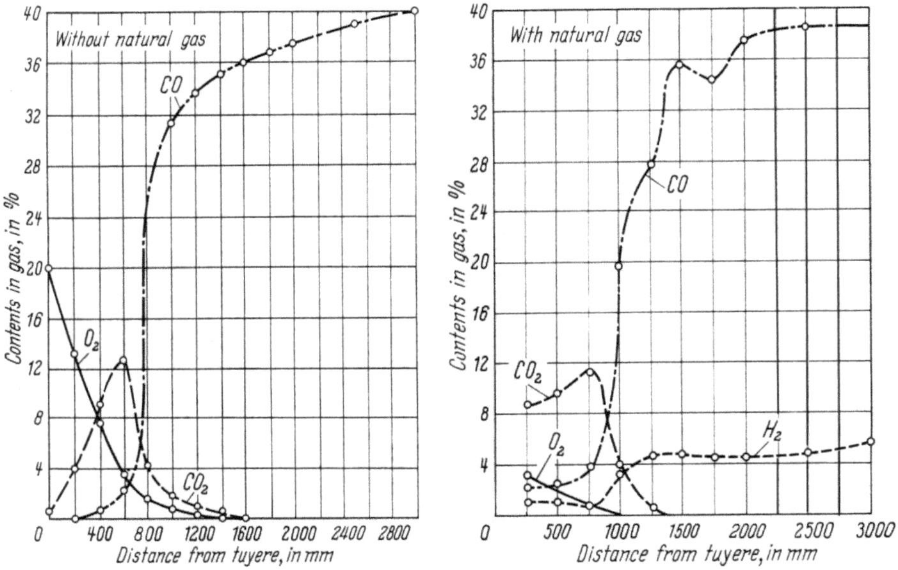

Fig. 354. Change in gas composition along the hearth radius without and with natural gas, according to NEKRASOV et al.[696a])

by STEINHAUER[660]) (Table 50) and European sources[696b–m]) (Table 50a). The replacement ratio increases with the hot blast temperature (Fig. 355a). With a rich-to-medium burden (case A to B in Table 48) and a blast consumption of some 1700 m³/ton iron at NTP a natural gas injection of 1% means a consumption of 17 m³/ton iron at NTP. When the hot blast temperature is raised by 5.0°C/m³ CH_4/ton iron at NTP (mean from 4.0 and 5.9) representing 85°C/% CH_4 in the blast, the heat content of the tuyere gas is kept constant according to the preceding equation. In these conditions, the theoretical replacement ratio of 1.9 kg coke/m³ CH_4 at NTP (Table 48) is obtained almost exactly (Fig. 355a). There is some variation in the observed changes in production. This may, however, be largely due to fluctuation in iron requirements from the furnaces.

If blast temperature is kept constant, the replacement ratio falls, according to Fig. 355a, to 1.10 kg coke/m³ natural gas at NTP, i.e. slightly below the calorific value ratio:

$$\frac{H_u(CH_4)}{H_u(\text{coke})} = \frac{8580}{7000} = 1.22.$$

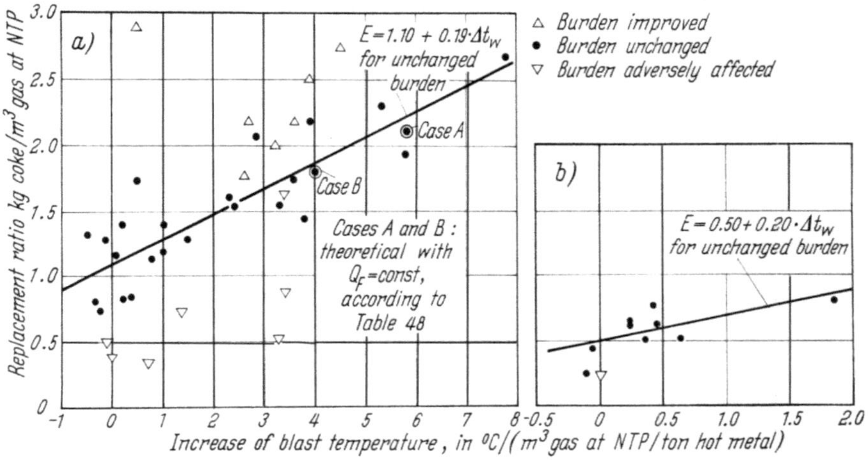

Fig. 355. Replacement ratio for natural gas and coke-oven gas in relation to increase in blast temperature

The calorific values of the natural gases used will, of course, vary. Equation of natural gas with CH_4 represents an ideal, only permissible in comparative calculations.

5.4.4.2. Coke-Oven Gas

Recent reports are available on the use of coke-oven gas[696m–s]. Table 50b shows the data for several blast furnaces working with coke-oven gas injection[696n–p]. The small range of data does not permit any definite statement as yet. The calorific value ratio indicates that, at constant blast temperature, a replacement ratio of

$$\frac{H_u(\text{coke-oven gas})}{H_u(\text{coke})} = \frac{4200}{7000} = 0.60 \frac{\text{kg coke}}{\text{m}^3 \text{ coke-oven gas at NTP}}$$

can be expected. The reading obtainable from Fig. 355b is 0.50 kg coke/m³ coke-oven gas at NTP.

5.4.4.3. Fuel Oil

A large number of operating data has been published on injection of fuel oil in the blast furnace[661, 662, 665a, 696c,s, 698–711, 711a–n]. Table 51 gives a broad survey of the plottable values.

Table 50. *Trial and full-scale results with*

Works	Furnace	Serial number	Hearth diameter m	Working volume m³	Number of tuyeres	Burden composition %				
						Pellets	Sinter	Screened lump ore	Unscreened ore	Mill-scale
Altos Hornos de Mexico, S. A.	2	1	6.40	806	14	—	51	49	—	—
		2				—	53	47	—	—
Bethlehem Steel Co., Johnstown	H	3	7.92	1290	20	40	30	15	10	5
		4				85	1	—	—	14
		5				88	1	—	—	11
		6				91	—	—	—	9
Bethlehem Steel Co., Lackawanna	J	7	9.12	1573	20	65	30	—	5	—
		8				65	30	—	5	—
		9				65	30	—	5	—
		10				65	30	—	5	—
Colorado Fuel & Iron Corp.	F	11	6.63	765	12	—	51	41	—	8
		12				—	65	23	—	12
		13				—	60	32	—	8
Great Lakes Steel Corp.	B	14	8.84	1340	24	—	55	45	—	—
		15				—	55	45	—	—
		16				—	55	45	—	—
		17				—	75	25	—	—
		18				—	69	31	—	—
Great Lakes Steel Corp.	A	19	9.22	1578	24	—	77	23	—	—
		20				—	85	15	—	—
Lone Star Steel Co.		21	8.23	1254	18	—	36	—	63	1
		22				—	38	—	58	4
Pittsburgh Coke Chemical Co.	B	23	6.70	860	12	~20	—	—	~80	—
		24				~20	—	—	~80	—
US-Steel Corp., Fairless	3	25	8.54	1340	20	—	60	40	—	—
		26				—	65	35	—	—

ıatural gas injection into American blast furnaces[660])

Scrap kg/ton iron	Blast volume $1000 m^3/h$ at NTP	Production tons/24 h	Coke consumption kg/ton iron	Natural gas injection m^3/ton iron at NTP	Replacement ratio kg coke/m^3 gas at NTP	Blast temperature °C	Blast humidity g/m^3 at NTP
—	95	951	833	—	—	684	32.3
—	104	995 + 4.6%	716 — 14.0%	54.0	2.17	810	29.7
72	144	1581	701	—	—	721	26.3
133	145	1736 + 9.8%	587 — 16.3%	52.6	2.17	810	15.0
141	142	1859 + 17.6%	553 — 21.1%	60.4	2.45	845	11.6
152	142	1888 + 19.4%	540 — 23.0%	59.0	2.73	875	11.8
—	165	1712	714	—	—	699	31.1
—	165	1834 + 7.1%	608 — 14.8%	61.0	1.74	785	17.8
—	158	1820 + 6.3%	577 — 19.2%	59.6	2.30	818	11.3
—	162	1832 + 7.0%	596 — 16.5%	61.1	1.93	820	8.0
—	74	674	707	—	—	597	9.6
—	76	693 + 2.8%	654 — 7.5%	59.5		628	10.1
				+16.7 litres oil/ton iron			
—	71	904 + 34.1%	575 — 18.7%	79.4		667	12.3
	(26% O_2)			+19.6 litres oil/ton iron			
				+122 m^3 O_2/ ton iron at NTP			
—	149	1737	661	—	—	965	42.2
—	148	1856 + 6.9%	585 — 11.5%	47.8	1.59	955	27.5
—	149	1870 + 7.7%	538 — 18.6%	80.6	1.54	1008	15.2
—	148	1878 + 8.1%	515 — 22.1%	73.1	2.01	995	16.4
—	141	1760 + 1.3%	549 — 16.9%	69.4	1.62	994	18.3
—	175	2080	647	—	—	750	28.9
—	168	1950(— 6.3%)	557 — 13.9%	41.6	2.16	748	16.4
225	131	958	1023	—	—	556	9.6
172	146	1200 + 25.3%	870 — 15.0%	87.0	1.76	693	8.0
—	82	766	853	—	—	757	23.2
—	75	678(— 11.5%)	751 — 12.0%	66.6	1.53	814	17.1
—	140	1800	625	—	—	877	38.6
—	137	2040 + 13.3%	499 + 20.2%	47.1	2.68	1080	12.5

Works	Reference	Furnace	Hearth diameter m	Working volume m³	Number of tuyeres	Burden composition: Sinter %	Ore %	Limestone kg/ton iron	Scrap kg/ton iron	Net burden weight kg/ton iron (% Fe in burden)	Blast volum 1000 m³/h at NTP
Hüttenwerk Oberhausen A.G.	696b)	3	7.0	802	16	65.9 65.3	34.1 34.7	52 54	63 58	1675 1732	124 121
		8	7.0	904	16	70.3 63.2	29.7 36.8	68 66	6 6	1600 1678	68 74
		8	7.0	904	16	84.9 86.1	15.1 13.9	52 58	15 10	1640 1759	62 66
Magnitogorsk	696c)	B	7.6			97.6 96.9 97.7	— — —		— — —	(55.3) (54.8) (53.6)	
	696d)	F				96 to 100 96 to 100 96 to 100 93.0	— — — 7.0		— — — —	(52.7 to 55.7) (52.7 to 55.7) (52.7 to 55.7) (52.7 to 55.7)	
	696e)		7.6			96 to 100 96 to 100	— —		— —	(52.7 to 55.7) (52.7 to 55.7)	
			8.0			96 to 100 96 to 100	— —		— —	(52.7 to 55.7) (52.7 to 55.7)	
Orsk-Khalilovo	696d)	C		1513		100 99.5 99.0 99.3	— — — —		— — — —	(48.7 to 51.8) (48.7 to 51.8) (48.7 to 51.8) (48.7 to 51.8)	
Chelyabinsk	696d)	C				65.7 74.7 69.4 57.6	34.3 25.3 30.6 42.4	30 52 112 72	22 — 6 65		 + 7 %
		E		1719		99.5 100 100 98.4 91.7	— — — —	2 1 1 9	1 —		
Gliwice	696i)		3.6	248	8	— — — —	100 100 100 100	674 705 724 731		2877* 2770* 2809* 2792*	
Dzerzhinsk	696f)					88.4 88.5 91.7 87.0 85.5 90.7	11.6 11.5 8.3 13.0 14.5 9.3			(51.1) (51.8) (51.3) (52.1) (52.4) (52.1)	171 + 0.2% 170 172 153 + 2.3% 155 + 2.4% 150 + 2.3%
Cherepovets	696f)	A				95.4 93.2 92.7	4.6 6.8 7.3				138 135 + 1.4% 133 + 2.5%
		B				98.1 96.3 97.3 96.8	1.9 3.7 2.7 3.2				135 132 + 1.5% 131 + 3.4% 125 + 6.1%
	696i)		7.35	930		58.7 100 100	41.3 — —				92 + 4.9% 94 + 10.1% 96 + 14.1%
Donetsk	696j)	B		700		18.4 32.5 52.6 61.5 68.3	81.6 67.5 47.4 38.5 31.7	802 601 517 405 374		2837* 2645* 2665* 2487* 2390*	

* Gross burden weight kg/ton iron.

Production	Coke consumption	Natural gas injection	Replacement ratio	Blast temperature	Blast humidity	Top gas CO_2	Top gas CO	Top gas H_2	High top pressure	Slag volume
tons/24h	kg/ton iron	m^3/ton iron at NTP	kg coke/ m^3 gas at NTP	°C	g/m^3 at NTP	%	%	%	atm gauge	kg/ton iron
1725	627 dry	—	—	858		13.9	28.6	0.9		~360
1645	610 dry	32.1	0.53	963		14.0	27.8	2.4		
1083	642 dry	—	—	856		13.7	28.1	1.2		~350
1251	608 dry	39.3	0.86	990		12.7	28.4	3.3		
1080	584 dry	—	—	978		15.8	26.2	1.1		305
1189	562 dry	30.4	0.72	1019		16.3	24.4	3.0		365
1.67**	609 raw	—	—	1100	32.0	14.4	27.1	2.0	0.8	480
1.76**	565 raw	60	0.73	1085	16.6	14.0	24.3	5.1		
1.74**	538 raw	90	0.79	1069	18.1	13.4	23.7	6.1		
100%	600	—	—	1060	30.9					
+0.6	547	64.3	0.83	1083	19.2					
to 1.0%	547	65.0	0.82	1073	22.6					
	576	61.8	0.39	1060	25.3					
100%	607 corr.	—	—	929	28.2	14.6	26.0	2.5	0.7	500
+0.6%	549 corr.	50	1.16	933	13.2	13.9	24.1	4.6		
100%	595 corr.	—	—	974	28.0	14.9	26.2	2.3	1.1	~490
+3.6%	527 corr.	60	1.13	1020	10.0	13.9	24.3	5.2		
100%	629	—	—	991	25	η_{CO} = 34%		2.4	1.4	570 to 680
+ 1.5%	610	58.4	0.33	1032	16	32		4.7		
+14.8%	572	48.4	1.18	1040	19	35		4.1		
+10.0%	563	38.0	1.73	1010	23	36		4.5		
100%	708	—	—	910	20.1	η_{CO} = 33.4%		2.0	1.3	} 940
+2.8%	658	39.2	1.27	904	16.9	38.0		2.9		
−3.6%	674	70.1	0.49	904	18.1	35.0		5.0		
+5.0%	630	61.0	(1.28)	882	10.9	36.1		5.1		840
100%	678	—	—	956	24.6	η_{CO} = 33.6%		1.9	1.5	916
+6.2%	621	19.7	2.89	957	15.0	34.2		2.1		788
+3.9%	625	40.5	1.31	936	17.3	34.3		4.0		} 860
+0.9%	613	46.3	1.40	1004	7.7	31.9		4.1		
+2.4%	621	44.5	1.28	1023	8.3	32.1		4.5		
248	1062 raw	—	—	612	25	11.5	28.8	2.3		
240	1017 raw	32.0	1.40	618	8	11.3	28.2	2.1		
260	991 raw	34.3	2.07	710	30	12.8	29.6	4.3		
270	992 raw	48.6	1.44	796	22	10.5	29.8	3.5		
	695	80.5	No comparison	864		13.6	25.6	5.2	0.6 to 1.0	620 to 670
	670	78.5		875		14.1	24.7	5.2		
	650	78.4		872		13.6	24.6	5.7		
	655	89.0		907		14.3	26.3	6.4		
	662	93.5		987		15.0	25.8	6.6		
	645	96.0		990		14.6	26.2	6.7		
2.15**	485 dry	84.0	No comparison	1032	12.4	16.3	22.2	5.6	1.5	~390
2.23**	468 dry	88.0		1036	8.7	17.1	23.0	6.0		
2.38**	472 dry	79.0		1035	6.9	17.6	23.8	6.1		
2.17**	483 dry	85.8	No comparison	1028	12.4	16.5	22.0	5.6	1.5	~390
2.24**	481 dry	85.5		1031	6.3	17.1	23.4	5.8		
2.38**	497 dry	82.8		1030	6.5	18.2	24.2	6.2		
2.51**	490 dry	82.0		1030	7.2	19.7	25.7	6.8		
~995	747	113	No comparison	848	21.2				0.2	550 to 660
~1510	544	115		925	18.5					
~1650	534	125		920	22.0					
833	804 raw	133	No comparison	880						913
854	763 raw	128		905						733
852	731 raw	126		929						799
840	684 raw	117		944						701
856	690 raw	105		944						713

** tons/(m^3 · 24 h).

Table 50b. *Operating results from blast furnaces*

Works	Ref-er-ence	Fur-nace	Hearth diam-eter m	Work-ing vol-ume m^3	Burden: Sinter %	Burden: Ore* %	Net burden weight kg/ton iron	Blast volume 1000 m^3/h at NTP	Production tons/24 h	Coke consumption kg/ton iron
Homécourt (Sidelor)	696n)	7	4.75	533	—	100	3164		253	957 dry
					—	100	3320		244	933 dry
Kuznetsk	696o)	A	8.09		82.1	17.9	1884	146	100%	618
					84.1	15.9	1895	145	—0.2%	603
					84.7	15.3	1867	144	100%	612
					80.6	19.4	1882	148	+2.0%	599
					81.2	18.8	1889	148	100%	607
					83.5	16.5	1866	148	+3.5%	562
					79.9	20.1	1869	148	+4.0%	549
					82.5	17.5	1888	156	100%	616
					77.4	22.6	1903	142	±0%	566
					78.6	21.4	1902	142	—5.0%	574
DHHU	696p)	3	6.25	675	28.7	71.3	1818	59	683	763 raw
					31.0	69.0	1874	64	769	722 raw
					46.3	53.7	1786	66	731	754 raw
					46.9	53.1	1763	64	801	685 raw
					38.3	61.7	1795	64	747	748 raw
					46.1	53.9	1750	59	729	680 raw

* Including additions.

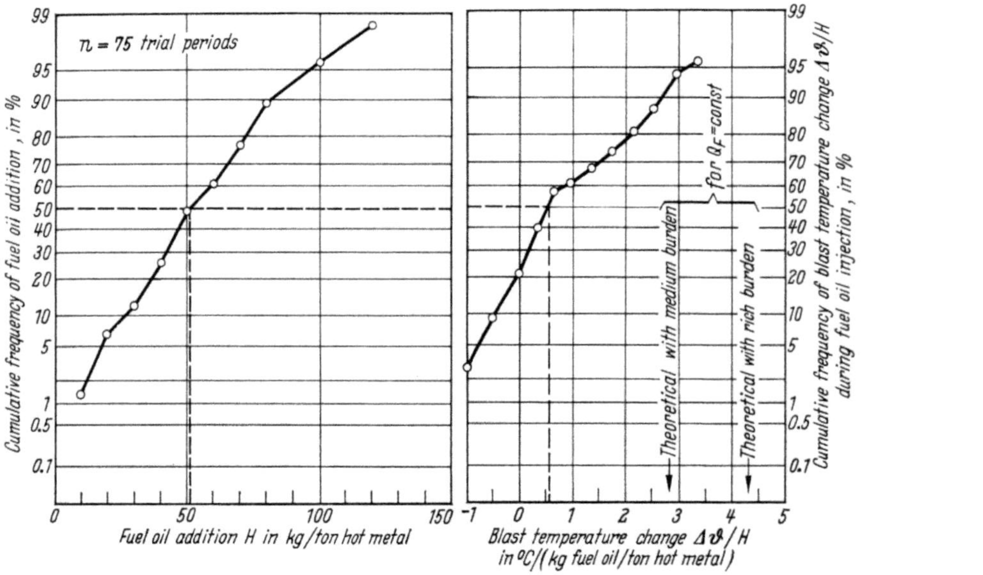

Fig. 356. Cumulative frequency of oil additions and change in blast temperature (from Table 51)

using injections of coke-oven gas

Coke-oven gas m^3/ton iron at NTP	Replacement ratio kg coke/m^3 gas at NTP	Blast temperature °C	Blast humidity g/m^3 at NTP	Top gas CO_2 %	Top gas CO %	Top gas H_2 %	Iron Si %	Iron S %	Slag volume kg/ton iron	Slag CaO/SiO_2
—	—	930							1135	
104	0.23	930							1080	
—	—	992	18.7	15.2	27.6	1.9	0.78	0.034	594	1.08
34.3	0.44	990	12.0	15.6	27.1	2.5	0.78	0.030	609	1.07
—	—	977	14.9	15.1	28.0	1.2	0.81	0.036	580	1.06
52.2	0.25	971	13.3	14.8	27.7	2.5	0.90	0.026	605	1.08
—	—	969	15.6	14.9	28.2	1.7	0.83	0.033	580	1.07
74.2	0.61	987	8.0	14.8	26.8	3.2	0.74	0.040	571	1.02
92.8	0.63	990	5.1	14.4	26.4	4.0	0.65	0.035	557	1.05
—	—	954	19.5	15.5	27.4	1.6	0.85	0.030	558	1.06
81.0	0.62	990	7.6	15.4	26.3	3.2	0.88	0.027	525	1.06
84.0	0.50	984	8.2	15.5	26.0	3.4	0.88	0.028	569	1.06
—	—	830	12.5	10.5	30.3	1.6			480	
80.1	0.51	881	11.1	10.1	29.7	3.8			493	
—	—	679	7.5	10.5	29.2	1.2			470	
87.4	0.80	841	10.2	11.2	27.6	4.1			424	
—	—	788	10.4	10.7	30.0	1.4			467	
88.1	0.77	825	10.7	11.0	27.6	4.0			411	

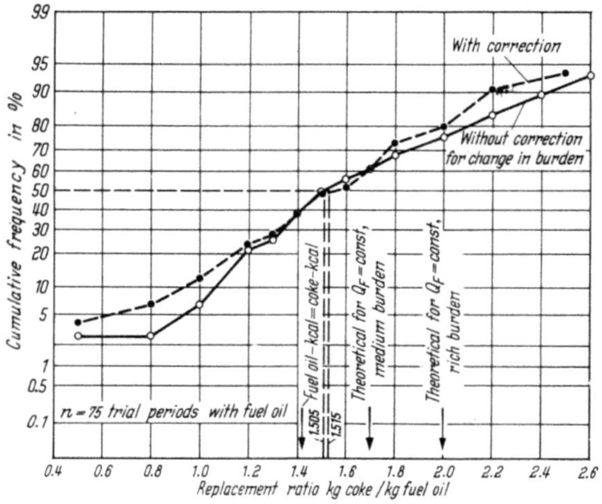

Fig. 357. Cumulative frequency of the replacement ratio E (from Table 51)

Table 51. *Tabulation of data for evaluation*

Works	Reference	Furnace	Serial number	Hearth diameter/area m/m²	Working volume m³	Net burden weight kg/ton iron	Burden composition: Sinter %	Pellets %	Lump ore* %	Flue dust kg/ton iron	Coke consumption kg/ton iron
Société des Aciéries de Pompey	662)	3	1	4.8/18.1	542.8	3022	—	—	100	202	1008
			2			2975	—	—	100	194	920
			3			2911	—	—	100	235	907
			4			2782	—	—	100	256	889
S.A. Cockerill-Ougrée	662)	4	5	3.8/11.3		2683	40.6	—	59.4		789
			6			2841	40.6	—	59.4		718
			7			2776	40.8	—	59.2		836
			8			2812	41.0	—	59.0		732
			9			2744	41.0	—	59.0		723
			10			2787	40.7	—	59.3		737
			11			2192	89.2	—	10.8		712
			12			2240	90.6	—	9.4		620
			13			2219	90.4	—	9.6		626
Phoenix Rheinrohr A.G., Duisburg-Ruhrort	661)	8	14	7.7/46.6	963	1877	48.4	—	51.6	74	648
			15			1855	48.4	—	51.6	84	626
			16			1826	48.2	—	51.8	68	595
			17			1877	48.4	—	51.6	76	582
			18			1889	49.8	—	50.2	90	560
			19			1841	48.8	—	51.2	101	620
			20			1887	48.8	—	51.2	70	562
			21			1834	48.3	—	51.7	85	552
			22			1867	48.6	—	51.4	103	565
			23			1883	48.9	—	51.1	78	536
			24			1855	49.8	—	50.2	96	510
			25			1879	48.7	—	51.3	97	616
			26			1828	49.6	—	50.4	83	536
			27			1866	48.2	—	51.8	62	508
			28			1905	50.0	—	50.0	106	500
Richard Thomas Baldwins Ltd., Ebbw Vale	698)	C	29	6.1/29.2	732.6	1790	—	—	100	49	923
			30			1773	—	—	100	58	808
			31			~1770	7.1	—	92.9	38	941
			32			1770	6.0	—	94.0	33	790
			33			1770	3.3	—	96.7	23	743
			34				30.2	—	69.8	87	670
ARBED, Terre Rouge	699)	1	35	6.0/28.3	841.5	3351	29.9	—	70.1	~40	1052
			36			3355	30.2	—	69.8		972
			37			3357	30.2	—	69.8		916
			38			3317	30.3	—	69.7		901
			39			3305	30.6	—	69.4		905
			40			3348	30.8	—	69.2		907
ARBED, Belval	700)	2	41	4.8/18.1	626	3584	29.5	—	70.5	110	1130
			42			3718	28.7	—	71.3	111	1095
			43			3731	29.3	—	70.7	138	1200
			44			3697	31.3	—	68.7	140	1217
			45			3740	22.7	—	77.3	105	1174
			46			3662	21.5	—	78.5	77	1126
Dofasco, Hamilton, Ontario	701, 711)	3	47	6.2/30.2		1567	—	61.8	38.2	23	617
			48			1560	—	57.8	42.2	19	564
Steel Company of Wales, Margam	702)	1	49	7.24/41.2	843	2028	48.9	—	51.1	43	719
			50			1938	50.0	—	50.0	65	613
			51			1962	51.2	—	48.8	64	578

* Lump ore: including nodules, remelting iron, slag- and limestone additions.

of oil injection trials on blast furnaces

Oil injection kg/ton iron	Iron production tons/24 h	Iron production tons/24 h m²	Blast temperature °C	Blast humidity g/m³ at NTP	Top gas CO %	Top gas CO_2 %	Top gas H_2 %	Pig iron Si %	Pig iron P %	Pig iron S 10^{-3}%	Slag volume kg/ton iron	CaO/SiO_2 in slag
—	258.2	14.3	781	5.35	26.7	12.8	3.7	0.55	1.74	59		1.47
51	271.3	15.0	869	4.46	27.0	13.3	4.4	0.59	1.83	35		1.47
43	292.0	16.1	903	6.64	27.3	13.1	4.3	0.67	1.80	30		1.52
48	291.3	16.1	910	6.78	26.9	13.9	4.5	0.63	1.77	41		1.46
—	202.8	17.9	710		26.6	13.8	1.7	0.32	1.85	71		1.50
61	200.6	17.8	810		25.7	14.1	3.6	0.35	1.84	65		1.42
—	225.4	19.9	697		27.2	13.0	1.9	0.38	1.86	64		1.50
62	222.0	19.6	817		24.5	13.0	3.7	0.30	1.80	80		1.45
63	230.8	20.4	821		26.1	13.6	3.8	0.35	1.81	43		1.37
58	229.3	20.0	823		27.0	12.9	3.8	0.37	1.80	55		1.43
—	266.4	23.6	707		26.3	12.7	1.3	0.67	1.78	114		1.22
50	269.4	23.8	822		25.6	13.0	2.6	0.47	1.82	115		1.17
49	293.0	25.9	795		26.0	12.7	2.4	0.40	1.80	128		1.22
—	1448	31.1	920	19	29.7	12.7	1.8	0.34	basic Bessemer iron	51	516	1.27
9	1574	33.8	934	20	29.7	12.4	2.1	0.32		47	500	1.31
20	1582	33.9	902	20	29.1	12.5	2.6	0.37		47	474	1.27
39	1610	34.5	943	20	28.6	12.4	3.4	0.37		52	489	1.26
92	1410	30.3	956	19	26.9	12.6	4.6	0.37		54	481	1.30
—	1525	32.7	998	20	29.4	13.2	1.8	0.30		55	518	1.29
42	1626	34.9	1003	20	27.9	13.3	3.0	0.37		52	474	1.31
45	1625	34.9	1010	20	28.0	13.1	3.1	0.33		55	468	1.29
53	1577	33.8	1029	20	28.0	12.8	3.4	0.34		52	498	1.30
78	1548	33.2	1003	18	27.6	13.1	4.1	0.32		56	493	1.29
115	1482	31.8	997	19	26.8	12.5	5.5	0.33		51	466	1.30
—	1644	35.3	1130	19	29.5	12.6	2.3	0.32		61	508	1.32
38	1700	36.5	1113	18	27.9	13.9	2.9	0.33		56	467	1.31
71	1693	36.3	1135	18	27.4	13.6	3.9	0.37		45	499	1.29
117	1573	33.8	1151	16	27.1	13.0	5.4	0.33		55	494	1.28
(7.5)	607	20.8	715	6.3	30.4	11.3	2.7	0.70	0.79	66	433.5	1.00**
69.0	654	22.4	653	7.5	28.3	12.0	3.9	0.62	0.75	30	423.0	1.09**
63.5	459	15.7	457	10.9	28.6	10.4	3.7	0.82	1.02	37	470	1.01**
52.5	595	20.4	516	7.1	27.2	12.5	3.7	0.89	0.76	35	390	0.99**
73.5	614	21.0	754	5.4	26.9	13.4	4.3	0.92	0.76	44	360	1.05**
56.0	670	22.9	676	5.1	26.1	14.1	6.0	0.86	0.30	35	300	1.06**
—	473.5	16.7	900	6.7	29.7	12.1	2.2	0.39	1.81	51	1341	1.38
59.2	500.2	17.7	900	6.4	29.5	12.2	3.3	0.25	1.77	53	1346	1.46
63.3	503.1	17.8	901	5.2	28.8	12.8	3.3	0.25	1.81	64	1331	1.37
78.0	500.9	17.7	904	3.6	28.5	12.6	3.6	0.27	1.80	63	1316	1.38
99.0	482.9	17.1	911	6.4	28.6	11.9	4.2	0.29	1.83	64	1336	1.38
80.4	496.3	17.5	902	7.8	28.5	12.3	3.5	0.34	1.84	53	1350	1.37
—	323.9	17.9	809	5.6	30.0	11.1	2.2	0.31	1.74	55	1422	1.36
22.5	311.9	17.2	847	7.6	29.8	11.5	2.6	0.25	1.68	53	1460	1.36
—	281.0	15.5	661	10.2	30.4	10.6	2.3	0.26	1.80	62	1473	1.35
39.2	287.0	15.9	672	10.5	29.9	10.1	2.8	0.18	1.74	70	1450	1.37
—	254.5	14.1	773	4.4	30.2	11.0	2.5	0.39	1.78	45	1457	1.35
37.5	258.0	14.3	814	4.8	28.8	11.4	2.9	0.26	1.73	70	1440	1.33
—	1458	48.3	863	41.4	25.8	16.8	2.3	1.11	0.10	26	304	1.04
29	1491	49.4	877	29.1	25.8	16.9	3.0	1.15	0.10	23	288	1.08
—	742	18.0	713	8.3	31.0	10.4	1.5	0.44	1.96	26	430	1.02**
76	809	19.6	748	8.3	29.3	10.7	3.4	0.44	2.02	24	378	1.06**
130	773	18.8	788	10.1	29.7	10.0	4.1	0.44	1.89	27	384	1.07**

$^{**}\ \dfrac{CaO + MgO}{SiO_2 + Al_2O_3}$

Table 51

Works	Reference	Furnace	Serial number	Hearth diameter/area m/m²	Working volume m³	Net burden weight kg/ton iron	Burden composition: Sinter %	Burden composition: Pellets %	Burden composition: Lump ore* %	Flue dust kg/ton iron	Coke consumption kg/ton iron
Hoesch A.G. Westfalenhütte, Dortmund		7	52	7.5/44.2	974.1	1854	74.7	—	25.3	59	620
			53			1689	76.8	—	23.2	51	496
			54			1845	68.2	—	31.8	107	548
			55			1722	66.0	—	34.0	65	521
Klöckner Werke A.G., Georgsmarienwerke	704)	3	56	5.28/21.9	537	1801	35.6	—	64.4	52	671
			57			1804	38.5	—	61.5	85	610
			58			1833	58.9	—	41.1	48	594
Fuji Iron and Steel Co., Ltd.***, Muroran Works		1	59	7.21/40.8	925	~1530	65.9	—	34.1		614
			60			~1510	61.4	—	38.6		588
			61			~1520	61.2	—	38.8		574
			62			~1500	60.1	—	39.9		561
			63			~1530	58.9	—	41.1		570
			64			~1490	59.2	—	40.8		553
			65			~1540	60.2	—	39.8		547
			66			~1520	76.5	—	23.5		546
Fuji Iron and Steel Co., Ltd.***, Kamaishi Works	705)	1	67	7.3/41.9	905	~1530	70.6	—	29.4		560
			68			~1570	63.8	—	36.2		515
			69			~1550	69.2	—	30.8		494
			70			~1490	65.2	—	34.8		479
			71			~1580	75.6	—	24.4		476
			72			~1490	67.2	—	32.8		456
			73			~1530	73.7	—	26.3		462
Hirohata Works		1	74	7.8/47.8	1140	~1580	71.5	—	28.5		577
			75			~1600	81.4	—	18.6		536
			76			~1610	68.4	—	31.6		506
			77			~1620	69.8	—	30.2		471
			78			~1630	65.8	—	24.2		482
		2	79	7.6/45.4	1128	~1620	59.7	—	40.3		583
			80			~1560	51.9	—	48.1		561
			81			~1580	60.8	—	39.2		513
			82			~1580	63.8	—	36.2		470
			83			~1620	60.3	—	39.7		475
Chusovoi Metallurgical Plant	706)		84			~2010	57.1	—	42.9	58	972
			85			~1910	48.9	—	51.1	64	903
			86			~1840	59.1	—	40.9	46	878
			87			~1880	58.6	—	41.4	87	835
			88			~1860	51.7	—	48.3	104	844
			89			~1890	57.9	—	42.1	83	809
Urals, USSR	707)		90	~5/~19.6	278	~2040	58.1	—	41.9		934
			91			~2160	60.8	—	39.2		833
			92			~1900	75.8	—	24.2		729
Société des Aciéries de Pompey	708)	4	93	4.8/18.1	542.8	~2900	—	—	100	202	1008
			94			~2900	—	—	100	194	908
USINOR, Louvroil	709)	1	95	4.50/15.9	488.5	2552	—	—	100	80	813
			96			2509	—	—	100	82	692
			97			2542	—	—	100	97	677
			98			2541	—	—	100	98	701
Société Hadir, Differdange	710)	10	99	6.0/28.3	748	3566	20.7	—	79.3	285	1127
			100			3712	20.4	—	79.6	272	1059
			101			3598	22.4	—	77.6	280.6	1034
			102			3480	25.6	—	74.4	255.8	1043
Dofasco, Hamilton, Ontario	711)	3	103	6.2/30.2		1618	—	9.4	90.6	21	752
			104			1670	—	0.4	99.6	43	664

* Lump ore: including nodules, remelting iron, slag- and limestone additions.
*** Net burden weight calculated from the coke consumption and the ore/coke ratio.

continuation)

Oil injection kg/ton iron	Iron production tons/24 h	tons/24 h m²	Blast temperature °C	Blast humidity g/m³ at NTP	Top gas CO %	CO_2 %	H_2 %	Pig iron Si %	P %	S 10^{-3}%	Slag volume kg/ton iron	CaO/SiO_2 in slag
—	1344	30.4	885	25.6	25.7	15.7	2.3	0.81	0.15	30	367†	1.09
50	1487	33.6	992	9.8	24.4	16.5	3.1	0.73	0.15	27	328	1.02
34	1468	33.2	868	4.0	24.0	16.8	2.5	0.56	0.14	39	309	1.12
44	1556	35.2	891	4.0	24.4	15.3	2.9	0.67	0.19	37	301	1.09
—	507	23.2	799	11.5	27.6	13.7	1.8	0.63	0.25	35	334	1.48
39.3	527	24.1	904	10.7	27.3	13.0	2.7	0.73	0.19	40	367	1.36
40.4	512	23.4	892	8.2	27.3	12.8	3.1	0.72	0.19	37	362	1.35
—	1158	28.4	770	26.6			2.1	0.71	basic iron	35	392	1.23
17.6	1283	31.4	830	18.6			2.3	0.84		27	378	1.27
27.8	1310	32.1	805	8.7			2.0	0.77		32	376	1.24
38.4	1303	31.9	785	2.8			2.1	0.73		30	376	1.28
39.4	1265	31.0	767	3.0			2.2	0.78		35	400	1.25
45.2	1271	31.1	820	2.8			2.2	0.76		24	384	1.31
48.8	1327	32.5	883	3.4			2.5	0.72		26	378	1.32
52.2	1192	29.2	908	5.0			2.7	0.82		24	431	1.27
—	1342	32.0	854	14.3			2.6	0.65		33	383	1.28
30.7	1393	33.2	824	14.3			3.4	0.64		36	389	1.29
45.4	1408	33.6	854	13.1			3.7	0.56		31	349	1.32
69.7	1362	32.5	884	12.0			4.9	0.56		35	379	1.32
71.4	1324	31.6	873	13.1			5.4	0.49		34	414	1.31
64.9	1390	33.2	891	11.0			4.9	0.47		33	349	1.36
60.7	1352	32.2	968	12.0			4.4	0.56		39	349	1.31
—	1574	32.9	944	54.7			2.4	0.66		24	349	1.30
18.0	1676	33.8	963	33.1			2.4	0.72		28	360	1.24
48.0	1524	31.9	977	33.6			3.0	0.68		32	364	1.25
81.0	1428	29.9	982	25.8			3.7	0.69		28	354	1.26
82.0	1468	30.7	958	34.3			4.3	0.65		33	336	1.27
—	1526	33.6	908	58.7			2.5	0.67		30	345	1.30
17.0	1612	35.5	967	36.8			2.5	0.70		29	345	1.26
34.0	1527	33.6	962	32.4			2.5	0.68		30	366	1.24
77.0	1351	29.8	971	25.8			3.0	0.66		27	350	1.27
80.0	1239	27.3	958	39.3			4.0	0.65		34	288	1.26
—			690		31.9	8.5	1.7				1384	
—			670		31.2	9.7	1.6				1353	
—			660		31.9	9.5	1.8				1349	
62			640		30.0	10.0	2.1				1382	
59			653		30.3	10.0	2.1				1342	
43			670		31.0	10.2	2.2				1313	
—	100%		695	22.8			1.5					
98	104.5%		732	18.8			3.5					
77	118.0%		745	17.7			—					
—	258.2	14.3	780		26.7	12.8	3.8					~1.50
49.2	283.4	15.7	910		27.0	13.3	4.5	0.63		40		1.46
—	362.2	22.8	709				2.4	0.40	1.77	97	~720	1.26
44	424.7	26.7	854				3.5	0.32	1.76	105	~720	1.33
59	457.4	28.8	917				4.1	0.45	1.81	92	~720	1.26
62	403	25.3	862				4.2	0.46	1.83	80	~720	1.28
—	417.6	14.8	790		30.8	11.2	2.3	0.53		46		1.43
50.5	419.4	14.8	800		30.4	10.8	3.7	0.45		63		1.37
57.5	451.3	15.9	796		29.8	11.1	3.6	0.38		91		1.36
56.9	478.5	16.9	808		29.3	10.8	4.1	0.28		85		1.40
—	1057	35.0	781	24.8				1.15		26	313	
66.5	1101	36.5	849	27.1				1.26		29	350	

† July 1962.

The cumulative frequency distribution of the observed injection rates and blast temperature changes is shown in Fig. 356. Replacement ratios with and without* correction for changes in burden are given in Fig. 357.

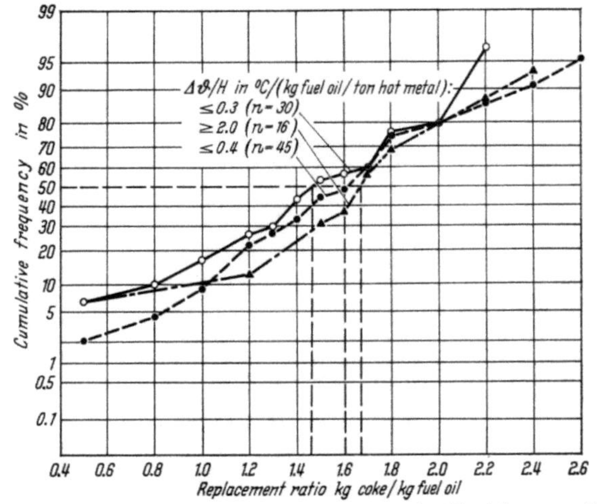

Fig. 358. Cumulative frequency of the replacement ratio E with correction for changes in burden and varying increases in blast temperature (from Table 51)

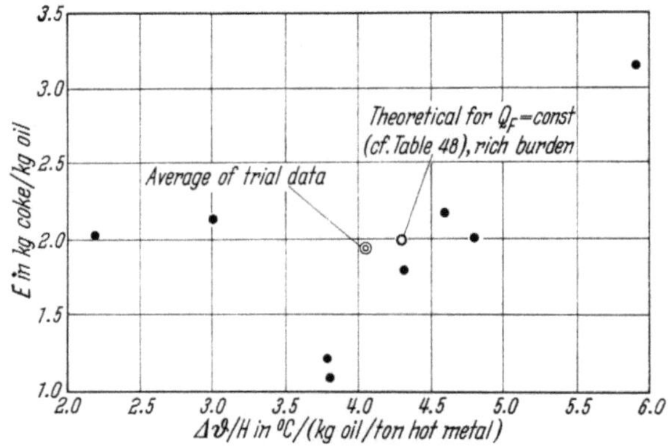

Fig. 359. Replacement ratio E in relation to blast temperature increase with different oil additions H, according to ISCHEBECK et al.[661])

The scatter is considerable. Mathematical/statistical evaluation shows, however, that the replacement ratio, as one might expect, increases with blast temperature (Fig. 358). The trial data obtained at Phoenix Rheinrohr A.G.[661]) (Fig. 359) for trial periods with an average temperature

* A change of ± 0.25 kg in coke consumption is assumed (according to section 5.2) for each ± 1 kg change in net burden weight.

increase of 4.1 °C/(kg oil/ton iron) show good agreement with the theoretical predictions in Table 48. Multi-dimensional regression analysis[689]) (Fig. 360) shows the influence of the quantities injected, related to constant net burden weight, blast temperature and humidity. The curve plot is adequately occupied statistically and shows good agreement with the theoretical calculation in the injection range of particular interest in ironmaking around 50 kg oil/ton iron. The results obtained with tar injection have proved to be comparable to those obtained with oil[711o,p]).

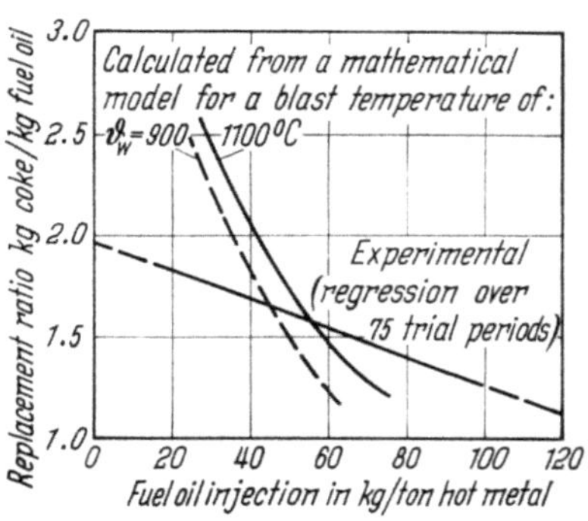

Fig. 360. Replacement ratio in relation to fuel oil addition with constant burden, blast temperature and humidity (from Table 51)[689])

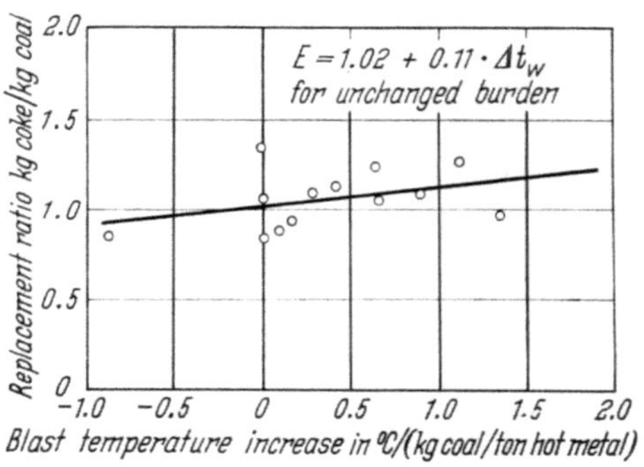

Fig. 361. Replacement ratio of pulverised coal in relation to blast temperature increase

5.4.4.4. Fine Coal

Table 52 gives a survey of the results obtained with fine coal injection[661a, 696s, 713, 714, 714a–e]). Statistical evaluation is not feasible because of the small number of data. According to Fig. 361, the replacement ratio is not so closely dependent on the blast temperature increase as in the case of fuel oil or natural gas. At constant blast temperature, a replacement ratio of 1.02 kg coke/kg coal is observed. The following value was determined from the calorific value ratio:

$$\frac{H_u(\text{coal})}{H_u(\text{coke})} = \frac{7600^*}{7000} = 1.08 \frac{\text{kg coke}}{\text{kg coal}}.$$

* Calorific value for water-free coal with 35% volatile matter and 8% ash (represents average of data in Table 52).

Table 52. *Replacement ratios and operating data for coal injection in the blast furnace* [661a, 713, 714])

Works	Stanton and Staveley			Hanna Furnace Co.				Cie des Hauts Fourneaux de Chasse				
Hearth diameter of furnace used	5.8 m			4.9 m				2.5 m				
Kind of fuel	Base coke	Colverton	Cadeby	Base coke	Lower-Kittanning	Base coke	Pittsburgh	Base coke	Coal with 34.5 % volatile matter		Coal with 6.6 % volatile matter	
1. *Fuel composition*												
Moisture . . . % by weight	4.9	9.1*	3.7*		max. 3		max. 3	1.8	1.9		1.1	
Ash . . . . . . % by weight	10.0	7.7	5.7	7.99	7.6	7.99	4.4	9.8	7.9		9.3	
Volatiles . . . % by weight	0.9	32.5	35.4	1.10	17.6	1.10	40.0	1.0	34.5		6.6	
S . . . . % by weight	1.11	0.70	1.47	0.74	0.94	0.74	1.14	0.90	0.80		0.60	
Size grading . . mm		0 to 4	0 to 4		0 to 4		0 to 4		0 to 0.2		0.05 to 2	
2. *Burden*												
Net burden wt. . kg/ton iron				1800	1782	1730	1715	1904	1948	1878	1946	1890
Proportion of sinter or pellets %	70.8	72.6	74.4	3.1	1.3	21.5	25.0	15.6	18.4	14.2	14.0	4.9
3. *Operating data*												
Test period. . . days	14	14	28	12	12	12	11	14	27	27	8	8
Iron make . . . tons/day	470	514	516	612	603	605	607	162	169	176	195	230
Blast temp. . . °C	850	910	950	790	803	795	816	750	871	795	820	855
Slag volume . . kg/ton iron	448	481	490	473	487	475	473	335	303	315	340	314
Coke consumpt. kg/ton iron	783	710	710	783	658	795	670	710**	579**	565**	578**	522**
Coal addition . kg/ton iron	0	67.5	74	—	142	—	132	0	108**	110**	110**	155**
Replacement ratio coke/coal kg/kg	0	1.08	0.98	—	0.88	—	0.94	—	1.27†	1.14†	1.24†	1.05†
η_{CO} . . . . %	23	23	24	39	38	38	37	30.3	32.8	34.1	33.7	35.5
Incr. in product. %	—	9.6	9.7	—	1.5	—	0.3	—	4.2	8.7	12.2	14.2
4. *Iron composition*												
C . . . . % by weight	3.77	3.72	3.72					4.22	4.06	4.18	4.19	4.24
Si . . . . % by weight	2.49	2.60	2.66	2.34	2.32	2.37	2.27	0.90	0.75	0.70	0.70	0.49
Mn . . . . % by weight	0.28	0.29	0.28	0.93	1.01	0.92	0.93	2.50	2.37	2.44	2.30	2.36
S . . . . % by weight	0.04	0.04	0.04	0.025	0.027	0.023	0.25					
P . . . . % by weight	0.62	0.64	0.65	0.36	0.99	1.06	1.01					

* Subsequently dried. ** Dry. † Corrected for change in burden.

There is, therefore, statistical agreement with the theoretical prediction. Occasional reports have been published on results obtained with industrial injection of pumpable coal/oil mixtures[661b, c, 714f–i]. Coal/oil mixtures with 30 to 62% of coal have been tried on full-scale furnaces.

It should again be emphasised that the theoretical calculation does not represent an absolute prediction but is based on data obtained from actual practice as far as the assumptions for the chemical and thermal efficiency of the stack gas are concerned. In so far as these data originated from the same industrial trials, as in the trial periods evaluated above, the agreement of the measured replacement ratios with the theory only proves that the balances were correct and the measuring error, small. The value of theoretical treatment lies chiefly in the fact that all the numerous factors involved can be systematically incorporated and reliable predictions can be made for future contingencies.

5.5. Mathematical Analysis, Prediction and Limitations of the Blast Furnace Process

5.5.1. Statement of the Problem

Prediction of blast furnace operating data according to the properties of the ores and fuels used and from particular operating parameters such as blast volume and blast temperature is of great theoretical and practical significance. Such prediction can provide the basis for numerous technical and economic decisions, for example, evaluation of ore, optimisation of driving conditions and, ultimately, for automation of the entire process. Determination of the minimum fuel consumption and maximum iron make for certain burden conditions is particularly important. A comprehensive assembly of the essential relationships describing the complex activity within the blast furnace is known as a mathematical model. Such models are best handled by electronic computers.

5.5.2. Traditional Blast-Furnace Models

Long-standing, traditional blast furnace models fall into three groups[717–726]:

thermodynamic models: energy balance models based on the first law of thermodynamics and energy balance models based on the first and second laws of thermodynamics;

refined energy balance models with allowance for chemical reaction and kinetics;

mathematical/statistical models.

There is considerable variation in the scope of the theoretical basis of these models and, therefore, in their value for forecasting but they can only be used within certain limitations for prediction.

5.5.2.1. Energy-Balance Models Based on the First Law of Thermodynamics

All reactions in the blast furnace involve reaction heats. According to the first law, the sum Q_R of all these reaction heats must equal the calorific value of the charged fuel plus the blast heat and minus the losses:

$$Q_R = K + W - H_g - F_g - V \text{ (kcal/ton iron).} \qquad (1)$$

In this equation,

K = coke charged × coke calorific value (kcal/ton iron),
W = blast heat (kcal/ton iron),
H_g = top gas calorific value × volume (kcal/ton iron),
F_g = top gas sensible heat × volume (kcal/ton iron),
V = cooling losses (kcal/ton iron),
Q_R = sum of the reaction heats of running reactions and the heat for heating the burden and coke and heat of fusion (kcal/ton iron).

The sum Q_R of the largely endothermic reaction heats, i.e. the useful heat, can be determined with great accuracy for each burden[718]). It is closely dependent on the type of ore and, particularly, the proportion of inert materials. The quantities of heat supplied and lost can be determined with good accuracy by weighing and analysing the coke and recording blast volume, blast temperature and top gas characteristics. Equation (1) is extremely useful for energy balance control, particularly in connection with material balances, as first shown by WESEMANN[717]). Equation (1) could be used to predict the principal operating parameters, especially, fuel consumption in the furnace, if the utilisation efficiency of the heat supplied,

$$\eta_W = \frac{Q_R}{K + W} = 1 - \frac{H_g + F_g + V}{K + W},$$

were near unity or, at least, constant. This is far from being the case, however. As shown in section 5.1, even with modern blast furnaces using prepared burdens, the heat supplied is considerably in excess of the heat required for reactions and melting. The ratio of heat used to heat supplied also varies widely. These variations are chiefly caused by fluctuating chemical utilisation of the stack gas rising up the furnace and, in particular, of the CO it contains (Fig. 381). Some 45 to 60% of the heat supplied in the coke is not used in the blast furnace and leaves in the form of unused CO in the top gas. Variation in the utilisation of the sensible heat in the stack gas is also significant. Genuine prediction of the required fuel input, the most important process parameter for the ore, is, therefore, only possible if a precise forecast can be made of the chemical utilisation of the CO from the coke or – which amounts to the same thing – of the degree of indirect reduction with each individual ore, which substantially affects the heat requirement. These data can only be obtained from measurements of reaction kinetics, however. They certainly cannot be considered as ore-dependent constants but depend extremely closely on

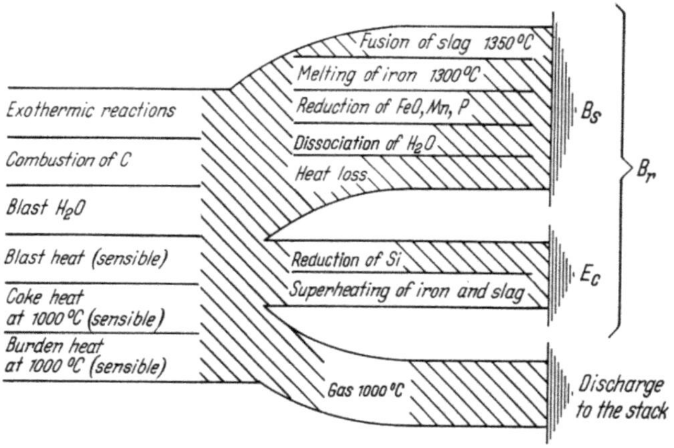

Fig. 362. Formation of the factor E_c, according to Poos et al.[731])

furnace driving practice. This inevitable theoretical deduction was confirmed only recently by Hupfer and Weidenmüller[727]) after probe measurements in the furnace.

Thermodynamic models of this type have, nevertheless, been used in interesting ways[728–731]). By continuous and accurate determination of blast volume and temperature, blast additives, top gas composition and temperature and the amounts and compositions of the materials charged, Eq. (1) can be used to obtain a continuous thermal balance. For constant cooling losses, the useful heat Q_R yielded to the burden can be calculated from the difference between the heat supplied (coke, hot blast and blast additives) and the heat removed in the top gas (quantity, calorific value and temperature). At constant burden composition, the iron temperature and silicon content will increase or decrease with Q_R. Continuous monitoring of the listed parameters, therefore, allows a constant prediction of iron temperature and silicon content which shows good agreement with practice[729]). It is possible to go one step further and counteract anticipated changes by deliberate control of blast temperature and injections. As Fig. 362 shows, the thermal balance can be made for the 1000°C zone of the furnace. The difference E_c between the heat B_r supplied to this zone and the minimum useful heat B_s is closely related to the Si-content of the iron (Fig. 363). Auto-control on this basis with the object of keeping

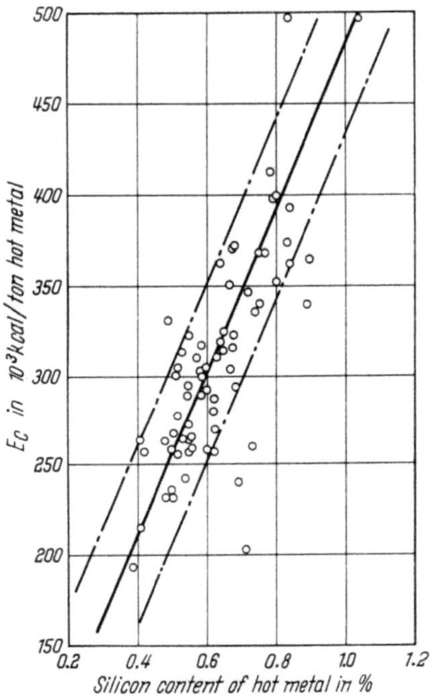

Fig. 363. Relationship between E_c and the silicon content of the hot metal, according to Poos et al.[731])

the Si-content of the iron constant would naturally require first-rate measuring techniques and application of automatic computers on the furnace. The most difficult operation in this event might well be the continuous and accurate determination of the chemical composition and other properties of the burden constituents. Mathematical/statistical evaluation of iron analyses with and without auto-control of the heat input has still not shown any convincing progress. For example, Horie[753]) obtained the following standard deviations in Si-contents in superlative furnace driving conditions:

without auto-control: $\pm 0.12\%$,
with auto-control: $\pm 0.10\%$.

5.5.2.2. Energy-Balance Models Based on the First and Second Laws of Thermodynamics*

One refinement, represented by REICHARDT's heat diagram[719]) is shown in Fig. 364**. This diagram starts from a recognition of the fact that, according to the second law of thermodynamics, the gas temperature in

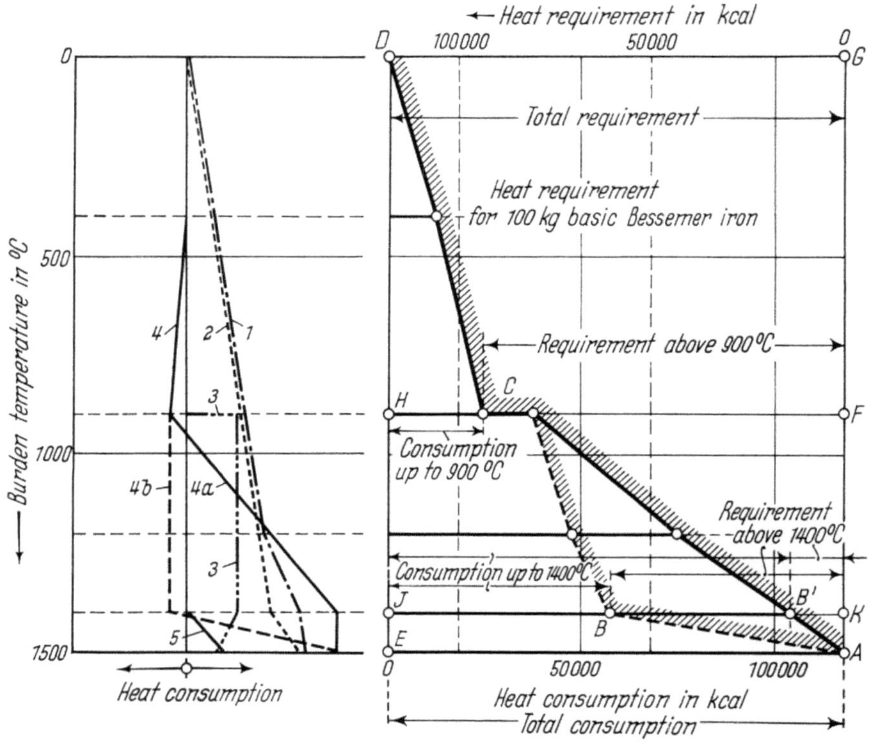

Fig. 364. Heat requirement for 100 kg basic Bessemer iron, according to REICHARDT[719])
1 Heating and melting of iron, *2* Heating and melting of the slag, *3* Decomposition of limestone and slag formation, *4* Reduction of Fe oxides, *4a* Gaseous reduction 900 to 1400°C, *4b* Direct reduction above 1400°C, *5* Reduction of Mn, P, Si

each temperature zone of the blast furnace must be higher than the burden temperature provided no exothermic reactions are in progress. The balance is elaborated and established separately for each temperature zone in the furnace. It is then necessary to know the temperature ranges of the chemical reactions in the furnace in order to insert the relevant reaction heats

* Energy balance models based on the first law retain their validity; additional statements become possible after introduction of the second law.

** This was preceded by the two-stage heat balance of MATHESIUS[726]), an improved form of which has recently been used by SCHÜRMANN and BÜLTER[720,721]).

in the correct zones. These temperature ranges are approximately known from laboratory and probe measurements.

Figure 365 summarises the results of several reports on the temperature ranges of the main blast furnace processes. Although these findings cannot be generalised, they do make it possible to draw up approximate heat

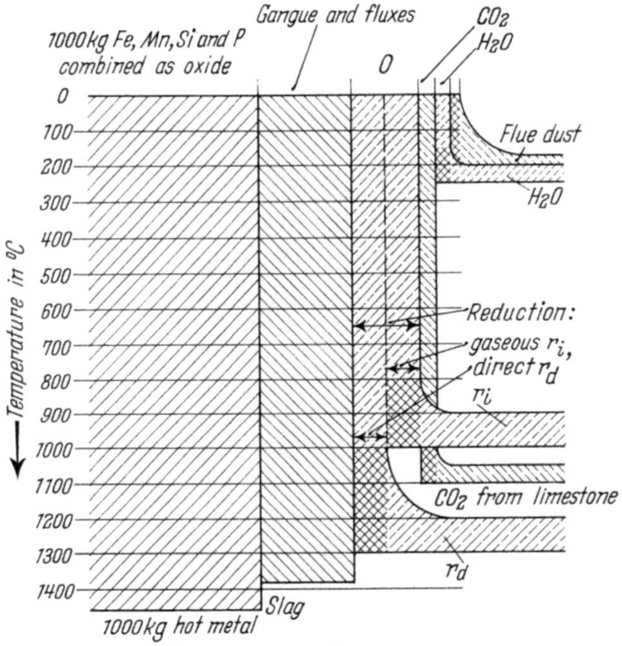

Fig. 365. Temperature ranges for reactions with the gas phase in the blast furnace[732] (schematic)

balances of the Reichardt type. This still does not solve the problem of real prediction, however, for the temperature ranges of the individual reactions can fluctuate several hundred degrees depending on the type of ore and the furnace driving rate. Examples are shown in Fig. 366. In this case, two different ores were reduced, as in the furnace, with an N_2/CO mixture and steadily rising temperature. The easily reducible Mano River ore was found to be reduced from 100°C onward whilst the less readily reducible Gellivara ore was only reduced from 400°C onward. Here too, these temperature ranges are, by no means, constants characterising the ore but depend on the nature of the burden as a whole and driving conditions. At identical temperature, for instance, the higher the CO-content of the gas phase, i.e. the higher the coke consumption rate, the earlier will reduction of the ore commence. The end result is, therefore, incorporated as a parameter in the bases of the calculation.

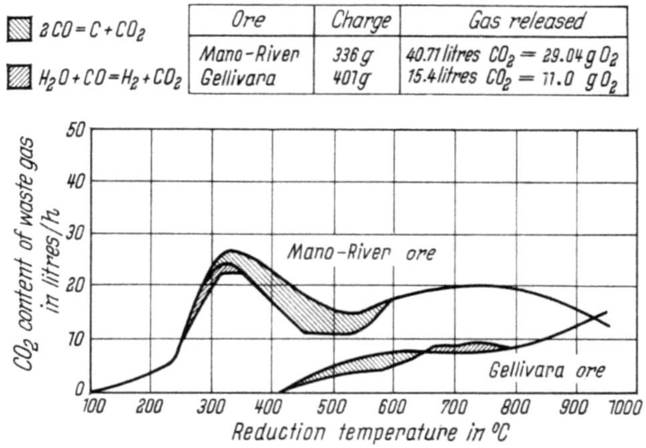

Fig. 366. Reduction test with Mano-River and Gellivara ore using the heating-up method[733a]

5.5.2.3. Refined Energy-Balance Models with Allowance for Chemical Reaction and Kinetics[722, 725, 734, 735]

WARTMANN[734] developed a blast furnace model in which corrective intervention in the zonal balances is effected on the basis of heat transfer from gas to burden and established laws of reduction kinetics and in which the ores were no longer characterised by parameters such as constant

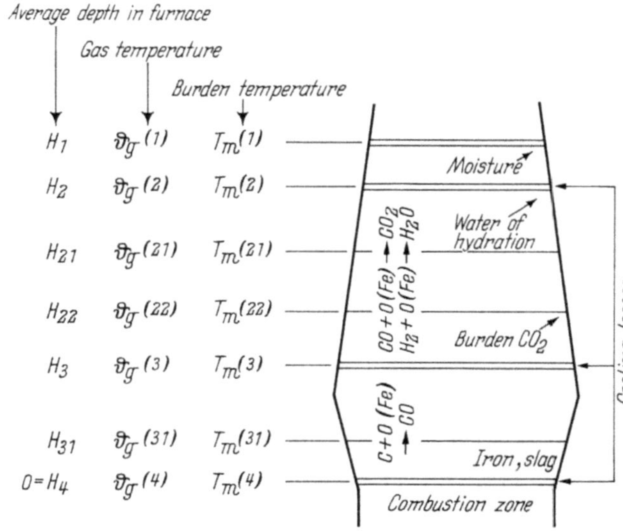

Fig. 367. Schematic outline of blast furnace processes for a mathematical model, according to WARTMANN[734]

indirect reduction but by a reactivity constant. The actual degree of indirect reduction of each ore is then calculated for each zone from the operating parameters (gas composition, rate of burden descent and so on) with the aid of this reactivity constant. The average value of these reactivity constants is chosen so that operational results match the results of calculations of blast furnace performance for the same mixed burden operation. For subsequent calculations with other burdens, the reactivity constants are scaled according to laboratory tests. Division of the reactions according to set temperature ranges is still retained however (Fig. 367), for example, allocation of 450 to 1100°C for indirect reduction and above 1100°C for direct reduction. Some results of this model are shown in Fig. 368.

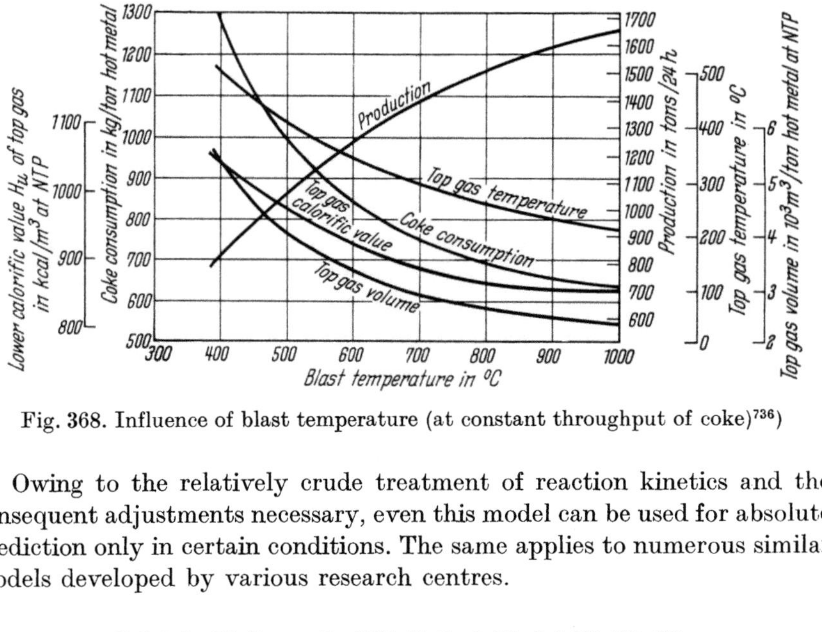

Fig. 368. Influence of blast temperature (at constant throughput of coke)[736]

Owing to the relatively crude treatment of reaction kinetics and the consequent adjustments necessary, even this model can be used for absolute prediction only in certain conditions. The same applies to numerous similar models developed by various research centres.

5.5.2.4. Mathematical/Statistical Models[732, 737–739]

Because of the requirements of practice and the long absence of a theoretical basis adequate for developing refined models, repeated attempts have been made (with remarkable results) to organize the abundant data from furnace observations in such a way that conclusions e.g. ore evaluations, can be established. Mathematical statistics can be used to organize numerical operating data into meaningful groups. The resulting regression equations can be regarded as mathematical/statistical blast furnace models. A very simple form of this type of model is, for example, the mathematical/statistical investigation of the relation between net burden weight and coke consumption rate according to whether a self-fluxing sinter burden or a fully prepared burden with non-self-fluxing sinter is used. The result is shown in Fig. 369 (cf. also section 5.2), the optimum values being obtained by VOICE and DIXON[739]). Even today, much of the ore evaluation in numerous iron and steel plants is based on relationships of this type.

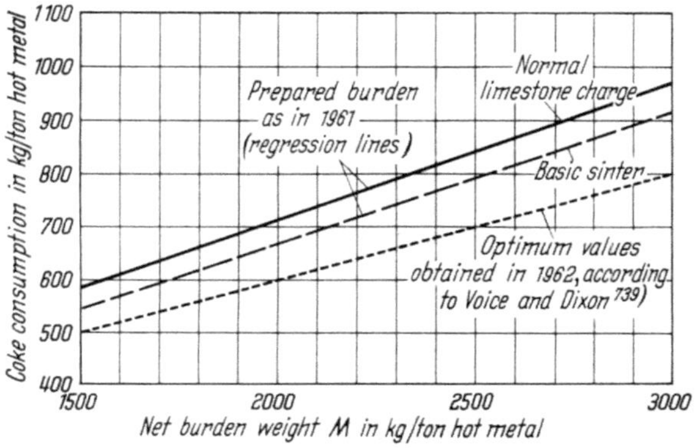

Fig. 369. Coke consumption (about 85% C) in relation to net burden weight (no fuel injection)

The basis of mathematical/statistical models can be refined with a much larger number of independent variables[738]).

The extent to which they can be applied, however, is always limited because the regression equations can be assumed valid only within the range of the evaluated data. Only the effects of changes within the range known and substantiated by operating trials can, therefore, be predicted.

5.5.3. Approaches to the Establishment of Condition-Free Models of the Blast Furnace Process

The usefulness of hitherto accepted classical blast furnace models was severely limited (as described, for example, in section 5.4 with reference to replacement fuels) by the necessity of incorporating the result in a more or less disguised form in the calculation e.g. as "degree of indirect reduction" or "chemical utilisation of CO or H_2" or "maximum possible gas throughput". Such models could not, therefore, give permanent satisfaction.

Another approach has now been gaining acceptance for some time based on analysis of the individual processes[736, 740–743]). The hydrogen-blown shaft furnace, the Purofer process, has provided an opportunity for using this approach with a simplified example and reliable measured data (section 4.6). In principle, the blast furnace can be treated in exactly the same way. Many other processes are, of course, going on at the same time for which measuring techniques are, in some cases, still at an early stage of development: e.g.

gasification of coke by O_2 and CO_2,
softening processes which affect both reduction rate and permeability,
liquid phase reactions,
slag formation.

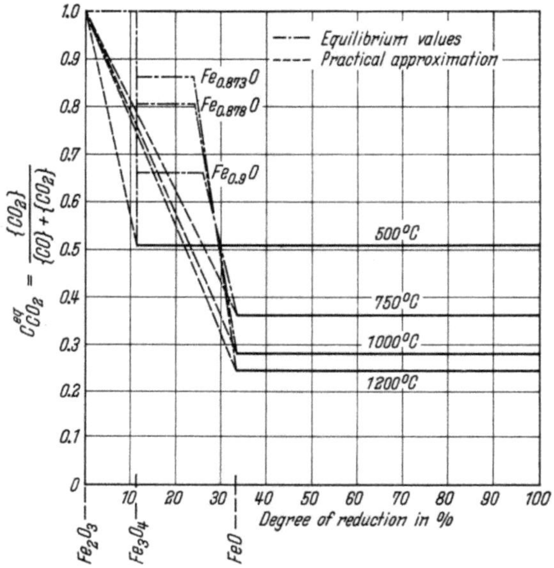

Fig. 370. Equilibria between iron oxides and CO/CO_2 mixtures at 500, 750, 1000, and 1200°C [743])

A reliable mathematical treatment of the blast furnace process on this basis is not yet even approximately feasible for lack of adequate data.

Only brief mention will, therefore, be made of the manner in which these problems have been initially approached.

It is first necessary to determine, for the full range of conditions conceivable in the blast furnace, the rates of the principal stack reactions, especially reduction of the ores:

$$Fe_nO_m + m\,CO = n\,Fe + m\,CO_2$$

and coke gasification by CO_2:

$$C + CO_2 = 2\,CO$$

in relation to temperature, gas composition, degree of reduction, previous history etc. Investigations of this type were made by PÜCKOFF[743]) on two ores between 500 and 1200°C using varying lump size and gas composition. As shown in section 4.4 using the example of reduction with hydrogen, the results can be summarised as follows.

The reduction rate $\frac{d\mathfrak{R}}{dt}$ measured in each case is split into two factors:

$$\frac{d\mathfrak{R}}{dt} = v(C^{eq}_{CO_2} - C^{0}_{CO_2}) \tag{3}$$

in which

$v = \imath(\mathfrak{R}, T)$ = rate constant (sec^{-1}),
$C^{eq}_{CO_2}$ and $C^{0}_{CO_2}$ = maximum and actual CO_2-content of the gas phase in volume fractions,
$\mathfrak{R}$ = degree of reduction.

$C^{eq}_{CO_2}$, as Fig. 370 shows, depends on the degree of reduction. For the reasons stated in section 4.4, linear interpolation is carried out between the individual maximum and minimum $C^{eq}_{CO_2}$-values for each temperature. The v-values are shown in Figs. 371 and 372 in relation to the degree

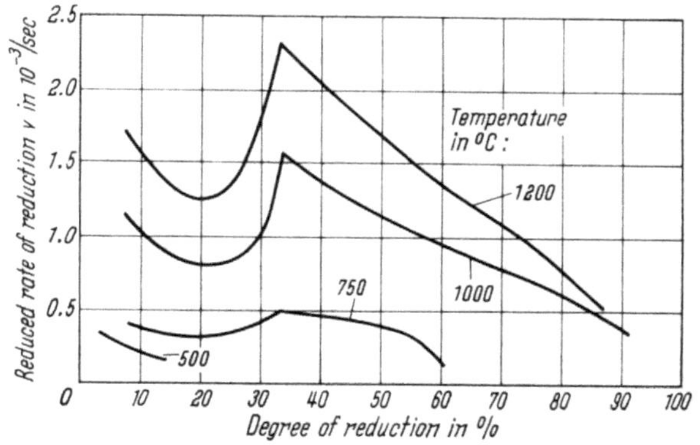

Fig. 371. Reduction rate in carbon monoxide, carbon dioxide and nitrogen gas mixtures ($p_{CO} + p_{CO_2} = 0.4$ atm absolute) in relation to temperature and degree of reduction (Venezuela ore, lump size 15 to 20 mm)[743]

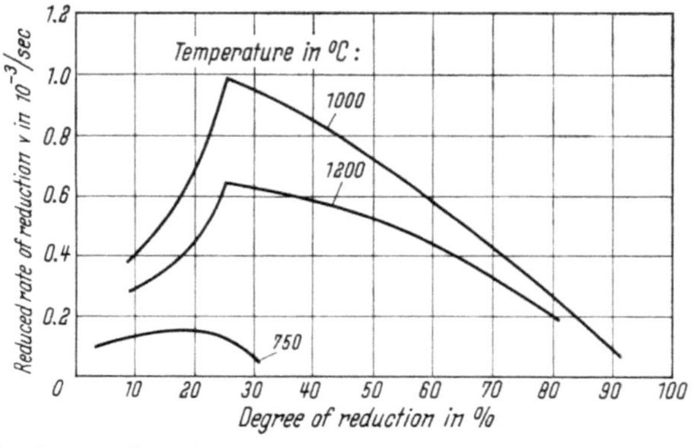

Fig. 372. Reduction rate in carbon monoxide, carbon dioxide and nitrogen gas mixtures ($p_{CO} + p_{CO_2} = 0.4$ atm absolute) in relation to temperature and degree of reduction (Kiruna-D ore, lump size 10 to 15 mm)[743]

of reduction and temperature. The reduction rate can be calculated from these data with Eq. (3) for any given combination of the principal conditions involved. The influence of previous history is not taken into account. Whether or not a certain degree of reduction was reached quickly or slowly at high or low temperature cannot be entirely disregarded, i.e. the relationships $v = v(\Re, T)$ and even $C^{eq}_{CO_2} = C^{eq}_{CO_2}(\Re, T)$ are not absolutely unequivocal. Fortunately, verification tests showed that the deviations in the range involved are not too wide and can, therefore, be disregarded initially[743]).

The reaction rates of coke with CO_2 were determined in the same way and summarised according to HEDDEN's formulae (cf. section 4.5).

JESCHAR's formulae are appropriate for calculating the heat transfer rate (section 2.2). As regards the remaining factors, e.g., rate of softening, melting, reduction of associated elements, etc., insufficient exploitable data are available. Probe measurements at tuyere level and three-dimensional ciné-pictures have so far provided valuable partial explanations[759]). The measured and existing data were sufficient for the development of a condition-free model for at least the "dry" part of the blast furnace. This model has already enabled statements to be made, e.g. on the manner in which coke consumption is affected by ore reducibility controllable by size range adjustment[743]) (see Table 53).

A really satisfactory mathematical blast furnace model, which must be based on reaction kinetics, will not be feasible until numerous missing constituents can be incorporated; one of the main tasks of our research centres will be to systematically determine these constituents.

The currently viable mathematical models will not be discussed further here because of the incomplete knowledge of reaction kinetics on which they are based. Further information can be found in the specialist literature[736, 741–743, 743a]).

Some important influencing factors will now be discussed with special reference to the limits of the blast furnace as regards output per unit time and fuel consumption per ton iron.

5.5.4. Output Limits

5.5.4.1. Theoretical Maximum Output

We will first consider the flow rate, i.e. the gas volume per unit time and unit furnace cross section in the most sensitive zone, namely, that in which the gas velocity is nearest the fluidization point. The coke charged per unit time and the iron make are intimately related to the gas volume.

As shown in section 3.1, gas velocity in packed beds is limited by the fluidization point. With uneven particle size distribution, the condition

Table 53. *Input data and results of calculation for a kinetic blast furnace model with a fixed temperature profile. Coke: Kaiserstuhl, 30 to 40 mm*

		Venezuela ore		Kiruna-D ore	
		15 to 20 mm	50 to 60 mm	10 to 15 mm	30 to 40 mm
a) Input data					
Ore bulk density	tons/m³	1.800	1.800	2.600	2.450
Coke bulk density	tons/m³	0.500	0.500	0.500	0.500
Fe-content of ore	%	51.50	51.60	59.00	59.00
O-content of ore	%	22.10	22.20	22.60	22.51
Slag forming materials	%	14.10	14.10	16.20	16.20
Ore moisture content	%	12.30	12.10	2.20	2.30
Slag volume	kg/ton iron	300	300	300	300
Furnace stack cross section	m²	50	50	50	50
Hearth diameter	m	7.5	7.5	7.5	7.5
Cold blast volume	m³/h at NTP	110000	110000	110000	110000
Blast temperature	°C	1000	1000	1000	1000
Cooling losses	10³ kcal/h	10000	10000	10000	10000
b) Results of calculation					
Burden temperature	°C	1250	1250	1250	1250
Stack gas temperature	°C	1413	1379	1310	1279
Ore consumption	tons/day	4302	3561	2773	2144
Degree of indirect reduction	%	61.0	51.4	55.7	32.4
Oxygen released	tons/h	24.705	17.106	14.789	6.585
Coke rate	kg/ton iron	445	534	545	724
Fuel coke	kg/ton iron	279	336	381	489
Coke for Boudouard reaction	kg/ton iron	132	164	134	202
Coke for Boudouard reaction	tons/h	0.537	0.268	0.326	0.122
Residence time in furnace stack	h	4.55	5.02	6.84	7.17
Iron production	tons/day	2382	1976	1759	1360
$\frac{CO_2}{CO + CO_2}$ top gas		0.49	0.34	0.32	0.14
Gas velocity at 950°C gas temperature (free cross-sectional velocity)	m/sec	1.9	1.9	1.8	1.9
Gas velocity at 300°C gas temperature (free cross-sectional velocity)	m/sec	3.4	3.4	2.9	2.9

known as "spouted bed" can arise even below the fluidization point; the characteristic curve of this phenomenon for a blast furnace using an unscreened burden has been clearly established by KEGEL (Fig. 373). Once

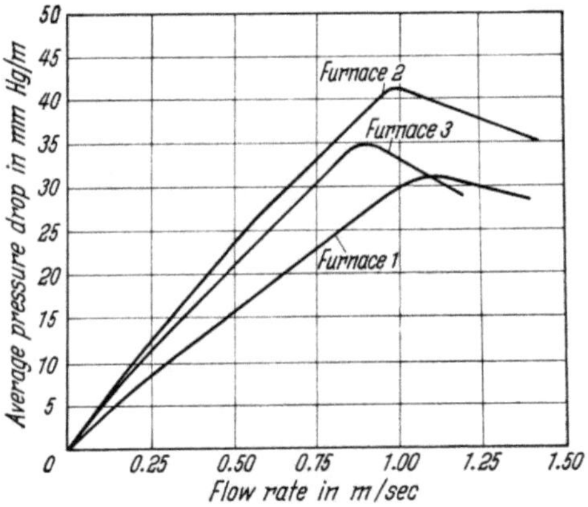

Fig. 373. Pressure drop in the blast furnace, according to KEGEL[744])

the maximum pressure is exceeded, the gas flow per unit time increases. Controlled driving of the furnace is then no longer possible.

In a burden screened to a narrow size range where the fine material, in particular, has been removed almost completely, the fluidization point can limit the gas velocity in the stack. As shown in Fig. 374, the fluidiza-

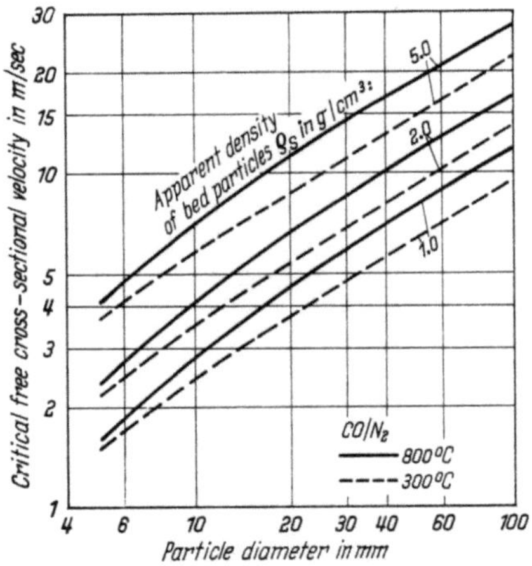

Fig. 374. Stability limit of beds of uniform particle size (fluidization point). Dependence on particle diameter (spheres)

tion point in uniformly-sized beds blown with CO/N_2 mixtures similar to those in the blast furnace is exceeded with much lower flow rates at 300°C than is the case at 800°C. The method of calculation was described in section 3.1. Special attention should, therefore, be given to gas velocity in the region of the throat. With a burden screened to 5 mm, the free cross-sectional gas velocity (defined as the gas velocity in the empty cross section in question) near the throat should not exceed 2.5 to 4 m/sec depending on the density of the ores (Fig. 374). Comparison with actual practice showed the following result: the gas velocities at the top and bottom of the stack, i.e. for temperatures of about 300°C and 1000°C respectively, were calculated from the published data[732, 746]) for the blast furnaces surveyed in Tables 39 and 54. It was found that there was a distinct parallel (Fig. 375a) between iron make and gas velocity and that the limiting velocities (fluidization point) for a burden screened to 5 mm* were partially attained.

It is clear from Figs. 375b to 375d that, for constant gas velocity, output can be increased appreciably by injection of gas or oil (enrichment of the tuyere gas with H_2) and by driving at high top pressure (increase in gas throughput with unchanged velocity). Figures 375a to 375d suggest that, in otherwise regular conditions, the furnace iron make is mainly determined by the volume of gas the furnace will accept per unit time and that the theoretical maximum gas velocities (and therefore outputs) are already being obtained with the currently used burden size ranges.

Further increases in maximum gas velocity can be achieved by the following measures:

a) An additional increase in gas pressure in the throat. According to section 3.1, the rate of gas throughput in the furnace is influenced only by the square root of the working pressure. The relatively small effect of high top pressure established in theory was also observed in practice[755]) as illustrated by the comparison in Fig. 376.

b) Further improvement in burden screening, especially removal of the −10 mm fractions. This should, according to Fig. 374, permit an appreciable rise in gas velocity. Whether or not the furnace output increases to the same extent will depend on the chemical utilisation of the gases and, therefore, the coke consumption rate and also on processes during softening and melting.

c) Charging of a partly or wholly reduced burden (cf. section 4.1).

---

* Where furnace inside diameter was not given in the literature, the free cross-sectional velocities were calculated on the assumption that bosh diameter $= 1.12\,D$ and throat diameter $= 0.85\,D$ (D = hearth diameter).

Table 54. *Characteristic operating data*

Works		Mannesmann A.G., Duisburg-Huckingen	Soviet Union	Phoenix Rheinrohr A.G., Duisburg-Ruhrort	Bethlehem Steel Corp., Bethlehem
Blast furnace				7	D
Period		20th May to 15th June 1963	Dec. 1962	1st to 15th Nov. 1963	Aug. 1963
Iron make	tons			30 533	91 291
Blowing time/shutdowns	h			295/17	701/43
Hearth diameter	m	7.0	9.75	7.5	8.76
Working volume	m^3	821	2002	981	1425
Ores	kg/ton iron	255		720	
Sinter	kg/ton iron		1890	677	660
Pellets	kg/ton iron	1320			762
Slags and other materials	kg/ton iron	135	4	196	80
Metallic charge	kg/ton iron	9		85	64
Limestone and dolomite	kg/ton iron	213	1	19	102
Gross burden	kg/ton iron	1932	1895	1697	1668
Flue dust	kg/ton iron	35	6	66	16
Net burden	kg/ton iron	1897	1889	1631	1652
Coke throughput (dry)	tons/24 h	1209	1847	1575	1706
Coke throughput (working volume)	tons/(m^3 · 24 h)	1.473	0.923	1.606	1.197
Coke consumption (dry)	kg/ton iron	578	525	634	546
Oil	kg/ton iron				
Natural gas	m^3/ton iron at NTP		35		
Coke throughput (corrected)*	tons/24 h		2009		
Coke consumption (corrected)	kg/ton iron		571		
Blast volume	m^3/h at NTP	143 000	213 420	162 000	201 000
Blast temperature	°C	882	1009	970	775
Blast pressure	kg/cm^2	1.61	3.07	1.35	1.79
Humidity	g/m^3 at NTP	14.6	10.7	6.7	34.4
Iron make	tons/24 h	2091	3518	2484	3125
Blast furnace coefficient	tons/(m^3 · 24 h)	2.547	1.757	2.532	2.193
Slag volume	kg/ton iron	386	585	368	274
Basicity CaO/SiO_2		1.17	1.17	1.29	0.99
Top gas temperature	°C	87	220	280	232
Top gas pressure	kg/cm^2		1.70		0.20
Calorific value	kcal/m^3 at NTP	770	761	1000	854
Hearth area	m^2	38.5	74.6	44.2	60.2
Coke throughput (corrected)	tons/(m^2 · 24 h)	31.4	26.9	35.6	28.3

* Conversion based on assumption that 1 kg fuel oil or 1 m^3 natural gas at NTP replaces 1.3 kg coke.

for various furnaces [512b, 746, 757, 758])

	S.A. Lorraine-Escaut, Mont-St-Martin	Bethlehem Steel Corp., Sparrows Point	Armco Steel Corp., Middletown	Armco Steel Corp., Middletown	Krivoi Rog	Nippon Kokan K. K., Mizue	Broken Hill Proprietary Co. Ltd., Port Kembla	Youngstown Sheet and Tube Co., Indiana Harbor
	3	J			Flagship	1	4	3
	July 1961	Oct. 1963	June 1961	March 1962	2nd decade of July 1961	April and May 1964	Sept. to Nov. 1962	Oct. 1961
		94742	80286	80741		188758	253882	75224
		715.8/28.2	704.5/15.5	721.5/22.5		1449/15	2110.5/73.5	—
	4.8	8.76	8.53	8.53	9.75	9.0	8.84	8.53
	640	1430	1529	1529	2002	1517	1522	1337
		11		140		580	161	
	2101	397			1890	993	1548	
		969	1540	1483				1565
	165	54	215	251	193	34	31	
		124		72				20
		71	220			70	16	302
	2266	1626	1975	1946	2083	1677	1756	1887
	79	56	94	37	60	22	18	29
	2187	1570	1881	1909	2023	1655	1738	1858
	500	1589	1729	1609	1574	1466	1816	1494
	0.781	1.111	1.131	1.052	0.786	0.966	1.193	1.117
	668	500	632	599	509	469	629	593
		44				47.8		
					76			
		1770			1881	1660		
		557			608	531		
	55500	218000	194500	177000	228000	164085	192102	185000
	852	900	704	788	950	1050	861	787
	0.82	2.45	2.07	1.71	2.90	1.84	2.00	1.85
	8.0		36.4	39.0		26.3	29.0	33.9
	760	3177	2735	2686	3094	3126	2887	2519
	1.188	2.222	1.789	1.757	1.545	2.061	1.897	1.884
	830	235	380	367	685	292	419	328
	1.20	1.04	1.03	1.04	1.13	1.15	1.15	
	278	188	182	145	430	217	201	140
		0.70	0.504	0.39	1.50	0.40	0.105	0.48
	824	844	764	764	878	810		
	18.1	60.2	56.8	56.8	74.6	63.6	61.3	56.8
	27.6	29.4	30.4	28.3	25.2	26.1	29.6	26.3

Table 54

Works		Inland Steel Co., Indiana Harbor	Dominion Foundries and Steel Co. Ltd. (DOFASCO), Hamilton	Phoenix Rheinrohr A.G., Duisburg-Ruhrort	Yawata Iron and Steel Co., Tobata
Blast furnace		4		6	3
Period		16th Oct. to 15th Nov. 1961	May 1962	July 1962	Dec. 1963
Iron make	tons	30808	50302	21831	98840
Blowing time/shutdowns	h	504.0/3.3		636/108	730.5/13.5
Hearth diameter	m	6.32	6.8	5.0	9.75
Working volume	m^3	764		635	1948
Ores	kg/ton iron	591	321	508	652
Sinter	kg/ton iron	1257		1209	918
Pellets	kg/ton iron		1441		
Slags and other materials	kg/ton iron		95	160	16
Metallic charge	kg/ton iron		125	30	
Limestone and dolomite	kg/ton iron	31		2	86
Gross burden	kg/ton iron	1879	1982	1909	1672
Flue dust	kg/ton iron	23	37	12	102
Net burden	kg/ton iron	1856	1945	1897	1570
Coke throughput (dry)	tons/24 h	876	761	509	1635
Coke throughput (working volume)	tons/($m^3 \cdot$ 24 h)	1.147		0.802	0.839
Coke consumption (dry)	kg/ton iron	594	469	618	500
Oil	kg/ton iron		70		29.4
Natural gas	m^3/ton iron at NTP				
Coke throughput (corrected)*	tons/24 h		909		1759
Coke consumption (corrected)	kg/ton iron		560		538
Blast volume	m^3/h at NTP	110400	117800	54610	184950
Blast temperature	°C	729	923	830	903
Blast pressure	kg/cm²	1.69	1.65	1.11	1.52
Humidity	g/m^3 at NTP	21.8	36.4	11.5	36.6
Iron make	tons/24 h	1474	1623	824	3269
Blast furnace coefficient	tons/($m^3 \cdot$ 24 h)	1.929		1.298	1.678
Slag volume	kg/ton iron	352	315	509	296
Basicity CaO/SiO_2				1.23	1.32
Top gas temperature	°C	141		157	199
Top gas pressure	kg/cm²	0.125	0.32		0.076
Calorific value	kcal/m^3 at NTP			848	781
Hearth area	m²	31.2	36.2	19.6	74.6
Coke throughput (corrected)	tons/(m² · 24 h)	28.1	25.0	26.0	23.6

* Conversion based on assumption that 1 kg fuel oil or 1 m^3 natural gas at NTP replaces 1.3 kg coke.

continuation)

Bethlehem Steel Co., Bethlehem	Fuji Iron and Steel Co., Muroran	Cherepovets	Steel Company of Canada (STELCO), Hamilton	Dominion Foundries and Steel Co. Ltd. (DOFASCO), Hamilton	Appleby-Frodingham	Dominion Foundries and Steel Co. Ltd. (DOFASCO), Hamilton	Youngstown Sheet and Tube Co., Indiana Harbor
B	3		2	3	Queen Victoria	3	1
May 1962	Dec. 1963	1959	1961/62	July 1961	28th Oct. to 3rd Nov. 1962	June 1961	June 1961
		33 days	6 days	18 days		17 days	
83574					14773		
					162.5/5.5		
8.76	7.61	7.20	5.49	6.78	9.45	6.78	7.92
1425	1088	1033	588	861	1448	861	1160
76				717		655	863
345	868	1916			2410		625
997			1441	983		1060	
113					38		196
86	9	27		61	12	70	15
86	160	115	120	209		234	212
1703	1742	2058	1561	1970	2460	2019	1911
38	12	101	27	20	20	22	36
1665	1730	1957	1534	1950	2440	1997	1875
1529	1013	1004	637	843	1545	898	1194
1.073	0.931	0.972	1.083	0.979	1.067	1.043	1.029
567	490	596	560	563	708	616	609
	47			29			
			34				
	1139		687	900			
	551		604	601			
189000	129540		74300	108600	182240	108500	151000
724	950	1027	787	876	954	865	630
1.79	1.84	2.21	1.24	1.74	1.74	1.66	1.57
45.3	11.8	12.9	16.2	30.0	8.4	43.7	21.6
2696	2067	1730	1138	1498	2182	1458	1960
1.892	1.900	1.675	1.935	1.740	1.507	1.693	1.690
268	321	693	168	288	1260	303	340
0.96		1.13	1.04	1.04	1.31		
238	190	148	185	134	202	140	
0.175	0.480	1.500	0.097	0.378	0.165	0.343	0.329
854	820	782	965	834	975	838	
60.2	45.6	40.7	23.7	36.3	70.1	36.3	49.2
25.4	25.0	24.7	29.0	24.8	22.0	24.7	24.3

Table 54

Works		Compañia de Acero del Pacifico S.A., Huachipato (Chile)	Phoenix Rheinrohr A.G., Duisburg-Ruhrort	Armco Steel Corp., Ashland	Yawata Iron and Steel Co., Tobata
Blast furnace		1	5	1	2
Period		June 1963	April 1964	1951 20 days	Dec. 1963
Iron make	tons	34541	27426		
Blowing time/shutdowns	h	669.6/	663/57		
Hearth diameter	m	6.00	5.5	4.18	8.90
Working volume	m^3	712	593	280	1458
Ores	kg/ton iron	1602	389	348	
Sinter	kg/ton iron		999		960
Pellets	kg/ton iron			1218	
Slags and other materials	kg/ton iron		327	174	
Metallic charge	kg/ton iron		79		
Limestone and dolomite	kg/ton iron	152	9		96
Gross burden	kg/ton iron	1754	1803	1740	1721
Flue dust	kg/ton iron	45	15	30	39
Net burden	kg/ton iron	1709	1788	1710	1682
Coke throughput (dry)	tons/24 h	677	593	338	1311
Coke throughput (working volume)	tons/(m^3 · 24 h)	0.951	1.000	1.207	0.899
Coke consumption (dry)	kg/ton iron	547	597	770	535
Oil	kg/ton iron	69			30
Natural gas	m^3/ton iron at NTP				
Coke throughput (corrected)*	tons/24 h	789			1406
Coke consumption (corrected)	kg/ton iron	637			574
Blast volume	m^3/h at NTP	93100	64110	40000	148200
Blast temperature	°C	985	778	614	890
Blast pressure	kg/cm^2	1.44	1.17	0.94	1.60
Humidity	g/m^3 at NTP	13.6	9.1	8.9	42.3
Iron make	tons/24 h	1238	993	439	2450
Blast furnace coefficient	tons/(m^3 · 24 h)	1.739	1.675	1.568	1.680
Slag volume	kg/ton iron	244	450	415	299
Basicity CaO/SiO_2		1.29	1.21	1.17	
Top gas temperature	°C	214	209	175	203
Top gas pressure	kg/cm^2	0.235		0.063	0.068
Calorific value	kcal/m^3 at NTP	868	878	878	821
Hearth area	m^2	28.3	23.7	13.7	62.2
Coke throughput (corrected)	tons/(m^2 · 24 h)	27.9	25.0	24.7	22.6

* Conversion based on assumption that 1 kg fuel oil or 1 m^3 natural gas at NTP replaces 1.3 kg coke.

(continuation)

Fuji Iron and Steel Co., Muroran	Kawasaki Steel Corp., Chiba	Great Lakes Steel Corp., Detroit	S.A. des Hauts-Fourneaux de la Providence, Rehon	Société Metallurgique de Normandie, Mondeville	Cherepovets	Nippon Kokan K.K., Mizue	Nippon Kokan K.K., Fukuyama
4	4			3			2
Dec. 1963	Dec. 1963	Sept. 1961	Oct. 1963	April 1964	1965	1 month	Nov. 1968
		17 days			31 days		
					131441		181 787
							708.75/11.25
8.80	8.80	8.84	5.30	6.70	9.75	9.0	11.20
1518	1498	1343	632	828	2000	1709	2255
				1205	3	37%	527
1244	589		100%	1061	1645	63%	1 060
	309	1790					
				87			
22	35			69			
42	34			203			
1637	1670	1790	2391	2625			1 587
24	35	19	111	27		25	35
1613	1635	1771	2280	2598			1 552
1158	1211	1126	396	620	1840	1709	2 900 raw
0.763	0.813	0.838	0.627	0.749	0.92	1.00	1.286
471	505	505	564	632	434	475	473 raw
40	33		39.8	42		50	27
		50			110		
1286	1314	1271	432	674	2446	1941	3 110 raw
523	548	570	616	687	577	540	508
140580	143460	169000	52000	87900	213700	195000	306 600
960	943	968	917	863	1135	1100	1153
1.68	1.51	1.74	0.88	0.96	3.08		2.63
14.6	24.8	21.4	9.5	11.1	13.2	30	23.8
2459	2398	2229	702	981	4240	3595	6154
1.620	1.610	1.660	1.111	1.185	2.12	2.1	2.73
274	271	307	1020	1130	361	290	263
			1.34	1.20			1.18
200	204		150	163	291		164
0.062	0.045	0.350	0.020	0.020	1.77	0.5	0.948
775	780	797	820	881	978		
60.8	60.8	61.3	22.1	35.2	74.6	63.6	98.5
21.2	21.6	20.7	19.5	19.1	26.3	26.6	31.5

HEYNERT and BEER[746]) propose a further partial process which could have the effect of limiting throughput: there is a counterflow movement within and above the hearth zone between the rising tuyere gas and the descending liquid slag and iron. When the gas velocity is too high, the

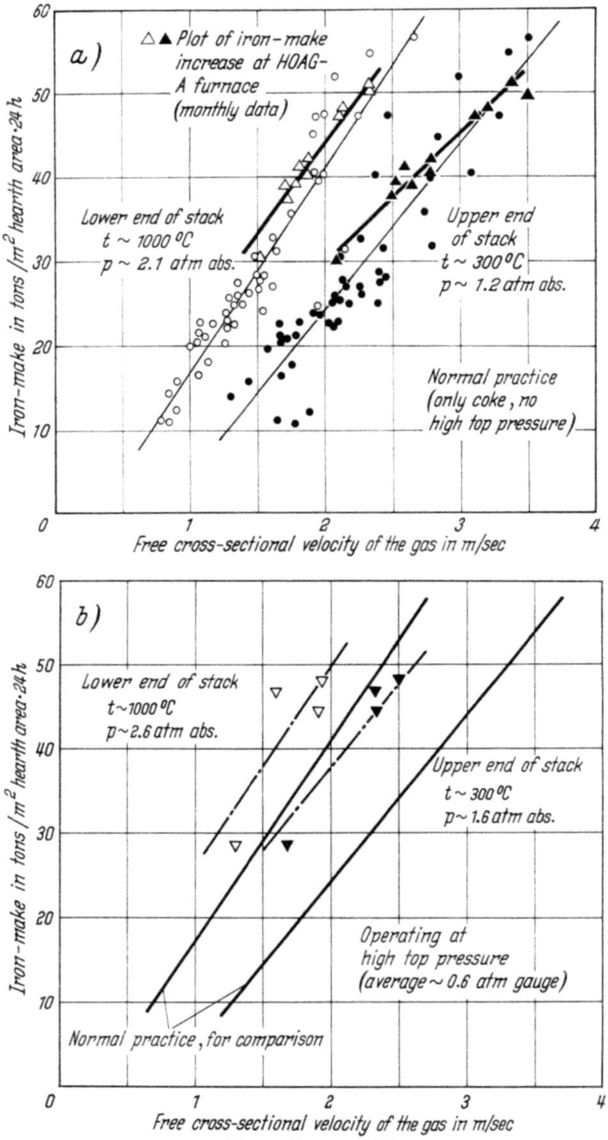

Figs. 375a and b

Fig. 375. Relationship between gas velocity and iron-make in blast furnaces. Gas velocity related to free cross-sectional velocity and working pressure and temperature. Iron-make related to hearth area flow of slag and metal is broken up and the burden column becomes blocked or "choked". If the operating parameters are used to form the counter flow factor

$$k = \frac{L}{G} \sqrt{\varrho_G/\varrho_L}$$

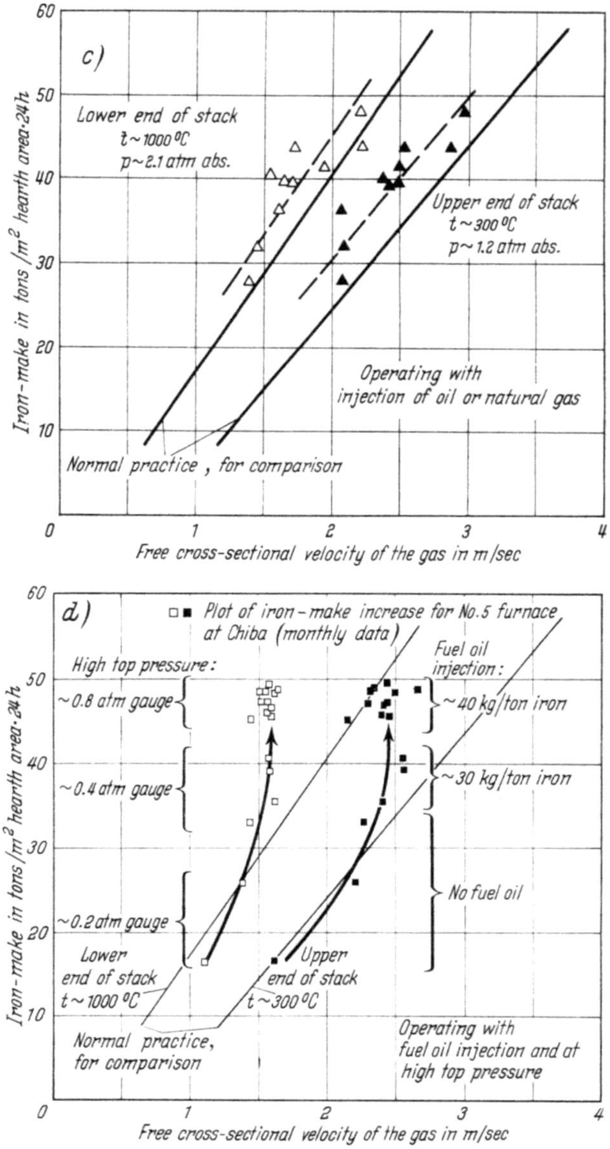

Figs. 375c and d

Fig. 375. Relationship between gas velocity and iron-make in blast furnaces. Gas velocity related to free cross-sectional velocity and working pressure and temperature. Iron-make related to hearth area

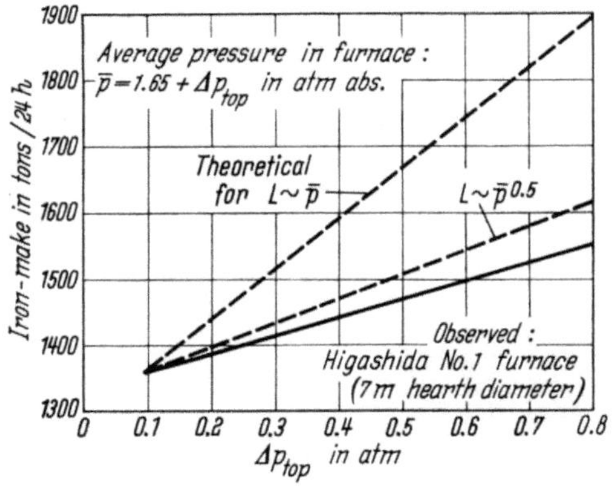

Fig. 376. Relationship between iron-make L and high top pressure Δp_{top}

in which

L = rate of slag flow (kg/m² h),
G = tuyere gas flowrate (kg/m² h),
ϱ_G = gas density (kg/m³),
ϱ_L = density of slag flow (kg/m³),

and also to form the choking factor

$$f = \frac{u^2}{g} \frac{F_s}{\varepsilon^3} \frac{\varrho_G}{\varrho_L} \eta^{0.2}$$

in which

u = gas flowrate (referred to working cross section of bosh) (m/sec),
g = acceleration due to gravity (m/sec²),
F_s = specific coke surface (m²/m³),
ε = voidage in coke charge (m³/m³),
η = viscosity of the fluid slag (cP),

then the degree of proximity to a limiting curve can be checked. JOURDE and HUGUET[754]) conducted similar investigations also taking into account the flow of liquid iron.

5.5.4.2. Maximum Outputs Achieved in Operating Blast Furnaces

Because of the varied size and dimensions of existing blast furnaces and the different burdens used, it is difficult to find an index which can be used for comparative assessment of output. The following productivity indices are normally used:

coke throughput per unit hearth area [tons coke/(m² · day)] or [kg coke/(m² · h)];

coke throughput per unit effective hearth area blown (assumed as up to 1.2 m wide) [tons coke/(m² · day)];

coke throughput per unit furnace working volume (tuyere level to stockline) [tons coke/(m³ · day)];

iron make per unit hearth area [tons iron/($m^2 \cdot$ day)];
iron make per unit furnace working volume [tons iron/($m^3 \cdot$ day)].

The reciprocal values are also used, e.g. the "kipo"* used in the USSR [m^3 furnace volume/(ton iron/day)]. From what has been said above, gas velocity in the upper furnace would also seem justified as an index of output. None of these methods of measuring productivity is convincingly superior to the rest. It seems expedient to virtually exclude the influence of different proportions of inert materials in the burden by using coke throughput as the measure of output instead of iron make. For blast furnaces of more than 5 m hearth diameter, the best coke throughputs so far achieved can be described by the equation:

$$K = 307\,(D - 3.15) \quad \text{(tons coke/day)}$$

in which D is the hearth diameter in metres. This equation was derived by mathematical/statistical evaluation of operating data from particularly smooth driving furnaces[748]). According to SCHENCK and KÜPPERSBUSCH[749]), the square of the hearth diameter should be incorporated and the coke throughput per day expressed in the form:

$$K = \gamma\, D^2 \quad \text{(tons coke/day)}$$

in which max. $\gamma = 25$.

According to Tables 34 and 54, with rich burdens, iron outputs per unit hearth area of up to some 62 tons iron/($m^2 \cdot$ day) have been achieved.

These data represent the current optimum outputs achieved and may, in the light of the statements in section 5.5.4.1, still fall somewhat short of the theoretically obtainable maximum outputs.

5.5.5. Limit of Fuel Requirement

5.5.5.1. Optimum Theoretical Rates

The following discussion refers to blast furnaces in which the only chemical energy source introduced is coke. Conditions during injection of gas, oil, etc. have already been described in section 5.4.

Attempts have often been made to calculate a theoretically based minimum limit for the coke consumption rate. Values quoted for this are in the region of 420 kg coke/ton iron[750, 751]).

As previously indicated, the following factors have a major influence on the coke consumption rate: proportion of inert material in the burden, blast temperature and chemical and thermal utilisation of the top gas.

* "kipo" is an abbreviation of the Russian expression "*k*oeffitsient *i*spol'zovaniya *p*olyeznovo *o*b'yoma" = "coefficient of utilisation of effective volume".

Whilst the proportion of inert materials and the blast temperature can be freely selected within certain limits, the chemical and thermal utilisation of the stack gases depends on a sequence of largely unknown reaction rates and heat transfer phenomena. It is, therefore, very important as regards chemical utilisation of the gases to have an ore and/or sinter of good reducibility which will at the same time permit a high degree of indirect reduction. Whether or not indirect reduction and/or chemical utilisation of the gases is a suitable measure of reduction efficiency, can be determined quite easily from the CO_2 balance of the top gas[751 a]). It can be deduced from this that

$$r_i = \frac{16}{12} \frac{C}{M_O} \eta_{CO}$$

in which

r_i = degree of indirect reduction,
C = reduction and combustion carbon (kg/ton iron),
M_O = dissociable oxygen in the burden (kg/ton iron),
$\eta_{CO} = \frac{CO_2}{CO + CO_2}$ = chemical utilisation of gas.

A CO_2-free burden and gases free from hydrogen and water vapour were assumed in this case. For a given burden and coke consumption, the indirect reduction can then be calculated from the chemical utilisation of the gas. It is shown later, moreover, that the carbon requirement can be represented as a function of the chemical utilisation of the gas. The chemical utilisation of the gas can therefore be considered a suitable criterion of the progress of reduction. The following simplified definitive equations are valid for determining coke consumption:

a) C balance $5.35 \cdot 10^{-3}\, V_G\, (CO + CO_2) = C$,

b) O balance $7.14 \cdot 10^{-3}\, V_G\, (CO + 2\, CO_2) = M_O + 0.3\, V_W$,

c) N_2 balance $12.5 \cdot 10^{-3}\, V_G\, (100 - CO - CO_2) = 0.988\, V_W$,

d) Thermal balance $C \cdot 7950 + 0.338\, V_W\, \vartheta_W = 2.11 \cdot 10^6 + S \cdot 350 + 30.2\, V_G \cdot CO + 0.33\, V_G\, \vartheta_G$,

e) $\eta_{CO} = \frac{CO_2}{CO + CO_2}$,

in which

V_G = top gas volume (m^3/ton iron at NTP),
V_W = blast volume (m^3/ton iron at NTP),
C = reduction and combustion carbon (kg/ton iron),
S = slag volume (kg/ton iron),
M_O = released oxygen (kg/ton iron),
ϑ_G = top gas temperature (°C),
ϑ_W = blast temperature (°C),
CO, CO_2 = volume concentration in top gas (%),
0.338 and 0.33 kcal/m^3 °C at NTP = specific heat of blast and top gas respectively,

$2.11 \cdot 10^6$ kcal/ton iron = heat yielded to burden including reduction heat requirement ($1.70 \cdot 10^6$ kcal/ton iron) + sensible heat of the iron ($0.29 \cdot 10^6$ kcal/ton iron) + cooling losses ($0.21 \cdot 10^6$ kcal/ton iron) — heat of solution and slag formation ($0.09 \cdot 10^6$ kcal/ton iron),
7950 kcal/kg = calorific value of the carbon,
350 kcal/kg = sensible heat of the slag.

ϑ_W, ϑ_G, M_O and S are known and/or accepted as parameters.

V_{CO}, V_G, CO, CO_2 and C are not known. The five definitive equations above are adequate for determining the five unknown factors as a function of the chemical utilisation of the gas η_{CO}.

M_O = 430 kg/ton iron,
ϑ_W = 1000°C,
ϑ_G = 100°C, 200°C,
S = 0, 200, 400, 600, 800 kg/ton iron.

The result of the calculation for 38 kg/ton iron of carbon for solution and coke with 87.5% C is given in Fig. 377. This diagram clearly shows the predominant influence of gas utilisation and, therefore, the importance of a high degree of ore reducibility for coke consumption. The objection could be raised here that the thermal balance contributing to the result has the disadvantage of working on the basis of the total heat requirement without indicating how much heat is required at which temperature. This disadvantage can be lessened with the aid of a multi-stage heat balance of the Reichardt type[719]), although this would involve new assumptions (cf. section 5.5.2.2). The result would, therefore, depend closely on these assumptions.

The chemical utilisation of the gas cannot, however, increase beyond a thermodynamic limit on account of the thermodynamic equilibria between the iron oxides and CO-CO_2 gas phase. This limit defines the minimum coke consumption.

One condition of its calculation is that reduction of the individual iron oxides proceeds in stages in the direction of the gas flow, i.e. first wüstite, then magnetite and, finally, hematite. A further condition is that hematite and magnetite should be reduced entirely by indirect reduction. This enables each η_{CO}-value to be broken down into a proportion for wüstite reduction, a proportion for magnetite reduction and a proportion for hematite reduction. The proportion for wüstite reduction at the relevant η_{CO}-value can be determined from the following balance:

$$V_G\,CO_2 \cdot 10^{-2}\frac{16}{22.4} - \frac{16\,(1.5-1.05)}{55.85}\,930 = V_G\,\Delta CO_2^{Fe/W} \cdot 10^{-2}\frac{16}{22.4}.$$

In this equation, the hematite composition was assumed to be 1.5 moles O/mole Fe, the wüstite, 1.05 moles O/mole Fe and the iron, 930 kg Fe/ton iron. Dividing by the balance equation a) (C balance), the result is:

$$\eta_{CO}\frac{16}{12} - \frac{16\,(0.45)}{55.85}\,\frac{930}{C} = \eta_{CO}^{Fe/W}\,\frac{16}{12}.$$

The proportion of magnetite and hematite reduction at the η_{CO}-value can be determined in the same way.

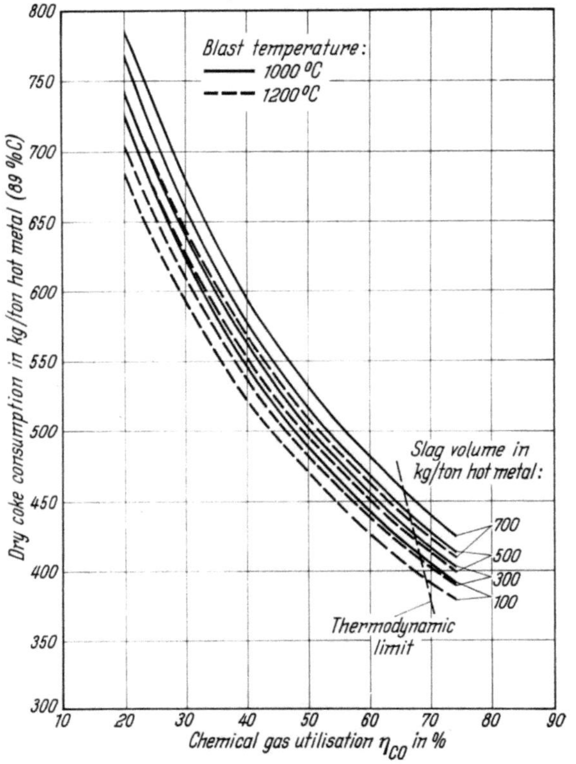

Fig. 377. Relationship between coke consumption and gas utilisation, slag volume and blast temperature

Reference point: Top gas temperature $\vartheta_{top} = 100°C$, releasable oxygen $M_O = 400$ kg/ton iron, coke temperature in front of tuyeres $\vartheta_C = 1500°C$, two-stage thermal balance

If the reduction of the wüstite is now assumed to give a $\eta_{CO}^{Fe/W}$-value corresponding in each case to the Fe/wüstite equilibrium, then the total of $\eta_{CO}^{Fe/W} + \eta_{CO}^{W/M} = \eta_{CO}^{Fe/M}$ exactly corresponding to equilibrium between wüstite and magnetite must be determined (cf. section 1.1). The η_{CO}-value for this total will then be the thermodynamic maximum and the corresponding coke consumption is the thermodynamic minimum. Figure 377 shows a curve plot of minimum calculated coke consumptions for the relevant parameters. The maximum η_{CO}-values can be seen to lie between 0.67 and 0.70. Figure 378 shows the minimum coke consumptions related to net burden weight. The minimum coke consumption data obtained by VOICE and DIXON[739]) have been entered for comparison.

The thermodynamic minimum coke consumption is, of course, only of theoretical interest since, although it will be approached, it will never be reached for kinetic reasons. Of practical interest and great economic importance are the kinetically attainable coke consumption and gas utilisation. Accurate information on this point can only be obtained with a condition-free mathematical blast furnace model. According to model calculations made by PÜCKOFF[743]), a coke consumption of 445 kg/ton iron and a η_{CO}-value of 49% are possible with a burden of optimum reducibility

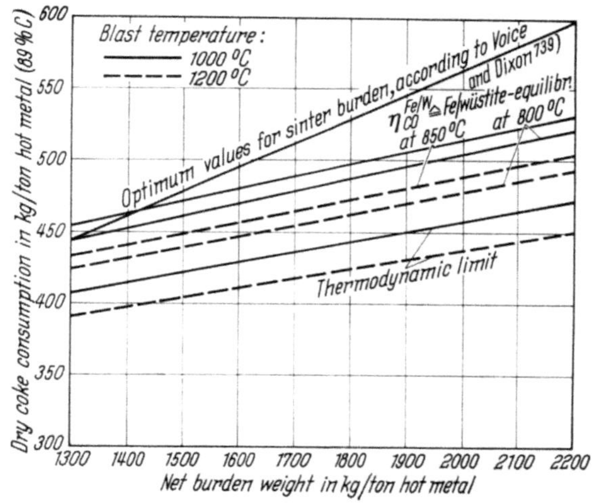

Fig. 378. Relationship between coke consumption and net burden weight. Theoretical and observed minimum values

(cf. Table 53). This result is of limited validity, however, because the calculation was made for a certain temperature profile in the furnace.

It is evident from Fig. 379 that the dry coke consumption, mathematically and kinetically calculated with a fixed temperature profile, can be further reduced with increasing ore reducibility. The lower limit is imposed by thermodynamics. The results obtained with the model were checked by assessment of full-scale blast furnace trials at two plants[562, 605]) where lump size and, therefore, reducibility of the particular characteristic burden constituent was varied (cf. section 5.2.2.2). As this constituent only represented 33.7 to 45% of the iron burden and the rest of the burden showed an average reducibility, the operationally-determined dry coke rate was necessarily above the values calculated with the mathematical/kinetic model. The curves for the operating trials run flatter than those for the model calculation for the same reason.

Figure 378 included the simplifying condition that reduction of wüstite results in a gas utilisation $\eta_{CO}^{W/Fe}$ corresponding to equilibrium concentration at 800°C and 900°C respectively and the coke consumption was determined from this. It is clear that, in this case, rapid reduction, especially at low temperatures, leads to an appreciable reduction in coke consumption. Reduction in the low temperature range, therefore, becomes increasingly important.

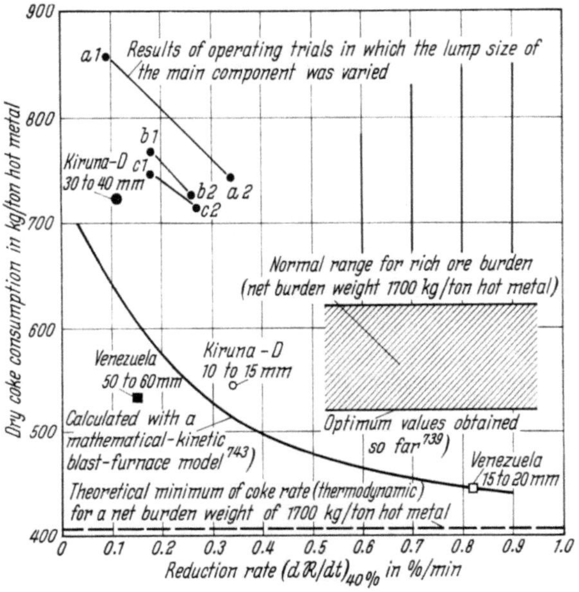

Fig. 379. Relationship between burden reducibility and coke rate in the blast furnace. Comparison of results obtained with a mathematical/kinetic blast-furnace model and actual operating practice. Reduction rate according to Stahleisen-Prüfblatt 1770-64

Trial periods *a 1* to *c 2* as in Fig. 284

The coke consumptions calculated from the balance equations for each of the η_{CO}-values can be used to determine the gain in output resulting from lower coke consumption. Figure 375 shows that output increases linearly with the free cross-sectional velocity and is, therefore, proportional to the top gas volume. This relationship can be easily deduced:

$$L_v = \frac{L}{V_N} \qquad \left(\frac{\text{tons iron}}{24\text{ h m}^3}\right),$$

$$\frac{u}{H_N} = \frac{\varepsilon}{\tau} = \frac{L_V\, V_G}{86400} \qquad (\text{sec}^{-1}),$$

$$L_v = \frac{\varepsilon \cdot 86400}{\tau\, V_G} \qquad \left(\frac{\text{tons iron}}{24\text{ h m}^3}\right)$$

where

L_v = specific output per m³ working volume,
L = output,

V_N = working volume,
H_N = effective height,
V_G = specific top gas volume,
u = free cross-sectional velocity,
τ = residence time,
ε = voidage.

In this instance, τ is a residence time related to the standard conditions. The actual residence time is determined from the blast furnace temperature profile. For any given operating state and with a constant gas residence time, however, the output gain can be calculated from

$$\frac{L}{L_0} = \frac{V_{G0}}{V_G}.$$

Figure 380 shows the result of one such calculation where the specific top gas volumes were determined from the η_{CO}-values and pertinent C-consumption with the aid of balance equations. Figures 377 and 380, therefore, show that a rise in the η_{CO}-levels is advantageous in two ways: coke consumption is reduced and output, increased.

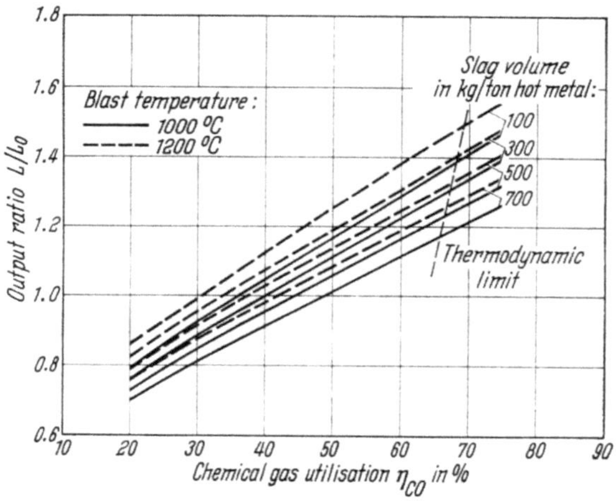

Fig. 380. Production increase resulting from reduced coke consumption at constant residence time in relation to gas utilisation, slag volume and blast temperature
Reference point: Gas utilisation η_{CO} = 40%, slag volume S = 300 kg/ton iron, dry coke consumption K = 568 kg/ton iron, blast temperature ϑ_W = 1000°C, top gas temperature ϑ_{top} = 100°C

5.5.5.2. Minimum Coke Consumptions Achieved in Practice

It was noted earlier that the optimum coke consumption rates achieved had been mathematically/statistically evaluated by VOICE and DIXON[739]). The resulting regression line has been entered for comparison with the theoretical minimum values plotted in Fig. 380. It is clear that, with increasing reducibility, further substantial improvements in coke consumption rate are possible.

It has often been feared that with maximum blast furnace throughput achieved with increasing gas velocity and decreasing gas residence time in the burden, chemical utilisation of the gas η_{CO} and, therefore, coke consumption rate, would deteriorate. This plausible conjecture was investigated by comparing the η_{CO}-values for normal and fast-driving fur-

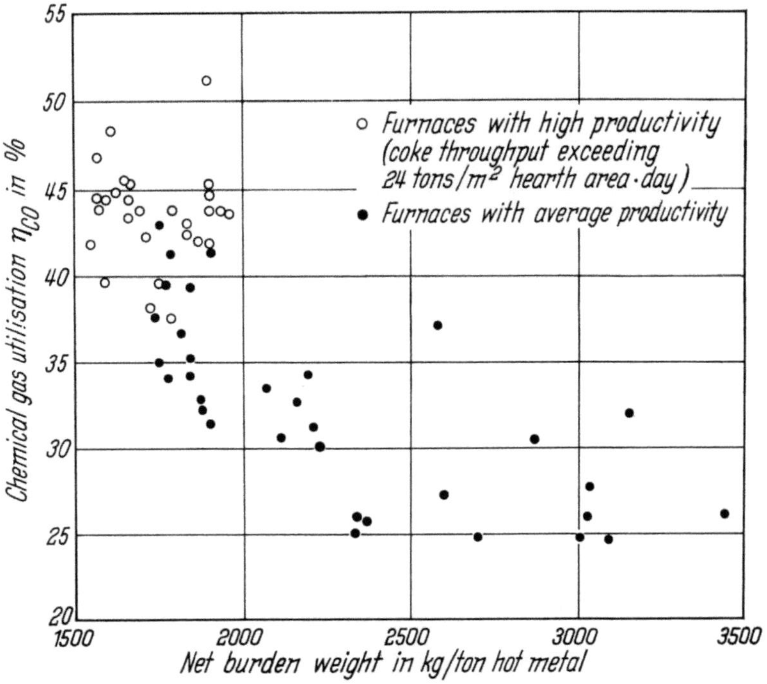

Fig. 381. Chemical utilisation of blast furnace gas η_{CO} in relation to net burden weight

naces. For normal operation furnaces with coke throughputs of 24 tons/(m^2 hearth area·day) and for fast-driving, furnaces with coke throughputs exceeding 24 tons/(m^2 hearth area·day), chosen from Tables 39 and 54, were used. The graph plot for CO utilisation (Fig. 381) shows that, in both groups of furnaces, the same net burden weight-dependent scatter band is obtained for η_{CO} and that the fast-driving furnaces are indeed somewhat better.

In conclusion, it is clear that further improvements in burden preparation, accompanied by a reduction in coke consumption rates, and by improved reducibility, will make it quite possible to effect a further substantial rise in blast furnace output.

Bibliography

1. Fundamentals

1.1. Equilibria

1. DARKEN, L. S., and R. W. GURRY: J. Amer. Chem. Soc. 67 (1945) p. 2; 68 (1946) p. 798–816; cf. Circ. Inform. Techn. 8 (1951) p. 1077–1117.
2. PHILLIPS, B., and A. MUAN: J. Phys. Chem. 64 (1960) p. 1451–1453.
3. SALMON, O. N.: J. Phys. Chem. 65 (1961) p. 550–556.
4. HANSEN, M.: The Constitution of Binary Alloys. 2nd ed. New York/Toronto/ London 1958.
5. ELLIOTT, J. F., and M. GLEISER: Thermochemistry for Steelmaking, Vol. 1. Reading (Mass.)/London 1960. — ELLIOTT, J. F., M. GLEISER, and V. RAMAKRISHNA: Thermochemistry for Steelmaking, Vol. 2, Thermodynamic and Transport Properties. Reading (Mass.)/London 1963.
6. SIFFERLEN, R.: C. R. Hebd. Séances Acad. Sci. 244 (1957) p. 1192–1193.

6a. STRANSKI, I. N.: Z. phys. Chem. 136 (1928) p. 259.

7. JETTE, E. R., and F. FOOTE: J. Chem. Phys. 1 (1933) p. 29–36.
8. ROTH, W. L.: Acta Crystallogr. 13 (1960) p. 140–149.
9. WAGNER, C.: Diffusion and high-temperature oxidation of metals. In: Atom Movements. Publ. by the American Soc. for Metals. Cleveland (Ohio) 1951, p. 153–173, esp. p. 165.
10. HIMMEL, L., R. F. MEHL, and C. E. BIRCHENALL: J. Metals Trans. 5 (1953) p. 827–843.
11. ENGELL, H.-J.: Acta Metallurg., New York 6 (1958) p. 439–445.
12. DESMARESCAUX, P., and P. LACOMBE: Mém. Sci. Rev. Métallurg. 61 (1964) p. 157–168.
13. WAGNER, C., and W. SCHOTTKY: Z. phys. Chem., Abt. B, 11 (1931) p. 163–210.
14. WAGNER, C.: Z. phys. Chem., Abt. B, 22 (1933) p. 181–194. — Thermodynamics of Alloys. Cambridge (Mass.) 1952, p. 54–66.
15. TAKEUCHI, S., and K. IGARI: Sci. Rep. Res. Inst. Tôhoku Univ., Ser. A, 4 (1952) p. 164–175.
16. BRYNESTAD, J., and H. FLOOD: Z. Elektrochem., Ber. Bunsenges. phys. Chem. 62 (1958) p. 953–958.

16a. HAUFFE, K., and H. PFEIFFER: Z. Metallkde. 44 (1953) p. 27–36.

16b. KATSURA, T., and A. MUAN: Trans. Metallurg. Soc. AIME 230 (1964) p. 77–84.

16c. DAVIES, M. W., and F. D. RICHARDSON: Trans. Faraday Soc. 55 (1959) p. 604 to 610.

16d. FUEKI, K., and J. B. WAGNER jr.: Electrochem. Soc. 112 (1965) p. 970–974.

17. ENGELL, H.-J.: Arch. Eisenhüttenwes. 28 (1957) p. 109–115. (Mitt. Max-Planck-Inst. Eisenforsch., Abh. 704.)
18. MARION, F.: Doc. Métallurg. No. 24 (1955) p. 87–136. — AUBRY, J., and F. MARION: C. R. Hebd. Séances Acad. Sci. 240 (1955) p. 1770–1771.

18a. BOHNENKAMP, K., and E. RIECKE: Arch. Eisenhüttenwes., in press.

19. CIRILLI, V., and C. BRISI: Ann. Chim. Appl. 41 (1951) p. 508—514.
20. FOSTER, F. K., and A. J. E. WELCH: Trans. Faraday Soc. 52 (1956) p. 1626 to 1635.
20a. GEIGER, G. H., R. L. LEVIN, and J. B. WAGNER jr.: J. Phys. Chem. Solids 27 (1966) p. 947—956.
20b. BRANSKY, I., and D. S. TANNHAUSER: Trans. Metallurg. Soc. AIME 239 (1967) p. 75—80.
20c. VALLET, P., and P. RACCAH: Mém. Sci. Rev. Métallurg. 62 (1965) p. 1—29.
21. FISCHER, W. A., A. HOFFMANN, and R. SHIMADA: Arch. Eisenhüttenwes. 27 (1956) p. 521—529. (Mitt. Max-Planck-Inst. Eisenforsch., Abh. 679.) — Also Dr. Ing. Diss. (abstract) by R. SHIMADA, TH Aachen. — FISCHER, W. A., and A. HOFFMANN: Arch. Eisenhüttenwes. 29 (1958) p. 107—113. (Mitt. Max-Planck-Inst. Eisenforsch., Abh. 755.)
21a. ILSCHNER, B., and E. MLITZKE: Acta Metallurg., New York 13 (1965) p. 855 to 867.
22. GORTER, E. W.: Philips Res. Rep. 9 (1954) p. 295—365, 403—443.
23. MORKEL, A., and H. SCHMALZRIED: Z. phys. Chem., N. F. 32 (1962) p. 76—90. — SCHMALZRIED, H.: Ber. Bunsenges. phys. Chem. 67 (1963) p. 93—96.
24. SCHMALZRIED, H.: Z. phys. Chem., N. F. 31 (1962) p. 184—197; 33 (1962) p. 111 to 128.
25. HIMMEL, L., R. F. MEHL, and C. E. BIRCHENALL: J. Metals Trans. 5 (1953) p. 827—843.
26. KRÖGER, F. A.: The Chemistry of Imperfect Crystals. Amsterdam 1964, p. 416.
27. VERWEY, E. J. W., and J. H. DE BOER: Rec. Trav. Chem. 55 (1936) p. 531—540. — HAUL, R., and T. SCHOON: Z. phys. Chem. 44 (1939), Section B, p. 216—226.
28. FERGUSON, G. A., jr., and M. HASS: Phys. Rev. 112 (1958) p. 1130—1131.
29. SCHMALZRIED, H., and C. WAGNER: Z. phys. Chem., N. F. 31 (1962) p. 198—221.
30. HÄGG, L. G., and G. SÖDERHOLM: Z. phys. Chem. 29 (1935) p. 88—94.
31. LUYKEN, W., and L. HELLER: Mitt. KWI Eisenforsch. 21 (1939) p. 271—288; cf. Stahl u. Eisen 59 (1939) p. 1138. — Also Dr. Ing. Diss. by L. HELLER, TH Aachen; Österr. Pat. 154368.
31a. PEPPERHOFF, W., and F. ZIRM: Arch. Eisenhüttenwes. 22 (1951) p. 295.
32. DE BOER, F. E., and P. W. SELWOOD: J. Amer. Chem. Soc. 76 (1954) p. 3365 to 3367.
33. DAVID, I., and A. J. E. WELCH: Trans. Faraday Soc. 52 (1956) p. 1642—1650.
34. VERWEY, E. J. W.: Z. Kristallogr. 91 (1935) p. 65—69.
35. KORDEN, E.: Z. Kristallogr. 91 (1935) p. 139—153.
36. FINCH, G. L., and K. P. SINHA: Proc. Roy. Soc. Lond., Ser. A, 241 (1957) p. 1—8.
37. SCHRADER, R., and G. BÜTTNER: Z. anorg. allg. Chem. 320 (1963) p. 220—234.
38. FLEISSNER, H.: Stahl u. Eisen 45 (1925) p. 1373—1379. — German Patent DRP 565672 of 11 Oct. 1938.
38a. FEITKNECHT, W.: Mém. Sci. Rev. Métallurg. 62 (1965), special number, p. 121 to 126.
38b. MEYER, K., H. RAUSCH, and M. OTTOW: Stahl u. Eisen 87 (1967) p. 654—660. (Hochofenaussch. 430.)
38c. WATANABE, S.: Annual Meeting of AIME, Los Angeles, 19—23 Feb. 1967. Reprint 67 B 9.
38d. GEIGER, H., and J. B. WAGNER jr.: Trans. Metallurg. Soc. AIME 233 (1965) p. 2092—2100.
38e. GORTON, A. T., G. BITSIANES, and T. L. JOSEPH: Trans. Metallurg. Soc. AIME 233 (1965) p. 1519—1525.

39. VALLET, P.: Recueil de données thermodynamiques. Publ. de l'Institut de Recherches de la Sidérurgie (IRSID), St. Germain-en-Laye, Serie B, No. 26, 1955.
40. KUBASCHEWSKI, O., and E. L. EVANS: Metallurgical Thermochemistry. 3rd ed. London 1958. (Metal Physics and Physical Metallurgy, Vol. 1.)
41. KELLEY, K. K.: U.S. Bur. Mines Bull. No. 584 (1960) p. 2325.
42. LANDOLT-BÖRNSTEIN: Zahlenwerte und Funktionen aus Physik, Chemie, Astronomie und Technik, 6th ed., Vol. 2, Part 4, Kalorische Zustandsgrößen. Edited by K. SCHÄFER and E. LAX. Berlin/Göttingen/Heidelberg 1961.
43. Hütte, Taschenbuch für Eisenhüttenleute, 5th ed. Edited by Akademischer Verein Hütte. Berlin/Düsseldorf 1961.
44. GIVAUDON, M. J.: Thermodynamique appliquée aux operations de la sidérurgie. Cahiers du CESSID (Centre d'Etudes Supérieures de la Sidérurgie) 1952, No. 1, p. 1–129.
45. LI, KUN: Proc. AIME, Blast Furnace, Coke Oven and Raw Materials Comm. 19 (1960) p. 153–169.
46. BAUR, L. E., and A. GLAESSNER: Z. phys. Chem. 43 (1903) p. 354–368.
47. ENGELL, H.-J.: Z. phys. Chem., N. F. 35 (1962) p. 192–195. — SCHMAHL, N. G., B. FRISCH, and G. STOCK: Arch. Eisenhüttenwes. 32 (1961) p. 297–302. — Also Dr. rer. nat. Diss. (abstract) by G. STOCK, Univ. d. Saarlandes, Saarbrücken. — HAHN, C. W., jr., and A. MUAN: Trans. Metallurg. Soc. AIME 224 (1962) p. 416 to 420.
48. ULRICH, K. H.: Über die Reduktion von Magnetit und Magnetit-Hausmannit-Mischkristallen. Diss., Clausthal 1964. — ULRICH, K. H., K. BOHNENKAMP, and H.-J. ENGELL: Arch. Eisenhüttenwes. 36 (1965) p. 611–618. (Aussch. metallurg. Grundlagen 17.)
49. RICHARDS, R. G., and J. WHITE: Trans. Brit. Ceram. Soc. 53 (1954) p. 233–270, esp. p. 253.
50. SCHMAHL, N. G., B. FRISCH, and G. STOCK: Arch. Eisenhüttenwes. 32 (1961) p. 297–302. — Also Dr. rer. nat. Diss. (abstract) by G. STOCK, Univ. d. Saarlandes, Saarbrücken. — PHILLIPS, B., S. SOMIYA, and A. MUAN: J. Amer. Ceram. Soc. 44 (1961) p. 167–169, esp. p. 169.
51. MUAN, A.: Amer. J. Sci. 256 (1958) p. 413–422.
52. KATSURA, T., and A. MUAN: Trans. Metallurg. Soc. AIME 230 (1964) p. 77–84, esp. p. 80.
53. WEBSTER, A. H., and N. F. H. BRIGHT: J. Amer. Ceram. Soc. 44 (1961) p. 110 to 116, esp. p. 115.
54. McCHESNEY, B., and A. MUAN: Amer. Mineral 44 (1959) p. 926–945.
55. Phase diagrams for ceramists. Edited by E. M. LEVIN, C. R. ROBBINS, and H. F. McMURDIE. Columbus 1964.
56. MUAN, A., and E. F. OSBORN: Phase Equilibria among Oxides in Steelmaking. Reading (Mass.) 1965.
57. BOWEN, N. L., J. F. SCHAIRER, and E. POSNJAK: Amer. Sci. J., 5th Ser., 26 (1933) p. 222–232.
58. SCHEEL, R.: Gleichgewichte zwischen flüssigem Eisen und gesättigten kalkreichen Schlacken des Systems CaO-SiO_{n2}-FeO unter Berücksichtigung der Schwefelverteilung. Diss., Bergakad. Clausthal 1965.
59. PHILLIPS, B., and A. MUAN: Trans. Metallurg. Soc. AIME 218 (1960) p. 1112 to 1118. — J. Amer. Ceram. Soc. 38 (1955) p. 264–280, esp. p. 274.
60. BRAUN, P. B., and W. KWESTROO: Philips Res. Rep. 15 (1960) p. 394–397.
61. NYQUIST, O.: In: Agglomeration. Edited by W. A. KNEPPER. New York/London 1961, esp. p. 809–864.

61a. HASS, K. P., G. BITSIANES, and T. L. JOSEPH: Proc. AIME, Blast Furnace, Coke Oven and Raw Materials Comm. 19 (1960) p. 429—453.
61b. EDSTRÖM, J. O.: Jernkont. Ann. 140 (1956) p. 101—113.
62. SCHENCK, R., H. FRANZ, and H. WILLEKE: Z. anorg. allg. Chem. 184 (1929) p. 1—38; cf. Stahl u. Eisen 50 (1930) p. 519—520. — SCHENCK, R., H. FRANZ, and A. LAYMANN: Z. anorg. allg. Chem. 206 (1932) p. 129—151; cf. Stahl u. Eisen 52 (1932) p. 731—732.
63. BOGAN, L. C., and H. K. WORNER: In: Agglomeration. Edited by W. A. KNEPPER. New York/London 1961, esp. p. 901—929.
64. KNEPPER, W. A., R. B. SNOW, and R. T. JOHNSON: In: Agglomeration. Edited by W. A. KNEPPER. New York/London 1961, esp. p. 787—807.
65. NYQUIST, O.: Jernkont. Ann. 146 (1962) p. 82—144.
66. BURDESE, A., and C. BRISI: Ric. Sci. Ital. 22 (1952) p. 1564—1567.
67. MEYER, G., and H. SERBENT: Techn. Mitt. Krupp, Forsch.-Ber. 22 (1964) p. 55—62.

1.2. Fundamentals of Reduction Kinetics

68. PRESENT, R. D.: Kinetic Theory of Gases. New York/Toronto/London 1958.
69. v. OBERMAYER, A.: Ber. d. kaiserl. Akad. zu Wien (math.-naturwiss. Klasse) 81 (II) (1880) p. 1102—1127; 85 (II) (1882) p. 147—168 and 748—761; 87 (II) (1883) p. 188—263.
70. LANDOLT-BÖRNSTEIN: Physikalisch-chemische Tabellen. 5th ed., Vol. I, 1923, suppl. vol. 1927. — LANDOLT-BÖRNSTEIN: Zahlenwerte und Funktionen aus Physik, Chemie, Astronomie, Geophysik und Technik. 6th ed., Vol. I, Part 1, Atome und Ionen. Edited by A. EUCKEN in collab. with K. H. HELLWEGE. Berlin/Göttingen/Heidelberg 1950, p. 325.
71. CHAPMAN, S., and T. G. COWLING: The Mathematical Theory of Non-Uniform Gases. London 1939.
71a. HIRSCHFELDER, J. O., R. B. BIRD, and E. L. SPOTZ: Chem. Rev. 44 (1949) p. 205—231; Ullmanns Enzyklopädie d. techn. Chemie. 3rd ed., Vol. 5. Munich 1954, esp. p. 848—850.
72. JOST, W.: Diffusion in Solids, Liquids, Gases. 3rd ed. New York 1960, p. 422 et seq., and p. A 44—53.
73. HELLUND, E. J.: Phys. Rev. 57 (1940) p. 33, 319—322, 737—742, and 743—744.
74. HAASE, R.: Thermodynamik der irreversiblen Prozesse (Fortschritte der phys. Chem., Vol. 8). Darmstadt 1963, p. 309—313.
75. BURCHARD, J. K., and H. L. TOOR: J. phys. Chem. 66 (1962) p. 2015—2022.
75a. SHERWOOD, T. K.: Absorption and Extraction. 1st ed. New York 1937.
75b. GILLILAND, E. R.: ibid., p. 11.
76. FRANK-KAMENETZKI, D. A.: Stoff- und Wärmeübertragung in der chemischen Kinetik. Berlin 1959. — GRASSMANN, P.: Physikalische Grundlagen der Chemie-Ingenieur-Technik (Grundlagen der chemischen Technik, Vol. 1). Aarau/Frankfurt a. M. 1962.
77. KRISCHER, O., and G. LOOS: Chem.-Ing.-Techn. (1958) p. 31—39 and 69—74. — KRISCHER, O.: Chem.-Ing.-Techn. 33 (1961) p. 155—162.
78. LANDOLT-BÖRNSTEIN: Physikalisch-chemische Tabellen, 5th ed., Vol. 1. Berlin 1923, p. 181.
78a. BIRD, R. B., J. O. HIRSCHFELDER, and C. F. CURTISS: In: Handbook of Physics. Edited by E. U. LONDON and H. ODISHAER. New York/Toronto/London 1958. p. 5—57.

78b. ANDRUSSOV, L.: Ber. Bunsenges. phys. Chem. 71 (1967) p. 528—534.
79. STAUDTE, H.: Physikalisch-chemisches Taschenbuch, Vol. 1. Leipzig 1945, p. 1050—1051.
80. TAMARU, K.: Trans. Faraday Soc. 55 (1959) p. 824—832.
81. MÜLLER, E. W.: private communication.
82. DE BOER, H. J.: Advances in Catalysis 9 (1957) p. 472—480.
83. TAYLOR, H. S., and C. O. STROTHER: J. Amer. Chem. Soc. 56 (1934) p. 586—590.
84. LANGMUIR, J.: J. Amer. Chem. Soc. 40 (1918) p. 1361.
85. DAMKÖHLER, G.: Z. phys. Chem. 174 (1935) p. 222—238.
86. LENNARD-JONES, J. E., and A. F. DEVONSHIRE: Proc. Roy. Soc. Lond., Ser. A, 156 (1936) p. 6—28 and 29—36.
87. CLAUSING, F.: Ann. Phys., Folge 5, 7 (1930) p. 489—520 and 521—568.
88. SUHRMANN, R.: Z. Elektrochem., Ber. Bunsenges. phys. Chem. 60 (1956) p. 804 to 815.
89. ENGELL, H.-J.: Z. Elektrochem., Ber. Bunsenges. phys. Chem. 66 (1962) p. 617 to 627.
90. EHRLICH, G.: J. chem. Phys. 31 (1959) p. 1111—1126.
91. TAMM, I.: Phys. Z. Sowjetunion 1 (1932) p. 733—746; cf. Chem. Zbl. 103 (1932) II, p. 3051.
92. HAUFFE, K., and H.-J. ENGELL: Z. Elektrochem., Ber. Bunsenges. phys. Chem. 56 (1952) p. 366—373. — ENGELL, H.-J., and K. HAUFFE: Z. Elektrochem., Ber. Bunsenges. phys. Chem. 57 (1953) p. 762—773. — ENGELL, H.-J.: In: Halbleiterprobleme, Vol. 1. Edited by W. SCHOTTKY. Braunschweig 1954, p. 249 to 274. — AIGRAIN, P., and C. DUGAS: Z. Elektrochem., Ber. Bunsenges. phys. Chem. 56 (1952) p. 363—366. — WEISS, P. B.: J. chem. Phys. 20 (1952) p. 1483 to 1484.
93. VOLKENSTEIN, F.: Zhurnal Fiz. Khim. 22 (1948) p. 311—330; 29 (1949) p. 917 to 930. — HORN, F., and L. KÜCHLER: Chem. Techn. 31 (1959) p. 1—11.
93a. HARTEN, H. V., and SCHULTZ: In: Halbleiterprobleme, Vol. 3. Edited by W. SCHOTTKY. Braunschweig 1956. — BOHNENKAMP, K., and H.-J. ENGELL: Z. Elektrochem., Ber. Bunsenges. phys. Chem. 61 (1957) p. 1184—1196. — HAUFFE, K., and E. G. SCHLOSSER: ibid. 61 (1957) p. 502—521.
94. SCHRIEFFER, J. R.: Phys. Rev. 94 (1954) p. 1420.
95. BRAUER, P.: Ann. Phys. (V) 25 (1936) p. 609—624. — BRAUER, P., and F. H. MÜLLER: Kolloid Z. 107 (1944) p. 129—131.
96. MILLER, P. H.: Semi-Conducting Materials. London 1951, p. 172. — HENISCH, H. K.: Z. phys. Chem. 198 (1951). — HOGARTH, C. A.: Z. phys. Chem. 198 (1951) p. 30—40.
97. HAHN, E. E.: J. Appl. Phys. 22 (1951) p. 853—863.
98. ELEY, D. D.: Trans. Faraday Soc. 49 (1953) p. 643—649.
99. TAYLOR, H. S.: Ann. N. Y. Acad. Sci. 58 (1954) p. 798—806.
100. BOHNENKAMP, K., and H.-J. ENGELL: Z. Elektrochem., Ber. Bunsenges. phys. Chem. 61 (1957) p. 1184—1196.
101. SHOCKLEY, W., and G. L. PEARSON: Phys. Rev. 74 (1948) p. 232—233.
102. KINGSTON, R. H.: Phys. Rev. 98 (1955) p. 1766—1775.
102a. EUCKEN, A.: Lehrbuch der chemischen Physik, 3rd ed., Vol. 2.2, Makrozustände der Materie, Part 2, Kondensierte Phasen und heterogene Systeme. Leipzig 1949, p. 1427.
102b. GARNER, W. E.: J. Chem. Soc. 1947, p. 1239—1244. — GARNER, W. E.: Chemistry of the Solid State. New York/London 1955.

102c. HAUFFE, K., and J.-H. ENGELL: Z. Elektrochem., Ber. Bunsenges. phys. Chem. 56 (1952) p. 336—373. — ENGELL, H.-J., and K. HAUFFE: ibid. 57 (1953) p. 762 to 773. — ENGELL, H.-J.: In: Halbleiterprobleme, Vol. 1. Edited by W. SCHOTTKY. Braunschweig 1954, p. 250—272. — HAUFFE, K.: Angew. Chem. 68 (1956) p. 776—784. — HAUFFE, K., and E. G. SCHLOSSER: Z. Elektrochem., Ber. Bunsenges. phys. Chem. 61 (1957) p. 506—521. — SCHLOSSER, E. G., and W. HERZOG: Ber. Bunsenges. phys. Chem. 71 (1967) p. 358—366.

102d. WINTER, E. R. S.: J. Chem. Soc. 1954, p. 1522—1527; 1955, p. 3824—2334. — WINTER, E. R. S.: In: Advances in Catalysis. Edited by D. D. ELEY, W. S. FRANKENBERG et al. New York 1958, p. 196—241.

103. GRABKE, H. J.: Ber. Bunsenges. phys. Chem. 69 (1965) p. 48—57.

103a. LANDLER, P. F. J., and K. L. KOMAREK: Trans. Metallurg. Soc. AIME 236 (1966) p. 138—148.

104. STOTZ, S.: Ber. Bunsenges. phys. Chem. 70 (1966) p. 37—44.

105. As in No. 48.

106. v. BOGDANDY, L., H. P. SCHULZ, I. N. STRANSKI, and B. WÜRZNER: Ber. Bunsenges. phys. Chem. 67 (1963) p. 958—964.

107. EISCHENS, R. P., W. A. PLISKIN, and M. J. D. LOW: J. Catalysis 1 (1962) p. 180 to 191.

108. WICKE, E.: Z. Elektrochem. angew. phys. Chem. 53 (1949) p. 279—285.

109. BÖTTICHER, H., L. v. BOGDANDY, E. FÖRSTER, and U. SCHIERLOH: Z. phys. Chem., N. F. 53 (1967) p. 240—255.

109a. RIECKE, E., K. BOHNENKAMP, and H.-J. ENGELL: Arch. Eisenhüttenwes. 38 (1967) p. 249—255. (Mitt. Max-Planck-Inst. Eisenforsch., Abh. 1087.)

109b. GRABKE, H. J.: unpublished.

109c. EDMISTON, W. A., and R. E. GRACE: Trans. Metallurg. Soc. AIME 236 (1966) p. 1547—1550.

110. McKEWAN, W. M.: Trans. Metallurg. Soc. AIME 221 (1961) p. 140—145.

111. ENDOM, A., K. HEDDEN, and G. LEHMANN: Arch. Eisenhüttenwes. 35 (1964) p. 577—584. (Aussch. metallurg. Grundlagen 4.)

111a. GEIGER, H., and J. B. WAGNER jr.: Trans. Metallurg. Soc. AIME 233 (1965) p. 2092—2100.

111b. GRAY, N. B., and J. HENDERSON: Trans. Metallurg. Soc. AIME 236 (1966) p. 1213—1217.

112. VOLMER, M., and A. WEBER: Z. phys. Chem. 119 (1926) p. 277—301. — VOLMER, M.: Die Kinetik der Phasenbildung. Dresden/Leipzig 1939.

113. BECKER, R., and W. DÖRING: Ann. Phys., Folge 5, 24 (1935) p. 719—752.

114. ZELDOVICH, B.: Zhurnal Exper. i Teor. Fiz. 12 (1942) p. 525. — Also REISS, H.: Ind. Engng. Chem. 44 (1952) p. 1284—1288, esp. p. 1286.

115. HIRTH, J. P., and G. M. POUND: Progress in Materials Sci., Vol. 11. New York 1963, esp. p. 47.

116. HOLLOMON, J. H., and D. TURNBULL: Progress Metal Phys. 4 (1953) p. 333—388, esp. p. 343.

117. STRANSKI, I. N.: Z. phys. Chem., Abt. B, 11 (1931) p. 342—349. — STRANSKI, I. N.. and R. KAISCHEW: Z. Kristallogr. 78 (1931) p. 373—385. — STRANSKI, I. N., and R. KAISCHEW: Phys. Z. 36 (1935) p. 393—403.

118. STRANSKI, I. N.: VDI-Ber. 20 (1957) p. 5—11.

119. MOSESMANN, M. A.: J. Amer. Chem. Soc. 73 (1951) p. 5635.

120. As in No. 48.

121. HIMMEL, L., R. F. MEHL, and C. E. BIRCHENALL: J. Metals Trans. 5 (1953) p. 827—843.

122. ENGELL, H.-J.: Acta Metallurg., New York 6 (1958) p. 439–445.
123. WAGNER, C.: Diffusion and high-temperature oxidation of metals. In: Atom Movements. Publ. by the American Society for Metals. Cleveland (Ohio) 1951, p. 153–173.
124. ENGELL, H.-J.: Z. Elektrochem., Ber. Bunsenges. phys. Chem. 60 (1956) p. 905 to 911.
124a. HAGEL, W. C.: Trans. Metallurg. Soc. AIME 236 (1966) p. 179–184.
125. v. BOGDANDY, L., H. P. SCHULZ, B. WÜRZNER, and I. N. STRANSKI: Arch. Eisenhüttenwes. 34 (1963) p. 401–409.
126. KOHL, H. K., and H.-J. ENGELL: Arch. Eisenhüttenwes. 34 (1963) p. 411–418. (Mitt. Max-Planck-Inst. Eisenforsch., Abh. 945 and Hochofenaussch. 377.)
127. WAGNER, C.: Z. Elektrochem., Ber. Bunsenges. phys. Chem. 63 (1959) p. 772 to 782.
128. WAGNER, C.: J. Electrochem. Soc. 103 (1956) p. 571–580.
129. KOHL, H. K.: Diss., Leoben 1962.
130. As in No. 48.
131. WAGNER, C.: private communication. — Also ZINTL, G.: Diss., Göttingen 1965.
132. ENGELL, H.-J., and H. K. KOHL: Z. Elektrochem., Ber. Bunsenges. phys. Chem. 66 (1962) p. 684–689, and unpublished researches.
132a. LIBANATI, N., J. PHILIBERT, and J. MANENC: C. R. Hebd. Séances Acad. Sci. 260 (1965) p. 1156–1159.
132b. WÜNSCH, J. B., and T. VASILOS: J. chem. Phys. 36 (1962) p. 2917–2922.
132c. MOORE, W. J., J. J. LANDER, S. R. LOGAN, M. O'KEEFE, J. S. CHOI, V. EBISUZAKI, S. BROWN, and D. MITCHELL: Physical Chemistry of the Solid State. Publ. by U. S. Atomic Energy Commission, Washington 1960, TID 11020.
133. MÜLLER, W., and H. SCHMALZRIED: Ber. Bunsenges. phys. Chem. 68 (1964) p. 270–276.
134. SCHMALZRIED, H., and C. WAGNER: Z. phys. Chem., N. F. 31 (1962) p. 198–221.
134a. LINDNER, R.: Acta chem. Scand. 6 (1952) p. 468.
134b. LINDNER, R.: Ark. Kemi 4 (1952) No. 27.
135. YUN, T. S.: Trans. Amer. Soc. Metals 54 (1961) p. 129–142.
136. JANDER, W.: Z. anorg. allg. Chem. 163 (1927) p. 1–30; 166 (1927) p. 31–52.
137. BAUKLOH, W., and R. DURRER: Arch. Eisenhüttenwes. 4 (1930/31) p. 455–460. — Also BAUKLOH, W.: Dr. Ing. Diss., TH Berlin.
138. BALDWIN, B. G.: J. Iron Steel Inst. 179 (1955) p. 306.
138a. KOHL, H. K., and B. MARINČEK: Arch. Eisenhüttenwes. 36 (1965) p. 851–859.
139. MAETZ, H., and H.-J. ENGELL: Arch. Eisenhüttenwes. 25 (1954) p. 397–404. (Mitt. Max-Planck-Inst. Eisenforsch., Abh. 608.)
140. EZZ, S. Y. M., and R. WILD: J. Iron Steel Inst. 194 (1960) p. 211–221.

2. Results of Experimental Investigations of the Kinetics of Reduction

2.1. Methods and Apparatus for Measurements of the Course of the Reaction

141. KNACKE, O.: Arch. Eisenhüttenwes. 30 (1959) p. 581–584.
142. GULBRANSEN, E. A.: Rev. Sci. Instrum. 15 (1944) p. 201–204. — RHODIN, T. N.: J. Amer. Chem. Soc. 72 (1950) p. 4343–4348.

143. MAURER, R. J.: J. chem. Phys. 13 (1945) p. 321—326. — DÜNNWALD, H., and C. WAGNER: Z. anorg. allg. Chem. 199 (1931) p. 321—346. — Also DUVAL, C.: Inorganic Thermogravimetric Analysis. Amsterdam/Houston/New York/London 1953.
144. GAST, T.: Dechema Monogr. 38 (1960) No. 579, p. 1—19.
145. ULRICH, K. H., K. BOHNENKAMP, and H.-J. ENGELL: Arch. Eisenhüttenwes. 36 (1965) p. 611—618. (Aussch. metallurg. Grundlagen 17.)
146. SCHMAHL, N. G., H. BAUMANN, and H. SCHENCK: Arch. Eisenhüttenwes. 29 (1958) p. 147—152.
147. MCKEWAN, W. M.: Trans. Metallurg. Soc. AIME 218 (1960) p. 2—6.
148. RAHMEL, A., W. JÄGER, and K. BECKER: Arch. Eisenhüttenwes. 30 (1959) p. 351—360. (Werkstoffaussch. 1192.) — Also BECKER, K.: Dr. rer. nat. Diss., TH Aachen.
149. HAUFFE, K., and H. PFEIFFER: Z. Metallkde. 44 (1953) p. 27—36.
150. QUETS, J. M., M. E. WADSWORTH, and J. R. LEWIS: Trans. Metallurg. Soc. AIME 218 (1960) p. 545—550.
151. DÜNNWALD, H., and C. WAGNER: Z. phys. Chem. 22 (1933) p. 212—225.
152. v. BOGDANDY, L., and W. JANKE: Z. Elektrochem., Ber. Bunsenges. phys. Chem. 61 (1957) p. 1146—1153.
153. SUBAT, G.: Techn. Mitt. Krupp, Forsch.-Ber. 21 (1963) p. 134—138.
154. HOCKINGS, W. A.: Proc. AIME, Blast Furnace, Coke Oven and Raw Materials Comm. 19 (1960) p. 170—184.
155. EZZ, S. Y. M., and R. WILD: J. Iron Steel Inst. 194 (1960) p. 211—221. — Also DALLA LANA, I. G., and N. R. AMUNDSON: Ind. Engng. Chem. 53 (1961) p. 22—26.
156. HAUFFE, K., and A. RAHMEL: Z. phys. Chem., N. F. 1 (1954) p. 105—128.
157. HEDDEN, K., and G. LEHMANN: Arch. Eisenhüttenwes. 34 (1963) p. 887—894. (Hochofenaussch. 386.)
158. KOHL, H. K., and H.-J. ENGELL: Arch. Eisenhüttenwes. 34 (1963) p. 411—418. (Mitt. Max-Plank-Inst. Eisenforsch., Abh. 945, and Hochofenaussch. 377.)
159. GRABKE, H. J.: Ber. Bunsenges. phys. Chem. 69 (1965) p. 48—57.
160. SCHMALZRIED, H., and C. WAGNER: Trans. Metallurg. Soc. AIME 227 (1963) p. 539—540.
161. STOTZ, S.: Max-Planck-Inst. f. phys. Chem., Göttingen, personal communication.
162. PLUSCHKELL, W.: Inst. f. theor. Hüttenkde., Bergakad. Clausthal, unpublished investigations.
163. VORONTSOV, E. S., and A. V. ERMAKOV: Ivz. Akad. Nauk USSR, OTN, Metallurg. i Gornoe Delo, 1963, p. 22—29.
164. BRILL-EDWARDS, H., B. L. DANIELL, and R. L. SAMUEL: J. Iron Steel Inst. 203 (1965) p. 361—368.
165. ZIMENS, K. E.: In: Handbuch der Katalyse 4, Heterogene Katalyse I. Prepared by R. A. BEEBE et al. Vienna 1943, p. 151—268. — WITZMANN, H.: Methoden der Strukturuntersuchung an hochdispersen und hochporösen Stoffen. Berlin 1961.
166. SCHENCK, H.: Stahl u. Eisen 75 (1955) p. 682—690. (Hochofenaussch. 286.)
167. BRUNAUER, S., P. H. EMMETT, and E. TELLER: J. Amer. Chem. Soc. 60 (1938) p. 309—319.
168. HAUL, R., and G. DÜMBGEN: Chem.-Ing.-Techn. 32 (1960) p. 349—354.
169. DE BOER, J. H.: The Structure and Properties of Porous Materials. London 1958, p. 68—94.

170. TROJER, F.: Die oxydischen Kristallphasen der anorganischen Industrieprodukte. Stuttgart 1963.
170a. SCHRAMM, B., and K. SCHÄFER: Ber. Bunsenges. phys. Chem. 69 (1965) p. 110 to 114.

2.2. Heat and Mass Transfer across the Gas-Flow Boundary-Layer

171. GRÖBER, H., S. ERK, and U. GRIGULL: Die Grundgesetze der Wärmeübertragung. 3rd ed. Berlin/Göttingen/Heidelberg 1963.
172. MCADAMS, W. H.: Heat Transmission. New York/London 1954.
173. FURNAS, C. C.: Bull. Bur. Mines No. 307 (1929) p. 1—144; cf. Stahl u. Eisen 51 (1931) p. 617—618.
174. JESCHAR, R.: Arch. Eisenhüttenwes. 34 (1964) p. 517—526. (Wärmestelle 544 and Aussch. metallurg. Grundlagen 3.)
174a. EUCKEN, A.: Die Wärmeleitfähigkeit keramischer feuerfester Stoffe. Ihre Berechnung aus der Wärmeleitfähigkeit der Bestandteile (VDI-Forschungsheft No. 353). Berlin 1932, p. 16.
174b. POLTHIER, K.: Arch. Eisenhüttenwes. 37 (1966) p. 365—374. (Aussch. metallurg. Grundlagen 23 and Wärmestelle 575.)
174c. WARTMANN, R., and H. MERTES: Arch. Eisenhüttenwes. 37 (1966) p. 201—207. (Wärmestelle 573 and Aussch. metallurg. Grundlagen 20.)
175. BEER, P.: Diss., Clausthal 1965. — BEER, P., and H. KRAINER: Techn. Mitt. Krupp, Forsch.-Ber. 24 (1966) p. 25—47.
175a. BEER, P.: Techn. Mitt. Krupp, Forsch.-Ber. 21 (1963) p. 120—122.
175b. TRAUSTEL, S.: Brennst. Wärme Kraft 15 (1963) p. 523—526.
175c. ANDRUSSOV, L.: Ber. Bunsenges. phys. Chem. 71 (1967) p. 528—534.
175d. ENGELL, H.-J., and P. BEER: Arch. Eisenhüttenwes. 37 (1966) p. 771—773.
176. CHAPMAN, S., and T. G. COWLING: The Mathematical Theory of Non-Uniform Gases. London 1939.
177. LANDOLT-BÖRNSTEIN: Physikalisch-chemische Tabellen. Edited by W. A. ROTH and K. SCHEEL. 5th ed., 2nd suppl. vol., 2nd part. Berlin 1931, p. 1279.
178. HANSEN, M.: Strömungsverhältnisse in der Hochofen-Schüttsäule, In: Troisièmes Journées Internationales de Sidérurgie, Luxembourg, 1—4 Oct. 1962. Publ. by Centre National de Recherches Métallurgiques, Liège 1962, p. 58—68.
179. MCKEWAN, W. M.: Trans. Metallurg. Soc. AIME 221 (1961) p. 140—145.
180. v. BOGDANDY, L., H. P. SCHULZ, I. N. STRANSKI, and B. WÜRZNER: Z. Elektrochem., Ber. Bunsenges. phys. Chem. 67 (1963) p. 958—964.
180a. WARNER, N. A.: Trans. Metallurg. Soc. AIME 230 (1964) p. 163—176.

2.3. The Diffusion of Gas and Solids in the Layer of Reaction Products

181. UDY, M. C., and C. H. LORIG: Trans. AIME 154 (1943) p. 162—181.
182. v. BOGDANDY, L., and W. JANKE: Z. Elektrochem., Ber. Bunsenges. phys. Chem. 61 (1957) p. 1146—1153.
183. SCHENCK, H., and H. P. SCHULZ: Arch. Eisenhüttenwes. 31 (1960) p. 691—702. (Hochofenaussch. 335.) — Also SCHULZ, H. P.: Dr. Ing. Diss. (extract), TH Aachen.
184. KAWASAKI, E., J. SANSKRAINTE, and T. J. WALSH: A. I. Ch. E. J. 8 (1962) p. 48—52.
185. LU, WEI KAO: Trans. Metallurg. Soc. AIME 227 (1963) p. 203—206.

186. Seth, B. B. L., and H. U. Ross: Trans. Metallurg. Soc. AIME 233 (1965) p. 180 to 185.
187. Woods, S. E.: Disc. Faraday Soc. No. 4, 1948, p. 184–193.
187a. Fischbeck, K.: Technische Reaktionsgeschwindigkeit. In: Der Chemie-Ingenieur. Edited by A. Eucken and M. Jakob. Vol. 3, Chemische Operationen, 1st part. Leipzig 1937, p. 242–359, esp. p. 307.
187b. Olsson, R. G., and W. M. McKewan: Trans. Metallurg. Soc. AIME 236 (1966) p. 1518–1522.
187c. Warner, N. A.: Trans. Metallurg. Soc. AIME 230 (1964) p. 163–176.
187d. Herr, H. D.: Dr. Diss., TH Aachen, 1967.
188. Wiberg, M.: Jernkont. Ann. 124 (1940) p. 172–212; cf. Stahl u. Eisen 61 (1941) p. 141–143.
189. Edström, J. O.: J. Iron Steel Inst. 175 (1953) p. 289–304 and 6 tables. — Edström, J. O.: Jernkont. Ann. 140 (1956) p. 116–129.
190. Riecke, E., K. Bohnenkamp, and H.-J. Engell: Arch. Eisenhüttenwes. 38 (1967) p. 249–255. (Mitt. Max-Planck-Inst. Eisenforsch., Abh. 1087.) — Kohl, H. K., and B. Marinček: Arch. Eisenhüttenwes. 36 (1965) p. 851–859.
191. Schürmann, E., W. Zischkale, P. Ischebeck, and G. Heynert: Stahl u. Eisen 70 (1960) p. 854–861. — Also Zischkale, W.: Dr. Ing. Diss. (abstract), Bergakad. Clausthal.
192. Hauffe, K., and A. Rahmel: Z. phys. Chem., N. F. 1 (1954) p. 104–128.
193. Kohl, H. K., and H.-J. Engell: Arch. Eisenhüttenwes. 34 (1963) p. 411–418. (Mitt. Max-Planck-Inst. Eisenforsch., Abh. 945, and Hochofenaussch. 377.)
194. Schenck, H., E. Schmidtmann, and H. Müller: Arch. Eisenhüttenwes. 31 (1960) p. 121–128.
195. Bohnenkamp, K., and H.-J. Engell: Arch. Eisenhüttenwes. 35 (1964) p. 1011 to 1018. (Mitt. Max-Planck-Inst. Eisenforsch., Abh. 998.)
196. Heynert, G., E. Schürmann, and J. Willems: Stahl u. Eisen 78 (1958) p. 1493 to 1505. (Hochofenaussch. 314.)
197. v. Bogdandy, L., H. P. Schulz, I. N. Stranski, and B. Würzner: Ber. Bunsenges. phys. Chem. 67 (1963) p. 958–964.
198. Brill-Edwards, H., B. L. Daniell, and R. L. Samuel: J. Iron Steel Inst. 203 (1965) p. 361–368.
199. Endom, A., K. Hedden, and G. Lehmann: Arch. Eisenhüttenwes. 35 (1964) p. 577–584. (Aussch. metallurg. Grundlagen 4.)

2.4. Investigations of the Phase-Boundary Reaction

200. McKewan, W. M.: Trans. Metallurg. Soc. AIME 224 (1962) p. 2–5.
201. v. Bogdandy, L., H. P. Schulz, B. Würzner, and I. N. Stranski: Arch. Eisenhüttenwes. 34 (1963) p. 401–409. (Hochofenaussch. 376.) — Also Würzner, B.: Dr. rer. nat. Diss. (abstract), TU Berlin.
202. Wilhelem, A. J., and G. R. St. Pierre: Trans. Metallurg. Soc. AIME 221 (1961) p. 1267–1269.
203. Knacke, O.: Arch. Eisenhüttenwes. 30 (1959) p. 581–584.
204. Krainer, H., M. Wahlster, and Th. Lettmann: Stahl u. Eisen 83 (1963) p. 578 to 585.
205. Kohl, H. K., and B. Marinček: Helv. Chim. Acta 48 (1965) p. 1857–1867.
206. Grabke, H. J.: Ber. Bunsenges. phys. Chem. 69 (1965) p. 48–57.
208. Quets, J. M.: M. E. Wadsworth, and J. R. Lewis: Trans. Metallurg. Soc. AIME 218 (1960) p. 545–550.

209. Kohl, H. K., and H.-J. Engell: Arch. Eisenhüttenwes. 34 (1963) p. 411–418.
209a. Jander, W.: Z. anorg. allg. Chem. 163, 164 (1927) p. 1–30.
210. McKewan, W. M.: Trans. Metallurg. Soc. AIME 212 (1958) S. 791–793.
211. v. Bogdandy, L.: Arch. Eisenhüttenwes. 32 (1961) p. 275–296, esp. p. 275–286.
212. Hockings, W. A.: Proc. AIME, Blast Furnace, Coke Oven and Raw Materials Comm. 19 (1960) p. 170–184.
213. McKewan, W. M.: Trans. Metallurg. Soc. AIME 221 (1961) p. 140–145.
214. Hedden, K., and G. Lehmann: Arch. Eisenhüttenwes. 34 (1963) p. 887–894. (Hochofenaussch. 386.) — Endom, A., K. Hedden, and G. Lehmann: Arch. Eisenhüttenwes. 35 (1964) p. 577–584. (Aussch. metallurg. Grundlagen 4.)
215. McKewan, W. M.: private communication.
216. Koch, H. J.: Gaswärme 10 (1961) p. 347–359.
217. Hauffe, K., and H. Pfeiffer: Z. Metallkde. 44 (1953) p. 27–36.
218. Pettit, F., R. Yinger, and J. B. Wagner: Acta Metallurg., New York 8 (1960) p. 617–623.

2.5. The Interaction of Diffusion and the Phase-Boundary Reaction

219. Lu, Wei Kao: Trans. Metallurg. Soc. AIME 227 (1963) p. 203–206.
220. As in No. 186.
221. Pluschkell, W.: Inst. f. theor. Hüttenkunde, Bergakad. Clausthal, unpublished results.
222. McKewan, W. M.: Trans. Metallurg. Soc. AIME 224 (1962) p. 2–5.
223. v. Bogdandy, L., and W. Janke: Z. Elektrochem., Ber. Bunsenges. phys. Chem. 61 (1957) p. 1146–1153.
224. v. Bogdandy, L.: Arch. Eisenhüttenwes. 32 (1961) p. 275–296.
225. Ulrich, K. H., K. Bohnenkamp, and H.-J. Engell: Arch. Eisenhüttenwes. 36 (1965) p. 611–618.
226. Engell, H.-J.: Z. Elektrochem., Ber. Bunsenges. phys. Chem. 60 (1956) p. 905 to 911.
227. Engell, H.-J.: Acta Metallurg., New York 6 (1958) p. 439–445.
228. Jeschar, R.: Arch. Eisenhüttenwes. 38 (1967) p. 663–667.
228a. Spitzer, R. H., F. S. Manning, and W. O. Philbrook: Trans. Metallurg. Soc. AIME 236 (1966) p. 726–742.
229. Thirle, E. W.: Ind. Engng. Chem. 31 (1939) p. 916–920.
230. Zeldovich, B.: Acta Physicochim. USSR 10 (1939) p. 583–592.
231. Jost, W.: Chem. Eng. Sci. 2 (1953) p. 199–202.
232. Morawietz, W.: Z. Elektrochem., Ber. Bunsenges. phys. Chem. 57 (1953) p. 539 to 548.
233. Wicke, E., and K. Hedden: Z. Elektrochem., Ber. Bunsenges. phys. Chem. 57 (1953) p. 636–641.
234. v. Bogdandy, L., and H. G. Riecke: Arch. Eisenhüttenwes. 29 (1958) p. 603 to 609. (Hochofenaussch. 313.)
235. McKewan, W. M.: Trans. Metallurg. Soc. AIME 224 (1962) p. 387–393.
236. Endom, A., K. Hedden, and G. Lehmann: Arch. Eisenhüttenwes. 35 (1964) p. 577–584. (Aussch. metallurg. Grundlagen 4.)
237. As in No. 201.
237a. Herr, H. D.: Dr. Diss., TH Aachen 1967.
238. Skodin, K. K.: Stal 23 (1963) p. 97–104; Stal in Deutsch 3 (1963) p. 745–752.
239. As in No. 211.
240. As in No. 180.

241. KRAINER, H.: Techn. Mitt. Krupp 19 (1961) p. 85–105.
242. KRAINER, H., M. WAHLSTER, and TH. LETTMANN: Stahl u. Eisen 83 (1963) p. 578–585.
243. PHILBROOK, W. O.: Blast Furn. Steel Plant 35 (1947) p. 893.

2.6. Observations on the Formation and Growth of Nuclei

245. As in No. 180.
246. WIBERG, M.: Jernkont. Ann. 124 (1940) p. 179–212; cf. Stahl u. Eisen 61 (1941) p. 141–143. — WIBERG, M.: Disc. Faraday Soc. No. 4, 1948, p. 231–233.
247. EDSTRÖM, J. O.: J. Iron Steel Inst. 175 (1953) p. 289–304 and 6 tables. — EDSTRÖM, J. O.: Jernkont. Ann. 140 (1956) p. 116–129.
248. KOHL, H. K., and H.-J. ENGELL: Arch. Eisenhüttenwes. 34 (1963) p. 411–418. (Mitt. Max-Planck-Inst. Eisenforsch., Abh. 945, and Hochofenaussch. 377.)
249. As in No. 145.
250. ENDOM, A., K. HEDDEN and G. LEHMANN: Arch. Eisenhüttenwes. 35 (1964) p. 577–584. (Aussch. metallurg. Grundlagen 4.)
250a. KUZNETSOV, A. N.: Zhurnal Fiz. Khim. 27 (1953) p. 1808–1815.
250b. GERLACH, H., H. PROBST, and D. NEUSCHÜTZ: Arch. Eisenhüttenwes. 35 (1965) p. 543–547.
251. GRABKE, H. J.: Ber. Bunsenges. phys. Chem. 68 (1965) p. 48–57.
251a. BÖTTICHER, H., L. v. BOGDANDY, E. FÖRSTER, and U. SCHIERLOH: Z. phys. Chem., N. F. 53 (1967) p. 240–255.
251b. ÖSTBERG, G.: Jernkont. Ann. 144 (1960) p. 46–76.
251c. ROSTOVTSEV, S. T., and A. P. EM: Doklady Akad. Nauk USSR 93 (1953) p. 329–332.
251d. WARNER, N. A.: Trans. Metallurg. Soc. AIME 230 (1964) p. 163–176.
251e. CECH, R. E.: Trans. Metallurg. Soc. AIME 215 (1959) p. 769–781.
251f. DIENES, G. J., and G. H. VINEYARD: Radiation Effects in Solids. New York 1957, esp. p. 203–264.
252. SCHAEFER, H. D.: Dr. Diss., TH Clausthal 1966. — MORAWIETZ, W., and H. D. SCHAEFER: Lecture at the 36e Congrès International de Chimie Industrielle, Bruxelles 1966.
253. MCKEWAN, W. M.: Trans. Metallurg. Soc. AIME 221 (1961) p. 140–145.
254. GELLNER, O. H., and F. D. RICHARDSON: Nature, London 168 (1951) p. 23–25.
254a. PLUSCHKELL, W., and H. YOSHIKOSHI: Arch. Eisenhüttenwes., to be published.
255. STOTZ, S.: Max-Planck-Inst. f. phys. Chem., Göttingen, unpublished results.
256. SCHMALZRIED, H., and C. WAGNER: Trans. Metallurg. Soc. AIME 227 (1963) p. 539–540.

2.7. The Reduction Kinetics of Mixed Oxides, Oxide Compounds, and Sinters

257. ENGELL, H.-J., and H. K. KOHL: Z. Elektrochem., Ber. Bunsenges. phys. Chem. 66 (1962) p. 684–689.
258. As in No. 145.
259. WATANABE, S.: In: Agglomeration. Edited by W. A. KNEPPER. New York/London 1962, p. 875.
260. MAZANEK, E., and S. JASIEŃSKA: J. Iron Steel Inst. 202 (1964) p. 319–324.
261. RUECKL, R. L.: Proc. AIME, Blast Furnace, Coke Oven and Raw Materials Comm. 21 (1962) p. 290–315.
262. SETH, B. B. L., and H. U. ROSS: Canad. Metallurg. Quarterly 2 (1963) p. 16–30.

263. NYQUIST, O.: In: Agglomeration. Edited by W. A. KNEPPER. New York/London 1962, p. 809–858.
263a. EDSTRÖM, J. O.: Jernkont. Ann. 142 (1958) p. 401–466.
263b. LIHL, F., and F. NEMEC: Ber. Dt. keram. Ges. 40 (1963) p. 365–372 and 467 to 477.
264. KNEPPER, W. A., R. B. SNOW, and R. T. JOHNSON: In: Agglomeration. Edited by W. A. KNEPPER. New York/London 1962, p. 787–804.
265. KOHL, H. K., and H.-J. ENGELL: Arch. Eisenhüttenwes. 34 (1963) p. 411–418. (Mitt. Max-Planck-Inst. Eisenforsch., Abh. 945, and Hochofenaussch. 377.)
265a. STRANGERAY, P. K., and H. U. ROSS: Canad. Metallurg. Quarterly 4 (1965) p. 97–111.
265b. SUBAT, G.: Techn. Mitt. Krupp, Forsch.-Ber., to be published.
265c. ÖSTBERG, G.: Jernkont. Ann. 144 (1960) p. 46–76.

2.8. Reduction of Slags Containing Iron Oxides

265d. KRAINER, H., H. P. BEER, and H. BRANDI: Techn. Mitt. Krupp, Forsch.-Ber. 24 (1966) p. 139–146.
265e. KONDAKOV, V. V., D. I. RYZHONKOV, and D. M. GOLENKO: Izv. Vyssh. Ucheb. Zav., Chern. Metallurg. 1960, No. 4, p. 23–28.
265f. GEL'D, P. V., V. G. VLASOV, and N. N. SEREBRENNIKOV: Doklady Akad. Nauk USSR 78 (1951) p. 693–696.
265g. KRAINER, H., and H. P. BEER: unpublished measurements.
265h. SHURYGIN, P. M., V. N. BORONENKOV, V. I. KRYUK, and V. V. REVERTSOV: Izv. Vyssh. Ucheb. Zav., Chern. Metallurg. 1965, No. 2, p. 23–28. — ORLOV, V. I., O. A. ESIN, and P. M. SHURYGIN: ibid. 1960, No. 11, p. 12–19. — BORONENKOV, V. N., O. A. ESIN, and P. M. SHURYGIN: Doklady Akad. Nauk USSR 160 (1965) p. 151–153.
265i. DANCY, T. E.: J. Iron Steel Inst. 169 (1951) p. 17–24.
265k. PHILBROOK, W. O., and L. D. KIRKBRIDE: J. Metals Trans. 8 (1956) p. 351–356.
265l. TURKDOGAN, E. T., P. GRIEVESON, and J. F. BEISLER: Trans. Metallurg. Soc. AIME 227 (1963) p. 1258–1274.
265m. GRIMBLE, M., R. G. WARD, and D. J. WILLIAMS: J. Iron Steel Inst. 203 (1965) p. 264–267.
265n. SHARMA, S. K., and R. G. WARD: J. Iron Steel Inst. 205 (1967) p. 196–202.
265o. GRIEVESON, P., and E. T. TURKDOGAN: Trans. Metallurg. Soc. AIME 230 (1964) p. 1609–1614.
265p. KOROS, P. J., and T. B. KING: Trans. Metallurg. Soc. AIME 224 (1962) p. 299 to 306.

2.9. Reaction Kinetics of the Gasification of Coal and Coke

266. BOUDOUARD, M. O.: C. R. Hebd. Séances Acad. Sci. 128 (1899) p. 824 – 825.
267. ILSCHNER, R.: In: GMELIN-DURRER: Metallurgie des Eisens, 4th ed., Vol. 1a, Geschichtliches. Weinheim/Bergstr. 1964, p. 163a–178a. (Supplementary volume to GMELIN's Handbuch System-No. 59, Eisen T. A., Lfg. 3–5.)
268. STRICKLAND-CONSTABLE, R. F.: Trans. Faraday Soc. 40 (1944) p. 333–343. — Chem. Ind. (1948) p. 771–774.
269. DUVAL, X.: J. Chim. phys. 47 (1950) p. 339–347.
270. EUCKEN, A.: Z. angew. Chem. 43 (1930) p. 986–993.

271. Tsukhanova, O. A.: Zhurnal Tekhn. Fiz. 9 (1930) p. 295–304.
272. Baetke, F. W.: Z. Elektrochem., Ber. Bunsenges. phys. Chem. 55 (1951) p. 655 to 657.
273. Meyer, L.: Z. phys. Chem., Abt. B, 17 (1932) p. 385–404.
274. Letort, M.: Rev. Univ. Mines, 9 Ser., 16 (1960) p. 255–271.
275. Blyholder, G., and H. Eyring: J. Phys. Coll. Chem. 63 (1959) p. 1004–1008.
276. Wicke, E., and K. Hedden: Z. Elektrochem., Ber. Bunsenges. phys. Chem. 57 (1953) p. 636–641.
277. Gulbransen, E. A., and K. F. Andrew: Ind. Engng. Chem. 44 (1952) p. 1034 to 1038 and 1039–1044. — Gulbransen, E. A.: ibid. 44 (1952) p. 1045–1051.
278. Henning, G. R.: Z. Elektrochem., Ber. Bunsenges. phys. Chem. 66 (1962) p. 629–635.
279. Rossberg, M.: Z. Elektrochem., Ber. Bunsenges. phys. Chem. 60 (1956) p. 952–956.
280. Walker, P. L., jr., F. Rusinko jr., and L. G. Austin: In: Advances in Catalysis. Edited by D. D. Eley, P. W. Selwood and P. B. Weisz. Vol. 11. New York 1959, p. 133–221.
281. Ergun, S.: Kinetics of the reactions of carbon dioxide and steam with coke. Bulletin Bureau of Mines, Dept. of the Interior, No. 598. Washington 1962.
282. Freundlich, J.: Z. Elektrochem., Ber. Bunsenges. phys. Chem. 66 (1962) p. 647–648.
283. Grabke, H. J.: Ber. Bunsenges. phys. Chem. 70 (1966) p. 664–674.
284. Gadsby, J., F. J. Long, P. Sleightholm, and K. W. Sykes: Proc. Roy. Soc. Lond., Ser. A, 193 (1948) p. 357–376. — Gadsby, J., C. N. Hinshelwood, and K. S. Sykes: Proc. Roy. Soc. Lond., Ser. A, 187 (1946) p. 129–151. — Wicke, E., and K. Hedden: Z. Elektrochem., Ber. Bunsenges. phys. Chem. 57 (1953) p. 636–641.
285. Hedden, K., H. H. Kopper, and V. Schulze: Z. phys. Chem., N. F. 2, 22 (1959) p. 23–62.
286. Boulangier, F., X. Duval, and M. Letort: The Kinetics of the Reactions of Carbon Filaments with Carbon Dioxide and Water Vapour at High Temperatures and Low Pressures. London/New York/Paris/Los Angeles.
287. Wicke, E., and M. Rossberg: Z. Elektrochem., Ber. Bunsenges. phys. Chem. 57 (1953) p. 641–645.
288. Hedden, K., and E. Wicke: The Kinetics of the Reactions of Carbon Filaments with Carbon Dioxide and Water Vapour at High Temperatures and Low Pressures. London/New York/Paris/Los Angeles, p. 249–256.
289. Kröger, C.: Angew. Chem. 52 (1939) p. 129–139.
290. Long, F. J., and K. W. Sykes: J. Chim. Phys. 47 (1950) p. 361–378.
291. Wicke, E.: Contributions to the Combustion Mechanism of Carbon. In: 5th Symposium (International) on Combustion, Pittsburgh (Pa.) 1954. Publ. by the Standing Committee on Combustion Symposia. New York/London 1955, p. 245–252.
292. Dahme, A., and H. J. Junker: Brennst.-Chem. 36 (1955) p. 193–199.
293. Peters, W.: Glückauf 96 (1960) p. 997–1006. (Kokereiaussch. 115.)
294. Hedden, K.: Brennst.-Chem. 41 (1960) p. 193–203.
295. Hunt, B. E., S. Mori, S. Katz, and R. E. Beck: Ind. Engng. Chem. 45 (1953) p. 677–680.
296. Baukloh, W., and G. Henke: Metallwirtsch. 19 (1940) p. 463–470. — Also Baukloh, W., and G. Hieber: Z. anorg. allg. Chem. 226 (1936) p. 321–332.
297. Sōma, T.: Tetsu to Hagané Overseas 4 (1964) p. 121–127.

298. Schenck, H., and W. Maschlanka: Arch. Eisenhüttenwes. 31 (1960) p. 271 to 277. — Also Maschlanka, W.: Dr. Ing. Diss. (extract), TU Aachen.
299. Baukloh, W., and E. Spetzler: Arch. Eisenhüttenwes. 13 (1939/40) p. 223 to 226.
300. Walker, P. L., J. F. Rakszawski, and G. R. Imperial: J. Phys. Chem. 63 (1959) p. 140—149.
301. Meyer, H. H.: Mitt. KWI Eisenforsch. 10 (1928) p. 107—116.
302. McLeod, J. M.: J. West Scotland Iron Steel Inst. 57 (1949/50) p. 242—253.
303. Baukloh, W., and F. Jaeger: Arch. Eisenhüttenwes. 13 (1939/40) p. 65—67. — Baukloh, W., and J. Hellbrügge: Arch. Eisenhüttenwes. 14 (1941/42) p. 163 to 166. — Baukloh, W., and B. Edwin: Arch. Eisenhüttenwes. 15 (1942/43) p. 197—200.
304. Baukloh, R., O. Knacke, and W. Löscher: Arch. Eisenhüttenwes. 27 (1956) p. 95—99.

2.10. The Water-Gas Reaction

304a. Wüst, F.: Stahl u. Eisen 48 (1928) p. 505—506.
304b. Royter, V. A., V. A. Yuza, and A. N. Kuznetsov: Zhurnal Fiz. Khim. 25 (1951) p. 960—970.
304c. Stålhane, B. S., and T. Malmberg: Jernkont. Ann. 114 (1930) p. 609—622; cf. Stahl u. Eisen 51 (1931) p. 716 — 717.
304d. Kulkova, N. V., and M. I. Temkin: Zhurnal Fiz. Khim. 23 (1949) p. 695—713. — Kulkova, N. V., Z. D. Kuznets, and M. I. Temkin: Doklady Akad. Nauk USSR 90 (1953) p. 1067—1070. — Temkin, M. I., M. L. Nakhmanovich, and N. M. Morozov: Kinetika i Kataliz 2 (1961) p. 650—654.
304e. Stotz, S.: Ber. Bunsenges. phys. Chem. 70 (1966) p. 37—44.
304f. Grabke, H. J.: unpublished measurements.
304g. Goldshtein, N. L., and N. S. Khromchenko: Stal 25 (1965) p. 197—200; Stal in Deutsch 5 (1965) p. 611—615.
304h. Grabke, H. J.: Ber. Bunsenges. phys. Chem., in press.

3. Gas Flow and Heat Transfer in Granular Materials

3.1. The Static Bed

305. Wicke, E., and W. Brötz: Chem.-Ing.-Technik 24 (1952) p. 58.
306. Furnas, C. C.: Fluid Flow through Beds of Broken Solids. Bull. Bur. Mines, No. 307, 1929.
307. Carman, P. C.: Trans. Inst. Chem. Eng. 15 (1937) p. 150.
308. Kozeny, J.: Ber. Akad. Wien 136a (1927) p. 271.
309. Jeschar, R.: Arch. Eisenhüttenwes. 35 (1964) p. 91—108. (Hochofenaussch. 391 and Wärmestelle 539.)
310. Brauer, H.: Dechema Monogr. 37 (1960) p. 7—77.
311. Doering, E.: Allg. Wärmetechn. 6 (1955) p. 82—89.
312. Ergun, S.: Chem. Engng. Progr. 48 (1952) p. 89—94.
313. Taschenbuch für Chemiker und Physiker. Edited by J. D'Ans and E. Lax. Berlin 1943, p. 1108.
314. Leva, M.: Fluidization. New York 1959, p. 42.
314a. Ergun, S.: Chem. Engng. Progr. 48 (1952) p. 89—94.

135. HANSEN, M.: Arch. Eisenhüttenwes. 34 (1963) p. 151—157. (Hochofenaussch. 374 and Wärmestelle 522.)
315a. As in No. 314, p. 169.
316. PERRY, J. H.: Chemical Engineers' Handbook. 4th ed. New York 1963, p. 20—41.
316a. MATHUR, K. B., and P. E. GISHLER: A. I. Ch. E. J. 1 (1955) p. 157.
317. Verfahrenstechnische Berichte. Publ. by Ing.-Wiss. Abt. der Farbenfabriken Bayer AG, Leverkusen. Weinheim/Bergstr. 1963, p. 561.
318. POLTHIER, K.: Arch. Eisenhüttenwes. 37 (1966) p. 365—374.
318a. JESCHAR, R.: Arch. Eisenhüttenwes. 37 (1966) p. 193—200. (Aussch. metallurg. Grundlagen 19 and Wärmestelle 572.)
318b. WARTMANN, R., and H. MERTES: Arch. Eisenhüttenwes. 37 (1966) p. 201—207. (Wärmestelle 573 and Aussch. metallurg. Grundlagen 20.)
318c. JESCHAR, R., R. WARTMANN, and H. MERTES: Arch. Eisenhüttenwes., forthcoming.
318d. MERTES, H.: Dr. Diss., TH Clausthal 1968.

3.2. The Fluidized Bed and the Transition to Particulate Fluidization

319. WICKE, E.: Chem.-Ing.-Technik 24 (1952) p. 82—91. — WICKE, E., and F. FETTING: Chem.-Ing.-Technik 26 (1954) p. 301—309.
320. BRÖTZ, W.: Chem.-Ing.-Technik 24 (1952) p. 60—81.
321. German Patent DRP 463772 of 3 Oct. 1922.
322. PERRY, J. H.: Chemical Engineers' Handbook. 4th ed. New York 1963.
323. LEVA, M.: Fluidization. New York 1959.
324. SCHYTIL, F.: Dechema Monogr. 38 (1960) p. 231—248.
325. GRASSMANN, P.: Physikalische Grundlagen der Chemie-Ingenieur-Technik. Aarau/Frankfurt a. M. 1961.
326. Fortschritte der Verfahrenstechnik. Weinheim/Bergstr. 1954.
327. v. BOGDANDY, L.: Stahl u. Eisen 79 (1959) p. 1064.
328. MEISSNER, H. P., and F. C. SCHORA: Trans. Metallurg. Soc. AIME 218 (1960) p. 12—21.
328a. DEEMTER, J. J. VAN: The counter-current flow model of a gas-solids fluidized bed. A lecture given at the International Symposium on Fluidization, 5—9 June 1967 in Eindhoven.
328b. BLOCK, F. R.: Counterflow heat exchange in fluidized beds. As in No. 328a.
328c. LAGO, E. C., A. R. OTERO, and J. J. S. ROF: Development of a countercurrent multistage fluidized bed reactor for solid-gas systems. As in No. 328a.
329. TRAWINSKI, H.: Chem.-Ing.-Technik 23 (1951) p. 416—419.
330. FETTING, F., and E. WICKE: Dechema Monogr. 24 (1955) p. 146—169.
331. NONNENMACHER, H.: Z. Elektrochem., Ber. Bunsenges. phys. Chem. 57 (1953) p. 512—518.

4. Techniques for the Reduction of Iron Ores, Apart from the Blast Furnace

4.1. Statement of the Problem

332. Stahleisen-Kalender 1966. Publ. by Verein Deutscher Eisenhüttenleute. Düsseldorf 1965.
333a. Eisen u. Stahl. Statistisches Vierteljahresheft der Eisen- u. Stahlindustrie. Publ. by Statist. Bundesamt, Außenstelle Düsseldorf. 3rd quarter 1963.

333b. HOFMANN, E. E.: In: Troisièmes Journées Internationales de Sidérurgie, Luxembourg 1–4 Oct. 1962. Publ. by Centre National de Recherches Metallurgiques, Liège 1962, p. 483–494.
333c. Statistisches Jahrbuch der Eisen- u. Stahlindustrie 1965, Düsseldorf, p. 296.
333d. Stahl u. Eisen 87 (1967) p. 696.
334. Bibliographical evaluation of methods for the direct reduction of iron ores. In: Eisenerzeugung außerhalb des Hochofens. 3rd ed. of "Verfahren der direkten Reduktion von Eisenerzen". Publ. by Europäische Gemeinschaft für Kohle u. Stahl, Hohe Behörde, Luxembourg 1967. (Copious references.)
335. J. Metals 15 (1963) p. 621.
336. HATCH, G. G., and N. H. CUKE: Canad. Min. Metallurg. Bull. 49 (1956) p. 619 to 622.
337. Steel 148 (1961) p. 122–123.
338. SIBAKIN, J. G.: Blast Furn. Steel Plant 50 (1962) p. 977–989.
338a. MEYER, K.: personal communication (1965).
339. LUCKE, F., H. SERBENT and G. MEYER: Stahl u. Eisen 82 (1962) p. 1222–1232.
339a. LUCKE, F.: personal communication (1965).
340. CELADA, J., C. K. MADER, and R. LAWRENCE: Iron Steel Eng. 37 (1960) p. 87 to 91.
340a. KOBRIN, C. L.: Iron Age 199 (1967) No. 11, p. 74–76.
341. LUBKER, R. A., and K. W. BRULAND: J. Metals 12 (1960) p. 321–324; also p. 440.
342. New Iron Ore Reduction Techniques. Publ. by Corp. Venez. Guyana, Caracas 1962.
342a. STOTLER, H. H.: Lecture at the Congrès International sur la Production et l'Utilisation des Minerais Réduits, 29–31 May 1967, Evian, France.
342b. ASTIER, J., U. KALLING, and J. STALHED: In: Ironmaking Tomorrow. Proc. of the conference held at the autumn general meeting of the Iron and Steel Institute on 22–23 Nov. 1966 in London (Spec. Rep. Iron Steel Inst., No. 102, 12935). London 1967.
342c. KALLING, B., and D. L. CAMPBELL: Iron and Steel Inst. 38 (1965) p. 651–654.
342d. WILD, R.: as in No. 342a.
343. Die Eisenerzwirtschaft der Welt in Zahlen. Publ. by Wirtschaftsvereinigung Eisen- u. Stahlindustrie, Düsseldorf 1961.
344. Anhaltszahlen für die Wärmewirtschaft in Eisenhüttenwerken. 5th ed. Publ. by Energie- u. Betriebswirtschaftsstelle Verein Deutscher Eisenhüttenleute. Düsseldorf 1957, p. 1.
344a. Ibid. p. 8.
345. Ibid. p. 10.
346. WUNSCH, W.: Stahl u. Eisen 82 (1962) p. 1213–1222. (Wärmestelle 513.)
347. WENZEL, W.: Arch. Eisenhüttenwes. 34 (1963) p. 75–82.
347a. WENZEL, W., and H. W. THEISGES: Klepzig Fachber. 73 (1965) p. 129–134.
348. MINTROP, R.: Stahl u. Eisen 78 (1958) p. 633–646. (Hochofenaussch. 305.)
348a. MINTROP, R.: personal communication.
349. Hoesch AG, Westfalenhütte Dortmund, unpublished research.
350. MASI, O., and P. CANNIZZO: J. Iron Steel Inst. 200 (1962) p. 199.
350a. LUCKE, F., H. SERBENT, and G. MEYER: Stahl u. Eisen 85 (1965), p. 1371 to 1378.
350b. BROWN, J. W., D. L. CAMPBELL, A. L. SAXTON, and J. W. CARR jr.: J. Metals 18 (1966) p. 237–242.
350c. CAMPBELL, D. L.: personal communication.

351. Pantke, H. D., and J. Wasmuht: Stahl u. Eisen 86 (1966) p. 1078.
351a. Archer, C., and J. L. Gatelais: as in No. 342a.
352. As in No. 342, p. 8.
352a. Meyer, K., G. Heitmann, and W. Janke: J. Metals 18 (1966) p. 748—752.
352b. Sibakin, J. G., P. H. Hookings, and B. A. Roeder: as in No. 342a.
352c. Antoine, J., and J. Dumont-Fillon: as in No. 342a.
353. Weaver, W. R.: Carbon Graphite News 7 (1960) No. 1; cf. Stahl u. Eisen 80 (1960) p. 1727—1728.
354. Plöckinger, E., and K. Borowski: Techn. Mitt. Krupp 18 (1960) p. 1—8.
354a. Rinesch, R. F.: In: Congrès International sur les Aciéries à l'Oxygène. Le Touquet, 23—25 Sept. 1963; Dunkerque, 26 Sept. 1963. Le Touquet 1964, p. 32—45.
354b. Borowski, K.: Techn. Mitt. Krupp, Werksber. 24 (1966) p. 33—38, esp. p. 36 to 38.
355. Gummeson, P. U.: Proc. AIME, Open-Hearth Comm. 41 (1958) p. 366—381.
355a. Ayala, J. L. P., J. G. Rodriguez, E. A. Bryan, and N. N. Hughes: Iron Steel Eng. 40 (1963), No. 8, p. 69—76.
355b. Celada, J. S.: as in No. 342a.
356. Chase, P. W., and D. L. McBride: Blast Furn. Steel Plant 51 (1963) p. 868 to 897
357. Schaefers, W.: Stahl u. Eisen 84 (1964) p. 482—484.
357a. Bernstein, N., J. L. Reuss, and P. L. Woolf: J. Metals 18 (1966) p. 652 to 656.
357b. As in No. 342b.
357c. Kalla, U., H. D. Pantke, and K. Pittel: Stahl u. Eisen 87 (1967) p. 534 to 542. (Aussch. metallurg. Grundlagen 34 and Aussch. Betriebswirtsch. 380.)
358. Keith, P. C., H. H. Stotler, and R. J. McMullan: Iron Steel Eng. 40 (1963) No. 11, p. 95—100.
359. Speith, K. G., and O. Steinhauer: Stahl u. Eisen 83 (1963) p. 961—971. (Stahlwerksaussch. 765.)
360. Iron Coal Trades Rev. 182 (1961) p. 477.
361. Cavanagh, P. E.: Rev. Techn. Luxemb. 50 (1958) p. 78—86.
362. Jolivet, H.: Rev. Métallurg. 61 (1964) p. 531—543.
363. Fine, M. M., P. L. Woolf, and N. Bernstein: U.S. Bureau of Mines Rep. 6523, 1964.
363a. Peart, J. A., and F. J. Pearce: J. Metals 17 (1965) p. 1396—1400.
363b. Woolf, P. L.: J. Metals 18 (1966) p. 243—247.
363c. Heynert, G., and H. Beer: Stahl u. Eisen 84 (1964) p. 1353—1365. (Hochofenaussch. 400.)

4.2. Reduction in Dispersed-Cloud Apparatus and Shower Furnaces

364. Knacke, O.: Arch. Eisenhüttenwes. 30 (1959) p. 581—584.
365. Moreau, J., J. Bardolle, and J. Bénard: Rev. Métallurg. Mém. 48 (1951) p. 486—494.
366. Ezz, S. Y. M., and R. Wild: J. Iron Steel Inst. 194 (1960) p. 211—221.
367. Cavanagh, R. L.: In: Iron Ore Reduction. Proceedings of a Symposium, Electrothermics and Metallurgy Division of the Electrochemical Society, held in Chicago, 3—5 May 1960. Oxford/London/New York/Paris 1962, p. 221—240.
368. Johnson, T. W., and J. Davison: J. Iron Steel Inst. 202 (1964) p. 406—419.
369. Edwards, J. A.: Iron and Steel 36 (1963) p. 608—609.

369a. RINESCH, R. F., and A. KRAUSZ: Berg- u. hüttenmänn. Mh. 111 (1966) p. 563 to 578.
370. SCHENCK, H., and W. WENZEL: Entwicklungsarbeiten auf dem Gebiete der Verhüttung von Erzstaub in Schmelzkammern (Forschungsberichte d. Wirtschafts- u. Verkehrsministeriums Nordrhein-Westfalen, No. 407). Köln/Opladen 1957.

4.3. Reduction in Fluidized Beds

371. WILLEMS, J., and G. QUADE: Stahl u. Eisen 79 (1959) p. 1058–1064. (Hochofenaussch. 322.) Discussion: v. BOGDANDY, L., esp. p. 1064.
372. MEISSNER, H. P., and F. C. SCHORA: Trans. Metallurg. Soc. AIME 218 (1960) p. 12–31.
373. YAMAMICHI, Y.: Tetsu to Hagané Overseas 3 (1963) p. 296–302; cf. Tetsu to Hagané 49 (1963) p. 747–752.
374. LUBKER, R. A., and K. W. BRULAND: J. Metals 12 (1960) p. 321–324.
375. VAVILOV, N. S., L. M. TSYLEV, and CHAO CH'UNG CH'I: Izv. Akad. Nauk USSR, OTN, Metallurgiya i Toplivo 1962, No. 1, p. 46–55.
376. WINZER, D.: Dipl.-Arbeit, TU Berlin 1956; cf. v. BOGDANDY, L.: Arch. Eisenhüttenwes. 32 (1961) p. 275–296.
377. SCHENCK, H., and W. MASCHLANKA: Arch. Eisenhüttenwes. 31 (1960) p. 271 to 277.
378. SCHENCK, H., W. WENZEL, and H. D. BUTZMANN: Arch. Eisenhüttenwes. 33 (1962) p. 211–216.
379. LANGSTON, B. G., and F. M. STEPHANS: as in No. 367, p. 207–220.
380. BRANT, H. H., and W. E. MARSHALL: In: Physical Chemistry of Process Metallurgical P. 2 and Petroleum Engineers, Pittsburgh, 27 Apr.–1 May, 1959. New York/London 1961, p. 647–669.
381. As No. 334.
382. STOTLER, H. H., and R. A. LUBKER: as in No. 367, p. 163.
383. KEITH, P. C., H. H. STOTLER, and R. J. MCMULLAN: Iron Steel Eng. 40 (1963) No. 11, p. 95–100.
384. CHASE, P. W., and D. L. MCBRIDE: Blast Furn. Steel Plant 51 (1963) p. 868 to 897.
385. REED, T. F., J. C. AGARWAL, and E. H. SHIPLEY: J. Metals 12 (1960) p. 317–320.
386. AGARWAL, J. C., P. W. CHASE, W. L. DAVIES, and T. F. REED: as in No. 342, p. 29.
387. NASONOV, P. YA., E. N. VASILEV, I. L. LURE, and V. F. KNYAZEV: Izv. Akad. Nauk USSR, OTN, Metallurgiya i Toplivo 1962, No. 5, p. 29–36.
388. HYDE, R. W.: as in No. 342, p. 93.
389. OLD, B. S., and R. W. HYDE: Chem. Eng. 64 (1957) p. 130.
389a. BROWN, J. W., D. L. CAMPBELL, A. L. SAXTON, and J. W. CARR: J. Metals 18 (1966) p. 237–242.
389b. CAMBON, T.: as in No. 342a.
389c. KOSKAS, R., M. ROUBY, and A. RIST: as in No. 342a.
389d. CAMBON, T.: personal communication.
390. As in No. 371.
391. WICKE, E., K. HEDDEN, and G. LÜTH: Stahl u. Eisen 79 (1959) p. 129–134. (Hochofenaussch. 318.)
392. STELLING, O., and I. PERESWETOFF-MORATH: Jernkont. Ann. 141 (1957) p. 237 to 260.

393. VAVILOV, N. S.: Izv. Vyssh. Ucheb. Zaved., Chern. Metallurgiya 1963, No. 5, p. 26–33.
394. FEINMANN, J.: Ind. Engng. Chem. Process Design Development 3 (1964) p. 241 to 247.

4.4. Reduction in Retorts

395. v. BOGDANDY, L., and W. JANKE: Z. Elektrochem., Ber. Bunsenges. phys. Chem. 61 (1957) p. 1146–1153.
396. HEDDEN, K., and G. SOMMER: Arch. Eisenhüttenwes. 35 (1964) p. 9–14. (Hochofenaussch. 390.)
397. BULL-SIMONSON, J.: Stahl u. Eisen 52 (1932) p. 457–461.
398. BROWN, W. E.: U. S. Bur. Mines Rep. Invest. No. 3925, 1948, p. 58.
399. CELADA, J., C. K. MADER, and R. LAWRENCE: Iron Steel Eng. 37 (1960) p. 87–91.
400. STARRATT, F. W.: J. Metals 11 (1959) p. 315–318.
401. Anon. in No. 432, p. 9.
401a. BROWN, D. C.: In: Recent Advances in Ferrous Metallurgy. Publ. by the American Institute of Chemical Engineers. New York 1963, p. 53–83.
401b. KOBRIN, C. L.: Iron Age 199 (1967) No. 11, p. 74–76.
401c. SKELLY, J. P.: as in No. 342a.
401d. The HyL-Process for Direct Reduction. Publ. by Hojalata y Lamina S.A. in Monterrey, Mexico.
401e. As in No. 334.
401f. JAEGER, F., and H. WINTERHOFF: Stahl u. Eisen 82 (1962) p. 290–293.
401g. Steelmaking with Gas. Publ. by M. W. Kellogg Company, New York.
403. v. BOGDANDY, L.: Lecture at the joint session of Ausschuß für metallurgische Grundlagen and Hochofenausschuß in Düsseldorf, 1964.
404. HATARESCU, O., G. ATANASIU, M. PETROSENEANU, and L. DIMITRIU: Metalurgia (Rumanian) 15 (1963) p. 298 (H. Brutcher Transl. No. 6427, 1964); ibid. 16 (1964) p. 425 (H. Brutcher Transl. No. 6479, 1965).
405. Hüttenwerk Oberhausen AG, unpublished research (1965).

4.5. Reduction in Rotary Kilns

406. LUCKE, F., H. SERBENT, and G. MEYER: Stahl u. Eisen 82 (1962) p. 1222–1232; 85 (1965) p. 1371.
407. WAHLSTER, M.: Techn. Mitt. Krupp 17 (1959) p. 330–342.
408. HEILIGENSTÄDT, W.: Wärmetechnische Rechnungen für Industrieöfen. 3rd ed. Düsseldorf 1951, p. 418 et seq.
409. JOHANNSEN, F.: Metall u. Erz 36 (1939) p. 325–333.
410. LUCKE, F., H. SERBENT, and G. MEYER: Techn. Mitt. Krupp 19 (1961) p. 189–195.
410a. SERBENT, H., and G. MEYER: Klepzig Fachber. 73 (1965) p. 452–457.
410b. SERBENT, H., and H. KRAINER: as in No. 342a.
411. HEDDEN, K.: Chem.-Ing.-Technik 30 (1958) p. 125–132.
412a. HEDDEN, K.: Brennst.-Chem. 41 (1960) p. 193–203.
412b. HEDDEN, K.: Chem. Eng. Sci. 14 (1961) p. 317. — HEYNERT, G., W. ZISCHKALE, and E. SCHÜRMANN: Stahl u. Eisen 80 (1960) p. 981–990. (Hochofenaussch. 333.)
413. Hoesch AG, Westfalenhütte Dortmund, unpublished research (1963).
414. KRAINER, H.: Techn. Mitt. Krupp 19 (1961) p. 86–106.
415. v. BOGDANDY, L., and W. JANKE: Z. Elektrochem., Ber. Bunsengesphys. Chem. 61 (1957) p. 1146–1153.

416. SIBAKIN, J. G.: Blast Furn. Steel Plant 50 (1962) p. 977–989.
417. As in No. 334.
418. WAHLSTER, M.: Techn. Mitt. Krupp, Forsch.-Ber. 20 (1962) p. 1 and 33–43.
419. v. BOGDANDY, L.: Stahl u. Eisen 82 (1962) p. 869–883. (Hochofenaussch. 361.) Discussion: JANKE, W., esp. p. 881.
419a. MEYER, K., G. HEITMANN, and W. JANKE: J. Metals 18 (1966) p. 748–752.
419b. MEYER, K., and G. HEITMANN: as in No. 342a.
419c. DAHL, E. Q.: Canad. Min. Metallurg. Bull. 59 (1966) p. 383–387.
420. HEISING, L. F., and R. C. BRIGGS: Preliminary Engineering Studies for Application of the R-N Process to Lake Superior Region Iron Ores. U.S. Dept. of Commerce, 1963.
421. BENGTSSON, E., K. ALMQUIST, and A. JOSEFSSON: J. Metals 16 (1964) p. 337–339.
421a. KALLING, B., and D. L. CAMPBELL: Iron and Steel 138 (1965) p. 651–654.
422. LEHMKÜHLER, H.: Stahl u. Eisen 59 (1939) p. 1281–1288. (Hochofenaussch. 189.)
423. JOHANNSEN, F.: Stahl u. Eisen 60 (1940) p. 910–912.
424. FRANK, Senior: as in No. 342, p. 66.
425. BRADDOCK, W. R.: Iron Steel Eng. 38 (1961) April, p. 123–125.
426. Ind. Heat 30 (1963) p. 1081.
427. MAURICE, H., and M. C. UDY: J. Metals 15 (1963) p. 630–631.
428. As in No. 367.
429. WAHLSTER, M., H. G. JOST, H. SERBENT, and G. MEYER: Techn. Mitt. Krupp, Forsch.-Ber. 21 (1963) p. 5–14.
430. JOSEFSSON, A.: personal communication (1964).
431. BARRETT, E. P.: Steel 105 (1939) No. 18, p. 48–52.
432. PAVLOVIC, P.: Entwicklung der Grundlagen eines Verfahrens für die Reduktion von Eisenerzen mit Öl. Dr. Ing. Diss., TH Aachen 1963.
433. PEARSON, T. F., C. G. ARMSTRONG, and D. R. CARTWRIGHT: J. Iron Steel Inst. 202 (1964) p. 1–10.
433a. ARCHER, C., and J. L. GATELAIS: as in No. 342a.
434. THEMELIS, N. J., J. W. DONALDSON, and M. C. UDY: Canad. Min. Metallurg. Bull. 57 (1964) p. 434–442.
435. v. BOGDANDY, L.: Lecture at the Metallkolloquium of the Technical University Berlin, July 1964.
435a. MEYER, K., H. RAUSCH, and M. OTTOW: Stahl u. Eisen 87 (1967) p. 654–660. (Hochofenaussch. 430.) — OTTOW, M.: Ursachen der Zerstörung von Hämatit-Pellets aus sehr reinem Magnetitkonzentrat während der Reduktion und Versuche zur Qualitätsverbesserung. Dr. Ing. Diss., TU Berlin 1966.

4.6. Reduction in Shaft Furnaces (Without Melting)

436. v. BOGDANDY, L., and W. JANKE: Z. Elektrochem., Ber. Bunsenges. phys. Chem. 61 (1957) p. 1146–1153.
437. v. BOGDANDY, L., and R. WARTMANN: Arch. Eisenhüttenwes. 36 (1965) p. 221 to 236.
438. As in No. 344, p. 358.
439. KUBASCHEWSKI, O., and E. L. EVANS: Metallurgical Thermochemistry. 2nd ed. (Metal Physics and Physical Metallurgy, Vol. 1). London 1956.
440. ELLIOTT, J. F., and M. GLEISER: Thermochemistry for Steelmaking, Vol. 1. Reading (Mass.)/London 1960.
441. As in No. 334, p. 83.
442. STÅLHED, J.: Stahl u. Eisen 72 (1952) p. 459–466. (Hochofenaussch. 258.)

443. Wiberg, M.: Jernkont. Ann. 142 (1958) p. 288—355.
444. As in No. 334, p. 95.
445. v. Bogdandy, L.: Stahl u. Eisen 82 (1962) p. 869—883. (Hochofenaussch. 361.)
446. v. Bogdandy, L.: Proc. United Nations Interregional Symposium on the Application of Modern Technical Practices in the Iron and Steel Industry of Developing Countries, 1963. Techn. Paper B 25.
446a. Hüttenwerk Oberhausen AG, unpublished research (1967).
447. v. Bogdandy, L., H. D. Pantke, and U. Pohl: J. Metals 18 (1966) p. 519.
448. Pehlke, B.: Über die Ursache des Dekrepitierens von Eisenerzen. Dipl.-Arbeit, Bergakad. Clausthal 1958.
449. Scortecci, A., and A. Palazzi: J. Iron Steel Inst. 195 (1960) p. 267—278.
450. Jeffries, C. F.: J. Metals 16 (1964) p. 805—808.
451. As in No. 441.
451a. Klein, L.: J. Metals 18 (1966) p. 923—931.

4.7. Electro-Reduction

452. Marinček, B.: Stoff- und Wärmeumsatz metallurgischer Vorgänge. Berlin/Göttingen/Heidelberg 1964, p. 67 et seq.
453. Wieder, H.: J. Metals 15 (1963) p. 628—629.
454. Colin, F. C.: as in No. 367, p. 98.
455. Hatch, G. G., and N. H. Cuke: Canad. Min. Metallurg. Bull. 49 (1956) p. 619 to 622.
456. Pecorari, A.: Radex-Rdsch. 1954, No. 4—5, p. 152—164.
457. Wenzel, W.: Stahl u. Eisen 79 (1959) p. 1183—1186. (Hochofenaussch. 324.)
458. As in No. 334.
459. Maurice, H., and M. C. Udy: J. Metals 15 (1963) p. 627—631.
460. As in No. 334, p. 13.
461. As in No. 367, p. 241.
462. Wimmer, A.: Stahl u. Eisen 80 (1960) p. 1469—1477.
463. As in No. 367, p. 134.
463a. Sato, Y., K. Murai, N. Kuboyama, A. G. Arnesen, and R. H. Friden: J. Iron Steel Inst. 203 (1965) p. 799—803.
464. Durrer, R., and G. Volkert: Die Metallurgie der Ferrolegierungen. Berlin/Göttingen/Heidelberg 1953.
465. Murakami, A., K. Takai, and Y. Sato: Tetsu to Haganè 46 (1960) p. 1140 to 1142.
466. Wenzel, W.: personal communication (1965).

4.8. Preparation of Fuels for the Reduction

467. Reerink, W.: Stahl u. Eisen 75 (1955) p. 322—335. (Kokereiaussch. 106.)
467a. Reerink, W., K. G. Beck, and W. Weskamp: Die Versuchskokerei des Steinkohlenbergbauvereins (Forschungsberichte des Landes Nordrhein-Westfalen, No. 1218). Köln/Opladen 1963.
468. Steinkohlenverkokung. In: Ullmanns Enzyklopädie der technischen Chemie. 3rd ed. Edited by W. Foerst. Vol. 10. Munich/Berlin 1958, p. 241—346.
468a. Handbuch des Kokereiwesens. Edited by O. Grosskinsky. Vols. 1 and 2. Düsseldorf 1955 and 1958.
468b. Simonis, W., and K. G. Beck: Glückauf, forthcoming.

469. Dickens, P., and W. Radmacher: Stahl u. Eisen 80 (1960) p. 129–136. (Hochofenaussch. 327, Chemikeraussch. 286 and Kokereiaussch. 113.)
470. Wesemann, F., U. Graf, and R. Wartmann: Stahl u. Eisen 76 (1956) p. 133 to 144. (Hochofenaussch. 291, Kokereiaussch. 104 and Wärmestelle 424.)
470a. Untersuchung über den Einfluß des Aschegehaltes von Koks im Hochofen. Publ. by Europäische Gemeinschaft für Kohle u. Stahl, Hohe Behörde, Luxembourg 1964.
470b. Scheidig, K., and G. Gruson: Report for the Internationale Tagung Koks im Eisenhüttenwesen, Charleroi 1966; cf. Steel Times 193 (1966) p. 612–613.— Also Neue Hütte 11 (1966) p. 385–390.
470c. White, R. H.: Blast Furn. Steel Plant 54 (1966) p. 241–245; cf. Stahl u. Eisen 86 (1966) p. 1606–1607.
470d. Knepper, W. A., and C. M. Sciulli: J. Metals 18 (1966) p. 228–230.
470e. VDEh-Hochofenausschuß, Unterausschuß Möllervorbereitung: Hillnhütter, F. W.: Reports of 24. 4. 65, 11. 11. 65, 19. 4. 67. — Winzer, G.: Report of 11.11. 65. — Zischkale, W.: Report of 19. 4. 67. — Schoppa, H.: Report of 19. 4. 67. — Schulz, D.: Report of 19. 4. 67.
470f. Wysocki, H., and F. W. Hillnhütter: as in No. 470b; cf. Steel Times 193 (1966) p. 613–614.
471. Heynert, G., W. Zischkale, and E. Schürmann: Stahl u. Eisen 80 (1960) p. 981–990. (Hochofenaussch. 333.)
472. Peters, W.: Glückauf 96 (1960) p. 997–1006.
473. Hedden, K.: Brennst.-Chem. 41 (1960) p. 193–203.
474. Pückoff, U.: Stahl u. Eisen 80 (1960) p. 989. Discussion: esp. p. 989–990.
474a. Schenck, H., and W. Wenzel: as in No. 470b; cf. Steel Times 193 (1966) p. 611–612.
475. Kahlhöfer, H., G. Pfrötschner, A. Send, and O. Steinhauer: Stahl u. Eisen 82 (1962) p. 547–556. (Hochofenaussch. 359.)
475a. Triska, A. A.: Blast Furn. Steel Plant 53 (1965) p. 508–515, 533, 539, and 543. — Also Triska, A. A., and C. D. Schubert: as in No. 470b; cf. Steel Times 193 (1966) p. 614–615.
476. Huebler, J.: as in No. 367, p. 24.
477. v. Gratkowski, H. W.: Stahl u. Eisen 80 (1960) p. 397–407. (Hochofenaussch. 329.)
478. v. Gratkowski, H. W.: as in No. 468, p. 360–458.
479. Mayer, M., and V. Altmayer: Ber. dtsch. Chem. Ges. 40 (1907) p. 2134–2144.
480. Cantelo, R. C.: J. phys. Chem. 28 (1924) p. 1036.
481. Cantelo, R. C.: J. phys. Chem. 30 (1926) p. 1641.
482. Schenck, R.: Z. anorg. allg. Chem. 164 (1927) p. 145–147 and 313–325.
483. Randall, M., and A. Mohammed: Ind. Engng. Chem. 21 (1929) p. 1048–1052.
484. Browning, L. C., and P. H. Emmett: J. Amer. Chem. Soc. 73 (1951) p. 581.
485. Rossini, F. D.: Selected Values of Properties of Hydrocarbons (Circ. Nat. Bur. Stand. C 461). Washington 1947.
486. Hiller, H.: Gaswärme 8 (1959) p. 151–154.
487. Peters, K., and E. Kappelmacher: Brennst.-Chem. 33 (1952) p. 296–307. — More recent data in: Cairns and Tevebaugh: J. Chem. Eng. Data 9 (1964) p. 453.
488. Montgomery, C. W., E. B. Weinberger, and D. S. Hoffmann: Ind. Engng. Chem. 40 (1948) p. 601–607.
489. Ulich, H., and W. Jost: Kurzes Lehrbuch der physikalischen Chemie. 12th and 13th eds. Darmstadt 1960.

490. FISCHER, F., and H. TROPSCH: Brennst.-Chem. 9 (1928) p. 39—46.
491. CELADA, J., C. K. MADER, and R. LAWRENCE: Iron Steel Engng. 37 (1960) p. 87 to 91.
492. v. BOGDANDY, L.: Stahl u. Eisen 82 (1962) p. 869—883. (Hochofenaussch. 361.)
493. v. BOGDANDY, L., W. RUTSCH, and I. N. STRANSKI: Z. Elektrochem., Ber. Bunsenges. phys. Chem. 66 (1962) p. 661—666.
494. CHOULAT, G.: Gaswärme 12 (1963) p. 180—185.
495. MEUNIER, J.: Vergasung fester Brennstoffe und oxydative Umwandlung von Kohlenwasserstoffen. Weinheim/Bergstr. 1962.
496. MÜLLER, L. D., R. W. BOUMANN, and M. C. CHANG: Blast Furn. Steel Plant 53 (1965) p. 381.
497. SCHMULDER, P.: Brennst.-Chem. 46 (1965) p. 23; cf. BARTHOLOMÉ, E., and H. NONNENMACHER: Erdöl u. Kohle (1959) p. 359.
498. v. GRATKOWSKI, H. W.: personal communication.
498a. KLEIN, L.: J. Metals 18 (1966) p. 923—931.
498b. KLEIN, L.: Trans. Metallurg. Soc. AIME 224 (1962) p. 121—129.

5. The Blast Furnace Process

5.1. An Approach to Physico-Chemical Concepts

499. SPRINGORUM, F. A.: Stahl u. Eisen 80 (1960) p. 1838—1851.
499a. Statist. Vierteljahreshefte. Publ. by Statist. Bundesamt, Außenstelle Düsseldorf.
500. Statistical Yearbook 1965. Publ. by Japan Iron and Steel Federation. Tokyo 1966.
501. WAGNER, A., A. HOLSCHUH, and W. BARTH: Arch. Eisenhüttenwes. 6 (1932) p. 129—136. (Hochofenaussch. 130.)
502. JOSEPH, T. L., P. H. ROYSTER, and S. P. KINNEY: Proc. Eng. Soc. Western Pennsylvania, Jan. 1926, p. 427.
503. OSANN, B.: Stahl u. Eisen 36 (1916) p. 477—530.
504. BANSEN, H.: Arch. Eisenhüttenwes. 1 (1927/28) p. 245—266. (Hochofenaussch. 86.)
505. REICHARDT, P.: Arch. Eisenhüttenwes. 1 (1927/28) p. 77—101. (Hochofenaussch. 83.)
506. WESEMANN, F.: Stahl u. Eisen 71 (1951) p. 873—877. (Hochofenaussch. 248 and Wärmestelle 374.)
507. GÖRGEN, R.: Arch. Eisenhüttenwes. 30 (1959) p. 381—390. (Hochofenaussch. 321 and Wärmestelle 475.)
508. SCHÜRMANN, E., and D. BÜLTER: Stahl u. Eisen 81 (1961) p. 1565—1574. (Hochofenaussch. 355.)
509. MINTROP, R.: Stahl u. Eisen 78 (1958) p. 633—646. (Hochofenaussch. 305.)
510. v. BOGDANDY, L.: Stahl u. Eisen 81 (1961) p. 12—22. (Hochofenaussch. 337.)
511. HEYNERT, G., P. ISCHEBECK, and W. v. SPEE: Stahl u. Eisen 81 (1961) p. 1—12. (Hochofenaussch. 336 and Maschinenaussch. 312.)
512. HORIE, S.: J. Metals 17 (1965) p. 243.
512a. Personal communication from the Works Management of Cherepovets 1966; cf. RAMM, A. N.: Stal in Deutsch 5 (1965) p. 106—118. — LEVIN, L. YA., et al.: Stal in Deutsch 4 (1964) p. 5—10.
512b. SUZUKI, G., M. HIGUCHI, and T. SHIBUJA: 6000-ton/day — Blast Furnace Practice. Nippon Kokan Kabushiki Kaisha, Dec. 1968.
512c. CROMWELL, D. P.: Blast Furn. Steel Plant 56 (1968) p. 309—313.

513. PETERSEN, U., H. KAHLHÖFER, and A. SEND: Stahl u. Eisen 83 (1963) p. 1397 to 1407. (Hochofenaussch. 382.)
514. BRANDI, H., G. HEYNERT, and H. BEER: Stahl u. Eisen 84 (1964) p. 1169–1174.
515. VOICE, E. W., and K. G. DIXON: J. Iron Steel Inst. 200 (1962) p. 438–443.
516. WARTMANN, R.: Stahl u. Eisen 83 (1963) p. 1414–1425. (Hochofenaussch. 383.)
517. PAVLOV, M. A.: Metallurgie des Roheisens, 4 vols. Moscow 1948. German transl., 2nd ed. Berlin 1953.
518. BRANDI, H. T., P. ISCHEBECK, and G. HEYNERT: Stahl u. Eisen 80 (1960) p. 65 to 73.
519. BEER, H., and G. HEYNERT: Stahl u. Eisen 84 (1964) p. 1353–1365. (Hochofenaussch. 400.)
520. As in No. 344, p. 87.

5.2. Preparation of the Burden

521. Arch. Eisenhüttenwes. 6 (1932/33), p. 129–136. (Hochofenaussch. 130.)
521a. KINNEY, S. P.: Bur. of Mines, Techn. Paper No. 459, 1930, p. 1–92.
522. LUYKEN, W.: Die Vorbereitung des Hochofenmöllers. Berlin 1953.
523. BIRNBAUM, H., and L. v. BOGDANDY: Stahl u. Eisen 82 (1962) p. 785–796. (Hochofenaussch. 360.)
524. MINTROP, R.: Stahl u. Eisen 78 (1958) p. 633–646. (Hochofenaussch. 305.)
525. v. BOGDANDY, L.: Stahl u. Eisen 81 (1961) p. 12–22. (Hochofenaussch. 337.)
526. As in No. 511.
526a. TROGNITZ, W. R.: Blast Furn. Steel Plant 52 (1964) p. 314–321 and 349, supplemented by a personal communication.
526b. WHITE, R. H.: Blast Furn. Steel Plant 54 (1966) p. 241–245; cf. Stahl u. Eisen 86 (1966) p. 1606–1610.
526c. GEBHARDT, K., H. SICHEL, C. HOLZAPFEL, and R. WALLACH: Stahl u. Eisen 83 (1963) p. 907–911.
526d. FRIEDETZKY, E.: Stahl u. Eisen 81 (1961) p. 1558–1564. (Hochofenaussch. 354.)
526e. TROJKA, D.: Neue Hütte 11 (1966) p. 520–526.
526f. KIKUCHI, T.: J. Iron Steel Inst. 205 (1967) p. 606–624.
527. STRASSBURGER, J. H.: J. Metals 8 (1956) p. 840–842.
528. STOWASSER, W. F.: Min. Engng. 7 (1955) p. 473–475.
529. MEYER, K.: Stahl u. Eisen 76 (1956) p. 588–595. (Hochofenaussch. 294 and Erzaussch. 72.)
530. PAQUET, J.: Rev. Techn. Luxemb. 47 (1955) p. 145–150.
531. VOICE, E. W., S. H. BROOKS, W. DAVIES, and B. L. ROBERTSON: J. Iron Steel Inst. 175 (1953) p. 97–152.
532. CARMAN, P. C.: Trans. Inst. Chem. Eng. 15 (1937) p. 150–166.
532a. HARADA, S.: J. Metals 17 (1965) p. 1305.
533. v. BOGDANDY, L., and R. SCHMOLKE: Stahl u. Eisen 77 (1957) p. 685–693. (Hochofenaussch. 299.)
534. HOLOWATY, M. O., and J. F. ELLIOTT: Proc. AIME, Blast Furnace, Coke Oven and Raw Materials Comm. 14 (1955) p. 26–42.
535. BAN, T. E., C. A. CZAKO, C. D. THOMPSON, and D. C. VIOLETTA: as in No. 61, p. 511–540.
536. WEILANDT, B., and W. STORSBERG: Stahl u. Eisen 76 (1956) p. 870–878. (Hochofenaussch. 296 and Erzaussch. 74.)
537. WOLF, W.: as in No. 536, p. 878.

538. KOSMIDER, H., E. BERTRAM, and H. SCHENCK: Stahl u. Eisen 76 (1956) p. 858 to 870. (Hochofenaussch. 295 and Erzaussch. 73.)
538a. STEINHAUER, O., and K. ZEHE: Stahl u. Eisen 81 (1961) p. 88–95. (Hochofenaussch. 338.)
538b. BRANDES, G., and H. RAUSCH: Proc. AIME, Blast Furnace, Coke Oven and Raw Materials Comm. 18 (1959) p. 232–249.
538c. STIRLING, H. T.: as in No. 259, p. 177–206.
538d. CHRISTMAS, R. B.: Broken Hill Prop. Rev. 37 (1960) p. 9–11.
538e. BENNETT, R. L., R. E. HAGEN, and M. V. MIELKE: J. Metals 8 (1956) p. 844 to 848.
538f. ELLIOTT, G. D., and N. D. MCDONALD: J. Iron Steel Inst. 167 (1951) p. 261 to 272. Discussion: 170 (1952) p. 123–128.
539. HAHN, R.: Mitteilung der Forschungs-Anstalt Gutehoffnungshütte-Konzern 9 (1942) p. 206–212.
540. RAUSCH, H., and F. CAPPEL: Stahl u. Eisen 81 (1961) p. 96–102. (Hochofenaussch. 339.)
540a. SCHENCK, H., W. WENZEL, and G. DIETRICH: Arch. Eisenhüttenwes. 34 (1963) p. 493–496.
541. SCHENCK, H., W. WENZEL, and G. DIETRICH: Forschungsberichte des Landes Nordrhein-Westfalen, No. 1122. Köln/Opladen 1962; cf. Stahl u. Eisen 81 (1961) p. 404–407. (Hochofenaussch. 344.)
541a. Ind. Heat 28 (1961) p. 1682–1688 and 1960.
542. ROHAN, T. M.: Iron Age 177 (1956) No. 15, p. 55–58.
543. HAVEN, W. A.: J. Iron Steel Inst. 180 (1955) p. 144–154.
543a. STARRATT, F. W.: J. Metals 14 (1962) p. 299–302.
544. SENGFELDER, G.: Stahl u. Eisen 70 (1950) p. 765–767 (Erzaussch. 54); 72 (1952) p. 1577–1579. (Erzaussch. 63.)
545. TIGERSCHIÖLD, M.: J. Iron Steel Inst. 177 (1954) p. 13–24.
545a. WINZER, G., and K. H. SCHMITZ: Stahl u. Eisen 87 (1967) p. 432–438. (Hochofenaussch. 428.)
546. MEYER, K.: Stahl u. Eisen 79 (1959) p. 222–225.
546a. BERKHAHN, R. W., and D. M. ULRICH: J. Metals 14 (1962) p. 303–305.
547. SEND, A.: Stahl u. Eisen 83 (1963) p. 972–978. (Hochofenaussch. 379.)
547a. MEYER, K.: Stahl u. Eisen 83 (1963) p. 1337–1345.
547b. ROHAN, T. M.: Iron Age 186 (1960) p. 159–161.
548. WOLF, W., and W. WYSOCKI: Stahl u. Eisen 81 (1961) p. 559–561. (Hochofenaussch. 349.)
548a. PETERSEN, W., and H. MINGENBACH: Z. Erzbergbau Metallhüttenwesen 16 (1963) p. 495–504.
549. RAMTHUN, H., and H. SCHEIWE: Arch. Eisenhüttenwes. 29 (1958) p. 165 to 167.
550. MOORE, J. E., and D. H. MARLIN: as in No. 259, p. 743–785.
550a. MCMULKIN, F. J., and N. G. THOMAS: J. Metals 16 (1964) p. 246–251.
550b. CAPPEL, F.: Stahl u. Eisen 84 (1964) p. 1062–1070.
551. WINZER, G., and E. PEETZ: Stahl u. Eisen 81 (1961) p. 1101–1107. (Hochofenaussch. 351.)
552. TOUSSAINT, F., G. HEYNERT, J. WILLEMS, and G. QUADE: Stahl u. Eisen 80 (1960) p. 473–483. (Hochofenaussch. 331.)
553. DOI, Y., and K. KASAI: Proc. AIME, Blast Furnace, Coke Oven and Raw Materials Comm. 18 (1959) p. 182–205; cf. J. Metals 11 (1959) p. 755–759.
554. DANIELSSON, C.: J. Iron Steel Inst. 175 (1953) p. 152–154.

555. Firket, A., and J. Molderez: Rev. Univ. Mines, 9 Sér., 102 (1959) p. 93—106; cf. J. Metals 11 (1959) p. 760—767.
556. Levin, L. Ya., I. A. Kusmin, V. D. Kailov, and A. B. Schur: Stal 18 (1958) p. 964—968; cf. Iron Coal Trades Rev. 178 (1959) p. 381—385.
557. Babarykin, N. N., and F. A. Yushin: Stal 18 (1959) p. 1057—1065, esp. p. 1058.
558. Möller, A.: Möglichkeiten zur Beeinflussung der Oxydationszone des Hochofens. Dr. Ing. Diss., Bergakad. Clausthal 1944. — Also as in No. 517.
559. Schürmann, E., W. Zischkale, P. Ischebeck, and G. Heynert: Stahl u. Eisen 80 (1960) p. 854—861.
560. Hupfer, K., and H. Weidenmüller: personal communication (1966).
561. Kitaico, B. I., R. Komar, and S. G. Mukha: Trans. Ind. Inst. Metals 11 (1958) No. 12, p. 3—20.
562. Birnbaum, H., L. v. Bogdandy, and W. Janke: Stahl u. Eisen 80 (1960) p. 781 to 788. (Hochofenaussch. 332.)
562a. Stahl-Eisen-Prüfblatt 1770-64. Verfahren zur Ermittlung der Reduzierbarkeit von Eisenerzen und Agglomeraten bei gleichbleibender Temperatur. 1st ed., Feb. 1964.
563. v. Bogdandy, L., P. Dickens, W. vor dem Esche, and J. Willems: Stahl u. Eisen 83 (1963) p. 129—139. (Hochofenaussch. 371.)
564. Klein, E.: Proc. AIME, Blast Furnace, Coke Oven and Raw Materials Comm. 17 (1958) p. 265—273.
565. Nitchie, C. M.: Paper presented at the Joint Meeting of the Eastern and Western States Blast Furnace and Coke Oven Association, Pittsburgh, Nov. 1963.
566. Ayres, H. S.: Blast furnace practice using sinter burdens. In: Deuxième Symposium International sur l'Agglomération des Minerais de Fer. Vol. II. Paris 1957, p. 612—613. — Figgis, K. J., and F. E. Charker: Blast Furn. Steel Plant 50 (1962) p. 1180—1191. — Stahl u. Eisen 84 (1964) p. 361—364.
567. Herr, H. D.: Untersuchung über den Brechvorgang an Erzen in der Möllervorbereitung. Dipl.-Arbeit, TU Berlin 1963.
568. Paschal, F.: Rev. Métallurg. 56 (1959) p. 285—301.
569. Elliott, G. D.: Ironmaking at the Appleby-Frodingham Works of the United Steel Companies Ltd. (Spec. Rep. Iron Steel Inst., No. 30). London 1944.
570. Dobscha, H. F.: Blast Furn. Steel Plant 34 (1946) p. 979—985.
571. Rueckel, W. C.: J. Metals 5 (1953) p. 509—514.
572. Kahlhöfer, H., A. Send, and G. Pfrötschner: Stahl u. Eisen 79 (1959) p. 1461 to 1471. (Hochofenaussch. 326.)
573. Wendeborn, H.: Saugzugsintern und -rösten. Berlin 1934.
574. Schluter, R., and F. Bitsianes: as in No. 259, p. 585—639.
575. Schenck, H., and H. P. Schulz: Arch. Eisenhüttenwes. 31 (1960) p. 691—702. (Hochofenaussch. 335.)
576. v. Bogdandy, L.: Arch. Eisenhüttenwes. 32 (1961) p. 275—296.
577. Schenck, R., O. Franz, and A. Laymann: Z. anorg. allg. Chem. 206 (1932) p. 192 et seq.; cf. Stahl u. Eisen 52 (1932) p. 731—732.
578. Schenck, R., H. Franz, and H. Willeke: Z. anorg. allg. Chem. 184 (1929) p. 1—38; cf. Stahl u. Eisen 50 (1930) p. 515—520.
579. vor dem Esche, W., and O. Steinhauer: Arch. Eisenhüttenwes. 30 (1959) p. 187—198. (Hochofenaussch. 319.)
580. Tigerschiöld, M.: Stahl u. Eisen 70 (1950) p. 397—403.
581. Danielsson, C.: J. Iron Steel Inst. 175 (1953) p. 152—154.
582. Notini, U.: In: Journées Internationales de Sidérurgie, 18—28 June 1958.

Comptes rendus. Publ. by Centre National de Recherches Métallurgiques, Liège 1958, p. 325–331.
583. Klein, E.: Iron Coal Trades Rev. 175 (1957) p. 549–552.
584. Cohen, E.: J. Iron Steel Inst. 175 (1953) p. 160–166.
585. Doi, Y., and K. Kasai: J. Metals 11 (1959) p. 755–759.
585a. Ikeno, E., and Y. Itoh: Tetsu to Haganė 48 (1962) p. 839.
585b. Leshchinskaya, E. I., and S. I. Rostovtsev: Izv. Vyssh. Ucheb. Zaved., Chern. Metallurgiya 6 (1962) No. 5, p. 5–15.
585c. Pokhvisnev, A. N.: as in No. 582, p. 295–296.
586. Elliott, G., J. A. Bond, and T. E. Mitchell: J. Iron Steel Inst. 175 (1963) p. 241–247.
587. Schürmann, E.: Arch. Eisenhüttenwes. 35 (1964) p. 484–486.
588. Kubaschewski, O., and E. L. Evans: Metallurgical Thermochemistry. 2nd ed. (Metal Physics and Physical Metallurgy, Vol. 1). London 1956.
589. Wuhrer, J., and G. Radermacher: Chem.-Ing.-Technik 28 (1956) p. 328–336.
590. Meyer, G.: Stahl u. Eisen 84 (1964) p. 479–482. Discussion of Schwabe, G., and H. Rellermeyer: Stahl u. Eisen 84 (1964) p. 327–349. (Hochofenaussch. 392.)
591. Baldwin, E. H., and I. M. Mathiesson: Proc. AIME, Blast Furnace, Coke Oven and Raw Materials Comm. 18 (1959) p. 218–231.
592. Raspopov, I. V., Ya. S. Gorbanev, and D. V. Gulyga: Stal 18 (1958) p. 676 to 682.
593. Schaefers, W., and O. Steinhauer: Stahl u. Eisen 83 (1963) p. 17–25. (Hochofenaussch. 369.)
594. Callender, W.: as in No. 259, p. 641–667.
595. Edström, J. O.: Jernkont. Ann. 141 (1957) p. 457–458.
595a. Cooke, S. R. B., and T. E. Ban: Min. Engng. 4 (1952) p. 1053–1058.
595b. Cooke, S. R. B., and W. F. Stowasser: Min. Engng. 4 (1952) p. 1223–1230.
595c. Furui, T.: Yawata Techn. Rep. No. 246, 1964, p. 4847–4854.
595d. Ottow, M.: Ursache der Zerstörung der Hämatit-Pellets aus sehr reinem Magnetitkonzentrat. Diss., TU Berlin 1966. — Ishimitsu, A.: Tetsu to Haganė Overseas 5 (1965) p. 214–216. — Koretata, K., et al.: Yawata Techn. Rep. No. 251, (1965), p. 5899–5909. — Meyer, K., H. Rausch, and M. Ottow: Stahl u. Eisen 87 (1967) p. 654–660. (Hochofenaussch. 430.)
595e. Grebe, K., R. Demuth, and K.-H. Schmitz: Report at the session of the Unterausschuß Erzeigenschaften des VDEh, 12 Dec. 1966.
595f. Kodama, K., A. Shigemi, K. Uno, T. Horrio, and T. Higashi: Yawata Techn. Rep. No. 251, 1965, p. 5899–5910.
595g. Watanabe, S., and Y. Otake: Tetsu to Haganė Overseas 5 (1965) p. 220–223.
595h. Stalhed, J., and R. Linder: Medd. Jernkont. No. 209, 1955.
595i. Wahlster, M.: Techn. Mitt. Krupp 19 (1961) p. 1–16.
595j. Vidal, R., N. Ponghis, and A. Poos: Centre National de Recherches Metallurgiques, No. 10, 1967, p. 3–6.
595k. Watanabe, S., and M. Yoshinaga: Report to the Annual Meeting of the American Institute of Mining and Metallurgical Engineers, Los Angeles, 1967.
595l. Ishimitsu, A.: Tetsu to Haganė Overseas 5 (1965) p. 214–216.
595m. Edström, J. O.: J. Iron Steel Inst. 179 (1953) p. 289–304.
596. v. Bogdandy, L., and H. G. Riecke: Arch. Eisenhüttenwes. 29 (1958) p. 603 to 609. (Hochofenaussch. 313.)
597. Hüttenwerk Oberhausen AG, not yet published.

597a. Kolesanov, F. F., and E. G. Gavrin: Stal 22 (1962) No. 4, p. 293—296; Stal in English 1962, No. 4, p. 247—250.
598. Marshall, W. E.: J. Metals 13 (1961) p. 308—313.
599. Heynert, G., W. Zischkale, and J. Willems: Stahl u. Eisen 82 (1962) p. 1641 to 1647. (Hochofenaussch. 365.)
600. Harada, S.: Report to the Metal Mining and Industrial Minerals Convention, American Mining Congress, Los Angeles, 15—18 Sept. 1963.
601. Petersen, U., H. Kahlhöfer, and A. Send: Stahl u. Eisen 83 (1963) p. 1397 to 1407. (Hochofenaussch. 382.)
602. Brandi, H., G. Heynert, and H. Beer: Stahl u. Eisen 84 (1964) p. 1169—1174.
603. Schenck, H.: Stahl u. Eisen 83 (1963) p. 1683—1690.
603a. Winzer, G.: Stahl u. Eisen 86 (1966) p. 74.
603b. Pohl, H., E. Tschierske, and K. Hagedorn: Stahl u. Eisen 86 (1966) p. 65 to 76. (Hochofenaussch. 410.)
604. Zudin, V. M., I. I. Sagaidak, A. P. Yakobson, N. N. Babarykin, V. G. Dorman, A. L. Galatanov, and P. V. Lekin: Stal 22 (1962) No. 8, p. 675—682; Stal in English 1962, No. 8, p. 578—582.
605. Send, A.: personal communication.
606. Wilburg, J. S.: Blast Furn. Steel Plant 52 (1964) p. 836—843, esp. p. 842.
606a. Burghardt, O.: personal communication.
607. Smedstum, J. A.: J. Metals 17 (1965) p. 465.

5.3. Testing Ores

608. Handbuch für das Eisenhüttenlaboratorium. Publ. by the Chemikerausschuß des Vereins Deutscher Eisenhüttenleute e.V., Vol. 3. Berlin/Düsseldorf 1956.
608a. German Standards: DIN 4187, Lochbleche für Prüfsiebe, July 1962. DIN 4188, Drahtgewebe für Prüfsiebe, Feb. 1957.
609. Birnbaum, H., L. v. Bogdandy, and W. Janke: Stahl u. Eisen 80 (1960) p. 781 to 788. (Hochofenaussch. 332.)
610. Hüttenwerk Oberhausen AG, not yet published.
611. Heynert, G., P. Ischebeck, and W. v. Spee: Stahl u. Eisen 81 (1961) p. 1—12. (Hochofenaussch. 336 and Maschinenaussch. 312.)
612. Janke, W.: Beitrag zur Prüfung von Stückerzen im Laboratoriums- und Betriebsversuch. Dr. Ing. Diss., TU Berlin 1960.
612a. ISO (International Standards Organization): Verfahren zur Trommelprüfung von Pellets, Sintern und unklassierten Eisenerzen. (Methods for trommel tests of pellets, sinters, and ungraded iron ores.)
613. Send, A., and B. Weilandt: Stahl u. Eisen 81 (1961) p. 303—310. (Hochofenaussch. 343.)
614. German Standard: DIN 51712, Steinkohlenkoks, Bestimmung der Trommelfestigkeit, 1950.
614a. Astier, J.: Circ. Inform. Techn. 21 (1964) p. 1493—1499.
615. v. Bogdandy, L., and R. Schmolke: Stahl u. Eisen 77 (1957) p. 685—693. Hochofenaussch. 299.)
616. Ramthun, H., and H. Scheiwe: Arch. Eisenhüttenwes. 29 (1958) p. 165—167.
617. Linder, R.: Iron Steel Inst. 189 (1958) p. 233—243.
617a. Inasaku, J.: Mines Exp. Station, Univ. of Minnesota, No. 8, July 1964.
618. Fortado, J.: Rev. Métallurg. Mém. 43 (1946) p. 219—228.
619. Debatty, L.: Circ. Inform. Techn. 10 (1953) p. 291—302.
620. As in No. 448.

621. VOLMER, M.: Kinetik der Phasenbildung. Dresden 1939.
622. KAISCHEW, R., and I. N. STRANSKI: Z. phys. Chem. 26 (1934) p. 317—326.
623. BAAKE, R.: Stahl u. Eisen 51 (1931) p. 1277—1283 and 1314—1319. (Hochofenaussch. 122.)
624. HARTMANN, F.: Stahl u. Eisen 63 (1943) p. 393—398. (Hochofenaussch. 214.)
625. DICKENS, P., W. VOR DEM ESCHE, and J. WILLEMS: Stahl u. Eisen 79 (1959) p. 905—917. (Hochofenaussch. 320 and Chemikeraussch. 271.)
626. POOS, A., and A. DECKER: Rev. Univ. Mines, 9 Sér., 102 (1959) p. 773—783.
627. GRIEVE, A.: J. Iron Steel Inst. 175 (1953) p. 1—4.
628. BABARYKIN, N. N., and F. A. YUSHIN: Stal 18 (1958) p. 1057—1065.
629. BRAUMANN, O.: Über das Erweichungsverhalten von Eisenerzen. Dr. Ing. Diss., TH Aachen 1964.
630. v. BOGDANDY, L., H. P. SCHULZ, and O. BRAUMANN: Stahl u. Eisen 84 (1964) p. 1561—1568. (Aussch. metallurg. Grundlagen 11.)
631. SCHENCK, H.: Stahl u. Eisen 75 (1955) p. 682—690. (Hochofenaussch. 286.)
632. MATHESIUS, L.: Stahl u. Eisen 34 (1914) p. 866—873.
633. SCHÜRMANN, E., and G. QUADE: Schrifttumsübersicht über die Reduktion von Eisenerzen. Parts 1 and 2. Publ. by Institut für Eisenhüttenkunde u. Gießereiwesen an der Bergakademie Clausthal, 1955.
633a. BOSLEY, J. J., N. B. MELCHER, and M. M. HARRIS: J. Metals 11 (1959) p. 610 to 615.
633b. REKAR, C.: Stahl u. Eisen 73 (1953) p. 1094—1101. (Hochofenaussch. 277.)
633c. POOS, A., and R. VIDAL: Etude théorique des phénomènes du haut fourneau basée sur les résultats du fourneau expérimental de Liège. Publ. by Centre National de Recherches Métallurgiques, Liège 1964.
634. v. BOGDANDY, L., P. DICKENS, W. VOR DEM ESCHE, and J. WILLEMS: Stahl u. Eisen 83 (1963) p. 129—139. (Hochofenaussch. 371.)
635. DICKENS, P., P. KÖNIG, and K. H. SCHMITZ: Stahl u. Eisen 84 (1964) p. 9—15. (Hochofenaussch. 387.)
636. SCHÜRMANN, E., J. WILLEMS, and G. SOMMER: Arch. Eisenhüttenwes. 35 (1964) p. 1—8 (Hochofenaussch. 389) and p. 169—171. (Hochofenaussch. 393.)
636a. HEDDEN, K., and G. SOMMER: Arch. Eisenhüttenwes. 35 (1964) p. 9—14. (Hochofenaussch. 390.)
637. SCHENCK, H., and W. WENZEL: Arch. Eisenhüttenwes. 37 (1966) p. 99—108.
638. WAHLSTER, M.: Techn. Mitt. Krupp 19 (1961) p. 1—16.
639. Stahl-Eisen-Prüfblatt 1770-64. Verfahren zur Ermittlung der Reduzierbarkeit von Eisenerzen und Agglomeraten bei gleichbleibender Temperatur. 1st ed., Feb. 1964.
640. SCHWABE, G., and H. RELLERMEYER: Stahl u. Eisen 84 (1964) p. 327—349. (Hochofenaussch. 392.)
641. MEYER, G.: Discussion of No. 640, p. 479—482.
642. SKODIN, K. K.: Stal 23 (1963) p. 57—104; Stal in Deutsch 3 (1963) p. 745.
643. KRAINER, H., T. LETTMANN, and M. WAHLSTER: Stahl u. Eisen 83 (1963) p. 578 to 585.
644. Hüttenwerk Oberhausen AG, not yet published.
645. POOS, A., and R. LINDER: In: Troisièmes Journées Internationales de Sidérurgie, Luxembourg, 1—4 Oct. 1962. Publ. by Centre National de Recherches Métallurgiques, Liège 1962, p. 117—126.
646. Hoesch AG, Westfalenhütte, not yet published.
647. BERG, B., J. HANCART, and A. POOS: Rev. Univ. Mines, 9 Sér., 105 (1962) p. 284 to 287.

648. Linder, R.: Jernkont. Ann. 148 (1964) p. 361—376.
648a. Pohl, H., E. Tschierske, and K. Hagedorn: Stahl u. Eisen 86 (1966) p. 73.
649. Barrett, E. P., and C. E. Wood: Ind. Eng. Chem. Anal. Ed. 18 (1946) p. 285.
650. Rist, A., and G. Bonnivard: Rev. Métallurg. 60 (1963) p. 23.

5.4. Injection of Fuels

651. British Patent 6207 (1831).
652. British Patent 7727 (1838).
652a. German Patent DRP 682 of 14 Sept. 1877.
653. Bertram, E.: Arch. Eisenhüttenwes. 1 (1927) p. 19—32. (Hochofenaussch. 82.)
654. German Patent DRP 350631 of 22 June 1918.
655. Kreide, R., and J. Roll: Stahl u. Eisen 56 (1936) p. 1177—1179. (Hochofenaussch. 155.)
656. Laurien, H., and G. Wedekind: Erdöl u. Kohle 14 (1961) p. 141—148.
656a. Gewerkschaftl. Rundschau, March 1964, p. 138—139.
657. Daten zur Entwicklung der Energiewirtschaft in der Bundesrepublik Deutschland. Bundesministerium für Wirtschaft, Jan. 1964.
658. Glückauf 100 (1964) p. 862—868.
658a. Segelken, L.: Gas- u. Wasserfach 104 (1963) p. 1293—1310.
658b. Statistische Unterlagen der Ruhrkohlenberatung GmbH, 1967.
659. As in No. 344.
660. Steinhauer, O.: Stahl u. Eisen 82 (1962) p. 1566—1572.
660a. Grimm, W.: Brennst. Wärme Kraft 15 (1963) p. 571—574.
661. Ischebeck, P., G. Heynert, and H. Beer: Stahl u. Eisen 82 (1962) p. 1476 to 1485. (Hochofenaussch. 364.)
661a. Strassburger, J. H., E. J. Ostrowski, and J. R. Dietz: J. Metals 14 (1962) p. 295—298.
661b. Anon.: Steel (1962) p. 67. — Crawford, C. W. J.: Steel and Coal 187 (1963) No. 4958, p. 174—175.
661c. Limpach, R.: Rapport Centre National de Recherches Métallurgiques 1964, No. 1, p. 3—13.
661d. Ostrowski, E. J., and J. R. Dietz: J. Metals 17 (1965) p. 1289.
661e. Uhl, U.: Diss., TH Aachen 1965.
662. v. Bogdandy, L., and W. Schaefers: Stahl u. Eisen 82 (1962) p. 1—18. (Hochofenaussch. 356.)
662a. Babarykin, N. N., and F. A. Yushin: Stal in Deutsch 6 (1966) p. 421—429.
662b. Schürmann, E., and D. Bülter: Arch. Eisenhüttenwes. 35 (1964) p. 677—698. (Hochofenaussch. 395.)
663. v. Bogdandy, L., and R. Görgen: Arch. Eisenhüttenwes. 35 (1964) p. 963—976. (Aussch. metallurg. Grundlagen 10 and Wärmestelle 550.)
663a. Beuthan, W.: Erdöl u. Kohle 18 (1965) p. 93.
664. v. Bogdandy, L.: Arch. Eisenhüttenwes. 32 (1961) p. 275—296.
664a. Skodin, K. K.: Stal 23 (1963) p. 97—104; Stal in Deutsch 3 (1963) p. 745.
664b. Krainer, H., M. Wahlster, and T. Lettmann: Stahl u. Eisen 83 (1963) p. 578—585.
665. Hofmann, E. E.: as in No. 178, p. 483—494, and personal communication.
665a. Schumacher, H.: Stahl u. Eisen 86 (1966) p. 309—321.
666. Melcher, N. B., J. P. Morris, E. J. Ostrowski, and P. L. Woolf: U.S. Bur. Mines Rep. Invest. No. 5621, 1960.

666a. Ostrowski, E. J., G. Kessler, and N. B. Melcher: Proc. AIME, Blast Furnace, Coke Oven and Raw Materials Comm. 19 (1960) p. 279—300.
667. Cordier, J. A.: as in No. 666a, p. 238—278.
667a. Cordier, J. A.: Techn. Mod. 55 (1963) p. 471—483.
668. Decker, A.: Stahl u. Eisen 81 (1961) p. 1264—1270.
668a. Brandi, H., P. Ischebeck, and H. Beer: Stahl u. Eisen 83 (1963) p. 1541 to 1546. (Hochofenaussch. 384.)
669. Joseph, T. L.: Blast Furn. Steel Plant 49 (1961) p. 239—246 and 324—328.
670. Hodge, A. L.: Blast Furn. Steel Plant 48 (1960) p. 665—675 and 689.
671. Ramm, A. N.: Use of enriched blast in blast-furnace operation. In: Contemporary Problems of Metallurgy. Edited by A. M. Samarin. Transl. from Russian. New York 1960, p. 51—79.
672. Ridgion, J. M.: J. Iron Steel Inst. 199 (1961) p. 135—143.
673. Melcher, N. B., W. M. Mahan, and P. L. Woolf: In: The Future of Ironmaking in the Blast-Furnace (Spec. Rep. Iron Steel Inst., No. 72). London 1962, p. 47—52.
674. Trense, R. V., and D. F. Rosborough: as in No. 673, p. 53—58.
675. Bell, H. B., and J. Taylor: J. Iron Steel Inst. 199 (1961) p. 285—287.
675a. Stephenson, R. L.: Iron Steel Eng. 39 (1962) No. 8, p. 91—98 — Blast Furn. Steel Plant 53 (1965) p. 389.
676. Krebs, E.: Stahl u. Eisen 70 (1950) p. 1290. — Kitaev, B. I.: Stal 14 (1954), No. 8, p. 684. — Zischkale, W., G. Heynert, and H. Beer: Stahl u. Eisen 83 (1963) p. 1117. — Babarykin, N. N.: Stal in Deutsch 5 (1965) Part 1, p. 10.
676a. Limpach, R.: Rev. Univ. Mines 19 (1963) p. 253—267.
677. Peters, W., and J. Lehmann: Erdöl u. Kohle 2 (1965) p. 112—118.
677a. Grimm, W.: Erdöl u. Kohle 16 (1963) p. 371.
677b. Fritsch, H. W.: Techn. Mitt. 55 (1962) p. 81.
677c. Masdin, E. G., and M. W. Thring: J. Inst. Fuel 35 (1962) p. 251.
678. Ostrowski, E. J., M. B. Royer, and L. J. Ropelewski: U.S. Bur. Mines Rep. Invest. No. 5648, 1960, p. 145.
678a. Nilles, P.: Rapport Centre National de Recherches Métallurgiques, Liége, No. 19, 1960.
679. Duftschmid, F., and F. Markert: Chem.-Ing.-Technik 32 (1960) p. 806—811.
680. Loginov, V. I., G. G. Oreshkin, J. G. Polovchenko, A. A. Sorokin, and I. N. Kardasevich: Stal 16 (1956) p. 675—682.
681. Grassmann, P.: Physikalische Grundlagen der Chemie-Ingenieur-Technik. Aarau/Frankfurt a. M. 1961. — Hansen, M.: Stahl u. Eisen 69 (1949) p. 526 to 528.
682. Gröber, H., S. Erk, and U. Grigull: Die Grundgesetze der Wärmeübertragung. Berlin/Göttingen/Heidelberg 1961.
683. Fritz, W., and H. Diemke: Feuerungstechn. 27 (1939) p. 129—136.
683a. Fritz, W.: Forsch. Ing.-Wes. 14 (1943) p. 1—10.
684. Gumz, W.: Kurzes Handbuch der Brennstoff- und Feuerungstechnik. 3rd ed. Berlin/Göttingen/Heidelberg 1962, p. 196.
685. Townend, D. T. A.: Iron Coal Trade Rev. 171 (1955) p. 1285—1292.
686. Krebs, E.: Stahl u. Eisen 70 (1950) p. 358—360. (Hochofenaussch. 238.) — Kitaev, B. I.: Stal 14 (1954) No. 8, p. 684—690. — Zischkale, W., G. Heynert, and H. Beer: Stahl u. Eisen 83 (1963) p. 1117—1125. — Babarykin, N. N.: Stal in Deutsch 5 (1965) Part 1, p. 10.
687. Schmitz, K. H.: Stahl u. Eisen 82 (1962) p. 14. (Hochofenaussch. 381.)
688. Wiethoff, G.: Stahl u. Eisen 84 (1964) p. 685—692.
689. v. Bogdandy, L., and R. Wartmann: Klepzig Fachber. 10 (1965) p. 458—467.

690. RAMIREZ, J.: Proc. AIME, Blast Furnace, Coke Oven and Raw Materials Comm· 20 (1961) p. 540—546.
691. KREGLOW, W. M., jr.: as in No. 690, p. 587—604.
692. CARLSON, J. W.: Iron Steel Eng. 38 (1961) p. 108—110. — CARLSON, J. W.: as in No. 690, p. 547—551.
693. COLLISON, W. H.: Iron Steel Eng. 39 (1962) p. 73—81.
694. MORELAND, J. C.: Blast Furn. Steel Plant 49 (1961) p. 317—323. — ANDERSON, S. G.: as in No. 690, p. 552—561.
695. DEAN, E. R., and R. A. POWELL: J. Metals 13 (1961) p. 49—50. — DEAN, E. R.: Blast Furn. Steel Plant 49 (1961) p. 417—423. — DEAN, E. R.: as in No. 690, p. 562—572.
696. WHITE, R. H., and C. M. SCIULLI: as in No. 690, p. 579—586.
696a. NEKRASOV, I. I., V. L. POKRYSKIN, A. V. ZAGREBA, and R. D. KRAMENEV: Stal 22 (1962) p. 199—205; Stal in Deutsch 2 (1962) p. 813.
696b. Hüttenwerk Oberhausen AG, unpublished results.
696c. STEFANOVICH, M. A., YU. V. FEDULOV, and L. YA. SPARBER: Stal in Deutsch 6 (1966) p. 1136—1143.
696d. VOLKOV, YU. P., L. YA. GAVRILYUK, I. A. KOPYRIN, I. A. LUBENETS, M. YA. OSTROUKHOV, and I. I. SAGAIDAK: Stal in Deutsch 6 (1966) p. 823—829.
696e. AGASHIN, A. A., N. N. BABARYKIN, YU. P. VOLKOV, A. I. GALATANOV, N. M. KRYUKOV, K. V. MALIKOV, M. YA. OSTROUKHOV, V. L. PISHANOV, A. N. TERNYATIN, and F. A. YUSHIN: Stal in Deutsch 5 (1965) p. 812—821.
696f. WANOT, J., W. WOŹNIACKI, and W. MATLINSKI: Neue Hütte 11 (1966) p. 591 to 593.
696g. LOGINOV, N. I., A. N. CHETSKURO, and V. M. DÓLGOV: Steel Times 191 (1965) p. 48—51. — MAKAEV, S. V.: Metallurgist 1965, No. 7, p. 4—7.
696h. LEVIN, L. YA., V. A. VANCHIKOV, V. D. KAILOV, A. B. SHUR, L. A. BYALYI, P. G. RUSAKOV, and A. P. KOTOV: Stal in Deutsch 5 (1965) p. 418—424.
696i. KORSHIKOV, G. V., YU. G. VORONOV, M. A. GEITLIN, YU. N. KIYASHKO, A. S. GOROKHOV, and M. A. SEKACHEV: Stal in Deutsch 6 (1966) p. 4—14.
696j. PANEV, G. A., A. G. KUZUB, P. P. CHUVPYLO, S. L. YAROSHEVSKII, and N. N. POPOV: Stal in Deutsch 6 (1966) p. 622—628.
696k. RUSAKOV, P. G., A. B. SHUR, and L. A. BYALYI: Stal in Deutsch 5 (1965) p. 1110 to 1114.
696l. LIZUNOV, G. I.: Iron and Steel of the USSR 6 (1965) p. 1118—1125.
696m. GOLDERMAN, G. D.: Metallurgist 1965, p. 311—312.
696n. Personal communication from "Sidelor", Homécourt (1966).
696o. ZHEREBIN, B. N., V. P. DEMBOVETSKII, M. S. KUDOYAROV, and P. P. MISHIN: Stal in Deutsch 5 (1965) p. 711—716.
696p. LOORZ, W., and D. WINZER: Stahl u. Eisen 87 (1967) p. 185—187. (Hochofenaussch. 424 and Wärmestelle 585.)
696q. GÖRGEN, R., and H. WEINECK: Stahl u. Eisen 87 (1967) p. 188—190. (Hochofenaussch. 425 and Wärmestelle 586.)
696r. DAUBENFELD, E., R. REILAND, and R. LIMPACH: In: Internationaler Kongreß Koks im Eisenhüttenwesen, Charleroi, 19—22 Sept. 1966. Liège 1967; cf. Steel Times 193 (1966) p. 676.
696s. HOWARTH, W. A., J. C. RYAN, and E. HOWELLS: as in No. 696r, p. 469—480; cf. Steel Times 193 (1966) p. 677.
697. Forschungsarbeiten über die Einspritzung von Kohlenwasserstoffen in Hochöfen. Publ. by Europäische Gemeinschaft für Kohle u. Stahl, Hohe Behörde, Luxembourg 1960. (Pompey blast furnace No. 3.)

697a. Bonnaure, E.: Stahl u. Eisen 79 (1959) p. 1105–1119. (Hochofenaussch. 323.) – J. Metals 13 (1961) p. 37–40.
697b. As in No. 697, Luxembourg 1961. (Seraing blast furnace No. 4.)
698. Penry, W. C. R., and J. Catlow: as in No. 645, p. 396–416.
699. Waxweiler, P., J. Lorang, R. Limpach, and N. Majerus: as in No. 645, p. 417–432.
700. Wagener, J., and N. Majerus: as in No. 645, p. 433–441.
701. Trense, R. V., and D. F. Rosborough: as in No. 673, p. 53–58.
702. Sharp, K. C.: J. Iron Steel Inst. 199 (1961) p. 69–75.
703. Wartmann, R.: unpublished analysis.
704. Schatz, E., S. Henkel, and W. Voss: unpublished report.
705. Hirase, M.: Blast Furn. Steel Plant 51 (1963) p. 21–31.
706. Lekontsev, Yu. A., and V. S. Nechev: Metallurgist No. 7 (1963) p. 8–9.
707. Borisov, Yu. S., and A. A. Fofanov: Stal 21 (1961) p. 492–498; cf. Stahl u. Eisen 82 (1962) p. 1751–1753.
708. Debruille, P.: Circ. Inform. Techn. 19 (1962) p. 1157–1170.
709. Guerbois, J.: Circ. Inform. Techn. 19 (1962) p. 1727–1754.
710. Erpelding, M.: Circ. Inform. Techn. 20 (1963) p. 1463–1470.
711. Taylor, H. C., and W. Rombough: Blast Furn. Steel Plant 50 (1962) p. 35–39.
711a. Walsh, J. P.: Regional Techn. Meet. Amer. Iron Steel Inst. 1964, p. 95–109. – Ind. Heat. 33 (1966) p. 485–486, 488, 490, 885–890, also p. 892, 894.
711b. Pavlik, O., and J. Hohn: Steel Times 192 (1966) p. 319–325.
711c. Beer, H.: Stahl u. Eisen 85 (1965) p. 872–875.
711d. Bakker, L.: Klepzig Fachber. 73 (1965) p. 468–480. – Bakker, L.: Erdöl u. Kohle 18 (1965) p. 106–112.
711e. Statz, H.: Neue Hütte 11 (1966) p. 538–540.
711f. Kotas, Z.: Neue Hütte 11 (1966) p. 536–537.
711g. Borisov, Yu. S., V. K. Kornev, I. I. Pushkash, B. D. Yantsen, A. E. Parenkov, and D. A. Zavarnitsyn: Stal in Deutsch 5 (1965) p. 922–928.
711h. Magazine of Metal Producing 4 (1966) p. 29.
711i. Opitz, W.: Stahl u. Eisen 86 (1966) p. 678.
711j. Brandi, H. T., G. Heynert, and F. Lobemeier: Stahl u. Eisen 86 (1966) p. 1345–1354. (Hochofenaussch. 418.)
711k. Schoppa, H., D. Winzer, and H. D. Bubner: Stahl u. Eisen 86 (1966) p. 1469 to 1473. (Hochofenaussch. 420.)
711l. Nagai, T., and K. Umegaki: as in No. 696r, p. 447–460; cf. Steel Times 193 (1966) p. 676.
711m. Egami, E.: as in No. 696r, p. 461–468; cf. Steel Times 193 (1966) p. 677.
711n. Miroshnichenko, A. A., Yu. A. Lekontsev, and G. G. Pashkeev: Metallurgist 1965, p. 244–246.
711o. Kobayashi, T., T. Izawa, and S. Matsui: Tetsu to Hagané 50 (1964) p. 369 to 371. – Hayashi, S., and T. Izawa: as in No. 696r, p. 502–508; cf. Steel Times 193 (1966) p. 678.
711p. Dowhaniuk, W. P., and G. S. Lucenti: Blast Furn. Steel Plant 53 (1965) p. 501–506.
712. Brandi, H., P. Ischebeck, and H. Beer: Stahl u. Eisen 83 (1963) p. 1541 to 1546. (Hochofenaussch. 384.)
713. Summers, E. M., L. McNaughton, and J. R. Monson: J. Iron Steel Inst. 201 (1963) p. 666–692.
714. Chauzat, P., C. Reveil, J. Cordier, A. Damgé, and J. C. Leclèrc: Rev. Métallurg. 60 (1963) p. 39–48.

714a. Ostrowski, J., and J. R. Dietz: J. Metals 17 (1965) p. 1289—1291.
714b. Tate, M., C. Nakane, C. W. Kim, and K. Suzuki: Tetsu to Hagané Overseas 5 (1965) p. 180—186.
714c. Voice, E. W.: Brit. Steelmaker 42 (1965) p. 282—288. — Voice, E. W.: Iron and Steel 138 (1965) p. 466—470.
714d. Summers, E. M., H. Williamson, and D. Morgan: as in No. 696r, p. 524 to 534; cf. Steel Times 193 (1966) p. 678.
714e. Uberoi, R.: Stahl u. Eisen 85 (1965) p. 1063—1064.
714f. Smith, P. G., G. P. Richard, and S. Harrison: Steel Times 193 (1966) p. 220 to 225.
714g. Limpach, R., P. Trousse, and A. Darimont: Rapport Centre National de Recherches Métallurgiques 1965, No. 4, p. 3—6. — Also as in No. 696r, p. 509 to 514; cf. Steel Times 193 (1966) p. 679.
714h. Steel Times 190 (1965) p. 882.
714i. Morinaga, K., and Y. Jomoto: as in No. 696r, p. 515—521; cf. Steel Times 193 (1966) p. 679.
715. Heynert, G.: Stahl u. Eisen 82 (1962) p. 1720—1726. (Hochofenaussch. 366.)
716. Mitsiu, K., M. Uchihara, H. Asai, and T. Yamada: Tetsu to Hagané 50 (1964) p. 366—369.

5.5. Mathematical Analysis, Prediction and Limitations of the Blast Furnace Process

717. Wesemann, F.: Stahl u. Eisen 68 (1948) p. 1—8. (Hochofenaussch. 230 and Wärmestelle 340.)
718. Görgen, R.: Arch. Eisenhüttenwes. 30 (1959) p. 381—390. (Hochofenaussch. 321 and Wärmestelle 475.)
719. Reichardt, P.: Arch. Eisenhüttenwes. 1 (1927/28) p. 77—101. (Hochofenaussch. 83.)
720. Schürmann, E., and D. Bülter: Stahl u. Eisen 81 (1961) p. 1565—1574. (Hochofenaussch. 355.)
721. Schürmann, E., and D. Bülter: Arch. Eisenhüttenwes. 35 (1964) p. 475—486.
722. Rist, A., and N. Meysson: Rev. Métallurg. 61 (1964) p. 121—145.
723. Dancoisne, P.: Canad. Metallurg. Quarterly 2 (1963) p. 197—223.
724. Hodge, A. L., and E. A. Wyczalek: A mathematical method for analysing and predicting changes in blast-furnace operation. Publ. by the Linde Co., Division of Union Carbide Corp., Newark (N.J.).
725. Hodge, A. L.: Iron and Steel 34 (1961) p. 232—238.
726. Mathesius, W.: Die physikalischen und chemischen Grundlagen des Eisenhüttenwesens. Leipzig 1916.
727. Hupfer, K., and H. Weidenmüller: personal communication (1966).
728. Staib, C., P. Dancoisne, and J. Michard: In: Internationale Eisenhüttentagung 1965: Automatisierung in Hüttenwerken, Berichtssammlung Tagungsteil Amsterdam.
729. Spallanzani, G. B., N. Andreotti, G. Sironi, and L. Pompilio: as in No. 728.
730. Katsura, K., and T. Yamamoto: as in No. 728.
731. Poos, A., R. Vidal, and J. Lückers: as in No. 728.
732. v. Bogdandy, L.: Stahl u. Eisen 81 (1961) p. 12—22. (Hochofenaussch. 337.)
733. Schürmann, E., W. Zischkale, P. Ischebeck, and G. Heynert: Stahl u. Eisen 80 (1960) p. 854—861.

733a. v. Bogdandy, L., P. Dickens, W. vor dem Esche, and J. Willems: Stahl u. Eisen 83 (1963) p. 129—139. (Hochofenaussch. 371.)
734. Wartmann, R.: Stahl u. Eisen 83 (1963) p. 1414—1425. (Hochofenaussch. 383.) — Arch. Eisenhüttenwes. 34 (1963) p. 879—885. (Hochofenaussch. 385 and Aussch. Betriebswirtsch. 352.)
734a. Lander, H. N., N. W. Meyer, and F. D. Delve: Proc. AIME, Blast Furnace, Coke Oven and Raw Materials Comm. 19 (1960) p. 219—232.
735. Koump, V., R. H. Tien, R. G. Olson, and T. F. Perzak: In: Symposium Simulation of Processes, Feb. 1964. Publ. by Amer. Inst. Min. Metallurg. and Petrol. Eng., forthcoming.
736. v. Bogdandy, L., and R. Wartmann: as in No. 728.
737. Sauer, K.: Techn. Mitt. Rheinhausen No. 4, 1955, p. 210—217.
738. Flint, R. V.: Blast Furn. Steel Plant 50 (1962) p. 46—47.
739. Voice, E. W., and K. G. Dixon: J. Iron Steel Inst. 200 (1962) p. 438—443.
740. v. Bogdandy, L., and R. Wartmann: Arch. Eisenhüttenwes. 36 (1965) p. 221 to 236. (Aussch. metallurg. Grundlagen 15.)
741. Beer, H., L. Diefenbach, and K. Hedden: as in No. 728.
742. Fliermann, G. A., and J. M. van Langen: as in No. 728.
743. v. Bogdandy, L., U. Pückoff, and R. Wartmann: Arch. Eisenhüttenwes. 37 (1966) p. 909—917. — Pückoff, U.: Diss., TH Aachen 1966.
743a. Muchi, I., J. Yagi, K. Tamura, and A. Muriyama: Nippon Kinzoku Gakkaishi 30 (1966) p. 826—831.
744. Kegel, H.: Stahl u. Eisen 70 (1950) p. 733—740. (Hochofenaussch. 240.)
745. Ridgion, J. M.: J. Iron Steel Inst. 188 (1958) p. 317—320.
746. Heynert, G., and H. Beer: Stahl u. Eisen 84 (1964) p. 1353—1365. (Hochofenaussch. 400.)
747. As in No. 344.
748. Petersen, U., H. Kahlhöfer, and A. Send: Stahl u. Eisen 83 (1963) p. 1397 to 1407. (Hochofenaussch. 382.)
749. Schenck, H., and H. Küppersbusch: Stahl u. Eisen 83 (1963) p. 1345—1348.
750. Heynert, G., and K. Hedden: Chem.-Ing.-Technik 33 (1961) p. 469—478.
751. Ramm, A. N.: Stal 24 (1964) p. 860—871; Stal in Deutsch 5 (1965) p. 106—118.
751a. Schierloh, U.: Stahl u. Eisen 90 (1970) p. 709—712.
752. Brandi, H., G. Heynert, and H. Beer: Stahl u. Eisen 84 (1964) p. 1169—1174.
753. Horie, S.: J. Metals 17 (1965) p. 243.
754. Jourde, P., and F. Huguet: Rev. Métallurg. 62 (1965) p. 197.
755. Tsujihata, K., Y. Shiraishi, K. Neshima, and A. Honda: Blast Furn. Steel Plant 53 (1965) p. 242.
756. Newton, W. M., T. B. Metcalf, and W. B. Mason: Petrol Refiner 32 (1953) p. 125. — Stage, H., and K. Bose: Die Belastungsverhältnisse in Füllkörpersäulen unter Destillationsbedingungen. Berlin/Göttingen/Heidelberg 1962.
757. Personal communication from the Works Management of Cherepovets (1966).
758. Schoppa, H., D. Winzer, and H.-D. Bubner: Stahl u. Eisen 86 (1966) p. 1469 to 1473.
759. Wysocki, H., and U. Pückoff: Stahl u. Eisen 86 (1966) p. 761—770.

Author Index

The numbers indicate the pages on which the authors' names or their work are referred to

Subject Index

721/24/70

MIX
Papier aus verantwortungsvollen Quellen
Paper from responsible sources
FSC® C105338

If you have any concerns about our products,
you can contact us on
ProductSafety@springernature.com

In case Publisher is established outside the EU,
the EU authorized representative is:
Springer Nature Customer Service Center GmbH
Europaplatz 3, 69115 Heidelberg, Germany

Printed by Libri Plureos GmbH
in Hamburg, Germany